Fundamentals of Analysis with Applications

Atul Kumar Razdan · V. Ravichandran

Fundamentals of Analysis with Applications

 Springer

Atul Kumar Razdan
MIET
Meerut, Uttar Pradesh, India

V. Ravichandran
Department of Mathematics
National Institute of Technology
Tiruchirappalli, India

ISBN 978-981-16-8385-5   ISBN 978-981-16-8383-1   (eBook)
https://doi.org/10.1007/978-981-16-8383-1

This Springer imprint is published by the registered company Springer Nature Singapore Pte Ltd.
The registered company address is: 152 Beach Road, #21-01/04 Gateway East, Singapore 189721,
Singapore

To my wife Anju and children, Anuja and Aman

—*Atul Kumar Razdan*

To Prof. Dato' Indera Rosihan M. Ali and my wife Kalaiselvi

—*V. Ravichandran*

Preface

The work on the book started some time ago with notes written by the second author to teach a semester course on *Real Analysis and Partial Differential Equations* to the second year B. Tech. engineering students at NIT, Tiruchirappalli. The mathematical courses students had, during previous semesters, focused mainly on the problem-solving techniques in single-variable calculus, linear algebra, ordinary differential equations, and the basics of complex analysis. In the first part, he introduced mainly those topics of analysis that are prescribed specifically as the prerequisites to deal with the intricacies of some problems in science and engineering, especially the ones modeled by partial differential equations. In the second part, he provided methods to solve some standard *initial boundary-value problems* concerning phenomena such as *vibrations, heat conduction, fluid flow*, and *electromagnetic waves*.

This textbook grew out of the need to provide a strong foundation in basic analysis and geometry to the reader interested in taking up a more rigorous course in the theory of partial differential equations. In particular, parts of the contents are suitable for a senior-level undergraduate course in physics and engineering, or a beginning-level graduate course in mathematics. For example, the single-variable analysis over $\mathbb{R}$ as presented in the first five chapters of the book is suitable for an early undergraduate semester course on *"Basic Real Analysis"*. The book assumes familiarity with elementary notions of the one-variable calculus, basic linear algebra techniques, and a working knowledge about some simple ordinary differential equations. Our choice for the topics in mathematical analysis as treated herein, and also the presentation style adopted, are not totally unaffected by the considerations that the analytical tools developed are ultimately needed to study certain important topics related to applications of differential equations.

In general, mathematical analysis studies the fundamental analytical properties such as continuity, differentiability, and integrability of some scalar- or vector-valued functions defined over *linear metric spaces* such as $\mathbb{R}^n$, $\ell_p(\mathbb{N})$ or $L^p(\mathbb{R})$, where it is assumed that only standard metrics are used to give topological and geometrical structures to such spaces. In this book, we shall be working with the Euclidean metric on $\mathbb{R}^n$, and, for the function spaces, the standard metrics such as induced by L^∞, L^1, and L^2- *norms*. We will discuss some important analytical properties of real-valued functions of the form $u = u(x, t)$ defined on a set $E \subseteq \mathbb{R}^n \times \mathbb{R}^+$, for some $n \geq 1$. The focus of the book is to prove some fundamental theorems related to advanced topics such as linear metric spaces, multivariable analysis, convergence of series of functions, Lebesgue's theory of integration, and to discuss some applications of Fourier series.

The concept of *convergence* is central to everything we do in modern mathematics, within and in applications. Some fundamental *types of convergence* used in the analysis are related to the *function spaces* defined over a set $E \subseteq \mathbb{R}^n, n \geq 1$. Considering a wider spectrum of engineering applications, the fundamental concept of the limit of a sequence in a metric space X is developed through various stages; of course, starting with the case when $X = \mathbb{R}$ and the convergence is discussed with respect to the *modulus metric* (Chap. 3). In general, for a given situation in science and engineering, we use a specific type of convergence considering the practical requirements of the physical problem in hand. We will prove some fundamental *convergence theorems* in the later part of the book.

Chapter 1 presents the fundamentals of sets and functions, with special emphasis on the *cardinality* aspects of sets. In Chap. 2, we discuss algebraic, order, and metric structures of the set $\mathbb{R}$ of real numbers. Analytical properties such as the continuity, differentiability, and integrability of single-variable real-valued functions are discussed in Chap. 4. Chapter 5 is about certain fundamental topological properties of the real line $\mathbb{R}$ such as compactness and connectedness, where the *modulus metric* is used to define a topology on $\mathbb{R}$.

In Chap. 6, we introduce some important linear metric spaces such as $\mathbb{R}^n$, and also discuss their analytical and topological properties, with special emphasis on certain important infinite-dimensional spaces defined over an interval $I \subseteq \mathbb{R}$. In Chap. 7, we introduce basic concepts of multivariable analysis and prove some standard theorems. The real emphasis here is to present a differential geometry viewpoint. In Chap. 8, we discuss *uniform convergence* of sequences (and hence series) of single-variable real-valued continuous, differentiable, and (Riemann) integrable functions. More generally, in Chap. 9, we discuss L^∞, L^1, and L^2-*convergence* in the space of *almost* real-valued Lebesgue integrable functions defined on $\mathbb{R}$. The chapter starts with an introduction to abstract measure spaces and then specializes to Lebesgue's theory of measure and integration over $\mathbb{R}$. Finally, in Chap. 10, we apply the ideas developed in previous chapters to discuss Fourier's idea of representing a nice (periodic) function as a *trigonometric series*.

We are thankful to two reviewers for their comments and to Dr. Lee See Keong, Universiti Sains Malaysia for pointing out typographic errors and for offering suggestions to improve the presentation. The research scholars of the second author at NIT-T have read several chapters of the book. We are also thankful to Dr. Shamim Ahmad, Publishing Editor, Mathematics, Springer India for his continued support; he first contacted the second author on February 27, 2012. We will be happy to receive comments for improvements including errors/misprints. The authors are thankful to the team at Springer, Mr. Nareshkumar and Mr. Lokeshwaran for their continuous support during the process.

New Delhi, India   Atul Kumar Razdan
Tiruchirappalli, India   V. Ravichandran
June 2021

Introduction

The book of nature is written in the language of mathematics.

Galileo Galilei (1564–1642)

Fundamental to modern analysis is our understanding of the algebraic, analytical, and geometric structures of sets consisting of objects such as *numbers, n-tuples of numbers, matrices, functions*, and their *higher-dimensional analogues*. A *set* as a collection of objects is viewed as a *single entity* under study, and a *function* is a rule to define correspondence among objects of sets, similar or different. The discussion in this text is limited to a study of standard analytical properties of real-valued functions defined over *analytical structures* constructed from the elements of the set $\mathbb{R}^n$, $n \geq 1$, and the sets of functions given on such types of structures. The foundation of some of the most important ideas in mathematics and physics is based on the concept of a *linear space over a field*.

A **field** $\mathbb{K}$ is a set with at least two *elements* $0 \neq 1$ such that it is possible to perform all the four fundamental arithmetic operations in $\mathbb{K}$. That is, more precisely, we have two *binary operations* on $\mathbb{K}$ given by

$$+ : \mathbb{K} \times \mathbb{K} \to \mathbb{K} \text{ and } \cdot : \mathbb{K} \times \mathbb{K} \to \mathbb{K},$$

so that we can *add* $x + y$ and *multiply* $x \cdot y$ elements $x, y \in \mathbb{K}$, and the following conditions hold:

1. (associative) For any $x, y, z \in \mathbb{K}$,

$$x + (y + z) = (x + y) + z \text{ and } x \cdot (y \cdot z) = (x \cdot y) \cdot z.$$

2. (commutative) For any $x, y \in \mathbb{K}$,

$$x + y = y + x \text{ and } x \cdot y = y \cdot x.$$

3.   (identity) For any $x \in \mathbb{K}$,

$$x + 0 = x \text{ and } x \cdot 1 = x.$$

4.   (inverse) For any $x \in \mathbb{K}$, there exist $y \in \mathbb{K}$ and $z \in \mathbb{K}$ such that

$$x + y = 0 \text{ and } x \cdot z = 1.$$

We write $y = -x$ and $z = x^{-1}$.

5.   (distributive) For any $x, y, z \in \mathbb{K}$,

$$x \cdot (y + z) = x \cdot y + x \cdot z \text{ and } (y + z) \cdot x = y \cdot x + z \cdot x.$$

In this case, we say that $(\mathbb{K}, +)$ is an *Abelian group* with the *additive identity* $0 \in \mathbb{K}$; and, also that $(\mathbb{K}, \cdot)$ is an *Abelian group* with the *multiplicative identity* $1 \in \mathbb{K}$. The condition (5) above is also called the *compatible condition*. For example, the set $\mathbb{R}$ of real numbers is a *field* with respect to *usual addition* and *usual multiplication*, and the real numbers 0 and 1 are, respectively, the additive and multiplicative identities. Moreover, as shown in Chapter 1, it is a unique *complete ordered field* with respect to the *natural order*.

Next, we say a set V is a **linear space** over a field $\mathbb{K}$ if there is a binary operation $+ : V \times V \to V$ so that $(V, +)$ is an *Abelian group*, where the additive identity $0 \in V$ is called the *zero vector*, and also there is an *action* of $\mathbb{K}$ on V given by $: \mathbb{K} \times V \to V$ such that the following conditions hold:

1.   For any $x, y \in V$ and $a \in \mathbb{K}$, $a \cdot (x + y) = a \cdot x + a \cdot y$.
2.   For any $x \in V$ and $a, b \in \mathbb{K}$, $(a + b) \cdot x = a \cdot x + b \cdot x$.
3.   For any $x \in V$ and $a, b \in \mathbb{K}$, $a \cdot (b \cdot x) = (ab) \cdot x$.
4.   For any $x \in V$ and $1 \in \mathbb{K}$, $1 \cdot x = x$.

Clearly then, the Abelian group $(\mathbb{R}, +)$ is a linear space over the field $\mathbb{R}$. In general, the set $\mathbb{R}^n$ of n-tuples $\mathbf{x} = (x_1, \ldots, x_n)$ with $x_i \in \mathbb{R}$ is also an Abelian group with respect to the usual coordinate-wise addition given by

$$\mathbf{x} + \mathbf{y} = (x_1 + y_1, \ldots, x_n + y_n), \text{ with } \mathbf{y} = (y_1, \ldots, y_n) \in \mathbb{R}^n.$$

In this case, the n-tuples $\mathbf{0} = (0, \ldots, 0)$ is the (additive) identity, and $-\mathbf{x} = (-x_1, \ldots, -x_n)$ is the (additive) inverse of $\mathbf{x}$. Also, the usual multiplication of an n-tuple $\mathbf{x} = (x_1, \ldots, x_n)$ by a real number $a \in \mathbb{R}$ given by

$$a \cdot \mathbf{x} = (ax_1, \ldots, ax_n)$$

is compatible with the coordinate-wise addition, which gives it a structure of a *linear space* over the field $\mathbb{R}$. More generally, given any set $X \neq \emptyset$, the collection $V := \mathbb{R}^X$

of real-valued functions f defined on the set X is a linear space over $\mathbb{R}$ with respect to the following two *pointwise operations*: For $f, g \in V$, and $a \in \mathbb{R}$,

$$(f + g)(x) := f(x) + g(x) \text{ and } (af)(x) := af(x), x \in X,$$

Notice that the *zero function* $0_X : x \mapsto 0$ is the (additive identity), and $(-f)(x) = -f(x)$ is the (additive) inverse of a function $f \in V$. Here, in this text, we are primarily interested in cases when $X = \mathbb{N}$ or X is an interval $I \subseteq \mathbb{R}$, bounded or unbounded. Taking $X = \mathbb{N}$, we conclude that the set $\mathbb{R}^{\mathbb{N}}$ of *sequences of real numbers* is a linear space over the field $\mathbb{R}$. Also, when X is an interval $I \subseteq \mathbb{R}$, we conclude that the set of *real-valued functions on I* is a linear space over the field $\mathbb{R}$. In this book, we will write $B(I)$ for the linear space of real-valued *bounded* functions, and $C(I)$ for the linear space of real-valued *continuous* functions, defined on the interval I. In fact, for a good part of the content, we will do an analysis with these two fundamental linear spaces.

We say, very roughly, a linear space V over $\mathbb{R}$ is *finite-dimensional* if the geometry of every object in V can be described (uniquely) in terms of finitely many real numbers, called the *coordinates* of the points. The minimal number of coordinates required to do so is called the *dimension* of V. Notice that $\mathbb{R}^n$ is n-dimensional linear space because each point $x \in \mathbb{R}^n$ can be described (uniquely) in terms of n real numbers $x_1, \ldots, x_n$ so that we write $x = (x_1, \ldots, x_n)$. However, the linear space $\mathbb{R}^{\mathbb{N}}$ of sequences of real numbers, and also both $B(I)$ and $C(I)$ are *infinite-dimensional* linear spaces over $\mathbb{R}$. Now, suppose $X = \mathbb{N}_n = \{1, \ldots, n\}$, and consider the n-functions $e_1, \ldots, e_n \in \mathbb{R}^{\mathbb{N}_n}$ given by

$$e_i(j) = \begin{pmatrix} 1, & \text{when } i = j \\ 0, & \text{when } i \neq j \end{pmatrix}, 1 \leq i, j \leq n.$$

Then, every $f \in V$ can be written (uniquely) as a *linear combination* of the form

$$f = a_1 e_1 + \cdots + a_n e_n, \text{ where } a_k = f(k), \text{ for } k = 1, \ldots, n.$$

In this case, we say the set $B = \{e_1, \ldots, e_n\}$ is a *basis* of the linear space $\mathbb{R}^{\mathbb{N}_n}$ over $\mathbb{R}$. More generally, if V is a linear space over $\mathbb{R}$, we say $B \subset V$ is a *spanning set* for V if each $v \in V$ is a finite linear combination of elements of B, with coefficients in $\mathbb{R}$. That is, there are elements $b_1, \ldots, b_n \in B$ and real numbers $a_1, \ldots, a_n$ such that

$$v = a_1 b_1 + a_2 b_2 + \cdots + a_n b_n.$$

Further, we say a set $B \subset V$ is *linearly independent* over $\mathbb{R}$ if, for every finite set $\{b_1, \ldots, b_m\} \subseteq B$,

$$a_1 b_1 + a_2 b_2 + \cdots + a_n b_n = 0 \Rightarrow a_1 = a_2 = \cdots = a_n = 0.$$

That is, the span of every finite subset of nonzero vectors in B is without a *nontrivial* linear combination. More formally, we say a set $B \subset V$ is a *basis* of V if B is a maximal linearly independent (and hence spanning) set for V. Then, the linear space V is called *finite-dimensional* if it has a basis B with a finite number of elements. Otherwise, V is said to be *infinite-dimensional*. It is a theorem in linear algebra that any two bases of a linear space have the same *cardinality*. Therefore, for a finite-dimensional linear space V, the *cardinality* of a basis of V is the *dimension* of V. Hence, more precisely, the dimension of the linear space $\mathbb{R}^n$ is n because the set B of the n-vectors

$$e_1 = (1, 0, , \ldots, 0), e_2 = (0, 1, 0, \ldots, 0), , \cdots, , e_n = (0, 0, \ldots, 1),$$

is a *basis* of $\mathbb{R}^n$. It is the *standard basis* of $\mathbb{R}^n$ in the sense that it automatically provides *n-unit vectors* along the positively oriented n mutually orthogonal axes. In this way, each point $x = (x_1, \ldots, x_n) \in \mathbb{R}^n$ is viewed as an n-dimensional *row vector* or a *column vector*. We choose one of the two representations depending upon the practical aspects of the problem under study. In actual applications, especially while dealing with matrices, we prefer to write the points of $\mathbb{R}^n$ as the *column vectors*. Linear spaces such as $\mathbb{R}^n$ are fundamental to many important concepts[1] in mathematics and physics.

Further, depending upon how we intend to extend the *analytical concepts* from a finite-dimensional linear space V such as $\mathbb{R}^n$ to an infinite-dimensional linear space such as $B(I)$ or $C(I)$, a key role is played by the **metric**, i.e., a function representing the *distance* between points of the space V, we choose to define the *convergence in V*. Most concepts related to convergence in a linear space V are defined in terms of sequences of elements in V. So, the fundamental idea of the *limit of a function* is of central significance to every concept developed in mathematical analysis, especially so while we attempt to extend certain characteristic properties of $\mathbb{R}^n$ to an infinite-dimensional space such as $B(I)$ or $C(I)$. Therefore, some of the metric spaces have always been an integral part of the evolution of diverse theories in mathematics and physics. We will discuss some simple topological and geometric properties of metric spaces in Chap. 6

Every metric on a linear space facilitates a mathematical description of intuitive idea of *nearness of points* in some appropriate sense. Most celebrated *convergence*

---

[1] For example, the *spectral theory* of certain nice operators defined over infinite-dimensional Hilbert spaces of *summable sequences* of complex numbers or *spaces of square integrable function* defined over $\mathbb{R}$. Some related ideas are developed in Chaps. 6 and 9.

theorems of modern analysis have been obtained as an answer to the following fundamental question: *If the terms of a sequence* $\langle x_n \rangle$ *in some linear space X have a property, and* $x_n \to x$ *with respect to a metric d on X, do we always have* $x \in X$? A *type of convergence* is *nice* if that is precisely the case. Such types of results are known as the *convergence theorems*. In general, the concept of *metric space* is a mathematical incarnation of the idea of the *distance between points* of a set, which is also used to discuss the geometric concepts such as the *size* and the *shape* of a metric space[2] [1]. In Chapter 6, we discuss some basic stuff related to metric spaces inasmuch as is needed to discuss certain core topological properties of linear spaces as mentioned above.

In abstract terms, a topology on a set is defined in terms of a *decorated collection* of subsets that are used to capture the fundamental idea of *neighbourhood* of a point. In a metric space, the set of all points that are at most $\varepsilon > 0$ distance away from a fixed point is a neighbourhood of the point, and hence a *topology* on a metric linear space (induced by the metric) provides for a mathematical framework to study the *proximity* of points. A similar construction leads to another key idea in mathematics, known as the *measure of a set*, which is fundamental to studies related to many different kinds of *physical measurements* given in terms of the integral of a function.

In abstract terms, the real purpose of mathematical analysis is to study the applications in science and engineering of analytical and geometrical properties of some *structure-preserving* mappings between certain infinite-dimensional linear metric spaces or, more generally, between *topological linear spaces*. Such a mapping defined between two similar types of spaces is called a *morphism*.[3] *Linear maps* are the morphisms of linear spaces, and the morphisms of metric spaces are known as *continuous maps*. Of particular interest is the interplay between analytical and geometrical properties of maps between linear metric spaces. For example, the derivative of a real-valued function at a point of an open set $U \subseteq \mathbb{R}^n$ is given by a linear map : $\mathbb{R}^n \to \mathbb{R}$. In this manner, the geometry of linear maps plays an important role in solving certain complex non-linear problems by a *local linearisation*. In Chap. 7, we discuss some fundamental theorems of multivariable analysis which are consequences of such an interplay.

In a typical practical scenario, a set represents quantities of some physical significance, and a function models the *cause-effect processes* among such types of quantities. In general, fundamental to our ability to use ideas of modern mathematics in science and engineering is our knowledge about the *mathematical structures* that a set has, and also about certain characteristic properties of the function(s) involved in the *modelling process*. That is a kind of general scheme how to solve complex problems related to diverse types of phenomena by applying certain important mathematical ideas. However, as the focus is to study the applications of differential equations in

---

[2] Typical examples of the size are *diameter* and *volume* of sets in a Riemannian manifold; and important invariant such as the *curvature* describes the local (and hence global) shape of a metric space.

[3] For example, a set of *bijective morphism* of a geometric configuration in the space $\mathbb{R}^n$ is a group under composition of maps, which is a very effective tool to study its *symmetries*.

science and engineering, the variables involved are *continuous*, i.e., they take values in some subset of $\mathbb{R}^n$. Moreover, since in most cases we are dealing with *evolutionary systems*, one of the independent variables represents the time. Therefore, in general, our concern is to study properties of functions of the form $u = u(x, t)$, where $x \in \mathbb{R}^n$ is the *position variable* and $t > 0$ represents the time.

The focus of the present course is to study the analytical/geometric properties of a function in some particular function space defined over a metric space. So, for a wider scope for applications, we also discuss the topology and geometry of the underlying metric spaces of vectors, sequences, continuous, and measurable functions. As said above, a working knowledge about the structure of such linear metric spaces is very crucial to solve some complex physical problems in physics and engineering. The treatment of most topics presented in this book is guided by our need to help the reader to reach to a certain acceptable level of understanding about basic concepts of real analysis so that s/he can meet the challenges offered by problems related to partial differential equations.

In Chap. 1, we review briefly concepts related to sets, functions, relations, and ordering, and fix notations for use in the sequel. A brief introduction to mathematical logic is provided as an appendix. To gain a contextual perspective about some concepts presented in later chapters, we present here historical developments related to the evolution of the concept of function. The chapter is concluded with a brief about Cantor's in-depth study of cardinal numbers of infinite sets.

For a stronger foundation in mathematical analysis, it is always good to start with a *single-variable analysis*. We will discuss the related concepts in the next four chapters. In Chap. 2, we use a constructive approach to introduce the set of real numbers $\mathbb{R}$ as a unique complete ordered field. The three standard methods of construction of real numbers due to Dedekind, Cantor, and Cauchy are described briefly. In Chap. 3, we discuss the convergence of sequences and series of real numbers. The standard concepts and theorems presented here are used in later chapters to study convergence in more general types of metric spaces. In Chap. 4, we discuss basic analytical properties such as continuity, differentiability, and integrability, for real-valued functions defined over an interval in $\mathbb{R}$. In Chap. 5, we describe topological concepts such as convergence, compactness, and connectedness of the set $\mathbb{R}$ with respect to *the usual metric*.

In Chap. 6, we introduce *metric spaces* and study some fundamental properties of the Euclidean space $\mathbb{R}^n$, the space ℓ_p of absolutely p-summable sequences of numbers, and certain important function spaces over a domain in $\mathbb{R}^n$. Also, *completeness* aspects of these spaces are discussed.

In Chap. 7, we discuss continuous functions between metric spaces, followed by a presentation covering various aspects of differentiability of functions $\mathbb{R}^m \to \mathbb{R}^n$. We extend here most important theorems of one variable analysis. We will also discuss briefly fundamental theorems like the Banach contraction principle, inverse function theorem, and the Picard-Lindelöff theorem.

In Chap. 8, we discuss pointwise and uniform convergence of series of functions defined over a finite interval. Basic theorems proved here are used later to discuss the convergence of some important power series and to discuss the convergence of Fourier series, presented in Chap. 10.

In Chap. 9, we discuss preliminary aspects of Lebesgue's theory of measure and integration. First, a crisp introduction to the abstract measure theory is given, and then some important theorems are reviewed. The basic concepts of Lebesgue measure and Lebesgue integration are then given followed by a discussion about the fundamental convergence theorems of Lebesgue integration. The chapter concludes with a brief discussion about L^p-spaces over $\mathbb{R}$.

In Chap. 10, we discuss the Fourier series representation of periodic functions in spaces such as $L^1([-\ell, \ell])$ or $L^1([0, \ell])$. For a better contextual understanding, we start with historical developments related to *vibrating string controversy*. We prove Dirichlet's theorem about the pointwise convergence of the Fourier series of a piecewise smooth periodic function. Some other issues related to the uniform convergence of the Fourier series are also discussed.

Reference

1.   M.M. Deza, E. Deza, *Encyclopedia of Distances* (4/e), Springer, Berlin, 2016.

Contents

List of Figures

Chapter 1
Sets, Functions, and Cardinality

The beginner should not be discouraged if he finds he does not
have the prerequisites for reading the prerequisites.

Paul R. Halmos (1916–2006)

A set is a collection of objects and a function is a rule to associate objects of two sets, where the terms *"collection"* and *"rule"* are *well-defined* in some appropriate sense. Every concept related to sets and functions is fundamental to the development of modern mathematics, and also for some theories in science and engineering. However, since a function itself is a special type of a set, most assertions in mathematics are actually logical formulation of statements in terms of sets, and a *"proof"* of an assertion uses certain axioms and the rules of deductive reasoning. It thus makes sense to say that the *mathematical logic* provides the language to communicate the mathematics. We refer the reader to Appendix A for a quick review of some basic concepts of mathematical logic.

The central idea of the set theory is the emphasis to view a set as a *single entity*. The *axiomatic system* of sets, as developed by *Ernst Zermelo* (1908) and *Abraham Fraenkel* (1922), is the foundation of the modern mathematics. It is usually referred to as the ZFC system,[1] which is a large and complicated subject in its own right. The interested reader may refer [1], for further details. We adopt here a naive approach to the concept of a set, as developed initially in 1870 by *George Cantor* (1845–1918) and *Richard Dedekind* (1831–1916).

On the other hand, in many practical situations, we study the properties of functional relationships between sets to express many different types of mathematical

---

[1] The axiomatic theory was enriched further by *Thoralf Skolem* (1922) and *Von Neumann* (1928). The letter C used here stands for the *axiom of choice*, as stated later.

© The Author(s), under exclusive license to Springer Nature Singapore Pte Ltd. 2022
A. K. Razdan and V. Ravichandran, *Fundamentals of Analysis with Applications*,
https://doi.org/10.1007/978-981-16-8383-1_1

structures in the simplest possible terms. For example, functions are used to study the size of a set in terms of its *cardinality*, control objects as in *control theory*, measure computational complexity of an *algorithm*, and so on. Fundamental concepts like *sequences* and *strings* are two special types of functions that have numerous applications in subjects such as computer science, economics, and financial modelling. Also, some *summation type functions* are used to study the limiting cases of many different types of infinite processes occurring in various dynamical systems. In certain important applications in physics and engineering, we also use some *integral type functions* such as *beta, gamma*, and *error functions*.

In this chapter, we discuss some important foundational aspects related to sets and functions inasmuch as we need in the sequel. The main purpose here is to introduce some basic terminology and fix notations. We refer the reader [2], for further details.

1.1 Naive Set Theory

As said before, a *set* is a *well-defined collection of objects of any type that may interest us*. An object in a set is called an *element* (or a *member*). The term *well-defined* used above means that, given any object, it is possible to decide in an *unambiguous way* if the object belongs to the collection. So, for example, the collection of *all English alphabets* and the collection of *all students* in a school do constitute a set. However, a collection of objects formed by using some *ambiguous* qualifying properties, such as *beautiful, dirty, intelligent*, etc., do not make a set in the usual sense.[2]

In the present course, we mostly have *numbers, vectors, matrices*, and also *functions*, for objects of a set. Each one of such types of sets is equipped with a certain kind of *natural* mathematical structures. For example, as seen later, relations on sets, graphs, polynomials, matrices, etc., given as abstract sets, have *discrete structures* that are defined in terms of their inherited properties. In computer science, *Boolean* refers to the set $\{0, 1\}$, together with three operators representing AND, OR, and NOT logic gates in usual sense of the electrical circuit theory.

The existence of a set with no element is assumed, which is denoted by $\emptyset$, and is called the *empty set*. In ZFC, this is known as the *axiom of the empty set*. In general, we use upper-case English letters

$$A, B, C, \ldots \quad \text{or} \quad X, Y, Z, \ldots,$$

to denote an arbitrary set, and their elements are usually denoted by lower-case English letters

$$a, b, c, \ldots \quad \text{or} \quad x, y, z, \ldots.$$

---

[2] However, we can deal with such collections as *fuzzy sets*.

For a nonempty set, we write a set by putting its elements in curly braces as { ... }. The elements of a set in the *tabular form* are listed without repetition, and without having any regard for the order of listing. For example, as a set, we can list 26 alphabets of English language in any order. However, we also come across applications when *ordering* of elements of a set will be significant. On the other hand, we may also write a set in *set-builder form*. That is, by stating a *characteristic property* shared by all its elements. So, if P is a property shared by a certain type of object, then the set X of elements x satisfying the property P is written as

$$X = \{x : x \text{ satisfies } P\}.$$

If X is a set, and x is an element of X, we write $x \in X$ (read $\in$ as *belong to* or *is a member of*). Otherwise, we write $x \notin X$ (read $\notin$ as *does not belong to*) when x is not an element of the set X. In this chapter, we assume that the reader has the working familiarity with basic arithmetic and order properties of the following *number sets*:

$$\mathbb{N} = \{1, 2, 3, \ldots\} \qquad (natural\ numbers)$$
$$\mathbb{Z} = \{\ldots, -2, -1, 0, 1, 2, \ldots\} \qquad (integers)$$
$$\mathbb{Q} = \{q = a/b : a, b \in \mathbb{Z}, b \neq 0\} \qquad (rational\ numbers)$$
$$\mathbb{R} = \mathbb{Q} \cup \{irrational\ numbers\} \qquad (real\ numbers).$$

Notice that $-2 \in \mathbb{Z}$, but $-2 \notin \mathbb{N}$. Also, $x \notin \mathbb{Q}$ is an *irrational number* of the form e, π, e^{π}, etc. A *constructive approach* to these number sets and a detailed description of their algebraic and order structures are given in the next chapter. We may write the set of even positive integers as

$$2\mathbb{N} = \{2, 4, 6, 8, \ldots\}.$$

Notice that this set in set-builder form can be written as

$$2\mathbb{N} = \{x : x = 2n, n \in \mathbb{N}\}.$$

Also, the set D of all *prime divisors* of 60 can be written as

$$D = \{2, 3, 5\} = \{n \in \mathbb{N} : n \text{ is a prime divisor of } 60\}.$$

We say X is a *finite set* if it has a finite number of elements. Otherwise, it is said to be an *infinite set*. For example, the set D and the set of all English alphabets E are finite, whereas all the number sets such as $\mathbb{N}, \mathbb{Z}, \mathbb{Q}$ are infinite.

Definition 1.1 If every element of a set A is also an element of the set B, we say that A is a *subset* of the set B (or B is a *superset* of the set A). We write $A \subseteq B$ (read as A is contained in B). We write $A \supseteq B$ if A is a superset of the set B.

So, if there is at least one element in A that is not in B, we say A is not a subset of B, and write $A \not\subseteq B$. If A is contained in B such that there is at least one element in B that is not in A, we say that A is a *proper subset* of the set B, and we write $A \subset B$. For example, $\mathbb{N} \subset \mathbb{Z}$ and $E \subset \mathbb{N}$, where E is the set of positive even integers. By definition, two sets A and B are *equal* if and only if $A \subseteq B$ and $B \subseteq A$. That is,

$$A = B \quad \Leftrightarrow \quad (x \in A \Rightarrow x \in B) \text{ and } (x \in B \Rightarrow x \in A).$$

Otherwise, we write $A \neq B$.

While dealing with a collection of sets, it is assumed that all sets in the collection are subsets of some *universal set* $\mathscr{U}$; we need this assumption to avoid paradoxes such as the one due to *Bernhard Russell*[3] (1872–1970). For our purpose, it suffices to assume that $\mathscr{U}$ is a family of sets, containing at least the set $\emptyset$, wherein all types of set theory operations are possible. In a particular case, the universal set refers to a collection containing everything currently under study. For example, the set of real numbers $\mathbb{R}$ would be the universal set for sets containing numbers. We may not mention the universal set explicitly in any given situation, but the context may always help.

Definition 1.2 The collection of all subsets of a set X, denoted by $\mathscr{P}(X)$, is called the *power set* of the set X.

Since every set is a subset of itself, by definition, $X \in \mathscr{P}(X)$ always. Also, it is *vacuously true* that the empty set $\emptyset$ is a subset of every set. So, $\emptyset \in \mathscr{P}(X)$. Indeed, the empty set $\emptyset$ and the set X itself are two extreme situations because

$$\emptyset \subset A \subset X, \quad \text{for any other subset } A \text{ of } X.$$

In particular, $\mathscr{P}(\emptyset) = \{\emptyset\}$. If a set X has n elements, then X has $\binom{n}{k}$ subsets with k elements, $k = 0, 1, 2, \cdots n$. Therefore, in total, a set X with n elements has

$$\binom{n}{0} + \binom{n}{1} + \binom{n}{2} + \cdots + \binom{n}{n-1} + \binom{n}{n} = (1+1)^n = 2^n$$

subsets, by using the *binomial theorem*. Hence, in this case, the set $\mathscr{P}(X)$ has $2^n = 2^{|X|}$ elements. It is an interesting fact that this simple assertion also holds for infinite sets (see Lemma 1.1).

We mainly have three basics *set operations* to form new sets from any given class of sets. Let A and B be any two sets, and x be some object. There are only following four possibilities to consider:

1. Suppose $x \in A$ <u>or</u> $x \in B$. This possibility suggests to consider the set

$$\{x : x \in A \text{ or } x \in B\},$$

---

[3] The barber shaves all people who do not shave themselves, but no one else. Who shaves the barber? More precisely, if $X = \{x : x \notin x\}$, then $X \in X$ if and only if $X \notin X$.

which is called the *union* of the sets A and B. It is denoted by $A \cup B$. That is, the set $A \cup B$ is a set having elements of both A and B, without repetition. Therefore, we always have

$$A \subseteq A \cup B \quad \text{and} \quad B \subseteq A \cup B.$$

For example, if $S = \{1, 2, 3\}$ and $T = \{3, a, b\}$, then $S \cup T = \{1, 2, 3, a, b\}$.

2. Suppose $x \in A$ <u>and</u> $x \in B$. This possibility suggests to consider the set of *common elements*, i.e., the set

$$\{x : x \in A \text{ and } x \in B\},$$

which is called the *intersection* of the sets A and B. It is denoted by $A \cap B$. In this case, by definition, we have

$$A \cap B \subseteq A \quad \text{and} \quad A \cap B \subseteq B.$$

For example, with sets S and T as above, $S \cap T = \{3\}$. We say A and B are *disjoint sets* if they do not have any common element, and we write $A \cap B = \emptyset$.

3. Suppose $x \in A$, but $x \notin B$; or $x \notin A$, but $x \in B$. Each one of these two possibilities leads to the concept of *relative complementation*. The *relative complement* of a set A relative to a set B, denoted by $A \setminus B$, is the set given by

$$A \setminus B = \{x : x \in A \text{ but } x \notin B\}.$$

Similarly, the *relative complement* of a set B relative to a set A, denoted by $B \setminus A$, is the set given by

$$B \setminus A = \{x : x \in B \text{ but } x \notin A\}.$$

The *absolute complement* of a set A is defined relative to the applicable universal set, say U. If U is clear from the context, we write $U \setminus A$ simply as A' (or A^c).

4. Suppose $x \notin A$ <u>and</u> $x \notin B$. For simplicity, we assume both the sets A, B are subsets of the same universal set, say U. Then, we have $x \in A'$ and $x \in B'$. That is, $x \in A' \cap B'$. As the next theorem shows, we have $x \notin (A \cup B)$.

The concept of relative complement leads to a yet another interesting set operation, denoted by $\triangle$ (or sometimes $\oplus$), which is given by as follows: For $A, B \subset X$, we define

$$A \triangle B = (A \setminus B) \cup (B \setminus A) = \{x \in X : x \in (A \cup B) \setminus (A \cap B)\}.$$

It is called the *symmetric difference* of the sets A and B. Notice that

$$A \triangle B = (A \cap B^c) \cup (A \cap B^c).$$

It follows directly from this definition that $\triangle$ satisfies the following properties:

$$X \triangle Y = Y \triangle X, \qquad X \triangle X = \emptyset,$$
$$X \triangle \emptyset = X, \quad X \triangle U = X^c, \quad X \triangle X^c = U.$$

What can be said about the sets A and B if $A \triangle B = A$?

The three types of set operations defined above satisfy the following fundamental properties, which makes the power set $\mathscr{P}(X)$ of a set X an important object in mathematics.

Theorem 1.1 *If A, B, C are any three sets, then*

1. Associativity Laws:

$$A \cap (B \cap C) = (A \cap B) \cap C; \quad A \cup (B \cup C) = (A \cup B) \cup C;$$

2. Commutative Laws :

$$A \cup B = B \cup A; \quad A \cap B = B \cap A;$$

3. Distributive Laws:

$$A \cup (B \cap C) = (A \cup B) \cap (A \cup C);$$
$$A \cap (B \cup C) = (A \cap B) \cup (A \cap C);$$

4. Identity Laws:
$$A \cup \emptyset = A, \qquad A \cap U = A;$$

5. Domination Laws: $A \cup U = U, \quad A \cap \emptyset = \emptyset;$
6. Idempotent Laws: $A \cup A = A, \quad A \cap A = A;$
7. Complementation Law: $(A^c)^c = A;$
8. Absorption Law:

$$A \cup (A \cap B) = A, \quad A \cap (A \cup B) = A;$$

9. de Morgan's Laws:

$$(A \cap B)^c = (A^c \cup B^c); \quad (A \cup B)^c = (A^c \cap B^c).$$

Proof Left to the reader as an easy exercise.   □

It should be clear that some of the concepts introduced above can be extended to a finite collection of sets. In fact, there are numerous situations wherein we also need to consider the set operations over an arbitrary indexed family of sets. Let X be some universal set, and I be an arbitrary nonempty set. Suppose $A_\alpha \subset X$, for each $\alpha \in I$. The collection

$$\mathscr{A} := \{A_\alpha \subset X \mid \alpha \in I\}$$

is called an *indexed family* of subsets of the set X, and the set I is called an *index set*. When I is a finite set, say with n elements, it is common to write

$$I = \mathbb{N}_n := \{1, 2 \ldots, n\}.$$

In this case, we write the collection of sets indexed by I as

$$\{A_1, A_2, \ldots, A_n\} := \{A_i : i \in I\},$$

and the associated union and intersection are called the *finite union* and the *finite intersection*, which we write as

$$\bigcup_{m=1}^{n} A_m := \{x \in X : x \in A_m \text{ for some } 1 \le m \le n\};$$

$$\bigcap_{m=1}^{n} A_m := \{x \in X : x \in A_m \text{ for all } 1 \le m \le n\}.$$

Further, when $I = \mathbb{N}$, the associated union and intersection of the indexed family are called the *countable union* and the *countable intersection*, which we write as

$$\bigcup_{m=1}^{\infty} A_m := \{x \in X : x \in A_m \text{ for some } m \in \mathbb{N}\};$$

$$\bigcap_{m=1}^{\infty} A_m := \{x \in X : x \in A_m \text{ for all } m \in \mathbb{N}\}.$$

We will see such considerations play an important role in subsequent chapters. In the general case, the union and intersection of the indexed family $\mathscr{A}$ are defined respectively by

$$\bigcup_{\alpha \in I} A_\alpha = \{x \in X : x \in A_\alpha \text{ for some } \alpha \in I\};$$

$$\bigcap_{\alpha \in I} A_\alpha = \{x \in X : x \in A_\alpha \text{ for every } \alpha \in I\}.$$

Theorem 1.2 *Let X be a universal set, and $\{A_\alpha : \alpha \in I\}$ be an indexed family of subsets of X. Then the following generalization of de Morgan's Laws hold:*

$$\left(\bigcup_{\alpha \in I} A_\alpha\right)^c = \bigcap_{\alpha \in I} A_\alpha^c \quad \text{and} \quad \left(\bigcap_{\alpha \in I} A_\alpha\right)^c = \bigcup_{\alpha \in I} A_\alpha^c.$$

Proof Left for the reader as an easy exercise. $\qquad\qquad\qquad\qquad\qquad\qquad$ □

The concept of the *Cartesian product* (or simply the *product*) of an indexed family of sets $\{X_\alpha : \alpha \in I\}$ is yet another fundamental set operation, which is useful in various important constructions in modern mathematics. We also need this concept in some interesting applications in science and engineering. The term *Cartesian* used here refers to the connection it has with the similar terminology applied in usual *coordinate geometry* of the plane or space. For example, a point in the plane is given by an *ordered pair* (x, y) and a point in the space is given by a *triplet* (x, y, z), for some $x, y, z \in \mathbb{R}$.

For a general discussion, as before, we start with the case when the index set I is finite, say with two elements. For a simpler motivation, suppose S is a set of students in an Institution and C is the set of courses on offer for some undergraduate science programme. The concept of the product of sets in this case is related to the question: *How can we use sets to write the statement such as the student s has opted for the course c* ? The simple answer is: Use the product of the sets S and C, *in that order*. That is, the set theoretic equivalent of the given statement is the expression $(s, c) \in S \times C$, where the latter set is defined as follows.

Definition 1.3 The *Cartesian product* of two sets A and B, denoted by $A \times B$, is the set given by
$$A \times B := \{(a, b) : a \in A \text{ and } b \in B\},$$

where $(a, b) = \{\{a\}, \{a, b\}\}$ is called an *ordered pair*, according to the set theoretic definition given by *Kazimierz Kuratowski* in 1921. We say a is the *first coordinate* of the ordered pair (a, b), and b is the *second coordinate*.

Next, let $I = \mathbb{N}_n := \{1, 2 \ldots, n\}$. The Cartesian product of nonempty sets $A_1, \ldots, A_n$, denoted by
$$\prod_{k=1}^n A_k = A_1 \times \cdots \times A_n,$$

is given by the set

$$\prod_{k=1}^n A_k := \left\{(a_1, \ldots, a_n) : a_k \in A_k, \text{ for } 1 \le k \le n\right\}.$$

The element $(a_1, \ldots, a_n)$ of this set is called an *n-tuple*, with a_k being its *k*th coordinate.

Example 1.1 (*Transportation Schedule*) For a simple practical application, let B be the set of buses in a *state terminus*, and C be the set of cities around the state. Then

$(b, c) \in B \times C$ implies that the destination of the bus b is the city c. Clearly, for a countrywide bus *transportation schedule*, we need to extend the notion of the product of sets to a finite collection of sets. A similar type of applications are important in some other more complex situations, including computer science.

Notice that, by definition, two n-tuples $(a_1, \ldots, a_n)$ and $(b_1, \ldots, b_n)$ are *equal* if and only if $a_k = b_k$ for each $k = 1, \ldots, n$. In particular, when $A_k = A$ for all k, we write

$$A^n = A \times \cdots \times A \quad (n\text{-copies}).$$

So, $\mathbb{R} \times \mathbb{R} = \mathbb{R}^2$ is the Cartesian plane, $\mathbb{R} \times \mathbb{R} \times \mathbb{R} = \mathbb{R}^3$ is the 3-space, and so on. The case of the Cartesian product of an arbitrary family of nonempty sets is more technical, which we will discuss at the end of the next section.

1.2 Relation and Ordering

Most applications of modern mathematics use diverse forms of *relations among sets*, possibly with a discrete or a continuous structure. We introduce here some simple notions such as *equivalence* and *order*, where both are based on the concept of *binary relations*. Extensions of these concepts in terms of *n- ary relations* have numerous applications in computer science.

Definition 1.4 Let X and Y be two nonempty sets. A *binary relation* from X to Y is a subset R of the product set $X \times Y$, which we usually write as $R : X \to Y$. The sets

$$\text{Dom}(R) := \{x \in X \mid (x, y) \in R, \text{for some } y \in Y\};$$
$$\text{Ran}(R) := \{y \in Y \mid (x, y) \in R, \text{for some } x \in X\},$$

are respectively called the *domain* and *range* of the relation R. We say a relation $R : X \to Y$ is *homogeneous* when $X = Y$.

By symmetry, a relation $S : Y \to X$ is any subset S of the set $Y \times X$. For $x \in X$ and $y \in Y$, we also write $x \sim y$ if x is *related to*, y i.e., $(x, y) \in R$, and $y \sim x$ if y is *related to* x, i.e., $(y, x) \in S$. The *inverse relation* $R^{-1} : Y \to X$ of a relation R is given by

$$R^{-1} := \{(y, x) \in Y \times X \mid (x, y) \in R\}.$$

The two functions $n \mapsto \pm n$ on the set of integers $\mathbb{Z}$ are given respectively by the relations

$$I = \{(n, n) \mid n \in \mathbb{Z}\} \quad \text{and} \quad R = \{(n, -n) \mid n \in \mathbb{Z}\}.$$

More generally, an *abstract definition* of a function between two sets is as given below. As said earlier, Dirichlet's interpretation of a function was not sufficiently adequate to meet the technical requirements of all applications, especially in computer science.

Definition 1.5 Let X and Y be two nonempty sets. A relation $f \subseteq X \times Y$ is called a *function* if $\text{Dom}(f) = X$ and, for any $x \in X$, $y_1, y_2 \in Y$,

$$(x, y_1), \ (x, y_2) \in f \ \Rightarrow \ y_1 = y_2.$$

Notice that all set theoretic operations apply to relations $R : X \to Y$ because each is a subset of a product set $X \times Y$. So, for relations $R_1, R_2 \subseteq X \times Y$, we have

$$R_1 \cup R_2 := \{(x, y) \in X \times Y \mid (x, y) \in R_1 \text{ or } (x, y) \in R_2\};$$
$$R_1 \cap R_2 := \{(x, y) \in X \times Y \mid (x, y) \in R_1 \text{ and } (x, y) \in R_2\};$$
$$R_1 \setminus R_2 := \{(x, y) \in X \times Y \mid (x, y) \in R_1 \text{ but } (x, y) \notin R_2\};$$
$$R_2 \setminus R_1 := \{(x, y) \in X \times Y \mid (x, y) \in R_2 \text{ but } (x, y) \notin R_1\};$$
$$R^c := \{(x, y) \in X \times Y \mid (x, y) \notin R\}.$$

Further, given two relations $R : X \to Y$ and $S : Y \to Z$, the *composition* of R and S is a relation $X \to Z$, which is written as $S \circ R$:

$$S \circ R := \{(x, z) \in X \times Z \mid (x, y) \in R \text{ and } (y, z) \in S, \text{ for some } y \in Y\}.$$

In particular, if R is a homogeneous relation on a set X, we obtain many new relations by applying set operations, composition, inverse operation, *in every possible combinations*. The homogeneous relation

$$I = \{(x, x) : x \in X\}$$

is called the *identity relation* (or the *diagonal relation*) on the set X. Whenever needed, we write I_X to emphasise X. Clearly, $I^{-1} = I$, and $R \circ I_Y = R$, for any relation $R : X \to Y$. Notice that

$$R^{-1} = \{(b, a) \in X \times X \mid (a, b) \in R\};$$
$$R^2 = R \circ R = \{(a, b) \in X \times X \mid (a, c), (c, b) \in R, \text{ for some } c \in X\}.$$

If $X = \{x_1, x_2, \ldots, x_n\}$ is a finite set, a computer understands a homogeneous relation $R : X \to X$ in terms of its *relation matrix*

$$M_R = (m_{ij}), \quad \text{where} \quad m_{ij} := \begin{cases} 1, & \text{if } (x_i, x_j) \in R \\ 0, & \text{otherwise} \end{cases}.$$

Also, in this case, we can *see* the relation $R : X \to X$ as the *digraph* $\Gamma_R = (V, E)$, where $V = X$ are the vertices and $E = R$ are the *directed edges*. These simple notions have numerous applications in computer science. Another two most important notions related to homogeneous relations are the concepts of *partition* and *ordering* of a set. We start with the following.

Definition 1.6 Let $\sim$ be relation on a set X. We say that $\sim$ is

1. *reflexive* if $a \sim a$, for all $a \in X$;
2. *symmetric* if $a \sim b \Rightarrow b \sim a$, for all $a, b \in X$;
3. *transitive* if $a \sim b$ and $b \sim c \Rightarrow a \sim c$, for all $a, b, c \in X$.

Notice that if we write $\sim$ in ordered pair notation as R, then (1) says $I \subseteq R$; (2) says $R^{-1} = R$; and (3) says $R^2 = R \circ R \subseteq R$. The following characterises the notion of *partitioning of a set*.

Definition 1.7 A relation $\sim$ on a nonempty set X is called an *equivalence relation* if it is reflexive, symmetric, and transitive. For $a \in X$, the *equivalence class* of a is the set

$$[a] := \{b \in X : b \sim a\}.$$

If $\sim$ is an equivalence relation on a nonempty set X, we write

$$\frac{X}{\sim} = \big\{ [x] : x \in X \big\}$$

for the set of all equivalence class. This is called the *quotient set* of X given by the equivalence relation $\sim$. Notice that there is a *natural* surjective function

$$\varphi : X \to \frac{X}{\sim} \quad \text{given by} \quad \varphi(x) = [x], \quad \text{for } x \in X.$$

Further, if $f : X \to Y$ is a function, then the relation $\sim'$ on X given by

$$x_1 \sim' x_2 \quad \Leftrightarrow \quad f(x_1) = f(x_2)$$

is an equivalence relation on X such that there is a unique injective function

$$\widehat{f} : \frac{X}{\sim'} \to Y, \quad \text{with} \quad f = \widehat{f} \circ \varphi.$$

Notice that, for any $x \in X$, $\widehat{f}([x])$ is the common value of the function f it takes elements of the equivalence class $[x]$. This simple construction is very helpful in number of important applications.

For example, consider the collection X of all lines ℓ in $\mathbb{R}^2$ passing through the origin, and take $\ell_1 \sim \ell_2$ if ℓ_1 and ℓ_2 are parallel. Then the relation $\sim$ is reflexive, symmetric, and transitive. Hence, it defines an equivalence relation on the set X. Can you see what are the equivalence classes of $\sim$? Does the conclusion holds if we replace *parallel* by *perpendicular* in our definition? Why the relation

$$\{(x, x) \mid x \in \mathbb{R}\} \setminus \{(0, 0)\}$$

is not an equivalence relation on $\mathbb{R}$?

Example 1.2 For any fixed natural number $n > 1$, we define a relation $\sim$ on the set of integers $\mathbb{Z}$ as follows:
$$a \sim b \quad \Leftrightarrow \quad n \mid (b - a).$$

We also write $a \equiv b \,(mod \, n)$ if $a \sim b$, and say a is congruent to b modulo n. It follows easily that $\sim$ is an equivalence relation, and that there are n distinct equivalence class given by

$$[a] = \{\ldots, -2n + a, -n + a, 0, n + a, 2n + a, \ldots, \}, \quad \text{for} \;\; 0 \leq a < n.$$

The second assertion follows from the Euclid's algorithm. For example, taking $n = 4$, we have $a \sim b$ if $b - a = 4k$, for some integer k. That is, if $b = a \pm 4k$, for some $k \geq 0$. In particular, we see that

$$[0] = \{\ldots, -8, -4, 0, 4, 8, \ldots \}.$$

In this case, we usually write $\mathbb{Z}/n\mathbb{Z}$ or $\mathbb{Z}_n$ for the quotient set of n congruence classes. The set $\mathbb{Z}/n\mathbb{Z}$ has some interesting algebraic structures, especially so when n is prime.

By a *partition* of a nonempty set X we mean an indexed family of sets $\{A_\alpha \subset X \mid \alpha \in I\}$ such that

$$X = \bigcup_{\alpha \in I} A_\alpha \;\; \text{and} \;\; A_\alpha \neq A_\beta \; \text{for} \; \alpha \neq \beta \in I.$$

Suppose $\{A_\alpha \subset X \mid \alpha \in I\}$ is a partition of a set X. Clearly then the relation on X given by
$$a \sim b \quad \Leftrightarrow \quad a, b \in A_\alpha, \;\; \text{for some} \;\; \alpha \in I,$$

defines an equivalence relation on the set X, and the equivalence classes with respect to this relation are precisely the sets A_α, $\alpha \in I$. The next theorem proves that the converse holds.

Theorem 1.3 *Every equivalence relation $\sim$ on a nonempty set defines a partition of the set X in terms of equivalence classes.*

Proof Let $\sim$ be an equivalence relation on a set X. By reflexive property, we have $x \in [x]$, for any $x \in X$. We next show that, for any $x, y \in X$,

$$[x] = [y] \;\; \text{or} \;\; [x] \cap [y] = \emptyset.$$

For, suppose $z \in [x] \cap [y]$. Then, $x \sim z$ and $z \sim y$ implies $x \sim y$, by transitivity of $\sim$. Therefore, $[x] = [y]$. It thus follows that the equivalence relation $\sim$ gives a partition of the set X. $\square$

Ordering

An *order* on a nonempty set X is a type of relation that is defined in terms of a *partial order* on the set X. We introduce first some terminologies. We say a relation $\sim$ on a set X is *antisymmetric* if

$$x \sim y \text{ and } y \sim x \Rightarrow x = y.$$

Notice that the identity relation I on the set X is both symmetric and antisymmetric relation. On the other hand, the relation on the set $\mathbb{N}$ given by

$$n \sim m \Leftrightarrow n \mid m$$

is antisymmetric. Notice that this relation is also reflexive and transitive.

Definition 1.8 Let X be a nonempty set. A reflexive and transitive relation $\sim$ on X is called a *pre-order*, and an antisymmetric pre-order $\sim$ on X is called a *partial order*. We usually write $\leq$ for a partial order $\sim$ on a set X, and the pair $(X, \leq)$ is called a *partial ordered set* (or simply a *poset*).

Let $(X, \leq)$ be a poset. We say two element $x, y \in X$ are *comparable* if $x \leq y$ or $y \leq x$; otherwise, we say x and y are *incomparable* with respect to the partial order $\leq$. The term partial used in above definition refers to the fact that not every pair of elements in a poset $(X, \leq)$ are comparable. More generally, we say a set C in a poset $(X, \leq)$ is a *chain* if every pair of elements in C are comparable.

Definition 1.9 A poset $(X, \leq)$ is said to be *linearly ordered* (or *totally ordered*) if every pair of elements in X are comparable with respect to the partial order $\leq$. In this case, we say $\leq$ is a *linear order* (or a *total order*) on the set X.

Example 1.3 The power set $X = \mathscr{P}(A)$ of a nonempty set A is a *poset* with respect to *inclusion relation* $\subseteq$. That is, for $S, T \in X$, we take $S \sim T$ if $S \subseteq T$. Since $S \subseteq S \Rightarrow S \sim S$ for any $S \in X$, the relation is reflexive. Also, $S \sim T$ and $T \sim S$ implies $S = T$. Thus, the relation is antisymmetric. Finally, since

$$S \subseteq T \text{ and } T \subseteq U \Rightarrow S \subseteq U,$$

it follows that $\sim$ is transitive. So, for $S, T \in X$, we may write $A \leq B$ if $A \subseteq B$. Clearly, not every pair of sets in X are *comparable*, but it contains many chains. Notice that replacing $\subseteq$ by $\supseteq$ would only reverse the ordering.

The following worked example describes the construction of an interesting linearly ordered set. It is very useful in computer science. For example, it applies in sorting the character data.

Example 1.4 Let $\mathscr{A}$ be a finite set, with elements $a_1, \ldots, a_n$, which we may call *alphabets*. By a *string* of length m over the *set of alphabets* $\mathscr{A}$ we mean a juxtaposed expression of the form

$$x_1 x_2 \cdots x_m, \quad \text{where each} \quad x_i \in X.$$

That is, we form strings of different lengths by putting together alphabets a_i in any order, possibly with repetitions. For instance, the following two strings are of length 5 and 7, respectively:

$$a_1 a_1 a_4 a_3 a_1 = a_1^2 a_4 a_3 a_1, \quad a_5 a_2 a_2 a_2 a_3 a_1 a_1 = a_5 a_2^3 a_3 a_1^2.$$

We write $\mathscr{A}^m$ for the set of strings of length m defined over the alphabets $\mathscr{A}$. Clearly then we have

$$\mathscr{S} = \mathscr{A}^1 \cup \mathscr{A}^2 \cup \cdots \cup \mathscr{A}^n \cup \cdots \tag{1.2.1}$$

is the collection of strings of all possible lengths defined over $\mathscr{A}$. Suppose there is a natural linear order $<$ defined on the set $\mathscr{A}$. We extend this linear order to the collection $\mathscr{S}$ by ordering two strings *lexicographically*. That is, in the *dictionary order*. For, if

$$s = x_1 x_2 \cdots x_p \quad \text{and} \quad t = y_1 y_2 \cdots y_q$$

are two strings of lengths p and q, then we say $s < t$ if

1. $x_1 < y_1$ in $(\mathscr{A}, <)$; or
2. $x_1 = y_1, x_2 = y_2, \ldots, x_k = y_k$, but $x_{k+1} < y_{k+1}$ in $(\mathscr{A}, <)$, for some $1 \le k < r = min(p, q)$.

It is then easy to verify that $<$ defines a linear order on $\mathscr{S}$. In algebraic terms, $\mathscr{S}$ is called a *free monoid* on the set $\mathscr{A}$, where binary operation defined by *juxtaposition* and *empty string* (of length 0) correspond to the identity element.

To understand the practical aspect of the above construction, recall that the human gene is a finite string of four *alphabets* that are actually initial letters of certain amino acids. Also, certain programming languages are written using strings based upon an unlimited number of alphabets (characters), with a limit on the number of characters actually used to determine if the two words are the same or different, as checked by a *compiler*. For example, in traditional C language, only the first eight characters of a string are checked by the compiler. So, if two words of any length agree in their first eight characters then the compiler treats the two same. This is where the concept of equivalence classes comes into play.

Definition 1.10 (*Order and Ordered Set*) An *order* on a nonempty set X is a transitive relation $<$ such that, for any $a, b \in X$, the following *trichotomy condition* holds:

$$a < b, \quad a = b, \quad \text{or} \quad b < a,$$

A set X with an order $<$ is called an *ordered set*.

When $a < b$, we say a is smaller than b or b is bigger than a. We write $a \leq b$ if either $a = b$ or $a < b$. We write $a > b$ when $b < a$ and $a \geq b$ if either $a = b$ or $a > b$. For example, the usual "less than" relation on the number sets $\mathbb{N}$, $\mathbb{Z}$, and $\mathbb{Q}$ is an order relation. Notice that the digraph of an order can never have a *closed loop*. Further, it follows easily that, for any *ordered set* $(X, <)$, we have

1. there is no $x \in X$ such that $x < x$; and
2. $x \neq y$ in $X \Rightarrow$ either $x \leq y$ or $y \leq x$.

In fact, it can be shown that for a transitive relation $<$ on a set X, (1) and (2) are equivalent to *trichotomy condition*. In general, a relation R on X is called *irreflexive* if it satisfies the condition (1). That is, $R \cap I = \emptyset$, where I is the identity relation on X. For example, on a collection of subsets of a set, the proper inclusion $\subset$ is an irreflexive relation (Example 1.3). Also, the usual order $<$ on the set of natural numbers $\mathbb{N}$ is an irreflexive relation. Notice that a relation that is not reflexive is not necessarily irreflexive, and vice versa. On the other hand, a relation R on X is called *connected* if it satisfies the condition (2).

Definition 1.11 Let $(X, <)$ be an ordered set, and $A \subset X$. An element $\alpha \in A$ is a *least element* of the set A if $\alpha \leq a$ for all $a \in A$. An element $\beta \in A$ is a *greatest element* of the set A if $a \leq \beta$ for all $a \in A$.

Example 1.5 The set $\mathbb{Z}$ has neither a greatest element nor a least element. The set $\mathbb{N}$ has 1 as its least element. It has no greatest element. The set $\{0, 1, 2, \ldots, 9\}$ has 0 as its least element and 9 as the greatest element. Also, the relation $<$ on $\mathbb{N}$ defined by $n < m \Leftrightarrow n|m$ is a partial order, and for any finite set $A = \{n_1, \ldots, n_k\} \subset \mathbb{N}$ we have

$$\alpha = gcd\{n_1, \ldots, n_k\} \quad \text{and} \quad \beta = lcm\{n_1, \ldots, n_k\}.$$

For any collection A of subsets of a set X, the elements α and β are respectively given by the *intersection* and *union* of the sets in the collection A.

Example 1.6 For any rational number $p > 1$, the set

$$A = \{x \in \mathbb{Q}^+ : x^2 < p\}$$

has no greatest element and the set

$$B = \{x \in \mathbb{Q}^+ : x^2 > p\}$$

has no least element. For a given $x \in \mathbb{Q}^+$, put $y = p(x + 1)/(x + p)$. Then $y \in \mathbb{Q}^+$, and

$$y - x = \frac{p(x+1)}{x+p} - x = \frac{p - x^2}{x+p};$$

$$(1.2.2)$$

$$y^2 - p = \frac{p(p-1)(x^2 - p)}{(x+p)^2}.$$

If $x \in A$, then $x^2 < p$ and hence it follows from the first (1.2.2) that $y > x$ and $y^2 < p$. Thus, $y \in A$ with $y > x$. Therefore, x is not a greatest element of A. Hence, no element of A can be a greatest element of A. Similarly, if $x \in B$, then $x^2 > p$ and hence it follows from the second (1.2.2) that $y < x$ and $y^2 > p$. Thus, $y \in B$ with $y < x$. Therefore, x is not a least element of B. Hence, no element of B can be a least element of B.

Definition 1.12 Let $(X, <)$ be an ordered set, and $A \subset X$. An element $u \in X$ is an *upper bound* of the set A if $a \le u$ for all $a \in A$. If a set $A \subset X$ has an upper bound u, we say that the subset A is *bounded above* (by u).

If A is bounded above, we may write

$$\sup A = \min\{u \in X : u \text{ is an upper bound of } A\},$$

called the *supremum* (or *least upper bound*, abbreviated as lub) of the set A. Notice that, by definition, $s = \sup A$ satisfies the property that, for each $\epsilon > 0$, there exists some element u of A such that $s - \epsilon \le u < s$.

Definition 1.13 Let $(X, <)$ be an ordered set, and $A \subset X$. An element $\ell \in X$ is a *lower bound* of the subset A if $\ell \le a$ for all $a \in A$. If a set $A \subset X$ has a lower bound ℓ, we say that the subset A is *bounded below* (by ℓ).

If A is bounded below, we may write

$$\inf A = \max\{\ell \in X : \ell \text{ is a lower bound of } A\},$$

called the *infimum* (or *greatest lower bound*, abbreviated as glb) of the set A. Notice that, by definition, $i = \inf A$ satisfies the property that, for each $\epsilon > 0$, there exists some element ℓ of A such that $i < \ell \le i + \epsilon$. Clearly, we always have $\inf A \le \sup A$, if the two exist.

Definition 1.14 (*LUB/GLB Property*) An ordered set X has the least *upper bound property* if every nonempty bounded above subset of X has a supremum in X. The ordered set X has the greatest *lower bound property* if every nonempty bounded below subset of X has an infimum in X.

Theorem 1.4 *If an ordered set has the least upper bound property, then it has the greatest lower bound property.*

Proof Let X be an ordered set and B a bounded below subset of X. Let A be the set of all lower bounds of B. Since B is bounded below, there is at least one lower bound and hence the set A is non-empty. We shall first show that sup A exists. Let $b \in B$ be an arbitrary element. Each element $a \in A$ is a lower bound of B and so $a \leq b$ for each $a \in A$. Thus, A is bounded above (by each element in B). Since A is a nonempty bounded above subset of A, by the least upper bound property of X, it follows that $s := \sup A$ exists.

We complete the proof by showing $s = \inf B$. Since $s = \sup A$, the element s is an upper bound of A and if $a < s$, then a is not an upper bound of A. Since every element of B is an upper bound of A, it follows that $a \notin B$ if $a < s$. Therefore, $s \leq a$ for all $a \in B$. Thus, s is a lower bound of B. Since $s = \sup A$, $a \leq s$ for all $a \in A$. If $a > s$, then $a \notin A$ and therefore a is not a lower bound of B. Thus, $s = \inf B$. Thus, every bounded below set has an infimum. □

Definition 1.15 Let $(X, <)$ be an ordered set, and $A \subset X$. We say A is a *bounded set* if it is both bounded above and bounded below.

1.3 Functions

In abstract terms, as defined in Definition 1.5, a *function* $f : X \to Y$ is a special type of a *relation* on the set $X \times Y$, denoted by $f : X \to Y$, which we write as

$$f = \{(x, y) \in X \times Y : y = f(x), \text{ for } x \in X\}.$$

We say X is the *domain* of the function f, and the set Y is called the *codomain* of f. We usually write $X := \mathrm{Dom}(f)$. In some situation, we also call an element $x \in \mathrm{Dom}(f)$ as an *argument* of the function f. The set $\mathrm{Ran}(f)$ given by

$$\mathrm{Ran}(f) := \{y = f(x) \in Y : x \in X\}$$

is called the *range* of the function f, and the element $y = f(x) \in Y$ is called the *image* of the element $x \in X$ (under the function f). For any fixed $y \in Y$,

$$f^{-1}(y) := \{x \in X : f(x) = y\}$$

is called the set of *preimages* of the element $y \in Y$. Notice that it is possible that $f^{-1}(y) = \emptyset$, for some $y \in Y$. The notation f^{-1} used later to write the *inverse of a function* f corresponds to the situation when, for every $y \in Y$, $f^{-1}(y)$ contains exactly one element of X. Further, the set

$$\Gamma_f := \{(x, f(x)) : x \in X\} \subset X \times Y$$

is called the *graph* of the function f. Finally, we say two functions $f : X \to Y$ and $g : U \to V$ are *equal* if $X = U$, $Y = V$, and

$$f(x) = g(x), \quad \text{for all} \ \ x \in X (= U).$$

Since a function $f : X \to Y$ in general is *a rule* that associates to every element of X a unique element of Y, we can define at least as many functions from X to Y as there are number of elements in Y. For, let $b \in Y$ be any fixed element, and take $C_b(x) = b$, for each $x \in X$. Then, $C_b : X \to Y$ is called a *constant function* at b defined on the set X. Notice that, for any $a \in \mathbb{R}$, the graph of a constant function $C_a : \mathbb{R} \to \mathbb{R}$ is the line $y = a$. In particular, the graph of the constant function C_0 is the x-axis.

The most basic types of functions of geometric significance are known as the *polynomial functions*. For example, if the independent variable x takes values in the set $\mathbb{R}$, the graph of a function of the form $y = f(x) = mx + c$ is a *line*; the graph of a function of the form $y = f(x) = ax^2 + bx + c$ is a *parabola*; and so on.

Definition 1.16 (*Polynomial Functions*) A *polynomial* (in a variable x) is an expression of the form
$$p(x) = a_0 x^n + a_1 x^{n-1} + \cdots + a_n,$$

where arbitrary constants $a_0, a_1, \ldots, a_n$ are called *coefficients* of $p(x)$. If $a_0 \neq 0$, we say that the polynomial $p(x)$ is of *degree n*.

When the coefficients $a_k \in \mathbb{Z}$, a polynomial $p(x)$ is called an *integer polynomial*, and we write $\mathbb{Z}[x]$ for the set of integer polynomials of any degree. A similar meaning stands for the set $\mathbb{Q}[x]$ of *rational polynomials*, and for the set $\mathbb{R}[x]$ of *real polynomials*.

The two related functions given in the next definition use Archimedean property of $\mathbb{R}$ (Remark 2.1). See also Exercise 2.30.

Definition 1.17 An integer-valued function $\lfloor \ \rfloor : \mathbb{R} \to \mathbb{Z}$ given by

$$\lfloor x \rfloor = n, \quad \text{if} \ \ x \geq n, \quad x \in \mathbb{R},$$

is called the *floor function*, and $\lceil \ \rceil : \mathbb{R} \to \mathbb{Z}$ given by

$$\lceil x \rceil = n, \quad \text{where} \ \ x \leq n, \quad x \in \mathbb{R},$$

is called the *ceiling function*. The floor function $\lfloor \ \rfloor$ is also known as the *greatest integer function*.

The graph of each one of these two functions is like a *stair* (see Fig. 1.1). So, such types of functions are also known as the *stair functions*. Any assertion concerning the floor functions $\lfloor \ \rfloor$ is dealt with by taking $x = n + \epsilon$, where $n = \lfloor x \rfloor \in \mathbb{Z}$ and $0 \leq \epsilon < 1$. Similarly, while dealing with any assertion concerning the *ceiling functions*, we take $x = n - \epsilon$, where $n = \lceil x \rceil \in \mathbb{Z}$ and $0 \leq \epsilon < 1$. For example, we can apply these arguments to prove some interesting properties of the *stair functions* such as given in the next theorem. These relations are used very extensively in computer science.

Fig. 1.1 Graph of the
greatest integer function
$f(x) = \lfloor x \rfloor$

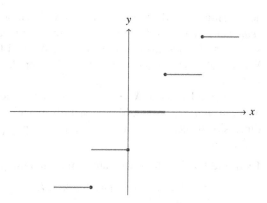

Theorem 1.5 *For any $x \in \mathbb{R}$,*

1. $\lfloor x \rfloor = n \Leftrightarrow n \leq x < n+1$; $\lceil x \rceil = n \Leftrightarrow n-1 < x \leq n$.
2. $\lfloor x \rfloor = n \Leftrightarrow x-1 < n \leq x$; $\lceil x \rceil = n \Leftrightarrow x \leq n < x+1$.
3. $x-1 < \lfloor x \rfloor \leq x \leq \lceil x \rceil < x+1$.
4. $\lfloor -x \rfloor = -\lceil x \rceil$ and $\lceil -x \rceil = -\lfloor x \rfloor$.
5. $\lfloor x+n \rfloor = \lfloor x \rfloor + n$ and $\lceil x+n \rceil = \lceil x \rceil + n$.

Proof Left to the reader as an exercise. $\qquad\qquad\qquad\qquad\qquad\qquad\square$

Next, we introduce an important class of functions given as an ordered list of elements of a nonempty set X, called *sequences*, where the set $\mathbb{N}$ is used to order the elements. The *strings* in finite number of symbols, as introduced in Example 1.4, are special types of sequences, which have applications in computer science and also in some other fields such as *genomics*. More generally, we come across *sequences* of many different types of mathematical objects such as numbers, vectors, matrices, functions, and even sets. In most cases, we use sequences as a tool to *approximate the reality*, by using the concept of *convergence*[4] as the central theme of the analysis.

Definition 1.18 (*Sequences*) Let X be a nonempty set. A function $s : \mathbb{N} \to X$ is called a **sequence** of elements of X. So, when

$$s(1) = x_1, \quad s(2) = x_2, \quad s(3) = x_3, \ldots$$

we write the *sequence s* simply as $\langle x_n \rangle$. The element $x_n = s(n)$ is called the *n*-term of the sequence $\langle x_n \rangle$.

Notice that two sequences $x = \langle x_n \rangle$ and $y = \langle y_n \rangle$ are *equal* if and only if $x = y$ as functions defined on $\mathbb{N}$. That is, $x_n = y_n$, for all $n \in \mathbb{N}$. We will also come

---

[4] It is a long journey to fully understand the *true meaning* of this phrase. In this book, we will try to elaborate it while playing around with the real numbers (Chap. 3), *n*-dimensional vectors (Chap. 6), and certain classes of real-valued functions defined on an interval (Chaps. 8 and 9).

across situations when, over some sets X, it is more appropriate to consider a *casual sequence* of the form $\langle x_0, x_1, x_2, \dots \rangle$. In general, a sequence in a set X is invariably seen as an infinite list of elements of X ordered by the set of natural numbers $\mathbb{N}$. We can talk about sequences of many different kinds of objects by taking the set X suitably.

In particular, when $X = \mathbb{R}$, we have the sequences of real numbers, which we also call *real sequences*. More generally, when $X = \mathbb{R}^2$, we obtain sequences of 2-dimensional vectors; when $X = \mathbb{R}^3$, we obtain sequences of 3-dimensional vectors; and so on.

Example 1.7 The following are some simple sequences in $\mathbb{R}$:

1. $s_n = 2n + 1$ gives the sequence $\langle 3, 5, 7, \dots \rangle$;
2. $s_n = (-1)^n \dfrac{n^2 - 1}{n}$ gives the sequence $\langle 0, -3/2, 8/3, \dots \rangle$;
3. $s_n = \begin{cases} \dfrac{1-n}{2}, & \text{if } n \text{ is odd} \\ \dfrac{n}{2}, & \text{if } n \text{ is even} \end{cases}$ gives the sequence $\langle 0, 1, -1, 2, -2, \dots \rangle$;
4. $s_n = \cos n\pi = (-1)^n$ gives the sequence $\langle -1, 1, -1, \dots \rangle$;
5. $s_n = \ln n$.

Notice that the sequence given by the function defined in (3) gives a *listing* of the set $\mathbb{Z}$ of integers as

$$0, 1, -1, 2, -2, \dots .$$

Also, the one given by the function defined in (4) picks the *discrete values* of the function $f(x) = \cos(\pi x)$, $x \in \mathbb{R}$, at the points $n = 1, 2, 3, \dots$. In certain applications, such a process is called *sampling*.

Next, we introduce some functions defined in terms of an *infinite series* of real numbers. In general, if X is an *additive set*,[5] then a *formal sum* of the form

$$\sum_{n=1}^{\infty} x_n = x_1 + x_2 + \cdots + x_n + \cdots \tag{1.3.1}$$

is called an **infinite series** over the set X. For example,

$$1 + 1 + \frac{1}{1 \cdot 2} + \frac{1}{1 \cdot 2 \cdot 3} + \frac{1}{1 \cdot 2 \cdot 3 \cdot 4} + \cdots$$

is an important infinite series of real numbers (see (1.5.3)). The infinite series expression for the function $f(x) = (1 + x)^q$, $q \in \mathbb{Q}$, with $|x| < 1$, given by

$$(1 + x)^q = 1 + \frac{q}{1}x + \frac{q(q-1)}{1 \cdot 2}x^2 + \frac{q(q-1)(q-2)}{1 \cdot 2 \cdot 3}x^3 + \cdots \tag{1.3.2}$$

---

[5] We say a nonempty set X is **additive** if, for each pair $x, y \in X$, we have $x + y \in X$, where $+$ is defined in some appropriate sense.

was first obtained by Sir Issac Newton, while extending the binomial theorem (Theorem 1.19) for fractional exponents.

We usually see an *infinite series* as the sum of terms of a sequence $\langle x_n \rangle$ over an additive set X. So, x_n is called the *nth term* of the infinite series. In general, to make sense out of such type of infinite sums, we use the sequence $\langle s_n \rangle$ given by

$$s_n = x_1 + x_2 + \cdots + x_n,$$

which is called the sequence of *partial sums* of the infinite series (1.3.1). To deal with the complexities involved with the *summing problems* associated with an infinite series, we need to analyse its sequence of partial sums $\langle s_n \rangle$. Notice that we can write the *n*th term of the infinite series (1.3.1) as

$$x_n = s_{n+1} - s_n, \qquad n \geq 1.$$

A large part of mathematics involves sequences and infinite series of various types of *numbers, vectors, matrices, functions, sets*, etc. For example, if $I \subseteq \mathbb{R}$ is an interval, and

$$X = \mathbb{R}^I = \{f : f : I \to \mathbb{R}\}, \quad I \subseteq \mathbb{R},$$

then a function $\sigma : \mathbb{N} \to X$ defines a sequence $\sigma = \langle f_n \rangle$ of real-valued functions defined on the interval I, and we also need to understand the function given by sum $\sum_{x \in I} f_n(x)$, if it exists. In this text, a good amount of time is spent discussing the possibility to sum an infinite series, and also ways to obtain the same in terms of the *limit* of the sequence $\langle s_n \rangle$, if it exists. We will consider sequences of real numbers in Chap. 3; of functions in Chap. 8; and sequences of some special kind of subsets of the set $\mathbb{R}$ in Chap. 9.

The function given in the next definition is not a formula-based function.

Definition 1.19 (*Characteristic Function*) Let X be a nonempty set and $A \subseteq X$. The *characteristic function* of the set A, denoted by χ_A, is a function $\chi_A : X \to \{0, 1\}$ given by

$$\chi_A(x) = \begin{cases} 1, & \text{if } x \in A \\ 0, & \text{if } x \in X \setminus A \end{cases}.$$

It is also known as the *indicator function* of the set A.

This type of function plays an important role in Lebesgue's theory of integration (see Chap. 9). Also, the relations given in the next theorem can be used to prove several set theoretic identities.

Theorem 1.6 *Let X be a nonempty set and $A, B \subseteq X$. The following relations hold:*

1. $\chi_{A \cap B} = \chi_A \cdot \chi_B$;
2. $\chi_{A \cup B} = \chi_A + \chi_B - \chi_A \cdot \chi_B$;
3. $\chi_{A^c} = 1 - \chi_A$;

4. $\chi_{A \oplus B} = \chi_A + \chi_B - 2\chi_A \cdot \chi_B$;
5. $a\chi_A + b\chi_B = a\chi_{A \setminus B} + (a + b)\chi_{A \cup B} + b\chi_{B \setminus A}$, for any real numbers a and b.

Proof Left to the reader as an exercise.   □

We are sure, as most would agree, it is not at all convenient for a beginner to grasp the *abstraction* of Definition 1.5 given earlier to introduce the function concept. Considering the fact that clarity about the *definition of a function* is fundamental to everything we do in mathematics, and also in most allied subjects such as physics and engineering, we decided to narrate the whole story about the evolution of the concept in Sect. 1.5.

Definition 1.20 Let $f : X \to Y$ be a function.

1. We say f is an *injective function* (or a *one-one function*) if every pair of distinct elements in X gets mapped to distinct elements in Y. That is, for any $x_1, x_2 \in X$,

$$x_1 \neq x_2 \quad \Rightarrow \quad f(x_1) \neq f(x_2).$$

2. We say f is a *surjective function* (or an *onto function*) if every element of Y has a preimage in X. Or, equivalently, if $\text{Ran}(f) = Y$. A function that is not an onto function is called an *into function*.
3. We say f is a *bijective function* (or simply a *bijection*) if f is both injective and surjective. In this case, we also say that f defines an one-to-one correspondence between the sets X and Y.

Notice that the condition in (1) is equivalent to the condition

$$f(x_1) = f(x_2), \quad \text{for} \quad x_1, x_2 \in X, \quad \Rightarrow \quad x_1 = x_2.$$

Further, not every injective function is surjective. For example, if A and B are two finite sets, with the set B having at least one element more than the set A, then an injective function $f : A \to B$ cannot be an surjective function (Why?).

Example 1.8 Let $X = \mathbb{N}$ and Y be the set of positive even numbers. That is, $Y = \{2n \mid n \in \mathbb{N}\}$. Let the function $\beta : \mathbb{N} \to Y$ be given by $\beta(n) = 2n$. Then, β is a bijection. For, to see that β is injective, let $\beta(n) = \beta(m)$, for some $n, m \in \mathbb{N}$. Then, by definition, we have $2n = 2m$, which implies $n = m$. So, β is an injective function. Next, to conclude, we show that β is surjective. For, let $y = 2n_0 \in Y$, for some $n_0 \in \mathbb{N}$. Clearly then, $\beta(n_0) = 2n_0 = y$. That is, β is surjective. Therefore, $\beta : \mathbb{N} \to Y$ is a bijective function, as asserted.

Example 1.9 Consider the function $f : \mathbb{N} \to \mathbb{Z}$ given by

$$f(n) = \begin{cases} -\dfrac{(n-1)}{2}, & \text{for } n \text{ is odd} \\ \dfrac{n}{2}, & \text{for } n \text{ is even} \end{cases}.$$

Then, f is a bijective function. We first show that f is surjective. For, let $m \in \mathbb{Z}$. Notice that if m is positive then $2m \in \mathbb{N}$, and so $f(2m) = m$, in this case. On the other hand, if m is zero or negative then $1 - 2m \in \mathbb{N}$, and so $f(1 - 2m) = m$ because $1 - 2m$ is an odd number. So, in this case also f is surjective. Next, to show that f is injective, note that

$$f(n) > 0 \text{ if } n \text{ is even; and } f(n) < 0 \text{ if } n \text{ is odd},$$

for each $1 \neq n \in \mathbb{N}$. Thus, if $f(r) = f(s) = a$, then s and r both must be 1 or both are even or odd, subjected to a is zero, positive, or negative, respectively. In case both s and r are even, we have

$$s = 2\left(\frac{s}{2}\right) = 2f(s) = 2f(r) = 2\left(\frac{r}{2}\right) = r.$$

And, if both s and r are odd, we have

$$s = (-2)\left[-\left(\frac{s-1}{2}\right)\right] + 1$$
$$= -2[f(s)] + 1$$
$$= -2[f(r)] + 1 = r.$$

Therefore, in each case, $f(r) = f(s) \Rightarrow r = s$. That is, f is an injective function. Hence, f is a bijective function, as asserted.

Let X be a nonempty set, and consider the set

$$2^X := \{f \mid f : X \to \{0, 1\}\}.$$

The next lemma shows that we can identify the class $\mathscr{P}(X)$ of all subsets of the set X with the set of functions 2^X.

Lemma 1.1 *Let X be a nonempty set. There is a bijective correspondence between $\mathscr{P}(X)$ and the set 2^X.*

Proof Clearly, for any $A \in \mathscr{P}(X)$, the characteristic function $\chi_A \in 2^X$. Conversely, given a $0 \neq f \in 2^X$, we have the set

$$A = \{x \in X : f(x) = 1\} \in \mathscr{P}(X).$$

Further, as $A^c = \{x \in X : f(x) = 0\}$, we have $\chi_\emptyset = 0$, the *zero function*. Also, $\operatorname{Ran}(\chi_A) = \{0, 1\}$ if and only if A is a proper subset of X. This completes the proof. □

Definition 1.21 Let $f : X \to Y$ and $g : Y \to Z$ be two functions. Then the function $g \circ f : X \to Z$ given by

$$(g \circ f)(x) = g\big(f(x)\big), \quad \text{for all } x \in X$$

is called the *composition* of the functions $f : X \to Y$ and $g : Y \to Z$.

For example, consider the function $f : \mathbb{N} \to \mathbb{Z}$ is given by $f(n) = n - 2$, and the function $g : \mathbb{Z} \to \mathbb{Z}$ is given by $g(n) = n^2$. Then their composition is given by

$$(g \circ f)(n) = g(f(n)) = g(n - 2) = (n - 2)^2 = n^2 - 4n + 4.$$

Notice that though the codomain of g is the set $\mathbb{Z}$, but as

$$\text{Ran}(g) = \{n^2 \mid n \in \mathbb{Z}\} \subset \mathbb{N},$$

we can define the composition $f \circ g$, where

$$(f \circ g)(n) = f(g(n)) = f(n^2) = n^2 - 2.$$

Here, $f \circ g \neq g \circ f$ due to both the following two reasons: (*a*) Their values are different; and (*b*) the domains of $f \circ g$ and $g \circ f$ are distinct. Notice that, in general, we need to have both types of equalities to ensure that the compositions of the two functions are the same.

Theorem 1.7 *Let $f : X \to Y$ and $g : Y \to Z$ be any two functions.*

1. *If both f and g are injective then so is the composition $g \circ f$.*
2. *If both f and g are surjective then so is the composition $g \circ f$.*
3. *If both f and g are bijections, then so is the composition $g \circ f$.*

Proof Left to the reader.   □

Let X be a nonempty set. A function $f : X \to X$ is also known as a *mapping* (or a *transformation*) of the set X. In many practical situations, we use class of mappings $f_t : X \to X$ indexed by a *time parameter* $t \geq 0$ to describe the time-dependent *dynamics* of the elements of X. We usually write $f_0 : X \to X$ for the *identity transformation* I_X so that $f_0(x) = x$, for all $x \in X$. In such cases, we say (X, f_t) defines a *dynamical system* on the set X, possibly with some additional structures. For example, we come across such systems wherein f_t represents *integral curves* of a differential equation. Such a geometric interpretation of functions $f_t : X \to X$ is also important in many other applications.

We consider here the following general question: *Given a function $f : X \to Y$, does there exists a function $g : Y \to X$ that will undo the* effect *of the function f?*

Definition 1.22 Let X and Y be two nonempty sets, and $f : X \to Y$ be a function. We say $g : Y \to X$ is an *inverse function* of f if

$$g \circ f = I_X \quad \text{and} \quad f \circ g = I_Y. \tag{1.3.3}$$

In this case, we write $g = f^{-1}$. We say a function $f : X \rightarrow Y$ is *invertible* if it has an inverse.

We remark that if a function $f : X \rightarrow Y$ is invertible, with $g : Y \rightarrow X$ as an inverse, then g is necessarily *unique*. For, suppose there are two functions $g_1, g_2 : Y \rightarrow X$ satisfying the conditions as in (1.3.3). That is,

$$f \circ g_1 = I_Y, \quad g_1 \circ f = I_X \quad \text{and} \quad f \circ g_2 = I_Y, \quad g_2 \circ f = I_X.$$

Then, we have

$$g_1 = g_1 \circ I_Y = g_1 \circ (f \circ g_2) = (g_1 \circ f) \circ g_2 = I_X \circ g_2 = g_2.$$

This proves our assertion.

Notice that the inverse of the function $f : \mathbb{N} \rightarrow \mathbb{Z}$ defined in Example 1.9 is the function $g : \mathbb{Z} \rightarrow \mathbb{N}$ defined by

$$g(n) = \begin{cases} 2n, & \text{if } n \geq 1; \\ -2n + 1, & \text{if } n \leq 0 \end{cases}.$$

In particular, it follows from (ii) of Exercise 1.13 that a function fails to be invertible when it is not injective or not surjective. The next theorem proves that, in general, a function is invertible if and only if it is bijective.

Theorem 1.8 *Let X and Y be two nonempty sets. A function $f : X \rightarrow Y$ has an inverse if and only if f is a bijective function.*

Proof First, let there exists a function $g : Y \rightarrow X$ such that $g \circ f = I_X$ and $f \circ g = I_Y$. It then follows that f is bijective (see (a) of Exercise 1.13). To prove the converse, suppose f is a bijective function. Then, for every $y \in Y$ there is a unique $x \in X$ such that $f(x) = y$. We define $g : Y \rightarrow X$ by $g(y) = x$. It follows from the definition that g satisfies the conditions as in (1.3.3). Hence, g is the inverse of the function f. $\square$

It follows easily that if $f : X \rightarrow Y$ is a bijective function then so is its inverse $f^{-1} : Y \rightarrow X$ (see (b) of Exercise 1.13). The next theorem proves that the set operations are more well behaved with respect to inverse functions.

Theorem 1.9 *Let $f : X \rightarrow Y$ be an invertible function. Then*

1. $f^{-1}(A^c) = [f^{-1}(A)]^c$; *for all $A \subseteq X$;*
2. $\bigcup_\alpha f^{-1}(A_\alpha) = f^{-1}\left[\bigcup_\alpha A_\alpha\right]$, *for every class $\{A_\alpha \subseteq X : \alpha \in I\}$;*
3. $\bigcap_\alpha f^{-1}(A_\alpha) = f^{-1}\left[\bigcap_\alpha A_\alpha\right]$, *for every class $\{A_\alpha \subseteq X : \alpha \in I\}$.*

Proof Left to the reader as an easy exercise. $\square$

We conclude our discussion here with a brief introduction to the *axiom of choice*, and some related concepts of ZF axiomatic set theory. In abbreviated form, the *axiom of choice* is usually written as AC. We start with an alternative description of Cartesian product of finite number of nonempty sets indexed by a set $I = \{1, \ldots, n\}$, say $A_1, \ldots, A_n$. Notice that, in this case, we can write

$$\prod_{k=1}^{n} A_k := \{f : I \to \cup_{k \in I} A_k : f(k) = a_k \in A_k, \text{ for } 1 \le k \le n\}.$$

More generally, given an arbitrary indexed family of nonempty sets $\mathscr{A} = \{A_\alpha : \alpha \in I\}$, we may define the *infinite product* of the sets A_α as

$$\prod_{\alpha \in I} A_\alpha := \{f : I \to \cup_{\alpha \in I} A_\alpha : f(\alpha) \in A_\alpha, \text{ for all } \alpha \in I\}.$$

The *axiom of choice* is the statement that $\prod_{\alpha \in I} A_\alpha \ne \emptyset$:

Axiom of Choice: *For any indexed family of nonempty sets* $\{A_\alpha : \alpha \in I\}$*, there exists a function* $f : I \to \bigcup_{\alpha \in I} A_\alpha$ *such that* $f(\alpha) \in A_\alpha$*, for each* $\alpha \in I$. Such an f is called a *choice function*.

The axiom says, no matter how many sets there may be in a family $\mathscr{A}$ of nonempty set A_α, it is always possible to *choose* one element from each A_α. In particular, if each A_α is a subset of some set X, then there exists a *choice function* $c : \mathscr{P}(X) \setminus \{\emptyset\} \to X$ such that $c(A) \in A$. Clearly, the converse holds.

In 1938, *Kurt Gòdel* (1906–1978) proved that AC is consistent with the ZF axioms of the set theory. And, in 1963, *Paul Cohen* (1934–2007) proved that AC is independent of the ZF axioms. Initially, *Ernest Zermelo* (1871–1953) formulated the *axiom of choice* in 1904 to prove the following important fact of axiomatic set theory. A partially ordered set is well ordered if every nonempty set has a least element.

Theorem 1.10 (Well-Ordering Theorem) *Every set can be well ordered.*

1.4 Cardinality

We introduce here the concept of *cardinality* of a set X, which we usually write as $|X|$ (or $\# X$). By definition, $|X|$ denotes the *size* of a set X in terms of the number of elements it contains, finite or infinite. The main point of discussion here is about how do we *compare* two sets on the basis of their cardinalities. Therefore, the following two questions define the focus of the section:

1. What is the cardinality of a set?
2. How to compare sizes of sets by using their cardinality?

To begin with, notice that the emptyset $\emptyset$ is the only set with $|\emptyset| = 0$. A *logical proof* of this fact uses the *axiom of extensionality* of ZFC [2]. Next, suppose X is a nonempty finite set. In this case, the size of X is determined by applying the *counting process* in the layman sense. That is, we can count the elements of X in succession as *one, two, three, ...*, and so on, assuming we have enough time to do so when the size of X is large. In mathematical terms, we say each $x \in X$ corresponds to a unique natural number $n = n(x)$, which is same as saying that there is an *injective* function $f : X \rightarrow \mathbb{N}$ such that

$$\text{Ran}(f) = \mathbb{N}_n := \{1, 2, 3, \ldots, n\}, \quad \text{for some} \quad n \geq 1.$$

Therefore, the elements of the set X can be written as a *finite list*, say $x_1, x_2, \ldots, x_n$, and we use numbers such as *five, fifty* or *two dozens*, etc., to specify its *cardinality*. For example, the set of *English alphabets* has the cardinality 26. More precisely, a set X has n elements if and only if there is a *bijective function* $f : X \rightarrow \mathbb{N}_n$, for some $n \in \mathbb{N}$.

Definition 1.23 We say X is a *finite set* if $X = \emptyset$ or there is a bijective function $f : A \rightarrow \mathbb{N}_n$, for some $n \in \mathbb{N}$.

Clearly then, given two nonempty finite sets X and Y, we have $|X| = |Y|$ if and only if there exists a bijective function $f : X \rightarrow Y$. For example, if $|X| = |Y| = n$, then both the sets are in bijective correspondence with the set $\mathbb{N}_n$. In general, if X and Y are two finite sets, and $f : X \rightarrow Y$ is an injective function that is *not* a surjective function, then we say Y has *more elements* than the set X. In this case, we write $|X| < |Y|$. We just concluded the introductory part of the story.

Now, suppose X is an *infinite set*. Recall that our aim is to introduce the concept of the *cardinality* of X, and then use the same to compare sizes of any two such sets. Here, at this stage, we must say that it was Georg Cantor's insight to use bijective functions between such types of sets to do the trick.

We first consider the case of infinite sets for which we can "*count the elements*" in some sense. For example, so is true for the set of natural numbers $\mathbb{N}$ because we can count its elements as 1, 2, 3, ... such that every $n \in \mathbb{N}$, howsoever large, is eventually counted. Therefore, in such cases, we can define the *cardinality* of an infinite set X by comparing its size with the set $\mathbb{N}$. This simple observation motivates the next definition.

Definition 1.24 We say X is a *denumerable set* (or a *countably infinite* set) if there exists a bijective function $: X \rightarrow \mathbb{N}$. That is, if it has the *cardinality* of the set $\mathbb{N}$. Further, we say X is a *countable set* if it is finite or a countably infinite set.

Example 1.10 The following are some simple denumerable sets.

1. The set $\mathbb{N}$ is countably infinite as the identity function $I : \mathbb{N} \rightarrow \mathbb{N}$ is a bijection.
2. The set $2\mathbb{N} = \{2, 4, 6, \ldots\}$ is countably infinite, because the function $f : \mathbb{N} \rightarrow 2\mathbb{N}$ defined by $f(n) = 2n$ is bijective.

3. Since the function $f : \mathbb{N} \to \mathbb{Z}$ defined in Example 1.9 is bijective, the set of integers $\mathbb{Z}$ is countably infinite.

Notice that Definition 1.24 in general provides a way to *count elements* of such type of infinite sets that have the cardinality of the set $\mathbb{N}$. For, if A is a *denumerable* set, and $f : \mathbb{N} \to A$ is a bijective function, then we can *enumerate* the elements of A as $a_1, a_2, a_3, \ldots$ That, we can write

$$A = \{a_n : a_n = f(n), \ n \in \mathbb{N}\},$$

where f is called an *enumeration function* of the set A. It justifies the terminology *enumerable* used occasionally for a countable set. For example, the enumeration $f : \mathbb{N} \to \mathbb{Z}$ as given in Example 1.9 *lists* the countably infinite set of integers $\mathbb{Z}$ as

$$0, \ 1, \ -1, \ 2, \ -2, \ 3, \ -3, \cdots.$$

In next few theorems, we prove some facts about countably infinite sets.

Theorem 1.11 *Every infinite subset of a countably infinite set is itself a countably infinite set.*

Proof Let S be a denumerable set, with an enumeration $f : \mathbb{N} \to S$ so that we may write

$$S = \{s_1, s_2, s_3, \ldots\}, \quad \text{where} \quad s_k = f(k), \ k \in \mathbb{N}.$$

Let $A \subset S$ be an infinite set so that it is a sub-collection of elements $s_1, s_2, s_3, \ldots$. Suppose n_1 is the *least* such that $s_{n_1} \in A$, with $s_{n_1} = s_j \in S$ for some $j \geq 1$. Having chosen $s_{n_1} \in A$, let n_2 be the *least* such that $s_{n_2} \in A$, with $s_{n_2} \neq s_{n_1}$, i.e., $s_{n_2} \in A \setminus \{s_{n_1}\}$. Continuing this process, for $k > 2$, let n_k be the *least* such that

$$s_{n_k} \in A \setminus \{s_{n_1}, s_{n_2}, \ldots, s_{n_{k-1}}\}.$$

Notice that, since A is infinite, it is possible to find infinite number of element $s_{n_k} \in A$ with desired property . That is,

$$A \setminus \{s_{n_1}, s_{n_2}, \ldots, s_{n_k}\} \neq \emptyset \quad \text{for all} \quad k \in \mathbb{N}.$$

So, we may write $A = \{s_{n_1}, s_{n_2}, \ldots, s_{n_k}, \ldots\}$. Then, the function $\alpha : \mathbb{N} \to A$ defined by $\alpha(k) = s_{n_k}$ is an enumeration. Clearly, α is a bijective function. Hence, A is denumerable. $\qquad\square$

The conditions stated in the next theorem are often used to determine the countability of a set.

Theorem 1.12 *Let A be a nonempty set. The following are equivalent:*

1. A is countable.

2. *There exists an injective function $f : A \to \mathbb{N}$.*
3. *There exists a surjective function $f : \mathbb{N} \to A$.*

In particular, every infinite set contains a denumerable set.

Proof The proof is completed by showing (1) $\Rightarrow$ (2) $\Rightarrow$ (3) $\Rightarrow$ (1). First, let A be a countable set. By definition, there is a bijective function $h : J \to A$, where $J = \mathbb{N}_k$ for some $k \in \mathbb{N}$, if A is finite; and $J = \mathbb{N}$, if A is denumerable. In both cases, $h^{-1} : A \to J$ is a bijective function. In particular, as $J \subseteq \mathbb{N}$, the function $h^{-1} : A \to \mathbb{N}$ is injective. This proves (1) $\Rightarrow$ (2).

Next, suppose there exists an injective function $f : A \to \mathbb{N}$ so that $f : A \to f(A)$ is a bijective function, with $f(A) \subseteq \mathbb{N}$. So, $f^{-1} : f(A) \to A$ is a bijective function. We shall use f^{-1} to define a function $g : \mathbb{N} \to A$ as follows: Fix an element $x \in A$, and define g by

$$g(n) = \begin{cases} f^{-1}(n), & \text{if } n \in f(A) \\ x, & \text{if } n \notin f(A) \end{cases}$$

It then follows easily that $g[f(A)] = f^{-1}[f(A)] = A$ and $g[\mathbb{N} \setminus f(A)] = \{x\}$. So, g is a surjective function. Therefore, (2) $\Rightarrow$ (3). Finally, suppose that there exists a surjective function $g : \mathbb{N} \to A$. Let the function $h : A \to \mathbb{N}$ be given by

$$h(s) = n, \quad \text{where } n \in \mathbb{N} \text{ is the least such that } g(n) = s.$$

Then, h is an injective function, so a bijection from A onto the set $h(A) \subseteq \mathbb{N}$. Since $\mathbb{N}$ is countable, so is $h(A)$, and hence $A \equiv h(A)$ is countable. $\qquad \square$

Corollary 1.1 *Every subset of a countable set is countable.*

Proof Let A be countable and $B \subset A$. If B is finite, then it is countable. If B is infinite, then Theorem 1.11 proves that it is countably infinite and hence countable. $\square$

Corollary 1.2 *The union of two countably infinite sets is countably infinite.*

Proof Let $A = \{a_1, a_2, a_3, \dots\}$ and $B = \{b_1, b_2, b_3, \dots\}$. If $A = B$, there is nothing to prove and so assume that $A \neq B$. First we consider the case when A and B are disjoint, that is, $A \cap B = \emptyset$. In this case, we write

$$A \cup B = \{a_1, b_1, a_2, b_2, a_3, b_3 \dots\}$$

and hence $A \cup B$ is countably infinite. If $B \setminus A$ is a finite set

$$\{b_{n_1}, b_{n_2}, \dots, b_{n_m}\},$$

then we have

$$A \cup B = \{b_{n_1}, b_{n_2}, \dots, b_{n_m}, a_1, a_2, a_3, \dots\}$$

Fig. 1.2 $\mathbb{N} \times \mathbb{N}$ is countable

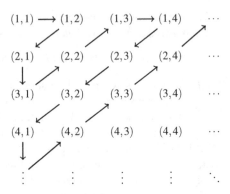

and hence $A \cup B$ is countably infinite. If $B \setminus A$ is infinite, then $A \cup B = A \cup (B \setminus A)$. Clearly, $A \cap (B \setminus A) = \emptyset$ as well as A and $B \setminus A$ are countably infinite and hence $A \cup B$ is countably infinite (by the first case). □

This result holds even for countable family of sets. We first use the following simple fact to prove that the product of two countable sets is a countable set.

Lemma 1.2 *The set $\mathbb{N} \times \mathbb{N}$ is countably infinite.*

Proof Arrange the elements of $\mathbb{N} \times \mathbb{N}$ as shown in Fig. 1.2, and enumerate all pairs by using *Cantor's diagonal process* as indicated by the arrows. Therefore, we can list elements of $\mathbb{N} \times \mathbb{N}$ as

$$(1, 1), \ (1, 2), \ (2, 1), \ (3, 1), \ (2, 2), \ (1, 3), \ (1, 4), \ (2, 3), \ \ldots .$$

Hence, $\mathbb{N} \times \mathbb{N}$ is countably infinite. □

Theorem 1.13 *If A_1 and A_2 are countably infinite, then $A_1 \times A_2$ is countably infinite. In particular, the set $\mathbb{Z} \times \mathbb{Z}$ is countably infinite.*

Proof Let $A = \langle a_1, a_2, a_3, \ldots \rangle$ and $B = \langle b_1, b_2, b_3, \ldots \rangle$ so that $A_1 \times A_2 = \{(a_i, b_j) \mid i, j \in \mathbb{N}\}$. Clearly, the mapping $f : A_1 \times A_2 \to \mathbb{N} \times \mathbb{N}$ given by $f((a_i, a_j)) = (i, j)$ is a bijection. Since $\mathbb{N} \times \mathbb{N}$ is countably infinite, by Lemma 1.2, there is a bijection g between $\mathbb{N} \times \mathbb{N}$ and $\mathbb{N}$. The composition $g \circ f$ is then a bijection between $A_1 \times A_2$ and $\mathbb{N}$. □

Theorem 1.14 *If, for each $n \in \mathbb{N}$, the set A_n is a countably infinite subset of a set X, then $\bigcup_{n=1}^{\infty} A_n$ is a countable infinite subset of X.*

Proof Since each A_n is countable, we can write $A_n = \langle x_{n1}, x_{n2}, \ldots \rangle$. The union $\bigcup_{n=1}^{\infty} A_n = \{x_{nm} \mid n, m \in \mathbb{Z}\}$. Assume that all these elements x_{mn} are distinct. Clearly, the function $f : \bigcup_{n=1}^{\infty} A_n \to \mathbb{Z} \times \mathbb{Z}$ given by $f(x_{mn}) = (m, n)$ is a bijection. Since $\mathbb{Z} \times \mathbb{Z}$ is countably infinite, by Theorem 1.13, there is a bijection $g : \mathbb{Z} \times \mathbb{Z} \to \mathbb{Z}$.

Fig. 1.3 Countable union of
countable sets is countable

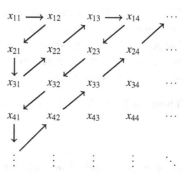

Then, $g \circ f : \bigcup_{n=1}^{\infty} A_n \to \mathbb{Z}$ is a bijection and hence $\bigcup_{n=1}^{\infty} A_n$ is countable. If some of the elements are equal, then $\bigcup_{n=1}^{\infty} A_n = \{x_{nm} | n, m \in \mathbb{Z}\}$ has a bijection with a subset of $\mathbb{Z} \times \mathbb{Z}$. Since any subset of $\mathbb{Z} \times \mathbb{Z}$ is countable, $\bigcup_{n=1}^{\infty} A_n = \{x_{nm} | n, m \in \mathbb{Z}\}$ is countable. Since A_1 is infinite, $\bigcup_{n=1}^{\infty} A_n = \{x_{nm} | n, m \in \mathbb{Z}\}$ is countably infinite.

Alternately, one can see that $\bigcup_{n=1}^{\infty} A_n$ is countable as its elements can be arranged in a sequence $\langle x_{11}, x_{12}, x_{21}, x_{31}, x_{22}, x_{13}, \ldots \rangle$. Elements are listed by following the arrows in Fig. 1.3. Some terms of the sequence may be missing if some of the elements x_{mn} are equal. $\square$

We can use the same argument to show that the countable union of finite sets is countable. However, in this latter case, the *row-sequences* are finite, which are treated as infinite sequences filling the remaining terms by a symbol $\square$. Of course, we can omit the symbols $\square$ while writing the sequence for the union $\bigcup_{n=1}^{\infty} A_n$.

We now consider the two questions raised earlier in the context of infinite sets. Notice that, for the second, we need a tool to determine how an infinite set is bigger (or has more elements) than some other similar type of a set. However, in this case, the situation is very different from what we found earlier about the finite sets. For example, think for a while about the following three simple questions:

1. Are there more even integers than odd integers?
2. Are there more even integers than the natural numbers?
3. Are there more rational numbers than negative integers?

As the function $n \mapsto 2n : \mathbb{N} \to 2\mathbb{N}$ defines a bijective function, the answer to the question in (2) above is NO. Notice that, however, the set $2\mathbb{N}$ is *half the size* of the set $\mathbb{N}$, where the other half is given by the set of positive odd integers $\{1, 3, 5, \ldots\}$. So, our common intuition fails while dealing with the infinite sets.

Definition 1.25 Let S and T be arbitrary sets. We say S is *equinumerous* to a subset of the set T if there is an injective function $f : S \to T$. In this case, we write $|S| \leq |T|$. Also, sets S and T are said to be *equinumerous* (or *equipotent sets*) if there exists an bijective function from S onto T. We then write $|S| = |T|$.

Clearly, two finite sets S and T are *equinumerous* if and only if they have the same *number of elements*. However, as said above, this no longer holds for infinite sets.

For example, by our discussion in Example 1.10, we have $|\mathbb{N}| = |\mathbb{Z}|$. Also, as the mapping $n \mapsto n + 1$ defines a bijection from the set of *odd integers* to the set of *even integers*, it follows that there are as many even integers as there are odd integers. The next theorem provides a general framework to introduce the concept of *cardinality* for a set, finite or infinite.

Theorem 1.15 *On the family $\mathscr{F}$ of all sets, define a relation $\sim$ by*

$$S \sim T \quad \Leftrightarrow \quad |S| = |T|, \;\; \text{for } S, T \in \mathscr{F}.$$

Then $\sim$ is an equivalence relation on the family[6] $\mathscr{F}$.

Proof For $S, T, U \in \mathscr{F}$, we have to prove that

1. $\sim$ is *reflexive*, i.e., $S \sim S$.
2. $\sim$ is *symmetric*, i.e., $S \sim T \;\Rightarrow\; T \sim S$.
3. $\sim$ is *transitive*, i.e., $S \sim T$ and $T \sim U \;\Rightarrow\; S \sim U$.

Notice that (1) is vacuously true if $S = \emptyset$; otherwise, the identity function $I : S \to S$ proves the assertion. Also, we know that if a function $f : S \to T$ is bijective then so is its inverse $f^{-1} : T \to S$. This proves (2). Finally, if $f : S \to T$ and $g : T \to U$ are two bijective functions, then so is their composition $f \circ g : S \to T$. This proves (3). $\square$

In particular, given any set $S \in \mathscr{F}$, we have $S \in [S]$. Further, two equivalence classes $[S]$ and $[T]$ are equal or $[S] \cap [T] = \emptyset$. For a proof of this assertion, let there exists $S, T \in \mathscr{F}$ such that $[S] \cap [T] \neq \emptyset$. Then, for $U \in [S] \cap [T]$, we have

$$U \in [S] \text{ and } U \in [T] \quad \Rightarrow \quad S \sim U \text{ and } U \sim T,$$

thus $S \sim T$, by (3). Therefore, we may write the family $\mathscr{F}$ as

$$\mathscr{F} = \bigcup_{S \in \mathscr{F}} [S], \quad \text{where} \quad [S] \cap [T] = \emptyset, \;\; \text{if } S \nsim T.$$

Since all the sets in a particular equivalence class are equinumerous, we assign the same *cardinality* to all the sets in the class. Further, by what we said above, each equivalence class has a *unique cardinality* representing the *size* of a set (and hence all the sets) in the equivalence class. This number is called the *cardinal number* of the equivalence class.

For example, since $[\emptyset] = \{\emptyset\}$, we have $|\emptyset| = 0$. So, there is a unique set with *cardinal number* 0. Next, the *cardinal number* of the equivalence class $[\{x\}]$ of singletons is 1, and so on. In general, the discussion above now provides a proper definition of the *cardinal number* of a finite set.

---

[6] As the collection $\mathscr{F}$ is not a set in usual sense, so the term *relation* used here is in a very rough sense.

Next, we consider the question: *What is the cardinal number of the equivalence class* [N]? By Definition 1.24, we know that $X \in [N]$ implies that X is a *denumerable set*. We usually assign all the countably infinite sets the cardinal number[7] $\aleph_0$, which is the cardinality of the set of natural numbers N. Therefore, the *cardinality* of all the countably infinite sets such as the set of integers $\mathbb{Z}$, the set of even integers, the set of odd integers, etc. is written as $\aleph_0$. A cardinal α is called a *transfinite number* if $\alpha \geq \aleph_0$. Notice that Theorem 1.14 and Theorem 1.13 imply the following two important facts of the *cardinal arithmetic*:

$$k \aleph_0 = \aleph_0 \quad \text{and} \quad \aleph_0^k = \aleph_0, \quad \text{for all } k \in N.$$

Finally, as it has been for Georg Cantor and some others, notice that the real issue arises when we try to determine the cardinality of an infinite set that is not countably infinite. Such an infinite set is called an *uncountable set*.

Example 1.11 It is possible to prove that, in general, a *countable product* of countably infinite sets such as N is an uncountable set. In particular, by extending the construction given earlier in Example 1.4, we obtain the set $\mathscr{S}^\infty$ of *strings of infinite length* over a set of k - alphabets, $k \geq 2$. Then, by the Cantor's *diagonal process*, it follows that $\mathscr{S}^\infty$ is an uncountable set. For, if possible, suppose $f : N \to \mathscr{S}^\infty$ is bijective, and let s_n be the nth symbol in the string $f(n)$. Define a string $\sigma \in \mathscr{S}^\infty$ such that the nth symbol of σ is $s_n + 1$ (*modulo* 2). Then, $\sigma \neq f(n)$, for all $n \in N$, proving that f is not surjective, a contradiction. The same argument applies to *infinite decimals* over the set $X = \{0, 1, \ldots, 9\}$, provided we work with respect to the base 10 in place of 2.

In this case, we start the general discussion with a formal definition of an infinite set, first proposed by Richard Dedekind. We say a set is **infinite** if it is equinumerous to a proper subset of its own. The next theorem proves that the condition stated in this definition is indeed a characteristic feature of every infinite set.

Theorem 1.16 (Dedekind Theorem) *A set S is infinite if and only if there exists a proper subset A of S that is equinumerous to S.*

Proof Since S is infinite, it is either denumerable or uncountable. In either case, we have to find a proper subset of S which is equivalent to S. We consider the following two cases:

Case I : Suppose S is denumerable. Then, there exists a bijective function $f : N \to S$, and we consider the function $g : N \to N$ given by $g(n) = n + 1$, which is a bijection from the set N to the set $N \setminus \{1\}$. So, the respective inverse functions f^{-1} and g^{-1} exist for both these two bijective functions. Thus, the function $h : S \to S$ given by

$$h(s) = (f \circ g \circ f^{-1})(s) = f(g(f^{-1}(s)))$$

---

[7] The symbol $\aleph_0$ is the first Hebrew letter, which is read as *aleph nought*.

is an injective function because so are functions f^{-1}, g, and f. But, h is not surjective, i.e., $h(S) \neq S$, because the element $s = h(1) \in S$ has no *pre-image* under the function h. Therefore, $h(S)$ is a proper subset of S. Hence, in this case, the set S has a proper subset that is equinumerous to S.

Case II : Suppose S is uncountable. Then, we can have a denumerable set $B \subset S$. So, by Case I, there is a bijective function h from B to a proper subset $A \subset B$. Now, consider a function $f : S \to S$ given by

$$f(s) = \begin{cases} s, & \text{if } s \in S \setminus B \\ h(s), & \text{if } s \in B \end{cases}.$$

We first show that f is injective. For, let $f(s_1) = f(s_2)$, where $s_1, s_2 \in S$. Now, if both s_1 and s_2 are in the set B, then

$$f(s_1) = f(s_2) \implies h(s_1) = h(s_2) \implies s_1 = s_2,$$

by injectivity of the function h. Similarly, if both s_1 and s_2 are in the set $S \setminus B$, then $s_1 = s_2$, by definition. Finally, consider the case when $s_1 \in B$ and $s_2 \in S \setminus B$. Here, we have

$$f(s_1) = f(s_2) \implies h(s_1) = s_2 \implies s_2 \in A \subseteq B,$$

which is a contradiction, because $s_2 \notin B$, by our choice. A similar argument applies when $s_2 \in B$ and $s_1 \in S \setminus B$. Therefore, we have $f(s_1) = f(s_2) \implies s_1 = s_2$, for all $s_1, s_2 \in S$. Hence, f is an injective function. Notice that, however, the function f cannot be surjective because A is a proper subset of the set B. It thus follows that, in this case as well, S is equivalent to one of its proper subset. This completes the proof of the direct part of the assertion.

For the converse part, let F be a finite set containing n elements. Then, any proper subset of F can have at most $n - 1$ elements. So, it is impossible to find a bijective function from F to any of its proper subset. Hence, a set which is equivalent to one of its proper subset can't be finite. Thus, a set S which has a proper subset equivalent to itself must be an infinite set.   □

The next theorem provides a far more useful tool to compare the *sizes* of infinite sets. It was proved independently by Georg Cantor (1887), Richard Dedekind (1888), Sergei Bernstein (1897), and some others, but never by Schröer, at least we have seen a proof by him that is correct. Cantor's argument used the following *trichotomy of cardinal numbers*: Given any two sets S, T, exactly one of the following three possibilities hold:

$$|S| < |T|, \quad |S| = |T|, \quad |S| > |T|.$$

It is known that this statement is equivalent to the well-ordering principle, and hence also to the axiom of choice. Peano's proof of the next theorem appeared in 1906, and that of Zermelo in 1908, who called it *Äquivalenzsatz*. Recall that, in general, $|S| \leq |T|$ implies that the set T has at least as many elements as the set S.

Theorem 1.17 (Cantor-Bernstein) *For any two sets X and Y,*

$$|X| \le |Y| \text{ and } |Y| \le |X| \quad \Rightarrow \quad |X| = |Y|.$$

Proof Let $f : X \to Y$ and $g : Y \to X$ be injective functions. The proof is complete if we are able to find a bijective function from X onto Y. Here, we present two different proofs.

Dedekind's Proof: The main argument given here rests upon the following fact: Let S, T, U be sets, with $S \subseteq T \subseteq U$. Then, $|U| = |S| \Rightarrow |U| = |T|$. For a proof of non-trivial part of the assertion, suppose $S \subset T \subset U$, and take $W = U \setminus T$ so that $U \setminus W = T$. Let $\alpha : U \to S$ be a bijective function, and consider the function $\beta : \mathscr{P}(U) \to \mathscr{P}(S)$ defined by

$$\beta(A) = \{\alpha(x) : x \in A\}, \quad \text{for any } A \subseteq U.$$

We can define a *recursive function* $\sigma : \mathbb{N}_0 \to \mathscr{U}$ such that

$$\sigma(0) = W \quad \text{and} \quad \sigma(n+1) = \alpha(\sigma(n)), \quad \text{for all } n \ge 0.$$

Let $\overline{W} = \cup_{n \ge 0} \sigma(n)$, and consider the function $\gamma : U \to U$ defined by

$$\gamma(x) = \begin{cases} \alpha(x), & \text{if } x \in \overline{W} \\ x, & \text{otherwise} \end{cases}.$$

Since α is bijective, it follows that $\gamma(\overline{W}) = \overline{W} \setminus W$, and that γ is injective. To prove the latter assertion, let $x \ne y \in U$, and consider the cases: (i) $x, y \in \overline{W}$; (ii) $x, y \in U \setminus \overline{W}$; and (iii) $x \in \overline{W}$, $y \in U \setminus \overline{W}$. Now, we know that

$$\gamma(\overline{W}) = \overline{W} \setminus W \quad \text{and} \quad \gamma(U \setminus \overline{W}) = U \setminus \overline{W}, \quad \text{by definition.}$$

Therefore, we obtain

$$\gamma(U) = (U \setminus \overline{W}) \bigcup (\overline{W} \setminus W) = U \setminus W = T,$$

which proves Dedekind's *Kettentheorie*. Finally, taking $S = (g \circ f)(X)$, $T = g(Y)$ and $U = X$, we have $S \subseteq T \subseteq U$, and $|U| = |S|$. So, we must have $|X| = |U| = |T|$. Also, since $|T| = |Y|$, the proof of the theorem follows.

Bernstein's Proof: We construct here sets A_k, $k \ge 0$, so that the function $h : X \to Y$ given by

$$h(x) = \begin{cases} f(x), & \text{if } x \in A \\ g^{-1}(x), & \text{if } x \notin A \end{cases}, \quad \text{where } A = \bigcup_{k \ge 0} A_k,$$

is a bijection. We may assume that g is not surjective; for, otherwise, we may take $h = g^{-1}$, with $A = \emptyset$. To start with, note that g^{-1} cannot be defined on the set $X \setminus g(Y)$, so we may take $A_0 = X \setminus g(Y)$. Next, if for some $a \in A_0$ we have $f(a) = g^{-1}(a')$, with $a' \in g(Y)$, then by *expected* injective property of h we must take $h(a') = f(a')$. But then

$$a' = g(f(a)) = (g \circ f)(a) \quad \Rightarrow \quad A_1 = (g \circ f)(A_0).$$

Continuing with this argument, we may define A_k *inductively* as

$$A_k = (g \circ f)(A_{k-1}) = (g \circ f)^k(X \setminus g(Y)), \quad \text{for } k \geq 1.$$

Observe that $A \supseteq X \setminus g(Y) \Rightarrow X \setminus A \subset g(Y)$, so h is defined for each $a \notin A$; indeed, it is defined on A, by our construction.

Now, for $x, x' \in S$, if both x, x' are in the set A or in the set $X \setminus A$, we have $h(x) = h(x') \Rightarrow x = x'$. If possible, suppose $x \in A$ and $x' \in X \setminus A$, with $h(x) = h(x')$. Then, $x \in A_k$ for some $k \geq 0$, and $x' \in g(Y)$ give

$$x' = g(h(x)) = g(f(x)) \quad \Rightarrow \quad x' \in S_{k+1} \quad \Rightarrow \quad x' \in A,$$

a contradiction. Therefore, we cannot find distinct x, x' with $x \in A$ and $x' \in X \setminus A$, or other way round such that $h(x) = h(x')$. Finally, to conclude the proof, let $y \in Y$, and consider the element $x = g(y) \in X$. Now, if $x \notin A$ then we have

$$h(x) = g^{-1}(x) = g^{-1}(g(y)) = y \quad \Rightarrow \quad h \text{ is surjective.}$$

And, if $x \in A$ so that $x \in A_k$ for some $k \geq 1$, then we have $x = (g \circ f)^k(a)$ for some $a \in A_0$; note that since $x = g(y) \in g(Y)$ we can't have $k = 0$. Take $a' = (g \circ f)^{k-1}(a) \in A_{k-1} \subseteq A$ so that

$$\begin{aligned} h(a') = f(a') &= g^{-1}(g(f(a'))) \\ &= g^{-1}((g \circ f)((g \circ f)^{k-1}(a))) \\ &= g^{-1}(x) = y. \end{aligned}$$

Hence, h is a bijective function. $\square$

Finally, let X be an *uncountable* set. In general, it is known that $k^{\aleph_0}$ is a transfinite cardinal number, for all $k \geq 2$. That is, $k^{\aleph_0} > \aleph_0$. The next theorem proves the assertion for $k = 2$. In fact, it also tells us how big a set can be.

Theorem 1.18 (Cantor, 1891) *For every set A, $A \leq \mathscr{P}(A)$ and $A \not\sim \mathscr{P}(A)$, where $\mathscr{P}(A)$ is the power set of the set A.*

Proof If $A = \emptyset$, the result is trivial because $\mathscr{P}(\emptyset) = \{\emptyset\}$. So, we assume $A \neq \emptyset$ and define a function $f : A \to \mathscr{P}(A)$, by $f(x) = \{x\}$, for $x \in A$. Then, f is injective as $\{x\} = \{y\} \Rightarrow x = y$. Hence, $A \leq \mathscr{P}(A)$. We next show that $A \not\sim \mathscr{P}(A)$, by way of contradiction. So, if possible, let us assume that $A \sim \mathscr{P}(A)$. Then, by definition, there exists a bijective function $g : A \to \mathscr{P}(A)$. Consider the subset B of A defined by

$$B = \{x \in A \mid x \notin g(x)\},$$

and then pick $a \in A$ such that $g(a) = B$. Now, notice that $a \in B \Leftrightarrow a \notin B$, which is impossible. This contradiction is reached because of our assumption that $A \sim \mathscr{P}$. This shows that our assumption is wrong, so the proof of the theorem is complete. $\square$

Finally, we discuss the cardinal number of the class $[\mathbb{R}]$. We anticipate the fact that the set $\mathbb{R}$ is uncountable, which we prove in the last section of Chap. 2. The cardinality of $\mathbb{R}$ is denoted by c, where c stands for the *continuum*. So, we take c as the *cardinal number* of $[\mathbb{R}]$. Notice that Theorem 1.18 implies that $\mathbb{N} \notin [\mathbb{R}]$. For, taking $A = \mathbb{N}$, it follows that there exists a set with cardinal number greater than $\aleph_0$; of course, it is the cardinal number of the set $\mathscr{P}(\mathbb{N})$. Thus, $\aleph_0 < $ c. Further, by Lemma 1.1, $\mathscr{P}(A)$ is in bijective correspondence with the set

$$2^A = \{f : f : A \to \{0, 1\} \text{ is a function}\},$$

and so, by a theorem of Bernstein, we have

$$|2^A| = 2^{|A|} \quad \Rightarrow \quad |\mathscr{P}(A)| = 2^{|A|},$$

for any set A. In particular, when $A = \mathbb{N}$, we obtain $|\mathscr{P}(\mathbb{N})| = 2^{\aleph_0}$. Therefore, $\aleph_0 < 2^{\aleph_0}$, by Theorem 1.18. On the other hand, by a deep result of Bernstein, $2^{\aleph_0} = $ c. There are many outstanding problems related to transfinite cardinals. One such problem asks: *Does there exists a transfinite cardinal α such that $\aleph_0 < \alpha < 2^{\aleph_0} = $ c ?* The following is a *conjecture* of Cantor:

Continuum Hypothesis. *There is no cardinal α, with $\aleph_0 < \alpha < $ c.*

The *continuum hypothesis* was the first of Hilbert's list of 23 problems presented in 1900. Notice that if the superset of a set is countable, all its subsets are countable and therefore by Corollary 1.1 we have that every superset of an uncountable set is uncountable. So, there are many sets of cardinality $> $ c.

1.5 Development of Function Concept—A Historical Note

As it was for *René Descartes* (1596–1650), *Pierre de Fermat* (1601–1665), Sir *Issac Newton* (1643–1727), *Gottfried Leibniz* (1646–1716), and many other mathematicians of the past, the term *function* is usually taken to be synonymous with a *formula* or an *analytical expression* involving a variable x (in 1-dimensional case), which gives a unique (output) value y, by plugging in a value for the variable x.

Of course, as the reader may be familiar with from their calculus course, that is precisely what we have while dealing with a *polynomial, rational, trigonometric, exponential,* or a *hyperbolic function.* A function of single variable formed by taking sums, products, and compositions of such type of functions is called an *elementary function.* However, as seen earlier, there are many other type of *functions* that do not fit into this *defn.* Some important ones in various engineering applications are *beta, gamma,* and *error functions.*

Since antiquity, mathematicians and scientists had been using such types of functions to study many different kinds of phenomena, but each understood a function in the sense of the next (reformulated) definition given by *Johann Bernoulli* (1667–1748). In the original text, he wrote ... *variable magnitude, a quantity that is composed in any possible manner of this variable magnitude and of constants. . . . .*

Definition 1.26 (*Johann Bernoulli, 1718*) A *function* is a mathematical entity used to study the dependencies among variables and constants defining a physical system, where the variables occurring in the (output) function are called *dependent variables* and (input) variables are called the *independent variables.*

The main purpose of taking the reader on a historical detour is to provide a proper perspective, and hence enhance the conceptual understanding about the concept of a function. It is a very illuminating experience to see how some of the *primitive notions* of the mathematical analysis evolved in the process. Various definitions for the *function concept* given in this section are written in exactly the way these were proposed by mathematicians in different time lines. Nonetheless, it is a very rewarding experience to read about the life of mathematicians whose work is referred to in this course. The main source for the exposition given here is the excellent book [3].

At the outset, it helps to have proper information about the terms and notions used in Definition 1.26. In fact, Newton was the first to use "*fluent*" for the independent variables, and "*relata quantités*" for the dependent variables. Also, notice that the "*correspondence between objects*" aspect of the concept had been known since antiquity. For example, it is there in *counting*; in defining the four *arithmetic operations*; in the *Babylonian's tables* of reciprocals, squares and square roots, cubics and cubic roots; and also in Ptolemy's *trigonometric computations* related to chords of a circle. However, there has never been a mention of the term "*function*". In 1673, Leibniz used the term *function* for the first time to denote any of several geometric concepts like *abscissa, ordinate, tangent,* etc., as derived from the *curve interpretation* of the modern definition of a function, as mentioned above.

Nicholas Oresme (1323–1382) could have foreseen the Definition 1.26 in 1350 while describing the nature as laws giving a dependence of one quantity on another. Some general ideas about independent and dependent variable quantities are seen to be present in his "*geometric theory of latitudes*". Galileo Galilei's (1564–1642) study of kinematics and astronomy is also regarded as an indication of the fact that he had a clear understanding about the relation among variable quantities. In 1638, he explained how to establish a one-to-one correspondence between two concentric circles, geometrically, with the larger one being of the circumference double than the smaller one.

We remark that mathematics before Descartes had only two aspects: *algebra* and *geometry*, each being used separately to study celestial phenomena, basic laws of physics, and economic problems faced by people during that era related to their daily needs. Starting from Euclid's *"Elements"* ($\sim 300\ BC$)—a compilation of rigorous proofs of coherent geometric results known during that time—it was only through translations produced by Arab writers that a part of the world came to know about the fundamental work of Greek astronomers and mathematicians like Aristotle, Plato, Archimedes, Apollonius, and Ptolemy. Western Europe discovered this treasure of knowledge only through translations of these Arabic manuscripts produced by Gerard of Cremona (1114–1187), Robert of Chester (twelfth century), Leonardo da Pisa (around 1200), and Regiomontanus (1436–1476).

The famous book *"Al-jabar w'al muqâbala"* by *Al-Khowârizmî* is the oldest known manuscript, dating back to 1342, that mainly has *geometric arguments* used to solve some *quadratic equation* like $x^2 + 21 = 10x$. Sometime between 1510 and 1515, Scipione del Ferro gave the first algebraic solution to the depressed cubic $x^3 + ax = b, a, b > 0$ and it was passed to one of his pupils, Antonio Maria Fior. Having got the formula for depressed cubic, in February 1535, Fior challenged Tartaglia to a public contest. But Tartaglia won the contest as could solve general cubic. Towards a solution of a *cubic equations*, Tartaglia wrote ... *I have discovered the general rule, but for the moment I want to keep it secret*... , which he ultimately had to tell Cardano (veiled in verses, without derivation). Cardano recreated the formula with great difficulty and got it published in his *"Ars Magna"* in 1545. Soon after, Ludovic Ferrari gave a method of solving a bi-quadratic equation. In 1826, Henry Abel gave a proof of *impossibility of solution by radicals* of a degree five polynomial.

The introduction of letters $A, B, C, X, \ldots$ by Franciscus Viéta in his *"Algebra Nova"* (1600) is viewed as the first milestone towards the formulation of *abstract definition* of a function, which he used to denote the known and unknown quantities while solving some geometric problems, algebraically. In 1637, Descartes used these ideas of Viéta to solve the *Pappus problem*, which was formulated initially around 350 BC. In 1637, the father of *analytical geometry*, René Descartes' *"La Géométrie"* introduced algebraic methods into the geometry of figures in 2- and 3-dimensional spaces. He was the first to write unknowns in an equation as variables x, y, z, etc., and coefficients by constants a, b, c, etc. His view of the equation as a dependence relation between variable quantities was in the sense of modern-day *graph interpretation* of a function. Soon after, *polynomial functions* (Definition 1.16) became natural entities to solve many algebraic problems such as *roots approximation* by using geometric argument.

An oldest record of use of logarithm (to base 2) is the following two-row series of M. Stifel (1544), which he called *in superiore ordine* and *in ineriore ordine*:

$$\cdots -3 \quad -2 \quad -1 \quad 0 \quad 1 \quad 2 \quad 3 \quad 4 \quad 5 \quad 6 \quad \cdots$$
$$\cdots 1/8 \quad 1/4 \quad 1/2 \quad 1 \quad 2 \quad 4 \quad 8 \quad 16 \quad 32 \quad 64 \quad \cdots,$$

where while moving from lower to upper row the product of two numbers converts to their sum. For Greeks, *logarithms* were only "*u*seful relations between numbers". The following definition of a *logarithmic function* is a culmination of works of *John Napier* (1550–1617), *Jobst Bürgi* (1552–1632), and *Henry Briggs* (1561–1631).

Definition 1.27 (*Logarithmic Function*) For $x > 0$, a number $\ell(x)$ is called the *logarithm* of x if, for all $x, y > 0$, the following relation holds:

$$\ell(x \cdot y) = \ell(x) + \ell(y).$$

If $\ell(a) = 1$, we say a is a *base of the logarithm* .

Replacing y by z/x gives $\ell(z/x) = \ell(z) - \ell(x)$ so that by taking $x = z = 1$, we obtain $\ell(1) = 0$. Similarly, we obtain $\ell(x^{-m}) = -m\ell(x)$, for all $m \in \mathbb{N}$. So, it follows easily that $\ell(x^{m/n}) = (m/n)\ell(x)$, for all $m \in \mathbb{Z}$ and $n \in \mathbb{N}$. In particular, we obtain $\ell(a^q) = q$, for any $q \in \mathbb{Q}$. As shown latter, this relation extends for all real number $x > 0$ by density of $\mathbb{Q}$ in $\mathbb{R}$, and the continuity of the function $y = \ell(x)$. Taking logarithm of $x > 0$ to base a, denoted by $\log_a x$, the inverse to the operation of forming exponentials, this follows that

$$y = \log_a x \quad \Leftrightarrow \quad x = a^y.$$

Euler's *golden rule* for base change is the following relation

$$\log_a x = \log_b x / \log_b a.$$

In early seventeenth century, problems related to *logarithm* and *maritime navigation* led to a formulation of the following *interpolation problem*:

Interpolation Problem *Given $n + 1$ points $(x_i, y_i)_{i=0}^n$ in the plane, with x_i evenly distributed, compute the (interpolation) polynomial $p(x)$ of degree n such that $y_i = p(x_i)$, for each $0 \le i \le n$.*

Briggs devised the method of *finite differences* in Chap. 8 of his book "*Arithmetica Logarithmica*", which he applied to logarithmic function $\log(1 + x)$ to construct tables of logarithm to the base $a = 10$. Use of Briggs' tables in scientific computations is more common owing to the fact that a shift of the decimal point just adds an integer to the logarithm. In 1676, Newton adopted ideas of Viéta's "*Ars Magna*" instead to write the unknown interpolation polynomial $p(x)$, say of degree 3, as

$$p(x) = a + b\,x + c\,x^2 + d\,x^3,$$

with undetermined constants a, b, c, d, where the arguments x_i were assumed to be

$$x_0 = 0, \quad x_1 = 1, \quad x_2 = 2, \quad \text{and } x_3 = 3.$$

Using first the condition $y_i = p(x_i)$, he transformed the problem to finding the solutions of the following system of algebraic equations:

$$y_0 = a, \quad y_1 = a + b + c + d,$$
$$y_2 = a + 2b + 4c + 8d, \quad y_3 = a + 3b + 9c + 27d.$$

So, eliminating a from these equations, we obtain

$$\Delta y_0 = y_1 - y_0 = b + c + d,$$
$$\Delta y_1 = y_2 - y_1 = b + 3c + 7d,$$
$$\Delta y_2 = y_3 - y_2 = b + 5c + 19d,$$

where $\Delta y_i = y_i - y_{i-1}$ are called *first forward differences*. Eliminating b from last three equations, we obtain

$$\Delta^2 y_0 = \Delta y_1 - \Delta y_0 = 2c + 6d,$$
$$\Delta^2 y_1 = \Delta y_2 - \Delta y_1 = 2c + 12d,$$

where $\Delta^2 y_i = \Delta y_i - \Delta y_{i-1}$ are called *second forward differences*. Finally, eliminating c from last two equations, we obtain

$$\Delta^3 y_0 = \Delta^2 y_1 - \Delta^2 y_0 = 6d.$$

Since y_i's are known, the desired polynomial is obtained. The general formula is derived similarly, using *methodus differentialis*. We remark that a form of this method to do computations efficiently has been in use since antiquity.

The *exponential notation* $a^{m/n}$, $m, n \in \mathbb{Z}$, emerged from the work of *Rafael Bombelli* (1526–1572), *Simen Stevin* (1548–1620), Descartes, and *Blaise Pascal* (1623–1662).

Theorem 1.19 (Pascal, 1654) *For real variables a, b, and $n = 0, 1, 2, \ldots$,*

$$(a + b)^n = a^n + na^{n-1}b + \frac{n(n-1)}{1 \cdot 2}a^{n-2}b^2 + \cdots + nab^{n-1} + b^n.$$

The first reader of *La Géométrie, I.F. Debeaune* (1601–1652) posed Descartes the following problem: *Find the locus of a point $P(x, y)$ such that the distances between the points where the vertical and the tangent line cut the x-axis are always equal to a given constant, say C.* The problem remained unresolved until, in 1684, Leibniz proposed the following solution: Increase x by a small increment a so that y is increased by ya/C, by similarity of the two involved triangles. Continuing with the scheme, one gets the sequence of values

$$y, \quad \left(1 + \frac{a}{C}\right)y, \quad \left(1 + \frac{a}{C}\right)^2 y, \quad \left(1 + \frac{a}{C}\right)^3 y, \ldots,$$

for the abscissae $x, x + b, x + 2b, x + 3b, \ldots$.

Similarity of the expressions in the last equation to the following *compound interest* formula is related to the practical needs of computing the total return A in t years, when an initial amount of money P is invested at an interest rate $x\%$ computed n times annually:

$$A(t) = P \left(1 + \frac{x}{n}\right)^{nt}, \quad t \geq 0. \tag{1.5.1}$$

In 1668, *Jacob Bernoulli* (1655–1705) was the first to study this formula for (continuous) compound interest. He used binomial theorem to prove that

$$2 < \lim_{n \to \infty} \left(1 + \frac{1}{n}\right)^n < 3.$$

The exact value of the limit is now known as the *Euler's number,*[8] in his honour. It is known that Leibniz mentioned this number as b in his 1690 letter to Huygens. The familiar notation e appeared for the first time in a letter of Euler to Goldbach, which he wrote in 1731.

The first indirect reference to the Euler number appeared in Napier's work on *logarithms* (1618). In 1624, Briggs obtained an approximation of $\log_{10} e$, with no explicit mention of the Euler number. In 1647, the Euler number appeared in Saint-Vincent's work on an area under a rectangular hyperbola. In 1661, Huygens was the first to describe the relation between logarithm and the rectangular hyperbola $yx = 1$, however, without any explicit mention of the fact that the area under the curve $yx = 1$ between the limits 1 and e equals 1. In fact, this property makes the Euler number e the base of the *natural logarithm* of a real number $x > 0$, usually denoted by $\ln x$.

Motivated by Fermat's 1636 work related to area under the curve $y = x^a$, the simple observation that $\ln(x) = Area(1 \to x)$, led *James Gregory* (1638–1675) and *Alfons Anton* (1618–1667) to discover the fact that *the area under the hyperbola* $yx = 1$ *is a logarithm*. *Nicolaus Mercator* (also known as *Kauffmann*) was the first to coin the term *natural logarithm* in his 1668 publication "*Logarithmo-technia*" who obtained the infinite series expression for the function $f(x) = \ln(1 + x)$ given by

$$\ln(1 + x) := x - \frac{1}{2}x^2 + \frac{1}{3}x^3 - \frac{1}{4}x^4 + \cdots \quad x > -1, \tag{1.5.2}$$

by shifting the origin to 1 so that $\ln(1 + x)$ is the area below $1/(1 + z)$ between 0 and x. For, using (1.3.2), substitute

$$\frac{1}{1 + z} = 1 - z + z^2 - z^3 + \cdots,$$

---

[8] The Euler number is also defined as the unique number a such that the slope of the *exponential function* $y = a^x$ has unit slope at $x = 0$. Its approximate value is $2.718281828459045 \cdots$, and we are counting.

and inserting for the areas under $1, z, z^2, \ldots$ from 0 to x, one obtains (1.5.2). The convergence of the series is discussed in a later chapter. Taking $x = 1$, one gets the following interesting series

$$\ln(2) = 1 - \frac{1}{2} + \frac{1}{3} - \frac{1}{4} + \frac{1}{5} - \frac{1}{6} + \cdots .$$

Replacing x by $-x$ in Mercator series, and subtracting, we obtain the following Gregory series (1668):

$$\ln\left(\frac{1+x}{1-x}\right) := 2\left(x + \frac{x^3}{3} + \frac{x^5}{5} + \frac{x^7}{7} + \cdots\right)$$

From this series, we obtain an approximation of $\ln(p)$, for any prime $p \geq 2$. Contrary to the common belief, Bernoulli's aforementioned work on *compound interest* (and not the logarithm) led him to the discovery of the constant e. Bernoulli's work of 1683 is the first authentic record of an explicit use of *limiting process* to approximate the number e. However, he could not see a connection between the limit $\lim_{n\to\infty}(1 + (1/n))^n$ and the logarithm.

In 1734, *Leonhard Euler* wrote ... if $f((x/a) + c)$ denotes an arbitrary function..., thereby introducing for the first time the notation $y = f(x)$. Euler in his "*Introductio in analysin infinitorum*" (1748) was led to consider the exponential expression $(1 + \epsilon)^n$, with ϵ small and n very large. For example, when $\epsilon = 1/n$, Theorem 1.19 gives

$$\left(1 + \frac{1}{n}\right)^n = 1 + \frac{n}{n} + \frac{n(n-1)}{1\cdot 2}\frac{1}{n^2} + \frac{n(n-1)(n-2)}{1\cdot 2\cdot 3}\frac{1}{n^3} + \cdots$$

$$= 1 + 1 + \frac{(1 - (1/n))}{1\cdot 2} + \frac{(n - (1/n))(1 - (2/n))}{1\cdot 2\cdot 3} + \cdots .$$

He thus concluded

$$e = 1 + 1 + \frac{1}{1\cdot 2} + \frac{1}{1\cdot 2\cdot 3} + \frac{1}{1\cdot 2\cdot 3\cdot 4} + \cdots , \quad \text{as } n \to \infty. \qquad (1.5.3)$$

In the later part of the book, we shall prove that (1.5.3) is true because the following two theorems of Euler hold:

1. $\lim_{n\to\infty}\left(1 + \frac{1}{n}\right)^n = e$ and
2. the infinite series on the right side of (1.5.3) converges to e.

It is an interesting exercise to find the *error* between *partial sums* of the series in (2) and the values of $(1 + (1/n))^n$ for some small values of n. In fact, to obtain e correct to 200 decimal places, one needs $n = 120$ terms of the series.

Definition 1.28 (*Euler, 1748*) A *function* of a variable quantity is an analytic expression composed in any way whatsoever of the variable quantity and numbers or constant quantities.

Euler said nothing about the term *analytic expression*, assuming the reader would understand that it refers to addition, multiplication, power, roots, etc., which immediately lead to inconsistencies. Taking $\epsilon = x/n$, Euler investigated further the situation when $n \to \infty$ and $\epsilon \to 0$ such that $n\epsilon$ is a constant x, say a rational number.

Theorem 1.20 (Euler, 1748) *As* $n \to \infty$,

$$\left(1 + \frac{x}{n}\right)^n \to e^x = 1 + x + \frac{x^2}{2!} + \frac{x^3}{3!} + \frac{x^4}{4!} + \cdots .$$

In particular, $\ln x$ *is the logarithm to base e.*

The next definition is from Joh. Bernoulli's "*Principia Calculi Exponentialium*".

Definition 1.29 (*Power Functions, 1697*) For any $a, b \in \mathbb{R}$, with $a > 0$,

$$a^b := \left(e^{\ln a}\right)^b = e^{b \ln a}.$$

Measurement of angles, mainly for problems related to astronomy, has been one of the oldest quests of mankind. Babylonian defined it in terms of *degrees* by dividing the circle into $360°$. Ptolemy, in his "*Almagest*" (AD 150) refined the measurements by including *minutes* and *seconds*. For logarithms, a natural measure is defined in terms of the *arc length* of a unit circle so that *one radian* equals $360°$. In 1719, de Lagny computed the *arc length* of the half of the unit circle with great precision. So, in terms of the notation π (*periphery*) introduced by W. Jones in 1706, we obtain the relation $2\pi = 360°$.

Leonhard Euler (1707–1783) is considered to be the most prolific writer of the eighteenth century. His collected works are in 92 volumes. Prior to Euler's work,[9] trigonometric quantities were regarded as lines connected with the circle rather than functions. For example, according to Brahmagupta (around 630),

$$\sin \alpha = (1/2)\,\text{chord}(2\alpha),$$

where α was used both for the *chord length* between the lines, and for the textslangle α subtended by these lines at the origin. The quantity $\sin \alpha$ was originally named *sinus rectus* (for vertical sine). Definition 1.28 led Euler to define *trigonometric functions* as follows.

Definition 1.30 (*Newton (1669), Leibniz (1691)*) For $x \in \mathbb{R}$,

---

[9] In words of Laplace, *Read Euler, read Euler, he is the master of us all.*

$$\cos x = 1 - \frac{x^2}{2!} + \frac{x^4}{4!} - \frac{x^6}{6!} + \frac{x^8}{8!} - \cdots \; ;$$

$$\sin x = x - \frac{x^3}{3!} + \frac{x^5}{5!} - \frac{x^7}{7!} + \frac{x^9}{9!} - \cdots .$$

Notice that these two functions are periodic: We say $f : \mathbb{R} \to \mathbb{C}$ is a *periodic function* if for some $p > 0$, $f(x + p) = f(x)$, for all $x \in \mathbb{R}$. The smallest such positive p is called the *fundamental period* (or simply *period*) of the function $f(x)$. The functions of period 1 are also known as $\mathbb{Z}$-periodic functions. Notice that, for a periodic function f of period p, it is suffice to work with the part of $f(x)$ with $x \in [0, p)$ or $x \in [-p/2, p/2)$. In particular, as both $\sin x$ and $\cos x$ are periodic of period 2π, we can always take $x \in [0, 2\pi)$ or $x \in [-\pi, \pi)$.

We will discuss *absolute convergence* of these series in the later part of the book. Since $\tan x$ is an odd function, and $\tan 0 = 0$, one may write

$$\tan x = a_1 x + a_3 x^3 + a_5 x^5 + \cdots .$$

Multiplying this equation by $\cos x$, and using series expression for $\sin x$, it follows by comparing the coefficients that

$$a_1 = 1, \quad -\frac{1}{6} = -\frac{a_1}{2} + a_3 \Rightarrow a_3 = \frac{1}{3}, \quad a_5 = \frac{2}{15}, \ldots .$$

So, we obtain the following series expression for $\tan x$

$$\tan x = x + \frac{x^3}{3} + \frac{2x^5}{15} + \frac{17x^5}{315} + \cdots .$$

Notice that these *trigonometric functions* are defined as a function of the *arc length* x on the unit circle. These definitions are extended to the set of real numbers $\mathbb{R}$ using *periodicity*, and the identification of the interval $[-1, 1]$ (or $(-\infty, \infty)$) with the unit circle being seen as the interval $[-\pi, \pi]$ or $[0, 2\pi]$ (or $(-\pi, \pi)$). So, conversely, *inverse trigonometric functions* define *arc* y as a function of $\sin x$, $\cos x$, and $\tan x$, which are denoted respectively by arcsin y, arccos y, and arctan y:

$$x = \arcsin y \quad \Leftrightarrow \quad y = \sin x, \quad \text{for} \quad -1 \le y \le 1, \ -\pi/2 \le x \le \pi/2;$$
$$x = \arccos y \quad \Leftrightarrow \quad y = \cos x, \quad \text{for} \quad -1 \le y \le 1, \ 0 \le x \le \pi;$$
$$x = \arctan y \quad \Leftrightarrow \quad y = \tan x, \quad \text{for} \quad -\infty < y < \infty, \ -\pi < x < \pi.$$

The series for arctan y was attributed to Gregory (1671), which got published by Leibniz in *Acta Eruditorum* (1682). However, the original discovery of this series from arcsin y had already been part of Newton's *"De Analysi"* (1669), which got published 40 years later. Alternatively, we can use the area of the corresponding circular sector. For, let x be the tangent of an angle whose *arc* $y = \arctan y$ we need to find. So, if A is the point (x, y) so that $OA = \sqrt{1 + x^2}$, by Pythagoras theorem, then for a point B on the y-axis we apply Thales theorem to write

$$OB = \frac{1}{\sqrt{1+x^2}} \quad \text{and also} \quad \Delta u = \frac{\Delta x}{\sqrt{1+x^2}}.$$

Consequently, the infinitesimal arc length Δy is given by

$$\Delta y = \frac{\Delta u}{\sqrt{1+x^2}} = \frac{\Delta x}{1+x^2}.$$

Thus, the desired *arc* y equals the total area between 0 and x of the function

$$\frac{1}{1+x^2} = 1 - x^2 + x^4 - x^6 + x^8 - \cdots,$$

which proves that

$$y = \arctan x = x - \frac{x^3}{3} + \frac{x^5}{5} - \frac{x^7}{7} + \frac{x^9}{9} - \cdots, \quad \text{for } |x| \leq 1.$$

The following infinite series was discovered by Newton (1669):

$$\arcsin x = x + \frac{1}{2}\frac{x^3}{3} + \frac{1 \cdot 3}{2 \cdot 4}\frac{x^5}{5} + \frac{1 \cdot 3 \cdot 5}{2 \cdot 4 \cdot 6}\frac{x^7}{7} + \cdots.$$

Newton used the binomial expression for $(1 + x^2)^{-1/2}$ to prove it.

Definition 1.31 (*Fonccnex, 1759; Lambert 1770*) For $x \in \mathbb{R}$, the **hyperbolic** cosine and sine functions of x are defined by

$$\cosh x = \frac{e^x + e^{-x}}{2} \quad \text{and} \quad \sinh x = \frac{e^x - e^{-x}}{2}.$$

Notice that, for $u = \cosh x$ and $v = \sinh x$, the point (u, v) lies on the hyperbola $u^2 - v^2 = 1$.

Calculus was developed initially to deal with the problem of finding at each point of a *curve* $y = f(x)$ its *slope, tangent line*, and the *normal line*. The early motivations were driven from Descartes' work on angle of intersection of curves; Johannes Kepler's laws of planetary motion (1609–1619); Galilei's work on linear motions (1638) and construction of telescope; Christian Huygens' mathematical analysis of *pendulums*[10] (1673); Pierre de Fermat's original work on tangents to curves, maxima/minima (1638), and quadrature; Leibniz's interpretation of *infinitely small* that led him to write $\Delta y / \Delta x$ as dy/dx (1684); Newton's laws of motion (1686) and law of universal gravitation (*The Principia*, 1687); etc.

---

[10] Huygens was the first to write the formula $T = 2\pi\sqrt{\ell/g}$ for the period of an ideal mathematical pendulum.

Leibniz's obscure publication of 1684 led Johann Bernoulli to reinvent differential calculus in 1691, assuming infinitely small quantities can be added to finite quantities without altering their values and interpreted curves to be a polygon with infinitely short sides. He and Jacob Bernoulli were among the first mathematicians to study, understand, and apply calculus to various physical problems. Besides, teaching his calculus to Marquis de L'Hospital, he also provided Euler weekend lessons on mathematics, while he was studying the philosophies of Descartes and Newton to write his Masters' dissertation. As a matter of fact, it was Johann Bernoulli who convinced Euler's father to allow him to take mathematics as a career.

A definitive breakthrough in discovering a new calculus appeared in 1696 as Guillaume L'Hospital's book "Analyse des infiniment petits pour l'Intelligence des Lignes Courbes", which is mainly a compilation of Johann Bernoulli's work. It faced no criticism related to infinitely small considerations (Nieventijt, 1694) or from Cartesians of France. In 1715, Taylor used derivative approach for finite differences to solve the problem of finding interpolation polynomial $p(x)$ for nodal points $x_t = x_0 + t\Delta x$, with $t = (x - x_0)/\Delta x$. By allowing nodal points to vary over an interval, he obtained the following formula:

$$f(x) = f(x_0) + (x - x_0)f'(x_0) + \frac{(x - x_0)^2}{2!} f'(x_0) + \cdots, \qquad (1.5.4)$$

with $\Delta^k y_0/\Delta x^k = f^{(k)}(x_0)$. The partial sums of the series on the right side are called the Taylor polynomials, in honour of Brook Taylor (1685–1731), which prove to be very helpful to approximate roots of an equation $f(x) = 0$, where f is some sufficiently smooth function. Notice that all the series derived earlier are special case of this series universalissima. Colin Maclaurin (1698–1746) in his "Treatise of Fluxions" (1742) considered . . . investigate the ratio which is the limit . . . . Maclaurin used the condition $p^{(k)}(x_0) = f^{(k)}(x_0)$ to obtain the series (1.5.4).

In 1754, D'Alembert was the first to introduce a clear notion of the limit: . . . This limit is the value which the ratio z/n approaches more and more . . . . In 1755, Euler published another influential two-volume book "Institutiones calculi differentialis" (i.e., the foundation of differential calculus) that he wrote in 1748. Here, he gave the following improved version of his earlier definition of a function.

Definition 1.32 (*Euler, 1755*) If some quantities so depend on other quantities that if the latter are changed the former undergoes change, then the former quantities are called functions of the latter. Here, he used the notation $f(x)$ for the first time to write the function f applied to the argument x.

This definition remained in use for the whole of the eighteenth century, and all modern treatises on the subject were based on this book of Euler. It is considered as the first textbook on differential calculus. In 1768, Euler published a three-volume textbook on "Institutionum calculi integralis" (on integral calculus), which also contains his discoveries about differential equations. In fact, due to its importance in computations related to areas, surfaces, and volumes, integral calculus has interested many great mathematicians since antiquity: Archimedes, Kepler, Cavalieri, Viviani,

Fermat, Gregory St. Vincent, Guldin, Gregory, and Barrow. It was the *fundamental of calculus*, independently discovered by Newton, Leibniz, and Johann Bernoulli, that was the real breakthrough. The symbol $\int$ is due to Leibniz (1686), and the term *integral* was first tossed by Johann Bernoulli.

Avoiding the notion of *infinitely small*, Lagrange based his analysis on power series: *"Théorie des fonctions analytiques"* (1797). He introduced the term *derivative*, and used for the first time the notation y' for $f'(x)$. Newton (1665) and Johann Bernoulli (1691/92) were the first to study the geometric interpretation of the second derivative:

$$f''(x) > 0 \;\Rightarrow\; f'(x) \text{ is increasing;}$$
$$f''(x) < 0 \;\Rightarrow\; f'(x) \text{ is decreasing,}$$

so respectively the curve $f(x)$ is *convex downward* and *convex upward*. We say a point a is *inflexion point* for $f(x)$ if $f''(a) = 0$.

In 1812, Joseph Fourier (1768–1830) published his celebrated book on the conduction of heat in solids. Soon after, certain issues related to the convergence of his proposed *series solution* of heat equations led to rigorous developments of some of the most fundamental ideas of modern mathematics. Such series are now known as the *Fourier series*, which we will discuss in Chap. 10.

In 1837, *Gustav Dirichlet* (1805–1859) formulated a definition of a function while investigating the behaviour of the Fourier series of a periodic function at points of discontinuity. At a later stage, in Chap. 10, we will have a lot more to add to this story.

Definition 1.33 (*Dedekind, 1888*) Let X and Y be any two nonempty sets. A *function* $f : X \to Y$ is a *rule* such that for each $x \in X$ there is a unique $y \in Y$ given by $f(x)$. We write $y = f(x)$. The set X is called the *domain of the function* f, which we usually write as $\mathrm{Dom}(f)$.

The set theoretic formulation, as given above, of Dirichlet's definition is due to *Richard Dedekind*. The real impetus for this modern definition came through the work of *Georg Cantor* on descriptive set theory.

The Taylor series (1.5.4) of a function, as discussed in Chap. 8, was generally taken to be valid until the discovery of the function $f(x) = e^{-1/x^2}$ by *Augustin-Louis Cauchy* (1789–1857), in 1823. As we show later, the Taylor series of $f(x)$ converges, but not to the function $f(x)$. We will also discuss some other examples of functions for which the Taylor series does not converge at any $x \neq a$, for some $a \in \mathbb{R}$. It further confirmed the importance of the idea of *infinitely small* as a limit, which ultimately provided the mathematical foundation to the *limit concept* first introduced by *Bernard Bulzano* (1781–1848), in 1817; followed by ϵ–δ definition due to Cauchy, in 1820; and further refined by *Karl Weierstrass* (1815–1897), in 1861, to formulate the definition of continuity of a function. Along with Eduard Heine, and his doctoral student Gorge Cantor, Weierstrass clarified during 1870–

1872 the fundamental notions such as the uniform convergence, uniform continuity, term-wise integration, and differentiation of infinite series.

Exercises 1

1.1 Supply a proof of Theorem 1.1.

1.2 Supply a proof of Theorem 1.2.

1.3 Find a sequence $\langle S_n \rangle$ of subsets of $\mathbb{N}$, with $S_n \supset S_{n+1}$ for all $n \geq 1$, such that $\bigcap_{n=1}^{\infty} S_n = \emptyset$.

1.4 Let A, X, and Y be three nonempty sets. Interpret the product $A \times X \times Y$, where A is the set of airlines and X, Y are sets of cities connected by air.

1.5 Let X, Y, and Z be three nonempty sets.

   a. What can be said about sets X and Y if $X \times Y = \emptyset$?

   b. For nonempty sets X and Y show that $X \times Y \neq Y \times X$ unless $X = Y$.

   c. Explain why $X \times Y \times Z$ and $(X \times Y) \times Z$ are not the same sets.

1.6 Supply a proof of Theorem 1.5.

1.7 Show that $\lfloor 2x \rfloor = \lfloor x \rfloor + \lfloor x + (1/2) \rfloor$, for all $x \in \mathbb{R}$.

1.8 (a) Prove or disprove: $\lceil x + y \rceil = \lceil x \rceil + \lceil y \rceil$, for all $x, y \in \mathbb{R}$.
(c) Show that, for any $x \in \mathbb{R}$, $\lceil x \rceil - \lfloor x \rfloor = 1$, if x is not an integer; and 0 otherwise.

1.9 Supply a proof of Theorem 1.6.

1.10 (Cantor) Show that there does not exist any surjective function $A :\rightarrow \mathcal{P}(A)$.

1.11 Give an example of a function $f : \mathbb{Z} \rightarrow \mathbb{Z}$ such that

   a. f is neither injective nor surjective;

   b. f is injective but not surjective; and

   c. f is surjective but not injective.

1.12 Let $f : X \rightarrow Y$ be a function, and $A \subseteq X, B \subseteq Y$.

   a. Show that $f^{-1}(Y \setminus B) = X \setminus f^{-1}(B)$, and

$$f(X \setminus A) \supset Y \setminus f(A) \quad \Leftrightarrow \quad f \text{ is surjective.}$$

   b. $f^{-1}(f(A)) = A$ if f is injective; and $f(f^{-1}(B)) = B$ if f is surjective.

1.13 Let $f : X \rightarrow Y$ and $g : Y \rightarrow Z$ be two functions.

   a. Show that if $f \circ g$ is bijective then g is injective and f is surjective;

   b. Show that if f is bijective then so is the inverse $f^{-1} : Y \rightarrow X$.

1.14 Supply a proof of Theorem 1.9.

1.15 Show that if there is an injective function $f : X \rightarrow Y$, then there exists a surjective function $g : Y \rightarrow X$.

1.16 Show that the polynomial function $f : \mathbb{Z}^+ \times \mathbb{Z}^+ \to \mathbb{Z}^+$ defined by

$$f(a, b) = \frac{(a + b - 2)(a + b - 1)}{b + 2}$$

is both injective and surjective.

1.17 Show that the total number of possible relations on a set X with n elements is 2^{n^2}. How many reflexive relations are possible on X ?

1.18 Show that a relation R on a nonempty set X is symmetric and transitive if and only if $R = R^{-1} \circ R$.

1.19 Prove the following statements.

   a. The relation $\sim$ on $\mathbb{Z}$ defined by $a \sim b$ if $a = b$ or $a = -b$ is an equivalence relation.
   b. The relation $\sim$ on $\mathbb{R}$ defined by $a \sim b$ if $b - a \in \mathbb{Z}$ is an equivalence relation.
   c. The relation $\sim$ on $\mathbb{R}$ defined by $r \sim s$ if $|r - s| < 1$ is not an equivalence relation.

1.20 If S and T are two nonempty subsets of $\mathbb{R}$ and $S \subset T$, show that $\inf T \leq \inf S$ if T is bounded below, and $\sup S \leq \sup T$ if T is bounded above.

1.21 If S and T are two bounded nonempty subsets of $\mathbb{R}$ and $S \subset T$, show that $\inf T \leq \inf S \leq \sup S \leq \sup T$.

1.22 If S and T are two subsets of $\mathbb{R}$ and $s \leq t$ for each $s \in S$ and $t \in T$, show that S is bounded above and T is bounded below. Further show that $\sup S \leq \inf T$.

1.23 If S and T are two subsets of $\mathbb{R}$ which are bounded above, find $\sup(S \cup T)$.

1.24 For two nonempty subsets S and T of $\mathbb{R}$, define $S + T = \{s + t : s \in S, t \in T\}$. If S and T are bounded below, find $\inf(S + T)$ in terms of $\inf S$ and $\inf T$. What can you say about $\sup(S + T)$? If $S = \{a\}$, write $S + T = a + T$. Give a suitable definition of $S - T$ for bounded subsets S and T of $\mathbb{R}$. Find $\inf(a + T)$, $\sup(a + T)$, $\sup(S - T)$, and $\inf(S - T)$ in terms of supremum and infimum of S and T.

1.25 If S is bounded above subset of $\mathbb{R}$, show that $kS = \{ks : s \in S\}$ is bounded above and $\sup(kS) = k \sup S$ for each $k > 0$. What happens if $k \leq 0$? Is $S + S = 2S$?

1.26 For two nonempty subsets S and T of positive real numbers in $\mathbb{R}$, define $ST = \{st : s \in S, t \in T\}$. Find $\inf(ST)$ in terms of $\inf S$ and $\inf T$. What can you say about $\sup(ST)$?

1.27 Let S be a subset of positive real numbers and $1/S = \{1/s : s \in S\}$. If S is bounded above, show that $1/S$ is bounded below and $\inf(1/S) = 1/ \sup S$.

1.28 Show that a set A is finite with n elements $\Leftrightarrow$ there is some surjective function $: \mathbb{N}_n \to A$ $\Leftrightarrow$ there is some injective function $: A \to \mathbb{N}_n$.

1.29 Show that a set A is infinite $\Leftrightarrow$ there is some injective function $: \mathbb{N} \to A \Leftrightarrow A$ is equinumerous to some proper subset.

1.30 Show that a set A is countable $\Leftrightarrow$ there is an injective function : $\mathbb{N} \to A$ $\Leftrightarrow$ there is a surjective function : $\mathbb{N} \to A$.
1.31 Prove that the product of a finite number of countably infinite set is countably infinite.

References

1. H.D. Ebbinghaus, *Ernst Zermelo—An Approach to His Life and Work* (Springer-Verlag, Heidelberg, 2007)
2. H.B. Enderton, *Elements of Set Theory* (Academic Press Inc, NY, 1977)
3. E. Hairer, G. Wanner, *Analysis by Its History* (Springer, Berlin, 2002)

Chapter 2
The Real Numbers

The ultimate nature of reality is numbers.

Pythagoras (570–495 BC)

Greeks used algebraic equations to solve some simple geometric problems. For example, to find the length a of sides of a square given that its perimeter is 8 units, the equation $4x = 8$ gives $x = 2$. So, the equation has an integer solution. Next, if the perimeter is 7 units, then the length of the side of the square is $7/4$ units. So, if we understand rational numbers $\mathbb{Q}$ as fractions, one obtains a concept of *fractional length*. To move a step further, suppose we need to find the length of a side of a square having area 2 units, then one needs to solve the equation $x^2 = 2$. In such situations, as the associated equation is not solvable in $\mathbb{Q}$, Pythagoreans failed to establish a connection between length and arithmetic numbers. It was one of many reasons that motivated the development of a number system larger than the set $\mathbb{Q}$, preferably without compromising geometry in favour of arithmetic. There enters the set of real numbers, $\mathbb{R}$!

We start with the set of natural numbers $\mathbb{N}$ and, gradually, construct the set of rational numbers $\mathbb{Q}$ as an *ordered field*. Once we are done constructing $\mathbb{Q}$, we call a number *irrational number* if it is not in the set $\mathbb{Q}$. Together they constitute the set of real numbers $\mathbb{R}$. The exposition given in the first section prepares ground to prove ultimately the fact that the set $\mathbb{R}$ is a *complete ordered field*, and it is *uniquely* so. We say a set F, with at least two elements $0 \neq 1$, is a *field* if it is possible to define addition and multiplication in F such that all the fundamental properties of four *arithmetic operations* hold in F. And, we say a field F is *complete* with respect to some *linear order* $<$ if there are no *gaps* in the sense that everything that is bounded above has a *least upper bound*. The discussion here is concluded with a word about some important countable and uncountable subsets of $\mathbb{R}$.

Most analytical structures, and all characteristic functional properties, that we will discuss in subsequent chapters of the book are derived directly from some of the

A. K. Razdan and V. Ravichandran, *Fundamentals of Analysis with Applications*,
https://doi.org/10.1007/978-981-16-8383-1_2

foundational theorems proved here. The prerequisites for the topics presented here are basic concepts about sets and functions as discussed in the previous chapter.

2.1 Ordered Field $\mathbb{Q}$

A *constructive approach* to the development of the set of real numbers $\mathbb{R}$ starts with the assumption that there exists a set $\mathbb{N}$ satisfying the following five properties:

1. $\mathbb{N}$ is nonempty (as $1 \in \mathbb{N}$).
2. Each $n \in \mathbb{N}$ has a *unique* successor, say $\hat{n}$.
3. There exists an $n^* \in \mathbb{N}$ that is not the successor of any $n \in \mathbb{N}$.
4. For $n_1 \neq n_2 \in \mathbb{N}$, $\hat{n}_1 \neq \hat{n}_2$.
5. The only subset of $\mathbb{N}$ that contains n^*, and successors of all its elements, is the set $\mathbb{N}$ itself.

These are known as the *Peano's axioms*, in honour of *Giuseppe Peano* (1858–1932), first published in 1889 as a pamphlet "*Arithmetices principia, nova methodo exposita*". The elements of the set $\mathbb{N}$ are called the *natural numbers*, which we usually write as

$$1, 2, 3, \ldots .$$

Though Peano later changed $Axiom - 1$ to include 0 as the *first element*, in this text, we write $\mathbb{N}_0$ for the set consisting of $0, 1, 2, 3, \ldots$, which are also called the set of *non-negative integers* (or *whole numbers*). The construction of the set $\mathbb{N}_0$ as an *axiomatic system*, using set theory axioms, is obtained in terms of von Neumann *ordinal numbers*:

$$0 = \emptyset, \quad 1 = \{\emptyset\}, \quad 2 = \{\emptyset, \{\emptyset\}\}, \quad \text{and so on.}$$

For obvious reason, the elements of $\mathbb{N}_0$ are also known as the *counting numbers*. According to famous German mathematician *Leopold Kronecker* (1823–1891), *the natural numbers are work of God, and all the rest is the work of mankind*.

By the Euclid's *division algorithm*, given $m, d \in \mathbb{N}_0$, with $d \neq 0$, there exist unique $q, r \in \mathbb{N}_0$ such that

$$m = qd + r, \quad \text{where} \quad 0 \leq r < |d|.$$

In particular, if $r = 0$ then $m = qd$ implies that d *divides* m. So, a natural number n divides a natural number m if $m = nk$ for some $k \in \mathbb{N}$. We also say that n *is a divisor of* m. A natural number $p \geq 2$ is called a *prime* if only divisors of p are 1 and p. Euclid's proof that he gave in 300 BC of the fact that there exist infinitely many prime numbers[1] is a true classic in mathematics. In fact, Euclid's argument can also

---

[1] The number $2^{82589933} - 1$ having 24862048 digits is the largest known prime as on January 2020.

be used to generate new primes from the known ones. A basic scheme is as follows: Start with $p_1 = 2$, and define inductively p_{k+1} to be the least prime not chosen earlier that divides $m + 1$, where m runs over all the divisors of the product $p_1 p_2 \cdots p_k$. It is an interesting theorem that the sequence of primes $\langle p_n \rangle$ constructed in this manner exhausts all primes. We refer the reader to [1] for more on prime numbers.

Peano's *fifth axiom*, also known as the principle of *mathematical induction* (PMI, in short), is an important *method of proof*. It applies when we need to *verify a conjectured statement* $p(n)$ for all $n \in K \subseteq \mathbb{N}_0$. For example, each one of the following is a statement $p(n)$:

1. $(1 + a)^n \geq 1 + na$, with $a \geq -1$, for $n \geq 0$.
2. $1 - na < (1 - a)^n < 1/(1 + na)$, with $0 < a < 1$, for $n \geq 2$.
3. $2n < n^2$, for $n \geq 3$;
4. the *closed form* of the *Fibonacci sequence* $\langle f_n \rangle$ defined *recursively* as $f_1 = f_2 = 1$, $f_{n+2} = f_n + f_{n+1}$, $n \in \mathbb{N}$, is given by

$$f_n = \frac{1}{\sqrt{5}} \left(\varphi^n + (-1)^{n-1} \frac{1}{\varphi^n} \right), \quad n \in \mathbb{N}, \qquad (2.1.1)$$

where $\varphi = (1 + \sqrt{5})/2$ is known as the *golden ratio*.

An argument used to prove any such conjectured statement $p(n)$ is known as the *proof by induction*. In 1654, Blaise Pascal used it for the first time to prove the binomial theorem for $n \geq 1$ (Theorem 1.19).

Lemma 2.1 (Principle of Mathematical Induction). *For $n \in \mathbb{N}$, let $p(n)$ be a conjectured proposition about n. Suppose*

1. *Base Step: $p(1)$ is true; and,*
2. *Induction Hypothesis: for $k \in \mathbb{N}$, $p(k) \Rightarrow p(k + 1)$.*

Then $p(n)$ holds for all $n \in \mathbb{N}$.

Proof Let $A = \{n \in \mathbb{N} : p(n) \text{ holds}\}$. By (1), we have $1 \in A$. Also, by (2), $k \in A \Rightarrow k + 1 \in A$. Therefore, by Peano's *fifth axiom*, we must have that $A = \mathbb{N}$. That is, $p(n)$ holds for all $n \in \mathbb{N}$. □

Notice that it is important to verify both (1) and (2) for an assertion $p(n)$ to hold for all $n \in \mathbb{N}$. For example, (2) holds if $p(n) : 2|2n - 1$, but $p(n)$ is not true for any $n \in \mathbb{N}$. On the other hand, $p(n) : n^2 = n$ is true only for $n = 1$. Also, the conjectured statement $p(n)$ that $n^2 - n + 41$ is a prime is true for $1 \leq n \leq 40$, but $p(41)$ fails to hold. However, we are free to start from any *base value*, other than the 1.

Lemma 2.2 (Principle of Mathematical Induction). *For any fixed $m \in \mathbb{Z}$, let each $p(n)$ be a proposition about an integer $n \geq m$. If*

1. *$p(m)$ is true;*
2. *for an integer $k \geq m$, $p(k) \Rightarrow p(k + 1)$,*

then $p(n)$ *holds for all integers* $n \geq m$.

Proof Apply Lemma 2.1 to $q(n) = p(n - m + 1)$, $n \in \mathbb{N}$. $\square$

Notice that there is a *natural order* on the set $\mathbb{N}$, given by Peano's second axiom, where 1 is (taken to be) the least element of the set $\mathbb{N}$, by Peano's third axiom. In general, every nonempty subset of $\mathbb{N}$ has the least element. This property of $\mathbb{N}$ is known as the *well ordering property* (WOP, in short).

The next theorem proves the well-ordering property of $\mathbb{N}$ is equivalent to the principle of mathematical induction. For proof, we anticipate the set of positive real numbers $\mathbb{R}^+$, and use the following concept: A subset $A \subset \mathbb{R}$ is called an *inductive set* if $1 \in S$, and $x \in S \Rightarrow x + 1 \in S$. In this setting, Peano's fifth axiom, and hence PMI, can be reformulated as follows: *The set* $\mathbb{N}$ *is the smallest inductive subset of* $\mathbb{R}^+$.

Theorem 2.1 $WOP \Leftrightarrow PMI$.

Proof First, suppose WOP holds for $\mathbb{N}$, and suppose $S \subseteq \mathbb{N}$ is an inductive set. If $S \neq \mathbb{N}$, then $\mathbb{N} \setminus S$ is a nonempty subset of $\mathbb{N}$. So, by WOP, $\mathbb{N} \setminus S$ has the least element, say ℓ. Since $1 \in S$, we have $\ell > 1$. Then, $\ell - 1$ is a natural number in S. But then, by inductive property of S, we have $\ell \in S$, a contradiction. Hence, $S = \mathbb{N}$. This proves PMI.

Conversely, suppose $\mathbb{N}$ is the smallest inductive set, and let S be a nonempty subset of $\mathbb{N}$. If possible, suppose S does not have the least element. Then, since $1 \in \mathbb{N}$ is the least element, we have $S \neq \mathbb{N}$. That is, $1 \in K = \mathbb{N} \setminus S$. We claim that K is an inductive set. For, let $k \in K$. We proceed to show that $k + 1 \in K$. Again, let us assume for a while that $k + 1 \notin K$. Then, $k + 1 \in S$ implies that $n_0 < k + 1$, for some $n_0 \in S$, because S has no least element. So, if $n_0 - 1 \in S$, we conclude from the previous inequality that $k \in S$, which is not the case by our choice of k. But then, we obtain $n_0 - 1 \in \mathbb{N} \setminus S = K$. This, in turn, implies that $n_0 \leq k$, using the relation $n_0 - 1 < k$, so $n_0 \in K$, which is a contradiction because $n_0 \in S$. Hence, we must have that $k + 1 \in K$. That is, K is an inductive set. Therefore, $K = \mathbb{N}$ by PMI, i.e., $S = \emptyset$, a contradiction to the fact that $S \neq \emptyset$. Hence, S has the least element. $\square$

Among all properties of the set $\mathbb{N}$, the following one is the simplest.

Theorem 2.2 (Fundamental Theorem of Arithmetic). *Every natural number* $n > 1$ *can be expressed uniquely as a finite product of prime numbers.*

Said differently, prime numbers constitute the (multiplicative) building blocks of the set $\mathbb{N}$. The following lemma, known as *Euclid's first theorem*, provides a proof of the existence part of Theorem 2.2. A proof of uniqueness part by the principle of mathematical induction is a bit involved.

Lemma 2.3 *The product* ab *of two natural numbers is divisible by a prime number* p *if and only if* p *divides* a *or* b.

Proof We prove it for the case when $a = b$. The general case follows on similar lines. If p does not divide a, then $a = p_1^{m_1} p_2^{m_2} \cdots p_k^{m_k}$ where none of these p_i's are equal to p. Then $a^2 = p_1^{2m_1} p_2^{2m_2} \cdots p_k^{2m_k}$ and p is not a factor of a^2 which is not possible as p divides a^2. □

The indirect proof given above is called the *method of contradiction*.

The modern decimal-based place value notation has its origin in *Aryabhatiya* of *Aryabhatta* (476–540 AD). *Brahmagupta* (598–668 AD) used zero[2] in 628 AD and gave rules to compute with zero in his work *Brāhmasphuṭasiddhānta*[3] (the Opening of the Universe) such as $n + 0 = n - 0 = n$ and $n \cdot 0 = 0$. He defined the number zero as the result of subtracting a number from itself.

Brahmagupta also gave arithmetical rules in terms of the positive and negative numbers like the product of a negative number and a positive number is a negative number. The numbers $\ldots, -3, -2, -1, 0, 1, 2, 3, \ldots$ are the *integers*, and the set of all integers is denoted by $\mathbb{Z}$ (the first letter of the German word "Zahlen" for numbers).

To construct integers from the set $\mathbb{N}$, define a relation on $\mathbb{N} \times \mathbb{N}$ as follows: For $m, n, m', n' \in \mathbb{N}$,

$$(m, n) \sim (m', n') \quad \Leftrightarrow \quad m + n' = m' + n.$$

It follows easily that $\sim$ is an equivalence relation. Let $\mathbb{Z}$ be the set of equivalence classes. Notice that, for any $(m, n) \in \mathbb{N} \times \mathbb{N}$,

$$[(m, n)] = \{(a, b) \in \mathbb{N} \times \mathbb{N} : m - n = a - b\}.$$

So, we may write

$$\ldots, -2 = [(n + 2, n)], \quad -1 = [(n + 1, n)], \quad 0 = [(n, n)],$$
$$1 = [(n, n + 1)], \quad 2 = [(n, n + 2)], \ldots.$$

Thus, the positive integers are given by the equivalence classes $[(n, n + m)]$, for $m = 0, 1, 2, \ldots$.

Definition 2.1 Let A be a nonempty set. A function $\star$ defined on the product set $A \times A$ is called a *binary operation* (on A) if $\star(a, b) \in A$, for all $a, b \in A$. For $a, b \in A$, we write $\star(a, b)$ simply as $a \star b$, and call it in general the *product* of elements a and b.

Every function $\star : A \times A \to A$ defines a binary operation on the set A. To continue with a general discussion about the properties of a binary operation on a set,

---

[2] Can you find the connection between "sifer" and "cyber"? The Sanskrit word "shunyam" or the Hindi equivalent "shunya" was translated to Arabic as al-sifer. Fibonacci called it as "cifra" and "cipher" came from it. From the Medieval Latin word "zephirum", the word "zero" originated.

[3] You can see this in many other old mathematics books at http://www.wilbourhall.org/index.html.

it is convenient to write the product $a \star b$ simply as $a \cdot b$ or ab. Notice that the term product used here is not necessarily means *multiplication* in the usual sense.

Clearly, for $m, n \in \mathbb{Z}$, the sum $m + n$ and the product mn are integers. That is, the usual addition and multiplication of integers are binary operations on the set $\mathbb{Z}$. Further, *addition* as a binary operation on the set $\mathbb{Z}$ satisfies the following properties: For all $m, n, p \in \mathbb{Z}$,

1. $(+)$ is *associative*, i.e., $(m + n) + p = m + (n + p)$;
2. $(+)$ has the *identity* 0, i.e., $0 + n = n = n + 0$;
3. $(+)$ is *invertible*, i.e., for each $n \in \mathbb{Z}$ there is some $m \in \mathbb{Z}$ such that $n + m = 0 = m + n$. Of course, $m = -n$; and,
4. $(+)$ is *commutative*, i.e., $n + m = m + n$.

All put together, we say the set of integers $\mathbb{Z}$ is an *Abelian group* with respect to the usual addition.

Definition 2.2 A nonempty set A with a binary operation $\cdot : A \times A \to A$ is called an *Abelian group* if

1. *Associativity* : $a \cdot (b \cdot c) = (a \cdot b) \cdot c$, for all $a, b, c \in A$;
2. *Existence of Identity* : There is some $e \in A$ such that $a \cdot e = a = e \cdot a$. The element e is called the *identity*;
3. *Existence of Inverse* : For each $a \in A$ there is some $b \in A$ such that $a \cdot b = e = b \cdot a$. We write $b = a^{-1}$, called it the *inverse* of a;
4. *Commutativity* : $a \cdot b = b \cdot a$, for all $a, b \in A$.

If $(A, \cdot)$ is an Abelian group, we prefer to write $a \cdot b$ as $a + b$, and the identity e as 0 so that with respect to additive notation the inverse of a is written as $-a$, i.e., $a + (-a) = 0 = (-a) + a$. So, in general, we write $(A, +)$ for an Abelian group A, with the (additive) identity 0.

Example 2.1 Clearly, usual addition and multiplication define binary operations on the set $\mathbb{N}$. But, usual subtraction is not a binary operation on $\mathbb{N}$. Notice that $(\mathbb{N}, +)$ is not an Abelian group because $0 \notin \mathbb{N}$ and $-n \notin \mathbb{N}$ for any $n \in \mathbb{N}$. Also, $(\mathbb{N}, \cdot)$ is not an Abelian group because $n^{-1} \notin \mathbb{N}$ for any $1 < n \in \mathbb{N}$, but it has the *multiplicative identity* 1.

Also, the *multiplication* as a binary operation on the set $\mathbb{Z}$ satisfies the following properties: For all $m, n, p \in \mathbb{Z}$,

1. $(\cdot)$ is *associative*, i.e., $(m \cdot n) \cdot p = m \cdot (n \cdot p)$;
2. $(\cdot)$ has the *identity* 1, i.e., $1 \cdot n = n = n \cdot 1$;
3. $(\cdot)$ *distributes* over the addition $(+)$, i.e., $m \cdot (n + p) = m \cdot n + m \cdot p$; and
4. $(\cdot)$ is *commutative*, i.e., $m \cdot n = n \cdot m$.

All put together, we say that $(\mathbb{Z}, +, \cdot)$ is a *commutative ring*, with (multiplicative) identity $1 \in \mathbb{Z}$.

Definition 2.3 An Abelian group $(A, +)$, with (additive) identity 0, is called a *commutative ring* if there is a binary operation $\cdot : A \times A \to A$ such that

5. *Associativity* : For all $a, b, c \in A$, $a \cdot (b \cdot c) = (a \cdot b) \cdot c$;
6. *Existence of Identity* : There is some $e \in A$ such that $a \cdot e = a = e \cdot a$. In this case, the element e is called the *multiplicative identity*, which is usually denoted by 1;
7. *Commutativity* : For all $a, b \in A$, $a \cdot b = b \cdot a$;
8. *Distributivity* : For all $a, b, c \in A$, $a \cdot (b + c) = a \cdot b + a \cdot c$.

Further, for $m, n \in \mathbb{Z}$, we say $m < n$ if $n - m \in \mathbb{N}$; and we write $m \leq n$ if either $m = n$ or $m < n$. This defines an *ordering* on $\mathbb{Z}$ in the following sense:

9. For any $m, n \in \mathbb{Z}$, exactly one of the following holds:

$$ m = n, \quad m < n, \quad \text{or} \quad n < m. $$

This is known as the *trichotomy law*;
10. If $m, n, p \in \mathbb{Z}$, $m < n$ and $n < p$, then $m < p$ (*transitivity*)
11. If $m, n, p \in \mathbb{Z}$, $m < n$, then $m + p < n + p$, for any p; and $mp < np$, for any $p > 0$ (*monotonicity* of addition and multiplication)

Therefore, we find that $(\mathbb{Z}, +, \cdot, 0, 1, <)$ is an *Ordered ring*. Notice that *division* is not possible in the commutative ring $(\mathbb{Z}, +, \cdot, 0, 1)$, because $a^{-1} \notin \mathbb{Z}$ for any $0 \neq a \in \mathbb{Z}$, unless $a = \pm 1$. Finally, we use the *ordered ring* $(\mathbb{Z}, +, \cdot, <)$ to construct the set of *rational numbers* ℚ.

Theorem 2.3 *The relation $\sim$ on the set $\mathbb{Z} \times \mathbb{N}$ defined by*

$$ (m, n) \sim (p, q) \quad \Leftrightarrow \quad mq = np, \quad \text{for } m, p \in \mathbb{Z} \text{ and } n, q \in \mathbb{N}, $$

is an equivalence relation. The set of equivalence classes is the set ℚ.

Proof It is easy to verify that, for all $(m, n), (p, q) \in \mathbb{Z} \times \mathbb{N}$,

1. $(m, n) \sim (m, n)$
2. $(m, n) \sim (p, q) \Rightarrow (p, q) \sim (m, n)$
3. $(m, n) \sim (p, q)$ and $(p, q) \sim (r, s) \Rightarrow (m, n) \sim (r, s)$.

Each equivalence class $[(m, n)]$ defines a *rational number*, and we write ℚ for the collection of all equivalence classes. The *positive* rational numbers are the equivalence classes $[(m, n)]$, with $(m, n) \in \mathbb{N} \times \mathbb{N}$. The *negative* rational numbers are equivalence classes $[(m, n)]$, where m is a negative integer and $n \in \mathbb{N}$. The set ℚ consists of the *zero* $0 = [(0, n)]$, $n \in \mathbb{N}$, the positive rational numbers and the negative rational numbers.

Since (m, n) and (m', n') belong to the same equivalence class if $mn' = m'n$, it is always possible to choose (m, n) such that m/n is a *reduced fraction*. That is, m and n

are *co-prime*, with $m \neq 0$. Thus, it is possible to write each *nonzero* rational number *uniquely* as a reduced fraction m/n, with $0 \neq m \in \mathbb{Z}$ and $n \in \mathbb{N}$. In particular, the set of *positive* rational numbers $\mathbb{Q}^+$ consists of reduced fractions m/n, with $m, n \in \mathbb{N}$. We say a rational number is in *reduced form* if it is represented by a reduced fraction.

In our quest to construct a (unique) *complete ordered field*, the set integer $\mathbb{Z}$ fails to meet an important requirement, viz., it is not a *field*; no integer other than 1 and -1 has the multiplicative inverse. However, the set of rational numbers $\mathbb{Q}$ constructed above is a *field* with respect to the addition and multiplication defined by

$$\frac{m}{n} + \frac{p}{q} = \frac{mq + np}{nq} \quad \text{and} \quad \frac{m}{n}\frac{p}{q} = \frac{mp}{nq}, \quad m, n, p, q \in \mathbb{Z} \times \mathbb{N}.$$

Definition 2.4 A set F with atleast two elements $0, 1 \in F$, $1 \neq 0$, is called a *field* if there are two binary operations $+$ and $\cdot$ such that both $(F, +, 0)$ and $(F^*, \cdot, 1)$ are Abelian groups, where $F^* = F \setminus \{0\}$, and the following *compatibility condition* holds for all $a, b, c \in F$:

$$a \cdot (b + c) = a \cdot b + a \cdot c \quad \text{and} \quad (b + c) \cdot a = b \cdot c + c \cdot a.$$

We say $K \subset F$ is a *subfield* if K itself is a field with respect to the induced binary operations.

As before, we will write ab for $a \cdot b$. Notice that elements 0 and 1 in F, in general, may not be the usual numbers. For any $a \in F$, the additive inverse of a is written as $-a$, and the multiplicative inverse of an element $a \neq 0$ is written as $a^{-1} := 1/a$. We need to prove some simple properties of fields for the later part of the discussion.

Lemma 2.4 (Cancellation Laws) *Let F be a field and $a, b, c \in F$. If $a + c = b + c$, then $a = b$. If $ac = bc$ and $c \neq 0$, then $a = b$.*

Proof Since $c \in F$, there is $-c \in F$ such that $c + (-c) = 0$. If $a + c = b + c$, then by adding $-c$ on both sides, we have

$$(a + c) + (-c) = (b + c) + (-c).$$

Using the associate law for addition, we have

$$a + (c + (-c)) = b + (c + (-c)).$$

Hence, using the definition of inverse element, we have

$$a + 0 = b + 0.$$

Since 0 is an additive identity, it follows that $a = b$. The other result follows by multiplying $ac = bc$ by c^{-1}. $\square$

Corollary 2.1 *The additive identity element in a field is unique or, equivalently, $a + b = a$ implies $b = 0$. The multiplication identity is also unique or, equivalently, $ab = a$ implies $b = 1$.*

Proof Since $a + b = a = a + 0$, it follows by cancellation law that $b = 0$. Thus, the additive identity of F is unique. If $ab = a$, then $ab = a = a1$ implies $b = 1$. $\square$

Lemma 2.5 *For each element a in a field F, we have $-(-a) = a$. If $a \neq 0$, then $(a^{-1})^{-1} = a$.*

Proof Since $b = -a$ if and only if $a + b = 0$, the result follows because $(-a) + a = a + (-a) = 0$. If $a \neq 0$, then $a^{-1} \in F$ and $aa^{-1} = 1$, so $a^{-1}a = 1$. Therefore, $(a^{-1})^{-1} = a$. $\square$

Lemma 2.6 *For every a, b in a field F, we have*

1. $0a = 0$.
2. *if $a \neq 0$, $b \neq 0$, then $ab \neq 0$.*
3. $(-a)b = a(-b) = -(ab)$.
4. $(-a)(-b) = ab$.

Proof For (1), since $0 + 0 = 0$, we have

$$0a + 0a = (0 + 0)a = 0a = 0a + 0$$

and, by cancellation law, $0a = 0$. For (2), suppose $a \neq 0$, $b \neq 0$ but $ab = 0$. Then, $1/a, 1/b \in F$ and, by using $0a = 0$ for each $a \in F$, we have

$$(ab)(b^{-1}a^{-1}) = 0(a^{-1}b^{-1}) = 0.$$

Also, by associativity of multiplication and properties of multiplicative inverse and multiplicative identity element, we see that

$$(ab)(b^{-1}a^{-1}) = a(b(b^{-1}a^{-1})) = a((bb^{-1})a^{-1})$$
$$= a(1a^{-1}) = aa^{-1} = 1.$$

Thus, we have $0 = 1$, a contradiction. Therefore, $ab \neq 0$. For (3), we show first that $(-a)b = -(ab)$. Since

$$ab + (-a)b = (a + (-a))b = 0b = 0$$

and $b = -a$ if and only if $a + b = 0$, we have $(-a)b = -(ab)$. Using $(-a)b = -(ab)$ and commutativity of multiplication, we see that

$$a(-b) = (-b)a = -(ba) = -(ab).$$

We can also give another proof by observing that the statement is equivalent to $ab = 0$, $a \neq 0$ implies $b = 0$. If $ab = 0$ and $a \neq 0$, we see that $0 = a^{-1}0 = a^{-1}(ab) = (a^{-1}a)b = 1b = b$. For (4), since $(-a)b = a(-b) = -(ab)$, we have

$$(-a)(-b) = -(a(-b)) = -(-(ab)) = ab.$$

The last equality follows by Lemma 2.5.   □

The *order relation* $<$ on $\mathbb{Q}$ is defined as follows: For m/n, $p/q \in \mathbb{Q}$, with $n, q \in \mathbb{N}$,

$$\frac{m}{n} < \frac{p}{q} \quad \Leftrightarrow \quad mq < pn.$$

In general, we apply first sign change to ensure that denominators are positive. Then, the field of all rational numbers $\mathbb{Q}$ is an *ordered field* in the following sense.

Definition 2.5 A field F is an *ordered field* if there is a relation $<$ defined on it such that, for any $a, b \in F$,

1. exactly one of the following holds: $a < b$, $a = b$, $b < a$;
2. $a < b$, $b < c \Rightarrow a < c$;
3. $a < b \Rightarrow a + c < b + c$; and $a > 0$, $b > 0 \Rightarrow ab > 0$.

The condition (1) says $<$ satisfies the *law of trichotomy*; condition (2) says $<$ is *transitive*; and condition (3) says both addition and multiplication are *monotone with respect to* $<$. Notice that a relation $<$ satisfying (1) and (2) is not an order relation *per se* because it is not reflexive. However, we may define

$$a \leq b \quad \Leftrightarrow \quad a < b \text{ or } a = b.$$

Also, $(F, \leq)$ is totally ordered by (1). The next lemma proves some simple properties implied directly by the above stated axioms.

Lemma 2.7 *Let F be an ordered field, and $a, b, c \in F$. Then the following holds:*

1. *$a > b$ if and only if $b - a < 0$. In particular, $a > 0$ if and only if $-a < 0$. Also $a > b$ if and only if $a - b > 0$.*
2. *If $b < c$, then $ab < ac$ for $a > 0$ and $ab > ac$ for $a < 0$. In particular, (i) $a > 0$, $b < 0$ implies $ab < 0$ and (ii) $a < 0$, $b < 0$ implies $ab > 0$.*
3. *$a^2 > 0$ for $a \neq 0$ and, in particular, $1 > 0$.*
4. *If $a > 0$, then $a^{-1} > 0$. Also, if $0 < a < b$, then $0 < b^{-1} < a^{-1}$.*

Proof In an ordered field F, (i) $a < b$ implies $a + c < b + c$ and (ii) $a > 0$, $b > 0$ implies $ab > 0$.

1. If $a > b$, then adding $-a$ and using (i), we get

$$0 = a + (-a) > b + (-a) = b - a.$$

This proves $b - a < 0$. If $b - a < 0$, then by adding a and using (i), we get

$$b = b + 0 = b + (-a + a) = (b - a) + a < 0 + a = a.$$

The particular case follows by taking $b = 0$. The other result follows as $-(b - a) = a - b$.

2. Let $b < c$. Then $c - b > 0$. If $a > 0$, then

$$ac - ab = a(c - b) > 0$$

and hence $ab < ac$. If $a < 0$, then $-a > 0$ and therefore

$$0 < (-a)(c - b) = (-a)c + (-a)(-b) = -ac + ab$$

or $ab > ac$.

3. If $a > 0$, then $a^2 = aa > 0$. If $a < 0$, then $-a > 0$ and so

$$a^2 = aa = (-a)(-a) > 0.$$

Since $1^2 = 1.1 = 1$ and $1 \neq 0$, it follows that $1 > 0$.

4. Let $a > 0$. If $a^{-1} = 0$, then $1 = aa^{-1} = 0$ which is absurd. Similarly, if $a^{-1} < 0$, then $1 = aa^{-1} < 0$ which contradicts $1 > 0$. Hence, $a^{-1} > 0$. Let $a > 0$ and $b > 0$. Then $a^{-1} > 0$ and $b^{-1} > 0$. By multiplying $a < b$ by a^{-1}, we get $1 = aa^{-1} < ba^{-1}$. Multiplying this by b^{-1}, we get

$$b^{-1} = b^{-1}1 < b^{-1}(ba^{-1}) = (b^{-1}b)a^{-1} = 1a^{-1} = a^{-1}. \qquad \square$$

An element a in an ordered field $(F, <)$ is called a *positive element* if $a > 0$; and it is called a *negative element* if $a < 0$. We write F^+ for the set of all positive elements of an ordered field F. Notice that the set F^+ has the following two properties:

1. $a, b \in F^+ \implies a + b, \, ab \in F^+$; and
2. for any $a \in F$, exactly one of the following holds:

$$a \in F^+ \quad \text{or} \quad a = 0 \quad \text{or} \quad a \notin F^+.$$

Conversely, suppose a nonempty subset P of a field F satisfies properties (1) and (2). Then, we say P is a *positive set* of the field F. In this formulation, we say a field F is *ordered* if it contains a positive set P. Notice that it gives an equivalent formulation of Definition 2.5.

Definition 2.6 Let F and K be two ordered fields. A *surjective* function $\phi : F \to K$ is called an (order-preserving) *isomorphism* if, for all $a, b \in F$, we have

1. ϕ is *additive*: $\phi(a + b) = \phi(a) + \phi(b)$;
2. ϕ is *multiplicative*: $\phi(ab) = \phi(a)\phi(b)$;
3. ϕ is *order-preserving*: $a < b \iff \phi(a) < \phi(b)$.

We say two ordered fields F and K are *order isomorphic* if there exists an isomorphism between them. In this case, we write $F \simeq K$.

Notice that (3) implies that ϕ is *injective*. We conclude the section by using the concept of *isomorphism* to prove that every ordered field $(F, <)$ contains a copy of $\mathbb{Q}$, and hence also of the sets $\mathbb{N}$ and $\mathbb{Z}$. To avoid the confusion, we write $0, 1 \in F$ as $0_F, 1_F$. Now, since the construction of $\mathbb{Q}$ started with the set of natural numbers $\mathbb{N}$, we may write

$$\underline{n} = n \, 1_F = 1_F + \ldots + 1_F \ (n \text{ terms}), \quad n \in \mathbb{N}.$$

The element $\underline{n} \in F$, for $n \in \mathbb{N}$, is called the nth *natural element* of the field F. We write $\mathbb{N}_F$ for the set of all natural elements of the ordered field F, and take $\underline{0} = 0_F$. Clearly, $\mathbb{N}_F$ is closed under addition and multiplication of F, and the function $\phi : \mathbb{N} \to \mathbb{N}_F$ given by

$$\phi(n) := \underline{n}, \quad n \in \mathbb{N},$$

is an isomorphism of ordered *sub-rings*. Notice that associativity of $+$ implies ϕ is an additive homomorphism; associativity of $\cdot$ and distributivity implies ϕ is a multiplicative homomorphism; and monotonicity of addition proves

$$n < m \quad \Leftrightarrow \quad \underline{n} < \underline{m}.$$

It thus follows that every ordered field $(F, <)$ contains a copy of $\mathbb{N}$. In particular, we conclude that the set of natural elements $\mathbb{N}_F \subset F$ has the well-ordering property.[4] We consider next the set of *integer elements* of an ordered field $(F, <)$ given by

$$\mathbb{Z}_F = \mathbb{N}_F \cup \{\underline{0}\} \cup \{-a \in F : a \in \mathbb{N}_F\},$$

and define a function $\phi : \mathbb{Z} \to \mathbb{Z}_F$ as follows:

$$\phi(n) = \begin{cases} \underline{n}, & \text{if } n \in \mathbb{Z}^+ \\ \underline{0}, & \text{if } n = 0 \\ -(\underline{-n}), & \text{if } n \in \mathbb{Z}^- \end{cases}.$$

With little extra effort, it follows that $\phi : \mathbb{Z} \to \mathbb{Z}_F$ is an isomorphism of ordered rings. Therefore, every ordered field $(F, <)$ contains a copy of the set of integers $\mathbb{Z}$. Finally, we consider the set of *rational elements* of an ordered field $(F, <)$ given by

$$\mathbb{Q}_F := \{mn^{-1} : m \in \mathbb{Z}_F, n \in \mathbb{N}_F\}.$$

Notice that, for $n < 0$, we have $\underline{n} = -(\underline{-n})$.

Theorem 2.4 *Every ordered field $(F, <)$ contains a copy of $\mathbb{Q}$.*

---

[4] Recall that the well-ordering property of $\mathbb{N}$ states that every nonempty subset of $\mathbb{N}$ has the least element with respect to the usual order.

Proof Using $\mathbb{Q}$ and $\mathbb{Q}_F$ as defined above, consider the function $\phi : \mathbb{Q} \to \mathbb{Q}_F$ given by

$$\phi(mn^{-1}) = \underline{mn}^{-1}, \quad \text{for } m \in \mathbb{Z} \text{ and } n \in \mathbb{N}.$$

We already know that the mapping $n \mapsto \underline{n}$ is an isomorphism from $\mathbb{N}$ onto $\mathbb{N}_F$ so that, for all $n \in \mathbb{N}$, $\underline{-n} = -\underline{n}$ and $\underline{n^{-1}} = \underline{n}^{-1}$. Thus, by associativity, commutativity, and distributivity, it follows that

$$
\begin{aligned}
\phi[(m_1 n_1^{-1})(m_2 n_2^{-1})] &= \phi[(m_1 m_2)(n_1 n_2)^{-1}] \\
&= \underline{(m_1 m_2)(n_1 n_2)}^{-1} = \underline{m_1}\,\underline{m_2}\,\underline{n_1}^{-1}\underline{n_2}^{-1} \\
&= (\underline{m_1}\,\underline{n_1}^{-1})(\underline{m_2}\,\underline{n_2}^{-1}) \\
&= \phi(m_1 n_1^{-1})\phi(m_2 n_2^{-1}); \\
\phi(m_1 n_1^{-1} + m_2 n_2^{-1}) &= \phi[(m_1 n_2 + m_2 n_1)(n_1 n_2)^{-1}] \\
&= \underline{[m_1 n_2 + m_2 n_1]}\,(\underline{n_1 n_2})^{-1} \\
&= [\underline{m_1}\,\underline{n_2} + \underline{m_2}\,\underline{n_1}]\underline{n_1}^{-1}\underline{n_2}^{-1} \\
&= \underline{m_1}\,\underline{n_1}^{-1} + \underline{m_2}\,\underline{n_2}^{-1} \\
&= \phi(m_1 n_1^{-1}) + \phi(m_2 n_2^{-1}).
\end{aligned}
$$

Finally, let $m_1 n_1^{-1} < m_2 n_2^{-1}$ so that $m_1 n_2 < m_2 n_1$, because $n_1 n_2 > 0$ and $m_1 n_2^{-1}$, $m_2 n_1^{-1} \in \mathbb{Z}$. Now, using the isomorphism $\mathbb{Z} \to \mathbb{Z}_F$ defined earlier, we have

$$n_2^{-1} n_1^{-1} > 0 \quad \Rightarrow \quad \underline{n_2}^{-1}\underline{n_1}^{-1} > 0.$$

It thus follows that

$$
\begin{aligned}
m_1 n_2 < m_2 n_1 &\Rightarrow \underline{m_1 n_2} < \underline{m_2 n_1} \\
&\Rightarrow \underline{m_1}\,\underline{n_2}\,\underline{n_2}^{-1}\underline{n_1}^{-1} < \underline{m_2}\,\underline{n_1}\,\underline{n_2}^{-1}\underline{n_1}^{-1} \\
&\Rightarrow \underline{m_1}\,\underline{n_1}^{-1} < \underline{m_2}\,\underline{n_2}^{-1}.
\end{aligned}
$$

Notice that ϕ is surjective, by definition. Therefore, ϕ is an isomorphism. $\qquad\square$

2.2 The Complete Ordered Field

To complete the construction of the set of real numbers, it remains to show that there exists a unique *complete ordered field* containing the ordered field $\mathbb{Q}$, which is written as $\mathbb{R}$. Since the three commonly known *methods of construction* of the set $\mathbb{R}$ are based on the existence of $\mathbb{Q}$, we start by discussing some of the deficiencies that the field $\mathbb{Q}$ has. We recommend [2], for some further finer details.

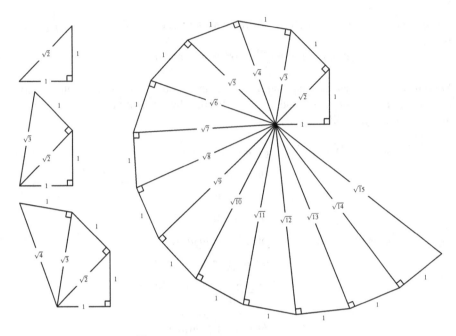

Fig. 2.1 Construction of $\sqrt{n}$

The first one is obvious because $\mathbb{Q}$ does not contain all the *conceivable numbers*. For example, in practical terms, there is no rational number that measures the length of diagonal of the square with unit side. In fact, as proved in the next theorem, there are infinitely many *numbers* that do not belong to $\mathbb{Q}$, which are called the **irrational numbers**.

Lemma 2.8 *For any prime number p, there is no rational number satisfying the equation $x^2 = p$. In particular, $\sqrt{2}$ is an irrational number.*

Proof For a contradiction, suppose there exists relatively prime positive integers a and b such that, for $x = a/b$, we have $x^2 = p$. Then, by squaring, we have $a^2 = pb^2$, which implies that p divides a^2. It then follows by Lemma 2.3 that p divides a. Let $a = pk$, for some integer k, so that

$$pb^2 = a^2 = (pk)^2 \quad \Rightarrow \quad b^2 = pk^2.$$

Therefore, we have p divides b^2. Once again, Lemma 2.3 implies that p divides b, i.e., $b = p\ell$, for some integer ℓ. We have gotten our contradiction because $a = pk$ and $b = p\ell$ implies p is a common factor of a and b. □

Figure 2.1 depicts a simple geometric procedure to construct the numbers $\sqrt{n}$, $n \in \mathbb{N}$. The next theorem proves that, for infinitely many $n \in \mathbb{N}$, $\sqrt{n} \notin \mathbb{Q}$. Clearly then, there must exist a number system larger than $\mathbb{Q}$ so that it is possible to measure such types of "*geometric lengths*".

Theorem 2.5 *For any $1 < n \in \mathbb{N}$ that is not a perfect square, $\sqrt{n}$ is not a rational number.*

Proof First, assume that n is a *square-free number*, i.e., it contains no square factor > 1. If possible, let $\sqrt{n} = a/b$, where a and b are natural numbers in lowest terms so that $nb^2 = a^2$. Let us write n as product of prime powers: $n = p_1^{k_1} p_2^{k_2} \cdots p_l^{k_l}$ where p_i's are prime numbers and k_i are some integers. Since n is not a perfect square, at least one k_i is odd. Let $p = p_i$ be the prime for which the power k_i in the prime factorisation of n is odd. Since p is a prime factor of a, it is factor of a^2 and hence a. But since there is no common factor between a and b, p is not a factor of b. Thus, p occurs to odd power in nb^2 and this contradicts the equation $nb^2 = a^2$ as p occurs in even power in a^2. □

There exist many other irrational numbers such as e, π, e^π, etc. Recall that e is called the *Euler constant*, which is usually defined as the infinite series given by

$$e = 1 + \frac{1}{1!} + \frac{1}{2!} + \frac{1}{3!} \cdots .$$

It is more familiar as the base of the *natural logarithm*. On the other hand, the number π is usually defined as the ratio of the circumference of any circle to its diameter. Archimedes was first to prove that the ratio is constant. He also obtained a fairly good estimate of π by applying two-sided *geometric limit process* and the *method of exhaustion*.

Example 2.2 We may also use arithmetic operations to find some other irrational numbers. 1. If r is a rational number and α is an irrational number, then $r + \alpha$ and $r\alpha$ are both irrational. For, if $r + \alpha$ is a rational number, then $\alpha = (r + \alpha) - r$, being difference of two rational numbers, is rational. Similarly, if $r\alpha$ is a rational number, then $\alpha = (r\alpha)/r$, being ratio of two rational numbers, is rational. 2. The numbers $\sqrt{8}$ and $\sqrt{12}$ are irrational. Since $\sqrt{2}$ and $\sqrt{3}$ are irrational, the irrationality of $\sqrt{8} = 2\sqrt{2}$ and $\sqrt{12} = 2\sqrt{3}$ follows.

We next introduce an important (topological) property of the ordered field $\mathbb{Q}$ as a subset of the set $\mathbb{R}$ of real numbers. This property has a very deep connection with a more delicate deficiency of $\mathbb{Q}$ that we will discuss a little later.

Definition 2.7 We say a subfield S of an ordered field $(F, <)$ is *dense* in F if, for any $a, b \in F$, with $a < b$, there is some $s \in S$ such that $a < s < b$.

Recall that between any two integers there are only finitely many integers. However, there exist infinitely many rational numbers between any two rational numbers. For, let $p \neq q \in \mathbb{Q}$. As $p < (p + q)/2 < q$ and $(p + q)/2$ is a rational number, it follows that there is at least one rational number between p and q. In general, we can define a sequence $\langle a_n \rangle$ in $\mathbb{Q}$ as follows:

$$a_1 = (p+q)/2, \quad \text{and} \quad a_{n+1} = \frac{a_n + q}{2}.$$

It is then easy to see that $p < a_1 < a_2 < \cdots < q$. It is also possible to find rational numbers between any two irrational numbers. The next example illustrates a procedure.

Example 2.3 We find some rational numbers between $\sqrt{2}$ and $\sqrt{3}$. Let $\sqrt{2} < p/q < \sqrt{3}$. Then $2q^2 < p^2 < 3q^2$. Choose $q = 2$. Then $8 < p^2 < 12$ and so $p = 3$. Thus, $3/2$ is a rational number between $\sqrt{2}$ and $\sqrt{3}$. If we take $q = 20$, then $800 < p^2 < 1200$. Since $29^2 = 841$ and $34^2 = 1156$, $p \in \{29, 30, 31, 32, 33, 34\}$. Hence the numbers $29/20, 30/20, \ldots, 34/20$ are rational numbers between $\sqrt{2}$ and $\sqrt{3}$.

Definition 2.8 We say an ordered field $(F, <)$ is **archimedean** if for each $a \in F$ there is some $n \in \mathbb{N}_F$ such that $n > a$. In this case, we also say $(F, <)$ is an **archimedean field**.

Recall that an element u of an ordered set $(S, \leq)$ is an *upper bound* of a subset $A \subset S$ if $a \leq u$, for all $a \in A$; and an element $\ell \in S$ is a *lower bound* of A if $\ell \leq a$, for all $a \in A$. We say A is *bounded above* if it has an upper bound; and A is said to be *bounded below* if it has a lower bound. Recall that a subset A of an ordered set S is said to be *bounded* if it is both bounded above and bounded below. In particular, it follows from Definition 2.8 that the set of *natural element* in every archimedean field is an *unbounded set*, as an ordered set.

Proposition 2.1 *Let $(F, <)$ be an archimedean field. Then, for any $a, b \in F$, with $a > 0$, there is some natural element $n \in \mathbb{N}_F$ such that $na > b$. The converse holds too.*

Proof Applying the above definition to the element $ba^{-1} \in F$, there is some $n \in \mathbb{N}_F$ such that $n > ba^{-1}$. So, since $a > 0$, we obtain $na > b$. Conversely, suppose an ordered field $(F, <)$ satisfies the given condition. Then, taking $a = 1_F$, it follows that the *archimedean property* holds for F. $\square$

Notice that Proposition 2.1, in layman sense, says that a person with a step size $a > 0$ can walk through any distance b if s/he takes sufficiently many number of steps, say n.

Remark 2.1 Notice that, for any real number $a > 0$, the set

$$S := \{n \in \mathbb{N} : \ | n > a\}$$

is nonempty by *archimedean property*, and it is bounded below by the number a. Hence, by well-ordering property of the set $\mathbb{N}$, the set S has the least element,[5] say n_0. Clearly then, we have $n_0 - 1 \leq a < n_0$. A similar argument applies to the case when $a < 0$. This simple observation justifies the definition of *floor functions* defined earlier in Chap. 1.

---

[5] Recall that an element l of an ordered set $(S, \leq)$ is the *least element* if $l \leq x$, for all $x \in S$.

The next lemma proves that the field $\mathbb{Q}_F$ of rational elements in any *archimedean field* $(F, <)$ is a dense set.

Lemma 2.9 (Density Lemma). *Let $(F, <)$ be an archimedean field, and $\mathbb{Q}_F$ be the set of rational elements in F. Then, $\mathbb{Q}_F$ is dense in F.*

Proof We may write the zero element and the identity of F as 0_F and 1_F, respectively. First, let $a > 0_F$ so that

$$b - a > 0_F \quad \Rightarrow \quad (b - a)^{-1} > 0_F.$$

Then, by the archimedean property, for some $m \in \mathbb{N}_F$ we have $m > (b - a)^{-1}$, i.e., $bm - am > 1_F$. Also, the set $\{k \in \mathbb{N}_F : k > ma\} \neq \emptyset$, by the archimedean property. So, as remarked earlier, this set has the least element, say n. Clearly, $n \neq 0_F$ because $ma > 0_F$. By minimality, we have $n - 1_F < ma$. And, since $bm - am > 1_F$, we obtain

$$(n - 1_F) + 1_F < (n - 1_F) + (bm - am) < ma + (bm - am).$$

That is, $n < bm$. Therefore,

$$ma < n < mb \quad \Rightarrow \quad a < nm^{-1} < b, \text{ where } m^{-1} > 0.$$

Observe that if $a = 0_F$ we may take $n \in \mathbb{N}_F$ such that $n > b^{-1}$. Then, $nb > 1_F \Rightarrow b > n^{-1} > 0_F = a$. Finally, if $a < 0_F$ then $-a > 0_F$ implies that $-a < k$ or $0_F < k + a$ for some $k \in \mathbb{N}_F$, by the archimedean property. But then, by previous discussion, $0_F < k + a < k + b$ gives a rational element $q \in \mathbb{Q}_F$ such that

$$k + a < q < k + b \quad \Rightarrow \quad a < q - k < b. \qquad \square$$

The most crucial property that $\mathbb{Q}$ lacks is known as the *completeness property*. To discuss this particular type of deficiency of the ordered field $\mathbb{Q}$, we need some preparation with the concepts introduced earlier in Chap. 1. Recall that, for a bounded above set A, we write

$$\sup A := \min \left\{ u \in F : u \text{ is an upper bound of } A \right\},$$

which is called the *least upper bound* (or the *supremum*) of the set A. Also, we write

$$\inf A := \max \left\{ \ell \in F : \ell \text{ is a lower bound of } A \right\},$$

which is called the the *greatest lower bound* (or the *infimum*) of the set A. Notice that we always have $\inf A \leq \sup A$, if the two exist.

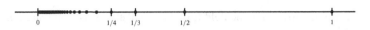

Fig. 2.2 The set $\{1/n : n \in \mathbb{N}\}$

Example 2.4 The subset $\mathbb{N} \subset \mathbb{Z}$ is bounded below but not bounded above. The integers $1, 0, -1, -2, \ldots$ are all lower bounds of $\mathbb{N}$ and inf $\mathbb{N} = 1$. The subset $\mathbb{Z}^- \subset \mathbb{Z}$ of all negative integers is bounded above but not bounded below. The integers $-1, 0, 1, 2, \cdots$ are all upper bound of $\mathbb{Z}^-$ and sup $\mathbb{Z}^- = -1$.

Notice that, in the above example, both the infimum and the supremum belong to the given subsets, and they are respectively the least and the greatest elements of the respective sets. However, that may not be the case, in general.

Example 2.5 Consider the set $A \subset \mathbb{Q}$, as depicted in Fig. 2.2, given by

$$A = \left\{ \frac{1}{n} : n \in \mathbb{N} \right\}.$$

Clearly, the set A is bounded, where any rational number ≥ 1 is an upper bound. Notice that, since $1 \in A$, no positive number < 1 is an upper bound of A. Therefore, we have sup $A = 1$, and sup A is the greatest element of A. Further, any rational number ≤ 0 is a lower bound, and no positive number is a lower bound of A (Corollary 2.4), and hence inf $A = 0$. Notice that, however, the set A has no least element, and sup $A \in A$, but inf $A \notin A$.

Recall that an order set S has the *upper bound property* (or the *supremum property*) if every bounded above subset of S has the supremum. For example, both $\mathbb{Z}$ and $\mathbb{N}$ have the supremum property, because the least upper bound of every bounded above subset of natural numbers or integers is its maximum. Notice that, however, that is not the case for the ordered field $\mathbb{Q}$.

Lemma 2.10 *The set $\mathbb{Q}$ of rational numbers does not satisfy the upper bound property.*

Proof For any prime number p, consider the set

$$A = \left\{ x \in \mathbb{Q} : x > 0, \ x^2 < p \right\}.$$

Clearly, A is bounded below by 0, and it is bounded above by p. Notice that inf $A = 0$. By Example 1.6, the set A has no greatest element and so none of its elements is an upper bound of A. But, every element of the set

$$B = \left\{ x \in \mathbb{Q} : x > 0, \ x^2 > p \right\}$$

is clearly an upper bound of A. None of the elements of B can be a supremum of A as the set B has no least element. Thus, sup A does not exist.   □

Definition 2.9 (*Completeness Axiom*). We say an ordered field $(F, <)$ is a **complete** if F satisfies the upper bound property.

So, it follows from Lemma 2.10 that $\mathbb{Q}$ is not a *complete ordered field*. An interesting feature of a complete ordered field is that the archimedean property holds by default. Therefore, the ordered subfield $\mathbb{Q}_F$ of rational elements in any complete ordered field $(F, <)$ is a dense set.

Lemma 2.11 *Every complete ordered field $(F, <)$ is archimedean.*

Proof If possible, suppose $(F, <)$ does not satisfy the archimedean property so that, for some $0 < a \in F$, we have $n > a$, for all natural elements $n \in \mathbb{N}_F$. Then, the element a is an upper bound of the set $\mathbb{N}_F$ of natural elements in F. Hence, by the completeness property of F, the set $\mathbb{N}_F$ has a least upper bound, say α. Therefore, for some $m \in \mathbb{N}_F$, we have $\alpha - 1 < m$, i.e., $\alpha < m + 1$. Since $m + 1 \in \mathbb{N}_F$, this contradicts the fact that α is the least upper bound of the set $\mathbb{N}_F$. $\square$

We next show that if there exists a complete ordered field, containing $\mathbb{Q}$ as a proper subfield, then such a field is necessarily *unique*.

Theorem 2.6 (Uniqueness). *Any two complete ordered fields, with the same subfield of rational elements, are (order) isomorphic.*

Proof Let $(F, 0_F, 1_F)$ and $(K, 0_K, 1_K)$ be two complete ordered fields. For both ordered fields, we may use a common symbol to write addition, multiplication, and order relation $<$; of course, the context will always help to figure out the difference. Let $\mathbb{Q}_F$ and $\mathbb{Q}_K$ respectively be the subfield of *rational elements* in F and K, and $\phi_1 : \mathbb{Q} \to \mathbb{Q}_F$ and $\phi_2 : \mathbb{Q} \to \mathbb{Q}_K$ be isomorphisms as provided by Theorem 2.4. Write $\phi = \phi_2 \circ \phi_1^{-1} : \mathbb{Q}_F \to \mathbb{Q}_K$ for the composite isomorphism. For $q \in \mathbb{Q}_F$, write $\phi(q) = \overline{q}$ so that the following hold for all $p, q \in \mathbb{Q}_F$:

$$\overline{p + q} = \overline{p} + \overline{q},$$
$$\overline{pq} = \overline{p}\,\overline{q}, \qquad (2.2.1)$$
$$p < q \Leftrightarrow \overline{p} < \overline{q}.$$

Now, since the *supremum* of the set

$$L = \{\overline{q} \in \mathbb{Q}_K \mid q < a\} \qquad (2.2.2)$$

exists for any $a \in F$, by the completeness of K, define $\varphi : F \to K$ by

$$\varphi(a) := \sup\{\overline{q} \in \mathbb{Q}_K : q < a\}, \quad \text{for each} \quad a \in F. \qquad (2.2.3)$$

First, we prove that $\varphi|_{\mathbb{Q}_F} = \phi$. Let $q \in \mathbb{Q}_F$. Then, for any $p \in \mathbb{Q}_F$ with $p < q$, $\overline{p} < \overline{q} \Rightarrow \varphi(q) \leq \overline{q}$, because $\overline{q}$ is an upper bound of the set $\{\overline{p} \in \mathbb{Q}_F : p < q\}$. However, $\varphi(q) < \overline{q}$ is not possible. For, otherwise, by the density lemma, there is some $\overline{r} \in \mathbb{Q}_K$

such that $\varphi(q) < \bar{r} < \bar{q}$, contradicting the fact that $\varphi(q)$ is the *supremum*. Therefore, $\varphi(q) = \bar{q} = \phi(q)$ for all $q \in \mathbb{Q}_F$. We complete the proof by verifying the following assertions:

1. $a < b$ in $F \;\Leftrightarrow\; \varphi(a) < \varphi(b)$;
2. For all $p, q \in \mathbb{Q}_F$,

$$\sup\{\bar{p} + \bar{q} : q < a\} = \sup\{\bar{q} : q < a\} + \bar{p},$$

and, for $a \in F^+$, $p \in \mathbb{Q}_F^+$ and $\mathbb{Q}_F$,

$$\sup\{\bar{p} \cdot \bar{q} : q < a\} = \sup\{\bar{q} : q < a\} \cdot \bar{p};$$

3. $\varphi(a + b) = \varphi(a) + \varphi(b)$;
4. $\varphi(a \cdot b) = \varphi(a) \cdot \varphi(b)$
5. φ is surjective.

For (1), suppose $a < b$ in F. By the density lemma, there are some $p, q \in \mathbb{Q}_F$ such that $a < p < q < b$. Now, since $\bar{p}$ is an upper bound of the set $\{\bar{p} : r < x\}$, we have $\varphi(a) \leq \bar{p}$. Similarly, it follows that $\bar{q} \leq \varphi(b)$. We already know that $p < q \Rightarrow \bar{p} < \bar{q}$, so $\varphi(a) < \varphi(b)$. Conversely, let $\varphi(a) < \varphi(b)$. If possible, suppose $a \geq b$ so that $\varphi(a) \geq \varphi(b)$, by previous case. This contradicts the trichotomy condition of $<$. Thus, (1) holds

For (2), observe that if $q < a$, then $\bar{p} < \sup\{\bar{q} : q < a\}$ so that $\bar{q} + \bar{p} < \sup\{\bar{q} : q < a\} + \bar{p}$. Thus,

$$\sup\{\bar{q} + \bar{p} : q < a\} \leq \sup\{\bar{q} : q < a\} + \bar{p}. \tag{2.2.4}$$

Now, if the inequality is strict, we can find by the density lemma a $\underline{r} \in \mathbb{Q}_K$ such that

$$\sup\{\bar{q} + \bar{p} : q < a\} < \underline{r} < \sup\{\bar{q} : q < a\} + \bar{p}.$$

In particular, we have

$$\underline{r} - \underline{p} = \underline{r - p} < \sup\{\underline{q} : q < a\},$$

which gives $r - p < a$, and so $\underline{r - p + p} \in \{\underline{q + p} : q < a\}$. Thus,

$$\underline{r} = \underline{r - p + p} \leq \sup\{\underline{q + p} : q < a\} < \underline{r},$$

a contradiction. So, the inclusion in (2.2.4) must be strict. For the second part, let a, p, q be as given in the hypothesis. Then, for $q < a$, we have $\underline{q} < \sup\{\underline{q} : q < a\}$ so that $\underline{qp} < \sup\{\underline{q} : q < a\} \cdot \bar{p}$. Thus,

$$\sup\{\bar{q} \cdot \bar{p} : q < a\} \leq \sup\{\bar{q} : q < a\} \cdot \bar{p}. \tag{2.2.5}$$

If possible, suppose the inequality is strict so that, by the density lemma, there is a $\bar{r} \in \mathbb{Q}_K$, with

$$\sup\{\bar{q} \cdot \bar{p} : q < a\} < \bar{r} < \sup\{\bar{q} : q < a\} \cdot \bar{p}.$$

In particular, we obtain $\bar{r}\,\bar{q}^{-1} < \sup\{\bar{q} : q < a\}$. Also,

$$\overline{p\,p^{-1}} = \overline{pp^{-1}} = 1_K \quad \Rightarrow \quad \bar{p}^{-1} = \overline{p^{-1}}.$$

Therefore, we conclude that

$$\overline{rp^{-1}} < \sup\{\bar{q} : q < a\}$$
$$\Rightarrow \quad rp^{-1} < a$$
$$\Rightarrow \quad \overline{rp^{-1}}\,\bar{p} \in \{\bar{q}\,\bar{p}\, q < a\} < \bar{r},$$

which is absurd. So, the inclusion in (2.2.5) is strict.
For (3), observe that since

$$\varphi(a + b) = \sup\{\bar{r} : r < a + b\}, \quad a, b \in F,$$

we have for $b = p \in \mathbb{Q}_F$

$$\{\bar{q} : q < a + p\} = \{\overline{q + p} : q < a\} = \{\bar{q} + \bar{p} : q < a\}.$$

So, taking supremum, it follows by (2) that

$$\sup\{\overline{q + p} : q < a\} = \sup\{\bar{q} : q < a\} + \bar{p}.$$

That is, $\varphi(a + p) = \varphi(a) + \bar{p}$. In general, for $p \in \mathbb{Q}_F$, suppose $p < b$ so that $a + p < a + b$ implies $\varphi(a + p) < \varphi(a + b)$, by (1), and so $\varphi(a) + \bar{p} < \varphi(a + b)$. Therefore,

$$p < b \Rightarrow \quad \bar{p} < \varphi(a + b) - \varphi(a)$$
$$\Rightarrow \quad \varphi(b) = \sup\{\bar{p} : p < b\} \leq \varphi(a + b) - \varphi(a).$$

We claim that strict inequality cannot hold. For, otherwise, by the density lemma there is a $\bar{r} \in \mathbb{Q}_K$ such that

$$\varphi(b) < \bar{r} < \varphi(a + b) - \varphi(a). \tag{2.2.6}$$

But then, by (1), $b < r$ implies that

$$\varphi(a + b) - \varphi(a) < \varphi(a + r) - \varphi(a) = \varphi(a) + \bar{r} - \varphi(a),$$

i.e., $\overline{r} < \overline{r}$, which is absurd. Hence, equality holds in (2.2.6).
For (4), since $\varphi\big|_{\mathbb{Q}_F} = \phi$, we have

$$\phi(0_F) = \overline{0}_F = 0_K \quad \text{and} \quad \phi(1_F) = \overline{1}_F = 1_K$$

so that by (3), we have $\varphi(-a) = -\varphi(a)$ for all $a \in F$. Now, first, let $a, b \in F^+$. Once again, using (1), $\varphi(0_F) = 0_K$ implies that $\varphi(a), \varphi(b) \in K^+$. To show that $\varphi(ab) = \varphi(a)\varphi(b)$, we may again start with $b = p \in \mathbb{Q}_F$ so that

$$\{\overline{q} : q < a\} = \{\overline{qp} : q < a\} = \{\overline{q}\,\overline{p} : q < a\}.$$

Thus, taking supremum and using (2), we obtain $\varphi(ap) = \varphi(a)\overline{p}$. Next, for $q \in \mathbb{Q}_F$,

$$0_F < b < q \;\Rightarrow\; aq < ab \;\Rightarrow\; \varphi(a)\overline{q} = \varphi aq < \varphi(ab),$$

by (3). Therefore, $q < b \Rightarrow \overline{q} < \varphi(ab)\varphi(a)^{-1}$. Now,

$$\varphi(b) = \sup\{\overline{q} : q < b\} \leq \varphi(ab)\varphi(a)^{-1}. \qquad (2.2.7)$$

For a contradiction, suppose strict inequality holds. Then, by density lemma there is a $\overline{r} \in \mathbb{Q}_K$ such that $\varphi(b) < \overline{r} < \varphi(b)\varphi(a)^{-1}$. Also, by (1), we have

$$b < r \;\Rightarrow\; \varphi(ab)\varphi(a)^{-1} < \varphi(ar)\varphi(a)^{-1} = \overline{r}$$
$$\Rightarrow\; \overline{r} < \overline{r},$$

a contradiction. This completes the proof for $a, b \in F^+$. To conclude, for $a, b \in F$, we may write $a = \epsilon_a |a|$ and $b = \epsilon_b |b|$, where $\epsilon_a, \epsilon_b = \pm 1_F$. By an earlier observation, we know that $\varphi(-a) = -\varphi(a)$ for all $a \in F$. Thus,

$$\begin{aligned}
\varphi(ab) &= \varphi(\epsilon_a |a| \epsilon_b |b|) \\
&= \epsilon_a \epsilon_b \varphi(|a||b|) \\
&= \epsilon_a \epsilon_b \varphi(|a|)\varphi(|b|) \\
&= \varphi(\epsilon_a |a|)\varphi(\epsilon_b |b|).
\end{aligned}$$

Hence, $\varphi(ab) = \varphi(a)\varphi(b)$ for all $a, b \in F$.

For (5), let $y \in K$, and consider the following two sets

$$\overline{L} := \{\overline{r} \in \mathbb{Q}_K : \overline{r} < y\} \quad \text{and} \quad L := \{r \in \mathbb{Q}_F : \overline{r} \in \overline{L}\}.$$

Then, we must have $y = \sup(\overline{L})$; for, otherwise, $y > \sup(\overline{L}) \Rightarrow \sup(\overline{L}) < \overline{r} < y$, for some $\overline{r} \in \mathbb{Q}_K$, by the density lemma, which is a contradiction. Observe that if $\overline{q} \in \mathbb{Q}_K$ is an upper bound of $\overline{L}$, then q is an upper bound of L, and so L is bounded above. By the completeness, $\sup(L) = x$ exists. We proceed to show that $\varphi(x) = y$.

For, let $\bar{r} \in \bar{L}$ so that $r \le x$. Then, by (1), we obtain $\bar{r} = \varphi(r) \le \varphi(x)$, i.e., $\varphi(x)$ is an upper bound of the set $\bar{L}$. But then, we must have $y \le \varphi(x)$. If possible, suppose $y < \varphi(x)$. Then, by the density lemma $y < \bar{q} < \varphi(x)$ for some $\bar{q} \in \mathbb{Q}_K$. This, in turn, implies that $q < x$ so that $q \in L$. Thus, $\bar{q} \in \bar{L} \Rightarrow \bar{q} < y$, a contradiction. Hence, $\varphi(x) = y$. □

There are a number of different methods of representing a *real number*, i.e., ways to construct the complete ordered field $\mathbb{R}$. Of course, by Theorem 2.6, all must be isomorphic as complete ordered fields. So, one is free to adopt any one of such constructions to write the set $\mathbb{R}$.

A brief description about the three more well-known *methods of construction* of the set $\mathbb{R}$, as given below, is included here only to complete the story, with a promise to provide the complete details for the second and the third one as soon as we are equipped with the essential *prerequisites*. In general, for most methods of construction of the set $\mathbb{R}$, the ordered field $\mathbb{Q}$ is used as the *base step* of the process.

1. $\mathbb{R}$ as the set of **Dedekind Cuts** : [Dedekind, 1872] A nonempty proper subset L of $\mathbb{Q}$ is called a *Dedekind cut* if for any $p \in L$, $q < p \Rightarrow q \in L$, and L has no maximal element. Clearly then, for any $q \notin L$, $q > p$, for every $p \in L$. For example, for any $p \in \mathbb{Q}$,

$$\bar{p} = L_p = \{q \in \mathbb{Q} : q < p\}$$

is a cut. Notice that the set L in (2.2.2) is a (Dedekind) cut. Two cuts are *equal* if they are so as sets. It then follows immediately that if two cuts L, L' are not equal then either $L \subset L'$ or $L' \subset L$. So, we may order cuts as follows: $L < L' \Leftrightarrow L \subset L'$. Notice that $\bar{p} = L_p < L \Rightarrow p \in L$; in particular, for $\bar{0} = L_0$, a cut is *positive* if $L > \bar{0}$, i.e., $0 \in L$. So, a cut L is *negative*, i.e., $L < \bar{0}$ if $0 \notin L$. The set of real numbers $\mathbb{R}$ is defined to be the collection of all cuts L, where a cut L is called a *rational number* if $\sup(L)$ exists, otherwise it is called an *irrational number*. Elementwise sum and product of two cuts define addition and multiplication operations on the set $\mathbb{R}$ so that it follows that $\mathbb{R}$ is a complete ordered field. Also, for $p \in \mathbb{Q}$, the mapping $p \mapsto \sup(L_p)$ defines an isomorphism of ordered fields of rational numbers. For further details, we refer to the text [3]. Notice that some related arguments are already being used in the proof of Theorem 2.6.

2. $\mathbb{R}$ as the set of **Limits of Cauchy Sequences** *over* $\mathbb{Q}$: [Cantor, 1872] A sequence of rational numbers $\langle q_n \rangle$ is called *Cauchy* if the terms q_n are *eventually very close* in the sense that, except for finitely many initial terms, the rest *cluster together*. For example, the sequence $q_n = 1/n$ is Cauchy (see Fig. 2.2). For two such sequences $r = \langle q_n \rangle$ and $r' = \langle q'_n \rangle$, we say $r \sim r'$ if $r - r' = \langle q_n - q'_n \rangle$ is a *null sequence*, i.e., $r = r'$, *eventually*. It then follows that the relation $\sim$ defines an equivalence on the collection $\mathscr{C}$ of Cauchy sequences over $\mathbb{Q}$. In this case, we take $\mathbb{R}$ to be the set of all equivalence classes obtained in this way. As we will see later, under usual addition and multiplication of representative Cauchy sequences, the set $\mathbb{R}$ turns into a field. Also, a natural *order relation* is defined between equivalence classes to give the set $\mathbb{R}$ structure of the complete ordered field. In historical terms,

Cauchy formulated this construction only to articulate his proof of uniqueness of trigonometric series. See Sect. 3.3 for the complete details.

3. $\mathbb{R}$ as the set of **Decimal Expansions** : A classical way of representing a (positive) real number $a \in \mathbb{R}$ is by using a *decimal expansion*, which is an expression of the form $a_0.a_1a_2a_3 \cdots$, where $a_0 = \lfloor a \rfloor \in \mathbb{Z}^+$ is called the *integral part* of the real number a, and $a_k \in \{0, 1, \ldots, 9\}$ $(k \geq 1)$ are the digits of the *fractional part* $f = 0.a_1a_2a_3 \cdots$ of the real number a. Clearly, we have $0 < f < 1$. For example,

$$\frac{1}{3} = 0.333 \cdots, \quad \frac{25}{4} = 6.25, \quad \sqrt{2} = 1.414213562 \cdots$$

We will prove that $f \in \mathbb{Q}$ if and only if either $a_k = 0$, for $k > n \geq 1$ (finite decimal), or it is an infinite decimal with a *repeating pattern* (Theorem 3.32). Also, we find a sequence of rational numbers

$$q_n = a_0.a_1a_2 \cdots a_n, \quad \text{with} \quad q_n \leq a < q_n + \frac{1}{10^n}, \tag{2.2.8}$$

such that the above decimal expansion of a can be obtained from the sequence $\langle q_n \rangle$, to any desired degree of accuracy (Lemma 3.2). The same procedure applies to $-a$ with $a > 0$. This particular *method of construction* of the set of real numbers $\mathbb{R}$ uses the concept of convergent infinite series in the sense that

$$a = \lim_{n \to \infty} \sum_{k=0}^{n} \frac{a_k}{10^k} = a_0.a_1a_2a_3 \cdots.$$

See Sect. 3.4.3 for the complete details.

Theorem 2.7 (Existence). *There is a complete ordered field containing a copy of the ordered field $\mathbb{Q}$.*

A proof of this theorem is given at the end of Sect. 3.3, using the second method of construction as introduced above. Here, we assume the *existence* of $\mathbb{R}$, and show that $\mathbb{R} \setminus \mathbb{Q} \neq \emptyset$. Recall that, for $0 < x \in \mathbb{R}$, if there exists some $0 < y \in \mathbb{R}$ such that $y^2 = x$, then we say y is the *square root* of x, which we denote by $y = \sqrt{x}$. The next theorem proves that every positive real number has a unique square root in $\mathbb{R}$.

Theorem 2.8 *For any real number $x > 0$, there exists a unique real number $y > 0$ such that $y^2 = x$.*

Proof Consider the set

$$A = \{t \in \mathbb{R} : t > 0 \text{ and } t^2 < x\}.$$

Let $t = x/(1 + x)$. Then, we have

$$0 < t < 1 \quad \text{and} \quad t^2 < t = \frac{x}{1+x} < x,$$

and so $t \in A$ implies that $A \neq \emptyset$. Also, the set A is bounded above by $1 + x$. For, if $t > 1 + x$ then $t > 1$, and so $t^2 > t > 1 + x > x$. Therefore, $t \in A \Rightarrow t \leq 1 + x$. Since $\emptyset \neq A \subset \mathbb{R}$ is bounded above, by the completeness property, $\sup A$ exist.

Suppose $y = \sup A$. Clearly, we have $y > 0$. We claim that $y^2 = x$. We prove our claim by showing that $y^2 < x$ and $y^2 > x$ lead to a contradiction. First, suppose $y^2 < x$. Notice that, as $1/n^2 < 1/n$ for any $1 < n \in \mathbb{N}$, we have

$$\left(y + \frac{1}{n}\right)^2 = y^2 + \frac{2y}{n} + \frac{1}{n^2} < y^2 + \frac{2y+1}{n}.$$

By the archimedean property, we choose m such that $m(x - y^2) > 2y + 1$. Then,

$$\left(y + \frac{1}{m}\right)^2 < y^2 + (x - y^2) = x$$

and hence $y + (1/m) \in A$. This is a contradiction because $y = \sup A$. Now, suppose $y^2 > x$. In this case, for any $1 < n \in \mathbb{N}$, we have

$$\left(y - \frac{1}{n}\right)^2 = y^2 - \frac{2y}{n} + \frac{1}{n^2} > y^2 - \frac{2y}{n}.$$

Again, by the archimedean property, we choose m such that $m(y^2 - x) > 2y$. Then,

$$\left(y - \frac{1}{m}\right)^2 > y^2 - (y^2 - x) = x.$$

We also have

$$my^2 > m(y^2 - x) > 2y \quad \Rightarrow \quad my > 2 \quad \Rightarrow \quad y > 2/m,$$

and so $y - 1/m > 1/m > 0$. As each positive real number t, with $t^2 > x$, is an upper bound of A, we got $y - (1/m)$ as an upper bound of A, smaller than y. This contradicts the fact $y = \sup A$. Therefore, we have $y^2 = x$. Finally, uniqueness of y follows easily. For, if $y_1^2 = y_2^2 = x$, for some real numbers $y_1 > 0$, $y_2 > 0$, then $y_1 = y_2$ or $y_1 = -y_2$. Notice that the second equality is not possible because $y_1 > 0$, $y_2 > 0$. □

2.3   Modulus Metric

We introduce here an important function, known as the *absolute value function*, which is fundamental to do analysis on the set $\mathbb{R}$. For a wider scope of applications, we define it over an ordered field $(F, <)$. Recall that, by (1) of Lemma 2.7,

$$0 < a \in F \quad \Leftrightarrow \quad -a < 0.$$

It thus follows that, for any $0 \neq a \in F$, $a \in F^+$ or $-a \in F^+$. Therefore, $\max\{a, -a\}$ is defined for any $0 \neq a \in F$.

Definition 2.10   Let $(F, <)$ be an ordered field, and $a \in F$. The *absolute value* (or *modulus*) of a, denoted by $|a|$, is given by

$$|a| = \begin{cases} \max\{a, -a\}, & \text{if } a \neq 0; \\ 0, & \text{if } a = 0. \end{cases}$$

The function $: F \to F^+$ given by $a \mapsto |a|$, $a \in F$, is called the **absolute value function** (or the *modulus function*).

The function $|\,| : F \to F^+$ has many interesting properties. For example, the first three properties stated in the next theorem show that it defines a *norm* on $(F, <)$, and the fourth says that the modulus function is *multiplicative*.

Theorem 2.9 (Norm Properties).   *Let $(F, <)$ be an ordered field. Then, for any $a, b \in F$,*

1. $|a| = 0 \Leftrightarrow a = 0$;
2. $|-a| = |a|$;
3. $|a + b| \leq |a| + |b|$; *and, the equality holds if and only if a and b have the same sign.*
4. $|ab| = |a|\,|b|$.

Proof   By definition, (1) holds. And, for (2), since $|a| = \max\{a, -a\}$, we have

$$|-a| = \max\{-a, -(-a)\} = \max\{a, -a\} = |a|.$$

The proof is completed by considering cases for $a, b \in F$. First, if a or b is 0, there is nothing to prove. Next, suppose $a > 0$, $b > 0$ so that $a > -a$ and $b > -b$. That is, $|a| = a$ and $|b| = b$. Then, $ab > 0$ implies that $|ab| = ab$. So, (4) holds in this case. Also, $a + b > 0 \Rightarrow |a + b| = |a| + |b|$. Now, consider the case of opposite signs, say $a > 0$ and $b < 0$. Then,

$$a > -a \Rightarrow |a| = a \quad \text{and} \quad b < -b \Rightarrow |b| = -b,$$

so that $|a| + |b| = a - b$ and $|a + b| = a + b$ or $-(a + b) = -a - b$. However, $-b > b \Rightarrow a - b > a + b$ and $-(a + b) = -a - b$. Thus, in both cases, we have $|a| + |b| > |a + b|$. Also, $b < 0 \Rightarrow ab < 0$, by (2) of Lemma 2.7 so that

$$|ab| = -(ab) = (-1)(ab) = a(-b) = |a||b|, \quad \text{for } a > 0.$$

Finally, suppose $a < 0$ and $b < 0$ so that $|a| = -a$ and $|b| = -b$, which by (2) of Lemma 2.7 implies that $|a||b| = (-a)(-b) = ab$. Also, $-a > 0$ and $b < 0$ implies $(-a)(b) = -(ab) < 0$, i.e., $ab > 0$, and so $|ab| = ab$. Thus, (4) holds for all $a, b \in F$. Next, by the monotone property of the operation $+$ with respect to $<$, we have $a + b < 0$ so that

$$|a + b| = -(a + b) = -a - b = |a| + |b|.$$

As shown above, if a and b have the same sign, the equality in (3) holds. Conversely, suppose $|a + b| = |a| + |b|$, for some $a, b \in F$. Look at the details given above to prove $|a + b| < |a| + |b|$. In the first case, $a + b > 0$, at least one of a or b is positive. Let us assume that $a > 0$. Then, $a = |a|$ and hence

$$|a| + |b| = |a + b| = a + b = |a| + b,$$

and so $b = |b|$ and hence $b > 0$. Therefore, a and b has the same sign. The proof in the case when $a + b < 0$ is similar. $\qquad\square$

Definition 2.11 Let X be a nonempty set. A function $d : X \times X \to \mathbb{R}$ is called a *distance function* (or a *metric*) on X if it satisfies the following three properties:

1. d is *positive*, i.e., $d(x, x) = 0$, and $d(x, y) > 0$, for all $x \neq y \in X$;
2. d is *symmetric*, i.e., $d(x, y) = d(y, x)$, for all $x, y \in X$;
3. d has the *triangle property*, i.e., $d(x, z) \leq d(x, y) + d(y, z)$, for all $x, y, z \in X$.

A nonempty set X with a metric d is called a *metric space*.

The next corollary provides examples of some interesting metric spaces.

Corollary 2.2 *The ordered field $\mathbb{R}$ is a metric space with respect to a distance function $m : \mathbb{R} \times \mathbb{R} \to \mathbb{R}^+$ given by*

$$m(a, b) = |a - b|, \quad \text{for } a, b \in \mathbb{R}. \tag{2.3.1}$$

Proof By assertions (1)–(3) of Theorem 2.9, it follows that the function m is positive, symmetric, and it satisfies the triangle property. That is, for all $a, b, c \in \mathbb{R}$, we have

$$|a - b| \leq |a - c| + |c - b|.$$

Hence, $(\mathbb{R}, m)$ is a metric space. $\qquad\square$

The metric defined in (2.3.1) is called the *modulus metric* on the set $\mathbb{R}$. The next theorem provides some properties of the modulus function $| \ | : F \to F^+$. A general discussion about topological properties of metric spaces is given later in Chap. 6.

Theorem 2.10 *Let $(F, <)$ be an ordered field. For $a, b \in F$,*

1. $|a| = a$, *for $a > 0$; and,* $|a| = -a$, *for $a < 0$;*
2. $a < |a|$, *for any $a \neq 0$;*
3. $\big| |a| - |b| \big| \leq |a + b|$;
4. $|a|^2 = a^2$;
5. $|a + b|^2 + |a - b|^2 = 2(|a|^2 + |b|^2)$ *(parallelogram law).*

Proof For (1), given any $a \in F$, we have by the trichotomy law: $a < 0$ or $a = 0$ or $a > 0$. If $a \neq 0$, and $a > 0$, then $|a| = a$ and so $\max\{a, -a\} = a$. Similarly, if $a < 0$, we have $-a > 0$ so that $|a| = -a$ and $\max\{a, -a\} = -a$. For (2), since $|a| = \max\{a, -a\}$, for $a \neq 0$, we have $a < \max\{a, -a\} = |a|$. For (3), note that since

$$|a| = |a + b - b| \leq |a + b| + |-b| = |a + b| + |b|,$$

we have $|a| - |b| \leq |a + b|$. Interchanging the roles of a and b, we obtain $|b| - |a| \leq |a + b|$. Therefore,

$$\big| |x| - |y| \big| = \max\{|x| - |y|, |y| - |x|\} \leq |x + y|.$$

For (4), if $a > 0$, we have $|a| = a$ and so $|a|^2 = a^2$; if $a < 0$, we have $|a| = -a$ and so $|a|^2 = (-a)^2 = a^2$. Finally, using (4), we have

$$|a + b|^2 + |a - b|^2 = (a + b)^2 + (a - b)^2$$
$$= 2(a^2 + b^2) = 2(|a|^2 + |b|^2). \qquad \square$$

Corollary 2.3 *Let $(F, <)$ be an ordered field. Then, for $a, \epsilon \in F$ with $\epsilon > 0$, $|a| < \epsilon$ if and only if $-\epsilon < a < \epsilon$.*

Proof Suppose $|a| < \epsilon$. Since $|a| = \max\{a, -a\}$, we get $a < \epsilon$ and $-a < \epsilon$. Since $-a < \epsilon$ implies $-\epsilon < a$, we get $-\epsilon < a < \epsilon$. For the converse, suppose that $-\epsilon < a < \epsilon$. Then $a < \epsilon$ and $-a < \epsilon$ and hence $|a| = \max\{a, -a\} < \epsilon$. The proof for the other part is similar. $\qquad \square$

Definition 2.12 Let $(F, <)$ be an ordered field, and $a < b$ in F. Then,

1. the set $(a, b) = \{x \in F : a < x < b\}$ is called an *open interval*;
2. the set $[a, b] = \{x \in F : a \leq x \leq b\}$ is called a *closed interval*;
3. the set $(a, b] = \{x \in F : a < x \leq b\}$ is called a left-open *semi-interval*;
4. the set $[a, b) = \{x \in F : a \leq x < b\}$ is called a right-open *semi-interval*.

For any $a \in F$, $\{a\} = [a, a]$ is called a *degenerate interval*. For any nondegenerate interval $I \subset F$ of the form (1) through (4), we write $\ell(I) = b - a$ for the *length* of the interval I. In the sequel, the next lemma referred to as the *epsilon argument*.

Lemma 2.12 (Epsilon Argument). *For any $a \in F$, with $|a| < \epsilon$, for each $0 < \epsilon \in F$, we have $a = 0$.*

Proof For, if $a \neq 0$, we may take $\epsilon = |a| > 0$ so that $|a| < \epsilon = |a|$ violates the trichotomy property of $<$.   □

2.4 Countable and Uncountable Sets in $\mathbb{R}$

We start with a quick review of concepts introduced earlier in Sect. 1.4. Recall that, for $n \in \mathbb{N}$, a set A has n elements if it is *equinumerous* to the set $\mathbb{N}_n = \{1, 2, \ldots, n\}$, which we denote by $A \sim \mathbb{N}_n$. That is, there is a bijective function : $A \to \mathbb{N}_n$. We say a set A is *finite* if $A = \emptyset$ or $A \sim \mathbb{N}_n$, for some $n \in \mathbb{N}$. A set A is *infinite* if it is not finite; or, equivalently, if A has a proper subset B such that there is a bijective function : $A \to B$ (Theorem 1.16).

There are mainly two types of infinite sets. We say a set A is *countably infinite* if there is a bijective function : $A \to \mathbb{N}$, and hence a bijective function $f : \mathbb{N} \to A$. A set is called *countable* if it is finite or countably infinite. By Example 1.10, we know that the set $\mathbb{Z}$ is countably infinite. In general, suppose A is a countably infinite set, and let $f : \mathbb{N} \to A$ be a bijective function. We write $f(n) = a_n$, for each $n \in \mathbb{N}$, so that the set A can be *enumerated* as

$$a_1, \ a_2, \ a_3, \ \ldots .$$

As before, we say f is an *enumeration* of the set A. Notice that, by using the enumeration as given in Example 1.9, we can enlist the set $\mathbb{Z}$ as

$$0, \ 1, \ -1, \ 2, \ -2, \ 3, \ -3, \cdots .$$

Also, we know that the countable union of countably infinite sets is countably infinite. For example, by arranging the elements of the product $\mathbb{N} \times \mathbb{N}$ as indicated by the arrows in Fig. 1.2 in Chap. 1, we see that the set $\mathbb{N} \times \mathbb{N}$ is a countable union of countably infinite sets, and hence is itself a countably infinite set. Notice that we can write

$$\mathbb{N} \times \mathbb{N} = \{(1, 1), (1, 2), (2, 1), (3, 1), (2, 2), (1, 3), (1, 4), (2, 3), \ldots\}.$$

In general, if A and B are two countably infinite sets so that we have

$$A = \{a_1, a_2, a_3, \ldots\} \quad \text{and} \quad B = \{b_1, b_2, b_3, \ldots\},$$

then the function $f : A \times B \to \mathbb{N} \times \mathbb{N}$ given by

$$f((a_i, b_j)) = (i, j), \quad i, j \geq 1,$$

is a bijection. As $\mathbb{N} \times \mathbb{N}$ is countably infinite, it follows that $A \times B$ is a countably infinite set. Clearly, this assertion holds for any finite product of countably infinite sets. In particular, as the set $\mathbb{Z}$ is countably infinite, we conclude that

$$\mathbb{Z}^n = \mathbb{Z} \times \cdots \times \mathbb{Z} \quad (n \text{ copies})$$

is a countably infinite set, for any $n \geq 1$. We use this simple observation to prove that the set of *algebraic numbers*[6] in the set $\mathbb{R}$ is a countably infinite set, and hence that the set of *transcendental* real numbers is an uncountable set. Let us start with the definition.

Definition 2.13 We say $\alpha \in \mathbb{R}$ is an *algebraic number* if there is an integral polynomial $p(x) \in \mathbb{Z}[x]$ of degree ≥ 1 such that $p(\alpha) = 0$. That is, for some integers $a_k \in \mathbb{Z}$, with $a_0 \neq 0$,

$$p(\alpha) = a_0\alpha^n + a_1\alpha^{n-1} + \cdots + a_n = 0. \tag{2.4.1}$$

We say a real number α is *transcendental* if it is not algebraic.

Every rational number is algebraic. For, if $\alpha = p/q \in \mathbb{Q}$, $q \neq 0$, we may write $q\alpha - p = 0$ so that the α satisfies the integral polynomial $p(x) = a_0x + a_1$, because $a_0 = q \neq 0$ and $a_1 = -p$ are integers. Also, $\alpha = \sqrt{2}$ is an algebraic number because it satisfies the equation $x^2 - 2 = 0$. Similarly, it follows that every real number of the form $\alpha = a^{p/q}$ is algebraic, where $p, 0 \neq q \in \mathbb{Z}$, and $a \in \mathbb{R}$ could be taken to be a finite combination of real *quadratic surds*. Therefore, every real number that is finite combination of cube root, fourth root, etc., is an algebraic number.

More generally, algebraic numbers are closed under field operations. In particular, we conclude that there are many irrational numbers that are algebraic. On the other hand, the real numbers such as π, e, e^π are transcendental. According to Hermite-Lindermann Theorem, for every nonzero algebraic number α, the real number e^α is transcendental. Hence so is any nonzero logarithm $\ln(\alpha)$. In fact, by a theorem of Gelfond and Schneider, if $\beta \notin \mathbb{Q}$ is algebraic then $e^{\beta \ln(\alpha)}$ is transcendental.

Theorem 2.11 *For $n \geq 0$, the set $\mathscr{P}_n$ of all integral polynomials of degree $\leq n$ is a countably infinite set. Hence, the set $\mathbb{Z}[x]$ of integral polynomials is countably infinite.*

Proof Let $\mathbb{Z}^* = \mathbb{Z} \setminus \{0\}$. Notice that for each *nonzero* polynomial $p(x) \in \mathscr{P}_n$, there is an element

$$(a_0, a_1, \ldots, a_n) \in \mathbb{Z}^* \times \mathbb{Z}^n,$$

and the converse is also true. As the elements of the set $\mathbb{Z}$ correspond to the *constant polynomials* over $\mathbb{Z}$, the set P_0 is a countably infinite set because so is the set $\mathbb{Z}$.

---

[6] In general, we say a complex number $a + ib$, $a, b \in \mathbb{R}$, is an *algebraic number* if it is a root of an integral polynomial. For example, every number of the form $p + iq \in \mathbb{C}$, $p, q \in \mathbb{Q}$, is an *algebraic number*. The concept plays an important role in *algebraic number theory* when $\mathbb{C}$ is replaced with an algebraically closed field.

So, we may assume that $n \geq 1$. Clearly then, for any fixed $n \geq 1$, the function $f : \mathscr{P}_n \to \mathbb{Z}^* \times \mathbb{Z}^n$ defined by

$$f\left(a_0 x^n + a_1 x^{n-1} + \cdots + a_n\right) = (a_0, a_1, \ldots, a_n), \quad \text{with} \quad a_0 \neq 0,$$

is a surjective function; and it is injective by *uniqueness of coefficients* of a polynomial. Therefore, for $n \geq 1$, $f : \mathscr{P}_n \to \mathbb{Z}^* \times \mathbb{Z}^n$ is a bijective function. Hence, for $n \geq 0$, the set $\mathscr{P}_n$ is countably infinite. Finally, as

$$\mathbb{Z}[x] = \bigcup_{n=0}^{\infty} P_n,$$

it follows that the set $\mathbb{Z}[x]$ is countably infinite, by our previous remark. □

Corollary 2.4 *The set of algebraic numbers $\mathscr{A} \subset \mathbb{R}$, and a fortiori the set $\mathbb{Q}$, is countably infinite.*

Proof This follows from the previous theorem because the set $\mathscr{A}$ is a proper subset of the set of roots of the polynomials in the countably infinite set $\mathbb{Z}[x]$. More precisely, for every positive $N \in \mathbb{N}$, there are only finitely many equations with

$$\deg(p(x)) + |a_0| + |a_1| + \cdots + |a_n| = N,$$

where $p(x)$ is as in Eq. (2.4.1). □

The fact that the set $\mathbb{Q}$ is countably infinite can also be proved directly.

Theorem 2.12 *The set of all rational numbers $\mathbb{Q}$ is countably infinite.*

Proof Let us write every rational number in *reduced form* as p/q so that $(p, q) \in \mathbb{Z} \times \mathbb{N}$, with p, q being coprime. Clearly then, there is a bijective function $f : \mathbb{Q} \to \mathbb{Z} \times \mathbb{N}$, where the latter set is known to be countably infinite. Alternatively, for each rational number $r \in \mathbb{Q}$ written in *reduced form* as p/q, associate a number $h = |p| + q \in \mathbb{N}$, known as the *height* of r. Clearly then, the set $\mathbb{Q}$ can be represented as a countable union of finite sets A_n, where A_n denotes the set of all rational numbers with height n. For example, A_1, A_2, and A_3 are given by

$$A_1 = \{0/1\}, \quad A_2 = \{-1/1, 1/1\},$$
$$A_3 = \{-2/1, -1/2, 1/2, 2/1\}$$
$$A_4 = \{-3, -2/2, -1/3, 1/3, 2/2, 3/1\}.$$

Hence, $\mathbb{Q}$ is countably infinite. Observe that, in general, A_n's are mutually disjoint finite subsets with $\mathbb{Q} = \bigcup_{n=1}^{\infty} A_n$. So, the conclusion also follows by an earlier remark. □

Recall that a set is said to be *uncountable* if it is not countable. Since a subset of a countable set is again a countable set, so superset of an uncountable set is uncountable. In particular, we infer the uncountability of $\mathbb{R}$, if we could show that it contains an uncountable set. In fact, the set $\mathbb{R}$ contains uncountably many uncountable sets. For example, as we will prove more conveniently in Chap. 9, every interval $I \subseteq \mathbb{R}$ of *positive length* is an uncountable set. Recall that, for $a < b$ in $\mathbb{R}$, an *open interval I* with end points a and b is a set of the form

$$I = (a, b) = \{x \in \mathbb{R} : a < x < b\}.$$

Also, $[a, b] = (a, b) \bigcup \{a, b\}$ is called a *closed interval*. Now, since the function $f : (-1, 1) \to \mathbb{R}$ given by

$$f(x) = \frac{x}{x^2 - 1}, \quad \text{for } x \in (-1, 1), \tag{2.4.2}$$

is a bijective function, it suffices to prove that the interval $(-1, 1)$ is an uncountable set. However, for any $c < d$ in $\mathbb{R}$, the function $g : (a, b) \to (c, d)$ given by

$$g(x) = \frac{d - c}{b - a}(x - a) + c, \quad \text{for } x \in (a, b),$$

is also a bijective function. Indeed, the above relation is the equation of the (open) line joining the *points* (a, c) and (b, d) in the plane $\mathbb{R}^2$. Therefore, we are done if it is shown that the *unit interval*

$$(0, 1) := \{x \in \mathbb{R} : 0 < x < 1\}$$

is an uncountable set. In fact, a more interesting fact is that the unit interval $I = [0, 1]$ itself contains a *thin*[7] uncountable set, known as the *Cantor ternary set*.

Example 2.6 (*Cantor Ternary Set*). Let $I = [0, 1]$. Take $C_1 \subset I$ to be the set obtained from the interval I by removing its middle-third open interval $(1/3, 2/3)$. That is,

$$C_1 := \left[0, \frac{1}{3}\right] \bigcup \left[\frac{2}{3}, 1\right].$$

Next, take $C_2 \subset I$ to be the set obtained from the interval I by removing the middle-third open intervals from each one of the closed intervals $[0, 1/3]$ and $[2/3, 1]$. That is,

$$C_2 := \left[0, \frac{1}{9}\right] \bigcup \left[\frac{2}{9}, \frac{1}{3}\right] \bigcup \left[\frac{2}{3}, \frac{7}{9}\right] \bigcup \left[\frac{8}{9}, 1\right].$$

---

[7] Here, the term *thin* is used in the sense that the Cantor ternary set C has the Lebesgue measure zero, as we shall prove later in Chap. 9.

Continuing in this way, in general, the set $C_n \subset I$ obtained at the nth stage of the process is a union of mutually disjoint 2^n closed intervals. Each one of these closed interval has the length $1/3^n$ so that the total length of the set C_n is $(2/3)^n$. By definition, the *Cantor ternary set* is the set given by

$$C := \bigcap_{n=1}^{\infty} C_n \subset [0, 1].$$

Notice that as the set C contains at least the points

$$0, 1, \frac{1}{3}, \frac{2}{3}, \frac{1}{9}, \frac{2}{9}, \frac{7}{9}, \frac{8}{9}, \dots,$$

which are the end points of its constituent intervals $C_n, n \geq 1$, we conclude that $C \neq \emptyset$.

In what follows, we use *decimal expansion* of real numbers in $(0, 1)$, as introduced in the concluding part of the previous section. A reader not fully comfortable with the concept may skip the remaining part of the chapter, and return to complete after reading the relevant details as provided in Sect. 3.4.3.

The main idea here is to apply the Cantor's ingenious *diagonal argument*, as described in the proof of the next theorem.

Theorem 2.13 *The set S of all binary expansions $0.a_1a_2 \cdots$ formed by random selection of digits $a_k \in \{0, 1\}$ is uncountable.*

Proof For a contradiction, suppose S is a countable set. Then, we can *list* all the elements of S, say as $s_1, s_2, s_3, \dots$. We write $s_j \in S$ as

$$s_1 = 0.a_{11}a_{12}a_{13}a_{14} \cdots$$
$$s_2 = 0.a_{21}a_{22}a_{23}a_{24} \cdots$$
$$s_3 = 0.a_{31}a_{32}a_{33}a_{34} \cdots$$
$$\vdots \quad \vdots$$

where each digit a_{jk} belongs to the set $\{0, 1\}$. Using the axiom of choice, we define a number $b \in S$ as follows:

$$b = 0.b_1b_2b_3 \cdots, \quad \text{with} \quad b_n \neq a_{nn}, \quad \text{for every } n \in \mathbb{N}.$$

As $b \neq s_k$ for any k, we obtained an element of S that is not *enlisted* yet. This contradiction proves the theorem. □

Notice that the set S in Theorem 2.13 can be replaced by any finite set. In particular, when $S = \{0, 1, \dots, 9\}$, we obtain *decimal expansions* of real numbers in the interval $(0, 1)$. Indeed, since each real number in $(0, 1)$ has a *decimal expansion* of the form

$0.a_1 a_2 \cdots$, with $a_k \in \{0, 1, \ldots, 9\}$, we can write a direct proof of the next corollary by applying the Cantor's *diagonal argument*.

Corollary 2.5 *The unit interval* $I = (0, 1)$ *is uncountable.*

Also, the *Cantor ternary set* C can be expressed in descriptive terms as the set

$$C = \left\{ \sum_{k=1}^{\infty} \frac{a_k}{3^k} : a_k \in \{0, 2\} \right\}. \qquad (2.4.3)$$

The details are left as an exercise for the reader (Exercise 3.4.3). So, by using Cantor's argument as given in the proof of Theorem 2.13, the next theorem follows.

Theorem 2.14 *The Cantor ternary set* C *is uncountable.*

Exercises 2

2.1 Prove uniqueness part of Theorem 2.2.

2.2 Show that there are infinitely many prime numbers.

2.3 Let a, b be two positive integers with no common factor such that $\sqrt{a/b} \in \mathbb{Q}$. Show that both a and b are perfect squares.

2.4 Give a direct proof of the fact that every natural number other than 1 is a product of prime numbers.

2.5 Let $p(n)$ be a statement about $n \in \mathbb{N}$. Suppose

   a. $p(n_0)$ is true for some $n_0 \in \mathbb{N}$; and,

   b. $p(n_0), p(n_0 + 1), \ldots, p(k)$ are true for each $k \geq n_0 \Rightarrow p(k+1)$ is true.

   Show that $p(n)$ is true for all $n \in \mathbb{N}$ with $n \geq n_0$.

2.6 Prove the following Bernoulli's inequalities:

   a. $(1 + a)^n \geq 1 + na$, for $a \geq -1$ and $n = 0, 1, 2, \ldots$.

   b. $1 - na < (1 - a)^n < 1/(1 + na)$, for $0 < a < 1$ and $n = 2, 3, \ldots$.

2.7 Prove relation (2.1.1) using mathematical induction.

2.8 Show that the field of complex numbers $\mathbb{C}$ is not an ordered field.

2.9 Suppose $f : \mathbb{R} \to \mathbb{R}$ is both additive and multiplicative map. Show that $f \equiv 0$ or f is the identity map.

2.10 Can sum and product of two irrational numbers be rational?

2.11 Show that the reciprocal of an irrational number is irrational.

2.12 Show that a number is irrational if its square is irrational.

2.13 If $n \in \mathbb{N}$, show that $\sqrt{n}$ is either an integer or an irrational number.

2.14 Show that there is no rational number whose cube is a prime number.

2.15 Show that $\sqrt{2} + \sqrt{3}$ and $\sqrt{6}$ are irrational.

2.16 Show that $\sqrt{n+1} \pm \sqrt{n}$ is irrational for each $n \in \mathbb{N}$.

2.17 Show that e, π, and $\sqrt{3}^{\sqrt{2}}$ are all irrational.

2.18 If $x, y \in \mathbb{R}$ and $x < y$, show that there is an irrational number $z \in \mathbb{R}$ such that $x < z < y$.

2.19 Is the set $\{\sqrt{n+1} - \sqrt{n} : n \in \mathbb{N}\}$ bounded? If so, find its infimum and supremum.

2.20 Suppose $\emptyset \neq S \subset \mathbb{R}$ is bounded below. Show that $\inf(-S) = -\sup S$.

2.21 For $a \in \mathbb{R}$, let $S_a = \{r \in \mathbb{Q} : r^2 < a\} \subset \mathbb{R}$. Find the supremum of S_a.

2.22 Let X be an ordered field, and $A \subset X$. Show that

    a. $\beta = \sup A$ if and only if for each $0 < \epsilon \in X$ there is an upper bound u of the set A such that $\beta - \epsilon < u \leq \beta$.

    b. $\alpha = \inf A$ if and only if for each $0 < \epsilon \in X$ there is a lower bound ℓ of the set A such that $\alpha \leq \ell < \alpha + \epsilon$.

2.23 For every pair of elements a, b in a field, prove that $-(a - b) = b - a$.

2.24 For each $x > 0$, show that there is a natural number n such that $0 < 1/n < x$.

2.25 If x is a positive real number and $n \in \mathbb{N}$, then there is a unique positive real number y such that $y^n = x$.

2.26 Show that $\sqrt{xy} \leq (x + y)/2$ for positive real numbers x and y. When does equality happen? Generalise the inequality for n real numbers.

2.27 Show that $\sqrt{x^2 + y^2} \leq |x| + |y|$ and $\sqrt{|x + y|} \leq \sqrt{|x|} + \sqrt{|y|}$ for any $x, y \in \mathbb{R}$.

2.28 For any $x, y \in \mathbb{R}$, show that $\max\{x, y\} = (x + y + |x - y|)/2$ and $\min\{x, y\} = (x + y - |x - y|)/2$.

2.29 If $A = \{1/n : n \in \mathbb{N}\} \subset \mathbb{Q}$, show that $\inf A = 0$.

2.30 For any real number $x \in \mathbb{R}$, there is an integer $m \in \mathbb{Z}$ such that $m - 1 \leq x < m$.

2.31 Show that the function f as defined in (2.4.2) is a bijective function.

References

1. R. Crandall, C. Pomerance, *Prime Numbers–A Computational Perspective* (2/e) (Springer, Berlin, 2005)
2. S.D. Abbott, *Understanding Analysis (2/e)* (UTM, Springer, New York, 2015)
3. J. Stillwell, *The Real Numbers—An Introduction to Set Theory and Analysis* (Springer, UTM, 2013)

Chapter 3
Sequences and Series of Numbers

As for methods I have sought to give them all the rigour that one requires in geometry, so as never to have recourse to the reasons drawn from the generality of algebra.

Augustin-Louis Cauchy (1789–1857)

Recall that a *sequence* in a set X is a function $s : \mathbb{N} \to X$. For $n \in \mathbb{N}$, we write the nth term of the sequence s as $x_n = s(n) \in X$. In particular, when $X = \mathbb{R}$, we have a sequence of real numbers (or simple a *real sequence*). For example, the real sequence $\langle t_n \rangle$ given by taking

$$t_n = 1 + 2 + 3 + \cdots + n, \quad \text{for} \quad n \geq 1,$$

counts the number of dots in a *triangular pattern*, and the tiling of squares with side lengths given by the *Fibonacci numbers*

$$a_0 = a_1 = 1, \quad \text{and} \quad a_{n+1} = a_n + a_{n-1}, \quad \text{for} \quad n \geq 1,$$

is an interesting real sequence $\langle a_n \rangle$, which is used to study many different kinds of *spiral patterns* in the nature.

Real sequences arise naturally in diverse types of problems in science and engineering. For example, *approximation* of some processes of a discrete system involves real sequences, which usually represent the *estimated values* obtained by applying an *iteration scheme* (such as in Example 5.6). Also, it is an everyday affair in financial institutions to conduct a quantitative analysis of a multi-layer market data to study the real sequences as *variations in values* of the stocks invested. In such cases, one usually applies *time series methods* to do analysis with the real sequences.

Further, as introduced earlier in Chap. 1, an *infinite series* over the set $\mathbb{R}$ is a formal sum of the form

$$\sum_{n=1}^{\infty} x_n = x_1 + x_2 + \cdots + x_n + \cdots, \quad \text{with} \quad x_n \in \mathbb{R},$$

which we usually see as the sum of the terms of a sequence $\langle x_n \rangle$. To make sense out of any such type of *formal sum*, we use the sequence $\langle s_n \rangle$ given by taking

$$s_n = x_1 + x_2 + \cdots + x_n,$$

which is called the sequence of *partial sums* of the *series*. Notice that we can then write the nth term x_n of the infinite series as

$$x_n = s_{n+1} - s_n, \quad \text{for} \quad n \geq 1.$$

We say an infinite series is *summable* if the associated *sequence of partial sums* $\langle s_n \rangle$ converges in the sense of Definition 3.2, and we write the **limit** of the sequence $\langle s_n \rangle$ as the **sum** of the infinite series. This *connection* between infinite series and sequences is the central idea of the topic. It helps to solve many practical problems in science and engineering involving approximation of an infinite process.

In historical terms, the beginning is attributed to the Greek mathematician *Archimedes* of Syracuse (287–212 *BC*) who used an infinite series to approximate the area under a parabolic arc. Since then, there had been many instances of infinite series being used to *define* various types of functions (see Sect. 1.3 for details). However, some of the most important results about infinite series, and the related fundamental theorems of analysis, were obtained during the eighteenth and nineteenth centuries by renowned mathematicians such as Maclaurin, Taylor, d'Alembert, Gauss, Einstein, Euler, Bernoulli brothers, Riemann, Fourier, Dirichlet, Weierstrass, Dedekind, Cantor, and many more. For example, a true classic is the proof of the following identity given by Euler in 1734:

$$\frac{1}{1^2} + \frac{1}{2^2} + \frac{1}{3^2} + \frac{1}{4^2} + \cdots = \frac{\pi}{6}.$$

In Chap. 10, we will give a proof by using the Fourier series.

We discuss here the convergence of a real sequence $\langle a_n \rangle$ with respect to the *modulus metric* (Corollary 2.2), where the concept of convergence is understood in the sense of Definition 3.2. Having introduced the fundamental idea of *the limit of a real sequence*, the main concern is to discuss certain important *convergence theorems*. Subsequently, we apply the concepts developed, and theorem proved, to derive some important *tests of convergence* for infinite series of real numbers, and also to discuss *convergence types* of some special types of series. In this chapter, the term *sequence* would always stand for a sequence of real numbers.

3.1 The Limit of a Sequence

Let us start with a list of some simple real sequences.

Example 3.1 We shall use the following sequences to illustrate the concepts introduced in the sequel.

1. The sequence $s(n) = 1$, $n \in \mathbb{N}$, is given by

$$s = \langle 1, 1, 1, \ldots \rangle.$$

2. The sequence $t(n) = (-1)^n$, $n \in \mathbb{N}$, is given by

$$t = \langle -1, 1, -1, \ldots \rangle.$$

3. The sequence $u(n) = n^2 - 1$, $n \in \mathbb{N}$, is given by

$$u = \langle 0, 3, 8, \ldots \rangle.$$

4. The sequence $v(n) = \dfrac{n}{2n + 1}$, $n \in \mathbb{N}$, is given by

$$v = \langle 1/3, 2/5, 3/7, \ldots \rangle.$$

5. The sequence $w(n) = 2^n$, $n \in \mathbb{N}$, is given by

$$w = \langle 2, 4, 8, \ldots \rangle.$$

Notice that most of the *set theoretic properties* of a sequence $x = \langle x_n \rangle$ are given in terms of the *range set* $\mathrm{Ran}(x)$ given by

$$\mathrm{Ran}(x) = \{ x_1, x_2, x_3, \ldots \}.$$

For example, the two sequences in (1) and (2) above are said to be *finite* because so are their range sets. And, for the same reason, the three sequences in (3) through (5) above are *infinite sequences*. Also, the three sequences in (1), (2), and (4) are *bounded* because so are their range sets, in the sense as defined earlier (Definition 1.15). In general, the concept of *boundedness* of a sequence is as given in the next definition.

Definition 3.1 We say $\langle x_n \rangle$ is a *bounded sequence* if, for some real number $M > 0$, we have $|x_n| \leq M$, for all $n \in \mathbb{N}$.

That is, in terms of the *modulus metric* $m : \mathbb{R} \times \mathbb{R} \to \mathbb{R}$ given by

$$m(x, y) = |x - y|, \quad \text{for } x, y \in \mathbb{R},$$

we have $x_n \in [-M, M]$, for all $n \geq 1$. Clearly then, the two sequences in (3) and (5) above are *unbounded sequences*. More generally, we say a sequence $\langle x_n \rangle$ is *bounded above* if

$$\sup(x) = \sup\{x_1, x_2, x_3, \ldots\} < \infty;$$

and, we say $\langle x_n \rangle$ is *bounded below* if

$$\inf(x) = \inf\{x_1, x_2, x_3, \ldots\} < \infty.$$

For example, the sequence $s(n) = (n + 1)/n$ is bounded above by 2 and bounded below by 1. As before, a sequence $\langle x_n \rangle$ is bounded if and only if it is both bounded above and bounded below. Therefore, the sequences $x(n) = n$ and $y(n) = 2^n$ are not bounded because both are bounded below, but not bounded above. Notice that, in view of next definition,

$$\lim_{n \to \infty} x_n = +\infty \quad \text{if } \langle x_n \rangle \text{ is not bounded above; and}$$

$$\lim_{n \to \infty} x_n = -\infty \quad \text{if } \langle x_n \rangle \text{ is not bounded below.}$$

In either situation, we say $\langle x_n \rangle$ is a *divergent sequence*.

Definition 3.2 Let $\langle x_n \rangle$ be a sequence. We say $\langle x_n \rangle$ is a **convergent sequence** if there is some $a \in \mathbb{R}$ such that, for any $\varepsilon > 0$, we can find a natural number $K = K(\varepsilon)$ satisfying the following condition:

$$n \geq K \quad \Rightarrow \quad |x_n - a| < \varepsilon. \tag{3.1.1}$$

In this case, $a \in \mathbb{R}$ is called a **limit** of the sequence $\langle x_n \rangle$, and write write

$$\lim_{n \to \infty} x_n = a \quad \text{or} \quad x_n \to a, \quad \text{as } n \to \infty.$$

We say $\langle x_n \rangle$ is a *divergent sequence* if no such $a \in \mathbb{R}$ can be found.

Notice that, for $n \geq K$, the equivalence

$$\lim_{n \to \infty} x_n = a \quad \Leftrightarrow \quad a - \varepsilon < x_n < a + \varepsilon, \tag{3.1.2}$$

implies that the sequence $\langle x_n \rangle$ is *eventually* close to the value a. Said differently, every interval centred around the value a contains all but finitely many terms of the sequence $\langle x_n \rangle$. We will use this topological interpretation of the condition (3.1.1) throughout the chapter. Clearly, if all terms of a *convergent sequence* are positive, then so is the *limit*.

Example 3.2 (*a*) The sequence $s_n = 1$, for all $n \in \mathbb{N}$, converges to 1; for, $|s_n - 1| = 0 < \varepsilon$, for any $\varepsilon > 0$ and for all $n \geq 1$;

(b) The sequence $t_n = (-1)^n$, $n \in \mathbb{N}$, is divergent. For, let $a \in \mathbb{R}$. Now, if $|a| \neq 1$, we may take $\varepsilon = \min\{|a - 1|, |a + 1|\}$ so that $|t_n - a| \geq \varepsilon$ for all n. And, if $|a| = 1$ then for $\varepsilon = 2$, we have

$$|t_n - a| \geq \varepsilon, \quad \text{for all } n \text{ odd, and} \quad |t_n - a| \geq \varepsilon, \quad \text{for all } n \text{ even.}$$

(c) The sequence $u_n = n/(2n + 1)$, $n \in \mathbb{N}$, is convergent, with limit $1/2$ because for any $\varepsilon > 0$ and for any natural number $K > 1/2(2\varepsilon) - 1$,

$$n > K \quad \Rightarrow \quad \left| u_n - \frac{1}{2} \right| = \frac{1}{2(2n + 1)} < \varepsilon.$$

(e) The sequence $v_n = n$, $n \in \mathbb{N}$, is divergent because for any $a \in \mathbb{R}$, we have $|v_n - a| \geq 1$, for each natural number $n > K > |a| + 1$.

Theorem 3.1 *Every convergent sequence has a unique limit.*

Proof Let $\langle x_n \rangle$ be a convergent sequence. If possible, let there be real numbers $a_1 \neq a_2$ satisfying the condition (3.1.1), for some $K_1, K_2 \in \mathbb{N}$, with $\varepsilon = (1/2)|a_1 - a_2|$. Let $K = \max\{K_1, K_2\}$. Then, for all $n \geq K$, we have

$$\begin{aligned} |a_1 - a_2| &= |(x_n - a_2) - (x_n - a_1)| \\ &\leq |x_n - a_2| + |x_n - a_1| \\ &< \varepsilon + \varepsilon = 2\varepsilon = |a_1 - a_2|, \end{aligned}$$

which is absurd. Hence, we conclude $a_1 = a_2$. □

Theorem 3.2 *Every convergent sequence is bounded. However, not every bounded sequence is convergent.*

Proof Let $\langle x_n \rangle$ be a convergent sequence, with the limit $a \in \mathbb{R}$. Take $\varepsilon = 1$. Then, for some $K \in \mathbb{N}$, we have

$$n > K \quad \Rightarrow \quad |x_n - a| < 1,$$

by condition (3.1.1). We write

$$M = \max \left\{ \max_{n,m \leq K} |x_m - x_n|, \ 1 + \max_{n \leq K} |x_n - a|, \ 2 \right\}.$$

It then follows that, for $m, n \leq K$,

$$|x_m - x_n| \leq \max_{i,j \leq K} |x_i - x_j| \leq M;$$

also, for $m \leq K < n$, we have

$$|x_m - x_n| \le |x_m - a| + |a - x_n| \le 1 + \max_{i \le K} |x_i - a| \le M;$$

and for $m, n > K$, we have

$$|x_m - x_n| \le |x_m - a| + |a - x_n| \le 2 \le M.$$

It thus follows that

$$\max_{m, n \ge 1} \{|x_n - x_m|\} \le M < \infty.$$

Therefore, in particular, $|x_n| \le |x_1| + M$ for all $n \ge 1$ and so the sequence $\langle x_n \rangle$ is bounded. For the converse, we know that the bounded sequence $\langle (-1)^n \rangle$ is not convergent (see the part (b) of Example 3.2). $\qquad\square$

Notice that, by taking

$$M = \max \{|x_1 - a|, \ldots, |x_K - a|, 1\},$$

it follows directly that every x_n is in the interval $(a - M, a + M)$. However, the proof given above suggests something more interesting: *If a bounded sequence $\langle x_n \rangle$ fails to converge, then it is possible because its terms x_n are not mutually arbitrarily close. More precisely, the sequence $\langle x_n \rangle$ is not a Cauchy sequence.* We will pursue this important aspect of the bounded sequences in the later part of the chapter.

Remark 3.1 The number $K \in \mathbb{N}$ in Definition 3.2 depends on ε in the sense that the smaller the value of ε the larger would be the value of K. In actual computations, if a sequence $\langle x_n \rangle$ is known to be convergent, then we need to apply some additional mathematical tools to make an intelligent guess about the possible value of a (see Example 3.3). It is then usually easy to find a suitable $K \in \mathbb{N}$ such that the condition (3.1.1) holds for any $\varepsilon > 0$.

The next theorem helps in a number of situations to find the limit of a convergent sequence. It is also known as the *pinching lemma*.

Theorem 3.3 (Sandwich Lemma). *Let $\langle u_n \rangle$, $\langle x_n \rangle$, and $\langle v_n \rangle$ be three sequences such that $u_n \le x_n \le v_n$, for all $n \ge k$, and*

$$\lim_{n \to \infty} u_n = \lim_{n \to \infty} v_n = \ell. \qquad (3.1.3)$$

Then, $\lim_{n \to \infty} x_n = \ell$.

Proof Let $\varepsilon > 0$. So, there are natural numbers N_1, N_2 such that

$$\ell - \varepsilon < u_n < \ell + \varepsilon \quad \text{and} \quad \ell - \varepsilon < v_n < \ell + \varepsilon,$$

for $n \ge N = \max\{N_1, N_2\}$. Then, for $K = \max\{k, N\}$, we have

$$\ell - \varepsilon < u_n \le x_n \le v_n < \ell + \varepsilon \quad \Rightarrow \quad \lim_{n \to \infty} x_n = \ell.$$

This completes the proof. The proof of the second assertion follows the same way. □

The next theorem is also very useful in evaluating the limit of a sequence.

Theorem 3.4 *If the sequences $\langle u_n \rangle$, $\langle x_n \rangle$, and $\langle v_n \rangle$ are convergent and satisfy $u_n \le x_n \le v_n$, for all $n \ge k$ for some $k \ge 1$, then*

$$\lim_{n \to \infty} u_n \le \lim_{n \to \infty} x_n \le \lim_{n \to \infty} v_n. \tag{3.1.4}$$

In particular, if $\langle x_n \rangle$ is convergent and $u \le x_n \le v$ for all $n \ge k$ for some $k \ge 1$, then

$$u \le \lim_{n \to \infty} x_n \le v.$$

Proof We first prove that $\lim_{n \to \infty} y_n \ge 0$ for each convergent sequence $\langle y_n \rangle$ of non-negative real numbers. Let $y = \lim_{n \to \infty} y_n$. We need to show that $y \ge 0$. On the contrary, assume that $y > 0$. Then $\varepsilon := -y > 0$ and there is a natural number N such that $|y_n - y| < \varepsilon$. Therefore, $y_n < y_n < y + \varepsilon = 0$ for all $n \ge N$, contradicting $y_n \ge 0$.

The general case now follows by applying this to the sequences $\langle v_n - x_n \rangle$ and $\langle x_n - u_n \rangle$. The particular case follows by using $u_n = u$ and $v_n = v$ for all n. □

Corollary 3.1 *Let $X \subset \mathbb{R}$. (i) If X is bounded above, with $\alpha = \sup X$, there exists a sequence $\langle x_n \rangle$ in X such that $\lim_{n \to \infty} x_n = \alpha$; and (ii) if X is bounded below, with $\beta = \inf X$, there exists a sequence $\langle y_n \rangle$ in X such that $\lim_{n \to \infty} x_n = \beta$.*

Proof By Exercise 2.4, there exists $x_n \in X$ such that

$$x_n > \alpha - \frac{1}{n}, \quad \text{for each } n \in \mathbb{N}.$$

As $x_n \le \alpha$, we also have

$$0 \le \alpha - x_n < \frac{1}{n}, \quad \text{for all } n \in \mathbb{N}.$$

Thus (i) follows from Theorem 3.3. For (ii), as $\beta = \inf X$, Exercise 2.4 implies that there exists $y_n \in X$ such that

$$y_n < \beta + \frac{1}{n}, \quad \text{for each } n \in \mathbb{N}.$$

As $y_n \ge \beta$, we also have

$$0 \le y_n - \beta < \frac{1}{n}, \quad \text{for all } n \in \mathbb{N}.$$

Therefore, once again, the assertion follows from Theorem 3.3. □

Example 3.3 Consider a sequence defined inductively as follows:

$$x_1 = \sqrt{2}, \quad x_{n+1} = \sqrt{2 + \sqrt{x_n}}, \quad \text{for } n \ge 1.$$

A simple inductive argument shows that $\sqrt{2} \le x_n < 2$. Also, since

$$x_{n+1}^2 - x_n^2 = \sqrt{x_n} - \sqrt{x_{n+1}},$$

a second application of induction shows that

$$\sqrt{2} = x_1 < x_2 < \cdots < x_n < \cdots < 2.$$

Assuming that $\lim_{n \to \infty} x_n = a$, we have, by Theorem 3.4, $\sqrt{2} < a \le 2$.
    We now proceed to find the value a. For, observe that

$$(x_{n+1}^2 - 2)^2 - x_n = 0$$

and this shows that

$$(a^2 - 2)^2 - a = 0.$$

Since $(a^2 - 2)^2 - a = (a - 1)(a^3 + a^2 - 3a - 4)$, and $a \ne 1$, the number a must
be a zero of the integral polynomial

$$f(x) = x^3 + x^2 - 3x - 4, \quad \text{for } x \in [\sqrt{2}, 2].$$

In fact, as $f(x)$ is a strictly increasing for $x \ge \sqrt{2}$ and $f(\sqrt{2}) < 0$ (see Fig. 3.1), it
has a unique root in the given interval. We can estimate the value of a numerically, to
any degree of accuracy, say by using Newton-Raphson method (see Example 5.6).

Fig. 3.1 The function
$f(x) = x^3 + x^2 - 3x - 4$,
$x \in [\sqrt{2}, 2]$

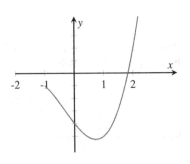

The limits obtained for three standard sequences in the next theorem are used frequently in the sequel. A proof given below assumes knowledge about some basic calculus facts related to continuity and differentiability, which we shall discuss more rigorously in Chap. 4.

Theorem 3.5 *1.* $\lim\limits_{n\to\infty} a^{1/n} = 1$, *for any positive number a;*

2. $\lim\limits_{n\to\infty} n \ln\left(1 + \dfrac{x}{n}\right) = x$;

3. $\lim\limits_{n\to\infty} \left(1 + \dfrac{x}{n}\right)^n = e^x$.

Proof For (1), we write $n^{1/n} = 1 + x_n$ where $x_n \geq 0$. Then, for $n \geq 2$, using binomial theorem, we have

$$n = (1 + x_n)^n > \frac{n(n-1)}{2} x_n^2$$

and so

$$0 < x_n < \sqrt{\frac{2}{n-1}}.$$

By Theorem 3.3, we have $\lim\limits_{n\to\infty} x_n = 0$. Therefore, we have $\lim\limits_{n\to\infty} n^{1/n} = 1$. Next, suppose that $a \geq 1$ is some real number. Now, if $n > a$, then $1 < a^{1/n} < n^{1/n}$ implies that $a^{1/n} \to 1$, by Theorem 3.3. And, if $0 < a < 1$, then $1/a > 1$. So, by the previous case, we have $(1/a)^{1/n} \to 1$. It thus follows that necessarily $a^{1/n} \to 1$.

For (2), we can assume that $x \neq 0$. Now, taking

$$t_n = n \ln\left(1 + \frac{x}{n}\right), \quad x \neq 0,$$

and since $h = x/n \to 0$ as $n \to \infty$, we have

$$\lim_{n\to\infty} t_n = x \lim_{n\to\infty} \frac{\ln(1 + (x/n)) - \ln 1}{(x/n)}$$

$$= x \lim_{h\to 0} \frac{\ln(1 + h)) - \ln 1}{h} = x \frac{d}{dy} \ln y \Big|_{y=1} = x.$$

Here, we have used the concept of derivative and the fact that the derivative of $\ln x$ is $1/x$.

Finally, the proof of (3) follows directly from (1) and (2) above. For, taking

$$w_n = \left(1 + \frac{x}{n}\right)^n, \quad x \neq 0,$$

we have

$$\ln\left(\lim_{n\to\infty} w_n\right) = \lim_{n\to\infty} \ln w_n = \lim_{n\to\infty} n \ln\left(1 + \frac{x}{n}\right) = x.$$

The first equality follows by the continuity of $\ln x$ and Theorem 4.4 discussed later in Chap. 4. Therefore, by exponentiating, we get $\lim\limits_{n\to\infty} w_n = e^x$. $\qquad\square$

The two assertions in the next theorem are known as Cauchy's first and second theorems on limits, which he gave in 1821. These hold importance to methods of *matrix summability*. We may also use them to describe the *Cesáro summability* of a series such as the *Grandi series*:

$$\sum_{n=0}^{\infty}(-1)^{n+1} = -1 + 1 - 1 + \cdots .$$

Theorem 3.6 (Cesáro Means). *If a sequence $\langle x_n \rangle$ converges to ℓ then*

$$\lim_{n\to\infty}\frac{x_1 + x_2 + \cdots + x_n}{n} = \ell.$$

In particular, if $x_n > 0$ for all n, then $\lim\limits_{n\to\infty}(x_1 x_2 \cdots x_n)^{1/n} = \ell.$

Proof For a proof of the first assertion, take $y_n = x_n - \ell,\ n \geq 1$, so that

$$\frac{y_1 + y_2 + \cdots + y_n}{n} = \frac{x_1 + x_2 + \cdots + x_n}{n} - \ell.$$

So, the first assertion follows if we show that

$$\lim_{n\to\infty}\frac{y_1 + y_2 + \cdots + y_n}{n} = 0.$$

For, let $\varepsilon > 0$ be arbitrary. Since $\lim_{n\to\infty} x_n = \ell$ we have $\lim_{n\to\infty} y_n = 0$ and therefore there is some $m \in \mathbb{N}$ such that

$$|y_n| < \frac{\varepsilon}{2}, \quad \text{whenever } n > m.$$

Since convergent sequences are always bounded, there exists some $M > 0$ such that $|y_n| \leq M$, for all n. Now,

$$\left| \frac{y_1 + \cdots + y_n}{n} \right| = \left| \frac{y_1 + \cdots + y_m + y_{m+1} + \cdots + y_n}{n} \right|$$

$$\leq \frac{|y_1 + \cdots + y_m|}{n} + \frac{|y_{m+1} + \cdots + y_n|}{n}$$

$$\leq \frac{|y_1| + \cdots + |y_m|}{n} + \frac{|y_{m+1}| + \cdots + |y_n|}{n}$$

$$\leq \frac{mM}{n} + \frac{(n-m)\varepsilon}{2n}$$

$$< \frac{mM}{n} + \frac{\varepsilon}{2} \qquad \left(\text{because } 0 < \frac{n-m}{n} < 1 \right)$$

$$< \frac{\varepsilon}{2} + \frac{\varepsilon}{2} = \varepsilon,$$

provided $mM/n < \varepsilon/2$ or, equivalently, $n > 2mM/\varepsilon$. So, taking a natural number $m' > 2mM/\varepsilon$, and $K = \max\{m, m'\}$, we obtain

$$\left| \frac{y_1 + y_2 + \cdots + y_n}{n} \right| < \varepsilon, \qquad \text{for each} \quad n > K,$$

which proves the first assertion of the theorem. The second assertion follows directly from the first by taking $y_n = \ln x_n$ so that

$$\lim_{n \to \infty} y_n = \lim_{n \to \infty} \ln x_n = \ln \left(\lim_{n \to \infty} x_n \right) = \ln \ell.$$

Therefore, we have

$$\ln \left(\lim_{n \to \infty} (x_1 x_2 \cdots x_n)^{1/n} \right) = \lim_{n \to \infty} \ln(x_1 x_2 \cdots x_n)^{1/n}$$

$$= \lim_{n \to \infty} \frac{\ln x_1 + \ln x_2 + \cdots + \ln x_n}{n}$$

$$= \lim_{n \to \infty} \frac{y_1 + y_2 + \cdots + y_n}{n} = \ln \ell$$

and hence $\lim_{n \to \infty} (x_1 x_2 \cdots x_n)^{1/n} = \ell$. We have used twice the continuity of $\ln x$ and Theorem 4.4 which are discussed later in Chap. 4. $\qquad \square$

Corollary 3.2 *For any sequence $\langle x_n \rangle$ of positive real numbers,*

$$\lim_{n \to \infty} \frac{x_{n+1}}{x_n} = \ell \quad \Rightarrow \quad \lim_{n \to \infty} x_n^{1/n} = \ell. \qquad (3.1.5)$$

Proof Clearly, we have $\ell > 0$. Take

$$y_1 = x_1, \quad y_2 = \frac{x_2}{x_1}, \quad y_3 = \frac{x_3}{x_2}, \quad \ldots, \quad y_n = \frac{x_n}{x_{n-1}}$$

so that

$$y_1 y_2 \cdots y_n = x_n \quad \Rightarrow \quad (y_1 y_2 \cdots y_n)^{1/n} = x_n^{1/n}.$$

Also, since

$$\lim_{n\to\infty} \frac{x_{n+1}}{x_n} = \ell \quad \Rightarrow \quad \lim_{n\to\infty} \frac{x_n}{x_{n-1}} = \ell \quad \Rightarrow \quad \lim_{n\to\infty} y_n = \ell,$$

the assertion follows from the second part of Theorem 3.6. For an alternative proof, let an arbitrary $\varepsilon > 0$, there exists some natural number $K = K(\varepsilon)$ such that

$$n > K \quad \Rightarrow \quad x_{n+1} < (\ell + \varepsilon)x_n \quad \Rightarrow \quad x_n < a(\ell + \varepsilon)^n,$$

where a depends on K, ℓ, and ε. Suppose $x_n^{1/n} > \ell + 2\varepsilon$. Then,

$$a(\ell + \varepsilon)^n > x_n \geq ((\ell + \varepsilon) + \varepsilon)^n$$
$$> n\varepsilon(\ell + \varepsilon)^{n-1}$$
$$\Rightarrow \quad n < \frac{a(\ell + \varepsilon)}{\varepsilon}.$$

We thus conclude that, for an arbitrary $\varepsilon > 0$,

$$n > \frac{a(\ell + \varepsilon)}{\varepsilon} \quad \Rightarrow \quad x_n^{1/n} < \ell + 2\varepsilon.$$

Now, if $\ell = 0$ then for these values of n,

$$0 < x_n^{1/n} < \ell + 2\varepsilon = 2\varepsilon,$$

and so the claim holds. Next, if $\ell > 0$ so that $u_n = 1/x_n$, the above procedure implies

$$u_n^{1/n} < \frac{1}{\ell} + \varepsilon = \frac{1 + \ell\varepsilon}{\ell} < \frac{1}{\ell(1 - \ell\varepsilon)},$$

for all sufficiently large n. But then, for these values of n, we have $x_n^{1/n} > \ell(1 - \ell\varepsilon)$, and so the proof is complete. $\square$

Example 3.4 For the sequence $x_n = n^n/n!$, we have

$$\frac{x_{n+1}}{x_n} = \left(\frac{n+1}{n}\right)^n \quad \Rightarrow \quad \lim_{n\to\infty} \frac{x_{n+1}}{x_n} = e,$$

by (3) of Theorem 3.5. Hence, by Corollary 3.2, we conclude that

$$x_n^{1/n} = \frac{n}{(n!)^{1/n}} \quad \Rightarrow \quad \lim_{n\to\infty} \frac{n}{(n!)^{1/n}} = e.$$

Likewise, to prove that

$$\lim_{n \to \infty} \frac{(n!)^{1/n}}{n} = \frac{1}{e},$$

we may apply Corollary 3.2 to the sequence $x_n = n!/n^n$.

Notice that the limit obtained in the above example is actually a weaker version of the following *Stirling's formula*:

$$n! \sim \left(\frac{n}{e}\right)^n \sqrt{2\pi n}, \quad \text{as} \quad n \to \infty,$$

where $s_n \sim t_n$ means that the two sequences are *asymptotic*. That is, $(s_n/t_n) \to 1$ as $n \to \infty$. We also have

$$\sqrt{2\pi}\, n^{n+(1/2)} e^{-n} \leq n! \leq e n^{n+(1/2)} e^{-n}$$

In general, we use the *Landau notations* introduced below to study the *asymptotic behaviour* of a sequence $\langle x_n \rangle$ with respect to some *majorant sequence* $\langle y_n \rangle$, with $y_n \geq 0$. We usually take n^a, $(\ln n)^b$, or e^{cn} for y_n. We write $x_n := O(y_n)$ if, for some $C > 0$,

$$|x_n| \leq C\, y_n, \quad \text{for all} \quad n \geq 1.$$

In particular, a sequence $\langle x_n \rangle$ is bounded if $x_n = O(1)$. For $y_n > 0$ for sufficiently large n, we write $x_n := o(y_n)$ if $x_n/y_n \to 0$ as $n \to \infty$. In particular, $x_n = o(1)$ implies that the sequence $\langle x_n \rangle$ is a null sequence.

Definition 3.3 If $\lim_{n \to \infty} x_n = 0$, we say $\langle x_n \rangle$ is a *null sequence*.

Further, as said above, we write

$$x_n \sim y_n \quad \text{if} \quad \frac{x_n}{y_n} \to 1 \quad \text{as} \quad n \to \infty.$$

In this case, $\sim$ defines an equivalence relation on the collection of all sequences, and any two sequences $\langle x_n \rangle$ and $\langle y_n \rangle$ in the same equivalence class are said to be *asymptotically equivalent* or the two are *asymptotic sequences*. Notice that

$$x_n \sim y_n \quad \Leftrightarrow \quad x_n = y_n + o(y_n).$$

We also say that the sequence $\langle y_n \rangle$ is the *principal part* of $\langle x_n \rangle$.

In general, *null sequences* play a significant role in some of the important developments to follow, particularly in the construction of the set $\mathbb{R}$, as described in a later part of this chapter. We will discuss here a little more about such sequences. We start with a simple observation. Suppose $\langle a_n \rangle$ is a convergent sequence, with $\lim_{n \to \infty} a_n = a$. Then, for any sequence $\langle x_n \rangle$,

$$|x_n| < a_n \quad \Rightarrow \quad -a < \lim_{n \to \infty} x_n < a, \tag{3.1.6}$$

by Theorem 3.3. Therefore, $\langle x_n \rangle$ is a null sequence if so is the sequence $\langle a_n \rangle$, i.e., if $a = 0$. In particular, it follows that if a sequence of *positive* real numbers $\langle x_n \rangle$ has a null sequence as a *majorant sequence* then $\langle x_n \rangle$ is null sequence.

Example 3.5 We show here that the sequence given by $x_n = a^n/n!$ is a null sequence, for any $a \in \mathbb{R}$. That is,

$$x_n = \frac{a^n}{n!} \to 0 \quad \text{as} \quad n \to \infty.$$

We first suppose $a > 0$. Let $\varepsilon > 0$ be arbitrary. By the archimedean property, there is a natural number N such that $N > a$, *i.e.*, $a/N < 1$. Then, a simple induction argument proves that

$$x_{N+k} \leq x_N \left(\frac{a}{N+1} \right)^k, \quad \text{for all} \ k \geq 1.$$

So, if $0 < a/(N+1) < 1$, then $a^k/(N+1)^k \to 0$ as $n \to \infty$. Thus, there is a natural number N_1 such that

$$k \geq N_1 \quad \Rightarrow \quad \left(\frac{a}{N+1} \right)^k < \frac{\varepsilon}{x_N}.$$

Thus, for $n > N + N_1$, by letting $k = n - N - 1$, it follows that $x_n < \varepsilon$. The case when $a < 0$ follows exactly the same way.

Theorem 3.7 *For any sequence $\langle x_n \rangle$ of nonzero real numbers,*

$$\lim_{n \to \infty} \left| \frac{x_{n+1}}{x_n} \right| = \ell, \ \text{with} \ |\ell| < 1, \quad \Rightarrow \quad \lim_{n \to \infty} x_n = 0. \tag{3.1.7}$$

Proof Since

$$|\ell| < 1 \ \Rightarrow \ 0 < \frac{1 - \ell}{2} < 1 \ \Rightarrow \ \lim_{n \to \infty} \left(\frac{1 - \ell}{2} \right)^n = 0,$$

so for $\varepsilon = (1 - \ell)/2$ there is some $K = K(\varepsilon) \in \mathbb{N}$ such that

$$n \geq K \Rightarrow \quad \ell - \varepsilon < \frac{x_{n+1}}{x_n} < \ell + \varepsilon$$

$$\Rightarrow \quad \ell - \frac{1 - \ell}{2} < \frac{x_{n+1}}{x_n} < \ell + \frac{1 - \ell}{2}$$

$$\Rightarrow \quad |x_{n+1}| < \frac{\ell + 1}{2}|x_n|$$

$$\Rightarrow \quad |x_{K+m}| < \left(\frac{\ell + 1}{2}\right)^m |x_K|, \quad \text{for all} \quad m \geq 1.$$

Therefore, $\langle x_n \rangle$ is a null sequence. $\qquad\square$

3.2 Algebra of Convergent Sequences

Suppose a nonempty set X has some algebraic operations such as *sum* and *product*, and let

$$X^{\mathbb{N}} := \{x = \langle x_n \rangle : x : \mathbb{N} \to X\}.$$

We can extend each algebraic operation on the set X to the collection $X^{\mathbb{N}}$, by using the concept of *pointwise operation*. In particular, we can have all the four arithmetic operations on the collection $\mathbb{R}^{\mathbb{N}}$ of real sequences. That is, for $x = \langle x_n \rangle$, $y = \langle y_n \rangle \in \mathbb{R}^{\mathbb{N}}$, we take

$$x + y := \langle x_n + y_n \rangle \quad \text{and} \quad x - y := \langle x_n - y_n \rangle;$$

$$x \cdot y := \langle x_n y_n \rangle \quad \text{and} \quad \frac{x}{y} := \langle x_n / y_n \rangle,$$

provided further, in the case of division, $y_n \neq 0$, for all $n \in \mathbb{N}$. Notice that as $a \in \mathbb{R}$ can be viewed as the *constant sequence* $\langle a_n \rangle$ given by $a_n = a$, for all $n \in \mathbb{N}$, we may define the *scalar multiplication* $a \cdot x$ as the usual product of two sequences.

Example 3.6 With notations as in Example 3.1, we have

1. $s_n + t_n = 1 + (-1)^n$, for $n \in \mathbb{N}$, implies that

$$s + t = \langle 0, 2, 0, 2, \ldots \rangle;$$

2. $u_n - t_n = (-1)^n - n^2 + 1$, for $n \in \mathbb{N}$, implies that

$$u - t = \langle -1, 6, -9, \ldots \rangle;$$

3. $t_n \cdot u_n = (-1)^n (n^2 - 1)$, for $n \in \mathbb{N}$, implies that

$$t \cdot u = \langle 0, 3, -8, 15, \ldots \rangle; \text{ and}$$

4. $v_n/w_n = n/2^n(2n+1)$, for $n \in \mathbb{N}$, implies that

$$\frac{v}{w} = \langle 1/6, \ 1/10, \ 3/56, \ \ldots \rangle.$$

It is possible to show that the subcollection of convergent sequences is an algebra with respect to the sum, product, and scalar multiplication, as defined above. We first prove that the subcollection is closed under *sum* and *product*.

Theorem 3.8 *Suppose $\langle u_n \rangle$ and $\langle t_n \rangle$ are two convergent sequences with limits u and t, respectively. Then, as $n \to \infty$,*

$$\begin{aligned}
u_n + t_n &\to u + t; \\
u_n t_n &\to ut.
\end{aligned}$$

(3.2.1)

In particular, for any $a \in \mathbb{R}$, the sequence $\langle au_n \rangle \to au$ as $n \to \infty$.

Proof Let $\varepsilon > 0$ be arbitrary. Then, for some natural numbers N_1, N_2, we have

$$n \geq N_1 \quad \Rightarrow \quad |u_n - u| < \frac{\varepsilon}{2}; \quad \text{and}$$

$$n \geq N_2 \quad \Rightarrow \quad |t_n - t| < \frac{\varepsilon}{2}.$$

Taking $N = \max\{N_1, N_2\}$, for $n \geq N$, we have

$$|(u_n + t_n) - (u + t)| \leq |u_n - u| + |t_n - t| < \frac{\varepsilon}{2} + \frac{\varepsilon}{2} = \varepsilon.$$

Therefore, $\lim\limits_{n \to \infty} (u_n + t_n) = u + t$. Next, for the product, we first consider the case when $\langle t_n \rangle$ is a *null sequence*, i.e., $t = 0$. Then, by Theorem 3.2, there exists $M > 0$ such that $|u_n| < M$, for all n. Further, for some $N \in \mathbb{N}$, we also have

$$n \geq N \quad \Rightarrow \quad |t_n| < \frac{\varepsilon}{M}.$$

It thus follows that

$$n \geq N \quad \Rightarrow \quad |u_n t_n| = |u_n||t_n| < \frac{\varepsilon}{M} M = \varepsilon.$$

Thus, in this case, $\lim\limits_{n \to \infty} (u_n \cdot t_n) = 0 = u \cdot 0$. Now, let $t \neq 0$. Then, as $u_n \to u$, there is some natural number K_1 such that

$$n \geq K_1 \quad \Rightarrow \quad |u_n - u| < \frac{\varepsilon}{2|t|}.$$

We also have a positive real number M such that $|x_n| < M$, for all n. Further, $t_n \to t$ implies that there is some natural number K_2 such that

$$n \geq K_2 \quad \Rightarrow \quad |t_n - t| < \frac{\varepsilon}{2M}.$$

Then, for $n \geq K = \max\{K_1, K_2\}$, we have

$$
\begin{aligned}
|u_n t_n - ut| &= |u_n(t_n - t) + (u_n - u)t| \\
&\leq |u_n|\,|t_n - t| + |u_n - u|\,|t| \\
&< M \frac{\varepsilon}{2M} + \frac{\varepsilon}{2|t|}\,|t| = \varepsilon.
\end{aligned}
$$

Therefore, $\lim_{n \to \infty}(u_n \cdot t_n) = u \cdot t$, in general. In particular, taking $t_n = a$, for all n, we obtain the last assertion. $\square$

This theorem implies, in particular, if $\langle u_n \rangle$ and $\langle t_n \rangle$ are convergent sequences with limits u and t, respectively, then

$$u_n - t_n = u_n + (-1)t_n \quad \Rightarrow \quad \lim_{n \to \infty}(u_n - t_n) = u - t.$$

Next, as the limit of the sequence $\langle x_n^{-1} \rangle$ does not exist if $\langle x_n \rangle$ is a *null sequence*, with $x_n \neq 0$ for all n. In all other cases, the sequence $\langle x_n^{-1} \rangle$ converges always.

Theorem 3.9 *Let $\langle x_n \rangle$ be a sequence, with $x_n \neq 0$ for all n, such that $\lim_{n \to \infty} x_n = a \neq 0$. Then, $x_n^{-1} \to a^{-1}$ as $n \to \infty$.*

Proof Let $\varepsilon > 0$ be arbitrary. By Exercise 3.15, we know that a convergent sequence that does not converge to a number stays *bounded away* from that number. Here, as $\langle x_n \rangle$ is not a null sequence, we can find some real number $M > 0$ and $N_1 \in \mathbb{N}$ such that

$$n > N_1 \quad \Rightarrow \quad |x_n| \geq M.$$

We also have some $N_2 \in \mathbb{N}$ such that

$$n > N_2 \quad \Rightarrow \quad |x_n - a| < \varepsilon M |a|.$$

Then, for $N = \max\{N_1, N_2\}$ and for $n > N$, we have

$$\left| x_n^{-1} - a^{-1} \right| = \frac{|x_n - a|}{|a|\,|x_n|} \leq \frac{|x_n - a|}{|a|\,M} < \frac{\varepsilon M |a|}{|a|\,M} = \varepsilon.$$

Therefore, $\lim_{n \to \infty} x_n^{-1} = a^{-1}$. $\square$

This theorem implies if $u_n \to u$ and $v_n \to v \neq 0$ with $v_n \neq 0$ for all n, then $\lim_{n \to \infty} u_n v_n^{-1} = u v^{-1}$. In particular, as $\lim_{n \to \infty} x_n = \lim_{n \to \infty} x_{n+1}$, we have

$$0 \neq \lim_{n \to \infty} x_n < \infty \quad \Rightarrow \quad \lim_{n \to \infty} \frac{x_n}{x_{n+1}} = 1.$$

Clearly, the assertion fails if $\lim_{n\to\infty} x_n = 0$. For example, consider the sequence

$$x_n = \begin{cases} 1/n, & \text{if } n \text{ is even} \\ 1/n^2, & \text{if } n \text{ is odd} \end{cases}.$$

Example 3.7 Let $P(x)$ and $Q(x) \neq 0$ be two polynomials with real coefficients of degrees p and q, respectively. From the discussion above, it follows that the sequence $\langle P(n)/Q(n) \rangle$ converges if and only if $q \geq p$. For $q > p$, the limit is zero. For $p = q$, if α and β respectively are the leading coefficients of the polynomials $P(n)$ and $Q(n)$, then

$$\frac{P(n)}{Q(n)} = \frac{n^p \left(\alpha + \text{a polynomial in } (1/n) \text{ of degree } \leq p-1\right)}{n^p \left(\beta + \text{a polynomial in } (1/n) \text{ of degree } \leq p-1\right)}$$

$$\to \frac{\alpha}{\beta}, \quad \text{as } n \to \infty.$$

For example, in particular,

$$\lim_{n\to\infty} \frac{n^3 + 3n - 1}{2n^3 - 4n^2 + 5} = \frac{1}{2}, \quad \lim_{n\to\infty} \frac{3n^2 + 5n + 7}{n^2 + 5} = 3,$$

and so on. The same argument gives $\lim_{n\to\infty} 4n/(3n - 7\sqrt{n}) = 4/3$.

We next prove that convergence preserves *order* and *root* operation.

Theorem 3.10 *Suppose $\langle u_n \rangle$ and $\langle v_n \rangle$ are convergent sequences with limits u and v, respectively. Then, $u_n \geq v_n$ for some $n \geq n_0$ implies that $u \geq v$. Conversely, $u < v$ implies that $u_n < v_n$ for some $n \geq n_0$. Further, when $u_n \geq 0$, then $u_n^{1/k} \to u^{1/k}$, for every $k \in \mathbb{N}$.*

Proof For the first assertion, let $u_n \geq v_n$ for some $n \geq n_0$. If possible, suppose $u < v$. Take $\varepsilon = (v - u)/2 > 0$. Then, for some natural numbers $n_1 > n_0$, we have

$$n \geq n_1 \Rightarrow |u_n - u| < \varepsilon$$

$$\Rightarrow v_n \leq u_n < u + \varepsilon = \frac{u+v}{2} = v - \varepsilon,$$

which contradicts the fact that $v_n \to v$. Therefore, $u \geq v$. Conversely, suppose $u < v$, and take $\varepsilon = (v - u)/2$. Then, for some natural numbers m_1 and m_2,

$$n \geq m_1 \Rightarrow u_n < u + \varepsilon;$$

$$n \geq m_2 \Rightarrow v_n > v - \varepsilon.$$

But then, for $n \geq n_0 = \max\{m_1, m_2\}$,

$$v - \varepsilon = \frac{u+v}{2} = u + \varepsilon \quad \Rightarrow \quad u_n < \frac{u+v}{2} < v_n.$$

For the second part, recall that $u_n \geq 0 \Rightarrow u \geq 0$. Let $\varepsilon > 0$ be arbitrary. Then, for some $m \in \mathbb{N}$,

$$n \geq m \quad \Rightarrow \quad |u_n - u| < \varepsilon^k$$
$$\Rightarrow \quad |u_n^{1/k} - u^{1/k}| < |u_n - u|^{1/k} < \varepsilon.$$

This completes the proof. $\qquad\qquad\qquad\qquad\qquad\qquad\qquad\qquad\qquad\qquad\qquad$ □

Notice that if $u_n \to u$ and $v_n \to v$, with $u_n < v_n$ for all $n \in \mathbb{N}$, then we may still have $u = v$. For, we may consider the following convergent sequences

$$u_n = \frac{n}{n^2+1} \quad \text{and} \quad v_n = \frac{1}{n}.$$

Further, as $\langle a_n^m \rangle$ converges to a^m, for any $m \in \mathbb{N}$, it follows that $a_n^q \to a^q$, for any positive rational number. And, when $a_n > 0$ for any n, the same holds for negative rational numbers as well.

3.3 Convergence Theorems

It is not possible in certain cases to do analysis without knowing the limit of a sequence. However, for many others, we are good if it is somehow known that a sequence $\langle x_n \rangle$ converges. Notice that, as convergent sequences are always bounded, we can always take $\langle x_n \rangle$ to be a bounded sequence. The focus here is to investigate some interesting properties of such types of sequences, which ensure their convergence.

We start with a property of a sequence as defined in the next definition.

Definition 3.4 We say a sequence $\langle x_n \rangle$ is *nondecreasing* if

$$x_1 \leq x_2 \leq x_3 \leq \cdots,$$

and it is *nonincreasing* if

$$x_1 \geq x_2 \geq x_3 \geq \cdots.$$

In case of strict inequalities, we use terminology such as *strictly increasing* for nondecreasing or *strictly decreasing* for nonincreasing. We say $\langle x_n \rangle$ is a *monotone sequence* if it is nondecreasing or nonincreasing.

In particular, constant sequences are monotone. Notice that every strictly increasing sequence is bound below, and every strictly decreasing sequence is bound above.

For example, both sequences $\langle n \rangle$ and $\langle 2^n \rangle$ are strictly increasing, with the lower bounds 1 and 2, respectively. On the other hand, both sequences $\langle 1 + (1/n) \rangle$ and $\langle 2^{-n} \rangle$ are strictly decreasing, with the upper bounds 2 and $1/2$, respectively.

A simple inspection of the terms of sequences given in the above paragraph is sufficient to determine their monotonicity. In general, we use *ratio of terms* or *difference of consecutive terms* of a sequence to verify the same. Also, for some sequences such as given in Example 3.3, we use the principle of mathematical induction.

Example 3.8 For any $a > 1$, consider the sequence $\langle x_n \rangle$ given by

$$x_1 = \sqrt{a}, \quad \text{and} \quad x_{n+1} = \sqrt{a\, x_n}, \quad \text{for} \quad n \geq 1.$$

We use here the mathematical induction to show that $x_{n+1} > x_n$, for all $n \geq 1$. For the base step, we have

$$a\sqrt{a} > a \;\Rightarrow\; \sqrt{a\sqrt{a}} > \sqrt{a} \;\Rightarrow\; x_2 > x_1.$$

That is, the assertion holds for $n = 1$. So, assume it holds for some $n \geq 1$. Then,

$$
\begin{aligned}
x_{n+1} > x_n &\Rightarrow & a x_{n+1} &> a x_n \\
&\Rightarrow & \sqrt{a x_{n+1}} &> \sqrt{a x_n} \\
&\Rightarrow & x_{n+2} &> x_{n+1},
\end{aligned}
$$

which shows that the assertion also holds for $n + 1$. Therefore, $\langle x_n \rangle$ is a nondecreasing sequence, with the lower bound $x_1 = \sqrt{a}$. Notice that, as $x_1 < a$ and $x_{n+1} = \sqrt{a\, x_n} < \sqrt{a^2} = a$ if $x_n < a$, it follows that a is an upper bound of the sequence $\langle x_n \rangle$. Further, if ℓ is the limit of the sequence $\langle x_n \rangle$, then

$$
\begin{aligned}
x_{n+1} = \sqrt{a\, x_n} &\Rightarrow & x_{n+1}^2 &= a\, x_n \\
&\Rightarrow & \lim_{n \to \infty} x_{n+1}^2 &= a \lim_{n \to \infty} x_n \\
&\Rightarrow & \ell^2 &= a\,\ell.
\end{aligned}
$$

So, $\ell = 0$ or $\ell = a$. However, as $\sqrt{a} \leq x_n < a$, we have $\displaystyle \lim_{n \to \infty} x_n = a$.

The next theorem, known as the *monotone convergence theorem*, is the first fundamental fact about the bounded sequences, which proves that the concluding remark given in the above-worked example holds in general.

Theorem 3.11 (Monotone Convergence Theorem) *Every bounded monotone sequence is convergent.*

Proof Suppose $\langle x_n \rangle$ is a bounded monotone sequence. *Without loss of generality*, we assume that $\langle x_n \rangle$ is a nondecreasing sequence. As the sequence $\langle x_n \rangle$ is bounded,

$$a = \sup\{x_1, x_2, x_3, \dots\}.$$

exists. Let $\varepsilon > 0$. Then, by the first part of Exercise 2.4, there is some $N \in \mathbb{N}$ such that $x_k \geq a - \varepsilon$, for all $k \geq N$. We also have $x_k \leq a + \varepsilon$, for each such x_k. Therefore, $x_n \to a$ as $n \to \infty$, as asserted. □

Example 3.9 Consider the sequence

$$x_n = \frac{1}{n+1} + \frac{1}{n+2} + \cdots + \frac{1}{n+n}.$$

Then, we can show that $|x_{n+1} - x_n| > 0$, for all $n \geq 1$. Also, $n + k > n$, for $k \geq 1$, implies that $x_n < 1$ for all $n \geq 1$. It thus follows by Theorem 3.11 that $\langle x_n \rangle$ is a convergent sequence.

Next, consider an arbitrary bounded sequence $\langle x_n \rangle$ such that

$$|x_n| \leq M, \quad \text{for some} \quad M > 0.$$

We define two new sequences given by

$$\ell_n = \inf_{k \geq n} \langle x_k \rangle = \inf\{x_n, x_{n+1}, \dots\}; \tag{3.3.1a}$$

$$u_n = \sup_{k \geq n} \langle x_k \rangle = \sup\{x_n, x_{n+1}, \dots\}. \tag{3.3.1b}$$

Clearly then, $\ell_n \leq u_n$ for all $n \in \mathbb{N}$, and

$$u_1 \geq u_2 \geq \dots \geq -M \quad \text{and} \quad \ell_1 \leq \ell_2 \leq \dots \leq M.$$

So, by Theorem 3.11, we can find real numbers[1] u and ℓ such that

$$u_n \to u \quad \text{and} \quad \ell_n \to \ell, \quad \text{as} \quad n \to \infty.$$

Further, by Exercise 1.5, notice that

$$\ell_n \leq \ell_{m+n} \leq u_{m+n} \leq u_n, \quad \text{for all} \quad m, n \quad \Rightarrow \quad \ell \leq u.$$

The limits ℓ and u respectively are called the *limit inferior* and *limit superior* of the sequence $\langle x_n \rangle$. We write

$$\ell = \liminf_{n \to \infty} x_n := \sup_{n \geq 1} \ell_n; \tag{3.3.2}$$

$$u = \limsup_{n \to \infty} x_n := \inf_{n \geq 1} u_n. \tag{3.3.3}$$

---

[1] It should be clear that $u = +\infty$ and $\ell = -\infty$ for a *divergent sequence*.

Notice that it follows from the definition of *supremum* that, for some $a \in \mathbb{R}$, if $a > u$ then only finitely many $x_n > a$; and, if $a < u$ then there are infinitely many $x_n < a$. Also, it follows from the definition of *infimum* that, for some $b \in \mathbb{R}$, if $b < \ell$ then only finitely many $x_n < b$; and if $b > \ell$ then there are infinitely many $x_n > \ell$. Details follow from the two assertions given in Exercise 2.4. We thus conclude that both ℓ and u exist for a bounded sequence.

For example, if $x_n = (5 + (-1)^n)/2$, then $\ell = 2$ and $u = 3$. Also, for $x_n = \sin n$, it can be seen that $\ell = -1$ and $u = 1$, using the fact that π is irrational. The next theorem provides a general criterion to decide when does a bounded sequence $\langle x_n \rangle$ converge.

Theorem 3.12 *A bounded sequence $\langle x_n \rangle$ converges if and only if $\ell = u$.*

Proof First, suppose $\lim\limits_{n \to \infty} x_n = a$. We prove that

$$\liminf_{n \to \infty} x_n = a = \limsup_{n \to \infty} x_n$$

For, let $\varepsilon > 0$ be arbitrary. Then, there is some $K \in \mathbb{N}$ such that

$$a - \varepsilon < x_n < a + \varepsilon, \quad \text{for all} \quad n > K. \tag{3.3.4}$$

But then, for $n > K$, we have

$$a - \varepsilon \leq \ell_n, \; u_n \leq a + \varepsilon, \tag{3.3.5}$$

by Eq. (3.3.1). Therefore,
$$0 \leq u - \ell \leq 2\varepsilon,$$

which implies that $\ell = u \, (= a)$. To prove the converse, suppose $\ell = u = a$. Then, since Eq. (3.3.5) holds for almost every n, so does Eq. (3.3.4). Hence, we obtain $\lim\limits_{n \to \infty} x_n = a$. $\qquad\qquad\qquad\qquad\qquad\qquad\qquad\qquad\qquad\qquad\qquad\qquad\qquad\qquad$ $\square$

This theorem is of little use in finding such properties of a bounded sequence which ensure its convergence. We thus continue with our quest to find some other properties for a bounded sequence with a desirable effect. For, we introduce the concept of *subsequences* of a sequence.

Definition 3.5 Let $x = \langle x_n \rangle$ be a sequence, and $s : \mathbb{N} \to \mathbb{N}$ be an injective function. We say the composition $x \circ s : \mathbb{N} \to \mathbb{R}$ defines a *subsequence* of the sequence $\langle x_n \rangle$ if s is a strictly increasing function.

Clearly, every subsequence of the sequence $\langle 1, 1, 1, \ldots \rangle$ is again a sequence of the same type. Also, the sequence $\langle 1, 1, 1, \ldots \rangle$ is a subsequence of the sequence $\langle -1, 1, -1, 1, \ldots \rangle$, for n even, and $\langle -1, -1, -1, \ldots, \rangle$ is a second subsequence, for n odd. Notice that these two subsequences of the sequence $\langle (-1)^n \rangle$ converge to 1 and -1, respectively. In general, if $s(k) = n_k \; (k \in \mathbb{N})$, with

$$1 \leq n_1 < n_2 < n_3 < \cdots ,$$

then

$$s_1 = x_{n_1}, \quad s_2 = x_{n_2}, \quad \ldots , s_k = x_{n_k}, \quad \ldots ,$$

is a subsequence of the sequence $\langle x_n \rangle$. In particular, by taking $s_1(n) = 2n$ or $s_2(n) = 2n - 1$, it follows that every sequence $\langle x_n \rangle$ has at least two subsequences $\langle x_{2n} \rangle$ and $\langle x_{2n-1} \rangle$.

Lemma 3.1 *If the two subsequences $\langle x_{2n-1} \rangle$ and $\langle x_{2n} \rangle$ of a sequence $\langle x_n \rangle$ have the same subsequential limit, then so does the sequence $\langle x_n \rangle$.*

Proof Suppose the given two subsequences share the same limit $a \in \mathbb{R}$, and let $\varepsilon > 0$ be arbitrary. Then, for some $N_1, N_2 \in \mathbb{N}$, we have

$$2n - 1 > N_1 \quad \Rightarrow \quad |x_{2n-1} - a| \leq \varepsilon;$$
$$2n > N_2 \quad \Rightarrow \quad |x_{2n} - a| \leq \varepsilon.$$

Clearly then, for $N = \max\{2N_1 - 1, 2N_2\}$, we have

$$n > N \quad \Rightarrow \quad |x_n - a| \leq \varepsilon.$$

Therefore, $x_n \to x$ as $n \to \infty$. $\qquad\square$

If a subsequence $\langle x_{n_k} \rangle$ of a sequence $\langle x_n \rangle$ converges to some $a \in \mathbb{R}$, we say a is a **subsequential limit** of the sequence $\langle x_n \rangle$. Clearly, by Theorem 3.1, every *subsequential limit* of a convergent sequence equals the limit of the sequence. As we will see shortly, the converse holds for a special type of sequence, known as the *Cauchy sequences* (Theorem 3.15).

We say $\langle x_n \rangle$ is an *oscillatory sequence* if it has at least two subsequential limits. To generalise the concept of *divergence* for sequences, we treat an oscillatory sequence as a divergent sequence. By Theorem 3.12, if a sequence $\langle x_n \rangle$ converges to a, then $\ell \leq a \leq u$. So, if $\langle y_n \rangle$ is a subsequence of a sequence $\langle x_n \rangle$, then

$$\ell = \liminf_{n \to \infty} x_n \leq \liminf_{n \to \infty} y_n \leq \limsup_{n \to \infty} y_n \leq \limsup_{n \to \infty} x_n = u.$$

The next theorem is useful in some interesting situations.

Theorem 3.13 *Every sequence contains a monotone subsequence.*

Proof Let $\langle x_n \rangle$ be an arbitrary sequence, with $n \geq 0$. Choose an increasing sequence of natural numbers $\langle n_k \rangle$ such that, given any k, either $x_{n_k} > x_{n_i}$, for all $k > i$, or $x_{n_k} \leq x_{n_i}$, for all $k > i$. Let $n_0 = 0$, and consider the following subsets of the set of natural numbers $\mathbb{N}$:

$$S_0 = \{n > 0 \mid x_n > x_0\} \quad \text{and} \quad T_0 = \{n > 0 \mid x_n \leq x_0\}.$$

Then, S_0 or T_0 is infinite. If it is S_0, we take only the subsequence consisting of x_0 and only those terms of the sequence $\langle x_n \rangle$ for which $n \in S_0$. And when S_0 is finite, we take only the subsequence consisting of x_0 and only those terms of the sequence $\langle x_n \rangle$ for which $n \in T_0$. Re-indexing, if necessary, we write the above subsequence as $\langle y_n \rangle$. Clearly, it has the property that either $y_n > y_0$, for all $n > 0$, or $y_n \leq y_0$, for all $n > 0$. Next, consider the following subsets of the set of natural numbers $\mathbb{N}$:

$$S_1 = \{n > 1 \mid y_n > y_1\} \quad \text{and} \quad T_1 = \{n > 1 \mid y_n \leq y_1\}.$$

Again, either S_1 or T_1 is infinite. If it is S_1, we take only the subsequence consisting of y_0, y_1, and only those terms of the sequence $\langle y_n \rangle$ for which $n \in S_1$. And, when S_1 is finite, we take only the subsequence consisting of y_0, y_1, and only those terms of the sequence $\langle y_n \rangle$ for which $n \in T_1$, and so on. Proceeding in this way, we obtain a subsequence (x_{n_k}) of the sequence $\langle x_n \rangle$ such that for any n, either $x_{n_i} > x_{n_j}$, for all $i > j$, or $x_{n_i} \leq x_{n_j}$, for all $i > j$. Hence, the claim holds. □

The next theorem proves a fundamental fact about bounded sequences.

Theorem 3.14 (Bolzano-Weierstrass Theorem) *Every bounded infinite sequence contains a convergent subsequence.*

Proof Let $\langle x_n \rangle$ be an infinite sequence, with $x_n \in I_0 = [a_0, b_0]$ for every $n \geq 1$. The proof given here uses bisection procedure. For, let c_0 be the midpoint of the interval I_0. Then, since the sequence $\langle x_n \rangle$ is infinite, at least one of the two *half-intervals* must contain infinitely many terms of $\langle x_n \rangle$. So, we obtain a subsequence $\langle x_{n_k} \rangle$ of the sequence $\langle x_n \rangle$. Write $x^0(k) = x_{n_k}$, $k \geq 1$. *Without loss of generality*, suppose $x^0(k) \in [a_0, c_0]$ for infinitely many values of k. We may write $a_0 = a_1$ and $c_0 = b_1$, and let c_1 be the midpoint of the interval $I_1 = [a_1, b_1]$. By the same reasoning, one of the two half-intervals of I_1 contains infinitely many terms of $\langle x_k^0 \rangle$, say $I_2 = [a_2, b_2]$, which defines a subsequence $\langle x_{k_p}^0 \rangle$. We may write $x^1(p) = x^0(k_p)$, $p \geq 1$. Continuing in this way, we obtain a sequence of intervals $I_n = [a_n, b_n]$, where each I_n contains infinitely many terms of the sequence $\langle x_n \rangle$. Clearly, $\langle a_n \rangle$ is a bounded above increasing sequence and $\langle b_n \rangle$ is a bounded below decreasing sequence. As shown later (Theorem 5.7), both sequences of endpoints converge to the same limit, say ℓ. We proceed to construct a subsequence of $\langle x_n \rangle$ that converges to ℓ.

Starting with I_1, we select an arbitrary value from the sequence $x^0(k)$, say $x_{n_1} \in I_1$. Next, among the infinitely many choices of k, there exists n_2 such that $n_2 > n_1$. Proceeding in this manner, we obtain a subsequence of $\langle x_n \rangle$, for $n = n_1 < n_2 < n_3, \ldots,$

$$x_{n_j} \in I_j = [a_j, b_j] \quad \Rightarrow \quad |x_{n_j} - \ell| < \frac{b_0 - a_0}{2^j}.$$

Clearly then, the subsequence $\langle x_{n_j} \rangle$ converges to ℓ. □

Notice that, by using Theorem 3.11, an alternative proof of the above theorem follows from Theorem 3.13.

3.3.1 Cauchy Sequences

We now discuss a very special type of property that some bounded sequences may have. The property we are referring to here requires that *almost all terms* of such types of sequences must be *packed together tightly*. A bounded sequence with this property is called a *Cauchy sequence*.

Definition 3.6 We say $\langle x_n \rangle$ is a **Cauchy sequence** if for each $\varepsilon > 0$ there is some positive integer N such that

$$n > m \geq N \quad \Rightarrow \quad |x_n - x_m| < \varepsilon. \tag{3.3.6}$$

Equivalently, a sequence $\langle x_n \rangle$ is a **Cauchy sequence** if, for all $p \in \mathbb{N}$,

$$|x_{m+p} - x_m| \to 0 \quad \text{as} \quad m \to \infty. \tag{3.3.7}$$

In particular, we have

$$|x_m - x_{m+1}| \to 0, \quad \text{as} \quad m \to \infty, \tag{3.3.8}$$

if $\langle x_n \rangle$ is a Cauchy sequence. Notice that, however, the converse of the last assertion is not true, in general. For example, the sequences $\langle x_n \rangle$ given by

$$x_n = 1 + \frac{1}{2} + \cdots + \frac{1}{n}, \quad \text{or } x_n = \ln n$$

satisfy the condition (3.3.8), but neither is Cauchy. Further, for $x_n = n$, we have

$$|x_{n+1} - x_n| = 1, \quad \text{for all} \quad n \geq 1,$$

it follows that $\langle n \rangle$ is not a Cauchy sequence. On the other hand, for $y_n = 1/n$, we have

$$|y_n - y_m| = \left| \frac{1}{n} - \frac{1}{m} \right| = \left| \frac{m-n}{mn} \right| < \frac{1}{n},$$

because $(m - n)/m < 1$, for $n \geq 1$, which implies that $|y_m - y_n|$ are very close for almost all $n, m \in \mathbb{N}$. Similarly, we can show that $\langle 1 + (1/n) \rangle$ is a Cauchy sequence.

Example 3.10 Consider the sequence

$$x_1 = a, \quad x_2 = b, \quad \text{and} \quad x_n = \frac{1}{2}(x_{n-1} + x_{n-2}), \quad \text{for} \quad n > 2.$$

First, note that if $a = b$ then $x_n = a$, for all $n \geq 1$, which implies $x_n \to a$, as $n \to \infty$. So, we assume that $a \neq b$. More specifically, we take $a < b$. Then, a simple inductive argument shows that $a < x_n < b$ for all $n \geq 3$, and that

$$\begin{aligned}
|x_{n+1} - x_n| &< \left| \frac{x_n}{2} + \frac{x_{n-1}}{2} - x_n \right| \\
&= \left| \frac{x_{n-1}}{2} - \frac{x_n}{2} \right| = \frac{1}{2}|x_n - x_{n-1}| \\
&< \frac{1}{2^2}|x_{n-1} - x_{n-2}|
\end{aligned}$$

$$\vdots$$

$$< \frac{1}{2^{n-1}}|x_2 - x_1| = \frac{1}{2^{n-1}}|b - a|, \quad \text{for all } n \geq 2,$$

which implies that, for $n > m$,

$$\begin{aligned}
|x_n - x_m| &\leq |x_m - x_{m+1}| + |x_{m+1} - x_{m+2}| + \cdots + |x_{n-1} - x_n| \\
&< \left[\frac{1}{2^{m-1}} + \frac{1}{2^m} + \cdots + \frac{1}{2^{n-2}} \right] |b - a| \\
&= \frac{1}{2^{n-2}} \left[1 + 2 + 2^2 + \cdots + 2^{n-m-1} \right] |b - a| \\
&= \frac{1}{2^{n-2}} (2^{n-m} - 1)|b - a| < \frac{1}{2^{m-2}}|b - a|.
\end{aligned}$$

For an arbitrary $\varepsilon > 0$, we may choose a natural number N larger than

$$m > 2 - \log_2 \left(\frac{\varepsilon}{|b - a|} \right) \qquad \Leftrightarrow \qquad \frac{1}{2^{m-2}}|b - a| < \varepsilon$$

so that

$$|x_n - x_m| < \frac{1}{2^{m-2}}|b - a| < \varepsilon, \quad \text{for all } n > m > N.$$

Therefore, $\langle x_n \rangle$ is a Cauchy sequence.

As almost all terms of a convergent sequence $\langle x_n \rangle$ are *very close* to its limit, it is quite natural to expect that almost all terms x_n are *eventually close*. The next theorem confirms the fact that almost all terms of a convergent sequence are packed tightly in every open interval around the limit.

Theorem 3.15 *Every convergent sequence $\langle x_n \rangle$ is Cauchy.*

Proof Notice that the convergence implies, for each $\varepsilon > 0$, there is some $N \in \mathbb{N}$ such that

$$n \geq N \quad \Rightarrow \quad |x_n - x| < \frac{\varepsilon}{2}.$$

But then, for $n, m \geq N$,

$$|x_n - x_m| \leq |x_n - x| + |x_m - x| < \frac{\varepsilon}{2} + \frac{\varepsilon}{2} = \varepsilon.$$

Hence, $\langle x_n \rangle$ is a Cauchy sequence. $\qquad\square$

The next theorem explains why we started with the bounded sequence, a priori.

Theorem 3.16 *Every Cauchy sequence $\langle x_n \rangle$ is bounded.*

Proof Since $\langle x_n \rangle$ is a Cauchy sequence, for $\varepsilon = 1$ there exists $K \in \mathbb{N}$ such that

$$n, m > K \implies |x_n - x_m| < 1$$
$$\implies |x_n - x_{K+1}| < 1, \quad \text{for } n > K.$$

But then, for $M = \max\{1, |x_1 - x_{K+1}| + 1, \ldots, |x_K - x_{K+1}| + 1\}$, we have

$$x_{K+1} - M < x_n < x_{K+1} + M, \quad \text{for all } n \geq 1.$$

So, the sequence $\langle x_n \rangle$ is bounded. $\qquad\square$

Remark 3.2 It is possible to construct many more examples of Cauchy sequences using the fact that the image of a Cauchy sequence under a uniformly continuous function is a Cauchy sequence (Exercise 7.5). In particular, it follows that $f(1/n)$ is a Cauchy sequence for any uniformly continuous function $f : [0, 1] \to \mathbb{R}$.

Recall that every *subsequential limit* of a convergent sequence equals the limit of the sequence. The next theorem proves that the converse holds for Cauchy sequences. Said differently, terms of a Cauchy sequence cannot go off in different directions.

Theorem 3.17 *If $\langle x_n \rangle$ is a Cauchy sequence, with a convergent subsequence $\langle x_{n_k} \rangle$, then $\langle x_n \rangle$ converges to $\lim_{k \to \infty} x_{n_k}$.*

Proof Let $\varepsilon > 0$. As $\langle x_n \rangle$ is Cauchy, and if $\lim_{k \to \infty} x_{n_k} = a$ for some $a \in \mathbb{R}$, then there are some $N_1, N_2 \in \mathbb{N}$ such that

$$|x_n - x_m| < \frac{\varepsilon}{2}, \ \forall\, n, m \geq N_1, \ \text{and} \ |x_{n_k} - a| < \frac{\varepsilon}{2}, \ \forall\, k \geq N_2.$$

Taking $N = \max\{N_1, N_2\}$, and using the triangle inequality, it follows that for arbitrary $k \geq N$

$$|x_n - a| \leq |x_n - x_{n_k}| + |x_{n_k} - a| < \frac{\varepsilon}{2} + \frac{\varepsilon}{2} = \varepsilon, \forall\, n \geq N.$$

Therefore, $x_n \to a$ as $n \to \infty$, as asserted. $\qquad\square$

More generally, if every subsequence of a sequence $\langle x_n \rangle$ contains a subsequence which converges to some $a \in \mathbb{R}$, then $\lim_{n \to \infty} x_n = a$, For, otherwise, we can find an interval I centred at a and a subsequence $\langle x_{n_k} \rangle$ not in I. But then, no subsequence of $\langle x_{n_k} \rangle$ would converge to a, which cannot be the case. This simple observation is very helpful to prove the convergence of a sequence in a space[2] if it is known that the related problem has a solution for *subsequential limit*, together with using the uniqueness of a solution of the problem.

Example 3.11 We show that the sequence $\langle s_n \rangle$ defined by

$$s_n = \sum_{k=1}^{n} \frac{1}{k^2} = \frac{1}{1^2} + \frac{1}{2^2} + \cdots + \frac{1}{n^2}$$

is convergent. First, note that a simple inductive argument gives $s_n \leq 2 - (1/n)$, for all $n \geq 1$. Also, since $k^2 \leq 2^k$, $k \geq 2$, we have

$$s_n \leq 1 + \frac{1}{4} + \frac{1}{8} + \cdots + \frac{1}{2^n}.$$

Thus, for n sufficiently large, $\langle s_n \rangle$ is bounded above by $3/2$. Therefore, $\langle s_n \rangle$ is convergent, and so is Cauchy. To prove that $\lim_{n \to \infty} s_n = \pi^2/6$, we need *Fourier series*, as shown in the last chapter of the text.

Since each Cauchy sequence is bounded, a monotonic subsequence of the Cauchy sequence $\langle x_n \rangle$ obtained by using Theorem 3.13 would still be bounded, which then must be convergent by Theorem 3.11. Hence, by Theorem 3.17, $\langle x_n \rangle$ is convergent. This gives the following *sequence version* of *general principle of convergence*.

Theorem 3.18 (General Principle of Convergence). *A sequence $\langle x_n \rangle$ converges if and only if for each $\varepsilon > 0$ there exists some positive integer N such that*

$$n > m \geq N \implies |x_n - x_m| < \varepsilon.$$

Proof If $\langle x_n \rangle$ converges, then the condition holds by Theorem 3.15. Conversely, suppose $\langle x_n \rangle$ is a Cauchy sequence, i.e., it satisfies the condition. Then, for $\varepsilon = 1$ and $N = N_1$, we select some $m \geq N_1$, and write

$$a_1 = x_m - 1 \quad \text{and} \quad b_1 = x_m + 1.$$

So, for $n \geq N_1$, $x_n \in [a_1, b_1]$, with $b_1 - a_1 = 2$. Again, for $\varepsilon = 2/3$ and $N = N_2 > N_1$, we select some $m \geq N_2$, and write

$$a_2 = x_m - (2/3) \quad \text{and} \quad b_2 = x_m + (2/3).$$

So, for $n \geq N_2$, $x_n \in [a_2, b_2]$, with $b_2 - a_2 = 4/3$. In this manner, we obtain a sequence of natural numbers N_k such that for $n \geq N_k$

---

[2] We use it specifically to show that not every notion of convergence in a space stems from a metric. For example, no metric on the space of continuous functions $C([a, b])$ induces the *pointwise convergence* of a sequence of functions in $C([a, b])$.

$$a_k = x_m - \frac{2^{k-1}}{3} \le x_n \le x_m - \frac{2^{k-1}}{3} = b_k, \quad k \ge 2.$$

Clearly then, we have

$$a_1 \le a_2 \le \cdots \le a_k \le \cdots \le b_k \le \cdots \le b_2 \le b_1,$$

and $\lim_{n \to \infty} a_n = \lim_{n \to \infty} b_n = \alpha$, say. Then for any $\varepsilon > 0$, there exists $K \in \mathbb{N}$ such that

$$\alpha - \varepsilon < a_n \le b_n < \alpha + \varepsilon \quad \Rightarrow \quad \lim_{n \to \infty} x_n = \alpha.$$

Hence, the proof is complete. $\qquad\qquad\qquad\qquad\qquad\qquad\qquad\qquad\qquad$ $\square$

Example 3.12 We consider, once again, the sequence given by

$$x_1 = a, \quad x_2 = b, \quad \text{and} \quad x_n = \frac{1}{2}(x_{n-1} + x_{n-2}), \quad \text{for} \quad n > 2.$$

As before, we assume that $a \ne b$. As seen in Example 3.10, the sequence $\langle x_n \rangle$ satisfies the condition of the above theorem, so it converges. Suppose $\lim_{n \to \infty} x_n = \ell$. Now, adding the equalities

$$x_n = \frac{1}{2}(x_{n-1} + x_{n-2}), \quad \text{for} \quad 3 \le n \le k,$$

we obtain the equation

$$x_k + \frac{1}{2}x_{k-1} = \frac{1}{2}(x_1 + 2x_2) = \frac{1}{2}(a + 2b).$$

So, taking the limit as $k \to \infty$, we have

$$\ell + \frac{1}{2}\ell = \frac{1}{2}(a + 2b) \quad \Rightarrow \quad \ell = \frac{1}{3}(a + 2b).$$

In particular, when $a = 0$ and $b = 1$, we get $\lim_{n \to \infty} x_n = 2/3$.

Notice that we may also use Theorem 3.14 to prove that every Cauchy sequence $\langle x_n \rangle$ is convergent. For, the boundedness of $\langle x_n \rangle$, together with Theorem 3.18, implies that $\langle x_n \rangle$ has a convergent subsequence, and so it converges, by Theorem 3.17.

Example 3.13 Let $I = [0, 1]$ and $a_n = 1/n$. Since $a_n = 1/n \to 0$, it is a Cauchy sequence in I. However, though it is a Cauchy sequence in the interval $J = (0, 1)$, it does not converge in J (Example 5.2). For, let $\varepsilon > 0$. Choose N such that $N > 2/\varepsilon$. Then, for $n, m \ge N$, we have

$$\left| \frac{1}{n} - \frac{1}{m} \right| \le \frac{1}{m} + \frac{1}{n} \le \frac{2}{N} < \varepsilon.$$

3.3.2   The Existence of the Set $\mathbb{R}$

We now have enough tools to deliver on our promised proof of the fact that there exists a complete ordered field.

Proof of Theorem 2.7 : Let $\mathscr{C}$ denotes the collection of all *rational* Cauchy sequences $\langle q_n \rangle$. That is, $q_n \in \mathbb{Q}$ for all $n \in \mathbb{N}$. For $r = \langle q_n \rangle$, $r' = \langle q_n' \rangle \in \mathscr{C}$, we define

$$r \sim r' \quad \Leftrightarrow \quad r - r' = \langle q_n - q_n' \rangle \;\to\; 0 \text{ as } n \to \infty.$$

That is, $r = r'$, eventually. It can be easily verified that $\sim$ is an equivalence relation on $\mathscr{C}$. We write $\mathbb{R}$ for the collection of all equivalence classes of $(\mathscr{C}, \sim)$. Let $\overline{0} \in \mathbb{R}$ be the equivalence class of *null* rational sequences. In general, we write $\overline{r} = \left[\langle r, r, r, \ldots \rangle \right]$ for the *rational elements* $r \in \mathbb{Q} \subset \mathbb{R}$. Notice that, in particular, as $\overline{1} \in \mathbb{R}$ is the equivalence class of rational Cauchy sequences converging to 1, we have $\overline{0} \neq \overline{1}$.

We define *addition* and *multiplication* of two equivalence classes in $\mathbb{R}$ in terms of sum and product of their representative rational Cauchy sequences, using the definition given earlier. That is, if

$$\overline{r} = \left[\langle p_n \rangle \right], \; \overline{s} = \left[\langle q_n \rangle \right] \in \mathbb{R}, \quad \text{with} \quad r = \langle p_n \rangle, \; s = \langle q_n \rangle \in \mathscr{C},$$

we take their sum and product respectively as

$$\overline{r} + \overline{s} := \left[\langle p_n + q_n \rangle \right]; \tag{3.3.9a}$$

$$\overline{r} \cdot \overline{s} := \left[\langle p_n q_n \rangle \right]. \tag{3.3.9b}$$

It is then easy to show that the two binary operations are well defined, and the set $\mathbb{R}$ is a field with respect to these two operations (Exercise 3.22). Notice that, for example,

$$r = \langle 1.4, 1.41, 1.414, \ldots \rangle \;\to\; \sqrt{2};$$

and

$$s = \langle 1.7, 1.73, 1.732, \ldots \rangle \;\to\; \sqrt{3}$$

implies that

$$r \cdot s = \langle 2.38, 2.4393, 2.449048, \ldots \rangle \to \sqrt{6},$$

though $\sqrt{6} = \left[\langle 2.4, 2.44, 2.449, \ldots \rangle \right]$. Next, for $\overline{r} = \left[\langle p_n \rangle \right]$, $\overline{s} = \left[\langle q_n \rangle \right] \in \mathbb{R}$, with $r = \langle p_n \rangle$, $s = \langle q_n \rangle \in \mathscr{C}$, we define

$$\overline{r} \leq \overline{s} \quad \Leftrightarrow \quad \overline{r} < \overline{s} \text{ or } \overline{r} = \overline{s}. \tag{3.3.10}$$

where we say $\bar{r} < \bar{s}$ if there exists $\varepsilon > 0$ and $M \geq 1$ such that

$$n \geq M \quad \Rightarrow \quad p_n \leq q_n - \varepsilon.$$

Notice that we must take a rational $\varepsilon > 0$ in Eq. (3.3.10). The condition $\bar{r} < \bar{s}$ means that the two rational Cauchy sequences $r, s \in \mathscr{C}$ are *properly separated*, which also makes the above definition of $<$ on $\mathbb{R}$ well-defined, and it follows easily that it is an *order relation* (Exercise 3.23). In fact, $\leq$ is a total order. We show that if $r \neq s$ then $r < s$ or $s < r$. For, as $r \neq s$, there is some $\varepsilon > 0$ such that for all $N \geq 1$,

$$\text{for some } n \geq N, \quad |p_n - q_n| \geq \varepsilon. \tag{3.3.11}$$

Let $\delta = \varepsilon/3$. Then, for some $N_1, N_2 \in \mathbb{N}$, and $k \geq 1$,

$$n \geq N_1 \Rightarrow |p_n - p_{n+k}| < \delta; \text{ and}$$
$$n \geq N_2 \Rightarrow |q_n - q_{n+k}| < \delta.$$

Taking $N = \max\{N_1, N_2\}$, and using (3.3.11), we find a natural number $n \geq N$ such that $|p_n - q_n| \geq \varepsilon$. Consider the following two cases:

$$p_n - q_n \geq \varepsilon \quad \text{or} \quad q_n - p_n \geq \varepsilon. \tag{3.3.12}$$

Since, for $k \geq 1$, the rational numbers p_{n+k} and q_{n+k} lie in the interval of length δ, it follows that (3.3.10) holds with $M = N$ and we have $r > s$ in the first case; and, $r < s$, for the second case.

Finally, we show that $\mathbb{R}$ is *complete*. For $i \geq 1$, let $\bar{r}_i = [\langle p_{in} \rangle]$ be a Cauchy sequence in $\mathbb{R}$. Since each rational sequence $\langle p_{in} \rangle$ is Cauchy, there is some $N_i \in \mathbb{N}$ such that

$$n \geq N_i \quad \Rightarrow \quad |p_{i,n} - p_{i,n+k}| < \frac{1}{2i}, \text{ for all } k \geq 1.$$

Put $q_i := p_{i,N_i}$ and consider the sequence $\langle q_i \rangle \in \mathscr{C}$. Observe that

$$|q_i - \bar{r}_i| = [\langle |q_i - p_{im}| \rangle],$$

implies that, for $m \geq N_i$,

$$|q_i - p_{im}| = |s_{i,N_i} - s_{i,m}| < \frac{1}{2i} = \frac{1}{i} - \frac{1}{i},$$

so by taking $\varepsilon = 1/2i > 0$ in (3.3.10) it follows that $|q_i - \bar{r}_i| < 1/i$. Also,

$$|q_i - q_{i+k}| = |q_i - \bar{r}_i + \bar{r}_i - \bar{r}_{i+k} + \bar{r}_{i+k} - q_{i+k}|$$
$$\leq |q_i - \bar{r}_i| + |\bar{r}_i - \bar{r}_{i+k}| + |\bar{r}_{i+k} - q_{i+k}|$$
$$< \frac{1}{i} + \varepsilon + \frac{1}{i+k} < 2\varepsilon,$$

for sufficiently large i and for $k \geq 1$. Thus, $[\langle q_n \rangle] \in \mathscr{C}$. We write $\bar{r} = [\langle q_n \rangle]$. Notice that $|q_i - \bar{r}| < 2\varepsilon$, for sufficiently large i, and so $q_i \to \bar{r}$. To conclude the proof, observe that

$$|\bar{r}_k - \bar{r}| \leq |\bar{r}_k - q_k| + |q_k - \bar{r}| < \frac{1}{k} + 3\varepsilon < 4\varepsilon,$$

for large enough k. Hence, $\bar{r}_i \to \bar{r}$. $\square$

3.4   Infinite Series

Recall that an *infinite series* over $\mathbb{R}$ is a *sum* of the form

$$\sum_{n=1}^{\infty} x_n = x_1 + x_2 + \cdots + x_n + \cdots . \qquad (3.4.1)$$

which we usually consider to be the sum of terms of an infinite sequence $\langle x_n \rangle$ in X. Once again, we call x_n the nth *term* of the infinite series. As said earlier, to make any sense out of such type of sums, we use the *sequence of partial sums* given by

$$s_n = x_1 + x_2 + \cdots + x_n. \qquad (3.4.2)$$

Our main purpose here is to demonstrate aforementioned *connection* between *series of real numbers* and the sequences $\langle s_n \rangle$. As there are two sequences into the play, we may write a series defined by a sequence $\langle x_n \rangle$ as the ordered pair $(\langle x_n \rangle, \langle s_k \rangle)$, subject to that the equation (3.4.2) holds. As before, the nth term of the series (3.4.1) is given by

$$x_n = s_{n+1} - s_n, \qquad n \geq 1. \qquad (3.4.3)$$

We also discuss certain standard *tests of convergence* for a series of real numbers.

Definition 3.7 We say (3.4.1) is a *convergent series* if the limit $s = \lim\limits_{k \to \infty} s_k$ exists. In this case, the limit s is the called the *sum of the series*, and we write

$$s = \sum_{n=1}^{\infty} x_n = x_1 + x_2 + \cdots + x_n + \cdots . \qquad (3.4.4)$$

In rest all situations, we say (3.4.1) is a *divergent series*.

We first remark that, as the convergence of a series is defined in terms of the convergence of its sequence of partial sums, Theorem 3.8 implies that the sum and difference of convergent series is again a convergent series, and

$$\sum_{n=1}^{\infty} u_n = u \quad \text{and} \quad \sum_{n=1}^{\infty} v_n = v \quad \Rightarrow \quad \sum_{n=1}^{\infty} [u_n \pm v_n] = u \pm v.$$

Also, multiplication by a nonzero constant gives again a convergent series. Notice that, however, the product of two convergent series in general may not be a convergent series. To illustrate these notions, we start with some familiar series.

Example 3.14 (*Arithmetic Series*). The nth term of the *arithmetic series*

$$a + (a + d) + (a + 3d) + \cdots$$

is $x_n = a + (n - 1)d$, with the first term $a \neq 0$ and common difference $d > 0$. The associated sequence of partial sums $\langle s_n \rangle$ is given by

$$s_n = a + (a + d) + (a + 2d) + \cdots + (a + (n - 1)d), \quad n \geq 1.$$

Rewriting s_n by reversing the order of the terms, and then adding the two expressions, we obtain

$$2s_n = n[2a + (n - 1)d]$$

and so

$$s_n = \frac{n}{2} [2a + (n - 1)d], \quad \text{for } n \geq 1.$$

In particular, it follows that

$$s_n = 1 + 2 + 3 + \cdots + n = n(n + 1)/2$$

defines the sequence of partial sums of series $\sum_{n=1}^{\infty} n = 1 + 2 + 3 + \cdots$. Clearly, $\langle s_n \rangle$ is a divergent sequence.

Example 3.15 (*Geometric Series*). Consider a *geometric series*

$$\sum_{n=1}^{\infty} ar^{n-1} = a + ar + ar^2 + \cdots + ar^{n-1} + \cdots,$$

defined by the sequence $x_n = ar^{n-1}$, with the first term $a \neq 0$ and a common ratio $r > 0$. As the sequence of partial sums $\langle s_n \rangle$ of this geometric series is given by

$$s_n = a + ar + ar^2 + \cdots + ar^{n-1}, \quad n \geq 1,$$

so that, we obtain

$$s_n - rs_n = s_n(1 - r) = a(1 - r^n)$$

and hence

$$s_n = \frac{a(1 - r^n)}{1 - r}.$$

Therefore, with $|r| < 1$, we have

$$s_n - \frac{a}{1 - r} = -\frac{a\,r^n}{1 - r}$$

and so

$$\left| s_n - \frac{a}{1 - r} \right| = \frac{a|r|^n}{|1 - r|} \rightarrow 0, \quad \text{as } n \rightarrow \infty,$$

So, in view of Definition 3.2, we conclude that

$$\sum_{n=1}^{\infty} ar^{n-1} = \lim_{n \to \infty} s_n = \frac{a}{1 - r}, \quad \text{provided } |r| < 1.$$

In particular, it follows that

$$(1 - r)^{-1} = 1 + r + r^2 + \cdots, \quad \text{for all } |r| < 1.$$

Likewise, we obtain the following famous geometric series (*Vite*, 1593):

$$(1 + r)^{-1} = 1 - r + r^2 - \cdots, \quad \text{for all } |r| < 1.$$

Notice that, if a series of the form (3.4.1) converges, then it follows from Eq. (3.4.3) that

$$\lim_{n \to \infty} x_n = \lim_{n \to \infty} s_{n+1} - \lim_{n \to \infty} s_n = s - s = 0. \tag{3.4.5}$$

That is, if a series $(\langle x_n \rangle, \langle s_k \rangle)$ is convergent, then $\langle x_n \rangle$ is a *null sequence*. Explicitly, as the sequence $\langle s_k \rangle$ is convergent, and so Cauchy, for any $\varepsilon > 0$ there is some $K \in \mathbb{N}$ such that

$$m, n > K \quad \Rightarrow \quad |s_n - s_m| < \varepsilon.$$

Taking $m = n + 1$, we obtain

$$|x_{n+1}| = |s_n - s_{n+1}| < \varepsilon, \quad \text{for all } n > K.$$

That is, for $n > K + 1$, $|x_n| < \varepsilon$. Therefore, $\lim\limits_{n \to \infty} x_n = 0$. The converse is not true in general. For example, consider the *harmonic series*

$$\sum_{n=1}^{\infty} \frac{1}{n} = 1 + \frac{1}{2} + \frac{1}{3} + \cdots . \tag{3.4.6}$$

Though its nth term converges to 0, but it is not a convergent series. For, the inequality

$$1 + \frac{1}{2} + \left(\frac{1}{3} + \frac{1}{4}\right) + \left(\frac{1}{5} + \cdots + \frac{1}{8}\right) + \cdots > 1 + \frac{1}{2} + \frac{2}{4} + \frac{4}{8} + \cdots ,$$

implies that the subsequence $\langle s_{2^n-1} \rangle$ of its sequence of partial sums is an *unbounded sequence*. More precisely, we have

$$s_{2^n - 1} > n/2, \quad \text{for all } n \geq 1.$$

Similarly, for the series

$$\sum_{n=1}^{\infty} \ln\left(1 + (1/n)\right),$$

we have $\lim\limits_{n \to \infty} x_n = \ln(1) = 0$, but

$$s_n = \sum_{k=1}^{n} \ln\left(\frac{k+1}{k}\right) = \ln\left(\frac{2}{1} \cdot \frac{3}{2} \cdots \frac{n+1}{n},\right) = \ln(n+1)$$

implies that $\lim\limits_{n \to \infty} s_n = \infty$. The simple assertion mentioned in Eq. (3.4.5) makes it easier to prove the divergence of some series. For example, all the following series are divergent:

1. The series $\sum\limits_{n=1}^{\infty} \left(1 + (1/n)\right)$ diverges because $\lim\limits_{n \to \infty} x_n = 1 \neq 0$;

2. The series $\sum\limits_{n=1}^{\infty} (1/n)^{1/n}$ diverges because $\lim\limits_{n \to \infty} x_n = 1 \neq 0$;

3. The series $\sum\limits_{n=1}^{\infty} \cos\left(1/n\right)$ diverges because $\lim\limits_{n \to \infty} x_n = 1 \neq 0$.

In general, we use some specific tests to decide the convergence of a series of the form (3.4.1). As described in the next section, some standard tests due to Cauchy, d'Alembert, etc., use the nth term x_n to give a necessary condition for the convergence (or divergence) of a series. We also use the following *series version* of Theorem 3.18.

Theorem 3.19 *A series $\sum_{n=1}^{\infty} x_n$ converges if and only if for each $\varepsilon > 0$ there exists some positive integer N such that*

$$n > m > N \quad \Rightarrow \quad \left| \sum_{i=m}^{n} x_i \right| < \varepsilon.$$

3.4.1 Tests of Convergence

Determination of *actual sum* of a convergent series is in general a difficult problem. In some cases, we apply an advanced method such as Fourier series to find the sum of a series (see Chap. 10). However, for several applications, it is enough to know that the sum of an infinite series exists, and so finding the *convergence type* of an infinite series is more significant. As said earlier, our main concern here is to discuss tests such as *comparison, ratio, root, and condensation*. Given an infinite series, we choose one of these tests according to their suitability considering the form of the nth term x_n of the series. The central idea in most situations is to use the *comparison test* in one of the two forms, as stated below.

We first consider infinite series of the form (3.4.1) with *positive terms*. That is, when $x_n > 0$ for all n. Clearly then, in this case, the sequence of partial sums given by

$$s_n = \sum_{k=1}^{n} x_k = x_1 + x_2 + \cdots + x_n,$$

is an increasing sequence. So, we can apply the following formulation of Theorem 3.19.

Theorem 3.20 *An infinite series $\sum_{n=1}^{\infty} x_n$, with $x_n \geq 0$, converges if and only if there exists some positive integer K such that*

$$s_n \leq K, \quad for\ each \quad n \geq 1.$$

Further, when this holds, we have $\sum_{n=1}^{\infty} x_n \leq K$.

Proof By definition, the convergence of a series $\sum_{n=1}^{\infty} x_n$ is equivalent to the convergence of the sequence of partial sums $\langle s_n \rangle$. So, if a series $\sum_{n=1}^{\infty} x_n$, with $x_n \geq 0$, converges, then $\langle s_n \rangle$ must be bounded, by Theorem 3.2. So, the condition holds. Conversely, if the condition holds for a positive terms series $\sum_{n=1}^{\infty} x_n$, then $\langle s_n \rangle$ is a bounded, increasing sequence. Therefore, the series converges, by Theorem 3.11. $\square$

In particular, it follows that the series of the form

$$a_0 + \sum_{n=1}^{\infty} \frac{a_n}{10^n}, \quad \text{with} \quad a_0 \in \mathbb{Z}, \quad \text{and} \quad a_n \in \{0, 1, \ldots, 9\},$$

converges. Notice that

$$a_0 + \frac{a_1}{10} + \cdots + \frac{a_n}{10^n} \leq a_0 + 9 \sum_{k=1}^{\infty} \frac{1}{10^k} = a_0 + 1,$$

and the sum of the series is the value assigned to the *decimal expansion* $a_0.a_1 a_2 \cdots$. More generally, we have the following *comparison test*.

Theorem 3.21 (Comparison Test). *Let* $\sum_{n=1}^{\infty} x_n$ *and* $\sum_{n=1}^{\infty} y_n$ *be two series, with* $0 \leq x_n \leq y_n$. *Then, the convergence of the latter series implies the convergence of the former series. And, the divergence of the former implies the divergence of the latter series.*

Proof We may write

$$s_n = x_1 + x_2 + \cdots + x_n \quad \text{and} \quad t_n = y_1 + y_2 + \cdots + y_n.$$

So, if $\sum_{n=1}^{\infty} y_n$ converges, then the sequence $\langle t_n \rangle$ is bounded. Thus, since $s_n \leq t_n$, the series $\sum_{n=1}^{\infty} x_n$ converges too. The second part follows from the same reasoning. $\square$

The next theorem is known as the Cauchy *condensation test*.

Theorem 3.22 *Let* $\sum_{n=1}^{\infty} x_n$ *be a positive term series, with* $x_{n+1} \leq x_n$. *Then the series* $\sum_{n=1}^{\infty} x_n$ *converges if and only if* $\sum_{n=1}^{\infty} 2^n x_{2^n}$ *converges.*

Proof As usual, $\langle s_n \rangle$ denote the sequence of partial sums of the series $\sum_{n=1}^{\infty} x_n$. We may write $\langle \sigma_n \rangle$ for the sequence of partial sums of the series $\sum_{n=1}^{\infty} 2^n x_{2^n}$ so that by the property $x_{n+1} \leq x_n$, we have

$$\sigma_n = 2x_2 + 4a_4 + \cdots + 2^n x_{2^n}, \quad \text{for all } n \geq 1.$$
$$\leq 2x_2 + (2x_3 + 2x_4) + \cdots + (2x_{2^{n-1}+1} + \cdots + 2x_{2^n})$$
$$\leq 2x_1 + 2x_2 + \cdots + 2x_{2^n} = 2s_{2^n}$$

Similarly, using the property $x_{n+1} \leq x_n$, we have

$$s_3 = x_1 + x_2 + x_3 \leq x_1 + 2x_2 = x_1 + \sigma_1$$
$$s_7 = x_1 + x_2 + \cdots + x_7 \leq x_1 + \sigma_1 + 4x_4 = x_1 + \sigma_2$$
$$\vdots$$
$$s_{2^{n+1}-1} \leq x_1 + \sigma_n, \quad n \geq 1,$$

where the last inequality follows from the identity

$$s_{2^{n+1}-1} - s_{2^n-1} = x_{2^n} + x_{2^n+1} + \cdots + x_{2^{n+1}-1} \le 2^n x_{2^n}.$$

Therefore, we obtain

$$s_{2^{n+1}-1} - x_1 \le \sigma_n \le 2x_{2^n}, \quad \text{for all } n \ge 1,$$

which proves that $\langle s_n \rangle$ is bounded if and only if $\langle \sigma_n \rangle$ is bounded. Hence, Theorem 3.20 completes the proof. □

Theorem 3.23 *Let* $\sum_{n=1}^{\infty} x_n$ *and* $\sum_{n=1}^{\infty} y_n$ *be two positive terms series, with* $\lim_{n \to \infty} (y_n/x_n) = 0$. *Then, the convergence of series* $\sum_{n=1}^{\infty} x_n$ *implies that of* $\sum_{n=1}^{\infty} y_n$. *More generally, the same conclusion holds if*

$$\lim_{n \to \infty} \frac{y_n}{x_n} = \ell, \quad \text{with } 0 < \ell < 1.$$

Further, the divergence of the series $\sum_{n=1}^{\infty} x_n$ *implies that of* $\sum_{n=1}^{\infty} y_n$ *if* $\lim_{n \to \infty} y_n/x_n = \ell$, *with* $\ell \ge 1$.

Proof The convergence of series $\sum_{n=1}^{\infty} x_n$ implies that, for each $\varepsilon > 0$, there exists an $N \in \mathbb{N}$ such that

$$\lim_{n \to \infty} \frac{y_n}{x_n} = 0 \quad \Rightarrow \quad y_n < \varepsilon x_n, \quad \text{for all } n > N.$$

Also, the sequence of partial sums of the series $\sum_{n=1}^{\infty} x_n$ is bounded above by Theorem 3.19. So, by above equation, the sequence of partial sums of the series $\sum_{n=1}^{\infty} y_n$ is bounded above. It thus converges by Theorem 3.19. □

Example 3.16 Consider the series

$$\sum_{n=1}^{\infty} n \left(\frac{1}{2} + (-1)^n \frac{1}{3} \right)^n = \frac{1}{6^n} + 2 \frac{5^n}{6^n} + \cdots .$$

If x_n denote the nth term of this series, we have

$$\frac{x_{n+1}}{x_n} = \frac{n+1}{n} \left(\frac{1}{2} + (-1)^{n+1} \frac{1}{3} \right) \le \frac{5(n+1)}{6n} \le \frac{6}{7} < 1,$$

provided $n \ge 35$. Hence, the convergence follows by the previous remark.

Example 3.17 (*p-Series*). Let p be a positive real number, and $n^p := e^{p \ln n}$. Consider the following *Riemann zeta series*

$$\sum_{n=1}^{\infty} \frac{1}{n^p} = \frac{1}{1^p} + \frac{1}{2^p} + \frac{1}{3^p} + \cdots , \tag{3.4.7}$$

also known as the *p-series*. To discuss about the convergence, we take cases when $p > 1$ or $p \le 1$.

1. <u>Case I</u> Let $0 < p \le 1$. For $p = 1$, we have

$$1 + \frac{1}{2} + \left(\frac{1}{3} + \frac{1}{4}\right) + \left(\frac{1}{5} + \cdots + \frac{1}{8}\right) + \cdots > 1 + \frac{1}{2} + \frac{2}{4} + \frac{4}{8} + \cdots,$$

   which implies that the sequence of partial sums of the series (3.4.7) for $p = 1$ has an unbounded subsequence because $S_{2^m} > m/2$, for all $m \ge 1$. So, the p-series diverges for $p = 1$. Next, for $0 < p < 1$, since $1/n^p > 1/n$, so the series diverges by *comparison test*.

2. <u>Case II</u> Let $p > 1$, we have

$$1 + \left(\frac{1}{2^p} + \frac{1}{3^p}\right) + \left(\frac{1}{4^p} + \frac{1}{5^p} + \cdots + \frac{1}{7^p}\right) + \cdots < 1 + \frac{2}{2^p} + \frac{4}{4^p} + \frac{8}{8^p} + \cdots,$$

   which is a geometric series with common ratio $1/2^{p-1}$. Hence, by *comparison test*, series converges for $p > 1$.

Hence, the p-series diverges, for $p \le 1$, and converges, for $p > 1$.

Example 3.18 Consider the positive terms series

$$\sum_{n=1}^{\infty} \frac{n!}{n^n} = 1 + \frac{1}{2} + \frac{2}{9} + \cdots .$$

If a_n denotes the nth term of the series, we have

$$0 < a_n \le \frac{2}{n^2}, \quad \text{for } n \ge 2.$$

By Example 3.4.7, we know that series $\sum_{n=1}^{\infty} \frac{1}{n^2}$ is convergent, so the convergence of the given series follows by comparison test.

The following is the convergence test in *ratio form*.

Theorem 3.24 *Let* $\sum_{n=1}^{\infty} x_n$ *and* $\sum_{n=1}^{\infty} y_n$ *be two positive terms series, with*

$$\frac{x_{n+1}}{x_n} \le \frac{y_{n+1}}{y_n}, \quad \text{for } n \ge n_0, \text{ say.}$$

Then, the convergence of the latter series implies the convergence of the former series. And, divergence of the former implies the divergence of the latter series.

Proof The inequities in the above equation for $n = n_0, \ldots, n-1$ are

$$\frac{x_{n_0+1}}{x_{n_0}} \leq \frac{y_{n_0+1}}{y_{n_0}}, \quad \frac{x_{n_0}}{x_{n_0-1}} \leq \frac{y_{n_0}}{y_{n_0-1}}, \quad \ldots, \quad \frac{x_n}{x_{n-1}} \leq \frac{y_n}{y_{n-1}}.$$

Multiplying all, we obtain for $n \geq n_0$,

$$\frac{x_n}{x_{n_0}} \leq \frac{y_n}{y_{n_0}}, \quad \text{i.e.,} \quad x_n \leq \varepsilon y_n, \quad \text{with} \quad \varepsilon = \frac{x_{n_0}}{y_{n_0}}.$$

So, the convergence of $\sum_{n=1}^{\infty} y_n$ implies that of the series $\sum_{n=1}^{\infty} x_n$. And, likewise, for the other part. □

The next theorem is known as the Cauchy's *ratio test*.

Theorem 3.25 (Cauchy Ratio Test). *Let $\sum_{n=1}^{\infty} x_n$ be a positive terms series. If* $\lim_{n \to \infty} (x_{n+1}/x_n) = \ell$, *then the series converges when $\ell < 1$, and it diverges when* $\ell > 1$.

Proof By definition, for each $\varepsilon > 0$ there is a natural number N such that

$$n \geq N \quad \Rightarrow \quad \ell - \varepsilon < \frac{x_{n+1}}{x_n} < \ell + \varepsilon.$$

Taking the product of $(n - N)$ inequalities obtained by using values $n - 1, n - 2, \ldots, N$ (in that order), we obtain

$$(\ell - \varepsilon)^{n-N} x_N < x_n < (\ell + \varepsilon)^{n-N} x_N$$

We first suppose $\ell < 1$. Choose $\varepsilon > 0$ such that $\ell + \varepsilon < 1$, and take

$$M_n := (\ell + \varepsilon)^{n-N} x_N.$$

Then $x_n \leq M_n$. Since the series $\sum_{k=1}^{\infty} M_k$ is convergent by comparison test, because $\ell + \varepsilon < 1$, it follows that $\sum_{n=1}^{\infty} x_n$ is convergent. Finally, assume $\ell > 1$. Choose $\varepsilon > 0$ such that $\ell - \varepsilon < 1$, and take

$$M_n := (\ell - \varepsilon)^{n-N} x_N.$$

Then, $M_n \leq x_n$ and the series $\sum_{k=1}^{\infty} M_k$ is divergent by comparison test, because $\ell - \varepsilon < 1$, it follows that $\sum_{n=1}^{\infty} x_n$ is divergent. □

Example 3.19 Once again, consider the positive terms series

$$\sum_{n=1}^{\infty} \frac{n!}{n^n} = 1 + \frac{1}{2} + \frac{2}{9} + \cdots .$$

Let x_n denotes the nth term of the series. Since

$$\lim_{n \to \infty} \frac{x_{n+1}}{x_n} = \lim_{n \to \infty} \left(1 + \frac{1}{n}\right)^{-n} = \left[\lim_{n \to \infty} \left(1 + \frac{1}{n}\right)^n\right]^{-1} = e^{-1} < 1,$$

the convergence of $\sum_{n=0}^{\infty} x_n$ follows by *ratio test*.

3.4.2 Arbitrary Terms Series

A general series $\sum_{n=0}^{\infty} x_n$ is called *absolutely convergent* if $\sum_{n=0}^{\infty} |x_n|$ is convergent. And, $\sum_{n=0}^{\infty} x_n$ is called *conditionally convergent* if it is convergent, but not absolutely. A series of the form

$$\sum_{k=1}^{\infty} (-1)^{k+1} a_n = a_1 - a_2 + a_3 - a_4 + \cdots \tag{3.4.8}$$

is called an **alternating series**. The next theorem gives a condition for the convergence of an *alternating series*.

Theorem 3.26 (Leibniz Theorem). *Suppose $\langle a_k \rangle$ is a monotonically decreasing sequence of positive numbers, with $\lim_{k \to \infty} a_k = 0$. Then the alternating series (3.4.8) is convergent.*

Proof Let $\langle s_n \rangle$ denote the sequence of partial sums of the given series. Notice that, as $a_{2n+1} \geq a_{2n+2}$, we have

$$s_{2n+2} = s_{2n} + \left(a_{2n+1} - a_{2n+2}\right) \geq s_{2n}, \quad \text{for all} \quad n \geq 1.$$

Similarly, we have

$$s_{2n+3} = s_{2n+1} + \left(a_{2n+2} - a_{2n+3}\right) \geq s_{2n+1}, \quad \text{for all} \quad n \geq 0.$$

That is, both the subsequences $\langle s_{2n} \rangle$ and $\langle s_{2n+1} \rangle$ of the sequence $\langle s_n \rangle$ are monotone. Clearly, both are bounded sequences. Therefore, by Theorem 3.11, the limits $a = \lim_{n \to \infty} s_{2n}$ and $b = \lim_{n \to \infty} s_{2n+1}$ exist. However, as $\lim_{k \to \infty} a_k = 0$, we obtain

$$b = \lim_{n \to \infty} s_{2n+1} = \lim_{n \to \infty} \left(s_{2n} + a_{2n+1}\right) = a + 0 = a.$$

Hence, $\langle s_n \rangle$ converges. This completes the proof. $\square$

For example, by Theorem 3.26, the alternating series

$$\sum_{k=1}^{\infty}(-1)^{k+1}\frac{1}{k} = 1 - \frac{1}{2} + \frac{1}{3} - \frac{1}{4} + \cdots$$

is convergent. As the series $\sum_{k=1}^{\infty}(-1)^{k+1}/k$ is convergent, but not absolutely, it is an example of a conditionally convergent series.

Theorem 3.27 *Every absolutely convergent series is convergent.*

Proof Suppose $\sum_{n=0}^{\infty} x_n$ is an absolutely convergent series. Let $\langle s_n \rangle$ and $\langle s_n' \rangle$ be respectively the sequences of partial sums of the series $\sum_{n=0}^{\infty} x_n$ and $\sum_{n=0}^{\infty} |x_n|$. Then, for $n \geq m$, we have

$$\begin{aligned}
|s_n - s_m| &= |x_{m+1} + x_{m+2} + \cdots + x_n| \\
&\leq |x_{m+1}| + |x_{m+2}| + \cdots + |x_n| \\
&= |s_n' - s_m'|.
\end{aligned}$$

So, the assertion follows by Theorem 3.19. □

Theorem 3.28 (Comparison Test). *Let $\sum_{n=0}^{\infty} x_n$ be a series such that*

$$|x_n| \leq M_n, \quad \text{for all } n \geq N_0 \geq 0.$$

If the series $\sum_{k=0}^{\infty} M_k$ is convergent then $\sum_{n=0}^{\infty} x_n$ is absolutely convergent.

Proof Let $\langle s_n \rangle$ and $\langle s_n' \rangle$ be respectively the sequences of partial sums of the series $\sum_{n=0}^{\infty} |x_n|$ and $\sum_{k=0}^{\infty} M_k$. Then, for $n \geq m \geq N_0$, we have

$$\begin{aligned}
|s_n - s_m| &= \left||x_{m+1}| + |x_{m+2}| + \cdots + |x_n|\right| \\
&= |x_{m+1}| + |x_{m+2}| + \cdots + |x_n| \\
&\leq M_{m+1} + M_{m+2} + \cdots + M_n \\
&= |s_n' - s_m'|.
\end{aligned}$$

So, the assertion follows by Theorem 3.19. □

Corollary 3.3 (Abel series). *Let $\sum_{n=0}^{\infty} x_n$ be a convergent series, and*

$$S(r) := \sum_{n=0}^{\infty} r^n x_n, \quad \text{for } 0 \leq r \leq 1.$$

Then $S(r)$ is absolutely convergent.

Theorem 3.29 (Cauchy Ratio Test). *Let $\sum_{n=0}^{\infty} x_n$ be a general terms series. If $\lim_{n \to \infty} |x_{n+1}/x_n| = \ell$ then the series converges absolutely when $\ell < 1$ and it diverges when $\ell > 1$.*

Proof By definition, for each $\varepsilon > 0$ there is a natural number N such that

$$n \geq N \quad \Rightarrow \quad \ell - \varepsilon < \frac{|x_{n+1}|}{|x_n|} < \ell + \varepsilon.$$

Taking the product of $(n - N)$ inequalities obtained by using values $n - 1, n - 2, \ldots, N$ (in that order), we obtain

$$(\ell - \varepsilon)^{n-N}|x_N| < |x_n| < (\ell + \varepsilon)^{n-N}|x_N|$$

We first suppose $\ell < 1$. Choose $\varepsilon > 0$ such that $\ell + \varepsilon < 1$. Take

$$M_n := (\ell + \varepsilon)^{n-N}|x_N|.$$

Then $x_n \leq M_n$. Since the series $\sum_{k=1}^{\infty} M_k$ is convergent by Theorem 3.28, because $\ell + \varepsilon < 1$, it follows that $\sum_{n=1}^{\infty} x_n$ is convergent. Finally, assume $\ell > 1$. Choose $\varepsilon > 0$ such that $\ell - \varepsilon < 1$. Take

$$M_n := (\ell - \varepsilon)^{n-N}|x_N|.$$

Then, $M_n \leq x_n$ and the series $\sum_{k=1}^{\infty} M_k$ is divergent by Theorem 3.28, because $\ell - \varepsilon < 1$, it follows that $\sum_{n=1}^{\infty} x_n$ is divergent. $\qquad\Box$

In particular, since

$$\ell = \lim_{n \to \infty} \left| \frac{x^{n+1}}{x^n} \right| = \lim_{n \to \infty} \frac{|x^{n+1}|}{|x^n|} = |x|,$$

it follows that the geometric series $\sum_{n=0}^{\infty} x^n$ converges absolutely for $|x| < 1$ and diverges for $|x| > 1$. The next theorem is known as the d'Alembert's *root test*.

Theorem 3.30 (d'Alembert's Root Test). *Let $\sum_{n=0}^{\infty} x_n$ be a general terms series. If $\lim_{n \to \infty} |x_n|^{1/n} = \ell$, then the series converges absolutely when $\ell < 1$ and it diverges when $\ell > 1$.*

Proof By definition, for each $\varepsilon > 0$, there is a natural number N such that

$$n \geq N \quad \Rightarrow \quad (\ell - \varepsilon)^n < |x_n| < (\ell + \varepsilon)^n.$$

First, suppose $\ell < 1$. Let $\varepsilon > 0$ be chosen so that $\ell + \varepsilon < 1$. Then the series $\sum_{k=N}^{\infty} M_k$ is convergent, by Theorem 3.28, and so the series $\sum_{n=0}^{\infty} x_n$ is absolutely convergent. Finally, assume $\ell > 1$. Let $\varepsilon > 0$ be chosen so that $\ell - \varepsilon < 1$. Then the series $\sum_{k=N}^{\infty} M_k$ is divergent, by Theorem 3.28, and so the series $\sum_{n=0}^{\infty} x_n$ is divergent. $\qquad\Box$

In particular, if $\langle b_n \rangle$ is a sequence with $\lim_{n \to \infty} |b_n| < 1$, then

$$\lim_{n\to\infty} |b_n^n|^{1/n} = \lim_{n\to\infty} |b_n| < 1$$

implies that the series $\sum_{k=0}^{\infty} b_k^k$ converges. So, taking $b_n = (n+1)/2n$, it follows that

$$\sum_{k=1}^{\infty} \left[\frac{(1+k)}{2k}\right]^k < \infty.$$

We conclude the chapter with a brief discussion about the connection between the concept of *conditional convergence* of a series and its rearrangements, where the meaning of the term *rearrangement of a series* is taken in the sense of the next definition.

Definition 3.8 Let $x = \langle x_n \rangle$ be a sequence, and $s : \mathbb{N} \to \mathbb{N}$ be a bijective function. Then the composition $x \circ s : \mathbb{N} \to \mathbb{R}$ defines a *rearrangement* of the sequence $\langle x_n \rangle$.

Notice that the convergence of a series is not affected by the addition or deletion of any finite number of terms. Further, it should be clear that it is always possible to introduce brackets over terms of a convergent series without changing its convergence type or altering the sum. Indeed, the introduction of brackets simply means that we are passing from a convergent sequence of partial sums to a sequence containing only some of the terms of the original sequence, and this will necessarily converge to the same value.

However, the matter gets complicated if we attempt to find the sum of a convergent series by insertion of brackets over some terms or by alteration of the number of terms in brackets. For example, consider the series

$$\sum_{n=1}^{\infty} (-1)^n = -1 + 1 - 1 + 1 - \cdots + (-1)^n + \cdots .$$

Clearly then, its sequence of partial sums is given by

$$s_n = \begin{cases} -1, & \text{when } n \text{ is odd} \\ 0, & \text{when } n \text{ is even} \end{cases},$$

which is a divergent sequence. However, if we write the same series as

$$[(-1) + 1] + [(-1) + 1] + [(-1) + 1] + \cdots ,$$

then this *rearranged series* converges to 0. Notice that the rearrangement

$$[(-1) + 1 + (-1)] + [1 + (-1) + 1] + [(-1) + 1 + (-1)] + \cdots ,$$

of this series is again divergent as, in this case, the sequence of partial sums is given by

$$\langle -1,\ 0,\ -1,\ 0,\ \dots \rangle.$$

Therefore, in general, it is not possible to insert or remove brackets over terms of a series without altering its convergence type. There are many mysterious facts about the *rearrangements of a series*, and the next theorem is the most unexpected of the lot.

Theorem 3.31 (Riemann). *Every conditionally convergent series can be rearranged to converge to any specified real number or diverge to $\pm\infty$.*

3.4.3 Decimal Expansion of Real Numbers

We discuss here the *method of construction* of the set $\mathbb{R}$ as *decimal expansion*, where each is seen as a convergent infinite series. We begin by proving that every real number can be approximated to any degree of accuracy by sequences of rational numbers expressed in terms of decimal expansions of some special type (Theorem 3.32). Trivially, the number 0 has the expansion 0, and since the decimal expansion of negative real number is by definition the negative of the expansion for its absolute value, so it suffices to consider only the positive real numbers.

Lemma 3.2 *Let $a > 0$ be a real number. Then, for every $n \in \mathbb{N}$, there is a finite decimal $r_n = a_0.a_1a_2 \cdots a_k$ such that*

$$r_n \leq a < r_n + \frac{1}{10^n}, \quad \text{for } n \in \mathbb{N}. \tag{3.4.9}$$

Proof Recall that, for any $x \in \mathbb{R}$, $n = \lfloor x \rfloor$ denotes the *largest integer* such that $n \leq x$. So, by Exercise 2.4, we may take $a_0 = \lfloor a \rfloor \in \mathbb{Z}^+$ so that $0 < a_0 \leq a < a_0 + 1$. Next, divide the interval $[a_0, a_0 + 1]$ into ten equal parts. If a is not any of these *subdivision points* then there is some digit $a_1 \in \{1, \dots, 9\}$ such that

$$a_0 + \frac{a_1}{10} \leq a < a_0 + \frac{a_1 + 1}{10}.$$

Continuing the process, we find digits $a_k \in \{0, 1, \dots, 9\}$ satisfying the inequality

$$a_0 + \frac{a_1}{10} + \cdots + \frac{a_k}{10^n} \leq a < a_0 + \frac{a_1}{10} + \cdots + \frac{a_k + 1}{10^n}, \quad n \geq 1.$$

$\square$

As each r_n is a positive rational number, Lemma 3.2 proves that there are two sequences of rational numbers, one increasing from below and the other decreasing from above, each converging to a. By repeating the process sufficiently many times, the decimal expansion of a can be obtained to any desired degree of accuracy. The integers $a_k \in \{0, 1, \ldots, 9\}$ ($k \geq 1$) obtained above are precisely the *digits* that one uses to define an infinite decimal expansion of the real number $0 < a < 1$.

Definition 3.9 A *decimal expansion* of a real number $a > 0$ is a series of the form

$$a := \lim_{n \to \infty} \sum_{k=0}^{n} \frac{a_k}{10^k} = \sum_{k=0}^{\infty} \frac{a_k}{10^k} = a_0.a_1a_2a_3 \cdots, \qquad (3.4.10)$$

where $a_0 = \lfloor a \rfloor$ is called the *integral part* of a, and $f = 0.a_1a_2a_3 \cdots$ is called its *fractional part*. The integers $0 \leq a_k \leq 9$ are called the decimal digits of a.

Notice that since the digits $a_k \in \{0, 1, \ldots, 9\}$, we have

$$\sum_{k=0}^{\infty} \frac{a_k}{10^k} \leq a_0 + 9 \sum_{k=1}^{\infty} \frac{1}{10^k} = a_0 + 1,$$

so the series $\sum_{k=0}^{\infty} \frac{a_k}{10^k}$ converges. Therefore, each real number $a > 0$ has a well defined decimal expansion of the form (3.4.10). Clearly, the same argument applies if $a < 0$.

In historical terms, the early traces of the concept of *decimal expansions* are found in the eighth-century algebra book by *Al-Khwarizmi* (780–850 *A D*). There had been some instances of occasional use of decimal notation during the twelfth century. In 1585, Flemish physicist and mathematician *Simon Stevin* (1548–1620) was first to introduce the *decimal point* and *decimal notation*. However, it was Scottish mathematician *John Napier* (1550–1617) who used the decimal expansions to separate the whole part of a positive real number from its fractional part, while writing his logarithmic tables in 1614.

Now, as it suffices to work with the *fractional part* of $a > 0$, we assume that $0 < a < 1$. We say that $a = 0.a_1a_2a_3 \cdots$ is a *finite decimal* (or a *terminating decimal*) if, for some n, $a_k = 0$, for $k > n \geq 1$; and, we say that a is an *infinite decimal* with *repeating decimals* if, except for finitely many initial digits, a *block of digits* keeps repeating infinitely many times. That is,

$$a = 0.a_1a_2 \cdots a_n b_1b_2 \cdots b_m b_1b_2 \cdots b_m b_1b_2 \cdots b_m \cdots.$$

In this case, we write $a = 0.a_1a_2 \cdots a_n \overline{b_1b_2 \cdots b_m}$, and call it a *periodic decimal*. Notice that a trivial way to view a terminating decimal as an infinite (non-terminating) decimal is by attaching an infinite string of 0's at the tail. However, apart from this obvious process, there is another interesting way that uses convergence of geometric series. For, since

$$0.\overline{9} = \sum_{k=1}^{\infty} \frac{9}{10^k} = 9 \sum_{k=1}^{\infty} \frac{1}{10^k} = 9 \frac{1/10}{1 - (1/10)} = 1,$$

the number 1 and the periodic decimal $0.\overline{9}$ are *taken to be* the same. For example, the decimal expansion 0.5 and $0.4\overline{9}$ represents the same rational number $1/2$. Also, multiplying by 3, we can write

$$\frac{1}{3} = 0.33333 \cdots \quad \text{as} \quad 1 = 0.99999 \cdots .$$

So, towards uniqueness of decimal expansion of the form (3.4.10), we remark that any exact decimal fraction $m10^n$ has two decimal expansions. Conversely, if $a > 0$ is a real number with more than one representation in a decimal expansion, then it has exactly two: one that has the expansion of the form $\cdots b999 \cdots$ or $\cdots (b+1)000 \cdots$. Opting for a non-terminating expansion of the first form avoids the issue of non-uniqueness. Decimal expansions are then ordered *lexicographically*.

Theorem 3.32 *The decimal expansion $a = 0.a_1a_2a_3 \cdots$ represents a rational number if and only if either $a_k = 0$ for $k > n \geq 1$ (finite decimal) or it is an infinite decimal with a block of digits repeating infinitely many times.*

Proof Clearly, for any choice of positive integers $0 \leq a_k \leq 9$, a *finite* decimal expansion of the form

$$a = \sum_{k=1}^{n} \frac{a_k}{10^k} = 0, a_1 \cdots a_n, \tag{3.4.11}$$

represents a rational number; indeed, we have $a = m/10^n$, where

$$m = a_1 10^{n-1} + a_2 10^{n-2} + \cdots + a_n.$$

Since there is no positive integer m such that $3m = 10^n$ for some $n \in \mathbb{N}$, we conclude that not every rational number can have a *terminating* decimal expansion. More generally, consider a decimal expansion of the form

$$a = 0.a_1a_2 \cdots a_n \overline{b_1b_2 \cdots b_m},$$

where the block of digits $b_1b_2 \cdots b_m$ repeats infinitely many times. Multiplying by 10^n, if necessary, we may assume that $a = 0.\overline{b_1b_2 \cdots b_m}$. Then

$$a = 0.\overline{b_1b_2 \cdots b_m}$$
$$= \frac{1}{10^m} \left[(b_1 10^{m-1} + b_2 10^{m-2} + \cdots + b_m) + a \right]$$
$$\Rightarrow \quad 10^m a = b_1 10^{m-1} + b_2 10^{m-2} + \cdots + b_m + a$$
$$\Rightarrow \quad a = \frac{b_1 10^{m-1} + b_2 10^{m-2} + \cdots + b_m}{10^m - 1}.$$

Thus, in both cases, a represents a rational number in $(0, 1)$. Conversely, suppose

$$a = 0.a_1a_2a_3 \cdots = \frac{m}{n}, \quad \text{where } m < n \text{ are positive integers.}$$

Using Euclid's *division algorithm*, repeatedly multiplying remainders by 10 and dividing by n, we obtain

$$10m = a_1n + r_1, \text{ where } r_1 = n[a_210^{-1} + a_310^{-2} + a_410^{-3} + \cdots]$$

$$10r_1 = a_2n + r_2, \text{ where } r_2 = n[a_310^{-1} + a_410^{-2} + a_510^{-3} + \cdots]$$

$$\vdots \qquad\qquad \vdots$$

$$10r_{k-1} = a_kn + r_k, \text{ where } r_k = n[a_{k+1}10^{-1} + a_{k+2}10^{-2} + \cdots]$$

$$\vdots \qquad\qquad \vdots$$

where each $r_k \in \{0, 1, \ldots, (n-1)\}$. If the process terminates, say at $k = n$, then we obtain a *finite* decimal expansion of the form $a = 0.a_1a_2 \cdots a_n$. Notice that a_k's are *quotients* obtained in the division process. If not, then $r_k \in \{1, \ldots, (n-1)\}$ for all $k \geq 1$. In this case, we use the *pigeon hole principle*[3] to complete the proof. For, observe that

$$10m = a_1n + r_1 \quad \Leftrightarrow \quad \frac{10m}{n} = a_1 + \frac{r_1}{n}$$

$$10r_1 = a_2n + r_2 \quad \Leftrightarrow \quad \frac{10r_1}{n} = a_2 + \frac{r_2}{n}$$

$$\vdots$$

$$10r_{k-1} = a_kn + r_k \quad \Leftrightarrow \quad \frac{10r_{k-1}}{n} = a_k + \frac{r_k}{n}$$

$$\vdots$$

By the pigeon hole principle, there must be a first pair of indices $\{j, k\}$ such that $r_j = r_k$, for some $1 \leq j < k \leq n$. Then,

$$10r_j = a_{j+1}n + r_{j+1} \quad \text{and} \quad 10r_k = a_{k+1}n + r_{k+1}$$

imply that $a_{j+1} = a_{k+1}$ and $r_{j+1} = r_{k+1}$. More generally, we have

$$a_{j+\ell} = a_{k+\ell} \quad \text{and} \quad r_{j+\ell} = r_{k+\ell}, \text{ for } 1 \leq \ell \leq k - j,$$

and therefore $m/n = 0.a_1a_2 \cdots a_j\overline{a_{j+1} \cdots a_k}$.   □

The following special case of Theorem 3.32 is often used directly.

---

[3] Dirichlet's *pigeon hole principle* states that if there are p pigeons and h holes, with $p > h$, then at least one hole must accommodate more than one pigeon.

Theorem 3.33 *A rational number a/b in its lowest term has a terminating decimal expansion if and only if $b = 2^m 5^n$, for some $m, n \in \mathbb{Z}^+$.*

Proof Suppose b has the given form. If $n \leq m$, multiply numerator and denominator of a/b by 5^{m-n} so that

$$\frac{a}{b} = \frac{a}{2^m 5^n} = \frac{a5^{m-n}}{10^m} = \frac{p}{10^m}, \quad \text{for some } p \in \mathbb{Z}^+.$$

Since division by 10^m simply requires inserting the decimal point at an appropriate place, we obtain a terminating decimal expansion. And, if $n > m$, multiplying numerator and denominator of a/b by 2^{n-m}, a similar argument applies. Rewriting a proof of the converse is left to the reader. $\qquad\square$

It follows from Theorem 3.32 that *irrational numbers* in the interval $(0, 1)$ have *non-terminating, non-repeating* decimal expansions. That is, $a \in (0, 1)$ is an irrational number if and only if a has an infinite decimal expansion of the form $0.a_1 a_2 a_3 \cdots$, where digits a_k are *random selections* from the set $\{0, 1, 2, \ldots, 9\}$. To illustrate, let $a = \sqrt{2}$ so that $1^2 < 2 < 2^2$. Considering squares of $1.1, 1.2, 1.3, 1.4$, and 1.5, we conclude that $1.4 < a < 1.5$. Next, consider squares of 1.41 and 1.42, we conclude that $1.41 < a < 1.42$. And so on, at each stage, doing a simple search through a smaller and smaller range of numbers to get one more digit.

In summary, if $\langle a_k \rangle$ is a real sequence such that $0 \leq a_k \leq 9$ (not all zero) then $\sum_{k=1}^{\infty} a_k 10^{-k}$ converges to a real number in $(0, 1]$. Conversely, each $a \in (0, 1]$ has a unique decimal expansion of the form $\sum_{k=1}^{\infty} a_k 10^{-k}$, with $a_k \in \{0, 1, \ldots, 9\}$, such that

$$a - 10^{-n} \leq \sum_{k=1}^{n} \frac{a_k}{10^k} < a, \quad \text{for each } n.$$

Exercises 3

3.1. If sequences $\langle u_n \rangle$ and $\langle v_n \rangle$ converge respectively to limits ℓ and ℓ', show that

$$\lim_{n \to \infty} \frac{u_1 v_n + u_2 v_{n-1} + \cdots + u_n v_1}{n} = \ell \ell'.$$

3.2. Show that, for $k \in \mathbb{N}$ and $0 < x < 1$, $n^k x^n \to 0$ as $n \to \infty$.

3.3. Show that, for $a, b > 0$, $x_n = \sqrt{n^2 + a} - \sqrt{n^2 + b}$ is a null sequence. Hence, find $\lim_{n \to \infty} (\sqrt{n^2 + n} - n)$.

3.4. Show that $\displaystyle\lim_{n \to \infty} \sum_{k=1}^{n} \frac{1}{k(k+1)} = 1$.

3.5. Show that the sequence $x_n = \displaystyle\sum_{k=1}^{n} \frac{1}{n+k}$ converges to a number a such that $1/2 < a \leq 1$.

3.6. Show that

$$\lim_{n \to \infty} x_n = a \quad \Rightarrow \quad \lim_{n \to \infty} \frac{x_1 + \cdots + x_n}{n} = a.$$

Also, give an example to show that the converse does not hold.

3.7. Suppose α, $x_1 > 0$, and $x_{n+1} = (1/2)(x_n + (\alpha/x_n))$. Show that $x_n \geq \sqrt{\alpha}$, and hence deduce that $\lim_{n \to \infty} x_n = \sqrt{\alpha}$.

3.8. Suppose a sequence $\langle x_n \rangle$ is such that $n^{-k} \leq x_n \leq n^k$, for some $k > 0$. Show that $\lim_{n \to \infty} x_n^{1/n} = 1$.

3.9. Show that $\lim_{n \to \infty} (n!)^{1/n^2} = 1$ and $\lim_{n \to \infty} n^{1/\sqrt{n}} = 1$.

3.10. Let $a > 0$ be a real number. Show that

$$\lim_{n \to \infty} \frac{1}{\sqrt{an^2 + 1}} + \frac{1}{\sqrt{an^2 + 2}} + \cdots + \frac{1}{\sqrt{an^2 + n}} = \frac{1}{\sqrt{a}}.$$

3.11. Prove that the following are *null sequences*

$$x_n = \frac{\cos n}{n}, \qquad t_n = \frac{(-1)^n}{n}, \qquad v_n = \frac{\ln n}{n}.$$

3.12. Show that the sequence $x_n = 1/(2n - 1)$, $n \geq 1$, is divergent.

3.13. Compute the limits of the following sequences

a. $x_n = \frac{1}{n} \left[1 + 2^{1/2} + 3^{1/3} + \cdots + n^{1/n} \right]$.

b. $x = \left[\frac{2}{1} \cdot \left(\frac{2}{3} \right)^2 \cdots \left(1 + \frac{1}{n} \right)^n \right]^{1/n}$.

3.14. Show that the following are *null sequences*:

$$(i) \quad u_n = \frac{n}{a^n} \ (a > 1); \qquad \text{and,} \qquad (ii) \quad t_n = \frac{n^p}{x^n}, \quad \text{for } |x| > 1.$$

3.15. Show that a convergent sequence $\langle x_n \rangle$ that does not converge to a number y is *bounded away from it*.

3.16. Reference [1] For any sequence $\langle x_n \rangle$, with $x_n > 0$ for all n, show that

$$\left(\frac{x_1 + x_{n+1}}{x_n} \right) > e$$

for infinitely many n's.

3.17. Reference [1] For $n \in \mathbb{N}$ and $0 < \theta < \pi$, show that

$$\sin \theta + \frac{\sin 2\theta}{2} + \cdots + \frac{\sin n\theta}{n} > 0.$$

3.18. Check the sequences $x_n = e^n/n$ and $u_n = a^n/n! \ (a > 0)$ for monotone property.

3.19. Check the sequences $x_n = (3n + 1)/2n - 2$, $u_n = a - (1/a^n)\,(a > 1)$, and $y_n = \sqrt{n}(\sqrt{n+1} - \sqrt{n})$ for monotone property.

3.20. Use Theorem 3.11 to prove that $\lim_{n \to \infty} (1 + (1/n))^n = e$.

3.21. Show that the sequence

$$x_n = 1 + \frac{1}{3} + \frac{1}{5} + \cdots + \frac{1}{2n - 1}, \quad n \geq 1.$$

is not a Cauchy sequence.

3.22. Show that the operations of addition and multiplication defined in (3.3.9) are well defined.

3.23. Show that $\leq$ defined in (3.3.10) is an order relation.

3.24. Prove that convergence of series $\sum_{n=1}^{\infty} n x_n$ gives convergence of $\sum_{n=1}^{\infty} x_n$.

3.25. Show that $(1/n) \sum_{n=1}^{k} x_n$ converges to 0 if the series $\sum_{n=1}^{\infty} (x_n/n)$ is convergent.

3.26. Suppose $\sum_{n=1}^{\infty} x_n = \infty$, where $x_n \geq 0$. Show that the series

$$\sum_{n=1}^{\infty} \frac{x_n}{(x_1 + \cdots + x_n)^a}$$

converges for $a > 1$ and diverges for $0 < a \leq 1$.

3.27. Suppose $\sum_{n=1}^{\infty} x_n y_n$ converges for every convergent sequence $\langle y_n \rangle$. Show that the series $\sum_{n=1}^{\infty} x_n$ converges.

3.28. This problem is about *base three* (ternary) decimal expansion of real numbers in the Cantor set C as defined in Example 2.6.

   a. Show that for every sequence $x_n \in \{0, 1, 2\}$, the series $\sum_{n=1}^{\infty} (x_n/3^n)$ converges in $[0, 1]$. Conversely, show that for every $x \in [0, 1]$ there is sequence $a_n \in \{0, 1, 2\}$ such that $x = \sum_{n=1}^{\infty} (a_n/3^n)$.

   b. If x is a rational number of the form $m/3^n$, with $m \equiv 1 \pmod 3$ and $n \in \mathbb{N}$, show that x has exactly one the following two ternary expansions:

$$0.a_1 \cdots a_{n-1} 1000 \cdots \qquad 0.a_1 \cdots a_{n-1} 0222 \cdots .$$

   c. Suppose x is a rational number of the form $m/3^n$, with $m \equiv 0$ or 2 (mod 3), or an irrational number. Show that if x has a unique ternary expansion that is different from the two given in the previous part.

   d. Show that $x \in [0, 1]$ if and only if x has at least one ternary expansion with ternary digits in the set $\{0, 2\}$.

3.29. The *Cauchy product* of two series $\sum_{n=0}^{\infty} x_n$ and $\sum_{n=0}^{\infty} y_n$ is the series having the nth term as $z_n := x_0 y_n + x_1 y_{n-1} + \cdots + x_n y_0$. Give examples of two convergent series such that their Cauchy product diverges. Further, show that

   a. if $\sum_{n=0}^{\infty} x_n$ and $\sum_{n=0}^{\infty} y_n$ converges respectively to X and Y, and their Cauchy product converges to Z, then $Z = XY$;

   b. if $\sum_{n=0}^{\infty} x_n$ converges absolutely to X, and $\sum_{n=0}^{\infty} y_n$ converges to Y, then their Cauchy product converges to XY.

   c. if $\sum_{n=0}^{\infty} x_n$ and $\sum_{n=0}^{\infty} y_n$ converges absolutely to X and Y, respectively, then their Cauchy product converges to XY.

3.30. Show that the series

$$\frac{1}{2} - \frac{1}{3} + 2\left(\frac{1}{4} - \frac{1}{5} + \frac{1}{6} - \frac{1}{7}\right) + \cdots$$

converges to *Euler constant*

$$\gamma := \lim_{n \to \infty} \left(1 + \frac{1}{2} + \cdots + \frac{1}{n} - \ln n\right).$$

Reference

1. M. Hata, *Problems and Solutions in Real Analysis* (World Scientific, 2007)

Chapter 4
Analysis on $\mathbb{R}$

> *Calculus required continuity, and continuity was supposed to require the infinitely little; but nobody could discover what the infinitely little might be.*

Bertrand Rusell (1872–1970)

Let $f : [a, b] \to \mathbb{R}$ be a function. We will discuss here the *continuity* and *differentiability* of the function $f(x)$ at a point $x = c \in [a, b]$, and also the *integrability* of f over the interval $[a, b]$. A thorough understanding about the concept of *a limit of a function at a point* is crucial to all the three notions. The standard *modulus metric* on $\mathbb{R}$, as introduced earlier in Chapter 2, provides a mathematical foundation to the intuitive notion such as *infinitely little* or *in near vicinity of* or *as close as possible*.

In 1817, *Bernard Bulzano* introduced the concept of a *limit* of a function at a point. In 1820, *Augustin-Louis Cauchy* formulated the (notorious) ε–δ definition of the limit concept. Later, in 1861, *Karl Weierstrass* refined the concept further to define the continuity of a function at a point. During 1870–1872, Weierstrass published his work related to fundamental notions such as uniform convergence, uniform continuity, termwise integration and differentiation of infinite series, etc., in collaboration with *Eduard Heine*, and his doctoral student *Gorge Cantor*.

In geometric terms, we say a function $f(x)$ is continuous at a point $x = c \in [a, b]$ if $f(x) \approx f(c)$ in near vicinity of the point $x = c$, which is usually written as $\lim_{x \to c} f(x) = f(c)$. Also, a function $f(x)$ is said to be *differentiable* at a point $x = c \in (a, b)$ if the curve $y = f(x)$ *is a line* in near vicinity of the point $x = c$. More specifically, f is differentiable at $x = c$ if, for some real number $f'(c)$,

$$f(x) \approx \ell(x) := f(c) + f'(c)(x - c),$$

© The Author(s), under exclusive license to Springer Nature Singapore Pte Ltd. 2022
A. K. Razdan and V. Ravichandran, *Fundamentals of Analysis with Applications*,
https://doi.org/10.1007/978-981-16-8383-1_4

Fig. 4.1 Area under the
curve $y = x$, bounded by the
lines $x = 0$ and $x = 1$, is the
integral $\int_0^1 y\,dx$

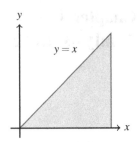

for all $x \in (a, b)$ very close to the point $x = c$. Then, the (linear) function $\ell(x) :=$
$f(c) + f'(c)(x - c)$ is called the *tangent line* to the function f at the point $x = c$.
In this case, we write

$$\lim_{x \to c} \frac{f(x) - f(c)}{x - c} = f'(c),$$

and $f'(c)$ is called the *derivative* of f at the point $x = c$. In fact, the value $f'(c) \in \mathbb{R}$
gives the *slope* of the (unique) line $\ell(x)$ passing through the point $(c, f(c))$. Finally,
the *integral* of a function f over $[a, b]$, denoted by $\int_a^b f\,dx$, is given by the *limit of
the sum of areas* of rectangular regions used to fill the area under the curve $y = f(x)$
between the lines $x = a$ and $x = b$, where the limit is taken over the decreasing
length of the bases of such rectangles. Clearly, no such *limiting process* is needed to
find the *triangular area* under the curve $y = x$ bounded by the lines $y = 0$ and $x = 1$,
as shown in Fig. 4.1.

Most aspects of the single-variable real analysis presented here, and theorems
proved, are prerequisites for the development of higher dimensional analysis as given
in later chapters, especially so for certain important infinite dimensional *function
spaces* defined over $\mathbb{R}$. We strongly recommend the reader to grasp the mathematics
involved with details given here.

4.1 Limit and Continuity

As required, here, we start with the concept of the *limit* of a function $f : [a, b] \to \mathbb{R}$
at a point $x = c \in [a, b]$. Subsequently, we shall use the same to define the continuity
of $f(x)$ at a point $x = c \in [a, b]$. First, we consider the case of *finite limits*. Let us
begin with an introduction of *one-sided limits*.

Definition 4.1 Let f be a real-valued function defined on an interval $[a, b]$, except
possibly at some $x = c \in [a, b]$. We say

1. $f(x)$ has a *left-hand limit* at $x = c \in (a, b]$ if, for any $\varepsilon > 0$, there is some
   $\delta = \delta(c, \varepsilon) > 0$ such that

$$x \in (c, c + \delta) \quad \Rightarrow \quad f(x) \in (\ell - \varepsilon, \ell + \varepsilon).$$

In this case, we write $\lim_{x \to c^-} f(x) = \ell$ or simply $f(c^-) = \ell$;

2. $f(x)$ has a *right-hand limit* at a point $c \in [a, b)$ if, for any $\varepsilon > 0$, there is some $\delta = \delta(c, \varepsilon) > 0$ such that

$$x \in (c - \delta, c) \quad \Rightarrow \quad f(x) \in (\ell - \varepsilon, \ell + \varepsilon).$$

In this case, we write $\lim_{x \to c^+} f(x) = \ell$ or simply $f(c^+) = \ell$.

Notice that, for any of these two limits to exist at a point $x = c$, we do not require the function $f(x)$ to be defined at $x = c$. Therefore, the concept of *limit of a function* $f(x)$ *at a point* $x = c$ is not about the value $f(c)$, it is rather about the *behaviour of a function in the near vicinity of the point* $x = c$. The next definition is a generalisation of the limit concept discussed earlier for sequences $x : \mathbb{N} \to \mathbb{R}$.

Definition 4.2 Let f be a real-valued function defined on an interval $[a, b]$, except possibly at some $x = c \in [a, b]$. We say a real number ℓ is a *limit* of the function $f(x)$ at $x = c$ if, for any $\varepsilon > 0$, there is some $\delta = \delta(c, \varepsilon) > 0$ such that

$$x \in (c - \delta, c + \delta) \setminus \{c\} \quad \Rightarrow \quad f(x) \in (\ell - \varepsilon, \ell + \varepsilon). \qquad (4.1.1)$$

In this case, we write $\lim_{x \to c} f(x) = \ell$.

The set $(c - \delta, c + \delta) \setminus \{c\}$ is called a *deleted interval*, with a puncture at the point $x = c$. The condition $x \neq c$ in (4.1.1), once again, signifies that $f(c)$ may not be defined. The next theorem proves that $\lim_{x \to c} f(x)$ exists precisely when both one-sided limits exist at $x = c$, and they have the same *limiting value*. This common value is the value of $\lim_{x \to c} f(x)$.

Theorem 4.1 *Let f be a real-valued function defined on an interval $[a, b]$, except possibly at some $x = c \in [a, b]$. Then*

$$\lim_{x \to c} f(x) = \ell \quad \Leftrightarrow \quad f(c^-) = f(c^+) = \ell.$$

Proof Left to the reader as a simple exercise. $\qquad \qquad \square$

Example 4.1 We first consider the greatest integer function $\lfloor \ \rfloor : \mathbb{R} \to \mathbb{Z}$ given by $\lfloor x \rfloor = n$, where $n \in \mathbb{Z}$ is the largest with $n \leq x$. Clearly then, at each integer point a, we have

$$\lim_{x \to c^-} \lfloor x \rfloor = a - 1 \quad \text{and} \quad \lim_{x \to c^+} \lfloor x \rfloor = a.$$

Hence, we conclude that $\lim_{x \to a} \lfloor x \rfloor$ does not exist at any $a \in \mathbb{Z}$ (see Fig. 1.1 in Chapter 1). Also, for the function $f : [0, 2] \to \mathbb{R}$ given by

$$f(x) = \begin{cases} x^2, & 0 \leq x \leq 1 \\ 3 - x^2, & 1 < x \leq 2 \end{cases},$$

Fig. 4.2 The graph of $f(x)$
as in Example 4.1

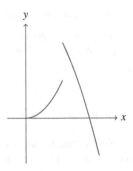

Fig. 4.3 The graph of the
function $g(x)$ as in
Example 4.1

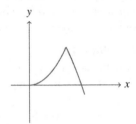

we have $\lim_{x \to 1^-} f(x) = 1$ and $\lim_{x \to 1^+} f(x) = 2$ (see Fig. 4.2). Hence, the limit of $\lim_{x \to 1} f(x)$ does not exist. However, in most such cases, a minor modification in the definition of a function $f(x)$ does the trick. Here, for the function $g : [0, 2] \to \mathbb{R}$ given by

$$g(x) = \begin{cases} x^2, & 0 \le x \le 1 \\ 2 - x^2, & 1 < x \le 2 \end{cases},$$

we have $\lim_{x \to 1} g(x) = 1$ (see Fig. 4.3).

The next theorem is the definition of *sequential limit* of a function.

Theorem 4.2 *Let $f : [a, b] \to \mathbb{R}$ be a function, and $c \in [a, b]$. Then, $\lim_{x \to c} f(x) = \ell$ if and only if, for every sequence $\langle a_n \rangle$ in I, with $a_n \ne c$,*

$$\lim_{n \to \infty} a_n = c \quad \Rightarrow \quad \lim_{n \to \infty} f(a_n) = \ell.$$

Proof First, let $\lim_{x \to c} f(x) = \ell$. Suppose $\langle a_n \rangle$ is a sequence in $[a, b]$, with $a_n \ne c$, such that $a_n \to c$ as $n \to \infty$. Let $\varepsilon > 0$ be arbitrary. By definition, we can find $\delta > 0$ such that $|f(x) - \ell| < \varepsilon$ whenever $x \in [a, b]$ and $0 < |x - c| < \delta$. For this $\delta > 0$, we choose $N \in \mathbb{N}$ such that

$$|a_n - c| < \delta, \quad \text{for all} \ n \ge N.$$

Fig. 4.4 Graph of
$f_1(x) = \sin(1/x)$ in
Example 4.2

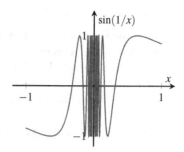

So, we have $|f(a_n) - \ell| < \varepsilon$. Thus, $\lim_{n \to \infty} f(a_n) = \ell$. To prove the converse, we use method of contradition. So, let $\lim_{x \to c} f(x) \neq \ell$. Then, for some $\varepsilon > 0$ and for every $\delta > 0$, there is an element $x \in [a, b]$ such that

$$|f(x) - \ell| \geq \varepsilon, \quad \text{but} \quad |x - c| < \delta.$$

For such an $\varepsilon > 0$, and $\delta = 1/n$, we find $a_n \in [a, b]$ such that $|f(a_n) - \ell| \geq \varepsilon$, with $|a_n - c| < 1/n$. Therefore, we have found a sequence $\langle a_n \rangle$ in $[a, b]$ such that $a_n \neq c$, $\lim_{n \to \infty} a_n = c$, and $\lim_{x \to c} f(x_n) \neq \ell$. $\qquad \square$

Example 4.2 We consider here the function $f : [-1, 1] \to \mathbb{R}$ given by

$$f(x) = \begin{cases} \sin(1/x), & x \neq 0 \\ b, & x = 0 \end{cases}.$$

Notice that the function $\sin(1/x)$ is not defined at $x = 0$, and the choice $f(0) = b \neq 0$ is arbitrary (see Fig. 4.4). This function is known as the *topologist's sine curve*. It oscillates infinitely often as $x \to 0$, has roots at the points $x_n = 1/(n\pi)$, $n \in \mathbb{Z}$, and $f(1/((n\pi + (\pi/2)))) = (-1)^n$. Therefore, applying an argument similar to the one used to prove the converse of the previous theorem, it follows easily that $\lim_{x \to 0} f(x)$ does not exist. Many interesting properties of this important function are discussed later in the last section of Chapter 6.

We are assuming that the reader is familiar with the following elementary calculus facts related to *algebra of limits* of functions defined on some interval $[a, b]$, where all operations are defined *pointwise*: Let $f, g : [a, b] \to \mathbb{R}$ be any two functions, and $x = c \in [a, b]$. Then

1. $\lim_{x \to c} f(x) = \ell$ and $\lim_{x \to c} g(x) = m \Rightarrow \lim_{x \to c} [f(x) + g(x)] = \ell + m$.
2. $\lim_{x \to c} f(x) = \ell$ and $\lim_{x \to c} g(x) = m \Rightarrow \lim_{x \to c} [f(x)g(x)] = \ell m$.
3. $\lim_{x \to c} f(x) = \ell$ and $\lim_{x \to c} g(x) = m \neq 0 \Rightarrow \lim_{x \to c} [f(x)/g(x)] = \ell/m$.

Further, the following two interesting facts can be proved easily. Notice that the property (3) holds in the sense of the property (4) below.

4. Let $g : [a, b] \to \mathbb{R}$ be a function, and $x = c \in [a, b]$. If $\lim_{x \to c} g(x) = m \neq 0$ then, for some $\delta > 0$, $g(x) \neq 0$ for all $x \in (c - \delta, c + \delta)$. In fact, in this interval, $g(x)$ has the same sign as m.

5. (*Sandwich Lemma*) Let $f, g : [a, b] \to \mathbb{R}$ be any two functions, and $x = c \in [a, b]$. Suppose $f(x), g(x) \to \ell$ over some interval $x \in (c - \delta, c + \delta)$. If $h : [a, b] \to \mathbb{R}$ is a function such that

$$f(x) \leq h(x) \leq g(x), \qquad \text{for all} \quad x \in (c - \delta, c + \delta),$$

then $\lim_{x \to c} h(x) = \ell$.

We also remark that the concept of *infinite limits* or *limits at* ∞ is defined in the most natural manner by using the *inversion operation*. That is, for any function $f : (a, b) \to \mathbb{R}$ and $x = c \in (a, b)$,

$$\lim_{x \to c} f(x) = \infty \qquad \Leftrightarrow \qquad \lim_{x \to c} \frac{1}{f(x)} = 0;$$

$$\lim_{x \to \infty} f(x) = \ell \qquad \Leftrightarrow \qquad \lim_{x \to 0} f\left(\frac{1}{x}\right) = \ell.$$

For example, notice that $\lim_{x \to \infty} f(x) = \infty$ if for any $M > 0$ there exists some $\delta > 0$ such that

$$0 < |x| < \delta \qquad \Rightarrow \qquad f\left(\frac{1}{x}\right) > M.$$

Similarly, $\lim_{x \to \infty} f(x) = -\infty$ if for any $M > 0$ there exists some $\delta > 0$ such that

$$0 < |x| < \delta \qquad \Rightarrow \qquad f\left(\frac{1}{x}\right) < -M.$$

We conclude here with a generalisation of the *Landau notations*, introduced earlier for sequences. These are used to study the *asymptotic behaviour* of a function $f(x)$ with respect to some *majorant function* $g(x) \geq 0$, as $x \to \ell$. Notice that, in this case, we may have $\ell = \pm\infty$. We usually take *powers, logarithms, exponentials*, etc., for the majorant function $g(x)$. As before, we write

$$f := O(g) \quad \text{if} \quad |f(x)| \leq C \, g(x), \quad \text{as } x \to \ell,$$

where C is some positive number. In particular, $f(x) = O(1)$ implies $f(x)$ is a *bounded function*. If $g(x) > 0$ for sufficiently large x, and

$$\frac{f(x)}{g(x)} \to 0 \quad \text{as} \quad x \to \ell,$$

we write $f := o(g)$. In particular, $f(x) = o(1)$ implies $f(x) \to 0$ as $x \to \ell$. Further, we write $f(x) \sim g(x)$ if

$$\frac{f(x)}{g(x)} \to 1 \quad \text{as} \quad x \to \ell.$$

Clearly, $\sim$ defines an equivalence relation among functions. Two functions $f(x)$ and $g(x)$ in the same equivalence class are said to be *asymptotically equivalent*. We also say that $f(x)$ *is asymptotic to* $g(x)$. Notice that

$$f(x) \sim g(x) \quad \Leftrightarrow \quad f = g + o(g).$$

In this case, the function $g(x)$ is called the *principal part* of $f(x)$.

Definition 4.3 (*Cauchy, 1820*) We say a function $f : [a, b] \to \mathbb{R}$ is *continuous* at a point $c \in (a, b)$ if $\lim_{x \to c} f(x) = f(c)$. That is, for any $\varepsilon > 0$ there is a $\delta = \delta(c, \varepsilon) > 0$ such that

$$x \in (c - \delta, c + \delta) \quad \Rightarrow \quad f(x) \in (f(c) - \varepsilon, f(c) + \varepsilon).$$

We say $f(x)$ is continuous at $x = a$ or $x = b$ if

$$\lim_{x \to a^+} f(x) = f(a) \quad \text{or} \quad \lim_{x \to b^-} f(x) = f(b),$$

respectively. We say $f(x)$ is continuous over $[a, b]$ if it is continuous at each point of $[a, b]$. Further, we say f is *uniformly continuous* over $[a, b]$ if $\delta > 0$ can be chosen to be independent of points $c \in [a, b]$.

As mentioned in the introduction of the chapter, in geometrical terms, a function $f : [a, b] \to \mathbb{R}$ is continuous at a point $c \in [a, b]$ if $f(x) \approx f(c)$ in the near vicinity of the point $x = c$. We will prove in a later chapter that, in general, every continuous function defined on a closed and bounded interval is uniformly continuous (Corollary 7.8).

We write $C(I)$ for the set of all real-valued continuous functions defined on an interval $I \subseteq \mathbb{R}$. The next theorem lists basic algebraic properties of functions in $C(I)$, whereas usual operations such as *sum*, *product*, and *scalar multiplication*, are defined *pointwise*. A proof in each case can be given using the corresponding property for the *algebra of limits*, as mentioned earlier.

Theorem 4.3 *Suppose $f, g \in C(I)$, and $\alpha \in \mathbb{R}$. Then, $f + g$, fg, αf, $|f| \in C(I)$. Further, if $g(c) \neq 0$ in some open interval in $[a, b]$ centred at the point $x = c \in (a, b)$, then f/g is continuous at $x = c$.*

Proof Left to the reader as an exercise. □

We also have that *composition* of two continuous functions is continuous. Notice that the *sandwich lemma* for the limits of functions as stated earlier implies the following simple fact: *If $f, g : I \to \mathbb{R}$ are continuous in some interval $J \subseteq I$, and a function $h : I \to \mathbb{R}$ is such that $f(x) \leq h(x) \leq g(x)$ for all $x \in J$, then $h \in C(J)$.*

Fig. 4.5 The graph of the
function $f(x) = 1/x$, $x > 0$

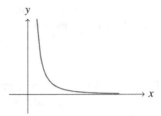

Example 4.3 Consider the function $f : \mathbb{R} \to \mathbb{R}$ given by $f(x) = x^2$. Let $c \in \mathbb{R}$. Choose $\delta = \min\{1, \varepsilon/(1 + 2|c|)\}$. If $|x - c| < \delta$, then, by using

$$|x + c| \le |x - c| + 2|c| \le \delta + 2|c| \le 1 + 2|c|,$$

we get

$$\left| x^2 - c^2 \right| = |x + c|\,|x - c| < (1 + 2|c|)\delta \le \varepsilon.$$

Therefore, the function f is continuous at $x = c$, and hence, it is continuous over $\mathbb{R}$.

Example 4.4 Consider the function $f : (0, \infty) \to \mathbb{R}$ given by $f(x) = 1/x$. Let $c > 0$. Choose $\delta = \min\{c/2, c^2\varepsilon/2\}$. If $|x - c| < \delta$, then $-\delta < x - c < \delta$ and so $x > c - \delta > c - c/2 = c/2 > 0$. Using this, we see that

$$\left| \frac{1}{x} - \frac{1}{c} \right| = \frac{|x - c|}{xc} \le \frac{2|x - c|}{c^2} < \frac{2\delta}{c^2} \le \varepsilon$$

whenever $|x - c| < \delta$. Therefore, the function f is continuous at $x = c$ and, hence, continuous on $(0, \infty)$ (see Fig. 4.5).

Let $f : [a, b] \to \mathbb{R}$ be a function, and $x = c \in [a, b]$. Notice that, in general, the equation $\lim_{x \to c} f(x) = f(c)$ fails to hold if any one of the following three conditions is true (in that order):

1. The function $f(x)$ is not defined at $x = c$.
2. (1) holds, but $\lim_{x \to c} f(x)$ does not exists.
3. (1) and (2) hold, but $\lim_{x \to c} f(x) \ne f(c)$.

We then say that the function $f : [a, b] \to \mathbb{R}$ is *discontinuous* at a point $x = c \in [a, b]$. For example, the function $f : [-1, 1] \to \mathbb{R}$ given by $f(x) = \sin(1/x)$ is discontinuous at $x = 0$ because it is not defined there (see Fig. 4.4). Also, the *greatest integer function* $\lfloor \ \rfloor : \mathbb{R} \to \mathbb{Z}$ given by $\lfloor x \rfloor = n$, where n is the largest integer $\le x$, is discontinuous at each integer point $a \in \mathbb{Z}$ because the limit does not exists at any of these points.

Further, we say that a function $f : [a, b] \to \mathbb{R}$ has a *discontinuity* of the *first kind* at a point $x = c \in [a, b]$ if (1) and (2) hold, but (3) fails. For example, the function $f : \mathbb{R} \to \mathbb{R}$ given by

Fig. 4.6 Graph of
$f(x) = \sin(x)/x$, $x \neq 0$;
$f(0) = 0.5$

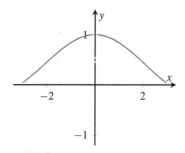

$$f(x) = \begin{cases} \sin(x)/x, & \text{for } x \neq 0 \\ 0.5, & \text{for } x = 0 \end{cases},$$

as shown in Fig. 4.6 has a discontinuity of the first kind at the point $x = 0$ because

$$\lim_{x \to 0} f(x) = 1 \neq 0.5 = f(0).$$

Recall that, by Theorem 3.8, $\lim_{n \to \infty} x_n^k = a^k$, for all $k \geq 1$, if $\lim_{n \to \infty} x_n = a$. Therefore, for any $a_0, a_1, \ldots, a_k \in \mathbb{R}$, we have

$$p(x_n) = a_0 + a_1 x_n + \cdots + a_k x_n^k \longrightarrow a_0 + a_1 a + \cdots + a_k a^k = p(a),$$

using Theorem 4.3. It thus follows that, for any polynomial $p(x) \in \mathbb{R}[x]$, we have

$$\lim_{n \to \infty} x_n = a \quad \Rightarrow \quad \lim_{n \to \infty} p(x_n) = p(a). \tag{4.1.2}$$

This simple fact suggests an interesting relationship between the concept of the continuity of a function $f : [a, b] \to \mathbb{R}$ and the convergence of sequences in $[a, b]$.

Definition 4.4 (*Sequential Continuity*). A function $f : [a, b] \to \mathbb{R}$ is *sequentially continuous* at $x \in [a, b]$ if for any sequence $\langle x_n \rangle$ in $[a, b]$,

$$x_n \to x \quad \Rightarrow \quad f(x_n) \to f(x), \quad \text{as } n \to \infty.$$

It is, in general, true that every continuous function between two spaces is sequentially continuous.[1] Here, let a function $f : [a, b] \to \mathbb{R}$ be continuous at $x = c \in [a, b]$. By Definition 4.3, for any $\varepsilon > 0$, there exists $\delta > 0$ such that for any $x \in [a, b]$ with $|x - c| < \delta$, we have $|f(x) - f(c)| < \varepsilon$. In particular, if $\langle x_n \rangle$ is a sequence in the interval $(x - c, x + c)$ such that $\lim_{n \to \infty} x_n = c$, then for some $N \in \mathbb{N}$

---

[1] For topological spaces, we use *nets* in place of sequences. The converse holds for *first countable* spaces such as metric spaces, as we shall discuss in Chapter 6.

$$f(x) = \begin{cases} 1, & x \in \mathbb{Q} \\ 0, & x \in \mathbb{R} \setminus \mathbb{Q} \end{cases}$$

Fig. 4.7 A nowhere continuous function

$$
\begin{aligned}
n \geq N \quad &\Rightarrow \quad |x_n - c| < \delta \\
&\Rightarrow \quad |f(x_n) - f(x)| < \varepsilon,
\end{aligned}
$$

by the continuity of $f(x)$ at $x = c$. More precisely, we take $N \in \mathbb{N}$ with $N\delta > 1$ so that

$$n \geq N \quad \Rightarrow \quad |x_n - c| < \frac{1}{n} \leq \frac{1}{N} < \delta.$$

So, $\lim_{n \to \infty} f(x_n) = f(x)$. Therefore, for any sequence $\langle x_n \rangle$ in $[a, b]$,

$$\lim_{n \to \infty} f(x_n) = f\left(\lim_{n \to \infty} x_n \right).$$

In the next theorem, we prove that the condition (4.1.1) implies that $f(x)$ is continuous if it is sequentially continuous. Hence, the two notions coincide, in this case.

Theorem 4.4 (Heine). *A function $f : [a, b] \to \mathbb{R}$ is continuous at $x = c \in [a, b]$ if, for every sequence $\langle a_n \rangle$ in $[a, b]$, with $a_n \neq c$,*

$$\lim_{n \to \infty} a_n = c \quad \Rightarrow \quad \lim_{n \to \infty} f(a_n) = f(c).$$

Proof Proof follows directly from Theorem 4.2.   □

In particular, it follows from (4.1.2) that polynomial functions are continuous, and hence so are *rational functions*. The best part of the sequential definition of continuity is that it helps to deal with functions that are not necessarily given by a *formula definition*.

Example 4.5 (*Dirichlet Function*). Let $f : \mathbb{R} \to \mathbb{R}$ be the *characteristic function* of the set $\mathbb{Q}$. That is,

$$f(x) = \begin{cases} 1 & \text{if } x \in \mathbb{Q}; \\ 0 & \text{if } x \in \mathbb{R} \setminus \mathbb{Q}. \end{cases}$$

Let $x = c$ be an arbitrary real number. Then, we can always find two sequences $\langle x_n \rangle$ and $\langle y_n \rangle$ converging to c, where $x_n \in \mathbb{Q}$ and $y_n \in \mathbb{R} \setminus \mathbb{Q}$. But then, we have $f(x_n) \to 1$ and $f(y_n) \to 0$, which shows that the function $f(x)$ is a *nowhere continuous* function (see Fig. 4.7).

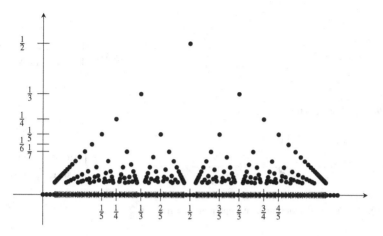

Fig. 4.8 Thomae function is continuous at every $x \in \mathbb{Q}$

In the next example, we now discuss *Thomae function* that is continuous precisely at irrational numbers. So, it is yet another example of a function with uncountably many discontinuities.

Example 4.6 (*Thomae Function*). Consider the function given by (see Fig. 4.8)

$$d(x) = \begin{cases} 1, & \text{for } x = 0 \\ \frac{1}{n}, & \text{for } 0 \neq x = \frac{m}{n} \in \mathbb{Q}, \text{ with } (m, n) = 1, \ n > 0 \ . \\ 0, & \text{for } x \notin \mathbb{Q} \end{cases} \qquad (4.1.3)$$

We show here that $d(x)$ is continuous precisely at irrational points. First, notice that $d(x)$ is a periodic function of period 1. It is thus enough to study it on the interval $(0, 1)$. As with increasing value of $n \in \mathbb{N}$, the number of rationals of the form $r = m/n$ increases, we conclude that the corresponding dots $(r, d(r))$ go down the x-axis. Therefore, clearly, the limit of the function is zero at every point $a \in \mathbb{R}$.

More precisely, let $\varepsilon > 0$, and choose an $n \in \mathbb{N}$ such that $n\varepsilon > 1$ (by the Archimedean property). Then, the set of all fractions m/n that lie within $x = 1$ from $x = a$ and whose denominators do not exceed n is a finite set. So, there is some $\delta > 0$ such that none of such fractions lies within the punctured δ-neighbourhood of the point a. Notice that it does not matter even if $x = a$ itself is one of these fractions, because the limit process ignores the limit point a.

Now, for any $x \in (0, 1)$ in this neighbourhood, if $x \in \mathbb{R} \setminus \mathbb{Q}$ then

$$d(x) = 0 \quad \Rightarrow \quad |d(x)| < \varepsilon.$$

And, if $x = p/q \in \mathbb{Q}$ then we must have $q > n$ because it cannot be one of the fractions that we excluded, therefore

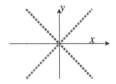

$$f(x) = \begin{cases} x, & x \in \mathbb{Q} \\ -x, & x \in \mathbb{R} \setminus \mathbb{Q} \end{cases}$$

Fig. 4.9 A function continuous only at $x = 0$

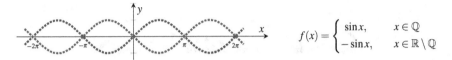

$$f(x) = \begin{cases} \sin x, & x \in \mathbb{Q} \\ -\sin x, & x \in \mathbb{R} \setminus \mathbb{Q} \end{cases}$$

Fig. 4.10 A function discontinuous except at all $x = n\pi, n \in \mathbb{Z}$

$$|d(x)| = \frac{1}{q} < \frac{1}{n} < \varepsilon.$$

In particular, $d(x)$ is continuous only at irrational numbers.

Example 4.7 Let $f : \mathbb{R} \to \mathbb{R}$ be given by

$$f(x) = \begin{cases} x & \text{if } x \in \mathbb{Q}; \\ -x & \text{if } x \in \mathbb{R} \setminus \mathbb{Q}. \end{cases}$$

This function f is continuous only at $x = 0$ (see Fig. 4.9).

Example 4.8 Let $f : \mathbb{R} \to \mathbb{R}$ be given by

$$f(x) = \begin{cases} \sin x & \text{if } x \in \mathbb{Q}; \\ -\sin x & \text{if } x \in \mathbb{R} \setminus \mathbb{Q}. \end{cases}$$

This function f is continuous only at $x = n\pi, n \in \mathbb{Z}$ (see Fig. 4.10).

Definition 4.5 We say a function $f : [a, b] \to \mathbb{R}$ is *piecewise continuous* if it is continuous everywhere in $[a, b]$ except perhaps on a finite set of points

$$a \leq t_1 < t_2 < \ldots < t_n \leq b,$$

where it may not even be defined, but the following limits

$$f(t_k^+) = \lim_{h \to 0} f(t_k + h) \quad \text{and} \quad f(t_{k+1}^-) = \lim_{h \to 0} f(t_{k+1} - h)$$

exist, for each $k = 1, 2, \ldots, n - 1$.

In particular, it follows that f is a *bounded function*. We write $PC[a, b]$ for the set of all real-valued piecewise continuous functions defined over the interval $[a, b]$. Notice that the function $f(t) = \sin(1/t)$, $t \neq 0$, is not a piecewise continuous function (Example 4.2). The significance of this class of functions would be clearer by Chapter 10.

4.2 Differentiability

We start with a formal definition of differentiability of a function at a point. We say a function $f : [a, b] \to \mathbb{R}$ is *differentiable* at $x = c \in (a, b)$ if the limit

$$\lim_{x \to c} \frac{f(x) - f(c)}{x - c} = \lim_{t \to 0} \frac{f(c + t) - f(c)}{t}$$

exists, and the limiting value is called the *derivative of the function* $f(x)$ *at the point* $x = c$. We usually write this value as $f'(c)$ or $Df(c)$. We say a function $f : [a, b] \to \mathbb{R}$ is differentiable over the interval (a, b) if it is differentiable at all points of (a, b). The derivative at the endpoint is defined using one-sided limits, as described in the previous section. The next theorem provides the first important fact.

Theorem 4.5 *Every differentiable function is continuous.*

Proof Suppose a function $f : [a, b] \to \mathbb{R}$ is differentiable at a point $x = c$. That is, $f'(c)$ exists, and we write

$$\lim_{x \to c} \frac{f(x) - f(c)}{x - c} = f'(c).$$

Once again, using product property of algebra of limits, we have

$$\lim_{x \to c}(f(x) - f(c)) = \lim_{x \to c}\left(\frac{f(x) - f(c)}{x - c}\right)\lim_{x \to c}(x - c)$$
$$= f'(c) \cdot 0 = 0.$$

Hence, $\lim_{x \to c} f(x) = f(c)$, which proves that $f(x)$ is continuous at $x = c$, as asserted. ☐

Example 4.9 The function $f : [-a, a] \to \mathbb{R}$ defined by $f(x) = x^n|x|$, for $n \in \mathbb{N}$, is differentiable. For $x > 0$, we have $f(x) = x^{n+1}$ and so $f'(x) = (n + 1)x^n$. Similarly, for $x < 0$, we have $f(x) = -x^{n+1}$ and so $f'(x) = -(n + 1)x^n$. For $x = 0$, it follows that $f'(x) = \lim_{x \to 0} x^{n-1}|x| = 0$. Hence, the derivative of f is given by $f'(x) = (n + 1)x^{n-1}|x|$ (see Fig. 4.11).

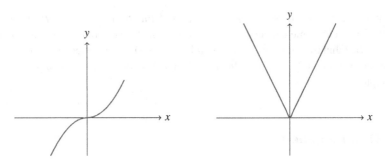

Fig. 4.11 Graphs of $f(x) = x|x|$ and its derivative $f'(x)$

Example 4.10 The function $f : [a, b] \to \mathbb{R}$ defined by $f(x) = x^n$, $n \in \mathbb{N}$, is differentiable. The identity

$$(x - c)(x^{n-1} + x^{n-2}c + \cdots + c^{n-1}) = x^n - c^n$$

is easily verified. Using this identity, we get

$$f'(c) = \lim_{x \to c} \frac{x^n - c^n}{x - c} = \lim_{x \to c}(x^{n-1} + x^{n-2}c + \cdots + c^{n-1}) = nc^{n-1}.$$

Example 4.11 The function $f : [0, 2n] \to \mathbb{R}$ defined by $f(x) = \sum_{k=1}^{n} |x - k|$ is non-differentiable exactly at $k = 1, 2, \ldots, n$. The function $f : [-a, a] \to \mathbb{R}$ given by $f(x) = x|x|$ is differentiable everywhere in $[-a, a]$, but its derivative $f'(x) = 2|x|$ is not differentiable at $x = 0$.

Example 4.12 The function $f : [-a, a] \to \mathbb{R}$ defined by $f(x) = x^2 \sin(1/x)$, for $x \neq 0$, with $f(0) = 0$, is differentiable on $[-a, a]$ and its derivative $f'(x)$ is given by

$$f'(x) = \begin{cases} 2x \sin(1/x) - \cos(1/x) & (x \neq 0) \\ 0 & (x = 0). \end{cases}$$

To see $f'(0) = 0$, we note that

$$\left| \frac{f(x) - f(0)}{x - 0} \right| = |x \sin(1/x)| \leq |x| \to 0$$

as $x \to 0$. Therefore,

$$f'(0) = \lim_{x \to 0} \frac{f(x) - f(0)}{x - 0} = 0.$$

This function f' is not continuous at $x = 0$ (Fig. 4.12).

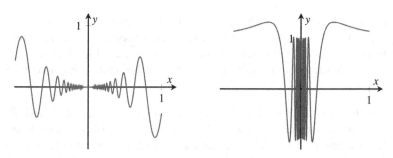

Fig. 4.12 Graphs of $f(x) = x^2 \sin(1/x)$ and its derivative $f'(x)$

The next theorem lists basic properties of *algebra of derivatives* for two real-valued functions defined on an interval. A proof in each case follows on similar lines as for the corresponding property for the algebra of limits.

Theorem 4.6 *Suppose $f, g : [a, b] \to \mathbb{R}$ are two functions, differentiable at a point $x = c \in (a, b)$. Then $f + g$, fg, αf are differentiable at the point c. If $g(c) \neq 0$ then f/g is differentiable at c. Further,*

1. $(f + g)'(c) = f'(c) + g'(c)$.
2. $(fg)'(c) = f'(c)g(c) + f(c)g'(c)$.
3. $(\alpha f)'(c) = \alpha f'(c)$.
4. $(f/g)'(c) = (f'(c)g(c) - f(c)g'(c))/(g(c))^2$.

Proof Left to the reader. □

Example 4.13 The function $f : [-a, a] \to \mathbb{R}$ defined by $f(x) = |x|$ is non-differentiable only at 0. For $x > 0$, we have $f(x) = x$ and so $f'(x) = 1$. Similarly, for $x < 0$, we have $f(x) = -x$ and so $f'(x) = -1$. Notice that $\lim_{x \to 0} |x|/x$ does not exist as $|x|/x \to 1$ when $x \to 0$ through positive values and $|x|/x \to -1$ when $x \to 0$ through negative values. Thus, the function f is not differentiable at $x = 0$.

Notice that, however, the function $f(x) = |x|$ is Lipschitz continuous.

Definition 4.6 We say a function $f : I \to \mathbb{R}$ is a *Lipschitz continuous* if, for some constant $C \geq 0$,

$$|f(x) - f(y)| \leq C|x - y|, \quad \text{for all} \quad x, y \in I. \tag{4.2.1}$$

In fact, the constant C can be taken to be the least positive number satisfying above condition. It is then called the *Lipschitz constant* of the function f. We say f is *short map* if $C = 1$; and, if $0 \leq C < 1$, then f is called a *contraction map*. As every Lipschitz continuous function is uniformly continuous, the constant C is also known as the *modulus of uniform continuity*. Lipschitz continuity is an important concept that is needed to prove many important theorems such as *Picard-Lindelöff* and *Banach contraction principle* (see Chapter 8).

Example 4.14 It follows from the condition (4.2.1) that every function $f : [a, b] \to$ $\mathbb{R}$ with $|f'(x)| \le C$, for some $C \ge 0$, is Lipschitz continuous. In particular, it follows that functions such as $f(x) = \sin x$ or $f(x) = \sqrt{x^2 + 3}$ are Lipschitz continuous. However, the function $f(x) = \sqrt{x}$, $x \in [0, 1]$ is not Lipschitz continuous because $f'(x) \to \infty$ as $x \to 0$.

We say function $f : I \to \mathbb{R}$ is *continuously differentiable* if both f and f' are continuous on the interval I. We write $C^1(I)$ for the set of all continuously differentiable functions $f : I \to \mathbb{R}$. Notice that the function $f(x)$ given in Example 4.11 is not continuously differentiable. Clearly, we can extend to idea to define k-*times continuously differentiable* functions $f : I \to \mathbb{R}$, and write $C^k(I)$ for the set of all k-times continuously differentiable functions $f : I \to \mathbb{R}$, for $k \ge 0$; of course, we take $C^0(I) := C(I)$. Further, we write $C^\infty(I)$ for the set of all *infinitely differentiable* (or *smooth*) functions $f : I \to \mathbb{R}$.

Definition 4.7 We say a function $f : [a, b] \to \mathbb{R}$ is *piecewise smooth* if both f and f' are piecewise continuous on the interval (a, b). We write $PS[a, b]$ for the set of all real-valued piecewise smooth functions defined over an interval $[a, b]$.

For example, the *periodic extension* to $\mathbb{R}$ of the function

$$f(x) = \begin{cases} x^2, & \text{for } -1 \le x \le 0 \\ -x, & \text{for } 0 < x \le 1 \end{cases} \tag{4.2.2}$$

obtained by setting $f(x + 2) = f(x)$, $x \in \mathbb{R}$, is is a piecewise smooth function. Notice that this function is discontinuous at each point $a \in \mathbb{Z}$. The significance of this class of functions would be clearer by Chapter 10.

Definition 4.8 A function $f : [a, b] \to \mathbb{R}$ is said to have a *local maximum* at $x = c \in (a, b)$ if there is some $\delta > 0$ such that

$$|x - c| < \delta \quad \Rightarrow \quad f(x) \le f(c).$$

Similarly, a function $f : [a, b] \to \mathbb{R}$ is said to have *local minimum* at $x = c \in (a, b)$ if there is some $\delta > 0$ such that

$$|x - c| < \delta \quad \Rightarrow \quad f(c) \le f(x).$$

Example 4.15 The function $f : [-a, a] \to \mathbb{R}$ defined by $f(x) = x^2$ has a local minimum at $x = 0$ as $f(0) = 0 \le x^2 = f(x)$ for all $x \in [-a, a]$. Similarly, the function $f : [-a, a] \to \mathbb{R}$ defined by $f(x) = |x|$ has a local minimum at $x = 0$ as $f(0) = 0 \le |x| = f(x)$ for all $x \in [-a, a]$. In both cases, we can choose any $\delta > 0$.

Fig. 4.13 Graphs of x^2 and $|x|$

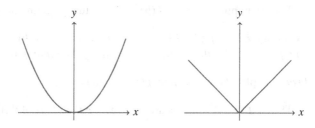

Theorem 4.7 (Fermat Theorem). *If $f : [a, b] \to \mathbb{R}$ has a local minimum or a local maximum at a point $x = c \in (a, b)$ and is differentiable at $x = c$, then $f'(c) = 0$.*

Proof We assume that f has a local maximum at $x = c$. By definition of local maximum, there is a $\delta > 0$, such that $f(x) \leq f(c)$ for all $x \in [a, b]$ with $|x - c| < \delta$. Thus, for all t with $c - \delta < t < c$, we have $f(t) \leq f(c)$ and so

$$\frac{f(t) - f(c)}{t - c} \geq 0.$$

By letting $t \to c$ through value less than c, we get $f'(c) \geq 0$. Similarly, for all t with $c < t < c + \delta$, we have again $f(t) \leq f(c)$ and so

$$\frac{f(t) - f(c)}{t - c} \leq 0.$$

By letting $t \to c$ through value less than c, we get $f'(c) \leq 0$. Therefore, it follows that $f'(c) = 0$. If f has a local minimum at $x = c$, then $-f$ has a local maximum at $x = c$ and so $f'(c) = 0$. $\qquad \square$

Notice that this theorem does not guarantee the differentiability at the point $x = c$ where it has a local minimum or maximum. For example, the function $f : [-a, a] \to \mathbb{R}$ given by $f(x) = |x|$ has a local minimum at $x = 0$, but it is non-differentiable at $x = 0$ (see Fig. 4.13). We prove later in Chapter 7 the existence of extreme points for a continuous function defined on a closed and bounded interval (see Corollary 7.3).

Theorem 4.8 (Rolle's Theorem). *If $f : [a, b] \to \mathbb{R}$ is continuous on $[a, b]$, is differentiable on (a, b), and $f(a) = 0 = f(b)$, then there is a $c \in (a, b)$ such that $f'(c) = 0$.*

Proof If f is constant, then $f'(x) = 0$ for all x. Hence, we may assume that f takes some positive or negative values. We first assume that f takes some positive values on $[a, b]$. By Corollary 7.3, $\sup\{f(x) : a \leq x \leq b\} = f(c)$ for some $c \in [a, b]$. Since f assumes some positive values, $f(c) > 0$ and so $c \neq a$ or b. Thus, $c \in (a, b)$. Also, $f(x) \leq f(c)$ for all x and so, f has a local maximum at $x = c$. By Theorem 4.7, $f'(c) = 0$. If f assumes only negative values, then $-f$ assumes all positive values on $[a, b]$. Therefore, there is an $c \in (a, b)$ such that $(-f)'(c) = 0$. Since $(-f)'(c) = -f'(c)$, it follows that $f'(c) = 0$. $\qquad \square$

The assumption $f(a) = f(b) = 0$ is actually redundant.

Corollary 4.1 *If $f : [a, b] \to \mathbb{R}$ is continuous on $[a, b]$, is differentiable on (a, b), and $f(a) = f(b)$, then there is a $c \in (a, b)$ such that $f'(c) = 0$.*

Proof Apply Theorem 4.8 to $f(x) - f(a)$. $\square$

The next two corollaries are *mean value theorems* of differential calculus.

Corollary 4.2 (Lagrange Mean Value Theorem). *If a function $f : [a, b] \to \mathbb{R}$ is continuous on $[a, b]$, and is differentiable on (a, b), then there is a $c \in (a, b)$ such that $f(b) - f(a) = (b - a)f'(c)$.*

Proof Define the function $g : [a, b] \to \mathbb{R}$ by

$$
g(x) = \begin{vmatrix} f(x) & x & 1 \\ f(a) & a & 1 \\ f(b) & b & 1 \end{vmatrix}
$$
$$
= (f(b) - f(a))(x - a) - (b - a)(f(x) - f(a)).
$$

Then g is continuous on $[a, b]$, is differentiable on (a, b), and $g(a) = 0 = g(b)$. Therefore, by Theorem 4.8, $g'(c) = 0$ for some $c \in (a, b)$. Since

$$
g'(x) = (f(b) - f(a)) - (b - a)f'(x),
$$

we have $f(b) - f(a) = (b - a)f'(c)$. $\square$

Corollary 4.3 (Cauchy Mean Value Theorem). *If $f, g : [a, b] \to \mathbb{R}$ are continuous on $[a, b]$ and differentiable on (a, b), then there is a $c \in (a, b)$ such that $(f(b) - f(a))g'(c) = (g(b) - g(a))f'(c)$.*

Proof Left to the reader as Exercise 4.14. $\square$

Definition 4.9 We say $f : [a, b] \to \mathbb{R}$ is an *increasing function* (or a *non-decreasing function*) over the interval $[a, b]$ if, for each $x_1, x_2 \in [a, b]$,

$$
x_1 < x_2 \quad \Rightarrow \quad f(x_1) \le f(x_2).
$$

We say it is *strictly increasing* if $x_1 < x_2 \Rightarrow f(x_1) < f(x_2)$. A similar definition holds for a *decreasing function* or a *strictly decreasing function*. Finally, we say $f : [a, b] \to \mathbb{R}$ is a *monotone function* if it is an increasing or a decreasing function.

Example 4.16 The function $f : [-1, 1] \to \mathbb{R}$ given by $f(x) = e^x$ is strictly increasing as $f'(x) = e^x > 0$. On the other hand, the function $f : (0, \infty) \to \mathbb{R}$ given by $f(x) = 1/x$ is strictly decreasing as $f'(x) = -1/x^2 < 0$. (see Fig. 4.14)

The criteria given in the next theorem are used frequently in the sequel.

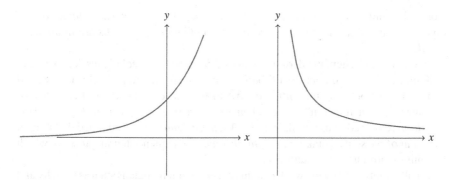

Fig. 4.14 Graph of the functions $y = e^x$ and $y = 1/x$

Theorem 4.9 *Let a function* $f : [a, b] \rightarrow \mathbb{R}$ *be continuous on* $[a, b]$, *and differentiable on* (a, b). *Then each one the following holds:*

1. f *is strictly increasing on* $[a, b]$ *if* $f'(x) > 0$ *for each* $x \in (a, b)$.
2. f *is increasing on* $[a, b]$ *if* $f'(x) \geq 0$ *for each* $x \in (a, b)$.
3. f *is constant on* $[a, b]$ *if* $f'(x) = 0$.
4. f *is strictly decreasing on* $[a, b]$ *if* $f'(x) < 0$ *for each* $x \in (a, b)$.
5. f *is decreasing on* $[a, b]$ *if* $f'(x) \leq 0$ *for each* $x \in (a, b)$.

Proof Let $x_1, x_2 \in [a, b]$ with $x_1 < x_2$. Since f is continuous on $[a, b]$ and differentiable on (a, b), it is continuous on $[x_1, x_2]$ and is differentiable on (x_1, x_2). By the Mean Value Theorem, there exists $c \in (x_1, x_2)$ such that $f(x_2) - f(x_1) = f'(c)(x_2 - x_1)$. We prove the first and the proofs of other four parts are similar. Since $f'(x) > 0$ for each $x \in (a, b)$, we have $f'(c) > 0$. Since $x_1 < x_2$, we have $x_2 - x_1 > 0$. Thus, $f'(c)(x_2 - x_1) > 0$ and hence $f(x_2) - f(x_1) = f'(c)(x_2 - x_1) > 0$ or $f(x_1) < f(x_2)$. $\qquad\square$

4.3 Riemann Integration

The *Rhind papyrus* is considered to be the first historical record of formulas, written purely in a recipe manner. Those formulas were used by the Egyptians used to approximate the area and volume of some regular and irregular geometric figures. Also, a form of *trapezoidal rule* was used by the Babylonians to do astronomical calculations. However, the realisation of the area and volume as the *limit* of a sum of infinitesimal parts of the whole is attributed to Eudoxus' *method of exhaustion* (408–355 BC). No meaningful progress on Eudoxus' idea appeared for a very long period of time until *Bonaventura Cavalieri* (1598–1647) formalised the method of exhaustion in terms of *indivisibles*. The work published in "*Geometria indivisibilibus continuorum nova quadam ratione promota*" (1635) was motivated by Galieo's early work. The main result is still known as the *Cavalieri's Theorem*: *The volumes of two objects are equal if the areas of their corresponding cross-sections are in all cases equal.*

Eventually, mathematical refinements of *infinitesimal techniques* led Newton, Leibnitz, and some others, to introduce the concept of integral of a real-valued bounded function f as the *antiderivative* in the sense of Theorem 4.22. That is, the integral of a function f was taken to a function F such that $F' = f$. This *definition* remained in use until, in 1821, *Augustin-Louis Cauchy* defined the integral as a *limit of the sums* by taking the *infinitesimal* as a sequence tending to zero, which led him to introduce the notation $\int$.

Finally, motivated by the work of some earlier mathematicians such as Cauchy and Jordan, *Bernhard Riemann* developed a *theory of integration* for real-valued bounded functions defined on a finite interval $I = [a, b]$. The main geometric argument that Riemann employed in the paper "*Über die Darstellbarkeit einer Function durch eine trigonometrische Reihe*" is based on the idea of approximating the *area under the curve* $y = f(x)$, $x \in [a, b]$, by using the areas of rectangles of *infinitesimal width* over tagged partitions of the interval $[a, b]$. He presented this work to the faculty at the University of Göttingen in 1854. His collected work on integral calculus was published in 1868. The main emphasis of Riemann's new theory of integration was to replace the *antiderivative approach* with a more powerful concept of *area under a curve*, and to drop the *continuity condition*, as assumed by Cauchy.

Let $I = [a, b]$ be a finite interval, and $f : I \to \mathbb{R}$ be a *bounded function*.[2] As said before, Riemann's *geometric argument* is based on the idea of approximating the area under the curve $y = f(x)$ using collections of rectangles formed on *non-overlapping*[3] subintervals $I_k := [x_{k-1}, x_k]$ given by *partitions* of the form

$$P : \quad a = x_0 \le x_1 \le x_2 \le \ldots \le x_{n-1} \le x_n = b. \tag{4.3.1}$$

Notice that it is always possible to arrange intervals I_k such that

$$\max I_k = \min I_{k+1}, \quad \text{for} \ = 1, \ldots, n. \tag{4.3.2}$$

As usual, the *length* of the kth subinterval I_k is denoted by $\ell(I_k) := x_k - x_{k-1}$, and the height of the rectangle on I_k is taken to be given by $t_k = f(x_k^*)$, for some $x_k^* \in I_k$, with $k = 1, 2, \ldots, n$. Let

$$T_P := T_P(I) = \{t_1, t_2, \ldots, t_n\}. \tag{4.3.3}$$

We write $R(f, P)$ for the *total area of rectangles* given by a *tagged partition* $P^* := (P, T_P)$, called the *Cauchy-Riemann sum* of the function f for a tagged partition P^*

---

[2] Since we are dealing only with bounded function all through this section, a function on an interval would be assumed to be bounded.

[3] We say two intervals are *non-overlapping* if they are disjoint or intersect in a single point, which is necessarily an endpoint.

of the interval I. That is,

$$R(f, P^*) := \sum_{k=1}^{n} t_k \, \ell(I_k). \tag{4.3.4}$$

The following number is called the *mesh* of the partition P:

$$\delta(P) := \max\{\ell(I_k) : 1 \leq k \leq n\}. \tag{4.3.5}$$

According to Riemann's original formulation, we say f is *Riemann integrable* over I if there is some real number $A_f(I)$ such that

$$A_f(I) = \lim_{\delta(P) \to 0} R(f, P^*). \tag{4.3.6}$$

The real number $A_f(I)$ is called the *Riemann integral* of the function f over the interval I, which we usually write as $\int_a^b f(x)\, dx$.

4.3.1 Darboux Integral

In what follows, we use *Gaston Darboux* (1842–1917) the formulation of Riemann integral, first published in 1875. It can be shown that the Darboux formulation is equivalent to Riemann's approach to integration (Corollary 4.4). In this case, the geometric process of computing the integral of $f(x)$ over I uses two collections of rectangles formed on *non-overlapping* subintervals $I_k := [x_{k-1}, x_k]$ given by *partitions* (4.3.1) of the interval I. Equivalently, we may also write a partition P of the form (4.3.1) as a set of *distinct points* $a = x_0, x_1, \ldots, x_n = b \in I$, and use the notation

$$P := \{a = x_0 < x_1 < x_2 < \cdots < x_n = b\}.$$

Here, we write the length of the kth subinterval I_k as

$$\Delta_k := x_k - x_{k-1}, \quad k = 1, 2, \ldots, n, \tag{4.3.7}$$

and the *heights* of rectangles are given by the *extreme values*

$$m_k = \min_{x \in I_k} f(x) \quad \text{and} \quad M_k = \max_{x \in I_k} f(x), \tag{4.3.8}$$

respectively. More precisely, notice that the boundedness of $f : I \to \mathbb{R}$ implies the existence of two real numbers m, M such that

$$m \leq f(x) \leq M, \quad \text{for all } x \in I. \tag{4.3.9}$$

Fig. 4.15 Approximation of area by lower rectangles

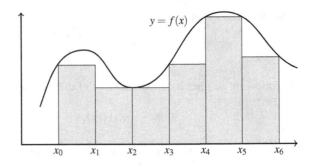

Let $\mathscr{P}[a, b]$ denotes the collection of all partitions of the interval I. As we will see later, it is safe to ignore finitely many values.[4]

$$f(a_1), \ldots, f(a_n), \quad a_k \in I.$$

For example, we can have $a_k \in I$ as points where the function f is not defined or it is discontinuous at these points.

The Darboux *lower sum* $L(f, P)$ and *upper sum* $U(f, P)$ of the function f relative to a partition P are defined, respectively, by

$$L(f, P) := \sum_{i=1}^{n} m_i \Delta x_i; \tag{4.3.10a}$$

$$U(f, P) := \sum_{i=1}^{n} M_i \Delta x_i. \tag{4.3.10b}$$

As $m \leq m_i \leq M_i \leq M$ for all $1 \leq i \leq n$, we conclude that

$$m(b - a) \leq L(f, P) = \sum_{i=1}^{n} m_i \Delta_i$$

$$\leq \sum_{i=1}^{n} M_i \Delta_i = U(f, P)$$

$$\leq M(b - a).$$

Notice that the last equation holds for each partition $P \in \mathscr{P}[a, b]$ (see Fig. 4.15).

Definition 4.10 Let P, Q be two partitions in $\in \mathscr{P}[a, b]$. We say P is *refinement* of Q if $Q \subset P$. Also, the partition $P \cup Q$ is called the *common refinement* of partitions P and Q.

---

[4] Using Lebesgue's theory of integration, as discussed in Chapter 9, we will discover that the value of the integral of f over a set $A \subset \mathbb{R}$ does not change even when the countably many *bad points* of f in I are *ignored*.

Lemma 4.1 *Let $f : [a, b] \to \mathbb{R}$ be a function, and $P, Q \in \mathscr{P}[a, b]$. If P is a refinement of Q, then*

$$L(f, Q) \leq L(f, P) \quad and \quad U(f, P) \leq U(f, Q).$$

Proof Proof left as an exercise for the reader. □

Let $f : [a, b] \to \mathbb{R}$ be a *positive function*, i.e., $f(x) \geq 0$ for all $x \in [a, b]$. The next lemma says that the total area of the *smaller rectangles $L(f, P)$* formed by a partition P underestimates the actual area under the curve, whereas the total area of the *larger rectangles $U(f, Q)$* formed by a partition Q overestimates it. Notice that, in both of these cases, the vertical line segments at partition points $x = x_i$ are not part of the total area. So, a geometric proof of the central idea of Riemann's theory is self-evident for *positive functions*.

Lemma 4.2 *If $f : [a, b] \to \mathbb{R}$ is a function and $P, Q \in \mathscr{P}[a, b]$, then $L(f, P) \leq U(f, Q)$.*

Proof Apply Lemma 4.1 to the refinement $R = P \cup Q$. □

The fact stated in Lemma 4.1 is the key idea of (Darboux) *integral*. It says that every refinement of partition *increases* the lower sums and *decreases* the upper sums. It, thus, makes sense to define the *lower (Darboux) integral* of a function f by

$$\underline{\int_a^b} f(x)dx = \sup_{P \in \mathscr{P}[a,b]} L(f, P); \tag{4.3.11a}$$

and the *upper (Darboux) integral* by

$$\overline{\int_a^b} f(x)dx = \inf_{P \in \mathscr{P}[a,b]} U(f, P). \tag{4.3.11b}$$

Observe that the *supremum* and *infimum* in these two equations are taken over all possible partition of the interval $[a, b]$, which may or may not exist for a given function. However, by Lemma 4.2, as

$$L(f, P) \leq U(f, Q), \quad \text{for any } P, Q \in \mathscr{P}[a, b],$$

it follows that

$$\sup_{P \in \mathscr{P}[a,b]} L(f, P) \leq U(f, Q), \quad \text{for a fixed } Q.$$

So, taking infimum over all partitions Q of $[a, b]$, we obtain

$$\sup_{P \in \mathscr{P}[a,b]} L(f, P) \leq \inf_{Q \in \mathscr{P}[a,b]} U(f, Q).$$

That is,

$$\int_{\underline{a}}^{b} f(x)dx \leq \int_{a}^{\overline{b}} f(x)dx. \qquad (4.3.12)$$

We say a function $f : [a, b] \to \mathbb{R}$ is *non-integrable* in Darboux sense if inequality in (4.3.12) is strict. That is, the function f is non-integrable if the lower and upper (Darboux) integrals of f are not equal. The function given in the next example provides a simple illustration.

Example 4.17 Let the function $f : [a, b] \to \mathbb{R}$ be given by

$$f(x) = \begin{cases} 1 & (x \in [a, b] \cap \mathbb{Q}) \\ 0 & (x \notin [a, b] \cap \mathbb{Q}). \end{cases}$$

As have seen before, this function is *nowhere continuous* on $[a, b]$. We assert that the function f is *not* (Darboux) integrable. For, let

$$P = \{a = x_0 < x_1 < x_2 < x_3 < \cdots < x_n = b\}$$

be a partition of $[a, b]$. Then, for each i,

$$m_i = \inf\{f(x) : x_{i-1} \leq x \leq x_i\} = 0$$

and

$$M_i = \sup\{f(x) : x_{i-1} \leq x \leq x_i\} = 1.$$

The lower (Darboux) sum is given by

$$L(f, P) = \sum_{i=1}^{n} m_i \Delta_i = \sum_{i=1}^{n} 0(x_{i-1} - x_i) = 0$$

and hence

$$\int_{\underline{a}}^{b} f(x)dx = 0.$$

Similarly, the upper (Darboux) sum is given by

$$U(f, P) = \sum_{i=1}^{n} M_i \Delta_i = \sum_{i=1}^{n} (x_{i-1} - x_i) = (b - a)$$

and hence

$$\int_{a}^{\overline{b}} f(x)dx = (b - a) \neq 0 = \int_{\underline{a}}^{b} f(x)dx,$$

which proves the assertion.

Definition 4.11 A function $f : (a, b) \to \mathbb{R}$ is called (Darboux) *integrable* if

$$\int_{\underline{a}}^{b} f(x)dx = \int_{a}^{\overline{b}} f(x)dx,$$

and the common value written as

$$\int_{a}^{b} f(x)dx$$

is called the (Darboux) *integral* of the function f. We write $\mathscr{R}(I)$ for the class of (Darboux) integrable functions defined over a bounded interval $I = [a, b]$.

Remark 4.1 The notion of Riemann integrability introduced above extends easily to define the *improper integral* of a function on any open *unbounded* interval $I \subseteq \mathbb{R}$. For, suppose $f : [a, b) \to \mathbb{R}$ is a function such that $f \in \mathscr{R}([a, b - \varepsilon])$ for any $\varepsilon \in (0, b - a)$, and

$$\int_{a}^{b-\varepsilon} f(x)dx \quad \text{converges, as} \quad \varepsilon \to 0^{+}. \tag{4.3.13}$$

Then, the **improper integral** of f over the interval $[a, b)$ is the limiting value, as given above. Similarly, we define the improper integral of a function defined over an interval $(a, b]$ or (a, b). Further, the *improper integral* of a function $f : [a, \infty) \to \mathbb{R}$ is given by

$$\int_{a}^{\infty} f(x)dx := \lim_{b \to \infty} \int_{a}^{b} f(x)dx, \tag{4.3.14}$$

provided $f \in \mathscr{R}([a, b))$, for all $b > a$, and the limit on the right side exists. A similar definition holds for the improper integral of a function defined over an interval $(-\infty, a]$ or $(-\infty, \infty)$.

Notice that the equation

$$\int_{-\infty}^{\infty} f(x)\, dx = \int_{-\infty}^{a} f(x)\, dx + \int_{a}^{\infty} f(x)\, dx$$

makes sense only when the two improper integrals on the right side converge. In this case, we say f is *integrable over* $\mathbb{R}$. It is known that the convergence of the improper integral of a function f needs not always imply the same for the function $|f|$. For example, that is the case for the *Fresnel integral* given by

$$\int_{0}^{\infty} \frac{\sin x}{\sqrt{x}} = \sqrt{\frac{\pi}{2}}. \tag{4.3.15}$$

We say a function $f : \mathbb{R} \to \mathbb{R}$ is *absolutely integrable* if the function $|f|$ is integrable over $\mathbb{R}$. It follows from the part (3) of the next theorem that *every absolutely integrable function is integrable*.

Theorem 4.10 *Suppose $f, g : [a, \infty) \to \mathbb{R}$ are integrable. Then, we have*

1. *If $c \in \mathbb{R}$, then $c + gf$ are integrable over $[a, \infty)$, and*

$$\int_a^\infty (cf + g)(x)\, dx = c \int_a^\infty f(x)\, dx + \int_a^\infty g(x)\, dx.$$

2. *If $f \geq 0$, and the set*

$$S = \left\{ \int_a^b f(x)\, dx : b \geq a \right\}$$

is bounded above, then the integral $\displaystyle\int_a^\infty f(x)\, dx$ converges to $\sup(S)$.

3. *If $|f| \leq g$, and the integral $\displaystyle\int_a^\infty g(x)\, dx$ converges, then the integral $\displaystyle\int_a^\infty f(x)\, dx$ also converges, and*

$$\left| \int_a^\infty f(x)\, dx \right| \leq \int_a^\infty g(x)\, dx.$$

Proof Left to the reader as an exercise. Notice that the proof of (3) follows by applying (2) to $f = f^+ - f^-$, where

$$f^+(x) := \max\left\{ f(x), 0 \right\} \quad \text{and} \quad f^-(x) := -\min\left\{ f(x), 0 \right\}$$

are as usual the *positive part* and the *negative part* of the function f.   □

The part (2) of Theorem 4.10 is a form of *comparison test*. Also, the *condensation test* given earlier in Theorem 3.22 is a special case of a more general statement, known as the *integral test*, as given in the next theorem.

Theorem 4.11 (Integral Test). *Suppose a continuous function $f : [1, \infty) \to \mathbb{R}$ is positive, monotonically decreasing, continuous function. Then*

$$\sum_{n=1}^{\infty} f(n) < \infty \quad \Leftrightarrow \quad \int_1^\infty f(x)\, dx < \infty.$$

Proof Left to the reader as an exercise.   □

4.3.2 Examples and Important Theorems

Let $I = [a, b]$. We first consider the function $f(x) = \alpha$, $\alpha \in \mathbb{R}$ for $x \in I$. Let $P = \{a = x_0 < x_1 < \cdots < x_n = b\}$ be a partition of I. Then, for each i, we have

$$m_i = \inf_{x \in [x_{i-1}, x_i]} f(x) = \alpha \quad \text{and} \quad M_i = \sup_{x \in [x_{i-1}, x_i]} f(x) = \alpha.$$

The lower sum is given by

$$L(f, P) = \sum_{i=1}^{n} m_i \Delta_i = \sum_{i=1}^{n} \alpha(x_{i-1} - x_i) = \alpha(x_n - x_0) = \alpha(b - a),$$

so that $\underline{\int_a^b} f(x)dx = \alpha(b - a)$. Similarly,

$$U(f, P) = \alpha(b - a) \quad \Rightarrow \quad \overline{\int_a^b} f(x)dx = \alpha(b - a).$$

Therefore, the constant function $f(x) = \alpha$ is integrable, with

$$\int_a^b f(x)dx = \alpha(b - a).$$

In certain situations, it is more convenient to work with a sequence of partitions $\langle P_n \rangle$ given by the *evenly spaced* points of the interval I. Such types of partitions are obtained by taking the *width of subintervals* h as a function of the number of partion points. That is, we take

$$h = h(n) := \frac{b - a}{n}, \quad n \geq 1.$$

So, an h-spaced partition $P_n = P_n(h)$ with n nodes is given by

$$a = x_0 < x_1 = a + h < x_2 = a + 2h < \cdots < x_n = a + nh = b.$$

Clearly then, $h \to 0$ as $n \to \infty$. Also, by (4.3.7), we have

$$\Delta x_i = x_i - x_{i-1} = \frac{b - a}{n} = h, \quad n \geq 1.$$

In this case, we write $\mathscr{P}'(I) := \{P_n : n \geq 1\}$ for the collection of all h-spaced partitions of the interval I. Notice that, as

$$\sup_{P_n \in \mathscr{P}'(I)} L(f, P_n) \leq \sup_{P \in \mathscr{P}[a,b]} L(f, P) := \int_a^b f(x)dx$$

$$\leq \int_{\underline{a}}^{\overline{b}} f(x)dx := \inf_{P \in \mathscr{P}[a,b]} U(f, P)$$

$$\leq \inf_{P_n \in \mathscr{P}'(I)} U(f, P_n),$$

it follows that a function $f : [a, b] \to \mathbb{R}$ is integrable if

$$\sup_{P_n \in \mathscr{P}'(I)} L(f, P_n) = \inf_{P_n \in \mathscr{P}'(I)} U(f, P_n)$$

and the common value is the Riemann integral of the function f.

To begin with this improved procedure, we first compute the integral $\int_a^b x\, dx$.

Example 4.18 Notice that the *identity function* $f : [a, b] \to [a, b]$ given by $f(x) = x$, for $x \in [a, b]$), is bounded as long as the interval $[a, b]$ is of finite length. For an h-spaced partition with n points

$$P_n = \{a < a + h < a + 2h < \cdots < a + nh = b\},$$

we have

$$m_i = \inf\{x : a + (i - 1)h \leq x \leq a + ih\}$$
$$= a + (i - 1)h;$$
$$M_i = \sup\{x : a + (i - 1)h \leq x \leq a + ih\}$$
$$= a + ih.$$

It, thus, follows that

$$L(f, P_n) = \sum_{i=1}^n m_i \Delta_i = \sum_{i=1}^n (a + (i - 1)h)h$$

$$= nah + h^2 \frac{n(n-1)}{2}$$

$$= a(b - a) + \frac{(b - a)^2}{n^2} \frac{n(n-1)}{2}$$

$$= \frac{b^2 - a^2}{2} - \frac{(b - a)^2}{2n}.$$

Now, as finding the supremum over all partitions $P_n \in \mathscr{P}'(I)$ is equivalent to taking the limit as $n \to \infty$, we have

$$\sup_{P \in \mathscr{P}'(I)} L(f, P) = \frac{b^2 - a^2}{2}.$$

Similarly, it follows that

$$U(f, P_n) = \sum_{i=1}^{n} M_i \Delta_i = \sum_{i=1}^{n} (a + ih)h$$

$$= nah + h^2 \frac{n(n+1)}{2}$$

$$= a(b-a) + \frac{(b-a)^2}{n^2} \frac{n(n+1)}{2}$$

$$= \frac{b^2 - a^2}{2} + \frac{(b-a)^2}{2n}.$$

As before, as finding the infimum over all partitions $P_n \in \mathscr{P}'(I)$ is equivalent to taking the limit as $n \to \infty$, we have

$$\inf_{P \in \mathscr{P}'(I)} U(f, P) = \frac{b^2 - a^2}{2}.$$

Hence, the function f is integrable over $[a, b]$, and

$$\int_a^b x\,dx = \frac{b^2 - a^2}{2}.$$

This approach to Riemann integration is also helpful in describing certain well known methods of *integral approximation* such as the *trapezoidal rule*. The basic idea is to estimate the integral $\int_a^b f(x)dx$ using a *polynomial interpolation* for the function $f(x)$ with respect to the nodal points $x_0, \ldots, x_n$ of a partition P_n of interval width $h = (b-a)/n$. In particular, considering a trapezoid over each one of these n subintervals, and using

$$m_k = M_k = \frac{f(x_{k-1} + f(x_k))}{2}, \quad k = 1, 2, \ldots, n,$$

it follows that

$$\int_a^b f(x)dx \approx h \sum_{k=1}^{n} \frac{f(x_{k-1} + f(x_k))}{2}$$

$$= h\left[\frac{f(x_0)}{2} + f(x_1) + \cdots + f(x_{n-1}) + \frac{f(x_n)}{2} \right]$$

$$:= T_n, \quad \text{say.}$$

The next theorem proves that *approximation error* for this method is least if $f \in C^2([a, b])$.

Theorem 4.12 *Suppose $f \in C^2([a, b])$, with*

$$\|f''\|_\infty := \sup\{|f''(x)| : x \in [a, b]\}.$$

Then the trapezoidal estimate T_n satisfies the condition

$$\left| \int_a^b f(x)dx - T_n \right| \leq \frac{(b-a)^3 \|f''\|_\infty}{12n^2}.$$

Proof Left to the reader as Exercise 4.14. $\square$

More generally, the integrability of a monotone function follows easily by using the class $\mathscr{P}'(I)$ of evenly spaced partitions. A proof given below is for the case when f is a monotonically decreasing function. A similar proof works when f is a monotonically increasing function.

Theorem 4.13 *Every monotone function $f : [a, b] \to \mathbb{R}$ is integrable.*

Proof Suppose f is monotonically decreasing so that $f(b) \leq f(x) \leq f(a)$ for all $x \in I = [a, b]$. Clearly, f is bounded on I. Let P_n denotes the h-spaced partition of $[a, b]$, with width

$$h = \Delta x_i = x_i - x_{i-1} = \frac{b-a}{n}, \quad n \geq 1,$$

so that we can write

$$P_n : a = x_0 < x_1 = a + h < \cdots < x_n = a + nh = b.$$

Also, in this case, we have

$$m_i = \inf\{f(x) : x_{i-1} \leq x \leq x_i\} = f(x_i); \quad \text{and,}$$
$$M_i = \sup\{f(x) : x_{i-1} \leq x \leq x_i\} = f(x_{i-1}).$$

Therefore, the lower and upper Riemann sums are given by

$$L(f, P_n) = \sum_{i=1}^n m_i \Delta_i = \frac{b-a}{n} \sum_{i=1}^n f(x_i);$$

$$U(f, P_n) = \sum_{i=1}^n M_i \Delta_i = \frac{b-a}{n} \sum_{i=1}^n f(x_{i-1}).$$

Hence, we obtain

$$U(f, P_n) - L(f, P_n) = \frac{b-a}{n} \sum_{i=1}^{n} [f(x_{i-1}) - f(x_i)]$$

$$= \frac{b-a}{n} \left[(f(x_0) - f(x_1)) + \cdots + (f(x_{n-1}) - f(x_n)) \right]$$

$$= \frac{b-a}{n} (f(b) - f(a)).$$

It, thus, follows that

$$\overline{\int_a^b} f(x)dx - \underline{\int_a^b} f(x)dx = \lim_{n\to\infty} [U(f, P_n) - L(f, P_n)] = 0,$$

which proves the assertion. $\square$

The general criteria stated in the next theorem prove helpful to check the integrability of some function. Notice that, however, this theorem does not help in identifying classes of functions that are integrable.

Theorem 4.14 *A function $f : [a, b] \to \mathbb{R}$ is integrable if and only if for each $\varepsilon > 0$ there is a partition P of $[a, b]$ satisfying*

$$U(f, P) - L(f, P) < \varepsilon. \tag{4.3.16}$$

Proof We first assume that, for each $\varepsilon > 0$, there is a partition P of $[a, b]$ such that (4.3.16) holds. It then follows that

$$\underline{\int_a^b} f(x)dx = \sup_{Q \in \mathscr{P}[a,b]} L(f, Q)$$

and

$$\overline{\int_a^b} f(x)dx = \inf_{Q \in \mathscr{P}[a,b]} U(f, Q),$$

imply that

$$L(f, P) \le \underline{\int_a^b} f(x)dx \le \overline{\int_a^b} f(x)dx \le U(f, P).$$

Therefore, we have

$$0 \le \overline{\int_a^b} f(x)dx - \underline{\int_a^b} f(x)dx \le U(f, P) - L(f, P) < \varepsilon,$$

which shows that

$$\int_a^{\overline{b}} f(x)dx - \int_{\underline{a}}^b f(x)dx = 0.$$

Hence, f is integrable. To prove the converse, suppose a function $f : [a, b] \to \mathbb{R}$ is integrable. That is,

$$\sup_{Q \in \mathscr{P}[a,b]} L(f, Q) = \int_a^b f(x)dx = \inf_{Q \in \mathscr{P}[a,b]} U(f, Q).$$

Let $\varepsilon > 0$ be given. Then, by the criterion of supremum and infimum as given in Exercise 2.4, we can find partitions P_1, P_2 of $[a, b]$ such that

$$L(f, P_1) > \int_a^b f(x)dx - \frac{\varepsilon}{2}; \tag{4.3.17}$$

$$U(f, P_2) < \int_a^b f(x)dx + \frac{\varepsilon}{2}. \tag{4.3.18}$$

Then, for $P = P_1 \cup P_2$, Lemma 4.2 and Lemma 4.1 imply that

$$\int_a^b f(x)dx - \frac{\varepsilon}{2} < L(f, P_1) \le L(f, P)$$
$$\le U(f, P) \le U(f, P_2)$$
$$< \int_a^b f(x)dx + \frac{\varepsilon}{2}.$$

Hence, $U(f, P) - L(f, P) < \varepsilon$, as required.   □

Corollary 4.4 *Suppose $f \in \mathscr{R}([a, b])$, and $\varepsilon > 0$. If*

$$P = \{a = x_0 < x_1 < \cdots < x_n = b\}$$

is a partition of $[a, b]$ such that $U(f, P) - L(f, P) < \varepsilon$, and $x_{i-1} \le x_i^* \le x_i$ are arbitrary, then

$$\left| \int_a^b f(x)dx - \sum_{i=1}^n f(x_i^*)\Delta_i \right| < \varepsilon.$$

Proof Clearly, $m_i \le f(x_i^*) \le M_i$ and hence

$$L(f, P) \le \sum_{i=1}^n f(x_i^*)\Delta_i \le U(f, P).$$

Since the function f is integrable, we have

$$L(f, P) \le \int_a^b f(x)dx \le U(f, P).$$

From the last two inequalities, we see that

$$\left| \int_a^b f(x)dx - \sum_{i=1}^n f(x_i^*)\Delta_i \right| \le U(f, P) - L(f, P) < \varepsilon.$$

$\square$

Example 4.19 Consider the function $f : [0, 1] \to \mathbb{R}$ given by $f(x) = \sqrt{x}$. By Theorem 4.14, for each $\varepsilon > 0$, we only need to find a suitable partition P_ε of the interval $[0, 1]$ that satisfies the condition of the theorem. For, since f is a *square root function*, we may consider the partition

$$P_\varepsilon : \quad 0 = x_0 < x_1 = \left(\frac{1}{n}\right)^2 < \cdots < x_{n-1} = \left(\frac{n-1}{n}\right)^2 < x_n = 1,$$

so that we have

$$U(P_\varepsilon, f) = \sum_{i=1}^n \left(\frac{i}{n}\right)\left[\left(\frac{i}{n}\right)^2 - \left(\frac{i-1}{n}\right)^2\right]$$

$$= \frac{1}{n^3}\sum_{i=1}^n [2i^2 - i];$$

$$L(P_\varepsilon, f) = \sum_{i=1}^n \left(\frac{i-1}{n}\right)\left[\left(\frac{i}{n}\right)^2 - \left(\frac{i-1}{n}\right)^2\right]$$

$$= \frac{1}{n^3}\sum_{i=1}^n [2i^2 - 3i + 1].$$

Therefore, we obtain

$$U(P_\varepsilon, f) - L(P_\varepsilon, f) = \frac{1}{n^3}\sum_{i=1}^n \left[2i - 1\right]$$

$$= \frac{1}{n^3}\left[n(n+1) - n\right] = \frac{1}{n},$$

implies that it is possible to make the difference smaller than ε as $n \to \infty$. Hence, f is integrable with value $2/3$, because both the sums approach to this value as $n \to \infty$.

Theorem 4.14 proves that every continuous function $f : [a, b] \to \mathbb{R}$ is integrable.

Theorem 4.15 *Let $[a, b]$ be a finite interval, and $f : [a, b] \to \mathbb{R}$ be continuous function. Then, f is integrable, except possibly at a or b.*

Proof A proof given here anticipates the fact that continuous function defined on a closed and bounded interval is always uniformly continuous (Theorem 5.19). So, we may assume that f is a uniformly continuous function. That is, for any $\varepsilon > 0$, we can find some $\delta > 0$ such that for any $x, y \in [a, b]$

$$|x - y| < \delta \quad \Rightarrow \quad |f(x) - f(y)| < \frac{\varepsilon}{2(a - b)}. \tag{4.3.19}$$

By the archimedean property, pick $n \in \mathbb{N}$ such that $(b - a)/\delta < n$. We write $h := (b - a)/n < \delta$. For such an n, consider the following partition of $[a, b]$:

$$P_n = \{a < a + h < a + 2h < \cdots < a + nh = b\}.$$

For $x, y \in [a + (i - 1)h, a + ih]$, by (4.3.19), we have

$$|f(x) - f(y)| < \varepsilon/(2(b - a)),$$

which implies that

$$
\begin{aligned}
M_i - m_i &= \sup\{f(x) : a + (i - 1)h \leq x \leq a + ih\} \\
&\quad - \inf\{f(y) : a + (i - 1)h \leq y \leq a + ih\} \\
&= \sup\{f(x) - f(y) : a + (i - 1)h \leq x, y \leq a + ih\} \\
&\leq \varepsilon/(2(b - a)).
\end{aligned}
$$

It, thus, follows that

$$
\begin{aligned}
U(f, P_n) - L(f, P_n) &= \sum_{i=1}^{n} (M_i - m_i)\Delta_i \\
&\leq \sum_{i=1}^{n} (\varepsilon/(2(b - a)))\Delta_i \\
&= \varepsilon/2 < \varepsilon.
\end{aligned}
$$

Hence, by Theorem 4.14, f is integrable. $\qquad\qquad\qquad\qquad\qquad\qquad\qquad\square$

Example 4.20 Let $c \in [a, b]$, and consider the function $f : [a, b] \to \mathbb{R}$ defined by

$$f(x) = \begin{cases} 1 & (x \neq c) \\ 0 & (x = c) \end{cases}.$$

Let $P = \{a = x_0 < x_1 < x_2 < x_3 < \cdots < x_n = b\}$ be a partition of $[a, b]$ such that $c \in [x_{j-1}, x_j]$, for some $1 \leq j < n$. Then, for each i, we have

$$M_i = \sup\{f(x) : x_{i-1} \leq x \leq x_i\} = 1.$$

Also, for $i \neq j$, we have

$$m_i = \inf\{f(x) : x_{i-1} \leq x \leq x_i\} = 1$$

and

$$m_j = \inf\{f(x) : x_{j-1} \leq x \leq x_j\} = 0.$$

The upper (Darboux) sum is given by

$$U(f, P) = \sum_{i=1}^{n} M_i \Delta_i = \sum_{i=1}^{n} (x_{i-1} - x_i) = x_n - x_0 = b - a$$

and hence

$$\int_{a}^{\overline{b}} f(x)dx = b - a.$$

Similarly, the lower (Darboux) sum is given by

$$L(f, P) = \sum_{i=1}^{n} m_i \Delta_i = \sum_{i=1}^{j-1}(x_{i-1} - x_i) + \sum_{i=j+1}^{n} (x_{i-1} - x_i)$$
$$= (x_n - x_0) - (x_j - x_{j-1}) = (b - a) - (x_j - x_{j-1}).$$

It thus follows that, for the partition P,

$$U(f, P) - L(f, P) = (b - a) - ((b - a) - (x_j - x_{j-1})) = x_j - x_{j-1}.$$

Given $\varepsilon > 0$, we choose a partition P with $x_j - x_{j-1} < \varepsilon$. For such a partition, we have $U(f, P) - L(f, P) < \varepsilon$ and hence the function f is integrable. The integral of the function f is given by

$$\int_{a}^{b} f(x)dx = \int_{a}^{\overline{b}} f(x)dx = b - a.$$

Notice that the argument used in the above-worked example could be extended easily to show that a function with a finite number of discontinuities is integrable. The next theorem proves a more general statement.

Theorem 4.16 *Suppose $f \in \mathcal{R}([a, b])$, and $a < c < b$. Then $f \in \mathcal{R}([a, c]) \cap \mathcal{R}([c, b])$, with*

$$\int_{a}^{b} f(x)dx = \int_{a}^{c} f(x)dx + \int_{c}^{b} f(x)dx.$$

The converse is also true.

Proof Let $\varepsilon > 0$ be given. Since the function $f : [a, b] \to \mathbb{R}$ is integrable, there is a partition Q of $[a, b]$ such that $U(f, Q) - L(f, Q) < \varepsilon$. Let P be the refinement of Q obtained by adding the point c, with $x_m = c$, so that

$$P = \{a = x_0 < x_1 < \cdots < x_m = c < x_{m+1} < \cdots < x_n = b\}.$$

Then, by Lemma 4.1,

$$L(f, Q) \leq L(f, P) \leq U(f, P) \leq U(f, Q).$$

Hence, $U(f, P) - L(f, P) \leq U(f, Q) - L(f, Q) < \varepsilon$. Let

$$P_1 = \{a = x_0 < x_1 < \cdots < x_m = c\},$$
$$P_2 = \{c = x_m < x_{m+1} < \cdots < x_n = b\}$$

so that $P = P_1 \cup P_2$. It is then easy to see that

$$L(f, P) = L(f, P_1) + L(f, P_2) \text{ and } U(f, P) = U(f, P_1) + U(f, P_2).$$

Therefore, we obtain

$$U(f, P_1) - L(f, P_1) + U(f, P_2) - L(f, P_2)$$
$$= U(f, P) - L(f, P) < \varepsilon.$$

Since $L(f, P_i) \leq U(f, P_i)$ for each i, it follows that

$$U(f, P_1) - L(f, P_1) < \varepsilon \quad \text{and} \quad U(f, P_2) - L(f, P_2) < \varepsilon.$$

This proves that f is integrable over both $[a, c]$ and $[c, b]$. So, given $\varepsilon > 0$, we can partitions P^* and Q^* respectively for $[a, c]$ and $[c, b]$ such that

$$U(f, P^*) - L(f, P^*) < \varepsilon/2 \text{ and } U(f, Q^*) - L(f, Q^*) < \varepsilon/2.$$

If $P = P^* \cup Q^*$, then

$$U(f, P) < L(f, P) + \varepsilon = L(f, P^*) + L(f, Q^*) + \varepsilon. \tag{4.3.20}$$

Using this, we have

$$\int_a^b f(x)dx = \inf_{P \in \mathscr{P}[a,b]} U(f, P) \le U(f, P)$$

$$< L(f, P) + \varepsilon$$

$$= L(f, P^*) + L(f, Q^*) + \varepsilon$$

$$\le \int_a^c f(x)dx + \int_c^b f(x)dx + \varepsilon.$$

Similarly,

$$\int_a^b f(x)dx = \sup_{P \in \mathscr{P}[a,b]} L(f, P) \ge L(f, P)$$

$$\ge U(f, P) - \varepsilon = U(f, P^*) + U(f, Q^*) - \varepsilon$$

$$\ge \int_a^c f(x)dx + \int_c^b f(x)dx - \varepsilon$$

and hence

$$\int_a^c f(x)dx + \int_c^b f(x)dx - \varepsilon \le \int_a^b f(x)dx$$

$$\le \int_a^c f(x)dx + \int_c^b f(x)dx - \varepsilon.$$

Since $\varepsilon > 0$ is arbitrary, it follows that

$$\int_a^b f(x)dx = \int_a^c f(x)dx + \int_c^b f(x)dx.$$

$\square$

A similar argument proves that, for every subinterval $[c, d] \subset [a, b]$,

$$f \in \mathscr{R}([a, b]) \quad \Rightarrow \quad f \in \mathscr{R}([c, d]),$$

In particular, it follows that every function $f : [a, b] \to \mathbb{R}$ with finite number of *discontinuities* is integrable. As mentioned earlier, Lebesgue's theory of integration proves that a function with countably many discontinuities is integrable (in Lebesgue sense). The *Dirichlet function* as given in Example 4.6 is one of many such functions that are Lebesgue integrable, but not Riemann integrable. We conclude here with a simpler formulation of Lebesgue's idea of the integral of a function that uses step functions.

Definition 4.12 A function $s : [a, b] \to \mathbb{R}$ is called a *step function* (or a *piecewise constant* function) if there is a partition

$$a = x_0 < x_1 < x_2 < \cdots < x_{n-1} < x_n = b,$$

and some numbers $a_k \in \mathbb{R}$ such that for $k = 1, \ldots, n$

$$s(x) = a_k, \quad \text{for } x \in I_k = (x_{k-1}, x_k). \tag{4.3.21}$$

In this case, we write $s \equiv (I_1, \ldots, I_n; a_1, \ldots, a_n)$.

Notice that the sum and product of two step functions are again a step function. For, if

$$s = (I_1, \ldots, I_n; a_1, \ldots, a_n) \text{ and } t = (J_1, \ldots, J_m; b_1, \ldots, b_m)$$

are two simple functions then the partition

$$P := \{ I_i \cap J_k : 1 \le i \le n, \ 1 \le i \le n \}$$

defines the sum and product of s and t. In this case, the (Darboux) lower and upper integrals of the function f over the interval $[a, b]$ are, respectively, given by

$$\underline{\int_a^b} f(x)dx = \sup_{s \le f} \left\{ \int_a^b s(x)dx : s \text{ is a simple function} \right\}; \tag{4.3.22a}$$

$$\overline{\int_a^b} f(x)dx = \inf_{f \le t} \left\{ \int_a^b t(x)dx : t \text{ is a simple function} \right\}. \tag{4.3.22b}$$

The next theorem proves that $\underline{\int_a^b} f(x)dx = \overline{\int_a^b} f(x)dx$, if $f \in \mathscr{R}([a, b])$.

Theorem 4.17 *Let* $f \in \mathscr{R}([a, b])$. *There exists an ascending sequence* $\langle s_n \rangle$ *and a descending sequence* $\langle t_n \rangle$ *of step functions such that* $s_n \le f \le t_n$ *for all* n *over* $[a, b]$, *and*

$$\int_a^b f(x)dx = \lim_{n \to \infty} \int_a^b s_n(x)dx = \lim_{n \to \infty} \int_a^b t_n(x)dx. \tag{4.3.23}$$

The converse is also true.

Proof Left as an exercise for the reader. Indeed, it is possible to show that

$$\lim_{n \to \infty} s_n = \lim_{n \to \infty} t_n = f \quad \text{over} \quad [a, b]. \qquad \square$$

Therefore, for any step function $s \in \mathscr{R}([a, b])$, we have

$$\int_a^b s(x)\, dx = \sum_{k=1}^n a_i (x_k - x_{k-1}). \tag{4.3.24}$$

Notice that all the properties of the class $\mathscr{R}([a, b])$ proved in the next sections remain valid for Darboux integrals. In particular, the following properties hold for every choice of step functions s, t defined over $[a, b]$:

$$\int_a^b (s + t)(x)dx = \int_a^b s(x)dx + \int_a^b t(x)dx; \tag{4.3.25}$$

$$\int_a^b (cs)(x)dx = c\int_a^b s(x)dx, \quad \text{where } c \text{ is a constant;} \tag{4.3.26}$$

$$s \leq t \quad \Rightarrow \quad \int_a^b s(x)dx \leq \int_a^b t(x)dx. \tag{4.3.27}$$

4.3.3 Properties of Riemann Integral

Let $I = [a, b]$ be a bounded interval. As before, we write $B(I)$ for the space of real-valued bounded functions defined over I, and $C(I) \subset B[a, b]$ is the subspace of continuous functions. Notice that, by Theorem 4.21, the integral of any $f \in B[a, b]$ belongs to the subspace $C[a, b]$. To begin with, we prove $\mathscr{R}([a, b])$ is an *additive set* with respect to pointwise addition, and use Theorem 4.14 to conclude that the (Riemann) integral operator

$$\int_I : B[a, b] \to C[a, b] \tag{4.3.28}$$

is an *additive map*. Theorem 4.21 makes the statement precise.

Theorem 4.18 *For any $f, g \in \mathscr{R}([a, b])$, the function $f + g \in \mathscr{R}([a, b])$, and*

$$\int_a^b (f(x) + g(x))dx = \int_a^b f(x)dx + \int_a^b g(x)dx.$$

Proof Here, for a given partition

$$P = \{a = x_0 < x_1 < \ldots < x_n = b\}$$

and a function $f : [a, b] \to \mathbb{R}$, we write

$$M_i(f) = \sup\{f(x) : x_{i-1} \leq x \leq x_i\}; \text{ and,}$$
$$m_i(f) = \inf\{f(x) : x_{i-1} \leq x \leq x_i\}.$$

We first show that

$$m_i(f) + m_i(g) \leq m_i(f + g).$$

For, clearly, $m_i(f) \leq f(x)$ for all $x_{i-1} \leq x \leq x_i$, and so

$$m_i(f) + m_i(g) \leq f(x) + g(x), \quad \text{for all} \quad x_{i-1} \leq x \leq x_i.$$

So, by taking the infimum over all $x \in [x_{i-1}, x_i]$, we obtain

$$m_i(f) + m_i(g) \leq m_i(f + g).$$

Therefore, we have

$$
\begin{aligned}
L(f, P) + L(g, P) &= \sum_{i=1}^{n} m_i(f)\Delta x_i + \sum_{i=1}^{n} m_i(g)\Delta x_i \\
&= \sum_{i=1}^{n} (m_i(f) + m_i(g))\Delta x_i \\
&\leq \sum_{i=1}^{n} m_i(f + g)\Delta x_i \\
&= L(f + g, P).
\end{aligned}
$$

Similarly, as $M_i(f + g) \leq M_i(f) + M_i(g)$, we have

$$U(f + g, P) \leq U(f, P) + U(g, P)$$

And hence,

$$
\begin{aligned}
L(f, P) + L(g, P) &\leq L(f + g, P) \\
&\leq U(f + g, P) \leq U(f, P) + U(g, P).
\end{aligned}
\tag{4.3.29}
$$

Now, since the functions f and g are integrable, for a given $\varepsilon > 0$ there are partitions P_1 and P_2 of $[a, b]$ such that

$$U(f, P_1) - L(f, P_1) < \frac{\varepsilon}{2} \quad \text{and} \quad U(g, P_2) - L(g, P_2) < \frac{\varepsilon}{2}.$$

Taking $P = P_1 \cup P_2$, we have

$$U(f, P) - L(f, P) < \frac{\varepsilon}{2} \quad \text{and} \quad U(g, P) - L(g, P) < \frac{\varepsilon}{2}. \tag{4.3.30}$$

Using this and inequality (4.3.29), we have

$$U(f + g, P) - L(f + g, P)$$
$$\leq U(f, P) - L(f, P) + U(g, P) - L(g, P)$$
$$< \frac{\varepsilon}{2} + \frac{\varepsilon}{2} = \varepsilon.$$

This proves the integrability of $f + g$. Further, using (4.3.29) and (4.3.30), we have

$$\int_a^b (f(x) + g(x))dx \leq U(f + g, P)$$
$$\leq U(f, P) + U(g, P)$$
$$\leq L(f, P) + L(g, P) + \varepsilon$$
$$\leq \int_a^b f(x)dx + \int_a^b g(x)dx + \varepsilon$$

and

$$\int_a^b (f(x) + g(x))dx \geq L(f + g, P)$$
$$\geq L(f, P) + L(g, P)$$
$$\geq U(f, P) + U(g, P) - \varepsilon$$
$$\geq \int_a^b f(x)dx + \int_a^b g(x)dx - \varepsilon.$$

It, therefore, follows that

$$-\varepsilon \leq \int_a^b (f(x) + g(x))dx - \left(\int_a^b f(x)dx + \int_a^b g(x)dx \right) \leq \varepsilon$$

or

$$\left| \int_a^b (f(x) + g(x))dx - \left(\int_a^b f(x)dx + \int_a^b g(x)dx \right) \right| < \varepsilon$$

showing that

$$\int_a^b (f(x) + g(x))dx = \int_a^b f(x)dx + \int_a^b g(x)dx.$$

$\square$

The next theorem proves that $\mathcal{R}([a, b])$ is closed under *scalar multiplication* with respect to pointwise operation and that the (Riemann) integral operator $\int_I :$ $B[a, b] \to C[a, b]$ is a *linear map*.

Theorem 4.19 *For any $f \in \mathcal{R}([a, b])$, and $\alpha \in \mathbb{R}$, the function $\alpha f \in \mathcal{R}([a, b])$ and*

$$\int_a^b \alpha f(x)dx = \alpha \int_a^b f(x)dx.$$

Proof For $\alpha = 0$, the result is trivial. So, we assume that $\alpha \neq 0$, and choose a partition P of $[a, b]$ such that

$$U(f, P) - L(f, P) < \varepsilon/|\alpha|.$$

For $\alpha > 0$, we have $M_i(\alpha f) = \alpha M_i(f)$, and so

$$U(\alpha f, P) = \sum_{i=1}^n M_i(\alpha f)\Delta_i = \alpha \sum_{i=1}^n M_i(f)\Delta_i = \alpha U(f, P).$$

Similarly, $L(\alpha f, P) = \alpha L(f, P)$ implies that

$$U(\alpha f, P) - L(\alpha f, P) = \alpha(U(f, P) - L(f, P)) < \varepsilon.$$

This proves that the function αf is integrable and

$$\int_a^b \alpha f(x)dx = \inf U(\alpha f, P)$$
$$= \alpha \inf U(f, P)$$
$$= \alpha \int_a^b f(x)dx.$$

Next, for $\alpha < 0$, we have $M_i(\alpha f) = \alpha m_i(f)$, and so

$$U(\alpha f, P) = \sum_{i=1}^n M_i(\alpha f)\Delta_i = \alpha \sum_{i=1}^n m_i(f)\Delta_i = \alpha L(f, P).$$

Similarly, $L(\alpha f, P) = \alpha U(f, P)$, and hence

$$U(\alpha f, P) - L(\alpha f, P) = -\alpha(U(f, P) - L(f, P))$$
$$= |\alpha|(U(f, P) - L(f, P)) < \varepsilon.$$

This proves that the function αf is integrable and

$$\int_a^b \alpha f(x)dx = \inf U(\alpha f, P)$$
$$= \inf(\alpha L(f, P))$$
$$= \alpha \sup L(f, P)$$
$$= \alpha \int_a^b f(x)dx. \qquad \square$$

The next theorem proves that the operator $\int_I$ in (4.3.28) is *monotone*.

Theorem 4.20 *Let* $f, g \in \mathcal{R}([a, b])$ *such that* $f(x) \leq g(x)$ *for all* $x \in [a, b]$. *Then*

$$\int_a^b f(x)dx \leq \int_a^b g(x)dx.$$

In particular, if $|f(x)| \leq M$ *for all* $x \in [a, b]$, *then*

$$\left| \int_a^b f(x)dx \right| \leq M(b - a).$$

Proof Let P be any partition of $[a, b]$. Since $f(x) \leq g(x)$ for all $x \in [a, b]$, we have $m_i(f) \leq m_i(g)$ and hence $L(f, P) \leq L(g, P)$. Using this, we see that

$$\int_a^b f(x)dx = \sup L(f, P)$$
$$\leq \sup L(g, P)$$
$$= \int_a^b g(x)dx.$$

If $|f(x)| \leq M$, then $-M \leq f(x) \leq M$ and hence

$$-M(b - a) = \int_a^b -Mdx \leq \int_a^b f(x)dx \leq \int_a^b Mdx = M(b - a),$$

or, equivalently,

$$\left| \int_a^b f(x)dx \right| \leq M(b - a).$$

$\square$

Theorem 4.21 (Fundamental Theorem of Calculus—I). *Suppose* $f \in \mathcal{R}([a, b])$, *and let the function* $F : [a, b] \to \mathbb{R}$ *be given by*

$$F(x) = \int_a^x f(x)dx.$$

Then $F \in C[a, b]$. *Further, if* f *is continuous at* $x = x_0 \in (a, b)$ *then the function* F *is differentiable at* $x = x_0$, *with* $F'(x_0) = f(x_0)$.

Proof By Theorem 4.16, we see for $a \leq x_0 < x$, and so for $a \leq x_0 \leq b$, that

$$F(x) - F(x_0) = \int_a^x f(x)dx - \int_a^{x_0} f(x)dx = \int_{x_0}^x f(x)dx. \tag{4.3.31}$$

Since an integrable function f is necessarily bounded (see (4.3.18)), we have $|f(x)| \leq M$, for all $x \in [a, b]$, and some $M > 0$. So, by applying Theorem 4.20 and Eq. (4.3.31), we have

$$|F(x) - F(x_0)| \leq \int_{x_0}^{x} |f(x)| dx \leq M|x - x_0|.$$

For a given $\varepsilon > 0$, let $\delta = \varepsilon/M$. Then $\delta > 0$ and

$$|x - x_0| < \delta \Rightarrow |F(x) - F(x_0)| < \varepsilon.$$

This shows that the function F is (uniformly) continuous on $[a, b]$. Let the function f be continuous at $x = x_0 \in [a, b]$. For a given $\varepsilon > 0$, choose $\delta > 0$ such that

$$|x - x_0| < \delta \Rightarrow |f(x) - f(x_0)| < \varepsilon.$$

Eq. (4.3.31) shows that

$$\frac{F(x) - F(x_0)}{x - x_0} - f(x_0) = \frac{1}{x - x_0} \int_{x_0}^{x} (f(x) - f(x_0)) dx.$$

Hence, by using Theorem 4.20, we see that

$$\left| \frac{F(x) - F(x_0)}{x - x_0} - f(x_0) \right| = \left| \frac{1}{x - x_0} \int_{x_0}^{x} (f(x) - f(x_0)) dx \right|$$

$$\leq \frac{1}{|x - x_0|} \int_{x_0}^{x} |f(x) - f(x_0)| dx$$

$$< \varepsilon$$

whenever $|x - x_0| < \delta$. This proves the function F is differentiable at x_0 and $F'(x_0) = f(x_0)$. $\square$

Theorem 4.22 (Fundamental Theorem of Calculus—II). *If a function $F : [a, b] \to \mathbb{R}$ is differentiable and its derivative $f(x) = F'(x)$ is continuous on $[a, b]$, then $f \in \mathscr{R}([a, b])$ and*

$$\int_{a}^{b} f(x) dx = \int_{a}^{b} F'(x) dx = F(b) - F(a).$$

Proof Since f is continuous, by Theorem 4.15, it is integrable. For a given $\varepsilon > 0$, choose a partition $P = \{a = x_0 < x_1 < \cdots < x_n = b\}$ of $[a, b]$ such that $U(f, P) - L(f, P) < \varepsilon$. Apply the Mean Value Theorem (Corollary 4.2) on F to get a point $x_{i-1} < t_i < x_i$ such that

$$F(x_i) - F(x_{i-1}) = F'(t_i)(x_i - x_{i-1}) = f(t_i)\Delta_i.$$

This show that

$$\sum_{i=1}^{n} f(t_i)\Delta_i = \sum_{i=1}^{n}(F(x_i) - F(x_{i-1}))$$
$$= F(x_n) - F(x_0) = F(b) - F(a).$$

This together with Corollary 4.4 shows that

$$\left| \int_a^b f(x)dx - (F(b) - F(a)) \right| < \varepsilon.$$

Since $\varepsilon > 0$ is arbitrary, we have

$$\int_a^b f(x)dx = F(b) - F(a).$$

□

Exercises 4

4.1. [1] Show that there exists no nonzero continuous functions $f, g, h : \mathbb{R} \to \mathbb{R}$ satisfying the condition

$$h(f(x) + g(y)) = xy, \quad \text{for all} \quad x, y \in \mathbb{R}.$$

4.2. Show that the function $f : \mathbb{R} \to \mathbb{R}$ given by

$$f(x) = \begin{cases} x, & \text{if } x \in \mathbb{Q} \\ -x, & \text{if } x \in \mathbb{R} \setminus \mathbb{Q} \end{cases}$$

is continuous only at $x = 0$.

4.3. Show that the function $f : \mathbb{R} \to \mathbb{R}$ given by

$$f(x) = \begin{cases} \sin x, & \text{if } x \in \mathbb{Q}; \\ -\sin x, & \text{if } x \in \mathbb{R} \setminus \mathbb{Q} \end{cases}$$

is continuous only at $x = n\pi, n \in \mathbb{Z}$.

4.4. (*Young's Inequality*) For a real number $1 < p < \infty$, we say $q \in \mathbb{R}$ is a *conjugate exponent* of p if

$$\frac{1}{p} + \frac{1}{q} = 1, \quad \text{with} \quad 1 < q < \infty.$$

For $p = 1$, we take $q = \infty$ as conjugate exponent. Show that, for any non-negative real numbers a, b,

$$ab \leq \frac{1}{p} a^p + \frac{1}{q} b^q;$$

and that the equality holds if and only if $a^p = b^q$.

4.5. Consider a function $f : [c, d] \to \mathbb{R}$ and $a \in [c, d]$. For $\delta > 0$, take

$$I_\delta(a) = \{ x \in [c, d] : |x - a| < \delta \}.$$

Then the number

$$M(a, f; \delta) = \sup_{x, y \in I_\delta(a)} \{ |f(x) - f(y)| \}$$

gives all oscillations of f on $I_\delta(a)$, and $o(f, a) := \lim_{\delta \to 0^+} M(a, f; \delta)$ is the *oscillation* of f at the point a. Show that

a. f is continuous at a if and only if $o(f, a) = 0$;
b. if f is monotone increasing, and x_i are n distinct partition points of $[c, d]$, then

$$\sum_{i=1}^{n} o(f, x_i) < f(d) - f(c).$$

4.6. For $1 \leq k \leq n$, find the kth derivative of the function $f : [-a, a] \to \mathbb{R}$ given by $f(x) = x^n |x|$, $n \in \mathbb{N}$. Does $(n + 1)$st derivative of f exist?

4.7. (Darboux) Let a function $f : [a, b] \to \mathbb{R}$ be differentiable, and $f'(a) < \lambda < f'(b)$. Show that there is an $c \in (a, b)$ such that $f'(c) = \lambda$.

4.8. Supply a proof of Corollary 4.3.

4.9. Suppose the three functions $f, g, h : \mathbb{R} \to \mathbb{R}$ are differentiable on (a, b), and continuous on $[a, b]$, show that there is some $c \in (a, b)$ such that the function $D : [a, b] \to \mathbb{R}$ given by

$$D(x) = \begin{vmatrix} f(x) & g(x) & h(x) \\ f(a) & g(a) & h(a) \\ f(b) & g(b) & h(b) \end{vmatrix}$$

satisfies $D'(c) = 0$. Hence deduce Cauchy's and Lagrange's Mean Value Theorems. (Use Rolle's theorem)

4.10. Suppose $f : \mathbb{R} \to \mathbb{R}$ be an *infinitely differentiable* function such that $f(0) f'(0) \geq 0$, and $f(x) \to 0$ as $x \to \infty$. Show that there exists a strictly increasing sequence $\langle x_n \rangle$ of positive numbers such that $f^{(n)}(x_n) = 0$ for all $n \in \mathbb{N}$.

4.11. Prove the equation (4.3.15). Further, show that the function $f : [0, \infty) \to \mathbb{R}$ given by $f(x) = \sin(x)/\sqrt{x}$ is not absolutely integrable.

4.12. Supply a complete proof for Theorem 4.10.

4.13. Supply a complete proof for Theorem 4.11, and hence show that integral test implies the condensation test.

4.14. Supply a complete proof of Theorem 4.12.

4.15. A function $f : [a, b] \to \mathbb{R}$ is said to be of *bounded variations* if

$$\sup_{P \in \mathscr{P}([a,b])} \left| f(x_{k+1} - f(x_k)) \right| < \infty.$$

Show that any such $f \in \mathscr{R}([a, b])$.

4.16. Show that, for any $f \in \mathscr{R}([a, b])$,

$$\lim_{n \to \infty} \frac{1}{n} \sum_{k=1}^{n} f\left(\frac{k}{n}\right) = \int_{a}^{b} f(x)dx.$$

4.17. Show that $f \in \mathscr{R}([a, b]) \Rightarrow |f| \in \mathscr{R}([a, b])$, and

$$\left| \int_{a}^{b} f(x)dx \right| \leq \int_{a}^{b} |f(x)|dx.$$

[You might need the inequality: $||x| - |y|| \leq |x - y|$ to prove $M_i(|f|) - m_i(|f|) \leq M_i(f) - m_i(f)$]

4.18. Show that $f \in \mathscr{R}([a, b]) \Rightarrow f^n \in \mathscr{R}([a, b])$ for each $n \in \mathbb{N}$. [Prove $M_i(f^n) - m_i(f^n) \leq nM^{n-1}(M_i(f) - m_i(f))$, where $|f(x)| \leq M$]

4.19. Show that $f, g \in \mathscr{R}([a, b]) \Rightarrow \max\{f, g\}$ and $\min\{f, g\} \in \mathscr{R}([a, b])$.

4.20. Prove the two assertions in equation (4.3.22), and hence Theorem 4.17.

4.21. Suppose $f \in \mathscr{R}([a, b])$, with $c \leq f(x) \leq d$, and $g \in C[c, d]$. Show that $g \circ f \in \mathscr{R}([a, b])$.

4.22. Suppose $f \in \mathscr{R}([a, b])$, and $a < c < b$. Show that the function $F : [a, b] \to \mathbb{R}$ given by $F(x) = \int_{a}^{x} f(x)dx$ has both left and right derivatives at $x = c$, and they are equal respectively to the left and right limit of f at $x = c$.

4.23. Suppose $f \in C[a, b]$ such that $f(x) \geq 0$ for all $x \in [a, b]$ and $\int_{a}^{b} f(x)dx = 0$. Show that $f(x) = 0$ for all $x \in [a, b]$. Can we relax the condition $f(x) \geq 0$?

4.24. [1] Suppose $f \in C([0, 1])$ and $g : \mathbb{R} \to \mathbb{R}$ is a periodic function of period 1. Show that

$$\lim_{n \to \infty} \int_{0}^{1} f(x)g(nx)dx = \int_{0}^{1} f(x)dx \int_{0}^{1} g(x)dx.$$

4.25. Suppose $f, g \in C([0, 1])$ are monotonically increasing functions. Show that

$$\int_{0}^{1} f(x)dx \int_{0}^{1} g(x)dx \leq \int_{0}^{1} f(x)g(x)dx.$$

4.26. Suppose $f : [0, 1] \to \mathbb{R}$ is a continuously differentiable function. Show that

$$\sum_{k=1}^{n} f\left(\frac{k}{n}\right) - n \int_0^1 f(x)dx \;\to\; \frac{f(1) - f(0)}{2}, \quad \text{as } n \to \infty.$$

4.27. Show that

$$\lim_{n \to \infty} \sqrt{n} \int_{-\infty}^{\infty} \frac{dx}{(1+x^2)^n} = \sqrt{\pi}.$$

4.28. Show that

$$\int_0^{\infty} \frac{e^{-x/a-(1/x)}}{x} \, dx \sim \ln a, \quad \text{as } a \to \infty.$$

4.29. [1] Suppose $f \in C(\mathbb{R})$ is such that the imporper integral $\int_{-\infty}^{\infty} |f(x)|dx$ converges. Show that

$$\lim_{n \to \infty} \int_{-\infty}^{\infty} f(x)| \sin nx|dx = \frac{2}{\pi} \int_{-\infty}^{\infty} f(x)dx$$

4.30. [1] Show that

$$\gamma = \int_0^1 \frac{1 - \cos x}{x} \, dx - \int_1^{\infty} \frac{\cos x}{x} \, dx,$$

where γ is the *Euler constant*. [Take $c_n = \int_0^{n\pi} \frac{1-\cos x}{x}dx - \ln(n\pi)$, $n \in \mathbb{N}$]

Reference

1. M. Hata, *Problems and Solutions in Real Analysis* (World Scientific, 2007)

Chapter 5
Topology of the Real Line

Mathematics is the art of giving the same name to different things.

Henri Poincaré (1854–1912)

In the previous chapter, we used *modulus metric* on $\mathbb{R}$ to discuss *locally* the three fundamental analytical properties of a function $f : [a, b] \to \mathbb{R}$. Most arguments applied therein used intervals of the form $I_\epsilon(c) := (c - \epsilon, c + \epsilon)$, called an ϵ-neighbourhood at $x = c \in (a, b)$. We say a set $N(c) \ni c$ is a neighbourhood of a point $c \in \mathbb{R}$ if it contains an ϵ-neighbourhood $I_\epsilon(c)$, for some $\epsilon > 0$. In general, a *system of neighbourhoods* $\mathcal{N}_x$ given at each point x of a set X provides a mathematical foundation to intuitive notion of *proximity of elements* in the set X. We say $U \subseteq X$ is an *open set* if, for each $x \in U$, we can find some $N \in \mathcal{N}_x$ such that $x \in N \subseteq U$. The collection $\mathscr{T}$ of all open sets in the set X is called a *topology* defined by the system of neighbourhoods $\mathcal{N}_x$, $x \in X$. In particular, a system of neighbourhoods $\mathcal{N}_x$ in a metric space (X, d) given by the sets of the form

$$B(x; r) := \{y \in X : d(x, y) < r\}, \qquad r > 0,$$

defines a topology on X, called the *metric topology*. We discuss here *metric topology* (also known as the *usual topology*) on the set $\mathbb{R}$ given by the modulus metric. The set $\mathbb{R}$ with usual topology is called the *real line*, where the intervals at a point form a natural choice for a system of neighbourhoods. The general case is discussed in the next chapter.

© The Author(s), under exclusive license to Springer Nature Singapore Pte Ltd. 2022
A. K. Razdan and V. Ravichandran, *Fundamentals of Analysis with Applications*,
https://doi.org/10.1007/978-981-16-8383-1_5

We say a property of a metric space is *topological* if it remains stable under *continuous deformations*.[1] Our main concern here is to discuss some important topological properties of the real line $\mathbb{R}$. The topological concepts such as convergence, compactness, and connectedness, discussed here are very helpful to understand the similar concepts developed in the next chapter for higher dimensional Euclidean spaces, and also for some infinite dimensional function spaces.

5.1   Open and Closed Set

As said above, we shall be working although with the *modulus metric* on $\mathbb{R}$, as introduced earlier in Corollary 2.2. The sets $\mathbb{N}$ and $\mathbb{Q}$ are also treated as metric spaces with respect to the *induced metric* obtained by restricting the modulus metric to these sets. Recall that, by Corollary 2.3, for any $a, x \in \mathbb{R}$ and $\epsilon > 0$,

$$|x - a| < \epsilon \quad \Leftrightarrow \quad a - \epsilon < x < a + \epsilon. \qquad (5.1.1)$$

So, for any $a \in \mathbb{R}$ and $\epsilon > 0$, we write

$$x \in (a - \epsilon, a + \epsilon) \quad \Leftrightarrow \quad |x - a| < \epsilon.$$

The interval $(a - \epsilon, a + \epsilon)$ is called an ϵ-*neighbourhood* centred at the point a, which we shall denote by $I_\epsilon(a)$. Notice that, we have

$$I_\epsilon(a) = \{x \in \mathbb{R} : |x - a| < \epsilon\}. \qquad (5.1.2)$$

More generally, we say a set $N(a) \ni a$ is a *neighborhood* of the point a if $I_\epsilon(a) \subseteq N(a)$, for some $\epsilon > 0$ (Hausdorff, 1914). For $a \in \mathbb{R}$, we write $\mathcal{N}_a$ for class of all neighbourhoods at the point a, and let

$$\mathcal{N} = \bigcup_{a \in \mathbb{R}} \mathcal{N}_a.$$

Trivially, the collection $\mathcal{N}$ satisfies the following properties:

1. For each $N \in \mathcal{N}_a$, we have $a \in N$;
2. If $N_1, N_2 \in \mathcal{N}_a$ then there is some $N \in \mathcal{N}_a$ such that $N \subseteq N_1 \cap N_2$;
3. If $N \in \mathcal{N}_a$ and $M \in \mathcal{N}_b$ then, for each $c \in N \cap M$, there is some $K \in \mathcal{N}_c$ such that $K \subset N \cap M$.

The collection $\mathcal{N}$ is a *neighborhood system* of $\mathbb{R}$, with the collections $\mathcal{N}_a$ as a *neighbourhood basis* at the point $a \in \mathbb{R}$. Notice that, by taking $a = b = c$ in (3)

---

[1] A *deformation* is a continuous map of a topological space that is independent of its *shape*. In geometrical terms, a deformation is given by bending, twisting, stretching, or even crumpling a space, but not by gluing together or tearing.

above, the property (2) follows. So, to show that an arbitrary collection of subsets of $\mathbb{R}$ is a neighbourhood system, it suffices to verify (1) and (3). Notice that the collection

$$\mathcal{N}_a = \left\{ \left[a - \frac{1}{n}, a + \frac{1}{n} \right] : n \in \mathbb{N} \right\}$$

is also a *basis* for a *neighborhood system* of $\mathbb{R}$. It can be shown that, for any *neighbourhood system* $\mathcal{N}$ of $\mathbb{R}$, the collection

$$\mathcal{T} = \left\{ U \subseteq \mathbb{R} : \text{for each } a \in U \text{ there is some } N \in \mathcal{N} \text{ with } N \subset U \right\}$$

defines a topology on $\mathbb{R}$ in the sense of the next definition.

Definition 5.1 (*Kuratowskii, 1922*) A collection $\mathcal{T}$ of subsets of a nonempty set X defines a *topology* if

1. $\emptyset, X \in \mathcal{T}$;
2. $\mathcal{T}$ is closed under arbitrary unions; and,
3. $\mathcal{T}$ is closed under finite intersections.

As said earlier, the topology defined by the neighbourhood system $\mathcal{N}$ of intervals of $\mathbb{R}$ is called the *usual topology* on $\mathbb{R}$. We understand the geometry of the usual topology on $\mathbb{R}$ by specifying the *relative position* of points in $\mathbb{R}$ with respect to a given set $A \subset \mathbb{R}$. We need to consider the following three possibilities.

Definition 5.2 Let $A \subseteq \mathbb{R}$. A point $a \in \mathbb{R}$ is called an *interior point* of the set A if there is some $I(a) \in \mathcal{N}$ entirely contained in A. And, a point $a \in \mathbb{R}$ is called an *exterior point* of A if there is some $I(a) \in \mathcal{N}$ entirely contained in $A^c = \mathbb{R} \setminus A$. Finally, a point $a \in \mathbb{R}$ is called a *boundary point* of A if every $I(a) \in \mathcal{N}$ meets both the sets A and A^c.

For example, taking $A = [a, b]$, the end points a and b are the boundary points of A; every point in the open interval (a, b) is an interior point of A; and, every point in the set

$$\mathbb{R} \setminus [a, b] = (-\infty, a) \cup (b, \infty)$$

is an interior point of A. In general, we write A^o for the set of interior points of a set $A \subseteq \mathbb{R}$. The next theorem proves the second assertion.

Theorem 5.1 *Every point of an open interval in $\mathbb{R}$ is an interior point.*

Proof Consider an open interval $I = (a, b)$, with $\ell(I) = b - a > 0$. For any $x \in I$, let $\epsilon = \min\{x - a, b - x\}$. We assert that the ϵ-neighbourhood $I_\epsilon(x)$ is contained in I. For, let $y \in I_\epsilon(x)$. Then,

$$|x - y| < \epsilon \leq |x - a| \quad \text{and} \quad |x - y| < \epsilon \leq |x - b|.$$

This implies that $I_\epsilon(x) \subset I$, i.e., x is an interior point of $I_\epsilon(a)$. $\qquad\square$

On the other hand, since every interval contains both rational and irrational numbers, the interior of the set $\mathbb{Q} \subset \mathbb{R}$ is empty. It thus follows that the boundary of the set $\mathbb{Q}$ in $\mathbb{R}$ is whole of the set $\mathbb{R}$. This simple illustration suggests that a common understanding about a topological notion may defy all our geometric imaginations.

Definition 5.3 (*Weierstrass, Hausdorff*) We say $U \subset \mathbb{R}$ is an *open set* if each $u \in U$ is an interior point of U. That is, for each $u \in U$, there is an ϵ-neighbourhood $I_\epsilon(u)$ entirely contained in U.

Clearly, the set $\mathbb{R}$ is open because every interval is a subset of $\mathbb{R}$. Also, the empty set $\emptyset$ is open; for, if not, then there is an $x \in \emptyset$ (this is not possible) which is not an interior point of $\emptyset$. By Theorem 5.1, every open interval is an *open set*.

Theorem 5.2 *Arbitrary unions and finite intersections of open subsets of $\mathbb{R}$ are open.*

Proof Let $\{U_\alpha : \alpha \in I\}$ be an indexed family of open sets. We need to show that $U = \cup_{\alpha \in I} U_\alpha$ is open. Let $a \in U$. Then $a \in U_\alpha$ for some $\alpha \in I$. Since U_α is open, there is an interval $(a - \epsilon, a + \epsilon) \subset U_\alpha$. Since $U_\alpha \subset \cup_{\alpha \in I} U_\alpha = U$, we have $(a - \epsilon, a + \epsilon) \subset U$ and hence U is open. Let $V_1, V_2, \ldots, V_m$ be open sets and let $V = \cap_{k=1}^{m} V_k$. Let $a \in V$. Then $a \in V_k$ for every k with $1 \leq k \leq m$. Since V_k is open, there are intervals $(a - \epsilon_k, a + \epsilon_k)$ such that $(a - \epsilon_k, a + \epsilon_k) \subset V_k$, $1 \leq k \leq m$. Let $\epsilon = \min\{\epsilon_1, \epsilon_2, \ldots, \epsilon_k\}$. So, $\epsilon > 0$, and $(a - \epsilon, a + \epsilon) \subset V_k$, for each k. Thus, $(a - \epsilon, a + \epsilon) \subset \cap_{k=1}^{m} V_k = V$. Hence, V is open. $\qquad\square$

However, the arbitrary intersection of open sets may not be open (Exercise 5.12). The discussion above proves that the collection of open sets in $\mathbb{R}$ forms a topology in the sense of Definition 5.1, which is known as the *usual topology* on $\mathbb{R}$. Notice that, using the argument as in the proof of Theorem 5.1, it follows easily that the usual topology on $\mathbb{R}$ is *Hausdorff*. That is, for every pair $a \neq b \in \mathbb{R}$, we can find neighbourhoods $I_a, I_b \in \mathcal{N}$ such that $I_a \cap I_b = \emptyset$.

Definition 5.4 A collection $\mathcal{B}$ of open subsets in a topological space $(X, \mathcal{T})$ is called a *base* for the topology $\mathcal{T}$ if every nonempty open set $U \subset X$ can be expressed as a union of some members of $\mathcal{B}$.

The next theorem proves that the collection $\mathcal{N}$ of all intervals (of every type) of $\mathbb{R}$ is a *base* for the usual topology on $\mathbb{R}$.

Theorem 5.3 *Every open set $U \subset \mathbb{R}$ can be expressed uniquely as a countable union of disjoint open intervals.*

Proof Consider the collection of intervals

$$\mathscr{I} = \bigcup_{x \in U} \{I_x \subset U : I_x \text{ is maximal}\}.$$

Notice that we can choose $I_x = (a_x, b_x)$, with $-\infty \leq a_x < b_x \leq \infty$, where

$$a_x = \inf\{a < x : (a, x) \subset U\}; \quad \text{and}$$
$$b_x = \sup\{b > x : (x, b) \subset U\}.$$

By maximality, $I_x \cap I_y \neq \emptyset$ implies $I_x = I_y$ and the countability of the collection $\mathscr{I}$ follows by considering only rational interior point for each interval. $\square$

Definition 5.5 A point $x \in \mathbb{R}$ is called a *limit point* of a set $S \subset \mathbb{R}$ if every ϵ-neighbourhood of the point x contains a point of S, other that x, in case $x \in S$. And, a point $x \in S$ is called an *isolated point* if it is not a limit point of S.

So, a point $x \notin S$ is a limit point of S if every ϵ-neighbourhood of x intersects S. That is,

$$S \cap I_\epsilon(x) \setminus \{x\} \neq \emptyset, \quad \text{for each} \quad \epsilon > 0.$$

Recall that the set $I_\epsilon(x) \setminus \{x\}$ is called a *punctured interval* centred at x. The set of all limit points of a set $S \subset X$ is called the *derived set* of S, usually denoted by S', and the *closure* $\overline{S}$ of a set S is the set $S \cup S'$. Notice that a limit point of S may or may not belong to S.

Example 5.1 Notice that every ϵ-neighbourhood centred at 0 is an interval of the form $(-\epsilon, \epsilon)$, which contains $1/n$, for some $n \in \mathbb{N}$, by the archimedean property of $\mathbb{R}$. It thus follows that 0 is a limit point of the set

$$S = \left\{ \frac{1}{n} : n \in \mathbb{N} \right\}.$$

Therefore, $S' = \{0\}$, and $\overline{S} = S \cup \{0\}$. On the other hand, for $I = (0, 1)$ and $\epsilon > 0$, every interval $(-\epsilon, \epsilon)$ contains infinitely many points of I. So, 0 is a limit point of I, which is not in I. Similarly, 1 is a limit point of I that does not belong to I. Every point of I is clearly a limit point of S. However, no point of the set $\mathbb{R} \setminus [0, 1]$ is a limit point of I. Therefore, in this case, $\overline{S} = S' = [0, 1]$.

Theorem 5.4 *Let $S \subset \mathbb{R}$. Then every ϵ-neighbourhood centred at a limit point of the set S contains infinitely many points of S. In particular, no finite subsets of $\mathbb{R}$ has a limit point.*

Proof Let a be a limit point of S and $I_\epsilon(a)$ be an ϵ-neighbourhood centred at a. If possible, suppose $I_\epsilon(a) \setminus \{a\}$ has only finitely many points of S, say, $b_1, b_2, \ldots, b_m$. Let

$$\rho = \min\{|a - b_1|, |a - b_2|, \ldots, |a - b_m|\}.$$

Then $\rho > 0$ and the ρ-neighbourhood $I_\rho(a) \setminus \{a\}$ has no point of S. Therefore, the point a cannot be a limit point of S. This contradiction proves that $I_\epsilon(a) \setminus \{a\}$ has infinitely many points of S, and hence so does $I_\epsilon(a)$. $\square$

Definition 5.6 We say $F \subset \mathbb{R}$ is a *closed set* if its complement $U = \mathbb{R} \setminus F$ is an open set.

Theorem 5.5 (Cantor, 1884) *A set $F \subseteq \mathbb{R}$ is closed if and only if F contains all its limit points, that is, $F' \subset F$.*

Proof Let F be closed so that $U = \mathbb{R} \setminus F$ is open. Let x be a limit point of F. To show that $x \in F$, it suffices to prove this by showing $x \notin U$. If possible, suppose $x \in U$. Then, as U is open, x is an interior point of U and hence there is an ϵ-neighbourhood $I_\epsilon(x)$ centred at x such that $I_\epsilon(x) \subset U$. The ϵ-neighbourhood $I_\epsilon(x)$ has no point in common with F and so x cannot be a limit point of F. This contradiction shows that $x \notin U$, and so $x \in F$. Conversely, suppose a set $F \subseteq \mathbb{R}$ contains all its limit points. To show that F is closed, it suffices to prove that $U = \mathbb{R} \setminus F$ is an open set. For, let $x \in U$. Since F contains all its limit points, so x cannot be a limit point of F and hence there is an ϵ-neighbourhood $I_\epsilon(x)$ such that $I_\epsilon(x) \cap F = \emptyset$. Therefore, $I_\epsilon(x) \subset U$ i.e., x is an interior point of U. Thus, U is open. Hence, F is closed. $\square$

For example, by Theorem 5.4, every finite subsets of $\mathbb{R}$ is a closed set. Also, by our discussion in Example 5.1, it follows that the set

$$F = \left\{ \frac{1}{n} : n \in \mathbb{N} \right\} \bigcup \{0\}.$$

is closed. In general, if F is a closed set and $u \notin F$, then there is some $\epsilon > 0$ such that $I_\epsilon(u) \cap F = \emptyset$, and hence $u \notin F'$.

Theorem 5.6 *For every set $F \subseteq \mathbb{R}$, its closure $\overline{F}$ is a closed set.*

Proof In view of Definition 5.6, it suffices to show that $U = \mathbb{R} \setminus \overline{F}$ is an open set. Recall that, by definition, $\overline{F} = F \cup F'$. So, by using de Morgan's law, we have

$$\mathbb{R} \setminus \overline{F} = \mathbb{R} \setminus (F \cup F') = (\mathbb{R} \setminus F) \cap (\mathbb{R} \setminus F').$$

Now, for $x \in U$, we have $x \in \mathbb{R} \setminus F$ and $x \in \mathbb{R} \setminus F'$ so that $x \notin F$ and $x \notin F'$. Thus, since $x \notin F'$, x is not a limit point of F, and so there is an ϵ-neighbourhood $I_\epsilon(x)$ of x that has no point of F other than x. Since $x \notin F$, $I_\epsilon(x)$ has no point of F and hence $I_\epsilon(x) \subset \mathbb{R} \setminus F$. We assert that $I_\epsilon(x) \subset \mathbb{R} \setminus F'$. For, if $y \in I_\epsilon(x)$, then ρ-neighbourhood $I_\rho(y)$, with $\rho = \epsilon - |x - y|$, is contained in $I_\epsilon(x)$ and hence the ρ-neighbourhood $I_\rho(y)$ does not intersect F. Thus, y is not a limit point of F. Therefore, $y \notin F'$ or $y \in \mathbb{R} \setminus F'$. Thus, we have shown that

$$I_\epsilon(x) \subset \mathbb{R} \setminus F \quad \text{and} \quad I_\epsilon(x) \subset \mathbb{R} \setminus F',$$

which shows that
$$I_\epsilon(x) \subset (\mathbb{R} \setminus F) \cap (\mathbb{R} \setminus F') = \mathbb{R} \setminus \overline{F}.$$

Thus, x is an interior point of $\mathbb{R} \setminus \overline{F}$ and hence $\mathbb{R} \setminus \overline{F}$ is an open set. Therefore, $\overline{F}$ is a closed set. $\square$

Corollary 5.1 *A set $F \subseteq \mathbb{R}$ is closed if and only if $F = \overline{F}$.*

Proof If $F = \overline{F}$, then F is a closed set by the previous theorem. On the other hand, if F is closed, then it contains all its limit points, i.e., $F' \subset F$. Therefore, we have $F = F \cup F' = \overline{F}$. □

In particular, we deduce that $\mathbb{R}$ and $\emptyset$ are closed sets. In general, a subset of a topological space is called a *clopen* if it is both open and closed.

Theorem 5.7 *Arbitrary intersections and finite unions of closed subsets in $\mathbb{R}$ are closed.*

Proof A proof follows directly from Theorem 5.2, by using de Morgan's laws. □

Notice that, however, the arbitrary union of closed sets is not closed, in general. For example, consider the sequence of closed sets $F_n = [0, 1/n]$, $n \in \mathbb{N}$. Clearly then, we have

$$\bigcup_{n=1}^{\infty} F_n = [0, 1),$$

which is not a closed set because 1 is a limit point of this union.

Example 5.2 Consider the unit interval $I = [0, 1]$, and let $x_n = 1/n$. Then, for any $\epsilon > 0$, we can find a natural number N such that $N > 1/\epsilon$, by using the archimedean property of $\mathbb{R}$. So, for $n \geq N$, we have

$$|x_n - 0| = 1/n \leq 1/N < \epsilon.$$

Therefore, $x_n = 1/n \to 0 \in I$. Now, let $J = (0, 1]$. If possible, suppose $x_n = 1/n \to \ell \in J$. Then, for any $0 < \epsilon < \ell$, there is a natural number N such that

$$|x_n - \ell| = |1/n - \ell| < \epsilon \quad \Leftrightarrow \quad \ell - \epsilon < 1/n < \ell + \epsilon.$$

In particular, we have $\ell - \epsilon < 1/n$, for all $n \geq N$. Since $\ell - \epsilon > 0$, by the archimedean property of $\mathbb{R}$, there is an integer m such that $(\ell - \epsilon)m > 1$. Take any $k \geq \max\{N, m\}$. Then, $k \geq N \Rightarrow \ell - \epsilon < 1/k$. And, $k \geq m$ implies that

$$(\ell - \epsilon)k \geq (\ell - \epsilon)m > 1 \quad \text{or} \quad \ell - \epsilon > 1/k,$$

which is impossible. Thus, $\langle 1/n \rangle$ does not converge in $J = (0, 1]$.

The next theorem provides the *sequential definition* of a limit point of a set $S \subseteq \mathbb{R}$.

Theorem 5.8 *Let $S \subseteq \mathbb{R}$. Then, $x \in S'$ if and only if $x = \lim_{n \to \infty} x_n$, for some sequence $\langle x_n \rangle$ of elements of S such that $x_n \neq x$, for all $n \in \mathbb{N}$.*

Proof First, let $x \in S'$. Then, the neighbourhood $I_{1/n}(x)$ intersects S, for each $n \in \mathbb{N}$. We may choose an element $x_n \in I_{1/n}(x) \cap S$, with $x_n \neq x$. Then, $x_n \in S$ such that $x_n \neq x$, for all $n \in \mathbb{N}$. We assert that $x_n \to x$. For, let $\epsilon > 0$. By the archimedean property, choose $N > 1/\epsilon$. Then, for all $n \geq N$, we have

$$|x_n - x| < 1/n \leq 1/N < \epsilon.$$

So, $x_n \to x$. Conversely, let $\langle x_n \rangle$ be a sequence of elements of S such that $\lim_{n \to \infty} x_n = x$, with $x_n \neq x$ for all $n \in \mathbb{N}$. Let $I_{\epsilon}(x)$ be a neighbourhood of x. Since $x_n \to x$, there is an natural number N such that $|x_n - x| < \epsilon$, for all $n \geq N$. Since $x_n \neq x$, there is some $n_0 \geq N$ such that $x_{n_0} \neq x$. Also, $x_{n_0} \in I_{\epsilon}(x)$. Therefore, x is a limit point of S. $\square$

Definition 5.7 A set $A \subseteq \mathbb{R}$ is said to be *dense* if $\overline{A} = \mathbb{R}$.

The next corollary is a topological version of the *density* of $\mathbb{Q}$ in $\mathbb{R}$.

Corollary 5.2 *For every $a \in \mathbb{R}$, there exists a sequence of rational numbers $\langle q_n \rangle$ such that $\lim_{n \to \infty} q_n = a$.*

Proof By Lemma 2.9, every ϵ-neighbourhood $I_{\epsilon}(a)$ centred at $a \in \mathbb{R}$ contains a rational number $q \neq a$. So, a is a limit point of $\mathbb{Q}$. Therefore, the proof is completed by applying Theorem 5.8. $\square$

Notice that we can use the same argument as above to conclude that, for every $a \in \mathbb{R}$, there exists a sequence of *irrational numbers* $\langle x_n \rangle$ such that $\lim_{n \to \infty} x_n = a$. However, we always prefer to work with $\mathbb{Q}$ mainly because it is countable.

Corollary 5.3 *A set $S \subseteq \mathbb{R}$ is closed if and only if for every Cauchy sequence $\langle x_n \rangle$ of elements of S, $x = \lim_{n \to \infty} x_n \in S$.*

Proof First, suppose S is a closed set, and $\langle x_n \rangle$ be a Cauchy sequence in S. Then, by Theorem 5.8, $x = \lim_{n \to \infty} x_n \in S' \subset \overline{S} = S$. Conversely, suppose every Cauchy sequence of elements of S converges in S itself. Let $x \in \mathbb{R}$ be a limit point of the set S. Once again, by Theorem 5.8, $x = \lim_{n \to \infty} x_n$, for some sequence $\langle x_n \rangle$ of elements of S such that $x_n \neq x$, for all $n \in \mathbb{N}$. Using the fact that every convergent sequence is Cauchy, it follows that $x \in S$. $\square$

We can use this corollary to prove a topological version of the *completeness* property of the ordered field $\mathbb{R}$. In general, we say a metric space (X, d) is *complete* if every Cauchy sequence in X converges to some $x \in X$ with respect to the metric d.

Theorem 5.9 *The real line $\mathbb{R}$ is a complete metric space.*

Proof Suppose $\langle x_n \rangle$ is a Cauchy sequence in $\mathbb{R}$. It is bounded, by Theorem 3.16, and so $|x_n| \leq M$, for some $M > 0$. Now, since $\langle x_n \rangle$ is a Cauchy sequence in the the interval $[-M, M]$, by Corollary 5.3, there is an $x \in [-M, M] \subset \mathbb{R}$ such that $x_n \to x$. Therefore, $\langle x_n \rangle$ converges in $\mathbb{R}$. Hence, $\mathbb{R}$ is complete. $\qquad \square$

Notice that the *completeness is not a topological property*. For example, the set $\mathbb{Q}$ is not complete with respect to the usual (modulus) metric. On the other hand, as shown in the next chapter, there exists two different types of metrics on the space $C(I)$ of real-valued continuous functions defined on an interval I such that one is complete, but the other is not.

The *local continuity* of a function $f : [a, b] \to \mathbb{R}$, as discussed in the previous chapter, is defined in terms of neighbourhood at a point, i.e., intervals. We introduce here the concept of *global continuity* given in terms of open sets. Notice that the topology on a set $I \subseteq \mathbb{R}$ is assumed to be induced by the usual topology, called the *subspace topology*. That is, a set $J \subseteq I$ is open if and only if $J = U \cap I$, for some open set U of $\mathbb{R}$.

Theorem 5.10 *Let $I \subseteq \mathbb{R}$. A function $f : I \to \mathbb{R}$ is continuous on I if and only if the inverse image $f^{-1}(U)$ of each open set $U \subseteq \mathbb{R}$ is an open set in I.*

Proof First, suppose f is continuous on I, and $U \subseteq \mathbb{R}$ be an open set. Let $J = f^{-1}(U)$. We need to show that J is open or, in other words, each point of J is an interior point. Let $c \in J$. Then $f(c) \in U$. Since U is open, there is an $\epsilon > 0$ such that the interval $I_\epsilon(f(c)) \subset U$. We may choose $\delta > 0$ such that

$$|x - c| < \delta \quad \Rightarrow \quad |f(x) - f(c)| < \epsilon.$$

We claim that $I_\delta(c) \subset J$. For, suppose $x \in I_\delta(c)$. Then, we have $|x - c| < \delta$, and so $|f(x) - f(c)| < \epsilon$. That is, $f(x) \in I_\epsilon(f(c)) \subset U$. Or, equivalently, $x \in f^{-1}(U) = J$. Therefore, $I_\delta(c) \subset J$. This proves that c is an interior point of J, and hence J is open in I.

Conversely, suppose $J = f^{-1}(U) \subseteq I$ is open for each open set $U \subseteq \mathbb{R}$. Let $x = c \in I$ be arbitrary, and $\epsilon > 0$. We write $U = I_\epsilon(f(c))$. We know that the set $U \subset \mathbb{R}$ is open and, by our assumption, $J = f^{-1}(U)$ is open in I. As $c \in J$, and J is open, there is some $\delta > 0$ such that $I_\delta(c) \subseteq J$. Now, if $x \in I_\delta(c)$, then $f(x) \in U = I_\epsilon(f(c))$, which shows that f is continuous at $x = c$. As $c \in I$ is chosen to be an arbitrary point, it follows that the function f is continuous on I. $\qquad \square$

A simple corollary of this topological version of continuity is that the composition of two continuous function is continuous.

Corollary 5.4 *Suppose $f : I \to \mathbb{R}$ is a continuous function, with $J = f(I)$, and $g : J \to \mathbb{R}$ is also a continuous function. Then, the composition $g \circ f : I \to \mathbb{R}$ is continuous on I.*

Proof Recall that, for any set $A \subseteq \mathbb{R}$, we have

$$(g \circ f)^{-1}(A) = f^{-1}\big(g^{-1}(A)\big).$$

In particular, as the function g is continuous, $g^{-1}(U)$ is open in J for any open set $U \subseteq \mathbb{R}$. Further, by the continuity of f, it follows that the set $f^{-1}(g^{-1}(V))$ is open in I. Therefore, the set $(g \circ f)^{-1}(V)$ is open in I. Hence, the composition $g \circ f$ is continuous on I.   □

Corollary 5.5 *A function $f : I \to \mathbb{R}$ is continuous on I if and only if the inverse image $f^{-1}(F)$ of each closed set $F \subseteq \mathbb{R}$ is a closed set in I.*

Proof First, notice that

$$\begin{aligned}
x \in I \setminus f^{-1}(F) &\Leftrightarrow x \notin f^{-1}(F) \\
&\Leftrightarrow f(x) \notin F \\
&\Leftrightarrow f(x) \in \mathbb{R} \setminus F \\
&\Leftrightarrow x \in f^{-1}(\mathbb{R} \setminus F).
\end{aligned}$$

That is, $I \setminus f^{-1}(F) = f^{-1}(\mathbb{R} \setminus F)$. Now, as $\mathbb{R} \setminus F$ is open in $\mathbb{R}$, it follows that the set $I \setminus f^{-1}(F) = f^{-1}(\mathbb{R} \setminus F)$ is open in I, and hence $f^{-1}(F)$ is closed in I. The proof of the converse is similar.   □

5.2   Compactness

We adopt here a topological definition of compactness, as formulated by *Eduard Heine* (1872) and *Émile Borel* (1895). Let $A \subseteq \mathbb{R}$. A collection

$$\mathscr{C} = \{A_\alpha \subseteq \mathbb{R} \mid \alpha \in I\}$$

is called a *cover* (or a *covering*) of the set A if $A \subset \cup_{\alpha \in I} A_\alpha$. If there is a finite set $J \subset I$ such that $A \subset \cup_{\alpha \in J} A_\alpha$, we say that the subcollection

$$\{A_\alpha : \alpha \in J\} \subset \mathscr{C},$$

is a *finite subcover* of the set A. That is, a finite subcover of a set A is a finite subset of a cover that also covers the set A. We say a covering $\mathscr{C}$ is an *open covering* if each A_α is open.

Definition 5.8 We say $K \subset \mathbb{R}$ is a *compact set* if every open covering of K contains a finite subcover for K.

Clearly, every finite set $K \subset \mathbb{R}$ is compact. For, given any open cover $\{U_\alpha\}$ of K, we can find finitely many indices $\alpha_1, \ldots, \alpha_m$ such that

$$K \subset U_{\alpha_1} \cup \ldots \cup U_{\alpha_m}.$$

Further, the set

$$K = \{0\} \cup \{1/n : n \in \mathbb{N}\}$$

is compact. For, given an open cover $\mathscr{C}$ of K, we choose $U \in \mathscr{C}$ such that $0 \in U$. As U is open, and $0 \in U$, there is a neighbourhood $I_\epsilon(0) \subset U$. By archimedean property, we have some $n_0 \in \mathbb{N}$ such that $1/n_0 < \epsilon$. Clearly then, $1/n \in U$, for all $n \geq n_0$. We now choose one open set U_n from the collection $\mathscr{C}$ for each $n = 1, 2, \ldots, n_0 - 1$ such that $1/n \in U_n$. It then follows that the finite collections $\{U, U_1, U_2, \ldots, U_{n_0-1}\}$ covers K.

In general, it is not an easy task to prove compactness of a set $K \subset \mathbb{R}$. For, in this case, we need to establish the existence of a finite subcover for every given open cover of the set K. However, on the other hand, to prove the non-compactness of a set $A \subseteq \mathbb{R}$, it suffices to find a suitable open cover of A such that no finite subcollection of this cover is able to cover for the set A.

Example 5.3 We give here examples of some non-compact sets in $\mathbb{R}$.

1. The open interval (a, b) is not a compact subset of $\mathbb{R}$. Consider $U_n = (a + 1/n, b - 1/n)$. Then, each U_n is open and $(a, b) = \cup_{n=1}^\infty U_n$. For any finite collection of integers $n_1, n_2, \ldots, n_m$, we have $\cup_{k=1}^m U_{n_k} = (a + 1/\ell, b - 1/\ell)$ where $\ell = \max\{n_i : 1 \leq i \leq m\}$. Therefore, no finite subcollection of $\{U_n\}$ can cover (a, b).
2. The real line $\mathbb{R}$ is not compact. Consider $U_n = (-n, n)$. Then, each U_n is open and $\mathbb{R} = \cup_{n=1}^\infty U_n$. For any finite collection of integers $n_1, n_2, \ldots, n_m$, we have $\cup_{k=1}^m U_{n_k} = (-l, l)$ where $l = \max\{n_i : 1 \leq i \leq m\}$. Therefore, no finite subcollection of $\{U_n\}$ can cover $\mathbb{R}$. The same open cover is also an open cover for $\mathbb{N}$, $\mathbb{Z}$, $\mathbb{Q}$, and $\mathbb{R} \setminus \mathbb{Q}$. These subsets are also not compact subsets of $\mathbb{R}$.
3. The half-open interval $L = [0, \infty)$ is not a compact subset of $\mathbb{R}$. Consider $U_n = (-1, n)$. Then, each U_n is open and $L \subset (-1, \infty) = \cup_{n=1}^\infty U_n$. For any finite collection of integers $n_1, n_2, \ldots, n_m$, we have $\cup_{k=1}^m U_{n_k} = (-1, l)$ where $l = \max\{n_1, n_2, \ldots, n_m\}$. Therefore, no finite subcollection of $\{U_n\}$ can cover L.

Theorem 5.11 *Every compact set in $\mathbb{R}$ is closed and bounded.*

Proof Suppose $K \subset \mathbb{R}$ is a compact set. Since K is closed if and only if $\mathbb{R} \setminus K$ is open, we show that $\mathbb{R} \setminus K$ is open. For, let $a \in \mathbb{R} \setminus K$ be an arbitrary point, and $x \in K$. Then, we must have $|a - x| > 0$. Take two real numbers $0 < r_x, s_x < |a - x|/2$, and consider the neighbourhoods $V_x = I_{r_x}(a)$ and $U_x = I_{s_x}(x)$. Then $V_x \cap U_x = \emptyset$ and $K \subset \cup_{x \in K} U_x$. So, $\{U_x : x \in K\}$ is an open cover of the compact set K. Thus, there is a finite set $\{U_{x_1}, \ldots, U_{x_m}\}$ that covers K. Put

$$V = \cap_{k=1}^m V_{x_k} \quad \text{and} \quad U = \cup_{k=1}^m U_{x_k}.$$

We know that both V and U are open, $x \in V$ and $K \subset U$. Since $V \subset V_{x_k}$, we have

$$V \cap U_{x_k} \subset V_{x_k} \cap U_{x_k} = \emptyset,$$

and hence $V \cap U_{x_k} = \emptyset$. Therefore, it follows that

$$V \cap U = V \cap (\cup_{k=1}^{m} U_{x_k}) = \cup_{k=1}^{m} (V \cap U_{x_k}) = \emptyset.$$

Since $K \subset U$ and $V \cap U = \emptyset$, we see that $V \cap K = \emptyset$ or $V \subset \mathbb{R} \setminus K$. Since $x \in V$, and V is open, there is an r-neighbourhood $I_r(x)$ of x contained in $V \subset \mathbb{R} \setminus K$. This proves that $\mathbb{R} \setminus K$ is open, and hence K is closed. Next, to show that K is bounded, fix a point $x \in K$. For each $a \in K$, $|x - a| < n$, for some $n \in \mathbb{N}$ and hence $a \in I_n(x)$. Therefore $\mathscr{C} = \{I_n(x) : n \in \mathbb{N}\}$ is an the open covering of K. Since K is compact, it contains a finite subcover $\{I_{n_1}(x), \ldots, I_{n_m}(x)\}$. Let $M = \max\{n_1, n_2, \ldots, N_m\}$. Then

$$K \subset \cup_{k=1}^{m} I_{n_k}(x) \subset I_M(x).$$

Thus, K is bounded.   □

In general, a set $S \subset \mathbb{R}$ is said to be *bounded* in metric sense[2] if there exists an element $a \in S$ and $M > 0$ such that $|x - a| < M$, for all $x \in S$. Notice that, as

$$I_M(b) \subseteq I_M(a) + |a - b|, \quad \text{for any } a, b \in S,$$

we may take *any* $a \in S$ to verify the *boundedness* of S. Said differently, a set $S \subset \mathbb{R}$ is bounded if S is contained in a δ-neighbourhood $I_\delta(a)$, for some $a \in S$ and $\delta > 0$. The interval $[a, \infty)$ is an *unbounded* closed subset of $\mathbb{R}$ because $\mathbb{R} \setminus [a, \infty) = (-\infty, a)$ is an open interval. As seen above, this interval is a non-compact set in $\mathbb{R}$. The next theorem proves that every closed and bounded interval is compact.

Theorem 5.12 *For any $a < b \in \mathbb{R}$, the closed interval $[a, b]$ is compact.*

Proof Suppose $\mathscr{C}$ is an arbitrary open cover of $[a, b]$. To complete the proof, we have to find a finite subcollection of $\mathscr{C}$ that covers the interval $[a, b]$. For, take K to be the set of all $x \in \mathbb{R}$ such that the interval $[a, x]$ is covered by a finitely subcollection of the open cover $\mathscr{C}$. We first prove that $b \in K$. Notice that, as $[a, a] = \{a\}$ can be covered by a single member of the cover $\mathscr{C}$, it follows that $K \neq \emptyset$. Now, let $x \in K$, and $a \leq y < x$. Then, as $[a, y] \subset [a, x]$, the finite subcollection of $\mathscr{C}$ that covers $[a, x]$ also covers $[a, y]$. Therefore, $y \in K$ whenever $x \in K$ and $a \leq y < x$.

In general, the set K is bounded above or it is unbounded. We first suppose K is not bounded above. It then follows that b is not an upper bound of K, and hence there is $x \in K$ such that $x > b$. In this case, as $x \in K$ and $b < x$, we see that $b \in K$. Suppose K is a bounded above set, with $\alpha = \sup K \in \mathbb{R}$. Here, we again have to consider two cases: $b < \alpha$ or $\alpha \leq b$. First, suppose $b < \alpha$. Then, by choosing $x \in K$ such that $b < x < \alpha$, we see that $b \in K$.

Finally, we show that $\alpha \leq b$ cannot hold. Suppose on the contrary that $\alpha \leq b$. Since $a \in K$, $\{a\} = [a, a]$ is covered by just an open set $U \in \mathscr{C}$, there is an open interval $[a - \epsilon, a + \epsilon] \subset U$. This shows that the interval $[a, a + \epsilon)$ is covered by U

---

[2] We will discuss this general type of boundedness of set in the next chapter.

and hence $[a, a + \epsilon) \subset K$. Therefore, $\alpha = \sup K \geq a + \epsilon > a$ and so $a < \alpha \leq b$. Since $\mathscr{C}$ is an open cover of $[a, b]$, there is an open set $U_\alpha \in \mathscr{C}$ such that $\alpha \in U_\alpha$. Since U_α is open, there is a neighbourhood $(\alpha - \epsilon, \alpha + \epsilon)$ of α contained in U_α. Since $\alpha = \sup K$, there is an $x \in K$ such that $x > \alpha - \epsilon$. Also $x \in K$ implies $a \leq x \leq \alpha$ and there is a finite subcollection $\mathscr{C}'$ of $\mathscr{C}$ that covers $[a, x]$. Hence, the finite collection $\mathscr{C}' \cup \{U_\alpha\}$ of $\mathscr{C}$ covers $[a, \alpha + \epsilon)$ and hence it covers $[a, \alpha + \epsilon/2]$. Thus, $\alpha + \epsilon/2 \in K$. This is a contradiction as $\alpha = \sup K$. Hence, $\alpha > b$. $\qquad\square$

Theorem 5.13 *Every compact set K in $\mathbb{R}$ is complete.*

Proof Suppose $\langle x_n \rangle$ is a Cauchy sequence in K, and $S = \{x_n : n \in \mathbb{N}\}$ be the range set. We know that S is a bounded set. If S is finite, then at least one element is repeating infinite number of times, say x_1. Then, the subsequence $\langle x_{n_k} \rangle$ of the sequence $\langle x_n \rangle$ formed by taking $x_{n_k} = x_1$ (for all $k \in \mathbb{N}$) converges in K. So, by Theorem 3.17, $\langle x_n \rangle$ converges in K. Next, suppose S is infinite. Then, by Theorem 3.14, S has a limit point $x \in K$. For $k \in \mathbb{N}$, we choose ascending natural numbers n_k such that $|x_{n_k} - x| < 1/k$. Notice that it is possible to do so because every neighbourhood of x contains infinitely many points of S. Now, given $\epsilon > 0$, we choose a natural number N such that $N > 1/\epsilon$. Then, we have

$$|x_{n_k} - x| < 1/k \leq 1/N < \epsilon, \quad \text{for all} \quad n \geq N.$$

Therefore, a subsequence $\langle x_{n_k} \rangle$ of the Cauchy sequence $\langle x_n \rangle$ converges in K. Hence, once again, by Theorem 3.17, $\langle x_n \rangle$ itself converges in K. $\qquad\square$

In particular, by Theorem 5.9, every closed and bounded interval $[a, b]$ is *complete*. In fact, every closed set in a compact metric space is complete. Notice that $\mathbb{R}$ is a non-compact complete metric space.

Theorem 5.14 *Let $F \subset K \subset \mathbb{R}$. If K is compact and F is closed in $\mathbb{R}$, then F is compact.*

Proof Suppose $\mathscr{C}$ is an open covering of F. Let $\mathscr{C}' = \mathscr{C} \cup \{\mathbb{R} \setminus F\}$. Clearly then, $\mathscr{C}'$ is an open covering of K (and indeed it cover whole of $\mathbb{R}$). Hence, there is a finite subcover $\mathscr{B}' \subset \mathscr{C}'$ that cover K. Let $\mathscr{B} = \mathscr{B}'$ if $\mathbb{R} \setminus K \notin \mathscr{B}'$ and $\mathscr{B} = \mathscr{B}' \setminus \{\mathbb{R} \setminus K\}$ if $\mathbb{R} \setminus K \in \mathscr{B}'$. Then the finite subcollection $\mathscr{B}$ covers K. Therefore, K is compact. $\square$

Definition 5.9 A collection $\mathscr{C}$ of subsets of $\mathbb{R}$ is said to have *finite intersection property* (FIP, in short) if, for each finite subcollection

$$\{C_1, C_2, \ldots, C_m\} \subset \mathscr{C},$$

the intersection $\cap_{k=1}^m C_k \neq \emptyset$. We also say that the collection $\mathscr{C}$ is *centred*.

Theorem 5.15 *For every collection of compact sets $\{K_\alpha\}$ in $\mathbb{R}$ having finite intersection property, $\cap_\alpha K_\alpha \neq \emptyset$.*

Proof For a contradiction, suppose $\cap_\alpha K_\alpha = \emptyset$. Since each K_α is compact, it is closed, and so each $U_\alpha = \mathbb{R} \setminus K_\alpha$ is an open set. Also,

$$\cap_\alpha K_\alpha = \emptyset \quad \Rightarrow \quad \cup_\alpha U_\alpha = \mathbb{R}, \quad \text{by taking complement.}$$

Fix one member of the collection $\{K_\alpha\}$, say K_β. Then $K_\beta \subset \mathbb{R} = \cup_\alpha U_\alpha$. Hence, the collection $\{U_\alpha\}$ forms an open covering of the compact set K_β. Therefore, there is a finite subcollection $\{U_{\alpha_1}, \ldots, U_{\alpha_m}\}$ that cover K_β. Thus,

$$K_\beta \subset \cup_{k=1}^m U_{\alpha_k} = \cup_{k=1}^m (X \setminus K_{\alpha_k}) = X \setminus (\cap_{k=1}^m K_{\alpha_k}).$$

Therefore, $K_\beta \cap (\cap_{k=1}^m K_{\alpha_k}) = \emptyset$, contradicting the finite intersection property of the collection $\{K_\alpha\}$. Hence the theorem holds. $\qquad\square$

Corollary 5.6 *For each $n \in \mathbb{N}$, let $K_n \subset \mathbb{R}$ be a nonempty compact set. If $K_k \supset K_{k+1}$ for each $k \in \mathbb{N}$, then $\cap_{k=1}^\infty K_k \neq \emptyset$.*

Proof Since $K_k \supset K_{k+1}$ for each $k \in \mathbb{N}$, we have $K_1 \supset K_2 \supset K_3 \cdots$. Thus, $K_{n_1} \cap K_{n_2} \cap \cdots \cap K_{n_m} = K_N \neq \emptyset$, where $N = \max\{n_1, n_2, \ldots, n_m\}$. Thus, the collection $\{K_n\}$ has finite intersection property. Therefore, by Theorem 5.15, we have $\cap_{k=1}^\infty K_k \neq \emptyset$. $\qquad\square$

The argument given in the next corollary is Cantor's original idea that he used to prove the completeness of the real line $\mathbb{R}$ (Theorem 5.9).

Corollary 5.7 *(Cantor) If $\{I_k : k \in \mathbb{N}\}$ is a collection of closed intervals in $\mathbb{R}$ satisfying $I_k \supset I_{k+1}$ for $k \in \mathbb{N}$, then $\cap_{k=1}^\infty I_k \neq \emptyset$.*

Proof Since each $I_k = [a_k, b_k]$ is compact, the required result follows from the previous corollary. A direct proof using *least upper bound property* of $\mathbb{R}$ is as follows. Since $I_k \supset I_{k+1}$, we have $a_1 \leq a_2 \leq \cdots \leq a_k \leq a_{k+1} \leq \cdots \leq b_{k+1} \leq b_k \leq \cdots b_2 \leq b_1$. Therefore, the set $\{a_k : k \in \mathbb{N}\}$ is bounded above by each b_k, $k \geq 1$, and the set $\{b_k : k \in \mathbb{N}\}$ is bounded below by each a_k, $k \geq 1$. Let $\alpha = \sup\{a_k : k \in \mathbb{N}\}$ and $\beta = \inf\{b_k : k \in \mathbb{N}\}$. Then $a_k \leq \alpha$ and $\beta \leq b_k$ for each $k \geq 1$. Since each a_k is a lower bound for $\{b_k : k \in \mathbb{N}\}$, it follows that $a_k \leq \beta$ and hence $\alpha = \sup\{a_k : k \in \mathbb{N}\} \leq \beta$. Hence, $a_k \leq \alpha \leq \beta \leq b_k$. Thus, $[\alpha, \beta] \subset I_k$ for each k and hence $\cap_{k=1}^\infty I_k = [\alpha, \beta] \neq \emptyset$. (When $\alpha = \beta$, $[\alpha, \beta] = \{\alpha\}$.) $\qquad\square$

More generally, the following holds.

Corollary 5.8 *If $K \subset \mathbb{R}$ is compact, then for every collection $\{F_\alpha\}$ of closed subsets of K having FIP, $\cap_\alpha F_\alpha \neq \emptyset$.*

Proof By Theorem 5.14, every closed subset of a compact set in $\mathbb{R}$ are compact, so the corollary follows immediately from Theorem 5.15. $\qquad\square$

The converse of this corollary is also true (Exercise 5.16).

Definition 5.10 We say a set $S \subset \mathbb{R}$ has the *Bolzano-Weierstrass property* if every infinite subset of S has a limit point in S. In this case, we also say the set S is *limit point compact* (or *Frechét compact*).

The next theorem proves compact sets in $\mathbb{R}$ are Frechét compact.

Theorem 5.16 *Every compact set $K \subset \mathbb{R}$ is limit point compact.*

Proof For a contradiction, suppose a compact set $K \subset \mathbb{R}$ is not limit point compact. Then, there is an infinite subset A of K that has no limit point (in K). Since A has no limit point, A is closed and therefore compact, by Theorem 5.14. Also, since A has no limit point, for each $a \in A$, we can find a neighbourhood I_a of a such that $A \cap I_a = \{a\}$. Clearly then, the collection $\{I_a : a \in A\}$ forms an open cover of A, which has no finite subcover of the set A. This contradiction to the compactness of A proves the theorem. $\qquad\square$

The next corollary is a topological version of Theorem 3.14.

Corollary 5.9 (Bolzano-Weierstrass Theorem) *Every bounded infinite subset $A \subset \mathbb{R}$ has a limit point.*

Proof Since $A \subset \mathbb{R}$ bounded, there is some $M > 0$ such that $A \subset I_M = [-M, M]$. Since I_M is compact, and A is an infinite subset, A has a limit in I_M, by Theorem 5.16, and hence in $\mathbb{R}$. $\qquad\square$

Example 5.4 The set $\{1/n : n \in \mathbb{N}\}$ is a bounded infinite subset of $\mathbb{R}$ and the number 0 is its limit point. The infinite set $\{n + (1/n) : n \in \mathbb{N}\}$ has no limit point. However, it is unbounded and so it does not contradict Bolzano-Weierstrass Theorem. The infinite set $\{n + (1/n) : n \in \mathbb{N}\}$ is unbounded but it has 0 as its limit point. Hence, the existence of limit points does not say anything about the boundedness of the set.

By Theorem 5.11, we know that every compact set $K \subset \mathbb{R}$ is closed and bounded. By Theorem 5.12, we know that the converse statement holds when K is an interval. The next celebrated theorem due to *Eduard Heine* and *Émile Borel* prove that the converse holds in general. The formulation given below is due to Borel (1895), though he proved it considering only *countable covers*.

Theorem 5.17 (Heine-Borel Theorem) *Every closed and bounded set $K \subset \mathbb{R}$ is compact.*

Proof Since K is bounded, there is $M > 0$ such that $K \subset I_M(0)$, and so $K \subset [-M, M]$, which is a compact set by Theorem 5.12. Since K is a closed subset of a compact set, it follows by Theorem 5.14 that K is compact. $\qquad\square$

Example 5.5 (*Cantor Set*) Let $C_n \subset [0, 1]$ be sets as defined in Example 2.6. By construction, the set C_n obtained at the nth stage is a union of 2^n disjoint closed intervals with each interval being of the length $1/3^n$. So, the total length of the set C_n is $(2/3)^n$. Then, by Theorem 5.7, the set

$$C = \bigcap_{k=1}^{\infty} C_k \subset [0, 1]$$

is closed, known as the *Cantor set*. Observe that $C \neq \emptyset$ because it contains end points of all intervals in C_n, $n \geq 1$. Being closed and bounded, the Cantor set C is compact. We also know that C is an uncountable set.

Definition 5.11 Two topological spaces X and Y are said to be *homeomorphic* if there exists a bijective continuous function $h : X \to Y$ such that $h^{-1} : Y \to X$ is also continuous. Such a function h is called a *homeomorphism*. Furthermore, we say a property $\mathscr{P}$ is a *topological invariant* (or simply an *invariant*) of a space X if every space Y that is homeomorphic to X has the property $\mathscr{P}$.

For example, it folows easily that the differentiable bijective function $f : (-1, 1) \to \mathbb{R}$ as defined in (2.4.2) is homeomorphism. The next theorem proves that the compactness is a topological property for the usual topology on the line $\mathbb{R}$.

Theorem 5.18 *Let $f : I \to \mathbb{R}$ be a continuous function, and $K \subset I$ be a compact set. Then, the set $f(K) \subset \mathbb{R}$ is compact. In particular, if the function f is bijective, with $J = f(K)$, then $g^{-1} : J \to K$ is also continuous.*

Proof Let $\{V_\alpha\}$ be an open covering of $f(K)$. As f is continuous, by Theorem 5.10, the open sets $U_\alpha = f^{-1}(V_\alpha)$ provides a covering of the set K. For, if $x \in K$, then $f(x) \in f(K) \subset \cup_\alpha V_\alpha$, and so $f(x) \in V_\beta$, for some β. Hence, $x \in U_\beta = f^{-1}(V_\beta)$. By compactness of K, we can find $\alpha_1, \alpha_2, \ldots, \alpha_n$ such that $K \subset \cup_{k=1}^{n} f^{-1}(V_{\alpha_k})$. Take $y \in f(K)$ so that $y = f(x)$, for some $x \in K$. Then, $x \in f^{-1}(V_{\alpha_k})$, for some $1 \leq k \leq n$, and hence $y = f(x) \in V_{\alpha_k}$. This shows that $f(K) \subset \cup_{k=1}^{n} V_{\alpha_k}$, which proves the compactness of the set $f(K)$. Next, to prove the second assertion, let $U \subset K$ be an open set. Then, $V = K \setminus U$, being a closed set of a compact space K, is it self compact, by Theorem 5.14. Therefore, by the first part, $f(V) \subset J$ is compact, and so closed, by Theorem 5.11. Also, since g is one-one, we have $f(V) = J \setminus f(U)$. Hence, $f(U) \subset J$ is open i.e., $f^{-1} : J \to K$ is continuous. $\square$

Theorem 5.19 (Heine-Cantor) *Every continuous function $f : [a, b] \to \mathbb{R}$ is uniformly continuous.*

Proof Let $\epsilon > 0$ and $a \in X$. By continuity of f, there exists $\delta_a > 0$ such that $d(x, a) < \delta_a$ implies that $d(f(x), f(a)) < \epsilon/2$. Then, the collection of open balls $\{B(a; \delta_a/2) : a \in X\}$ covers X, and so there must be finitely many δ_{a_i}, say $1 \leq i \leq n$, such that

$$X \subseteq \bigcup_{i=1}^{n} B(a_i; \delta_{a_i}/2).$$

Take $\delta = (1/2) \min_{1 \leq i \leq n} \delta_{a_i}$. Now, for $x, y \in X$ with $d(x, y) < \delta$, $x \in B(a_i; \delta_{a_i}/2)$ implies that $d(x, a_i) < \delta_{a_i}/2$. So, by triangle inequality, it follows that

$$d(y, a_i) \le d(y, x) + d(x, a_i) < \delta + \frac{\delta_{a_i}}{2} \le \delta_{a_i}.$$

Thus, we have

$$d(f(x), f(y)) \le d(f(x), f(a_i)) + d(f(a_i), f(y)) < \frac{\epsilon}{2} + \frac{\epsilon}{2} = \epsilon.$$

Hence, f is uniformly continuous. $\square$

Corollary 5.10 (Boundedness Theorem) *Let $f : I \to \mathbb{R}$ be a continuous function, where I is a closed and bounded interval. Then, the set $J := f(I) \subset \mathbb{R}$ is a closed and bounded set.*

Proof By Theorem 7.9, $f(I) \subset \mathbb{R}$ is compact. We know that, by Heine-Borel theorem, compact sets in $\mathbb{R}$ are exactly the closed and bounded sets, and hence J is a closed and bounded set. $\square$

More precisely, the next theorem proves that every real-valued continuous function defined on a compact interval attains its maximum and minimum. It was first proved by *Bernard Bolzano* in 1830. However, the formulation given below is due to *Karl Weierstrass*.

Theorem 5.20 (Extreme Value Theorem) *Let $f : I \to \mathbb{R}$ be a continuous function, where I is a compact interval. Then, there exist elements a, $b \in I$ such that*

$$f(a) = \inf\{f(x) : x \in X\} \quad and \quad f(b) = \sup\{f(x) : x \in X\}.$$

Proof By previous corollary, $f(I) \subset \mathbb{R}$ is a closed and bounded set. So, $s := \sup f(I)$ exists. By Exercise 5.15, $s \in \overline{f(I)} = f(I)$, and so there is some $b \in I$ such that $f(b) = s = \sup f(I)$. The second assertion follows similarly by applying the previous argument to the function $(-f)$. $\square$

5.3 Connectedness

The real line $\mathbb{R}$ as the infinite interval $(-\infty, \infty)$ is *connected* because it is one single entity. In topological terms, we say there exists no *separation* for the real line $\mathbb{R}$. However, for any $a \in \mathbb{R}$, the set $\mathbb{R} \setminus \{a\}$ is *not connected*. For, in this case, the open intervals $U = (-\infty, a)$ and $V = (a, \infty)$ form a *separation* of the set $\mathbb{R} \setminus \{a\}$ (into two pieces).

In general, we say a pair (U, V) of nonempty disjoint open sets in $\mathbb{R}$ forms a *separation* of the real line $\mathbb{R}$ if we can write $\mathbb{R} = U \cup V$. Notice that, in this definition, we can replace *open sets* by *closed sets*. Therefore, the *connectedness* of the line $\mathbb{R}$ is equivalent to the fact that the only *clopens*[3] in $\mathbb{R}$ are $\emptyset$ and $\mathbb{R}$. We now introduce

---

[3] We say a set A in a topological space is a *clopen* if it is both open and closed.

the concept of connectedness of a nonempty sets $A \subseteq \mathbb{R}$. For any $B \subset A \subset \mathbb{R}$, we write $Cl_A(B)$ for the *closure* of the set B relative to the set A. As before, $\overline{B}$ is the closure of the set B in $\mathbb{R}$.

Definition 5.12 We say two sets $A, B \subset \mathbb{R}$ are *separated* if none contains a limit point of the other. That is,

$$\overline{A} \cap B = \emptyset = A \cap \overline{B}.$$

Clearly then, $A \cap B = \emptyset$.

Notice that disjoint sets need not be separated. For $x \neq y \in \mathbb{R}$, let $\epsilon = |x - y|/2 > 0$. Then, the open intervals $I = (x - \epsilon, x + \epsilon)$ and $J = (y - \epsilon, y + \epsilon)$ in X are separated because

$$\overline{I} \cap J = \emptyset = I \cap \overline{J}.$$

Notice that, for $B, C \subseteq A \subseteq \mathbb{R}$,

$$B \cap Cl_A(C) = \emptyset \quad \Leftrightarrow \quad A \cap \overline{B} = \emptyset; \text{ and,}$$
$$C \cap Cl_A(B) = \emptyset \quad \Leftrightarrow \quad B \cap \overline{A} = \emptyset,$$

which imply that subsets of separated sets are separated, and conversely. Thus, a separation in a subspace A is same as the separation in $\mathbb{R}$.

Definition 5.13 A nonempty set $A \subset \mathbb{R}$ is called *disconnected* if there exists nonempty sets B, C such that $A = B \cup C$ and

$$Cl_A(B) \cap C = \emptyset = B \cap Cl_A(C).$$

Such a pair (B, C) is called a *separation* of the subspace A.

Since $A \cap \overline{B} = Cl_A(B)$, a subspace $A \subset \mathbb{R}$ has a separation (B, C) if

$$A = B \cup C \quad \Rightarrow \quad B, C \text{ are closed in } A,$$

and so these are also open sets in A. The intervals $I = [0, 1/2)$ and $J = (1/2, 1]$ are separated in $A = [0, 1] \setminus \{1/2\}$ and also in $\mathbb{R}$, but I, J are not open in $\mathbb{R}$. The only *nontrivial* connected sets in $\mathbb{R}$ are intervals.

Theorem 5.21 *A nonempty subset $A \subseteq \mathbb{R}$ is connected if and only if A is an interval, unless it is a singleton.*

Proof First, suppose $A \subset \mathbb{R}$ is connected with $a \neq b \in A$. Now, for any $a < x < b$, if $x \notin A$ then the open sets $(-\infty, x)$ and (x, ∞) in $\mathbb{R}$ produce a separation of A, contradicting the connectedness of A. Hence, $x \in A$, and so A is an interval. Conversely, let $A = [a, b]$ be an interval, with $-\infty \leq a < b \leq \infty$. If possible, suppose (J, K) is a separation of $[a, b]$, i.e.,

$$[a, b] = J \cup K, \quad \text{with} \quad J \cap Cl_A(K) = \emptyset = Cl_A(J) \cap K.$$

We may assume that $b \in J$. Then, the nonempty open set $[a, b] \cap K$ in the interval $[a, b]$ is bounded above by b, so it has a least upper bound $c \leq b$, which belongs to J or K. If $c \in J$, it also contains an ϵ-ball around c. This, however, can not be the case, for otherwise $c - \epsilon$ would be a smaller upper bound. On the other hand, for $c \in K$, there exists an ϵ-ball around c that lies completely inside K, and so there would be points in K larger than c. Thus, we must have that $[a, b]$ is connected. $\square$

The next theorem proves that *connectedness* is a topological invariant of the real line $\mathbb{R}$.

Theorem 5.22 *Let $f : \mathbb{R} \to \mathbb{R}$ be a (nonconstant) continuous function. Then, for any connected set $A \subseteq \mathbb{R}$, the image $f(A) \subseteq \mathbb{R}$ is connected.*

Proof We can assume that A is a nondegenerate interval, say $A = [a, b]$. If possible, suppose $f(A)$ is disconnected, and (B, C) be a separation. Then, there are open sets U and V in $\mathbb{R}$ such that

$$f(A) = B \cup C \subseteq U \cup V, \quad \text{with} \quad B \cap V = C \cap U = \emptyset.$$

But then, for $B_1 = f^{-1}(B), C_1 = f^{-1}(C), U_1 = f^{-1}(U)$, and $V_1 = f^{-1}(V)$, we can express A as a union of two nonempty disjoint subsets B_1, C_1 covered by open sets U_1 and V_1 in A. This contradicts the connectedness of A. Hence, $f(A)$ is connected.$\square$

The next theorem is a very useful result of real analysis.

Theorem 5.23 (Intermediate Value Theorem) *Let $f : A \to \mathbb{R}$ be a (nonconstant) continuous function, and $A \subseteq \mathbb{R}$ be a connected set. For any $[x, y] \subset A$, if $f(x) < b < f(y)$, then there is some $a \in [x, y]$ such that $f(a) = b$.*

Proof By Theorem 5.22, we know that $f(A) \subset \mathbb{R}$ is a connected set so that $f([x, y])$ is an interval, by Theorem 5.21. Suppose $f([x, y]) = [c, d]$. Further, as f is a nonconstant function, we have $c < d$. Clearly then, for any $b \in (c, d)$ there is some $a \in [x, y]$ such that $f(a) = b$. $\square$

In particular, for the function $f(x) = x - \cos(x)$, $x \in [0, \pi]$, we find that $\cos(x_0) = x_0$, for some $x_0 \in (0, \pi)$. That is, the function $f : [0, \pi] \to [-1, \pi - 1]$, has a *fixed-point*. Clearly, this property holds for any continuous function defined on a closed and bounded interval $I \subset \mathbb{R}$. To conclude our discussion here, we give some simple applications of Theorem 5.23. The next theorem is also known as the *weighted mean value theorem*.

Theorem 5.24 (First Integral Mean Value Theorem) *Suppose $f \in C([a, b])$ and $g \in \mathscr{R}([a, b])$, with $g(x) \geq 0$ for all $x \in [a, b]$. Then there exists a number $c \in (a, b)$ such that*

$$\int_a^b f(x)g(x)dx = f(c) \int_a^b g(x)dx.$$

Proof By Theorem 5.22, there exists real numbers m, M such that $m \leq f(x) \leq M$. So, we have

$$m\,g(x) \leq f(x)g(x) \leq M\,g(x), \quad \text{for all} \quad x \in [a, b].$$

Thus, by Theorem 4.20, it follows that

$$m \int_a^b g(x)dx \leq \int_a^b f(x)g(x)dx \leq M \int_a^b g(x)dx.$$

There is nothing to prove if $\int_a^b g(x)dx = 0$. So, we may assume that $\int_a^b g(x)dx \neq 0$. In this case, we have

$$m \leq \frac{\int_a^b f(x)g(x)dx}{\int_a^b g(x)dx} \leq M.$$

Therefore, by Theorem 5.23, it follows that there is some $c \in (a, b)$ such that

$$f(c) = \frac{\int_a^b f(x)g(x)dx}{\int_a^b g(x)dx}.$$

This completes the proof. $\square$

Theorem 5.25 (Second Integral Mean Value Theorem) *Suppose $f : [a, b] \to \mathbb{R}$ is a positive decreasing function, and $g \in \mathscr{R}([a, b])$. Then there exists a number $c \in (a, b)$ such that*

$$\int_a^b f(x)g(x)dx = f(a^+) \int_a^c g(x)dx.$$

Proof Left to the reader as Exercise 5.21. $\square$

An application of Theorem 5.23 given in the next worked example is about an important numerical method, called *Newton-Raphson method*, which is usually applied to approximate a root of an equation $f(x) = 0$, where f is a sufficiently smooth function.

Example 5.6 (*Newton-Raphson Method*) Suppose $f : [a, b] \to \mathbb{R}$ is an arbitrary differentiable function such that $f(c)f(d) < 0$ for some interval $[c, d] \subseteq [a, b]$. Then, by Theorem 5.23, there exists a (simple) root α of the equation $f(x) = 0$ in the interval $[c, d]$. Before starting with the process of estimation of $\alpha \in [c, d]$, we apply *intermediate value theorem* ample number of times to ensure that such an interval $[c, d]$ is of *very small length*. Let $x_0 \in [c, d]$ be a near vicinity *initial estimate* of the root α such that

$$\alpha = x_0 + \epsilon, \quad \text{for some small} \quad \epsilon > 0.$$

Then, by Taylor's theorem, we have

$$0 = f(\alpha) = f(x_0 + \epsilon) \approx f(x_0) + \epsilon f'(x_0).$$

Hence, if $\langle x_n \rangle$ is the sequence of subsequent estimates of α in the interval $[c, d]$, then according to Newton-Raphson method x_{n+1} is given by the *iteration formula*

$$x_{n+1} = x_n - \frac{f(x_n)}{f'(x_n)}, \qquad n \geq 0.$$

This scheme gives a very efficient approximation of the root α in the sense that the associate error sequence $\langle e_n \rangle$ given by $e_n = \alpha - x_n$ $(n \geq 1)$ has quadratic *rate of convergence*. That is, $e_{n+1} = O(e_n^2)$ as $n \to \infty$.

Exercises 5

5.1 Define $d : \mathbb{N} \times \mathbb{N} \to \mathbb{R}$ by $d(m, n) = |m^{-1} - n^{-1}|$. Show that d is a metric. Is $\langle n^2 \rangle$ a Cauchy sequence? Does it converge in $\mathbb{N}$? Is $\mathbb{N}$ a complete metric space with this metric? Can this definition of d be extended to the set of positive real numbers or to the set of all irrational numbers to get a metric on them?

5.2 Let $\mathbb{R}^+$ denote the set of all positive real numbers. Show that the function $d : \mathbb{R}^+ \times \mathbb{R}^+ \to \mathbb{R}$ defined by $d(x, y) = |\log(y/x)|$ is a metric on $\mathbb{R}^+$.

5.3 Let $A, B \subseteq \mathbb{R}$.

    a. Suppose $x \in \mathbb{R}$ lies in $A^o \cap B^o$. Is it also true that $x \in (A \cap B)^o$?
    b. Show that $A \subseteq B \Rightarrow A^0 \subseteq B^0$.
    c. Show that $(A \cap B)^o = A^o \cap B^o$.
    d. Is $(A \cup B)^o = A^o \cup B^o$? If not, give a suitable example.

5.4 Show that, for any $A \subseteq \mathbb{R}$, A^o is the largest open set of $\mathbb{R}$ contained in A.

5.5 Find all limit points of $\mathbb{N}$ in $\mathbb{R}$. Do the same for $\mathbb{Z}$, $\mathbb{Q}^+$. and $\mathbb{Q}$.

5.6 Find the limit points of the sets $\{2^n + 2^{-n} : n \in \mathbb{N}\}$, and $\{m^{-1} + n^{-1} : m, n \in \mathbb{N}\}$ in $\mathbb{R}$.

5.7 For fixed $a, b \in (0, 1)$, find the limit points of the set $\{a^n + b^m : n, m \in \mathbb{N}\}$.

5.8 Let $A, B \subseteq \mathbb{R}$.

    a. Suppose $x \in \mathbb{R}$ lies in $\overline{A \cap B}$. Is it also true that $x \in \overline{A} \cap \overline{B}$?
    b. Show that $A \subseteq B \Rightarrow \overline{A} \subseteq \overline{B}$.
    c. Show that $\overline{A \cup B} = \overline{A} \cup \overline{B}$.
    d. Is $\overline{A \cap B} = \overline{A} \cap \overline{B}$? If not, give a suitable example.

5.9 Show that, for any $A \subseteq \mathbb{R}$, $\mathbb{R} \setminus \overline{A} = (\mathbb{R} \setminus A)^o$.

5.10 Show that, for any $A \subseteq \mathbb{R}$, $\overline{A}$ is the smallest closed set of $\mathbb{R}$ containing A.

5.11 Give a topological proof of the fact that the set $\mathbb{Q}$ is not a complete ordered field.

5.12 Show that the arbitrary intersection of open sets may not be open.

5.13 Show that every point of the set

$$S = \{1/n : n \in \mathbb{N}\} \subset \mathbb{R}$$

   is an *isolated point*.

5.14 Prove Corollary 5.10 using Bolzano-Weierstrass property.

5.15 If $S \subset \mathbb{R}$ be bounded above, with $s = \sup S$, show that $s \in \overline{S}$.

5.16 Suppose for every collection $\{F_\alpha\}$ of closed subsets of a set $K \subset \mathbb{R}$ that has finite intersection property, we have $\cap_\alpha F_\alpha \neq \emptyset$. Show that K is compact.

5.17 With notations as in Exercise 4.3.3, show that, for any $\epsilon > 0$, the set $\{x : o(f, x) \geq \epsilon\}$ is a closed set.

5.18 Show that, for any connected set $A \subseteq \mathbb{R}$, every continuous function $f : A \to \{0, 1\}$ is constant.

5.19 If $I \subset \mathbb{R}$ is an interval, and $f : I \to \mathbb{R}$ is monotone, prove that f is continuous if and only if $\text{Ran}(f)$ is an interval. Hence conclude that if f is strictly monotone then f^{-1} defined on the interval $[\inf f, \sup f]$ is continuous. How about the converse of the second assertion?

5.20 Suppose $f \in C([a, b])$. Show that for some $c \in (a, b)$ we have

$$\int_a^b f(x)dx = (b - a)f(c).$$

5.21 Supply a proof of Theorem 5.25.

5.22 Apply Newton-Raphson method to estimate the smallest positive root of the equation $f(x) = x - e^{-x}\cos(x)$, $x \in [0, 1]$.

5.23 Apply Newton-Raphson method to estimate the smallest positive root of the equation $f(x) = x - \tan x$, $x \in [0, 3]$.

Chapter 6
Metric Spaces

Therefore, either the reality on which our space is based must form a discrete manifold or else the reason for the metric relationships must be sought for, externally, in the binding forces acting on it.

Bernhard Riemann (1826–1866)

A metric structure on a (nonempty) set is a manifestation of the fundamental concept of a *distance function*, as introduced earlier in Chap. 2 (Definition 2.11). Recall that a *metric d* on a set X is a positive, symmetric function $d : X \times X \to \mathbb{R}$ satisfying the triangle inequality. The pair (X, d) is then called a *metric space*, and elements of the set X are called *points*. When we consider more than one metrics on a set X, we refer to X as the *carrier set* of each one of the metric spaces so defined. In any practical application, the problem in hand guides our choice for a particular metric on the carrier set.

We discuss here metric structures of certain important real linear spaces such as the space of n-dimensional vectors $\mathbb{R}^n$, the space $\ell_p(\mathbb{R})$ of p-summable sequences of real numbers, and various types of subspaces of the space $C(I)$ of continuous real-valued functions defined on an interval $I \subseteq \mathbb{R}$. After having introduced in Chap. 9 the concept of (Lebesgue) *p-integrability* of a function $f : \mathbb{R} \to \mathbb{R}$, we will also discuss the metric of the space $L^p(\mathbb{R})$ of *p-integrable* real-valued functions defined on $\mathbb{R}$.

In general, a metric on a set facilitates the idea of *proximity* among the points, and hence defines a topology on the underlying set. It is then possible to talk about the *convergence* of sequences and series of elements of a linear space, with respect to some assigned metric. For abstract spaces such as *Riemannian manifolds*, metric geometry defined in terms of *geodesics* is used to study the *shape* of the space in

A. K. Razdan and V. Ravichandran, *Fundamentals of Analysis with Applications*, https://doi.org/10.1007/978-981-16-8383-1_6

terms of its *curvature*, and also the *size* of some special types of subsets in terms of their *diameter* and *volume*.

In this chapter, we discuss topological properties of some of the above-mentioned metric spaces inasmuch as required in the next chapter to discuss the analytical properties such as continuity and differentiability of multivariable vector-valued functions. We also discuss certain interesting *types of convergence* in such spaces. Some important completeness theorems are also proved. The emphasis all through is on the problem of extending analytical properties of functions from the space $\mathbb{R}$ to some infinite dimensional space such as $\ell_p(\mathbb{R})$ or $C(\mathbb{R})$.

Considering the involved complexities of some archetypical *partial differential equation* models of phenomena in science and engineering, it is very important that the reader has a firm grasp over the concepts introduced here. Also, a good understanding about the *content* of the theorems proved here is crucial to our comfort in dealing with the geometry of the associated *integral surfaces*.

6.1   Some Important Metric Spaces

Let X be a nonempty set. As defined earlier in Chap. 2, a function $d : X \times X \to \mathbb{R}$ is called a *distance function* (or a *metric*) on the set X if it is *positive, symmetric*, and satisfies the *triangle inequality*.[1] That is, for all $x, y, z \in X$,

1. $d(x, y) = 0$ if and only if $x = y$;
2. $d(x, y) = d(y, x)$;
3. $d(x, z) \leq d(x, y) + d(y, z)$.

Notice that, by using properties (2), (3) and (1) (in that order), we have

$$d(x, y) = \frac{1}{2}\left(d(x, y) + d(y, x)\right) \geq \frac{1}{2}d(x, x) = 0, \ \forall \, x, y \in X.$$

Therefore, every metric is a *nonnegative* function. On the other hand, if a *nonnegative* function $\rho : X \times X \to \mathbb{R}$ has properties (1) and (3) as above, then the function $d : X \times X \to \mathbb{R}$ given by

$$d(x, y) = \rho(x, y) + \rho(y, x), \ \ x, y \in X,$$

is a metric on X. More generally, we say a function $s : X \times X \to \mathbb{R}$ is a *semimetric* (or a *pseudometric*) on a set X if it satisfies properties (2) and (3) as above, but we may have $s(x, y) = 0$, for some pair $x \neq y \in X$. As before, every semimetric is a nonnegative function. For example, if

$$\boldsymbol{x} = (x_1, y_1), \quad \boldsymbol{y} = (x_2, y_2) \in \mathbb{R}^2,$$

---

[1] The terminology corresponds to the geometry fact: *The sum of lengths of two sides of a triangle is always greater than or equal to the length of the third side.*

then the function $s : \mathbb{R}^2 \times \dot{\mathbb{R}}^2 \to \mathbb{R}$ given by

$$s(\boldsymbol{x}, \boldsymbol{y}) = |(x_1 - x_2) + (y_1 - y_2)|, \qquad (6.1.1)$$

is a semimetric. Later, in Chap. 9, we will come across an interesting semimetric on the space $L^p(\mathbb{R})$ of p-integrable real-valued functions defined on $\mathbb{R}$.

Definition 6.1 (*Metric Space*) A nonempty set X with a metric d is called a *metric space*, which we write as (X, d). Similarly, a set X with a semimetric s is called a *semimetric space*, which we write as (X, s).

Let (X, s) be a semimetric space, and $\sim$ be a relation on X given by

$$x \sim y \quad \Leftrightarrow \quad s(x, y) = 0, \text{ for } x, y \in X.$$

Clearly then, $\sim$ is an equivalence relation on the set X. Consider the collection of equivalence classes

$$X' = \{[x] : x \in X\}.$$

It then follows easily that X' is a metric space with respect to the induced function $d' : X' \times X' \to \mathbb{R}$ given by

$$d'([x], [y]) = d(x, y), \quad \text{for } x, y \in X.$$

It is called the *quotient space* of the semimetric space (X, s), which we may write as X/s. In certain situations, this way of constructing a metric space from a semimetric space proves to be a useful tool.

Lemma 6.1 *Every nonempty set is a metric space.*

Proof Let X a nonempty set, and define

$$\delta(x, y) = \begin{cases} 0, & \text{if } x = y \\ 1, & \text{if } x \neq y \end{cases}, \quad \text{for } x, y \in X. \qquad (6.1.2)$$

Then, with this metric, the set X is a metric space. Clearly, $\delta(x, y) \geq 0$. Also, by definition, we have $\delta(x, y) = 0$ if and only if $x = y$ and $\delta(y, x) = \delta(x, y)$. Since $\delta(x, z) = 0$ or $\delta(x, z) = 1$, we verify the triangle inequality

$$\delta(x, z) \leq \delta(x, y) + \delta(y, z)$$

in each case separately. In the first case, $x = z$, and so

$$\delta(x, y) + \delta(y, z) = 2\,\delta(y, z).$$

Clearly then,

$$\delta(x, z) = 0 \le 2\delta(y, z) = \delta(x, y) + \delta(y, z).$$

The next case is $x \ne z$. If $x = y$ and $y = z$, then $x = z$. Since $x \ne z$, we must have either $x \ne y$ or $y \ne z$ and so $\delta(x, y) = 1$ or $\delta(y, z) = 1$ and hence $\delta(x, y) + \delta(y, z) \ge 1$. Finally, we have

$$\delta(x, z) = 1 \le \delta(x, y) + \delta(y, z).$$

Hence, δ is a metric on X.

The metric δ defined above is known as the *discrete metric*, and the pair (X, δ) is called the *discrete metric space*. There is no practical use of the metric δ. However, in some situations, it helps to construct *non-examples* for some concepts of theoretic significance.

To continue with the general discussion, we next prove that there are certain easy ways to construct new metrics from a known metric defined on a (carrier) set.

Lemma 6.2 *Let (X, d) be a metric space, and define $\rho : X \times X \to \mathbb{R}$ by*

$$\rho(x, y) = \min\{1, d(x, y)\}, \quad x, y \in X.$$

Then, (X, ρ) is metric space.

Proof Clearly, $\rho(x, y) \ge 0$, as so is $d(x, y)$. If $d(x, y) = 0$, then

$$\rho(x, y) = \min\{1, d(x, y)\} = \min\{1, 0\} = 0.$$

If $\rho(x, y) = 0$, then

$$\rho(x, y) = \min\{1, d(x, y)\} = 0$$

and so $d(x, y) = 0$. Hence, $x = y$. Since $d(x, y) = d(y, x)$, we have

$$\rho(x, y) = \min\{1, d(x, y)\} = \min\{1, d(y, x)\} = \rho(y, x).$$

To prove the triangle inequality for ρ, we consider two cases. First, if either $d(x, y) \ge 1$ or $d(y, z) \ge 1$, then $\rho(x, y) = 1$ or $\rho(y, z) = 1$, and hence $\rho(x, y) + \rho(y, z) \ge 1$. Using this, we see that

$$\rho(x, z) \le 1 \le \rho(x, y) + \rho(y, z).$$

Now, consider the case when $d(x, y) \le 1$ and $d(y, z) \le 1$. In this case, $\rho(x, y) = d(x, y)$ and $\rho(y, z) = d(y, z)$. Using the triangle inequality satisfied by d: $d(x, z) \le d(x, y) + d(y, z)$, we get

$$\rho(x, z) \le d(x, z) \le d(x, y) + d(y, z) = \rho(x, y) + \rho(y, z).$$

Therefore, the function ρ is a metric on X.

Lemma 6.3 (Bounded Metric) *Let d be a metric on a set X. Define*

$$\rho(x, y) = \frac{d(x, y)}{1 + d(x, y)}, \quad x, y \in X.$$

Then, (X, ρ) is metric space.

Proof As $d(x, y) \geq 0$, we have $\rho(x, y) \geq 0$. Also, $\rho(x, y) = 0$ if and only if $d(x, y) = 0$ and this holds if and only if $x = y$. Also, $\rho(x, y) = \rho(y, x)$. We prove the triangle inequality for ρ using the triangle inequality for d: $d(x, z) \leq d(x, y) + d(y, z)$. Using this, we have

$$\begin{aligned}
\rho(x, z) &= \frac{d(x, z)}{1 + d(x, z)} = 1 - \frac{1}{1 + d(x, z)} \\
&\leq 1 - \frac{1}{1 + d(x, y) + d(y, z)} \\
&= \frac{d(x, y)}{1 + d(x, y) + d(y, z)} + \frac{d(y, z)}{1 + d(x, y) + d(y, z)} \\
&\leq \frac{d(x, y)}{1 + d(x, y)} + \frac{d(y, z)}{1 + d(y, z)} = \rho(x, y) + \rho(y, z)
\end{aligned}$$

Therefore, the function ρ is also a metric on X.

Lemma 6.4 (Product Spaces) *Let (X_i, d_i) be metric spaces, for $1 \leq i \leq n$, and let $\underline{X} := X_1 \times \cdots \times X_n$. For brevity, we write $\underline{x} = (x_1, \ldots, x_n) \in \underline{X}$. For $p \in [1, \infty)$, let $d_p : \underline{X} \to \mathbb{R}$ be given by*

$$d_p(\underline{x}, \underline{y}) = \left(\sum_{i=1}^{n} [d_i(x_i, y_i)]^p \right)^{1/p}, \quad \text{for } \underline{x}, \underline{y} \in \underline{X}.$$

Then, $(\underline{X}, d_p)$ is a metric space, for every $p \in [1, \infty)$.

Proof Clearly, d_p satisfies the first two properties of a distance function. The verification of the triangle inequality for d_p is left to the reader (Exercise 6.10). So, d_p is a metric, for each $1 \leq p < \infty$. Further, by taking the limit as $p \to \infty$, we have that

$$d_\infty(\underline{x}, \underline{y}) = \max_{1 \leq i \leq n} \{d_i(x_i, y_i)\},$$

is also a metric on the set $\underline{X}$. Again, the verification of the triangle inequality for d_∞ is left to the reader (Exercise 6.11). An interesting special case of this construction is when each $X_i = \mathbb{R}$, and n is allowed to be infinite (see Example 6.4).

In general, given any metric space (X, d), there are numerous ways to construct new metrics on the (carrier) set X. For example, given any $\alpha > 0$, it follows easily that *dilated metric* (or a *scaled metric*) $d_\alpha = \alpha d$ defines a metric on X, which is called a *dilation* of the metric d by a *factor* α. Similarly, given any two metrics on the set X, the *sum metric* is again a metric on the set X. Some other types of metric spaces, and the related properties, are discussed in the subsequent sections of the chapter.

6.1.1 The Euclidean Space $\mathbb{R}^n$

In Chap. 2, we introduced algebraic, order, and metric structures on the set $\mathbb{R}$ of real numbers. The interplay between the algebraic and order structures helped us to prove that the set $\mathbb{R}$ is a unique *complete ordered field*. We used *modulus metric* on $\mathbb{R}$, in Chap. 3, to discuss the convergence of sequences and infinite series of real numbers; and, in Chap. 4, to study the three fundamental analytical properties of a single-variable real-valued function defined on an interval. Again, in Chap. 5, we discussed topological properties of the linear space $\mathbb{R}$ with respect to the *modulus metric*. The part of *real analysis* discussed so far provides enough mathematical tools to deal with many diverse types of 1-*dimensional problems* concerning various phenomena important to science and engineering.

For example, with little extra effort, we can deal with the *Cauchy Problem* related to finding a continuously differentiable function $x = x(t) : I \to \mathbb{R}$ such that it satisfies the following *initial-value problem*

$$\frac{dx}{dt} = f(t, x), \quad \text{with} \quad x(t_0) = x_0, \tag{6.1.3}$$

where $f(t, x)$ is a real-valued continuous function defined on an open set $D \subset (t_0 - \epsilon, t_0 + \epsilon) \times \mathbb{R}$, with $(t_0, x_0) \in D$ being an *interior point*. Notice that, we must have that

$$\Gamma_x := \{(t, x(t)) : t \in I\} \subset D,$$

where Γ_x is the *graph* of the function $x = x(t)$. To solve the problem, assuming that $f(t, x)$ is bounded over D, and satisfies the Lipschitz condition

$$\|f(t, x) - f(t, y)\| \leq C|x - y|, \quad \text{over} \quad D,$$

the basic idea is to apply the method of *successive approximations* to the integral equation of the form

$$x(t) = x_0 + \int_{t_0}^{t} f(t, x(t))dt, \quad \text{for} \quad t \in I.$$

In the beginning of the twentieth century, the work of *Ivar Fredholm* (1866–1927), *David Hilbert* (1862–1943), and *Erhard Schmidt* (1876–1959), on the general theory of integral equations laid foundation of *abstract analysis*. Fredholm and

Hilbert had similar approach to solve the main problem, namely, to solve first the associated eigenvalue problem for finite dimensional spaces and then extend the ideas to the infinite dimensional setting by using routine convergence techniques with respect to some suitable metric. The focus of the later part of the study was to use a (unique) *eigenfunctions expansion of a square integrable* real-valued function in terms of an *orthogonal system* of some simpler types of functions. Therefore, the issues related to *convergence* of expansions needed serious attention. We will explain some of the terms used above in this chapter, and others in the subsequent chapters. The point here is to take the gist of the story.

In 1906, *Maurice Fréchet* (1878–1973) investigated the relationship between the metric and geometric structures of such types of infinite dimensional function spaces in his doctoral dissertation *Sur quelques points du calcul functionnel*. He introduced the concept of metric spaces[2] (*E-class* for him), and important notions such as completeness, compactness, etc. His main purpose was to develop a unifying theory for various types of convergence that were in use for different kind of function spaces. The abstract theory of metric spaces for arbitrary sets was developed by Hausdorff in 1914. His fundamental work on measure spaces and abstract topological spaces appeared in the famous book *Grundzüge der Mengenlehre*. The interested reader may refer to the text [1] for more details.

Further, motivated by the work of Fredholm and Hilbert, *Frigyes Riesz* (1880–1956) and *Maurice Fréchet* (1878–1973) introduced the concept of *inner product*[3] on the space $L^2(\mathbb{R})$, and established independently in 1907 an important result of functional analysis, which is now known as the *Riesz Representation Theorem*. In 1929, *John von Neumann* (1903–1957) developed the abstract theory of Hilbert spaces to *model* and *solve* problems of quantum mechanics. In fact, the theory proposed by von Neumann culminated from the above mentioned work of Fredholm, Hilbert, and Schmidt.

In general, to study multivariable problems related to various types of physical phenomena, we need a suitable space to do analysis in dimension $n \geq 2$. In what follows, we show that the set

$$\mathbb{R}^n = \left\{ x = (x_1, \ldots, x_n) : x_k \in \mathbb{R} \right\}$$

of *n-tuples* of real numbers has all the requisite properties to obtain a *mathematical formulation* of some significant problems in science and engineering. More precisely, we prove that $\mathbb{R}^n$ is a *Hilbert space*. To start with, we first discuss the linear structure of $\mathbb{R}^n$. Recall the definition of an abelian group from Chap. 2.

Definition 6.2 An abelian group $(V, +)$ is said to be a *linear space* over a field $(F, +, \cdot, 0, 1)$ if there is a mapping $(a, v) \mapsto a \cdot v : F \times V \to V$, called the *scalar multiplication*, such that the following conditions hold, for all $a, b \in F$ and $u, v \in V$:

1. $a \cdot (u + v) = a \cdot u + a \cdot v$;

---

[2] Fréchet used *écart* for metric. The terminology *metric space* is due to Felix Hausdorff.

[3] The first usage is attributed to *Giuseppe Peano* who used it in his 1888 work on linear spaces.

2. $(a + b) \cdot u = a \cdot u + b \cdot u$;
3. $(a \cdot b) \cdot u = a \cdot (b \cdot u)$.
4. $1 \cdot u = u$.

We call elements of V as *vectors*, and of the field F as scalars.

   In particular, a linear space over the field $\mathbb{R}$ is called a *real linear space*. All linear spaces we consider in this text are over the field $\mathbb{R}$, so we may simply write *linear spaces*. For example, we can show that $V = \mathbb{R}$ itself is a linear space. By Definition 2.2, $(\mathbb{R}, +)$ is an abelian group, where $+$ is the usual addition of the real numbers. Also, since $V = F = \mathbb{R}$ in this case, we may take the usual multiplication of real numbers as the *scalar multiplication*. Notice that the conditions (1) and (2) of the above definition are same as the distributivity property of $\mathbb{R}$, and the condition (3) is the associativity of multiplication of real numbers. Similarly, we can see that the abelian group $(\mathbb{R}, +)$ is a linear space over the field $F = \mathbb{Q}$ of rational numbers, by using the fact that $ra \in \mathbb{R}$, for all $r \in \mathbb{Q}$ and $a \in \mathbb{R}$.
   Next, for a real number $a \in \mathbb{R}$, and *n-tuples*

$$\boldsymbol{x} = (x_1, \ldots, x_n), \quad \boldsymbol{y} = (y_1, \ldots, y_n) \in \mathbb{R}^n,$$

define the addition and scalar multiplication as follows:

$$\boldsymbol{x} + \boldsymbol{y} = (x_1 + y_1, x_2 + y_2, \ldots, x_n + y_n);$$
$$a \cdot \boldsymbol{x} = (ax_1, ax_2, \ldots, ax_n).$$

It then follows easily that the set $\mathbb{R}^n$ of *n-tuples* is a linear space. The related verifications of conditions are left for the reader as a simple exercise. Notice that the *n-tuple* of zeros $\boldsymbol{0} = (0, \ldots, 0)$ is the additive identity of this linear space, which we will call the *zero vector*. The next example illustrate a general procedure to produce examples of many interesting linear spaces.

Example 6.1 Let X be a nonempty set, and consider the set

$$V = \mathbb{R}^X := \{f : f : X \to \mathbb{R}\}.$$

We define the *sum* and *scalar multiplication* on V *pointwise*. That is, for $f, g \in V$ and $a \in \mathbb{R}$, we write

$$(f + g)(x) = f(x) + g(x), \quad x \in X;$$
$$(af)(x) = a f(x), \quad x \in X.$$

It is then easy to see that $(V, +)$ is an abelian group, where the additive identity is given by the *zero function* $0(x) = 0$, for all $x \in X$. Also, the conditions (1) through (4) of Definition 6.2 hold for the abelian group $(V, +)$. So, $V = \mathbb{R}^X$ is a linear space. Notice that, to make the set $V = \mathbb{R}^X$ a linear space, we only need the fact that $\mathbb{R}$ itself is a linear space. Therefore, we can use any linear space in place of $\mathbb{R}$.

Notice that, when $X = \mathbb{N}$, $V = \mathbb{R}^{\mathbb{N}}$ is the linear space of *real sequences*. Also, by taking $X = \mathbb{N}_m \times \mathbb{N}_n$, it follows that the set $V = M_{m,n}(\mathbb{R})$ of $m \times n$ real matrices is a linear space. Further, for $X = [a, b] \subset \mathbb{R}$, we conclude that a collection of real-valued functions defined on the interval X such as $C^1([a, b])$ is a linear space.

Now, suppose $X = \mathbb{N}_m = \{1, \ldots, m\}$, and consider the m-functions $e_1, \ldots, e_m$ in the real linear space $V = \mathbb{R}^X$ given by

$$e_i(j) = \begin{cases} 1, & \text{when } i = j \\ 0, & \text{when } i \neq j \end{cases}, \quad 1 \leq i, j \leq m.$$

So, every function $f \in V$ can be written (uniquely) as a *linear combination* of the form

$$f = a_1 e_1 + a_2 e_2 + \cdots + a_m e_m,$$

where $a_k = f(k)$, for $k = 1, \ldots, m$. We then say that the set $B = \{e_1, \ldots, e_m\}$ is a *basis* of the linear space V over $\mathbb{R}$. Notice that, if we identify the function e_k with $k \in \mathbb{N}_m$, then the set $X = \{1, \ldots, m\}$ itself can be viewed as a *basis* of the real linear space $V = \mathbb{R}^X$. More generally, if V is a linear space over a field F, we say $B \subset V$ is a *spanning set* for the linear space V if every $v \in V$ is a linear combination of elements of B. That is, there are some elements $b_1, \ldots, b_n \in B$ and scalars $a_1, \ldots, a_n \in F$ such that

$$v = a_1 b_1 + a_2 b_2 + \cdots + a_n b_n.$$

In this case, we write $V = \langle B \rangle_F$. Further, we say $B \subset V$ is *linearly independent* over the field F if, for every finite set $\{b_1, \ldots, b_m\} \subseteq B$,

$$a_1 b_1 + a_2 b_2 + \cdots + a_n b_n = 0 \quad \Rightarrow \quad a_1 = a_2 = \cdots = a_n = 0.$$

That is, the span $\langle B \rangle_F$ contains no *non-trivial* linear combination. In particular, for any $v \neq 0$ in a linear space V over F, the set $B = \{v\}$ is linearly independent. For $v \neq 0$, the set

$$L(v) := \{a\,v : a \in \mathbb{R}\}$$

is called a *line* in the space V passing through the vector $v \in V$. Also, for any $w \notin L(v)$, the set

$$P(v, w) := \{a\,v + b\,w : a, b \in \mathbb{R}\}$$

is called a *plane* in V passing through the vectors $v, w \in V$. We may continue this process *ad infinitum* unless V is a *finite dimensional* space in the following sense.

Definition 6.3 Let V be a linear space over a field F. A set $B \subset V$ is called a *basis* of V if B is a linearly independent, spanning set of V. We say V is a *finite dimensional* space over the field F if it has a basis B with finite number of elements. Otherwise, we say V is an *infinite dimensional* linear space.

It is a theorem in linear algebra that any two bases of a *finite dimensional* linear space have the same *cardinality*. So, in a finite dimensional linear space V over a field F, the *cardinality* of any basis of V is the *dimension* of the linear space V. For example, the dimension of the linear space $\mathbb{R}^n$ is n because the set B of the n-vectors

$$e_1 = (1, 0, , \ldots, 0),\, e_2 = (0, 1, 0, \ldots, 0),\, , \cdots ,\, , e_n = (0, 0, \ldots, 1),$$

form a *basis* of the linear space $\mathbb{R}^n$. This is called the *standard basis* of the linear space $\mathbb{R}^n$ in the sense that it automatically provides the *positively oriented n-unit vectors* along the n mutually orthogonal axes. Each point $x = (x_1, \ldots, x_n) \in \mathbb{R}^n$ is viewed as an n-dimensional *row vector* or a *column vector*. Depending upon the practical aspects of a problem under study, we choose one of the two representations. In actual applications, while dealing with matrices, we prefer to write points of $\mathbb{R}^n$ as *column vectors*.

Definition 6.4 A (real) linear space V is called an *inner product space* if there exists a mapping $(u, v) \mapsto \langle u, v \rangle : V \times V \to \mathbb{R}$ satisfying the following conditions: For all $a, b \in \mathbb{R}$ and $u, v \in V$,

1. $\langle u, v \rangle = \langle v, u \rangle$;
2. $\langle a\, u + b\, v, w \rangle = a \langle u, w \rangle + b \langle v, w \rangle$;
3. $\langle u, u \rangle > 0$, for $u \neq 0$.

The function $\langle\,,\,\rangle : V \times V \to \mathbb{R}$ is called an *inner product* on the space V.

For example, the usual *dot product* of two vectors $x = (x_1, \ldots, x_n)$, $y = (y_1, \ldots, y_n) \in \mathbb{R}^n$ given by

$$\langle x, y \rangle := x \cdot y = x_1 y_1 + \cdots + x_n y_n, \tag{6.1.4}$$

satisfies all the three conditions of Definition 6.4. It thus follows that the linear space $\mathbb{R}^n$ is an *inner product space* with respect to usual *dot product* of vectors. Further, the mapping $x \mapsto \|x\| : \mathbb{R}^n \to \mathbb{R}$ given by

$$\|x\| := +\sqrt{\langle x, x \rangle} = \sqrt{\sum_{k=1}^{n} x_k^2}, \quad x = (x_1, \ldots, x_n) \in \mathbb{R}^n, \tag{6.1.5}$$

is a *positive, homogeneous* function such that Theorem 6.1 holds. For $x \in \mathbb{R}^n$, $\|x\|$ is also called the *radial length* of the vector x. Notice that, the second assertion in above theorem says that the function $\|\ \| : \mathbb{R}^n \to \mathbb{R}$ given by (6.1.5) is a *subadditive function*. That is, it defines a *norm* on the linear space $\mathbb{R}^n$ in the sense of the next definition.

Definition 6.5 Let V be a (real) linear space. A function $n : V \to \mathbb{R}$ is called a *norm* if it satisfies the following conditions: For all $u, v \in V$ and $a \in F$:

4. n is *positive*, i.e., $n(u) > 0$, for $u \neq \mathbf{0}$.
5. n is *homogeneous*, i.e., $n(a\,u) = |a|\,n(u)$;
6. n is *subadditive*, i.e., $n(u + v) \leq n(u) + n(v)$;

A linear space V with a norm function $n : V \to \mathbb{R}$ is called a *normed space*.

Theorem 6.1 *If $x, y \in \mathbb{R}^n$, then*

$$\left| \langle x, y \rangle \right| \leq \|x\| \, \|y\| \quad and \quad \|x + y\| \leq \|x\| + \|y\|.$$

Proof Let $a = \|y\|^2$, $b = \langle x, y \rangle$ and $c = \|x\|^2$. Notice that $a \geq 0$. If $a = 0$, then $y = 0$ and in this case, $\langle x, y \rangle = 0$ and $\|y\| = 0$. Therefore, the inequality $\langle x, y \rangle \leq \|x\| \, \|y\|$ holds, trivially. So, we may assume that $a > 0$. Now, using $\langle x, x \rangle = \|x\|^2$, we have

$$\begin{aligned}
0 \leq \|x + \lambda y\|^2 &= \langle x + \lambda y, x + \lambda y \rangle \\
&= \langle x, x \rangle + 2\lambda \langle x, y \rangle + \lambda^2 \langle y, y \rangle \\
&= \|x\|^2 + 2\lambda \langle x, y \rangle + \lambda^2 \|y\|^2 \\
&= a\lambda^2 + 2b\lambda + c
\end{aligned}$$

for all λ. By taking $\lambda = -b/a$, we see that

$$-\frac{b^2}{a} + c \geq 0.$$

Since $a > 0$, we obtain $b^2 \leq ac$ or, equivalently,

$$\left| \langle x, y \rangle \right|^2 \leq \|x\|^2 \|y\|^2.$$

Taking square roots on both side, we obtain

$$\left| \langle x, y \rangle \right| \leq \|x\| \, \|y\|, \tag{6.1.6}$$

as asserted. Next, for $x, y \in \mathbb{R}^n$, we have

$$\begin{aligned}
\|x + y\|^2 &= \langle x + y, x + y \rangle \\
&= \|x\|^2 + \|y\|^2 + 2\langle x, y \rangle \\
&\leq \|x\|^2 + \|y\|^2 + 2\|x\| \, \|y\| \\
&= (\|x\| + \|y\|)^2.
\end{aligned}$$

The proof is completed by using the fact that the square root function is an increasing function on the interval $[0, \infty)$.

The inequality (6.1.6) in the proof of Theorem 6.1 is known as the *Cauchy-Schwarz inequality*, which we can use to define the concept of the *angle* θ between two vectors x, $y \in \mathbb{R}^n$:

$$\cos \theta := \frac{\langle x, y \rangle}{\|x\| \, \|y\|}; \tag{6.1.7}$$

Notice that, for all x, $y \in \mathbb{R}^n$, we have $|\cos \theta| \leq 1$. The argument used to prove Theorem 6.1 also implies the following important identity, which is known as the *parallelogram law*:

$$\|x + y\|^2 + \|x - y\|^2 = 2(\|x\|^2 + \|y\|^2). \tag{6.1.8}$$

For, since

$$\|x + y\|^2 = \|x\|^2 + \|y\|^2 + 2x \cdot y,$$

replacing y by $-y$, we get

$$\|x - y\|^2 = \|x\|^2 + \|y\|^2 - 2x \cdot y.$$

Adding the last two equations, we obtain the desired identity.

Notice that it follows from the definition of angle as in (6.1.7) that two vectors x, $y \in \mathbb{R}^n$ are *orthogonal* if $\langle x, y \rangle = 0$. Clearly then, we have

$$\langle e_i, e_j \rangle = 0, \quad \text{for} \quad 1 \leq i \neq j \leq n;$$
$$\langle e_i, e_i \rangle = \|e_i\|^2 = 1, \quad \text{for all} \quad i = 1, \ldots, n.$$

We say the *standard basis* $B = \{e_1, \ldots, e_n\}$ is an *orthonormal basis* for the inner product space $\mathbb{R}^n$ because each vector $x = (x_1, \ldots, x_n) \in \mathbb{R}^n$ can be written *uniquely* as

$$x = \langle x, e_1 \rangle e_1 + \langle x, e_2 \rangle e_2 + \ldots + \langle x, e_n \rangle e_n. \tag{6.1.9}$$

This property characterises every *orthonormal basis* for the linear space $\mathbb{R}^n$, and Definition 6.1.5 is independent of our choice for any orthonormal basis of the inner product space $\mathbb{R}^n$ (Exercise 6.8).

Definition 6.6 We say two vectors u, v in an inner product space $(V, \langle , \rangle)$ are *orthogonal* if $\langle u, v \rangle = 0$. In this case, we usually write $u \perp v$.

Finally, we introduce on $\mathbb{R}^n$ a metric that makes it most appropriate inner product space to study the problems of the real world.

Theorem 6.2 *For $x = (x_1, \ldots, x_n)$, $y = (y_1, \ldots, y_n) \in \mathbb{R}^n$, the* Pythagorean distance $d_2 : \mathbb{R}^n \times \mathbb{R}^n \to \mathbb{R}$ *given by*

$$d_2(x, y) = \|x - y\| = \sqrt{\sum_{k=1}^{n}(x_k - y_k)^2},\qquad(6.1.10)$$

is a metric on $\mathbb{R}^n$.

Proof Clearly, d_2 is a positive, symmetric function. To complete the proof, we show that d_2 satisfies the triangle inequality. For, let $x, y, z \in \mathbb{R}^n$. Then, by the second assertion of Theorem 6.1, we have

$$\begin{aligned} d_2(x, z) &= \|x - z\| \\ &= \|(x - y) + (y - z)\| \\ &\leq \|x - y\| + \|y - z\| \\ &= d_2(x, y) + d_2(y, z). \end{aligned}$$

Hence, $(\mathbb{R}^n, d_2)$ is a metric space.

The linear metric space $(\mathbb{R}^n, d_2)$ is called the *n*-dimensional *Euclidean space*, and the Pythagorean distance d_2 is called the *Euclidean metric*. It follows from the above discussion that, in the Euclidean space $\mathbb{R}^n$, we have all the fundamental geometric concepts such as the *length, angle,* and the *distance*. Moreover, as we will see in the next section, every Cauchy sequence in $(\mathbb{R}^n, d_2)$ converges. That is, the linear space $\mathbb{R}^n$ is *complete* with respect to the metric d_2. Therefore, as said earlier, $\mathbb{R}^n$ is a *Hilbert space*. Hence, we can do geometry and analysis over $\mathbb{R}^n$.

The inner product space $\mathbb{R}^n$ has many other types of metrics, where each hold a special significance to some practical problem. For example, in certain applications, it is more meaningful to use the Euclidean distance between *weighted vectors*. That is, if the *k*th values of vectors $x, y \in \mathbb{R}^n$ has the *weight attribute* given by $\omega_k \in \mathbb{R}$, then we use the Euclidean distance given by

$$d_2^w(x, y) = \sqrt{\sum_{k=1}^{n}\omega_k(x_k - y_k)^2}.\qquad(6.1.11)$$

The metric d on $\mathbb{R}^2$, as given in the next example, is called the *rail metric*. It has a simple practical interpretation. Suppose the *origin* $\mathbf{0} = (0, 0) \in \mathbb{R}^2$ represents the starting point of all metro trains of a city. To commute between places x and y, one may take a direct metro, if the two stations are on the same line (i.e., $x = \alpha y$); otherwise, one moves from x to the source station $\mathbf{0}$ and then to the station y.

Example 6.2 (*Rail Metric*) For $X = \mathbb{R}^2$, let $d : X \times X \to \mathbb{R}$ be defined as follows: For $x = (x_1, x_2), y = (y_1, y_2) \in X$, and some $0 \neq \alpha \in \mathbb{R}$, take

$$d(x, y) = \begin{cases} d_2(x, y), & \text{if } x = \alpha y \\ d_2(x, 0) + d_2(y, 0), & \text{otherwise} \end{cases}.$$

It follows easily that d is a metric on $\mathbb{R}^2$. The verification of details is left to the reader.

Notice that the *shortest distance* between two points x, $y \in \mathbb{R}^n$ with respect to the metric d_2 is given by the *length of the line joining the two points*. Said differently, in the metric space $(\mathbb{R}^n, d_2)$, the *geodesics* are given by the straight lines. Notice that only birds benefit from this *characterising property* of the Euclidean spaces $\mathbb{R}^n$. For rest all of us, what matters is the *taxicab metric* $d_1 : \mathbb{R}^n \times \mathbb{R}^n \to \mathbb{R}$ given by

$$d_1(x, y) = \sum_{k=1}^{n} |x_k - y_k|, \tag{6.1.12}$$

where $x = (x_1, \ldots, x_n)$, $y = (y_1, \ldots, y_n) \in \mathbb{R}^n$.

Example 6.3 (*Taxicab Metric*) Let d_1 be as defined in (6.1.12). As $|x_k - y_k| \geq 0$, for $k = 1, \ldots, n$, we have $d_1(x, y) \geq 0$. Clearly, $d_1(x, x) = 0$. Next,

$$d_1(x, y) = 0 \quad \Rightarrow \quad |x_k - y_k| = 0, \text{ for } k = 1, \ldots, n.$$

Or, equivalently, $x_k = y_k$, for all $k = 1, \ldots, n$. Therefore, $d_1(x, y) = 0 \Rightarrow x = y$. Also, since $|-x| = |x|$ for any $x \in \mathbb{R}$, we have

$$d_1(y, x) = \sum_{k=1}^{n} |y_k - x_k| = \sum_{k=1}^{n} |-(x_k - y_k)| = \sum_{k=1}^{n} |x_k - y_k| = d_1(x, y).$$

Finally, by triangle inequality property of modulus metric, we have

$$|x_k - z_k| \leq |x_k - y_k| + |y_k - z_k|, \quad \text{for all } k = 1, \ldots, n,$$

which implies that

$$\begin{aligned} d_1(x, z) &= \sum_{k=1}^{n} |x_k - z_k| \\ &\leq \sum_{k=1}^{n} (|x_k - y_k| + |y_k - z_k|) \\ &\leq \sum_{k=1}^{n} |x_k - y_k| + \sum_{k=1}^{n} |y_k - z_k| \\ &= d_1(x, y) + d_1(y, z). \end{aligned}$$

Hence, $(\mathbb{R}^n, d_1)$ is a metric space. The metric d_1 is also known as the *Manhattan metric*. This particular metric holds special significance because, for many practical applications,[4] it is more logical to work with the metric space $(\mathbb{R}^n, d_1)$. Notice that d_1 defines a *non-Euclidean geometry* on $\mathbb{R}^n$ in the sense that there could be more than one *geodesics* between points of the space $\mathbb{R}^n$.

More generally, the Euclidean metric d_2 and the taxicab metric d_1 are two special cases of a family of metrics $d_p : \mathbb{R}^n \times \mathbb{R}^n \to \mathbb{R}$ introduced by the German physicist *Hermann Minkowski* (1864–1909), which are now known as the *Minkowski metrics*. For $1 \leq p < \infty$, we define

$$d_p(x, y) = \left(\sum_{k=1}^{n} |x_k - y_k|^p \right)^{1/p}, \qquad (6.1.13)$$

where $x = (x_1, \ldots, x_n)$, $y = (y_1, \ldots, y_n) \in \mathbb{R}^n$. Clearly, each d_p is a positive, symmetric function. The triangle inequality of d_p follows from a generalisation of Theorem 6.1, which is known as the *Minkowski-Hölder inequality* (Exercise 6.13). Also, the *supremum metric* $d_\infty : \mathbb{R}^n \times \mathbb{R}^n \to \mathbb{R}$ is given by

$$d_\infty(x, y) = \max_{1 \leq k \leq n} |x_k - y_k|, \qquad (6.1.14)$$

where $x = (x_1, \ldots, x_n)$, $y = (y_1, \ldots, y_n) \in \mathbb{R}^n$. It is also known as the *Chebyshev metric*, named after the Russian mathematician *Pafnuty Chebyshev* (1821–1894). Notice that the triangle property for d_p fails when $p < 1$. For example, in $\mathbb{R}^2$, the point $(0, 1)$ is at a unit distance from the points $(0, 0)$ and $(1, 1)$, but

$$d_p\big((0, 0), (1, 1)\big) = 2^{1/p} > 2.$$

The metric d_∞ is also known as the *uniform metric* because, as we will see, it defines the uniform convergence of sequences. Notice that the function $d_\infty : \mathbb{R}^n \times \mathbb{R}^n \to \mathbb{R}$ has the property

$$d_\infty(x, y) =: \lim_{p \to \infty} d_p(x, y) = \max_{1 \leq k \leq n} |x_k - y_k|. \qquad (6.1.15)$$

6.1.2 Function Spaces

We introduce here some important classes of *infinite dimensional function spaces*. Like $\mathbb{R}^n$, there are more than one metrics defined on the carrier set. We begin with the space of real-valued functions defined on the set $\mathbb{N}$ of natural numbers, i.e., the spaces of real sequences.

---

[4] E. F. Krause, Taxicab Geometry, Dover Publications, New York, 1975.

Example 6.4 (*Sequence Spaces*) Let $1 \leq p < \infty$. We write $\ell_p := \ell_p(\mathbb{R})$ for the set of all *p-summable sequences* $\langle x_n \rangle$ of real numbers. That is, we have

$$\langle x_n \rangle \in \ell_p \quad \Leftrightarrow \quad \sum_{n=1}^{\infty} |x_n|^p < \infty.$$

Notice that, in particular, the set ℓ_1 consists of the sequences defining *absolutely convergent series* of real numbers. Recall that each ℓ_p is a real linear space with respect to *pointwise* addition and scalar multiplication. Further, for $1 \leq p < \infty$, the space ℓ_p is infinite dimensional because it contains the linearly independent p-summable sequences of the form $e_k = \langle \delta_{kn} \rangle$, for $k \geq 1$, where δ_{kn} is the *Kronecker delta*. Let

$$x = (x_1, x_2, \dots), \quad y = (y_1, y_2, \dots) \in \ell_p,$$

and a function $d_p : \ell_p \times \ell_p \to \mathbb{R}$ be given by

$$d_p(x, y) = \left(\sum_{i=1}^{\infty} |x_i - y_i|^p \right)^{1/p}. \tag{6.1.16}$$

Clearly, d_p is a positive, symmetric function, and the triangle inequality follows from the discrete version of Hölder-Minkowski's inequality (Exercise 6.13). So, (ℓ_p, d_p) is a metric space, for each $1 \leq p < \infty$, which are usually referred to as the **ℓ_p-spaces**. In particular, the metric linear space (ℓ_2, d_2) plays an important role in many parts of mathematics, including Fourier series expansion of periodic functions, as discussed in Chapter 10. Next, consider the linear space of ℓ_∞ of *bounded sequences* of real numbers. In this case, it is natural to define a metric $d_\infty : \ell_\infty \times \ell_\infty \to \mathbb{R}$ as follows:

$$d_\infty(x, y) = \sup_{n \in \mathbb{N}} |x_n - y_n|. \tag{6.1.17}$$

It follows easily that (ℓ_∞, d_∞) is a metric space.

We need to explain some notations before introducing our next class of function spaces. Recall that $C(\mathbb{R})$ is the linear space of real-valued continuous functions defined on the line $\mathbb{R}$. We write $C_c(\mathbb{R})$ for the space of functions $f \in C(\mathbb{R})$ with *compact support*. That is,

$$f \in C_c(\mathbb{R}) \quad \Leftrightarrow \quad K = \overline{\{a \in \mathbb{R} : f(a) \neq 0\}}$$

is a compact set in $\mathbb{R}$. And, we write $C_0(\mathbb{R})$ for the linear space of functions $f \in C_c(\mathbb{R})$ *vanishing at infinity*. That is, for each $\epsilon > 0$, there is some compact set $K \subset \mathbb{R}$ such that $|f(x)| < \epsilon$ for $x \notin K$. Clearly then, $C_c(\mathbb{R}) \subset C_0(\mathbb{R})$. Since

$$f(x) = \frac{1}{1 + x^2} \in C_0(\mathbb{R}) \setminus C_c(\mathbb{R}),$$

Fig. 6.1 Graph of
$f(x) = 1/(1 + x^2)$

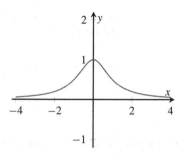

the inclusion is strict (see Fig. 6.1). We also need the continuous version of Hölder-Minkowski's inequality, as given in the next theorem. Notice that it is assumed that the involved integrals exist in Riemann sense.

Theorem 6.3 (Hölder-Minkowski's Inequality) *For any real number $p \geq 1$, and $f, g \in C_0(\mathbb{R})$,*

$$\left(\int_{\mathbb{R}} |f(t) + g(t)|^p dt \right)^{\frac{1}{p}} \leq \left(\int_{\mathbb{R}} |f(t)|^p dt \right)^{\frac{1}{p}} + \left(\int_{\mathbb{R}} |g(t)|^p dt \right)^{\frac{1}{p}}. \quad (6.1.18)$$

Proof For any $f \in C_0(\mathbb{R})$, we write

$$I_p(f) = \left(\int_{\mathbb{R}} |f(t)|^p dt \right)^{\frac{1}{p}}.$$

We first prove the following continuous version of Hölder's inequality: *Let $0 < p, q \in \mathbb{R}$ be a pair of conjugate exponents, i.e., $1/p + 1/q = 1$, with $1 < q < \infty$. Then, for any $f, g \in C_0(\mathbb{R})$,*

$$\int_{\mathbb{R}} |f(t)g(t)| dt \leq I_p(f) I_q(g).$$

Of course, there is nothing to prove if f or g is identically zero. So, we may assume that both the integrals $I_p(f)$ and $I_q(g)$ are *positive* real numbers. Then, the following holds (Exercise 4.4):

$$\frac{|f(x)g(x)|}{I_p(f)I_q(g)} = \frac{|f(x)|}{I_p(f)} \frac{|g(x)|}{I_q(g)} \leq \frac{1}{p} \frac{|f(x)|^p}{[I_p(f)]^p} + \frac{1}{q} \frac{|g(x)|^q}{[I_q(g)]^q}.$$

Integrating over $\mathbb{R}$, we have

$$\int_{\mathbb{R}} \frac{|f(x)g(x)|}{I_p(f)I_q(g)}dx \leq \frac{1}{p}\int_{\mathbb{R}}\frac{|f(x)|^p}{[I_p(f)]^p}dx + \frac{1}{q}\int_{\mathbb{R}}\frac{|g(x)|^p}{[I_q(g)]^q}dx$$

$$= \frac{1}{p}\frac{[I_p(f)]^p}{[I_p(f)]^p} + \frac{1}{q}\frac{[I_q(g)]^q}{[I_q(g)]^q}$$

$$= \frac{1}{p} + \frac{1}{q} = 1,$$

which proves the Hölder inequality, as asserted. Finally, to prove (6.1.18), observe that a simple computation takes care of the case $p = 1$. So, we may assume that $p > 1$, and take $q \in \mathbb{R}$ such that (p, q) forms a pair of *conjugate exponents*. Now, since

$$I = \int_{\mathbb{R}}|f(x)+g(x)|^p dx$$

$$= \int_{\mathbb{R}}|f(x)+g(x)||f(x)+g(x)|^{p-1}dx$$

$$\leq \int_{\mathbb{R}}|f(x)||f(x)+g(x)|^{p-1}dx + \int_{\mathbb{R}}|g(x)||f(x)+g(x)|^{p-1}dx$$

$$= \int_{\mathbb{R}}|f(x)||[f(x)+g(x)]^{p-1}|dx + \int_{\mathbb{R}}|g(x)||[f(x)+g(x)]^{p-1}|dx,$$

by applying Hölder inequality to integrals on the right side, we have

$$I \leq I_p(f)I_q([f+g]^{p-1}) + I_p(g)I_q([f+g]^{p-1})$$

$$= (I_p(f)+I_p(g))I_q([f+g]^{p-1})$$

$$= (I_p(f)+I_p(g))\left(\int_{\mathbb{R}}|f(x)+g(x)|^{(p-1)q}dx\right)^{1/q}$$

$$= (I_p(f)+I_p(g))\left(\int_{\mathbb{R}}|f(x)+g(x)|^p dx\right)^{1-1/p}.$$

So, if $\int_{\mathbb{R}}|f(x)+g(x)|^p dx = 0$, then we have $I_p(f+g) = 0$, and hence the inequality (6.1.18) holds trivially. And, when $\int_{\mathbb{R}}|f(x)+g(x)|^p dx \neq 0$, we divide the last inequality by $[\int_{\mathbb{R}}|f(x)+g(x)|^p dx]^{1-1/p}$ to obtain the Minkowski inequality, as asserted.

Example 6.5 (*Function spaces*) Consider the linear space $C[a, b]$ of continuous real-valued functions defined on a compact interval $[a, b]$ ($a < b$), where addition and scalar multiplication are given by *pointwise operations*. Since the polynomial functions $p(x)$ defined by $1, x, x^2, x^3, \ldots$ lie in $C[a, b]$, we conclude that it is an infinite dimensional space.

1. We first use the metric properties of the usual metric on $\mathbb{R}$ to conclude that the function

$$d_\infty(f, g) = \sup\{|f(x) - g(x)| \mid x \in [a, b]\}, \quad f, g \in C[a, b], \quad (6.1.19)$$

defines a metric on the set $C[a, b]$. The metric d_∞ is called the *uniform metric* because, as we will see in Chap. 8, it defines *uniform convergence* in the space $C[a, b]$. Further, as it measures the largest vertical distance between graphs of two functions, the metric d_∞ is frequently used to assess the error in approximating elements of $C[a, b]$ by polynomials (Theorem 8.12). Notice that these assertions hold even for the *larger* linear space $B[a, b]$ of bounded real-valued functions defined over $[a, b]$.

2. Next, it follows easily that the function

$$d_1(f, g) = \int_a^b |f(t) - g(t)| \, dt, \quad f, g \in C[a, b], \quad (6.1.20)$$

defines a metric on the space X. The only nontrivial part is to show that $d_1(f, g) = 0 \Rightarrow f \equiv g$. For, observe that if for some $h \in X$, $a_0 = h(t_0) > 0$ for $t_0 \in [a, b]$, then by the continuity of h, there exists a positive number $\delta < 1$ such that $|h(t) - a_0| < a_0/2$ for all $t \in [a, b] \cap [a_0 - \delta, a_0 + \delta]$. But then, we have

$$h(t) \geq a_0/2, \quad \text{for all} \quad t \in [a, b] \cap [a_0 - \delta, a_0 + \delta],$$

$$\Rightarrow \int_a^b h(t)dt \geq \int_{a_0-\delta}^{a_0+\delta} h(t)dt \geq \frac{\delta a_0}{2} > 0.$$

3. Finally, the function d_2 given by

$$d_2(f, g) = \left(\int_a^b [f(t) - g(t)]^2 dt \right)^{1/2}, \quad f, g \in C[a, b], \quad (6.1.21)$$

defines yet another interesting type of metric on the space $C[a, b]$. Verification of metric properties is left for the reader. Notice that the triangle inequality follows from (6.1.18).

We remark that the two metrics given by (6.1.19) and (6.1.20) do not satisfy the identity (6.1.8), and so none can give the space $C[a, b]$ a Hilbert space structure. However, the metric given by (6.1.21) satisfies the identity (6.1.8), and hence $(C[a, b], d_2)$ is a Hilbert space. The significance of this remark is discussed in a later chapter.

6.1.3 Metric Subspace

Let (X, d) be a metric space, and $\emptyset \neq Y \subset X$. Clearly then, the function $d_Y = d|_Y :$ $Y \times Y \to \mathbb{R}$ given by

$$d_Y(y_1, y_2) = d(y_1, y_2) \quad y_1, y_2 \in Y,$$

defines a metric on the set Y. We call (Y, d_Y) a *subspace* of the metric space (X, d), and d_Y is called the *induced metric*. However, since any $\emptyset \neq Y \subset X$ can be seen as a *subspace* of X in this way, we may continue writing d for d_Y. By using this simple scheme, we can obtain examples of many other interesting metric spaces. The next example illustrates the fact that $\mathbb{R}^k \subset \mathbb{R}^n$, for all $1 \leq k < n$, as a subspace of dimension k.

Example 6.6 By fixing any one coordinate of the points in the Euclidean space $(\mathbb{R}^n, d_2)$, say kth for $k = 1, \ldots, n$, we obtain a subset which we may write as Y_k. Clearly then, Y_k is a subspace of $\mathbb{R}^n$ of dimension $(n - 1)$ with a metric induced by the metric d_2. For example, when $n = 2$, the usual distance between points $x_1 = (x_1, y_1)$ and $x_2 = (x_2, y_2)$ in the plane $\mathbb{R}^2$ is given by the formula

$$d_2(x_1, x_2) = \sqrt{(x_1 - x_2)^2 + (y_1 - y_2)^2}.$$

We may identify $x \in \mathbb{R}$ with the point $x = (x, 0) \in \mathbb{R}^2$ on the x-axis so that restriction of d_2 to $\mathbb{R}$ gives

$$d_2(x, y) = \sqrt{(x - y)^2 + (0 - 0)^2} = |x - y|.$$

Therefore, the usual modulus metric on $\mathbb{R}$ is a particular case of the *Pythagorean distance* on $\mathbb{R}^2$.

The next example lists some important subspaces of (ℓ_∞, d_∞).

Example 6.7 The metric space (ℓ_∞, d_∞) contains the following three important subspaces, where the metric d_∞ is given by (6.1.17).

1. The space c of *convergent sequences* of real numbers. Notice that the inclusion is proper because $(-1, 1, -1, \ldots) \in \ell_\infty \setminus c$.
2. The space $c_0 \subset c$ of *null sequences*. Clearly, the inclusion is proper.
3. The space $c_{00} \subset c_0$ of null sequences having only *finitely many nonzero entries*.

Notice that $\ell_p \subset c_0$, for $1 \leq p < \infty$, and

$$\ell_1 \subset \ell_p \subset \ell_q \subset \ell_\infty,$$

for $1 < p < q < \infty$.

The induced metric may not always be a useful way of considering the distance between points of a subset of a metric space. For example, the induced Euclidean metric on the *unit circle*

$$\mathbb{S}^1 = \{(x, y) : x^2 + y^2 = 1\} \subset \mathbb{R}^2,$$

gives length of segments. Similarly, we are less interested to work with the Euclidean metric on the *surface of the earth*, which we may view as the sphere of radius $R > 0$ given by

$$S^2(\mathbf{0}; R) = \{x \in \mathbb{R}^3 : \|x\| = R\}.$$

In all such cases, it is more meaningful to consider an *intrinsic metric* on a subset of a metric space, which is usually *isometrically different* from the induced metric (Definition 6.10). For example, define the distance between two points on the unit circle $\mathbb{S}^1$ as the *arc length* of the shorter path pointing the two points. More generally, consider the *n-dimensional unit sphere* given by

$$\mathbb{S}^n = \{x \in \mathbb{R}^{n+1} : \|x\| = 1\}, \quad n \geq 1.$$

Notice that it has a metric structure as a subspace of $\mathbb{R}^{n+1}$ with respect to any one of the metrics we defined earlier.

Example 6.8 We use here the concept of the *angle* as in (6.1.7) (and not the length) to define the *angular metric* α on $\mathbb{S}^n$ as follows: By (6.1.6), we have

$$\langle x, y \rangle^2 \leq \|x\|^2 \|y\|^2 = 1, \quad \text{for } x, y \in \mathbb{S}^n.$$

So, there exists $0 \leq \alpha \leq \pi$ such that $\cos \alpha = \langle x, y \rangle$. Thus, we define

$$\alpha(x, y) = \cos^{-1}(\langle x, y \rangle) = 2 \sin^{-1}\left(\frac{\|x - y\|}{2}\right). \tag{6.1.22}$$

That is, α is the Euclidean angle between unit vectors at points x and y on the *great circle*, which is the length of the *shortest arc* of the great circle connecting these points. Clearly, α is positive and symmetric. However, the proof of the triangle inequality is bit involved (Exercise 6.14).

6.2 Topology of Metric Spaces

In historical terms, the first satisfactory definition of a *topology on a set* appeared in the famous book *Grundzüge der Mengenlehre* published by *Felix Hausdorff* (1868–1943) in 1914. However, the abstract definition of a *topology* in terms of a collection of subsets satisfying certain properties was given by *Kazimierz Kuratowski*(1896–1980) in 1922.

As said earlier in Chap. 5, neighbourhoods of points in an abstract topological space are the basic building blocks to define any notion in topology. The concept is also fundamental to the study of important topological properties such as the convergence, completeness, and compactness of metric spaces.

Definition 6.7 Let (X, d) be a metric space, and $a \in X$. By a basic *neighbourhood* at a we mean the set

$$B(a; r) := \{x \in X : d(a, x) < r\}, \quad \text{for some } r > 0.$$

It is called an *open ball* at a of radius $r > 0$. When emphasis on the metric d is needed, or there are several metrics in play, we write $B_d(a; r)$.

Our first example explains why the *discrete metric* δ on a set X is of little significance to the real world. For $x \in X$, the open ball $B_\delta(x; r)$ at x of radius $r > 0$ is given by

$$B_\delta(x; r) = \begin{cases} \{x\}, & \text{if } r < 1 \\ X, & \text{if } r \geq 1 \end{cases}$$

In general, for $a \in X$ and $r > 0$, the set

$$\overline{B}(a; r) := \{x \in X : d(a, x) \leq r\}.$$

is called the *closed ball* $\overline{B}(a; r)$ at a of radius $r > 0$. Also, the set

$$S(a; r) =: \{x \in X : d(x, a) = r\} = \overline{B}(a; r) \setminus B(a; r),$$

is called the *circle* at a of radius r. While dealing with some particular metric on a set X, we may also call $B(a; r)$ an open r-ball, etc.

Example 6.9 In the metric space $(\mathbb{R}^2, d_2)$, the open ball at a point $a = (a_1, a_2) \in \mathbb{R}^2$ of radius $r > 0$ is the *open disk* $D(a; r)$ of points $x = (x_1, x_2) \in \mathbb{R}^2$ such that $\|x - a\| < r$. That is,

$$D(a; r) = \left\{(x_1, x_2) \in \mathbb{R}^2 : \sqrt{(x_1 - a_1)^2 + (x_2 - a_2)^2} < r\right\}.$$

Similarly, the closed ball $\overline{B}(a; r)$ is the *closed disk* $\overline{D}(a; r)$ of points $x = (x_1, x_2) \in \mathbb{R}^2$ such that $\|x - a\| \leq r$. That is,

$$\overline{D}(a; r) = \left\{(x_1, x_2) \in \mathbb{R}^2 : \sqrt{(x_1 - a_1)^2 + (x_2 - a_2)^2} \leq r\right\}.$$

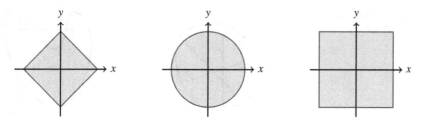

Fig. 6.2 Open unit balls in $\mathbb{R}^2$ with metric d_1, d_2, and d_∞

Also, the set

$$S^1(a; r) = \left\{(x_1, x_2) \in \mathbb{R}^2 : (x_1 - a_1)^2 + (x_2 - a_2)^2 = r^2\right\}$$

is the circle at a of radius r.

In general, the *geometry* of balls with respect to different metrics may not be as expected. For example, in the metric space $\mathbb{R}^2$, the unit ball $B_{d_1}(\mathbf{0}; 1)$ is a region with the boundary given by the square having vertices at points $(-1, 0)$, $(0, 1)$, $(1, 0)$, and $(0, -1)$; the unit ball $B_{d_2}(\mathbf{0}; 1)$ is an open disk with the boundary given by the unit circle $\mathbb{S}^1$; and, the open unit ball $B_{d_\infty}(\mathbf{0}; 1)$ is a region with the boundary given by the square having vertices at points $(-1, 1)$, $(1, 1)$, $(1, -1)$, and $(-1, -1)$ (see Fig. 6.2).

Lemma 6.5 *Let (X, d) be a metric space, and r, s be positive real numbers. Then*

$$d(x, y) \geq r + s \quad \Rightarrow \quad B(x; r) \cap B(y; s) = \emptyset;$$
$$d(x, y) \leq r - s \quad \Rightarrow \quad B(x; s) \subset B(y; r).$$

The converse of these implications may not hold, in general.

Proof Left to the reader as an exercise. One may use the *discrete metric* δ to show that the reverse implications may not hold in general.

Recall that a set A in the metric space $(\mathbb{R}^n, d_2)$ is said to be *convex* if, for every pair of points $x_1, x_2 \in A$, the line segment given by

$$L(x_1, x_2) = \left\{x = x_1 + \lambda (x_2 - x_1) \in \mathbb{R}^n : 0 \leq \lambda \leq 1\right\},$$

is contained in the set A. Otherwise, A is called a *non-convex* set. For example, the first region in Fig. 6.3 is convex, but the second is not. We show that every ball in the Euclidean space $(\mathbb{R}^n, d_2)$ is a convex set.

Fig. 6.3 A convex set and a
non-convex set

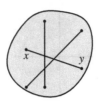
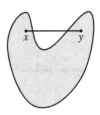

Theorem 6.4 *Consider the Euclidean space* $(\mathbb{R}^n, d_2)$. *For any* $\mathbf{a} \in \mathbb{R}^n$ *and* $r > 0$,
the open ball

$$B(\mathbf{a}; r) = \{x \in \mathbb{R}^n : \|x - \mathbf{a}\| < r\}$$

is a convex set.

Proof To show that $B(\mathbf{a}; r)$ is convex, let $x_1, x_2 \in B(\mathbf{a}, r)$ so that

$$\|x_1 - \mathbf{a}\| < r \quad \text{and} \quad \|x_2 - \mathbf{a}\| < r.$$

Then, for $0 \le \lambda \le 1$ and $x = \lambda x_1 + (1 - \lambda)x_2$, we have

$$\|x - \mathbf{a}\| = \|\lambda x_1 + (1 - \lambda)x_2 - \mathbf{a}\| = \|\lambda(x_1 - \mathbf{a}) + (1 - \lambda)(x_2 - \mathbf{a})\|$$
$$\le \lambda\|x_1 - \mathbf{a}\| + (1 - \lambda)\|x_2 - \mathbf{a}\| \le \lambda r + (1 - \lambda)r = r.$$

This shows that the line joining the points x_1 and x_2 lies inside the ball $B(\mathbf{a}; r)$.
Therefore, the open ball $B(\mathbf{a}; r)$ is convex. The closed ball $\overline{B}(\mathbf{a}; r) = \{x \in \mathbb{R}^n : \|x - \mathbf{a}\| \le r\}$ in $\mathbb{R}^n$ is also convex.

6.2.1 Open Sets

Each metric d defines a *topology* on a set X given in terms of *open* sets, which
are constructed from balls in the metric space (X, d) (Theorem 6.6). In general, a
point $u \in U \subset X$ is called an *interior point* of U if there is an open ball $B(u; r)$
contained entirely in U (See Fig. 6.4). Notice that, since every interval of $\mathbb{R}$ contains
both rational and irrational numbers, the interior of the set $\mathbb{Q} \subset \mathbb{R}$ with respect to
usual topology is an empty set. On the other hand, Theorem 6.5 prove that every
point of an open ball in a metric space is an interior point.

Definition 6.8 Let (X, d) be a metric space. A set $U \subset X$ is said to be **open** if each
$u \in U$ is an interior point of U.

Theorem 6.5 *Every open ball in a metric space is an open set.*

Fig. 6.4 Interior points of set U

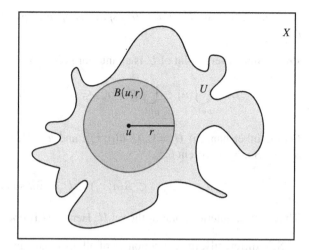

Fig. 6.5 A ball inside an open ball

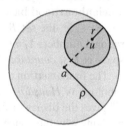

Proof Let (X, d) be a metric space, and consider an open ball $U = B(a; \rho) \subset X$, for some $\rho > 0$. Then, for $u \in U$, we have $r = \rho - d(a, u) > 0$ because $d(a, u) < \rho$ (See Fig. 6.5). We need to show that $B(u; r) \subset B(a; \rho)$. For, if $x \in B(u; r)$, then $d(u, x) < r$ implies that

$$d(a, x) \le d(a, u) + d(u, x) < d(a, u) + r = \rho.$$

Therefore, $B(u; r) \subset B(a; \rho)$, which proves that $u \in U$ is an interior point. As u is arbitrary, we conclude that U is an open set.

More generally, in geometrical terms, an open set in a metric space specifies the *relative position* of its points with respect to the given metric. For a nonempty subset $Y \subset X$, an interior point of the set $X \setminus Y$ is called an *exterior point* of the set Y. And, a point $y \in Y$ is called a *boundary point* if each open ball $B(y, r)$ intersects both Y and $X \setminus Y$.

Clearly, the set X itself is open, because $B(x, r) \subset X$ for each $x \in X$ and for every $r > 0$. Also, the empty set $\emptyset$ is open; for, if not, then there is an $x \in \emptyset$ (this is not possible) which is not an interior point of $\emptyset$. Since balls in $\mathbb{R}$ are intervals, the next theorem generalises Theorem 5.3.

Theorem 6.6 *A set U in a metric space is open if and only if U is union of open balls.*

Proof Since every point of U is an interior point, so

$$U = \bigcup_{u \in U} \{u\} \subset \bigcup_{u \in U} B(u; r_u) \subseteq U \quad \Rightarrow \quad U = \bigcup_{u \in U} B(u; r_u).$$

On the other hand, if $U = \bigcup_{a \in U} B(a; r_a)$, and $u \in U$, then $u \in B(a; r_a)$ for some $a \in U$. So, by Theorem 6.5,

$$u \in B(u; \epsilon) \subset B(a; r_a) \subset U, \quad \text{for some } \epsilon > 0.$$

Thus, u is an interior point of the set U. Hence, U is open.

Said differently, the collection $\mathscr{B}$ of all open balls in a metric space X constitute a **neighbourhood base** for the *metric topology*. That is, $X = \bigcup_{B \in \mathscr{B}} B$ and, for all $B_1, B_2 \in \mathscr{B}$, the set $B_1 \cap B_2$ is a union of sets in $\mathscr{B}$. Moreover, since we can take rational numbers $1/n$ ($n \in \mathbb{N}$) for the radii of these balls, it follows that every metric space is *first countable*.[5]

The first assertion of the next theorem proves that we always have that the metric topology is *Hausdorff*, and the second asserts that the linear structure of the space $\mathbb{R}^n$ gives the liberty to study its topology in terms of the unit balls at the origin, which explains why the concept of the *radial vector* is used so frequently.

Theorem 6.7 *Let (X, d) be a metric space. Then, (a) distinct points can be separated by open balls; and, (b) for any $\mathbf{x} \in \mathbb{R}^n$ and $r > 0$, $B(\mathbf{x}; r) = \mathbf{x} + r B(0; 1)$.*

Proof For (a), let $x \neq y \in X$, and take $r = d(x, y) > 0$. Then, the disjoint open balls $B(x; r/2)$ and $B(y; r/2)$ separate the points x and y. For (b), notice that

$$\mathbf{x} + r B(0; 1) := \{ \mathbf{x} + r\mathbf{a} : \mathbf{a} \in B(0; 1) \}.$$

It thus follows that, for any $\mathbf{y} \in \mathbf{x} + r B(0; 1)$, we have

$$\mathbf{y} = \mathbf{x} + r\mathbf{a}, \quad \text{for some } \mathbf{a} \in B(0; 1).$$

Then, $\|\mathbf{a}\| < 1 \Rightarrow \|\mathbf{y} - \mathbf{x}\| = r\|\mathbf{a}\| < r \Rightarrow \mathbf{y} \in B(\mathbf{x}; r)$. On the other hand, we have

---

[5] A topological space is called **first countable** if each point has a countable neighbourhood base.

$$y \in B(x; r) \quad \Rightarrow \quad \|(y - x) - 0\| < r$$
$$\Rightarrow \quad \frac{\|y - x\|}{r} < 1$$
$$\Rightarrow \quad \frac{1}{r}(y - x) \in B(0; 1)$$
$$\Rightarrow \quad y = x + ra, \quad \text{for some } a \in B(0; 1).$$

This completes the proof.

The next theorem proves that the collection of open sets in a metric space defines a topology, which is called the *metric topology*.

Theorem 6.8 *Arbitrary unions and finite intersections of open subsets of a metric space are open.*

Proof Let $\{U_\alpha : \alpha \in I\}$ be an indexed family of open sets. We need to show that $U = \cup_{\alpha \in I} U_\alpha$ is open. Let $x \in U$. Then $x \in U_\alpha$ for some $\alpha \in I$. Since U_α is open, there is a neighbourhood $B(x; r)$ of x such that $B(x; r) \subset U_\alpha$. Since $U_\alpha \subset \cup_{\alpha \in I} U_\alpha = U$, we have $B(x; r) \subset U$ and hence U is open. Let $V_1, V_2, \ldots, V_m$ be open sets and let $V = \cap_{k=1}^m V_k$. Let $x \in V$. Then $x \in V_k$ for every k with $1 \le k \le m$. Since V_k is open, there is a neighbourhood $B(x; r_k)$ of x such that $B(x; r_k) \subset V_k$. Let $r = \min\{r_1, r_2, \ldots, r_k\}$. Then $r > 0$ and $B(x; r) \subset B(x; r_k) \subset V_k$ for each k and hence $B(x; r) \subset \cap_{k=1}^m V_k = V$. Hence, V is open.

Notice that, however, arbitrary intersection of open sets need not be open. The reader may construct some examples to prove the assertion.

Next, we introduce the concept of *product topology* on a finite product of metric spaces. For brevity, we consider here the case of two metric spaces, say (X, d) and (Y, ρ). Recall that, by Example 6.4, the Cartesian product $Z = X \times Y$ is a metric space with respect to the metric $d_\infty : Z \times Z \to \mathbb{R}$ given by

$$d_\infty(x_1, x_2) = \max\{d(x_1, x_2), \rho(y_1, y_2)\},$$

where $x_1 = (x_1, y_1), x_2 = (x_2, y_2) \in Z$.

Lemma 6.6 *The metric d_∞ defines the product topology on the set Z.*

Proof Notice that it suffices to show that every open ball in the product space (Z, d_∞) is the product of open balls in (X, d) and (Y, ρ). For, let $(a, b) \in Z$ and $\epsilon > 0$. Then, we have

$$B_{d_\infty}((a, b); \epsilon) = \{(x, y) \in X \times Y : d_\infty((x, y), (a, b)) < \epsilon\}$$
$$= \{(x, y) \in X \times Y : d(x, a) < \epsilon \text{ and } \rho(y, b) < \epsilon\}$$
$$= B_d(a; \epsilon) \times B_\rho(b; \epsilon).$$

This completes the proof.

In particular, the metric d_2 on the set Z defines the *Euclidean topology*. A similar remark stands for the metric d_1. In fact, all Minkowski metrics d_p $(1 \leq p < \infty)$ define the same topology on Z, but their metric geometries may not be the same.

Example 6.10 Viewing the space $\mathbb{R}^n$ as a product space, with respect to the metric d_∞, so that each $\mathbb{R}$ has the usual topology. Then, as seen in Lemma 6.6, open balls in $(\mathbb{R}^n, d_\infty)$ are precisely the product of open balls in $(\mathbb{R}, m)$ i.e., open intervals. Also, we see that the collection

$$\mathcal{B} = \{U_1 \times \cdots \times U_n : \text{ each } U_i \text{ is an open set in } \mathbb{R}\}$$

is a *base* for the product topology on $\mathbb{R}^n$. Notice that, for $n = 2$, if $B_1 = U_1 \times V_1$ and $B_2 = U_2 \times V_2$ then

$$B_1 \cap B_2 = (U_1 \cap U_2) \times (V_1 \cap V_2).$$

So, by Theorem 5.3, every open set in $\mathbb{R}^n$ $(n > 1)$ is a union of *almost disjoint* rectangles. Notice that, here, the term *almost* is used in the sense that these rectangles can only intersect along boundaries. Such considerations are important to deal with practical problems in many situations. In this text, we will use th same in Chap. 10 to introduce the concept of Riemann integration of multivariable bounded functions defined on some bounded open set $U \subset \mathbb{R}^n$.

Notice that the three *types* of balls in Fig. 6.2 defined by metrics d_1, d_2, and d_∞ on $\mathbb{R}^2$ have the property that, at each point $x \in \mathbb{R}^2$, a ball with respect to one metric contains an open ball with respect to the other two metrics. In other words, the three metrics on $\mathbb{R}^2$ are *equivalent*.

Definition 6.9 Two metrics d and ρ on a nonempty set X are said to be *equivalent* if, for some positive constants c_1, c_2, we have

$$c_1 \rho(x, y) \leq d(x, y) \leq c_2 \rho(x, y), \quad \text{for all} \ \ x, y \in X.$$

It follows from the above condition that two equivalent metrics on a set X define the same *topology* on X. However, it is important to notice that the *metric geometry* defined by the three metrics d_1, d_2 and d_∞ on $\mathbb{R}^n$ $(n > 2)$ may not be the same.

Definition 6.10 Let (X, d) and (Y, ρ) be two metric spaces. A function $f : X \to Y$ is called an *isometry* if

$$\rho(f(x_1), f(x_2)) = d(x_1, x_2), \quad \text{for all} \ \ x_1, x_2 \in X.$$

We say X and Y are *isometric* if there is a surjective isometry[6] $f : X \to Y$.

---

[6] In metric geometry, the term **motion** is used synonymously for a surjective isometry.

Fig. 6.6 Open interval
$(a - \epsilon, a + \epsilon) =$
$B(a; \epsilon) \cap \mathbb{R}$

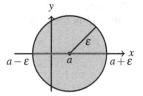

Clearly, every isometry f is an injective function. Therefore, *a surjective isometry between metric spaces is a bijective, distance preserving function*. Clearly, we can form partitions of the family of metric spaces by taking in an *equivalence class* all metric spaces of a particular *isometry type*.

In particular, we say two metrics d, d' on a (carrier) set X define the same *metric geometry* if there is an isometry $f : (X, d) \to (X, d')$. For example, the semimetric s on $\mathbb{R}^2$ as defined in (6.1.1) gives a metric space $\mathbb{R}^2/s$ that is isometric to the line $(\mathbb{R}, m)$. Furthermore, the metric spaces $(\mathbb{R}^2, d_1)$ and $(\mathbb{R}^2, d_2)$ are isometric, but $(\mathbb{R}^n, d_1)$ and $(\mathbb{R}^n, d_\infty)$ are not isometric metric spaces, for $n > 2$ (Why?).

The next theorem is about the *subspace topology* of a set Y in a metric space, which relates the open sets of Y to the open sets of X. For early motivation, notice that no open interval $(a, b) \subseteq \mathbb{R}$ is open in $\mathbb{R}^2$ because every open set of $\mathbb{R}^2$ must contain an open disk of some positive radius around each point in it (See Fig. 6.6).

Theorem 6.9 *Let X be a metric space and Y be a subset of X. A subset $V \subset Y$ is open in Y if and only if $V = U \cap Y$ for some open subset U of X.*

Proof We use $B_X(x; r)$ for a ball in X centred at x with radius r. Let $V = U \cap Y$ for some open subset U of X. Let $x \in V$. Then $x \in U$. Since U is open in X, there is a $r > 0$ such that $B_X(x; r) \subset U$ and hence $B_X(x, r) \cap Y \subset U \cap Y$. But $B_X(x, r) \cap Y = B_Y(x; r)$ (see Fig. 6.7) and therefore $B_Y(x; r) \subset U \cap Y$. This proves that $U \cap Y$ is open in Y.

For the converse, let $V \subset Y$ be open in Y. Then, for each $x \in V$, there is a real number $r_x > 0$ such that $B_Y(x; r_x) \subset V$. This shows that $\cup_{x \in V} B_Y(x; r_x) \subset V$. Since $x \in V$, we have $x \in B_Y(x; r_x) \subset \cup_{x \in V} B_Y(x; r_x)$, So, we have $V \subset \cup_{x \in V} B_Y(x; r_x)$.

Fig. 6.7 $B_Y(x; r) = B_X(x; r) \cap Y$ is the shaded region

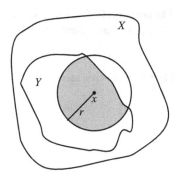

This shows that $V = \cup_{x \in V} B_Y(x; r_x)$. Let $U = \cup_{x \in V} B_X(x; r_x)$. Since each $B_X(x; r_x)$ is open, U is open and

$$
\begin{aligned}
U \cap Y &= (\cup_{x \in V} B_X(x; r_x)) \cap Y \\
&= \cup_{x \in V} (B_X(x; r_x) \cap Y) \\
&= \cup_{x \in V} B_Y(x; r_x) = V.
\end{aligned}
$$

$\square$

6.2.2   Limit Points and Closed Sets

The concept of limit point of a set S in a metric space X is very important. We say a point $a \in X$ is a *limit point* (or an *accumulation point*) of S if every ball $B(a; r)$ contains a point $b \in S$, with $b \neq a$, in case $a \in S$. Said differently, $a \notin S$ is a limit point of S if every ball $B(a; r)$ intersects S. The set of all limit points of a set $S \subset X$ is called the *derived set* of S, denoted by S', and the *closure* $\overline{S}$ of S is the union $S \cup S'$. Therefore,

$$
a \in S' \quad \Leftrightarrow \quad S \cap (B(a; \epsilon) \setminus \{a\}) \neq \emptyset, \quad \text{for any } \epsilon > 0.
$$

As before, the set $B(a; r) \setminus \{a\}$ is called a *punctured ball* at a with radius r. Notice that a limit point of S may or may not belong to S.

Example 6.11   Consider the *top-open* square

$$
S = \{(x, y) \in \mathbb{R}^2 : 0 \leq x \leq 1 \text{ and } 0 \leq y < 1\},
$$

as shown in Fig. 6.8. Then, every point of S is a limit point of S. The points $(x, 1)$ with $0 \leq x \leq 1$ are limit point of S but they are not in S. In this case, we have

$$
\overline{S} = S' = \{(x, y) : 0 \leq x \leq 1, 0 \leq y \leq 1\}.
$$

Theorem 6.10   *Let S be a subset of a metric space X. Then every ball around a limit point of the set S contains infinitely many points of S.*

Fig. 6.8   A *top-open* square in $\mathbb{R}^2$

Proof Let a be a limit point of S and $B(a, r)$ be some ball around a. Suppose that $B(a; r) \setminus \{a\}$ has only finitely many points of S, say, $b_1, b_2, \ldots, b_m$. Let

$$\rho = \min\{d(a, b_1), d(a, b_2), \ldots, d(a, b_m)\}.$$

Then $\rho > 0$ and the neighbourhood $B(a; \rho) \setminus \{a\}$ has no point of S. Therefore, the point a cannot be a limit point of S. This contradiction shows that $B(a; r) \setminus \{a\}$, and hence $B(a; r)$, has infinitely many points of S.

In particular, finite set in a metric space can not have a limit point.

Corollary 6.1 *Every finite set in a metric space is without any limit point.*

Definition 6.11 Let (X, d) be a metric space. A set $F \subset X$ is said to be *closed* if its complement $U = X \setminus F$ is open in X.

For example, by Corollary 6.1, every finite set in a metric space is a closed set. In general, for each $u \notin F$, there exists an $r > 0$ such that $B(u, r) \cap F = \emptyset$. Notice that we may take

$$r = \inf \{d(u, x) : x \in F\}.$$

So, an element $u \in U = X \setminus F$ is not in the derived set of F.

Theorem 6.11 *Let X be a metric space. A set $F \subseteq X$ is closed if and only if it contains all its limit points, i.e., $F' \subset F$.*

Proof Let F be closed so that $U = X \setminus F$ is open, by definition. Let x be a limit point of F. We need to show that $x \in F$. Equivalently, we prove $x \notin U$. For a contradiction, suppose $x \in U$. Then, as U is open, we have $x \in U^\circ$, and hence $B(x; r) \subset U$, for some $r > 0$. Since this ball $B(x; r)$ has no point in common with F, we got our contradiction, because x is assumed to be a limit point of F. It thus follows that $x \notin U$, and so $x \in F$. To prove the converse, suppose F contains all its limit points. Again, to show that F is closed, it suffices to prove that the set U is open. For, let $x \in U$ be arbitrary. Since F contains all its limit points, x can not be a limit point of F, and hence we can find some ball $B(x; r)$ such that $F \cap B(x; r) = \emptyset$. Equivalently, $B(x; r) \subset U$. This implies that x is an interior point of U. That is, U is open. Hence, F is closed.

Theorem 6.12 *For every set F in a metric space X, its closure $\overline{F}$ is a closed subset of X.*

Proof We first show that $U = X \setminus \overline{F}$ is open. By definition, $\overline{F} = F \cup F'$. By using de Morgan's law, we have

$$X \setminus \overline{F} = X \setminus (F \cup F') = (X \setminus F) \cap (X \setminus F').$$

Fig. 6.9 $B(y, s) \cap F = \emptyset$

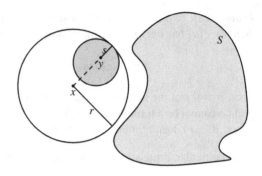

Now, for $x \in U$, we have $x \in X \setminus F$ and $x \in X \setminus F'$ so that $x \notin F$ and $x \notin F'$. Since $x \notin F'$, x is not a limit point of F and hence there is a ball $B(x; r)$ of x that has no point of F other than x. Since $x \notin F$, $B(x; r)$ has no point of F and hence $B(x; r) \subset X \setminus F$. We claim that $B(x; r) \subset X \setminus F'$. For each point $y \in B(x; r)$, the ball $B(y; s)$, with $s = r - d(x, y)$, is contained in $B(x; r)$ and hence the ball $B(y; s)$ does not intersect F (Fig. 6.9). Thus, y is not a limit point of F. Therefore, $y \notin F'$, that is, $y \in X \setminus F'$. Thus, we have shown that

$$B(x; r) \subset X \setminus F \quad \text{and} \quad B(x; r) \subset X \setminus F'.$$

This immediately shows that

$$B(x; r) \subset (X \setminus F) \cap (X \setminus F') = X \setminus \overline{F}.$$

Thus, x is an interior point of $X \setminus \overline{F}$ and hence $X \setminus \overline{F}$ is open. Therefore, $\overline{F}$ is closed.

Corollary 6.2 *Let F be a subset of a metric space X. Then F is closed if and only if $F = \overline{F}$.*

Proof If $F = \overline{F}$, then F is closed, by Theorem 6.12. On the other hand, if F is closed, it contains all its limit points and hence $F' \subset F$. Therefore, we have $F = F \cup F' = \overline{F}$.

So, both X and $\emptyset$ are closed sets as well. In general, a set in a metric space is called *clopen* if it is both open and closed. As closed sets are important subsets of a metric space, it is of interest to construct new closed sets from known closed sets.

Theorem 6.13 *Arbitrary intersection and finite union of closed subsets of a metric space are closed.*

Proof Follows from Theorem 6.8, by using the de Morgan's laws for sets. See Exercise 6.18

Notice that, however, an arbitrary union of closed sets may not be a closed set. The reader may construct some examples to prove the assertion.

6.3 Convergence and Completeness

The *convergence problems* related to sequences and series of points in some linear metric space arise naturally while dealing with practical problems in physics, computer science, engineering, economics, finance, and even in medical sciences. For example, a typical problem concerning design and analysis of a discrete system[7] takes *discrete-time* signals as inputs, and we need to understand the behaviour of the outputs. In most such types of problems, we need to know the following.

Does a sequence $\langle x_n \rangle$ in a metric space (X, d) converges ? That is, does there exist some x such that $d(x_n, x)$ is as small as we please ? Is $x \in X$? Such x is called the limit of the sequence $\langle x_n \rangle$, if it exists. In some case, we also need to find the limit of a convergent sequences.

In Chap. 3, we discussed the situation when the underlying linear space consists of the real numbers, and we studied the convergence of sequences and series with respect to *modulus metric*. In Chap. 8, we will discuss above stated questions for the linear metric spaces of real-valued bounded functions defined on some interval $I \subseteq \mathbb{R}$, with respect to the *uniform metric* (Definition 6.13). In Chap. 9, we consider sequences and series of real-valued Lebesgue integrable functions defined on certain *nice subsets* of the set $\mathbb{R}$ with respect to L^p-*metric*, as defined therein.

Clearly then, in each case, the central idea is to use some metric to deal with the convergence problems of sequences in the respective linear space. Recall that the convergence problems related to the *summability* of an infinite series are studied using the associated sequences of *partial sums*. To solve a practical problem, say concerning a discrete system, the focus is always to study the convergence of a sequence[8] with respect to a metric. Therefore, in general, we need to understand how a metric defined on a set (that is mostly a linear space) is used to talk about the closeness of terms of a sequence to some point, which may or may not be in a set.

Having set the goal for the section, let us start with some definitions. Recall that a sequence $\langle a_n \rangle$ in the set $\mathbb{R}$ is said to converge with respect to the *modulus metric* if there is some real number α such that, for any $\epsilon > 0$ and some $N \in \mathbb{N}$, we have

$$n \geq N \quad \Rightarrow \quad |a_n - a| < \epsilon. \tag{6.3.1}$$

In this case, we say α is the *limit* of the sequence $\langle a_n \rangle$, and we write $\lim_{n \to \infty} a_n = \alpha$. The concept of convergence of a sequence in a metric space (X, d), as given in the next definition, is a natural extension of the above idea, provided we replace the *modulus metric* by the metric d.

---

[7] A *discrete system* is usually represented by some *difference equations* involving *casual sequences* or a block diagram. The concept of Fourier series, as we shall discuss later in Chap. 10, is a special type of a discrete system, which takes *discrete-time* periodic signals as the inputs and returns the *frequency spectrum* of the signal as outputs.

[8] In certain situations, we also need to analyse the *rate of convergence* of a sequence. However, we will not pursue the topic in this text.

Definition 6.12 Let (X, d) be a metric space, and $\langle x_n \rangle$ be a sequence in the set X. We say $\langle x_n \rangle$ *converges in X with respect to the metric d* if, for some $x \in X$, we have $\lim_{n \to \infty} d(x_n, x) = 0$. The point $x \in X$ is called the *limit* of the sequence $\langle x_n \rangle$. We also write

$$\lim_{n \to \infty} x_n = x \quad \text{or} \quad x_n \to x \quad \text{as} \quad n \to \infty.$$

So, a sequence $\langle x_n \rangle$ in a metric space (X, d) converges to some $x \in X$ if and only if the sequence of real numbers given by $\alpha_n = d(x_n, x)$ converges to 0 in the sense of the Eq. (6.3.1). In topological terms, we say a sequence $\langle x_n \rangle$ converges to some $x \in X$ if every ball $B(x; \epsilon)$ contains all the terms x_n, *eventually*. Therefore, by the Hausdorff property of the metric topology, limit of a convergent sequence in every metric space is unique (Exercise 6.22).

Example 6.12 (*Convergence in the Euclidean Spaces*) The concept of the convergence in $(\mathbb{R}^n, d_2)$, and in some other similar spaces, is a straightforward extension of the idea of convergence in $\mathbb{R}$. For, consider a sequence of n-*vectors*

$$x^m = \langle x_1^m, \ldots, x_n^m \rangle \in \mathbb{R}^n, \quad m \geq 1.$$

Suppose $\langle x^m \rangle$ converges to a point $x = (x_1, \ldots, x_n) \in \mathbb{R}^n$. Then, for any $\epsilon > 0$ and some $N \in \mathbb{N}$, we have

$$m \geq N \quad \Rightarrow \quad d_2(x^m, x) = \left(\sum_{k=1}^{n} |x_k^m - x_k|^2 \right)^{1/2} < \epsilon.$$

In particular, for $m \geq N$ and for each $1 \leq k \leq n$, we have

$$|x_k^m - x_k|^2 \leq \sum_{k=1}^{n} |x_k^m - x_k|^2 = \left[d_2(x^m, x) \right]^2 < \epsilon^2$$
$$\Rightarrow \quad |x_k^m - x_k| < \epsilon.$$

This implies that, for $m \geq N$, the real sequence of kth *coordinates* $\langle x_k^m \rangle$ of the sequence $\langle x^k \rangle$ converge to x_k, the kth coordinate of the point $x = (x_1, \ldots, x_n)$. A similar argument implies that the converse holds.

Example 6.13 (*Convergence in ℓ_p Spacess*) Recall the construction of the metric spaces (ℓ_p, d_p), as given earlier in Example 6.4. Consider the following three sequences in ℓ_p:

$$x_n = \langle \ldots, 1/n, \ldots \rangle;$$
$$e_n = \langle \ldots, 1, \ldots \rangle;$$
$$y_n = \langle 1/n, \ldots, 1/n, 0, \ldots \rangle,$$

where the only nonzero entry in sequences x_n and e_n is in their nth term, with value $1/n$ and 1 respectively and the sequence y_n has $1/n$ in the first n terms, with rest all entries 0. Then, by definition of the metric d_p, we have

$$d_p(x_n, \mathbf{0}) = \frac{1}{n} \;\Rightarrow\; d_p(x_n, \mathbf{0}) \to 0, \quad \text{as } n \to \infty.$$

So, $x_n \to \mathbf{0}$. The sequence e_n does not converge in any ℓ_p. And,

$$d_\infty(y_n, \mathbf{0}) = \frac{1}{n} \;\Rightarrow\; d_\infty(y_n, \mathbf{0}) \to 0, \quad \text{as } n \to \infty,$$

shows that y_n converges to $\mathbf{0}$ with respect to the metric d_∞. However, as $d_1(y_n, \mathbf{0}) = 1$, it follows that $\langle y_n \rangle$ is not convergent with respect to the metric d_1. What can be said about the convergence of $\langle y_n \rangle$, for $1 < p < \infty$?

As said before, in general, the convergence (or *non-convergence*) of a sequence $\langle x_n \rangle$ in a (carrier) set X depends on our choice for some particular metric d given on the set X. So, by working with different kinds of metrics d given on a (carrier) set X, one introduces various *types of convergence* for a sequence $\langle x_n \rangle$ in the metric space (X, d). The next definition illustrates the point.

We write $X = B(I)$ for the linear space of real-valued bounded functions defined on a set $I \subset \mathbb{R}$. Let $\langle f_n \rangle$ be a sequence in $B(I)$. We say $f : I \to \mathbb{R}$ is the *limit function* (or a *pointwise limit*) of the sequence $\langle f_n \rangle$ if the sequence of real numbers $\langle f_n(a) \rangle$ converges to the real number $f(a)$, for each $a \in I$. That is,

$$f(a) := \lim_{n \to \infty} f_n(a), \quad \text{for } a \in I.$$

In this case, we also say that the sequence $\langle f_n \rangle$ *converges pointwise* on the set I (or simply $f_n \to f$ *pointwise* on the set I).

Definition 6.13 Let $\langle f_n \rangle$ be a sequence in $B(I)$. We say $\langle f_n \rangle$ *converges uniformly* (or *globally*) to a function $f : I \to \mathbb{R}$ if

$$d_\infty(f_n, f) = \sup_{a \in I} \{|f_n(a) - f(a)|\} \;\to\; 0, \quad \text{as } n \to \infty. \tag{6.3.2}$$

In this case, the function f is called the *uniform limit* of the sequence $\langle f_n \rangle$, and we write $f_n \rightrightarrows f$ as $n \to \infty$. The metric d_∞ is called the *uniform metric* (or the *supremum metric*).

Notice that, however, not all type of convergences are metric based, i.e., some types of convergence may not stem from a metric. For example, there is no metric on the linear space $C[0, 1]$ that induces the concept *pointwise convergence* of sequences. However, as have been said earlier, the concept of *uniform convergence* of sequences in $B(I)$ is the convergence with respect to the *supremum metric* $d_\infty(f, g)$, with

$f, g \in B(I)$. As we will see later in Chap. 8, the concept of *uniform convergence* is far more superior than the *pointwise convergence*.

Example 6.14 Consider the sequence $\langle f_n \rangle$ in $C[0, 1]$ given by

$$f_n(x) = \max \left(1 - |nx - 1|, 0\right), \qquad x \in I.$$

Then, the pointwise limit function of $\langle f_n \rangle$ is the *zero function* $f \equiv 0$. Indeed, for all $n \geq 2$, we have

$$f_n(0) = 0 \quad \text{and} \quad f_n(x) = 0, \quad \text{on the interval } \left[2/n, 1\right].$$

Notice that, however, as $f_n(1/n) = 1$, the sequence fails to converge with respect to the uniform metric as in (6.3.2).

The next theorem generalises Theorem 5.8 for metric spaces.

Theorem 6.14 *Let S be a subset of a metric space (X, d). If x is a limit point of S, then there is a sequence $\langle x_n \rangle$ of elements of S such that $x_n \to x$. Converse holds, if the terms of the sequence are distinct.*

Proof Since x is a limit point of S, the neighbourhood $B(x; 1/n)$ intersects S for each $n \in \mathbb{N}$. Choose an element $x_n \in B(x; 1/n) \cap S$. Then $x_n \in S$. We claim $x_n \to x$. Let $\epsilon > 0$ be given. Choose $N > 1/\epsilon$. Then, for all $n \geq N$, we have

$$d(x_n, x) < 1/n \leq 1/N < \epsilon.$$

Thus, $x_n \to x$. Conversely, let $\langle x_n \rangle$ be a sequence of distinct elements of S, and $x_n \to x$, for some $x \in X$. Let $B(x; \epsilon)$ be a neighbourhood of x. Since $x_n \to x$, there is a natural number N such that $d(x_n, x) < \epsilon$ for all $n \geq N$. Since x_n's are distinct, there is some $n_0 \geq N$ such that $x_{n_0} \neq x$. Also, $x_{n_0} \in B(x; \epsilon)$. Therefore, x is a limit point of S.

Theorem 6.15 *In a metric space X, a sequence $\langle x_n \rangle$ converges to $x \in X$ if and only if every subsequence $\langle x_{n_k} \rangle$ converges to x.*

Proof If every subsequence $\langle x_{n_k} \rangle$ converges to x, $\langle x_n \rangle$, being a subsequence of itself, also converges to x. Conversely, if $\langle x_n \rangle$ converges to x, then for a given $\epsilon > 0$ there is an N such that $d(x_n, x) < \epsilon$ for all $n \geq N$. Now, for $n_k \geq N$, we have $d(x_{n_k}, x) < \epsilon$ and hence the subsequence $\langle x_{n_k} \rangle$ converges to x.

To test when a sequence converges, we need to know the limit of the sequence first. So, we need to find a criterion for convergence that does not involve the limit value. Roughly speaking, a sequence $\langle x_n \rangle$ converges to ℓ if every term x_n is close to ℓ when n is sufficiently large. In such case, any two terms x_n and x_m of the sequence are also close to each other for large values of n and m. This motivates the following definition.

Definition 6.14 (*Cauchy Sequence*) A sequence $\langle x_n \rangle$ in a metric space X is called a *Cauchy sequence* if, for a given $\epsilon > 0$, there is a natural number $N = N(\epsilon)$ such that

$$d(x_n, x_m) < \epsilon, \quad \text{for all } n, m \geq N.$$

The next theorem prove that every convergent sequence is Cauchy.

Theorem 6.16 *If* $\langle x_n \rangle$ *is a convergent sequence in a metric space* X, *then it is a Cauchy sequence.*

Proof Suppose $\lim_{n \to \infty} x_n = x$, and let $\epsilon > 0$. Then, for some $N = N(\epsilon) \in \mathbb{N}$, we have

$$n \geq N \quad \Rightarrow \quad d(x_n, x) < \frac{\epsilon}{2}.$$

In particular, for $n, m \geq N$, we have

$$\begin{aligned} d(x_n, x_m) &\leq d(x_n, x) + d(x, x_m) \\ &= d(x_n, x) + d(x_m, x) \\ &< \frac{\epsilon}{2} + \frac{\epsilon}{2} = \epsilon, \end{aligned}$$

which shows $\langle x_n \rangle$ is Cauchy.

Definition 6.15 Let (X, d) be a metric space, and $A \subseteq X$. We say A is a *bounded set* if, for some $a \in A$ and $M > 0$, we have $d(x, a) < M$, for all $x \in A$. Or, equivalently, if A lies in a ball at some point in A.

Notice that, since

$$B(b; M) \subseteq B(a; r) + d(a, b), \quad \text{for any } a, b \in A, \tag{6.3.3}$$

we can take any point $a \in A$ in the above definition. Alternatively, we say a set $A \subseteq X$ is *bounded* if it has a *finite diameter*, where

$$\text{diam}(A) := \sup \{ d(x, y) : x, y \in A \}$$

is called the *diameter* of the set A. In particular, we say a sequence $\langle x_n \rangle$ in X is *bounded* if the set $A = \{ x_n : n \in \mathbb{N} \}$ is bounded in above sense. That is, for some $x_m \in A$ and $M > 0$, we have $d(x_n, x_m) \leq M$, for all $n \in \mathbb{N}$. As $x_m \in A$ can be chosen arbitrarily, equivalently, $\langle x_n \rangle$ is bounded if $d(x_n, x_m) \leq M$, for some $M > 0$ and for all $m, n \in \mathbb{N}$.

Theorem 6.17 *Every Cauchy sequence, and hence every convergent sequences, is bounded. The converse is not true, in general.*

Proof Let $\langle x_n \rangle$ be a Cauchy sequence in a metric space X. Since $\langle x_n \rangle$ is Cauchy, there is a natural number N such that

$$d(x_n, x_m) \leq 1 \quad \text{for all } n, m \geq N.$$

Let $r = \max\{1; d(x_i, x_N), 1 \leq i \leq N - 1\}$. Then $d(x_n, x_N) \leq r$ for all $n \in \mathbb{N}$. This shows that $\langle x_n \rangle$ is bounded. Notice that the sequence $\langle (-1)^n \rangle$ is bounded, but it is not convergent.

As said before, the converse of Theorem 6.16 is not true in general. In fact, convergence of a Cauchy sequence depends upon the topology of the ambient space (Example 3.13). So, convergence of a Cauchy sequence in a space is a *topological property*. Thus, Cauchy condition separates topological aspects of a metric space from its geometric properly induced by the metric defined on it (Example 5.2). Further, if Y is a subset of a metric space X, and a sequence $\langle x_n \rangle$ in Y is Cauchy in X, then it would remain so in Y. But, the converse need not be true (Example 3.13).

Theorem 6.18 *In a metric space X, a Cauchy sequence $\langle x_n \rangle$ converges to $x \in X$ if it has a subsequence $\langle x_{n_k} \rangle$ that converges to x.*

Proof Let the subsequence $\langle x_{n_k} \rangle$ converges to x. Then, for a given $\epsilon > 0$ there is an N_1 such that $d(x_{n_k}, x) < \epsilon/2$ for all $n_k \geq N_1$. Since $\langle x_n \rangle$ is a Cauchy sequence, there is an N_2 such that $d(x_n, x_m) < \epsilon/2$, for $n, m \geq N_2$. Now, for $n \geq N = \max\{N_1, N_2\}$, choose some $n_i \geq N$. Then, we have

$$d(x_n, x) \leq d(x_n, x_{n_i}) + d(x_{n_i}, x) < \epsilon/2 + \epsilon/2 = \epsilon.$$

This proves that $\langle x_n \rangle$ converges to x.

More generally, if every subsequence of a sequence $\langle x_n \rangle$ converges to some $x \in X$, then $\lim_{n \to \infty} x_n = x$; for, otherwise, there exist a ball $B(x; \epsilon)$, and a subsequence $x_{n_k} \notin B(x; \epsilon)$, which contradicts the fact every subsequence of $\langle x_n \rangle$ converges to x. This simple observation helps to establish convergence of a sequence in situations wherein it is known that the limit of a subsequence solves a given problem together with *uniqueness* of a solution of the problem.

Definition 6.16 A metric space X is said to be *complete* if every Cauchy sequence of elements of X converges in X.

By our discussion following Definition 6.12, we find that the completeness of $\mathbb{R}^p$ ($p \geq 2$) follows from the completeness of $\mathbb{R}$ (Theorem 5.9). For, if $\mathbf{x}^m = \langle x_n^m \rangle$ is a Cauchy sequence in $\mathbb{R}^p$, then for each $\epsilon > 0$ here is some natural number N such that by the Cauchy criterion,

$$d_2(\mathbf{x}^N, \mathbf{x}^{N+s}) = \left(\sum_{n=1}^{\infty} |x_n^N - x_n^{N+s}|^2 \right)^{1/2} < \epsilon, \quad \text{for all } s \geq 1.$$

Since, for $m \geq N$ and for all $s \geq 1$, we have

$$|x_n^m - x_n^{m+s}|^2 \leq \sum_{k=1}^{\infty} |x_k^m - x_k^{m+s}|^2 = [d_2(x^m, x^{m+s})]^2 < \epsilon^2$$
$$\Rightarrow \quad |x_n^m - x_n^{m+s}| < \epsilon,$$

which implies that, for $m \geq M$, real sequence of kth coordinate

$$y^k = \langle x_k^m, x_k^{m+1}, x_k^{m+2}, \ldots \rangle, \quad \text{for } 1 \leq k \leq p,$$

of the sequence $\langle x^m \rangle$ is Cauchy. So, we can find real numbers $x^k \in \mathbb{R}$ such that $y^k \to x^k$ as $k \to \infty$, for $1 \leq k \leq p$. It is then easy to show that $x^m \to (x^1, x^2, \ldots, x^p)$ as $m \to \infty$.

The same idea proves that the space ℓ_p, $1 \leq p < \infty$, of *absolute p-summable* real (or complex) sequences introduced in Example 6.4 is complete. As said before, this important property of ℓ_2 plays a very significant role in the theory of Fourier series discussed in Chap. 10.

Example 6.15 For $1 \leq p < \infty$ and $k \geq 1$, let $x^k = \langle x_n^k \rangle$ be a Cauchy sequence in (ℓ_p, d_p). Let $\epsilon > 0$. Then there is some natural number N such that

$$[d_p(x^N, x^{N+s})]^p = \sum_{n=1}^{\infty} |x_n^N - x_n^{N+s}|^p < \epsilon^p, \quad \text{for all } s \geq 1. \qquad (6.3.4)$$

Then, for $k \geq N$ and for all $s \geq 1$, we have

$$|x_n^k - x_n^{k+s}|^p \leq \sum_{n=1}^{\infty} |x_n^N - x_n^{N+s}|^p = [d_p(x^N, x^{N+s})]^p < \epsilon^p$$
$$\Rightarrow \quad |x_n^k - x_n^{k+s}| < \epsilon,$$

which implies that real sequence of mth coordinate

$$y^m = \langle x_m^k, x_m^{k+1}, x_m^{k+2}, \ldots \rangle, \quad \text{for } k \geq N,$$

is Cauchy. So, we can find real numbers $x^m \in \mathbb{R}$ such that $y^m \to x^m$, for every $t \geq 1$. We claim that $x = \langle x^1, x^2, x^3, \ldots \rangle \in \ell_p$ and $x^k \to x$ with respect to metric d_p. For, note that it follows from (6.3.4) that

$$\sum_{m=1}^{q} |x_m^N - x_m^{N+s}|^p < \epsilon^p, \quad \text{for all } m, s \geq 1. \qquad (6.3.5)$$

This proves the claim $x^k \to x$ as $k \to \infty$. To conclude, note that

$$\sum_{m=1}^{\infty} |x^m|^p = \sum_{m=1}^{\infty} |x^m - x_m^k + x_m^k|^p$$

$$\leq \sum_{m=1}^{\infty} |x^m - x_m^k|^p + \sum_{m=1}^{\infty} |x_m^k|^p$$

$$< \epsilon^p + \sum_{m=1}^{\infty} |x_m^k|^p < \infty.$$

Hence, (ℓ_p, d_p) is a complete space. A similar argument proves that (ℓ_∞, d_∞) is also a complete metric space.

The following theorem, together with Theorem 6.26, help to find more examples of complete metric spaces.

Theorem 6.19 *Closed sets in a complete metric space are complete.*

Proof Let X be a complete metric space, and $S \subseteq X$ be closed. Let $\langle x_n \rangle$ be a Cauchy sequence in S. Since X is complete, and the sequence $\langle x_n \rangle$ is Cauchy in X, there is some $x \in X$ such that $x_n \to x$ as $n \to \infty$. Now, since $x_n \in S$, we must have $x \in S'$, by Theorem 6.14. So, $x \in S$, because S is closed. Hence, S is complete.

We will prove in a later chapter that the space $C[0, 1]$ of real valued continuous function defined on the compact set $[0, 1]$ is complete with respect to *uniform metric* (Theorem 8.3). However, the metric spaces as in next two examples are *not complete*.

Example 6.16 Consider the metric space $(C[0, 1], d_2)$, where the metric d_2 is as defined in (6.1.21). For $n \geq 3$, let the sequence $f_n : [0, 1] \to \mathbb{R}$ be given by

$$f_n(t) = \begin{cases} 1, & \text{when } 0 \leq t \leq \frac{1}{2} \\ -nt + \frac{n}{2} + 1, & \text{when } \frac{1}{2} \leq t \leq \frac{1}{2} + \frac{1}{n}, \\ 0, & \text{when } \frac{1}{2} + \frac{1}{n} \leq t \leq b \end{cases} \quad n \geq 3.$$

Clearly, each $f_n \in C([0, 1])$. Moreover, for $m, n \geq 3$, we have

$$[d_2(f_m, f_n)]^2 = \int_0^{1/2} |f_m(t) - f_n(t)|^2 dt + \int_{1/2}^1 |f_m(t) - f_n(t)|^2 dt$$

$$= \int_{1/2}^1 |f_m(t) - f_n(t)|^2 dt,$$

which implies that

$$d_2(f_m, f_n) = \left(\int_{1/2}^{1} |f_m(t) - f_n(t)|^2 dt \right)^{1/2}$$

$$\leq \left(\int_{1/2}^{1} |f_m(t)|^2 dt \right)^{1/2} + \left(\int_{1/2}^{1} |f_n(t)|^2 dt \right)^{1/2}$$

$$\leq \left(\int_{1/2}^{1} |f_m(t)|^2 dt + \int_{1/2}^{1} |f_n(t)|^2 dt \right)^{1/2},$$

and so

$$[d_2(f_m, f_n)]^2 \leq \int_{1/2}^{1} |f_m(t)|^2 dt + \int_{1/2}^{1} |f_n(t)|^2 dt. \tag{6.3.6}$$

However, since

$$\int_{1/2}^{1} |f_m(t)|^2 dt = \int_{1/2}^{1/2+1/m} |f_m(t)|^2 dt + \int_{1/2+1/m}^{1} |f_m(t)|^2 dt$$

$$= \int_{1/2}^{1/2+1/m} (-mt + m/2 + 1)^2 dt + 0$$

$$\leq \left(\int_{1/2}^{1/2+1/m} (-mt + m/2 + 1) dt \right)^2 = \frac{1}{(2m)^2};$$

and, likewise, $\int_{1/2}^{1} |f_n(t)|^2 dt \leq \frac{1}{(2n)^2}$, it follows from (6.3.6) that

$$[d(f_m, f_n)]^2 \leq \frac{1}{(2m)^2} + \frac{1}{(2n)^2} \leq \left(\frac{1}{2m} + \frac{1}{2n} \right)^2,$$

and so

$$d(f_m, f_n) \leq \frac{1}{2m} + \frac{1}{2n} \to 0, \quad \text{as} \quad m, n \to \infty.$$

That is, $\langle f_n \rangle$ is a Cauchy sequence in $C[0, 1]$. However, if for some $g \in C[0, 1]$, $\lim_{n \to \infty} f_n = g$, then the definition of f_n implies that

$$g(t) = \begin{cases} 1, & \text{when } 0 \leq t \leq 1/2 \\ 0, & \text{when } 1/2 \leq t \leq 1 \end{cases},$$

which is discontinuous at $t = 1/2$. Thus, $\{f_n\}$ is not convergent in X. Hence, the metric space $(C[0, 1], d_2)$ is *not complete*.

Example 6.17 Consider the subspace $C^1[0, 1] \subset C[0, 1]$ of continuously differentiable functions defined on the interval $[0, 1]$, and let d_∞ be the *supremum metric* as defined in (6.1.19). Let a sequence $f_n : [0, 1] \to \mathbb{R}$ be given by

$$f_n(x) = \sqrt{x + (1/n)}, \quad x \in [0, 1].$$

Then, for $f(x) = \sqrt{x}$, we have

$$
\begin{aligned}
d_\infty(f_n, f) &= \sup_{x \in [0,1]} \left| \sqrt{x + (1/n)} - \sqrt{x} \right| \\
&= \sup_{x \in [0,1]} \frac{1/n}{\sqrt{x + (1/n)} + \sqrt{x}} \\
&= \frac{1}{\sqrt{n}} \to 0, \quad \text{as} \quad n \to \infty.
\end{aligned}
$$

As $f(x) = \sqrt{x} \notin C^1[0, 1]$, we conclude that $(C^1[0, 1], d_\infty)$ is not a complete metric space.

As said before, in most cases of interest to us, the *convergence* (or *non-convergence*) of a sequence $\langle x_n \rangle$ in a (carrier) set X is defined with respect to some specified metric d given on the set X. Each metric defines a *convergence type* for the sequence $\langle x_n \rangle$ in the metric space (X, d). The terminology adopted for a convergence type in a particular situation is *somehow* related to our preference for having such a concept in the first place.

Notice that, however, the convergence is not a topological property. That is, a sequence may converges with respect to one metric defined on a set X, whereas it may diverge with respect to some other. We say a metric d on a set X is *nice* if the convergence of a sequence $\langle x_n \rangle$ in the metric space (X, d) satisfies the conditions that the limit $x := \lim_{n \to \infty} x_n$ enjoys most of the properties that the terms x_n have.

6.4 Compactness

Let X be a metric space, and $K \subset X$. A collection

$$\mathscr{C} = \left\{ A_\alpha \subset X : \alpha \in I \right\}$$

is called a *cover* (or a *covering*) of K if $K \subset \bigcup_{\alpha \in I} A_\alpha$. When the index set I is finite, $\mathscr{C}$ is called a *finite cover*. A cover $\mathscr{C}$ of K is said to admit a *finite subcover* for K if, for some finite set $J \subset I$, we have $K \subset \bigcup_{\alpha \in J} A_\alpha$. Said differently, only a finite subcollection of the cover $\mathscr{C}$ is enough to cover the set K. In this case, we also say that the cover $\mathscr{C}$ contains a *finite subcover* for the set K. A collection $\mathscr{C}$ of subsets of X is called an *open cover* of K if each member of $\mathscr{C}$ is an open set.

Definition 6.17 Let X be a metric space. We say a set $K \subset X$ is *compact* if every open cover of K contains a finite subcover (for K).

That is, K is compact if from *any given collection* of open sets $\{U_\alpha\}$, with $K \subset \bigcup U_\alpha$, we can find finitely many indices $\alpha_1, \ldots, \alpha_m$ such that

$$K \subset U_{\alpha_1} \cup \ldots \cup U_{\alpha_m}.$$

Clearly then, every finite set $K \subset X$ is compact. For, if K has n elements, and $\mathscr{C} = \{U_\alpha\}$ is any open covering of K, then we need at most n members of the collection $\mathscr{C}$ to cover K. In particular, every metric space on a finite set X is compact. On the other hand, if K is an infinite set, and we consider the *discrete metric* on the set X, then K can not be compact. For, in this case, each singleton $\{x\}$ is an open set, and hence no finite subcollection of the open covering $\mathscr{C} = \{\{a\} : a \in K\}$ of K can cover the set K.

Now, let $K \subset Y \subset X$. Then, we can talk about open covering of K using open sets in X or open sets in Y. In this case, we use the terminology that K is compact relative to X or relative to Y. The next theorem prove that K is compact relative to X if and only if it is compact relative to Y. Therefore, in virtue of this result, we do not have to mention underlying space, and so we can simply say K is compact.

Theorem 6.20 *Let X be a metric space, and $K \subset Y \subset X$. Then, K is compact relative to X if and only if K is compact relative to Y.*

Proof First, suppose K is compact relative to X, and let $\{V_\alpha : \alpha \in I\}$ be an open covering of K by open sets $V_\alpha \subseteq Y$. Then, by Theorem 6.9, there are open sets $U_\alpha \subseteq X$ such that $V_\alpha = U_\alpha \cap Y$, for each $\alpha \in I$. Since $V_\alpha \subset U_\alpha$, we have $K \subset \bigcup V_\alpha \subset \bigcup U_\alpha$. Therefore, $\{U_\alpha : \alpha \in I\}$ is an open covering of K by open sets in X. Hence, we can find a finite subcollection $\{U_{\alpha_1}, \ldots, U_{\alpha_m}\}$ that covers K. That is, $K \subset \bigcup_{k=1}^m U_{\alpha_k}$, and hence

$$K = K \cap Y \subset \left(\bigcup_{k=1}^m U_{\alpha_k}\right) \cap Y = \bigcup_{k=1}^m (U_{\alpha_k} \cap Y) = \bigcup_{k=1}^m V_{\alpha_k}.$$

It thus follows that $\{V_{\alpha_1}, \ldots, V_{\alpha_m}\}$ covers K.

To prove the converse, suppose K is compact relative to Y, and let $\{U_\alpha : \alpha \in J\}$ be an open covering of K by open sets U_α in X. Then, by Theorem 6.9, the sets $V_\alpha = U_\alpha \cap Y$ are open in Y, for each $\alpha \in J$. Since $K \subset \bigcup U_\alpha$, we have

$$K = K \cap Y \subset \left(\bigcup U_\alpha\right) \cap Y = \bigcup (U_\alpha \cap Y) = \bigcup V_\alpha.$$

Therefore, $\{V_\alpha : \alpha \in J\}$ is an open covering of K by open sets in X. Hence, we can find a finite subcollection $\{V_{\alpha_1}, \ldots, V_{\alpha_m}\}$ that covers K and hence

$$K \subset \bigcup_{k=1}^m V_{\alpha_k} \subset \bigcup_{k=1}^m U_{\alpha_k}.$$

Therefore, $\{U_{\alpha_1}, \ldots, U_{\alpha_m}\}$ covers K. This completes the proof.

The next theorem proves closed set of a compact space is compact.

Theorem 6.21 *Let X be a metric space and $F \subset K \subset X$. If K is compact and F is a closed set in X, then F is compact.*

Proof Let $\mathscr{C}$ be an open covering of F by sets open in X. Let $\mathscr{C}' = \mathscr{C} \bigcup \{X \setminus F\}$. Then $\mathscr{C}'$ is an open covering of K (and indeed it covers all of X). Hence, there is a finite subcover $\mathscr{B}' \subset \mathscr{C}'$ that covers K. Let $\mathscr{B} = \mathscr{B}'$ if $X \setminus K \notin \mathscr{B}'$ and $\mathscr{B} = \mathscr{B}' \setminus \{X \setminus K\}$ if $X \setminus K \in \mathscr{B}'$. Then the finite subcollection $\mathscr{B}$ covers K. Therefore, K is compact.

Theorem 6.22 *Let X be a metric space, and $K \subset X$ be a compact set. Then, K is a closed and bounded set.*

Proof First, suppose K is compact. To prove that K is closed, we show that $X \setminus K$ is open. For, let $a \in X \setminus K$ be an arbitrary point. Since $x \in K$ and $a \in X \setminus K$, $d(a, x) > 0$. For each $x \in K$, let $0 < r_x, s_x < d(a, x)/2$. Consider the neighbourhoods $V_x = B(a; r_x)$ and $U_x = B(x; s_x)$. Then $V_x \bigcap U_x = \emptyset$ and $K \subset \bigcup_{x \in K} U_x$. Therefore, $\{U_x : x \in K\}$ is an open cover of K. Since K is compact, there is a finite subset $\{U_{x_1}, \ldots, U_{x_m}\}$ that covers K. Put

$$V = \bigcap_{k=1}^{m} V_{x_k} \quad \text{and} \quad U = \bigcup_{k=1}^{m} U_{x_k}.$$

Then, V and U are open, $a \in V$ and $K \subset U$. Since $V \subset V_{x_k}$, we have $V \bigcap U_{x_k} \subset V_{x_k} \bigcap U_{x_k} = \emptyset$ and hence $V \bigcap U_{x_k} = \emptyset$. Therefore, it follows that

$$V \bigcap U = V \bigcap \left(\bigcup_{k=1}^{m} U_{x_k} \right) = \bigcup_{k=1}^{m} (V \bigcap U_{x_k}) = \emptyset.$$

(See Fig. 6.10). Since $K \subset U$ and $V \bigcap U = \emptyset$, we see that $V \bigcap K = \emptyset$ or $V \subset X \setminus K$. Since $a \in V$, and V is open, there is some ball $B(a; r)$ contained in V, and so in the set $X \setminus K$. Therefore, $X \setminus K$ is open, and hence K is closed.

Finally, to show that K is bounded, fix a point $x \in K$. For each $a \in K$, we can find some $n \in \mathbb{N}$ such that $d(x, a) < n$, by archimedean property. That is, $a \in B(x; n)$. So, the collection $\mathscr{C} = \{B(x; n) : n \in \mathbb{N}\}$ is an the open cover of the set K. By compactness of K, we find $n_1, \ldots, n_m \in \mathbb{N}$ such that $\{B(x; n_1), \ldots, B(x; n_m)\}$ is a finite subcover of K. Take $M = \max\{n_1, n_2, \ldots, n_m\}$. Clearly then,

$$K \subset \bigcup_{k=1}^{m} B(x; n_k) \subset B(x; M).$$

Hence, K is bounded.

A collection $\mathscr{C}$ of subsets of a metric space X is said to have finite intersection property if, for each finite subcollection $\{C_1, C_2, \ldots, C_m\}$ of $\mathscr{C}$, the intersection $\bigcap_{k=1}^{m} C_k \neq \emptyset$.

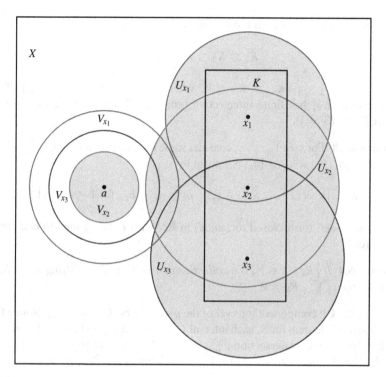

Fig. 6.10 Compact sets are closed

Theorem 6.23 *If* $\{K_\alpha\}$ *is a collection of compact subsets of a metric space X having finite intersection property, then the intersection* $\bigcap_\alpha K_\alpha \neq \emptyset$.

Proof Proof is by contradiction. So, we may assume that $\bigcap_\alpha K_\alpha = \emptyset$. Let $U_\alpha = X \setminus K_\alpha$. As each K_α is compact, and so closed, we have that U_α is open, for each α. Also, by taking complement of $\bigcap_\alpha K_\alpha = \emptyset$, we obtain $\bigcup_\alpha U_\alpha = X$. Fix one member of the collection $\{K_\alpha\}$, say K_β. Then, $K_\beta \subset X = \bigcup_\alpha U_\alpha$ implies that the collection $\{U_\alpha\}$ is an open cover of the compact set K_β. So, there is a finite subcollection $\{U_{\alpha_1}, \dots, U_{\alpha_m}\}$ that covers K_β. Therefore,

$$K_\beta \subset \bigcup_{k=1}^{m} U_{\alpha_k} = \bigcup_{k=1}^{m} (X \setminus K_{\alpha_k}) = X \setminus \left(\bigcap_{k=1}^{m} K_{\alpha_k} \right).$$

Hence, $K_\beta \cap \left(\bigcap_{k=1}^{m} K_{\alpha_k} \right) = \emptyset$, which contradicts the finite intersection property of the collection $\{K_\alpha\}$. This completes the proof.

Corollary 6.3 *For* $n \in \mathbb{N}$, *let* K_n *be a nonempty compact subset of a metric space. If* $K_k \supset K_{k+1}$, *for each* $k \in \mathbb{N}$, *then* $\bigcap_{k=1}^{\infty} K_k \neq \emptyset$.

Proof As $K_k \supset K_{k+1}$ for each $k \in \mathbb{N}$, we have

$$K_1 \supset K_2 \supset K_3 \supset \cdots .$$

So, $K_{n_1} \cap K_{n_2} \cap \cdots \cap K_{n_m} = K_N \neq \emptyset$, where $N = \max\{n_1, n_2, \ldots, n_m\}$. Thus, the collection $\{K_n\}$ has finite intersection property. Therefore, by Theorem 6.23, we have $\bigcap_{k=1}^{\infty} K_k \neq \emptyset$.

Definition 6.18 For $k = 1, \ldots, n$, consider some intervals $I_k = [a_k, b_k] \subset \mathbb{R}$. Then, the set $R = [a_1, b_1] \times \cdots \times [a_n, b_n]$ given by

$$R = \left\{ x = (x_1, \ldots, x_n) \in \mathbb{R}^n : a_k \leq x_k \leq b_k, \text{ for } 1 \leq k \leq n \right\}$$

is called an *n-cell* (or a *closed rectangle*) in $\mathbb{R}^n$, with I_k being the *k*th *component interval* for $k = 1, \ldots, n$.

Corollary 6.4 *If* $\left\{ R_k : k \in \mathbb{N} \right\}$ *is a collection of n-cells in* $\mathbb{R}^n$ *satisfying* $R_k \supset R_{k+1}$, *for* $k \in \mathbb{N}$, *then* $\bigcap_{k=1}^{\infty} R_k \neq \emptyset$.

Proof Let the *k*th component interval of the *n*-cell R_k be $I_k = [a_k, b_k]$. Notice that, by Heine-Borel theorem for $\mathbb{R}$, each interval I_k satisfies the hypothesis of the previous corollary, and so their intersection $\bigcap_{k=1}^{\infty} I_k \neq \emptyset$. Take a point γ_k in the intersection $\bigcap_{k=1}^{\infty} I_k$. Clearly then, $(\gamma_1, \ldots, \gamma_n) \in \bigcap_{k=1}^{\infty} R_k$. Therefore, $\bigcap_{k=1}^{\infty} R_k \neq \emptyset$.

Corollary 6.5 *Every n-cell in* $\mathbb{R}^n$ *is compact.*

Proof The argument is similar to the proof given above.

Corollary 6.6 *For any collection* $\{F_\alpha\}$ *of closed sets in a compact metric space X, with finite intersection property, the intersection* $\bigcap_\alpha F_\alpha \neq \emptyset$.

Proof Since closed sets in a compact space are compact, the corollary follows directly from Theorem 6.23.

The converse of this corollary holds too.

Theorem 6.24 *A metric space X is compact if, for every collection* $\{F_\alpha\}$ *of closed sets in X, with finite intersection property, the intersection* $\bigcap_\alpha F_\alpha \neq \emptyset$.

Proof Suppose the condition holds. We prove that X is compact. For, let $\mathscr{C} = \{U_\alpha\}$ be an open cover of X so that $X = \bigcup U_\alpha$. For each α, we write $F_\alpha = X \setminus U_\alpha$. Then, $\mathscr{B} = \{F_\alpha\}$ is a collection of closed sets in X. Further, as $X = \bigcup U_\alpha$, we have

$$\bigcap_\alpha F_\alpha = \bigcap_\alpha (X \setminus U_\alpha) = \emptyset.$$

So, the collection $\mathscr{B}$ cannot have finite intersection property. Thus, there is some finite subcollection $\{F_1, F_2, \ldots, F_m\} \subset \mathscr{B}$ such that $\bigcap_{k=1}^{m} F_k = \emptyset$. Therefore,

$$\bigcup_{k=1}^{m} U_k = \bigcup_{k=1}^{m} (X \setminus F_k) = X \setminus \left(\bigcap_{k=1}^{m} F_k \right) = X \setminus \emptyset = X.$$

Hence, $\{U_1, U_2, \ldots, U_m\}$ is a finite subcollection of the cover $\mathscr{C}$ that covers X. This proves that X is compact.

In general, to prove that a set is not compact, it suffices to construct a suitable open cover that has no finite subcover for the set. We have already illustrated this point in Chap. 5. However, it is usually difficult to prove that a set $K \subset X$ is compact, because we have to deal with every open cover of K. So, it makes sense to seek some easy-to-apply equivalent conditions. The condition stated in the next definition is about the fundamental questions: *When does a sequence has a convergent subsequence ?*

Definition 6.19 Let X be a metric space, and $A \subset X$. We say A has the *Bolzano-Weierstrass property* (or simply *BWP*) if every infinite subset of A has a limit point in A. In this case, we also say A is *limit point compact*.

We prove that compact sets are always *limit point compact*.

Theorem 6.25 *Let X be a metric space, and $K \subseteq X$ be a compact set. Then K has the Bolzano-Weierstrass property.*

Proof For a contradiction, suppose there is an infinite compact set $K \subset X$ that fails the Bolzano-Weierstrass property. Then, by definition, K contains an infinite set A such that A has no limit point (in K). Notice that, since A has no limit point, A is closed, and hence compact (Theorem 6.21). Also, for each $a \in A$, there is an open ball $U_a := B(a; r_a)$ $(r_a > 0)$ such that a is the only point it has in common with A. Clearly then, the collection $\mathscr{C} := \{U_a : a \in A\}$ is an open cover of A that admits no finite subcover for A. This contradicts the compactness of A, and hence completes the proof.

We next prove that every compact metric space is complete. Notice that, however, the converse statement may not hold. For example, the set $\mathbb{R}$ with respect the usual metric is complete, but not compact.

Theorem 6.26 *If (X, d) is a compact metric space, then it is complete.*

Proof Let $\langle x_n \rangle$ be a Cauchy sequence in X. If the set $S = \{x_n\}$ is finite, then some x_n is already repeating itself infinitely many times. Clearly then, by using such an element, we obtain a subsequence that converges in X, and hence the sequence $\langle x_n \rangle$ converges in X, by Theorem 6.18. So, we assume that the set $S = \{x_n\}$ is infinite. Then, by Theorem 6.25, S has a limit point $x \in X$. Take $n_1 \in \mathbb{N}$ such that $d(x_{n_1}, x) < 1$. After having chosen $n_1, n_2, \ldots, n_{k-1}$, we take $n_k > n_{k-1}$ such that $d(x, x_{n_k}) < 1/k$. Notice that it is possible to do so because every ball at x contains infinitely many points of S. Now, let $\epsilon > 0$, and take $N \in \mathbb{N}$ such that $N > 1/\epsilon$, by the archimedean property. We then have

$$d\left(x, x_{n_k}\right) < \frac{1}{k} \leq \frac{1}{N} < \epsilon, \quad \text{for all} \quad n \geq N.$$

Therefore, the subsequence $\langle x_{n_k} \rangle$ of the Cauchy sequence $\langle x_n \rangle$ converge in X. Hence, by Theorem 6.18, the sequence $\langle x_n \rangle$ is itself convergent in X.

Recall that, in view of the second assertion of Theorem 6.7, a set $A \subset \mathbb{R}^n$ is *bounded* if $\|a\| < M$, for all $a \in A$. Notice that, if $\|a\| < M$ for all $a \in A$, then $d_2(x, 0) = \|x\| < M$, and hence it bounded in metric sense (Definition 6.15). On the other hand, if a set $A \subset \mathbb{R}^n$ is bounded in the metric sense, then we can find some $a \in \mathbb{R}^n$ such that $\|x - a\| < M$, for all $x \in A$, and hence

$$\|x\| \leq \|x - a\| + \|a\| < M + \|a\|, \quad \text{for all} \quad x \in A.$$

Therefore, as before, the two definitions are equivalent.

Theorem 6.27 *(Heine-Borel Theorem)* *A set $K \subset \mathbb{R}^n$ is compact if and only if it is closed and bounded.*

Proof We know that every compact set $K \subset \mathbb{R}^n$ is closed and bounded, by Theorem 6.22. To complete the proof, suppose $K \subset \mathbb{R}^n$ is a closed and bounded set. Then, we have $K \subset B(\mathbf{0}; M)$, for some $M > 0$. So,

$$K \subset [-M, M] \times \cdots \times [-M, M].$$

Now, as n-cells are compact, by Corollary 6.5, and K is a closed set of a compact set, we conclude that K is compact, by Theorem 6.21.

Theorem 6.28 *(Bolzano-Weierstrass Theorem)* *Every bounded infinite subset $F \subset \mathbb{R}^n$ has a limit point in $\mathbb{R}^n$.*

Proof Since $F \subset \mathbb{R}^n$ is bounded, it is contained in some n-cell. Also, by Corollary 6.5, as n-cells are compact, so F is an infinite subset of a compact set. Therefore, by Theorem 6.25, F has a limit in the n-cell, and hence in $\mathbb{R}^n$.

6.5   Connectedness

We say a space X is *connected* if it is not topologically separated. That is, it is not possible to find a pair of nonempty open sets (U, V) such that

$$X = U \cup V \quad \text{and} \quad U \cap V = \emptyset. \tag{6.5.1}$$

We also say that (U, V) is a *separation* of X. Therefore, a space X is *disconnected* if it has a separation. Notice that we can write the condition (6.5.1) in terms of the closed sets $F = X \setminus U$ and $S = X \setminus V$. Therefore, equivalently, a space X is connected if $\emptyset$ and X are the only *clopen sets* in X.[9] For, if $\emptyset \neq S \neq X$ is a clopen, then the pair $(S, X \setminus S)$ provides a *separation* of the space X. Clearly, a *one-point* space is connected.

Notice that, in particular, *discrete space* (X, δ) is disconnected unless X is a *singleton*, because every set in (X, δ) is a clopen. We have shown in the previous chapter that the real line $\mathbb{R} = (-\infty, \infty)$ is connected. However, for any $a \in \mathbb{R}$, the set $\mathbb{R} \setminus \{a\}$ is disconnected with respect to induced usual topology, because the open sets $U = (-\infty, a)$ and $V = (a, \infty)$ provide a separation.

The next theorem provides a simple criterion of connectedness of a space X in terms of continuity of the *indicator function* $\chi : \mathscr{P}(X) \to \{0, 1\}$ given by

$$\chi(A) = \begin{cases} 1, & \text{if } x \in A \\ 0, & \text{if } x \in X \setminus A \end{cases}, \quad \text{for } A \in \mathscr{P}(X).$$

We anticipate here the following topological definition of continuity of a function between spaces X and Y: *A function $f : X \to Y$ is continuous if the inverse image of every open set of Y under f is an open set of X* (Theorem 7.5). We will also use this definition in some other situations.

Theorem 6.29 *A space X is connected if and only if every continuous function $f : X \to \{0, 1\}$ is constant.*

Proof First, suppose there exists a *nonconstant* continuous function $f : X \to \{0, 1\}$. Clearly then, the set $\emptyset \neq A = f^{-1}(\{0\}) \neq X$ is a clopen. Therefore, X is disconnected. Conversely, if X is disconnected, and $\emptyset \neq A \neq X$ is a clopen, then the *indicator function* $\chi : \mathscr{P}(X) \to \{0, 1\}$ is a *nonconstant* continuous function.

We now introduce the concept of *connectedness* for a nonempty set $Y \subset X$ with respect to the induced metric topology. For a $C \subset Y$, we write $Cl_Y(C)$ for the *relative closure* of C with respect to the induced topology on Y, and reserve the notation $\overline{C}$ for the *closure* in the space X. In this case, we say a pair of sets (A, B) is *separated* if neither one of the two contains a limit point of the other. That is,

$$\overline{A} \cap B = \emptyset = A \cap \overline{B}.$$

Clearly then, $A \cap B = \emptyset$. Notice that disjoint sets need not be separated. By Hausdorff property of a metric space X, for $x \neq y$, open sets $A = B(x; r)$ and $B = B(y; r)$ in X, with $r = |x - y|/2 > 0$, are separated because

$$\overline{A} \cap B = \emptyset = A \cap \overline{B}.$$

---

[9] A set S in a space is called a *clopen* if S is both open and closed set in X.

Notice that, in general, for $A, B \subseteq Y \subseteq X$,

$$A \cap Cl_Y(B) = \emptyset \quad \Leftrightarrow \quad A \cap \overline{B} = \emptyset; \text{ and,}$$
$$B \cap Cl_Y(A) = \emptyset \quad \Leftrightarrow \quad B \cap \overline{A} = \emptyset,$$

imply that subsets of separated sets are separated, and conversely. Thus, a *separation in Y* is the same as a *separation in X*.

Definition 6.20 A nonempty set $Y \subset X$ is *disconnected* if there exists nonempty sets A and B such that $Y = A \bigcup B$ and

$$Cl_Y(A) \cap B = \emptyset = A \cap Cl_Y(B).$$

We say (A, B) is a *separation* of the space Y.

Since $Y \cap \overline{C} = Cl_Y(C)$, if (A, B) is a separation of $Y \subset X$, then

$$Y = A \cup B \quad \Rightarrow \quad A, B \text{ are closed in } Y,$$

and so A and B are also open **in** Y. Notice that, when $X = \mathbb{R}$, the sets $A = [0, 1/2)$ and $B = (1/2, 1]$ are separated both in $Y = [0, 1] \setminus \{1/2\}$ and X, but A, B are not open in X. Recall that the only *nontrivial* connected sets in $\mathbb{R}$ are intervals.

We digress here to prove some results that are required to continue with the general discussion. The next theorem proves an interesting fact about a *continuous image* of a connected set in a space.

Theorem 6.30 *Let X and Y be two spaces, and $A \subseteq X$. If $f : X \to Y$ is a continuous function, and A is a connected set, then $f(A) \subset Y$ is a connected set.*

Proof For a possible contradiction, suppose $f(A) \subset Y$ is a disconnected set. Then, we can express it as a union of two disjoint nonempty subsets of Y, say B and C, such that these sets are covered by open sets in Y, say U and V, where

$$f(A) = B \cup C \subseteq U \cup V, \text{ with } B \cap V = C \cap U = \emptyset.$$

But then, for the sets $B_1 = f^{-1}(B), C_1 = f^{-1}(C), U_1 = f^{-1}(U)$, and $V_1 = f^{-1}(V)$, we can express A as a union of disjoint nonempty subsets B_1 and C_1 respectively exclusively covered by the open sets U_1 and V_1 in Y. This contradicts A is connected. Hence, $f(A)$ is connected.

The next theorem generalises Theorem 5.23.

Theorem 6.31 (Intermediate Value Theorem) *Let X be a connected space, and $f : X \to \mathbb{R}$ be a (nonconstant) continuous function. Then, for any pair of elements $x, x' \in X$, if r is any real number between $f(x)$ and $f(x')$ then there exists a point $x_0 \in X$ such that $f(x_0) = r$.*

Proof It follows from Theorem 6.30 that $f(X)$ is connected. But the, since $f(X) \subset \mathbb{R}$ is a nonempty connected set, Theorem 6.30 and our assumption $f(x) < r < f(x')$ (say) implies that $r \in f(X)$. Hence, there exists an $x_0 \in f^{-1}(X)$ such that $r = f(x_0)$, as asserted.

Remark 6.1 Applying Theorem 6.31 to a continuous function $f : [-1, 1] \to [-1, 1]$, it follows that $f(x_0) = x_0$, for some $x_0 \in [-1, 1]$. That is, the function f has a *fixed-point* over the set $[-1, 1]$. This assertion holds in general for any continuous function $f : \mathbb{D}^n \to \mathbb{D}^n$, where $\mathbb{D}^n$ is the closed n-disk in $\mathbb{R}^n$, $n \geq 2$. However, it requires sophistication of ideas from algebraic topology to prove this intuitively obvious fact.

By Theorem 6.30, if $f : X \to Y$ is a continuous function, with X connected, then the *graph* $\Gamma_f \subset X \times Y$ is a connected set, because it is *homeomorphic* to the space X (Definition 7.3). For example, since the function $f : (0, \infty) \to \mathbb{R}$ given by $y = f(x) = 1/x$ is continuous, it follows that the graph of f given by

$$\Gamma_f = \{(x, y) \in \mathbb{R}^2 : x > 0 \text{ and } y = x^{-1}\}$$

is a connected set in $\mathbb{R}^2$ (see Fig. 6.11). Notice that the converse may not hold, in general. For example, the function $f : [0, 1] \to \mathbb{R}$ given by

$$f(x) = \begin{cases} \sin(\frac{\pi}{x}), & 0 < x \leq 1 \\ 0, & x = 0 \end{cases}$$

oscillates between -1 and 1 more rapidly near $x = 0$, where it is discontinuous, but the graph Γ_g of the continuous function $g = f|_{(0,1]}$ is connected (see Fig. 6.12). In fact, since

$$\overline{\Gamma}_g = \Gamma_g \cup (\{0\} \times [-1, 1]),$$

Fig. 6.11 The graph Γ_f of $y = 1/x$, $x > 0$, is connected

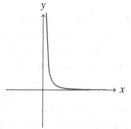

Fig. 6.12 Discontinuous
function with connected
graph

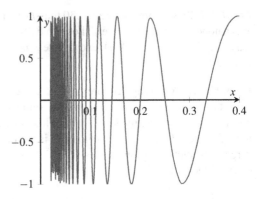

Corollary 6.9 implies that $\Gamma_g \bigcup Z$ is connected for any $Z \subseteq \{0\} \times [-1, 1]$. In partic-ular, the graph $\Gamma_f = \Gamma_g \bigcup \{(0, 0)\}$ of the function f is connected. The only failure is due to the fact that Γ_f is not *closed*.[10]

Let us continue with the general discussion. Notice that the two connected sets A and B given by

$$A = \{(x, y) \in \mathbb{R}^2 : y = 0\};$$
$$B = \{(x, y) \in \mathbb{R}^2 : x > 0 \text{ and } y = 1/x\}$$

satisfy the condition that $\overline{A} \bigcap B = \emptyset = A \bigcap \overline{B}$, and so $Y = A \bigcup B$ is *disconnected* in $\mathbb{R}^2$. So, we conclude that the *union of connected sets may not be connected*. How-ever, as a separation (U, V) of a space X produces a separation for every nonempty set $Y \subset X$ given by the pair $(U \cap Y, V \cap Y)$, if Y is a connected we must have that one of the two separating sets is empty, i.e., $Y \subset U$ or $Y \subset V$. Thus, it is reasonable to expect that a union of overlapping connected subsets is connected.

Theorem 6.32 *Arbitrary union* $X = \bigcup_{\lambda \in \Lambda} X_\lambda$ *of connected spaces is connected if there is some* $x_0 \in \bigcap X_\lambda$.

Proof For, if (S, T) is a separation of X, then the point x_0 is in one of the sets S or T. Suppose $x_0 \in S$. Since X_λ is a connected set, it lies entirely in S or T; indeed, since $x_0 \in S$, we must have that $X_\lambda \subset S$. But then, $X \subset S$, forcing $T = \emptyset$, a contradiction. Whence X is connected.

For example, since $\mathbb{R}^n$ can be written as a union of lines (each seen as $\mathbb{R}$) passing through the origin, it follows that $\mathbb{R}^n$ is connected. In fact, unlike the real line $\mathbb{R}$, the n-dimensional space $\mathbb{R}^n$ ($n > 1$) stays connected even when *countably many* points are being removed from it (Exercise 6.28). In particular, the set $\mathbb{R}^n \setminus \mathbb{Q}^n$ is connected,

---

[10] C.E. Burgess, *Continuous Functions and Connected Graphs*, The American Mathematical Monthly, April 1990, 337–339.

for $n > 1$. To prove this assertion, we need the following corollary of the previous theorem.

Corollary 6.7 *Let X be a space such that each pair of points in X lie in some connected set of X. Then, the space X itself is connected.*

Proof Suppose every pair of points $x, y \in X$ lies in some connected set C_{xy}. Then, for any fixed $a \in X$, we can write

$$X = \bigcup \{C_{ay} : y \in X\}, \quad \text{where each } C_{ay} \text{ is connected,}$$

and so X is connected, by Theorem 6.32.

We can use Theorem 6.29 to prove a generalised version of Theorem 6.32.

Theorem 6.33 *Let $\{X_\lambda : \lambda \in \Lambda\}$ be a family of connected spaces such that $X_\lambda \cap X_\nu \neq \emptyset$ for all $\lambda \neq \nu$. Then, $X = \bigcup_{\lambda \in \Lambda} X_\lambda$ is connected.*

Proof Since X_λ is connected, by Theorem 6.29, any continuous function $f_\lambda : X_\lambda \to \{0, 1\}$ is constant, say $f_\lambda(X_\lambda) = a_\lambda \in \{0, 1\}$. Now, to prove that X is connected, let $f : X \to \{0, 1\}$ be a continuous function. Then,

$$f\Big|_{X_\lambda}(X_\lambda) = a_\lambda \text{ and } f\Big|_{X_\nu}(X_\nu) = a_\nu \Rightarrow a_\lambda = a_\nu,$$

becasue $X_\lambda \cap X_\nu \neq \emptyset$. As this is true for all $\lambda \neq \nu$, it follows that f itself is a constant function. Hence, X is connected.

For two spaces X and Y, consider the product space $Z = X \times Y$. Suppose Z is connected. Then, by Theorem 6.30, continuity of the *natural projections* $\pi_1 : Z \to X$, $\pi_2 : Z \to Y$ implies that both X and Y are connected. The converse follows from the previous theorem.

Corollary 6.8 *For two connected spaces X and Y, the product $Z = X \times Y$ is connected.*

Proof For fixed $a \in X$ and $y \in Y$, consider the *vertical line* $V_a = \{a\} \times Y$ and the *horizontal lines* $H_y = X \times \{b\}$. Clearly, these lines are connected sets in the product space Z. Since $H_y \cap V_a = \{(a, b)\} \neq \emptyset$, so the set $L_y = V_a \cup H_y$ is connected for all $y \in Y$, by Theorem 6.33. But then, $L_y \cap L_{y'} \neq \emptyset$ for $y \neq y'$, which implies that the set

$$Z = X \times Y = \bigcup_{y \in Y} L_y$$

is connected.

In particular, $\mathbb{R}^n$ is connected, as a product space.

Remark 6.2 The argument used to prove Theorem 6.33 can be applied in a more abstract situation. For, let X be a connected space, and $f : X \to Y$ be a continuous function that is *locally constant*, i.e., for each $x \in X$ there is some ball $B(x : \epsilon)$ such that the restriction of the function f to the ball $B(x; \epsilon)$ is constant. Let $f(x) = y_0$, for all $x \in B(x; \epsilon)$. We claim that the function f is identically constant on X. For, the sets

$$\emptyset \neq A = \{x : f(x) = y_0\} \quad \text{and} \quad B = \{x : f(x) \neq y_0\}$$

are disjoint open sets, with $X = A \bigcup B$, and so $A = X$, by the connectedness of X. In particular, when Y is a two-point set such as $\{1, 0\}$, and if we wish to prove that all points of X have a particular topological property $\mathscr{P}$, it is then sufficient to prove the following three assertions:

1. There is at least one point with property $\mathscr{P}$.
2. If x has property $\mathscr{P}$, the same applies to all points in a sufficiently small ball.
3. If x does not has property $\mathscr{P}$, then the same applies to all points in a small ball.

This observation applies to numerous situations in analysis and geometry.

The following is yet another interesting special case of Theorem 6.32.

Corollary 6.9 *A set $Y \subset X$ obtained by adding some or all limit points of a connected space C results into a connected space. In particular, closure of a connected space is always connected.*

Proof We are given that $C \subseteq Y \subseteq Cl_X(C)$. For each $y \in Y$, $\{y\}$ and C are not separated. So, by Theorem 6.32,

$$Y = C \cup \left(\cup_{y \in Y} \{y\} \right)$$

is a connected set.

Notice that, however, the interior of a connected set may not be a connected.

Example 6.18 Since any point of the set

$$X = \left\{(x, y) \in \mathbb{R}^2 : xy \geq 0\right\}$$

can be joined to the *origin* by a *stright line*, it follows that X is connected. But, the interior of X given by the set

$$X^o = \left\{(x, y) \in \mathbb{R}^2 : xy > 0\right\}$$

is not *connected*. For, notice that it is the union of the first and third quadrants, without the axes.

The *path-connectivity* property used in Example 6.18 is far more stronger condition than the connectedness.

6.5.1 Path-Connectedness

Definition 6.21 Let X be a metric space, and $x, y \in X$. By a *path* (or an *arc*) in X from the point x to y we mean a continuous (and *injective*) function $\alpha : [0, 1] \to X$ such that $\alpha(0) = x$ and $\alpha(1) = y$, which we may write as $\alpha(x; y)$. We say X is *path connected* (or *arcwise connected*) if any two points $x, y \in X$ can be joined by a path (or an arc).

Clearly, for any fixed $a \in X$, the continuous function $e_a : [0, 1] \to X$ given by $e_a(t) = a$, for all $t \in [0, 1]$, is a path. It is called the *constant path* at the point a. Next, given two paths $\alpha(x; y)$ and $\beta(y; z)$, we define their *product* $\gamma := \alpha \cdot \beta$ as the path $\gamma(x; z)$ obtained by *concatenation*. That is,

$$\gamma(t) := \begin{cases} \alpha(2t) , & \text{for } 0 \le t \le (1/2) \\ \beta(2t - 1), & \text{for } (1/2) \le t \le 1 \end{cases}, \quad t \in [0, 1].$$

Notice that the continuity of γ follows from the *pasting lemma* (Theorem 7.7). By a *polygonal path* $\gamma : [0, 1] \to X$ we mean a concatenation of paths of the form

$$\gamma = \alpha_1 \cdots \alpha_n, \quad \text{for } n \ge 2,$$

where $\alpha_{k+1}(1) = \alpha_k(0)$, for all $1 \le k < n$. Also, the *reverse path* of a path $\alpha(x; y)$ is the path $\beta(y; x)$ given by

$$\beta(t) = \alpha(1 - t), \quad \text{for } t \in [0, 1].$$

In this case, we write $\beta = \alpha^{-1}$. Notice that

$$\alpha \cdot \alpha^{-1} = e_x \quad \text{and} \quad \alpha^{-1} \cdot \alpha = e_y.$$

Definition 6.22 We say a set $C \subset \mathbb{R}^n$ is *star-convex* if there exists a point $a \in C$ such that, for any $x \in C$, the *line segment* joining x to a given by

$$L(x; a) := \left\{ y \in \mathbb{R}^n : y = x + t(a - x) \text{ for } t \in [0, 1] \right\}$$

lies completely in the set C. We call a a *star point* of the set C.

Example 6.19 Notice that, if a set $C \subset \mathbb{R}^n$ is star-convex with respect to a *star point* $a \in C$, then we can join any two points $x, y \in C$ by a (polygonal) path $\alpha \cdot \beta$, where $\alpha(x, a)$ and $\beta(a; x)$ are the paths given by

$$\begin{cases} \alpha(t) = x + t(a - x) \\ \beta(t) = a + t(y - a) \end{cases}, \quad \text{for } t \in [0, 1].$$

So, each star-convex set in $\mathbb{R}^n$ is path connected. In particular, each convex set in $\mathbb{R}^n$ is path connected. Also, by the above argument, the *punctured space* $\mathbb{R}^n \setminus \{0\}$, $n > 1$, is path connected, because we can use a *polygonal path* to join the two points, if the line joining these points passes through the origin. Therefore, using the fact that the image of the connected set $\mathbb{R}^{n+1} \setminus \{0\}$ under the continuous function given by

$$x \mapsto \frac{x}{\|x\|}, \quad x \in \mathbb{R}^{n+1} \setminus \{0\},$$

is the *n-dimensional sphere* $\mathbb{S}^n$ given by

$$\mathbb{S}^n = \left\{ x \in \mathbb{R}^{n+1} : \|x\| = 1 \right\}, \quad n \geq 1,$$

we conclude that $\mathbb{S}^n$ is path connected, by Theorem 6.30.

As said earlier, the path connectedness of a space is a stronger condition than the connectedness. For example, we have proved already a stronger statement in Corollary 6.7. The next theorem completes the argument.

Theorem 6.34 *Every path connected space is connected.*

Proof For, if (U, V) is a separation of a path connected space X, then no point in U can be joined to any point in V since the connected set as continuous image of the connected set $I = [0, 1]$ has necessarily to lie entirely in U or V. Alternatively, fix $a \in X$. For each $x \in X$, let $\alpha_x : [0, 1] \to X$ be a path from a to x, and let $C_x = \alpha_x([0, 1])$. So, as each C_x is connected and $X = \bigcup_{x \in X} C_x$, we conclude that X is connected, by Theorem 6.32.

The converse of the above theorem is not true, in general. The famous *topologist sine curve* (see Fig. 6.13)

$$T = \left\{ \left(x, \sin(\pi/x) \right) : x \in (0, 1] \right\} \cup \{(0, 0)\} \tag{6.5.2}$$

Fig. 6.13 Topologist sine curve $= \sin(\pi/x)$, $x \neq 0$

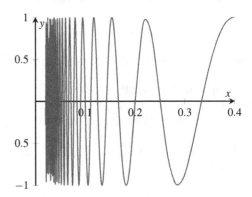

provides an example of a connected space that is not path connected. We write $f(x) = \sin\left(\pi/x\right)$, for $x \in (0, 1]$. More generally, it is possible to prove that no point on the *vertical segment*

$$L = \{0\} \times [-1, 1] = \{(0, y) : -1 \leq y \leq 1\}$$

can be joined by a path to any point of the *graph*

$$\Gamma_g = \{(x, \sin(\pi/x)) : 0 < x \leq 1\},$$

where $g = f|_{(0,1]}$ (Exercise 6.29).

Theorem 6.35 *There exists a connected space that is not path connected.*

Proof We prove here the space T as in (6.5.2) is such a space. For convenience, we prove the assertion for the points $a = (0, 0) \in L$ and $b = (1, 0) \in \Gamma_g$, where notations are as introduced above. If possible, suppose $\gamma : [0, 1] \to T$ is a path from $\gamma(0) = (0, 0)$ to $\gamma(1) = (1, 0)$. Then, for the continuous function $F = \pi_1 \circ \gamma : [0, 1] \to \mathbb{R}$, we have $F(0) = 0$ and $F(1) = 1$. Thus, by Theorem 6.31, $F(t_1) = 2/3$, for some $t_1 \in (0, 1)$. For the same reason, there exists a $t_2 \in (0, t_1)$ such that $F(t_2) = 2/5$, and so on. So, there exists a *monotone* sequence $\langle t_n \rangle$ in $(0, 1)$ such that

$$t_{n+1} \in (0, t_n) \quad \text{and} \quad F(t_n) = \frac{2}{2n + 1}, \quad n \geq 1.$$

Let $t^* = \lim\limits_{n \to \infty} t_n$ so that $0 \leq t^* < 1$. Also, since $\gamma : [0, 1] \to T$ is continuous, we have $\gamma(t^*) = \lim\limits_{n \to \infty} \gamma(t_n)$. However, since the sequence

$$\gamma(t_n) = \left(\frac{2}{2n + 1}, (-1)^n \right)$$

is not convergent, we arrived at a contradiction.

A similar argument proves that, for any set $Y \subset L$, the set $\Gamma_g \bigcup Y$ is not path connected. In particular, it follows that $\overline{\Gamma}_g = \Gamma_g \bigcup L$ is not path connected. Thus, the *closure of a path connected set may not be path connected*, in general. Notice that Γ_g, being homeomorphic to $(0, 1]$, is a path connected space.

6.5.2 Components

When a space X is disconnected, we need to study the extent of the disconnection in terms of its (maximal) connected subsets. Of course, the extreme situation arises when the only connected sets in a space X are *singletons*. Such a space is called *totally disconnected*. For example, all the sets $\mathbb{N}$, $\mathbb{Z}$, $\mathbb{Q}$ are totally disconnected spaces with respect to the induced modulus metric. And, so is the *Cantor ternary set*.

Example 6.20 The Cantor set C, as defined in Example 2.6, is a *totally disconnected* set. For, if $x \neq y \in C$ are such that $|x - y| > 3^{-k}$, it follows by construction of C that x and y belong to distinct C_k's. So, we conclude that there exists $z \notin C$, with $x < z < y$. Yet C has no isolated points because for some end point $y_k \in C_k$, $x \neq y_k$, and $|x - y_k| > 3^{-k}$.

It thus follows that, for a totally disconnected space, the *maximal connected subsets* are one-point set. On the other extreme, a connected space has only one component, just itself.

Definition 6.23 A maximal connected set $C \subset X$ is called a *component*.

Equivalently, a connected set $C \subset X$ is a *component* if for any $K \subset X$, with $K \not\subset C$, the set $C \bigcup K$ is disconnected. This suggests the following construction of *components* of a space X: For a fixed $x \in X$, consider the set

$$C_x = \bigcup \{C \subseteq X : C \text{ is connected, and } x \in C\}.$$

For example, $C_x = X$ for all x, if X is connected; and $C_x = \{x\}$ for each $x \in X$, if X is *totally disconnected*. Notice that, in general, $\{x\} \in C_x \Rightarrow C_x \neq \emptyset$. So, by Theorem 6.32, C_x is connected. Trivially, it is a *maximal connected set* subject to $x \in C_x$. That is, C_x is a *component* containing x. In general, for any $x, y \in X$,

$$C_x \neq C_y \quad \Rightarrow \quad C_x \cap C_y = \emptyset.$$

To prove this, suppose $z \in C_x \cap C_y$. Then, $C_x \subseteq C_z \Rightarrow x \in C_z$, and so $C_x = C_z$. Similarly, $C_y = C_z$. Hence, $C_x = C_y$. Also, for any nonempty connected set $Y \subseteq X$, $y \in Y \Rightarrow Y \subset C_y$. Notice that, by previous observation, C_y is unique to contain Y. Thus, every nonempty connected set in X is contained in precisely one C_x. Thus, $\{C_x : x \in X\}$ provides a partition of the space X. Conversely, consider the relation on X given by

$$x \sim y \quad \Leftrightarrow \quad x \text{ and } y \text{ lie in some connected set in } X. \qquad (6.5.3)$$

Clearly, it is reflexive and symmetric. And, transitivity follows by using Theorem 6.32. That is, it defines an equivalence relation on X, and so partitions X into equivalence classes, which are precisely the *components* C_x, $x \in X$.

Theorem 6.36 *Every space X is a disjoint union of its components, which are closed connected sets.*

Proof Clearly, the first part of the statement follows from the above discussion. And, Corollary 6.9 proves that components are always closed sets.

Notice that if a space has only finitely many components then all these components are necessarily open. Hence, for a space with finite number of components, these are *clopen sets*. But, in general, components of a space need not be open sets. For example, all components of $\mathbb{Q}$ are one-point sets. Hence, in general, a component and its complement do not provide a *disconnection* of X.

The next theorem proves that the *number of components* of a space X is a topological invariant. As is customary in algebraic topology, let $\pi_0(X)$ denote the collection of components of a space X.

Theorem 6.37 *For two homeomorphic spaces X and Y, $\pi_0(X) \approx \pi_0(Y)$.*

Proof The main argument is to show that every continuous bijective function $f : X \to Y$ gives a bijection $f_* : \pi_0(X) \to \pi_0(Y)$. For, if $C \subseteq X$ is a component and $x \in C$, we have $C = C_x$. Consider a component $D \subseteq Y$ containing $y = f(x)$ so that $D = D_y$. Since $f(C_x)$ is a connected set in Y containing y, we must have $f(C_x) = D_y$. Thus, by Theorem 6.36, the function $f_* : \pi_0(X) \to \pi_0(Y)$ given by $f_*(C) = D$ is a well-defined bijection.

Using the construct as given above to find the components of a space X, we can introduce the notion of a *path component* $P_x \subseteq X$ containing a point $x \in X$, by replacing the term "*connected*" with the term "*path connected*" all through the discussion leading to the relation (6.5.3). The same change applies to the discussion following it. It then follows that the equivalence classes are precisely the *path components* P_x, $x \in X$, and any path connected set is contained in precisely one path component of X. In this case, we write,

$$P_x = \cup\{P \subseteq X : P \text{ is path connected, and } x \in P\} \qquad (6.5.4)$$
$$= \{y \in X : x \text{ and } y \text{ lie in some path connected set in } X\} \qquad (6.5.5)$$

However, unlike in connected case, *path components* may neither be open nor closed. For example, the *topologist sine curve* has exactly two path components: the *vertical segment* $L = \{0\} \times [-1, 1]$ and the graph

$$\Gamma_g = \left\{\left(x, \sin\frac{1}{x}\right) : x \in (0, 1]\right\}$$

of the restriction $g = f|_{(0,1]}$. The graph Γ_g is not closed. If, however, we consider a subspace $Y \subset T$ dropping all rational points from the vertical segment L, then the space Y has a single component, but uncountably many path components.

By Theorem 6.34, we have $P_x \subseteq C_x$, $x \in X$. As above, topologist sine curve shows that the equality $P_x = C_x$ ($x \in X$) may not hold, in general. The next theorem proves that equality holds in *locally* path connected spaces. It is a generalisation of Theorem 6.34.

Definition 6.24 A space X is called *locally path connected* if every point in X has a path connected neighbourhood.

For example, since ball in $\mathbb{R}^n$ are always path connected, so $\mathbb{R}^n$ is a *locally* path connected space.

Theorem 6.38 *Let X be a locally path connected space. Then, P_x is open, and $P_x = C_x$, for all $x \in X$.*

Proof Since each $x \in X$ has a path connected neighbourhood U_x, we have $U_x \subseteq P_x$, by relation (6.5.4). And, if $y \in P_x$ is arbitrary then $P_x = P_y$ implies that P_x also contains a neighbourhood of y. Thus, P_x is open. In particular, it follows that the union of all path components other that P_x is an open set. So, the path component P_x as the complement of this union is a closed set. That is, $P_x \subseteq C_x$ is a *clopen*. Hence, we must have $P_x = C_x$, because C_x is connected.

Exercises

6.1 Let (X, d) be a metric space. Show that

$$d(x_1, x_n) \leq d(x_1, x_2) + d(x_2, x_3) + \cdots + d(x_{n-1}, x_n),$$

   for all $x_1, x_2, \ldots, x_n \in X$.

6.2 Let (X, d) be a metric space. Show that

$$\left| d(x, y) - d(y, z) \right| \leq d(x, z), \quad \text{for } x, y, z \in X.$$

6.3 Let (X, d) be a metric space, and suppose $g : [0, \infty) \to [0, \infty)$ is a monotone increasing function, with $g(t) > 0 = g(0)$, for $t > 0$, and

$$g(s + t) \leq g(s) + g(t), \quad \text{for all } s, t \geq 0.$$

   Show that $\rho = g \circ d$ is a metric on the set X. Do the functions $g(t) = t^p$, for $0 < p \leq 1$; $g(t) = \min\{1, t\}$, for $t \geq 0$; and, $g(t) = t/(1 + t)$, for $t \geq 0$, satisfy the above stated conditions?

6.4 Let (X, d) be a metric space. Show that $\sqrt{d(x, y)}$ and $\max\{1; d(x, y)\}$ are also metrics on the set X.

6.5 If d_1 and d_2 are two metrics on a set X, and $\alpha > 0$, show that $\alpha d_1, d_1 + d_2$ and $\max\{d_1, d_2\}$ are also metrics on the set X.

6.6 Let (X, d) be a metric space, Z be an arbitrary nonempty set, and $f : Z \to X$ be a bijective function. Show that the function $\rho : Z \times Z \to \mathbb{R}$ given by $\rho(x, y) = d(f(x), f(y))$ defines a metric on the set Z.

6.7 For any finite set Y, let $X = \mathscr{P}(Y)$, the *power set* of the set Y. Take $d(A, B)$ to be the number of elements in the *symmetric difference* $(A \setminus B) \bigcup (B \setminus A)$, for $A, B \in X$. Show that d is a metric on X.

6.8 A set $L = \{v_1, \ldots, v_m\}$ in $\mathbb{R}^n$ is *orthonomal* if $\langle v_i, v_j \rangle = \delta_{ij}$, for $1 \leq i, j \leq m$, where δ_{ij} is the usual Kronecker delta. Prove that L is linearly independent and, for any $a_1, \ldots, a_m \in \mathbb{R}$,

$$\|\sum_{k=1}^{m} a_k v_k\| = \left(\sum_{k=1}^{m} |a_k|^2\right)^{1/2}.$$

6.9 Prove that the function $d : \mathbb{R}^2 \times \mathbb{R}^2 \to \mathbb{R}$ defined by

$$d(x, y) = \begin{cases} |x - y| & \text{if } x, y \text{ are linearly dependent;} \\ |x| + |y| & \text{otherwise} \end{cases}$$

is a metric on $\mathbb{R}^2$.

6.10 For each $m \in \mathbb{N}$, the function $d : \mathbb{R}^n \times \mathbb{R}^n \to \mathbb{R}$ defined by $d(x, y) = (\sum_{k=1}^{n} |x_k - y_k|^m)^{1/m}$ where $x = (x_1, x_2, \ldots, x_n)$ and $y = (y_1, y_2, \ldots, y_n)$, is a metric on $\mathbb{R}^n$.

6.11 Prove that the function $d : \mathbb{R}^n \times \mathbb{R}^n \to \mathbb{R}$ defined by $d(x, y) = \max\{|x_k - y_k| : 1 \le k \le n\}$ where $x = (x_1, x_2, \ldots, x_n)$ and $y = (y_1, y_2, \ldots, y_n)$ is a metric on $\mathbb{R}^n$.

6.12 (*Hölder's Inequality*) For $1 < p < \infty$, let $1 < q < \infty$ be the *conjugate exponent* of p, i.e., $(1/p) + (1/q) = 1$. We may take $q = \infty$ for $p = 1$, and $q = 1$ for $p = \infty$. Show that, for any sequences $x = \langle x_n \rangle$, $y = \langle y_n \rangle \in \ell_p$,

$$\sum_{n=1}^{\infty} |x_n y_n| \le [S_p(x)]^{1/p}[S_q(y)]^{1/q},$$

where $S_p(x) = \sum_{n=1}^{\infty} |x_n|^p$ and $S_q(y) = \sum_{n=1}^{\infty} |y_n|^q$.

6.13 (*Minkowski's Inequality*) Show that, for any real number $p \ge 1$, $x = \langle x_n \rangle$, $y = \langle y_n \rangle \in \ell_p \Rightarrow x + y \in \ell_p$, and

$$\left(S_p(x + y)\right)^{1/p} \le \left(S_p(x)\right)^{1/p} + \left(S_p(y)\right)^{1/p}.$$

6.14 Prove that the metric defined in (6.1.22) satisfies the triangle inequality.

6.15 Let (X, d) be a metric space, and $x \in X$. Show that the set $U = X \setminus \overline{B}(a; r)$ is open, for every $r > 0$.

6.16 Prove that the intersection of an arbitrary collection of convex sets in $\mathbb{R}^n$ is a convex set.

6.17 For $k = 1, 2, \ldots, n$, let $-\infty < a_k < b_k < \infty$, and $I_k = [a_k, b_k]$. Show that the open n-cell $R = I_1 \times \cdots \times I_n$ given by

$$R = \{x = (x_1, \ldots, x_n) \in \mathbb{R}^n : a_k < x_{ki} < b_k \text{ for } 1 \le k \le n\}$$

is a convex set of the space $\mathbb{R}^n$.

6.18 Show that a finite union of closed sets is a closed set, and a finite intersection of open sets is an open set. Give an example, in each case, to show that these assertions fail if the term "*finite*" is dropped.

6.19 Let S and T be two subsets of a metric space, with $S \subset T$. Show that $S' \subset T'$.

6.20 Let T be closed subset of a metric space. Show that $\overline{S} \subset T$, for any $S \subset T$.

6.21 Show that a set S in a metric space X is closed if and only if for each sequence $\langle x_n \rangle$ in S, with $x_n \to x$, we have $x \in S$.

6.22 If a sequence $\langle x_n \rangle$ in a metric space converges to two points x, $y \in X$, then $x = y$. [*Hint*: Use Theorem 6.7]

6.23 Let X be a metric space and x_n, $x \in X$. Show that $x_n \to x$ if and only if every ball around x contains all but finitely many terms of $\langle x_n \rangle$.

6.24 Let X be a complete metric space, and $Y \subset X$. Show that Y is complete if and only if it is closed.

6.25 Show that a metric space X is complete if and only if for every decreasing sequence of closed subsets $\{F_n\}$ of the space X, with *diameter* $\mathrm{diam}(F_n) \to 0$ as $n \to \infty$, the intersection $\bigcap_{n=1}^{\infty} F_n$ contains a single element.

6.26 Let F be a closed set, and K a compact set, in a metric space. Show that $F \cap K$ is compact.

6.27 Let $\mathscr{F} = \{F_n : n \in \mathbb{N}\}$ be a collection of nonempty closed sets in a compact metric space, with $F_k \supset F_{k+1}$, for each $k \in \mathbb{N}$. Show that $\bigcap_{k=1}^{\infty} F_k \neq \emptyset$.

6.28 Show that, for any countable set $C \subset \mathbb{R}^n$ $(n > 1)$, the set $\mathbb{R}^n \setminus C$ is connected.

6.29 With notations as in Theorem 6.35, prove that no point on the *vertical segment* L can be joined by a path to any point of the *graph* Γ_g.

6.30 Show that every open connected set in $\mathbb{R}^n$ is path connected. Give an example to illustrate the fact that the assertion fails if we replace the term *open* by *closed*.

Reference

1. M.M. Deza, E. Deza, *Encyclopedia of Distances (4/e)* (Springer, Berlin, 2016)

Chapter 7
Multivariable Analysis

> *Practical application is found by not looking for it, and one can*
> *say that the whole progress of civilization rests on that principle.*

Jacques Hadamard (1865–1963)

We start with the concept of the continuity of a function defined between metric spaces. In the special case when $E \subseteq \mathbb{R}^m$ is an open set, the continuity of a multivariable vector-valued function $f : E \to \mathbb{R}^n$ at the point $x = c \in E$ is equivalent to the geometrical condition that $f(x) \approx f(c)$, for points x *very close* to the point c. Further, f is *differentiable* at c if, as $\|h\| \to 0$, the vectors $\big(f(c + h) - f(c)\big)$ approximate the *tangent plane* to the *level surface* of f at the point c given by a linear map $L_c : \mathbb{R}^n \to \mathbb{R}^n$. Said differently, f is differentiable at $x = c$ if the *vectors $f(x)$* for points x *very close* to c lie in a plane. The linear map L_c is called the derivative $Df(c)$ of f at c. Some related concepts developed in this chapter, and theorems proved, are significant to the analysis of certain important linear partial differential equations.

In a typical practical situation, we have $E \subseteq \mathbb{R}^n$, and a differentiable function $f : E \to \mathbb{R}^n$ represents a solution of some explicit or implicit non-linear problem. Then, the derivative $Df(c)$ given in terms of the linear map L_c provides a tool to *linearise* the problem locally, if $\det(L_c) \neq 0$. Therefore, if the *linearised problem* has a solution in some class of functions, then a unique (local) solution of the non-linear problem is obtained by considering a particular such solution that is *nearest to f* in some sense. For many applications in science and engineering, such a *linearisation principle* helps to apply tools from linear algebra to do *geometry* on surfaces embedded nicely in the Euclidean spaces $\mathbb{R}^n$.

© The Author(s), under exclusive license to Springer Nature Singapore Pte Ltd. 2022   273
A. K. Razdan and V. Ravichandran, *Fundamentals of Analysis with Applications*,
https://doi.org/10.1007/978-981-16-8383-1_7

Our main focus here is to discuss certain interesting geometric consequences of *linearisation principle.*[1] Most of the analytical and geometrical concepts introduced in this chapter are needed later to study the types of smooth surfaces that arise as the *integral surfaces* of the vector field defined by the function f. We conclude with a brief discussion about the *inverse function theorem, implicit function theorem,* and a standard theorem related to existence and uniqueness of a *local solution* of a non-linear ordinary differential equation. The reader may refer to excellent text on the topic by Munkres [1], for missing details or any unfinished argument.

7.1 Limit of Functions

Let (X, d_X) and (Y, d_Y) be two arbitrary metric spaces, and $E \subseteq X$. In general, the limit of a function $f : E \to Y$ at a point $x = c \in E$ is a *local property* in the sense that we are mainly concerned about how f changes in a very small ball around $x = c$, even when $f(c)$ is not defined. So, we may take $c \in E'$, where E' the set of all *limit points* of the set E with respect to metric d_X. Further, since a metric on a set in any given situation is always specified, it is convenient to write d for both d_X and d_Y.

We say that $y = l \in Y$ is a *limit* of a function $f : E \to Y$ at a point $x = c \in E'$ if, for any $\epsilon > 0$, there is some $\delta = \delta(c, \epsilon) > 0$, such that

$$0 < d(x, c) < \delta \quad \Rightarrow \quad d(f(x), l) < \epsilon. \tag{7.1.1}$$

In this case, we write $\lim_{x \to c} f(x) = l$. Recall that the inverse image of a set $S \subset Y$ with respect to the function $f : E \to Y$ is the set

$$f^{-1}(S) = \{x \in X : f(x) \in S\}.$$

So, a topological version of the condition (7.1.1) is given by

$$\lim_{x \to c} f(x) = l \quad \Leftrightarrow \quad B(c; \delta) \setminus \{c\} \subset f^{-1}\big(B(l; \epsilon)\big), \tag{7.1.2}$$

where $B(c; \delta) \setminus \{c\}$ is a *punctured ball* at $x = c$ of radius δ. A formal definition is as given next.

Definition 7.1 Let X and Y be metric spaces, and $E \subset X$. A function $f : E \to Y$ is said to have a *limit* $l \in Y$ at $x = c \in E'$ if, for any $\epsilon > 0$ and some $\delta = \delta(c, \epsilon) > 0$, we have

$$x \in E, \ 0 < d(x, c) < \delta \Rightarrow d(f(x), l) < \epsilon. \tag{7.1.3}$$

We write $\lim_{x \to c} f(x) = l$ or $f(x) \to l$, as $x \to c$.

---

[1] The reader may notice that some aspects of the multivariable analysis discussed here are straightforward generalisations of the ideas introduced earlier in Chap. 4.

Notice that $x = c \in X$ need not be in E. However, the assumption $c \in E'$ ensures that for any given $\delta > 0$ there are enough points $x \in E$ such that $d(x, c) < \delta$. Also, the limit $l \in Y$ does not depend on the value of the function f at the point $x = c$, even when $f(c)$ is defined. In general, limit of a function at a point may not exist. However, it is necessarily unique, if it exists.

Theorem 7.1 *Let X and Y be metric spaces, and $E \subset X$. The limit of a function $f : E \to Y$ at $c \in E'$ is unique, if it exists.*

Proof The main argument is that every metric defines a Hausdorff topological space. More precisely, suppose f has two limits l_1 and l_2 as $x \to c$. Let $\epsilon > 0$. Then, for some $\delta_1 > 0$ and $\delta_2 > 0$, we have

$$x \in E, \ 0 < d(x, c) < \delta_1 \quad \Rightarrow \quad d(f(x), l_1) < \epsilon/2,$$

and

$$x \in E, \ 0 < d(x, c) < \delta_2 \quad \Rightarrow \quad d(f(x), l_2) < \epsilon/2.$$

Take $\delta = \min\{\delta_1, \delta_2\}$. Then, for all $x \in E$ with $0 < d(x, c) < \delta$, we have $0 < d(x, c) < \delta_1$ and $0 < d(x, c) < \delta_2$, and hence

$$\begin{aligned} d(l_1, l_2) &\leq d(l_1, f(x)) + d(f(x), l_2) \\ &= d(f(x), l_1) + d(f(x), l_2) \\ &< \epsilon/2 + \epsilon/2 = \epsilon. \end{aligned}$$

Therefore, by usual argument, $l_1 = l_2$.

The next theorem gives definition of *sequential limit* of a function.

Theorem 7.2 *Let X and Y be metric spaces, and $E \subset X$. Let $f : E \to Y$ be a function, and $c \in E'$. Then, $\lim_{x \to c} f(x) = l$ if and only if $\lim_{n \to \infty} f(x_n) = l$, for every sequence $\langle x_n \rangle$ in E, with $x_n \neq c$, and $\lim_{n \to \infty} x_n = c$.*

Proof Let $\langle x_n \rangle$ be a sequence in E with $x_n \neq c$ and $\lim_{n \to \infty} x_n = c$. Let $\epsilon > 0$ be given. Choose $\delta > 0$ such that $d(f(x), l) < \epsilon$ whenever $x \in E$ and $0 < d(x, c) < \delta$. For this $\delta > 0$, choose a natural number N such that $d(x_n, c) < \delta$ for all $n \geq N$. Since $x \in E$ and $d(x_n, c) < \delta$, we have $d(f(x_n), l) < \epsilon$ or $\lim_{n \to \infty} f(x_n) = l$.

To prove the converse, assume that $\lim_{x \to c} f(x) \neq l$. Then there is an $\epsilon > 0$ such that, for every $\delta > 0$, there is an element $x \in E$ such that $d(f(x), l) \geq \epsilon$ but $d(x, c) < \delta$. Using this, choose $x_n \in E$ such that $d(f(x_n), l) \geq \epsilon$ but $d(x_n, c) < 1/n$. Thus, we have a sequence $\langle x_n \rangle$ with $\lim_{n \to \infty} x_n = c$, but $\lim_{x \to c} f(x_n) \neq l$.

We now consider the case when both X and Y are Euclidean spaces of some dimension. As before, we follow the convention of writing the vector quantities in *boldface*. Let $E \subset \mathbb{R}^m$ be an open set, and $x = c \in E'$. If $c \in E$, then there is some $\epsilon > 0$ such that

$$\|x - c\| < \epsilon \quad \Rightarrow \quad x \in E.$$

Otherwise, for a sequence $\langle x_n \rangle$ in E, with $x_n \neq c$, we have $\lim_{n \to \infty} x_n = c$.

Suppose the metric space Y is the real line $\mathbb{R}$, and consider a multivariable real-valued function $\varphi : E \to \mathbb{R}$. By definition, φ has the limit $a \in \mathbb{R}$ at a point $x = c \in E'$ if the condition (7.1.3) holds, for every $\epsilon > 0$ and some $\delta > 0$. That is,

$$0 < \|x - c\| < \delta \quad \Rightarrow \quad |\varphi(x) - a| < \epsilon. \tag{7.1.4}$$

Example 7.1 Consider the function $f : \mathbb{R}^2 \to \mathbb{R}$ given by

$$f(x) = \begin{cases} \dfrac{xy}{x^2 + y^2}, & \text{for } x = (x, y) \neq 0 \\ 0, & \text{for } x = 0 \end{cases}$$

Then, along the *line* $y = m\,x$, we have

$$f(x) = \frac{xy}{x^2 + y^2} = \frac{m}{1 + m^2}.$$

So, as $x \to 0$, the function $f(x)$ approaches different values. Therefore, $\lim_{x \to 0} f(x)$ does not exist at $x = 0$.

Example 7.2 Consider the function $f : \mathbb{R}^2 \to \mathbb{R}$ given by

$$f(x) = \begin{cases} \dfrac{x^2 y}{x^4 + y^2}, & \text{for } x = (x, y) \neq 0 \\ 0, & \text{for } x = 0 \end{cases}$$

In this case, along the *parabola* $y = m\,x^2$, we have

$$f(x) = \frac{x^2 y}{x^4 + y^2} = \frac{m}{1 + m^2}.$$

So, as $x \to 0$, the function $f(x)$ approaches different values. Therefore, $\lim_{x \to 0} f(x)$ does not exist at $x = 0$.

More generally, consider a multivariable *vector-valued function* $f : E \to \mathbb{R}^n$. Since $f(x) \in \mathbb{R}^n$, for each $x \in E$, we write

$$f(x) = \big(f_1(x), \ldots, f_n(x)\big), \quad \text{for } x \in E, \tag{7.1.5}$$

where the n multivariable real-valued functions $f_k : E \to \mathbb{R}$ $(1 \leq k \leq n)$ are called the *component functions* of the function f. The next theorem proves that it suffices to study the limit of functions f_k, for $k = 1, \ldots, n$.

Theorem 7.3 *Let X be a metric space, $E \subset X$, and $c \in E'$. A function $f : E \to \mathbb{R}^n$ has the limit $l = (l_1, \ldots, l_n) \in \mathbb{R}^n$ as $x \to c$ if and only if the component function $f_k : E \to \mathbb{R}$ has the limit l_k, as $x \to c$, for each $k = 1, \ldots, n$.*

Proof For $k = 1, \ldots, n$, let $\pi_k : \mathbb{R}^n \to \mathbb{R}$ denote the usual *projection maps* given by $\pi_k(x) = x_k$, for $vx = (x_1, \ldots, x_n) \in \mathbb{R}^n$. Then, it follows easily that $l = (l_1, \ldots, l_n)$ is the limit of f at $x = c$ if and only if the function $\pi_k \circ f : X \to \mathbb{R}$ has the limit l_k at $x = c$, for each $k = 1, \ldots, n$. The details are left to the reader. For example, a simple adaption of the proof given for Theorem 7.10 may prove to be helpful.

Therefore, a multivariable vector-valued function $f : E \to \mathbb{R}^n$ has the limit $l = (l_1, \ldots, l_n) \in \mathbb{R}^n$ at a point $c \in E$ if, for every $\epsilon > 0$ and some $\delta - \delta(c, \epsilon) > 0$, we have

$$0 < \|x - c\| < \delta \Rightarrow |f_k(x) - l_k| < \epsilon, \quad \text{for} \quad k = 1, \ldots, n. \tag{7.1.6}$$

In this case, we write $\lim\limits_{x \to c} f(x) = l$.

Sometimes, we need to assume that $E \subseteq \mathbb{R}^m$ is a *region*, i.e., E is an open connected set. In particular, when $m = 1$, we may take $E = (a, b)$, for some $a < b$, because the connected sets in $\mathbb{R}$ are only intervals. A function $\gamma : (a, b) \to \mathbb{R}^n$ is called a *vector function*. If the *component functions* of γ are given by the single variable real-valued functions $\gamma_k : (a, b) \to \mathbb{R}$, for $k = 1, \ldots, n$, then we write

$$\gamma = (\gamma_1, \gamma_2, \ldots, \gamma_n).$$

By Theorem 7.3, a vector function $\gamma : (a, b) \to \mathbb{R}^n$ has the limit $l = (l_1, \ldots, l_n) \in \mathbb{R}^n$ at $t = c \in (a, b)$ if and only if, for each $k = 1, \ldots, n$, the component function $\gamma_k : (a, b) \to \mathbb{R}$ has the limit l_k at $t = c$.

7.2 Continuity of a Function

We start with the following general definition.

Definition 7.2 Let X and Y are two metric spaces, and $E \subset X$ be an open set. We say a function $f : E \to Y$ is *continuous* at $x = c \in E$ if, for any $\epsilon > 0$ and some $\delta = \delta(c, \epsilon) > 0$, we have

$$x \in E, \ d(x, c) < \delta \quad \Rightarrow \quad d(f(x), f(c)) < \epsilon. \tag{7.2.1}$$

The function f is *continuous* on E if it is continuous at each point of E. Further, if $\delta > 0$ can be chosen to be independent of a point $x = c \in E$ then f is said to be *uniformly continuous* over E.

Let X and Y be metric spaces, and $y = l \in Y$. Consider the *constant function* $f_l : X \to Y$ given by $f_l(x) = l$, for all $x \in X$. Then, for $x_1, x_2 \in X$, we have

$$d\big(f_l(x_1),\, f_l(x_2)\big) = d(l, l) = 0 < \epsilon,$$

for each $\epsilon > 0$, implies that the condition (7.2.1) holds, for any $\delta > 0$. Therefore, every constant function on X is uniformly continuous. Next, consider the *identity function* $e : X \to X$ given by $e(x) = x$, for $x \in X$. In this case, we have

$$d(e(x_1), e(x_2)) = d(x_1, x_2), \quad \text{for all} \quad x_1, x_2 \in X.$$

So, given any $\epsilon > 0$, we may take $\delta = \epsilon$. Therefore, the identity function $e : X \to X$ is uniformly continuous.

By condition (7.2.1), if a function $f : E \to Y$ is continuous at $x = c \in E$ then, for any $\epsilon > 0$ and some $\delta = \delta(c, \epsilon) > 0$, we have

$$0 < d(x, c) < \delta \quad \Rightarrow \quad d(f(x), f(c)) < \epsilon.$$

In particular, we can find a sequence $\langle x_n \rangle$ in $B(c; \delta) \setminus \{c\}$ such that $\lim_{n \to \infty} x_n = c$, and so $d(f(x_n), f(c)) < \epsilon$. For, by archimedean property, there is some $N \in \mathbb{N}$ with $N\delta > 1$, so that

$$n \geq N \quad \Rightarrow \quad d(x_n, c) < \frac{1}{n} \leq \frac{1}{N} < \delta.$$

Therefore, $f : X \to Y$ is *sequentially continuous* at $x = c \in X$. In general, a similar argument proves that *every continuous function between arbitrary topological spaces is sequentially continuous.*[2] However, not every sequentially continuous function between topological spaces is continuous, unless the space X is *first countable*. As said before, metric spaces are indeed first countable, so the converse holds.

Theorem 7.4 (Sequential continuity) *Let X and Y be metric spaces, and $E \subset X$. A function $f : E \to Y$ is continuous at $x = c \in E$ if and only if for every sequence $\langle x_n \rangle$ in E, with $x_n \neq c$ and $\lim_{n \to \infty} x_n = c$, we have $\lim_{n \to \infty} f(x_n) = f(c)$.*

Proof It follows directly from Theorem 7.2.

The definition of the continuity of a function at a point given earlier is called a *local version* of the concept. The next theorem gives a *global version* in terms of open sets.

Theorem 7.5 *Let X and Y be metric spaces. A function $f : X \to Y$ is continuous on X if and only if, for each open set $V \subseteq Y$, the inverse image $f^{-1}(V) \subseteq X$ is an open set.*

Proof First, suppose f is continuous, and $V \subseteq Y$ be an open set. To prove that the set

$$U = f^{-1}(V) = \{x \in X : f(x) \in V\}$$

---

[2] For arbitrary topological spaces, we use *nets* in place of sequences.

is open, we show each point $c \in U$ is an interior point. For, since $f(c)$ is in the open set V, there is some $\epsilon > 0$ such that $B(f(c); \epsilon) \subset V$. So, by the continuity of f at the point c, we can find some $\delta > 0$ such that

$$d(x, c) < \delta \quad \Rightarrow \quad d(f(x), f(c)) < \epsilon.$$

Now, if $x \in B(c; \delta)$, then $d(x, c) < \delta$ implies that $d(f(x), f(c)) < \epsilon$. That is, $f(x) \in B(f(c); \epsilon) \subset V$. Therefore, $x \in f^{-1}(V) = U$, which shows $B(c, \delta) \subset U$. Hence, $x = c$ is an interior point of U, and so U is an open set.

To prove the converse, suppose the condition holds, and $x = c \in X$ be an arbitrary point. Let $\epsilon > 0$. Then, by the given condition, for the open set $V = B(f(c); \epsilon) \subset Y$, we have that the set $U = f^{-1}(V) \subseteq X$ is an open set. In particular, as $c \in U$ is an interior point, we can find some $\delta > 0$ such that $B(c; \delta) \subset U$. Therefore, to conclude the proof, it remains to verify the $\epsilon - \delta$ condition. For, suppose $x \in X$ is such that $d(x, c) < \delta$. Then,

$$x \in B(c; \delta) \subset U \quad \Rightarrow \quad f(x) \in V \quad \Rightarrow \quad d(f(x), f(c)) < \epsilon.$$

Hence, f is continuous at $x = c$. As $c \in X$ is arbitrary, we conclude that the function f is continuous on X.

In particular, it follows that the composition of continuous functions is a continuous function.

Corollary 7.1 *Let X, Y and Z be metric spaces, and $f : X \to Y$ and $g : Y \to Z$ be two continuous functions. Then, the composition $g \circ f : X \to Z$ is continuous on X.*

Proof Notice that $(g \circ f)^{-1}(A) = f^{-1}(g^{-1}(A))$ for any set $A \subset Z$. Let V be an open subset of Z. Since g is continuous, the inverse image $g^{-1}(V)$ is open in Y. The continuity of f now implies that the set $f^{-1}(g^{-1}(V))$ is open in X. Therefore, the set $(g \circ f)^{-1}(V)$ is open in X and hence the composition $g \circ f$ is continuous on X.

Since closed sets are complement of open sets, the next theorem follows from Theorem 7.5.

Theorem 7.6 *Let X and Y be metric spaces. A function $f : X \to Y$ is continuous on X if and only if the inverse image $f^{-1}(F)$ of each closed subset $F \subseteq Y$ is closed in X.*

Proof First, notice that

$$\begin{aligned}
x \in X \setminus f^{-1}(F) \quad &\Leftrightarrow \quad x \notin f^{-1}(F) \\
&\Leftrightarrow \quad f(x) \notin F \\
&\Leftrightarrow \quad f(x) \in Y \setminus F \\
&\Leftrightarrow \quad x \in f^{-1}(Y \setminus F).
\end{aligned}$$

It thus follows that, for any subset $F \subseteq Y$,

$$X \setminus f^{-1}(F) = f^{-1}(Y \setminus F).$$

Now, suppose $f : X \to Y$ is continuous, and $F \subseteq Y$ is closed. Then, $Y \setminus F$ is an open set in Y, and so $X \setminus f^{-1}(F) = f^{-1}(Y \setminus F)$ is open in X, by Theorem 7.5. Thus, $f^{-1}(F)$ is closed in X. The proof of the converse is similar.

The next theorem helps to prove the continuity of a function obtained as the *concatenation* of continuous functions.

Theorem 7.7 *[Pasting Lemma] Let A and B be two closed sets (or open sets) in a space X such that $X = A \bigcup B$, and Y be an arbitrary space. If the two restrictions of a function $f : X \to Y$ to sets A and B are continuous, then the function f is continuous.*

Proof We write $f_A = f|_A$ and $f_B = f|_B$. Consider a closed set $F \subseteq Y$. Then, by Theorem 7.6, $f_A^{-1}(F)$ is a closed set of A because the function $f_A : A \to Y$ is continuous. Therefore, $f_A^{-1}(F)$ is closed in X. Similarly, $f_B^{-1}(F)$ is a closed set of X because $f_B : B \to Y$ is also a continuous function. Finally, since

$$f^{-1}(F) = f_A^{-1}(F) \cup f_B^{-1}(F)$$

is a closed closed set of X, it follows from Theorem 7.6 that f is continuous. This completes the proof.   □

The next theorem is a fundamental fact about continuous functions.

Theorem 7.8 (Heine-Cantor) *Every continuous function on a compact space is uniformly continuous.*

Proof Let X and Y be metric spaces, with X be compact. Suppose $f : X \to Y$ is a continuous function. We take an arbitrary but fixed $\epsilon > 0$. Then, for every $a \in X$, there is some $\delta_a > 0$ such that

$$d(x, a) < \delta_a \quad \Rightarrow \quad d(f(x), f(a)) < \frac{\epsilon}{2}.$$

We consider the collection of balls $\{B(a; \delta_a/2) : a \in X\}$. Clearly then, this collection forms an open covering of the compact space X, and so we can find finitely many δ_{a_k}, say for $1 \leq k \leq n$, such that

$$X \subseteq \bigcup_{k=1}^{n} B(a_k; \delta_{a_k}/2).$$

We take $\delta = (1/2) \min_{1 \leq k \leq n} \delta_{a_k}$. So, for $x, y \in X$ with $d(x, y) < \delta$, we have $x \in B(a_k; \delta_{a_i}/2)$, which implies that $d(x, a_k) < \delta_{a_k}/2$. Thus, by the triangle inequality, we obtain

$$d(y, a_k) \le d(y, x) + d(x, a_k) < \delta + \frac{\delta_{a_k}}{2} \le \delta_{a_k}.$$

It therefore, follows that

$$d\big(f(x), f(y)\big) \le d\big(f(x), f(a_k)\big) + d\big(f(a_k), f(y)\big)$$
$$< \frac{\epsilon}{2} + \frac{\epsilon}{2} = \epsilon.$$

Hence, f is a uniformly continuous function.

Definition 7.3 Let X and Y be two metric spaces. We say a *bijective* function $f : X \to Y$ is a *Homeomorphism* if both f and $f^{-1} : Y \to X$ are continuous functions. We say two metric spaces X and Y are *homeomorphic* if there exists a homeomorphism $f : X \to Y$.

We say a property $\mathscr{P}$ of a metric space X is a *topological invariant* if every metric space *homeomorphic* to X has the property $\mathscr{P}$. In the previous chapter, we have shown that *connectedness* is a topological invariant of a metric space (Theorem 6.30). The next theorem proves that the compactness is also a topological invariant.

Theorem 7.9 *Let X and Y be metric spaces, where X is compact. Then, for any continuous function $f : X \to Y$, $f(X)$ is compact in Y. In particular, if f is a bijective, then X and Y are homeomorphic spaces.*

Proof Let $\{V_\alpha : \alpha \in I\}$ be a cover of $f(X)$ by open sets in Y. Since f is continuous, $f^{-1}(V_\alpha)$ is open in X. For each $x \in X$, $f(x) \in f(X) \subset \bigcup_\alpha V_\alpha$, and so $f(x) \in V_\alpha$, for some α. Thus, $x \in f^{-1}(V_\alpha)$, and $X \subset \bigcup_\alpha f^{-1}(V_\alpha)$. Therefore, $\{f^{-1}(V_\alpha)\}$ is an open cover of X. Since X is compact, we can find $\alpha_1, \alpha_2, \ldots, \alpha_n \in I$ such that $X \subset \bigcup_{k=1}^n f^{-1}(V_{\alpha_k})$. Let $y \in f(X)$ so that $y = f(x)$ for some $x \in X$. Then $x \in f^{-1}(V_{\alpha_k})$, for some $1 \le k \le n$, and hence $y = f(x) \in V_{\alpha_k}$. This shows that $f(X) \subset \bigcup_{k=1}^n V_{\alpha_k}$, and hence $f(X)$ is compact.

Next, to prove the second assertion, let $U \subset X$ be open. Then, $V = X \setminus U$ is a closed set in the compact space X, and so V itself compact. But then, by the first part, $f(V)$ is a compact set, and so it is a closed set in Y (Theorem 6.21). Also, since f is bijective, we have $f(V) = Y \setminus f(U)$. Thus, $f(U) \subset Y$ is open, which prove that the function $f^{-1} : Y \to X$ is also continuous. Hence, f is a homeomorphism. This completes the proof.

Corollary 7.2 *Let X be a compact metric space, and $f : X \to \mathbb{R}^n$ be a continuous function. Then, the set $f(X) \subset \mathbb{R}^n$ is closed and bounded.*

Proof By Theorem 7.9, $f(X)$ is a compact set of $\mathbb{R}^n$. By Theorem 6.27, we also know that compact sets in $\mathbb{R}^n$ are precisely the closed and bounded sets. It thus follows that $f(X)$ is a closed and bounded set.

The next corollary proves that every real valued continuous function defined on a compact space attains its maxima and minima. This result was first proved by *Bernard Bolzano* in 1830. However, the formulation given below is due to *Karl Weierstrass*.

Corollary 7.3 (Extreme Value Theorem) *Let X be a compact metric space, and $f : X \to \mathbb{R}$ be a continuous function. Then, there exist elements $x_1, x_2 \in X$ such that*

$$f(x_1) = \inf\{f(x) : x \in X\} = \inf f(X)$$

and

$$f(x_2) = \sup\{f(x) : x \in X\} = \sup f(X).$$

Proof By Corollary 7.2, $f(X) \subset \mathbb{R}$ is closed and bounded, and so $s = \sup f(X)$ exists, and $s \in \overline{f(X)} = f(X)$ implies that $f(x_2) = \sup f(X)$, for some $x_2 \in X$. The second assertion follows similarly by applying above conclusion to the function $-f$.

Let $E \subseteq \mathbb{R}^m$ be an open set, and consider a multivariable vector-valued function $f : E \to \mathbb{R}^n$. For $k = 1, \ldots, n$, let $f_k : E \to \mathbb{R}$ be the functions such that (7.1.5) holds. We may assume that f is continuous on the set $\overline{E} = E \cup E'$. More generally, for any metric space X, the next theorem prove that the continuity of the *component functions* $f_k : X \to \mathbb{R}$ implies the continuity of a vector-valued functions $f : X \to \mathbb{R}^n$.

Theorem 7.10 *Let X be a metric space. A vector-valued function $f : X \to \mathbb{R}^n$ is continuous over X if and only if each component function $f_k : X \to \mathbb{R}$ is continuous over X, for $k = 1, \ldots, n$.*

Proof Let $x = c \in X$ and $\epsilon > 0$ be arbitrary. First, suppose the function f is continuous at $x = c$. Then, for some $\delta > 0$, we have

$$d(x, c) < \delta \Rightarrow |f(x) - f(c)| < \epsilon.$$

As $f = (f_1, \ldots, f_n(x))$, we have

$$\|f(x) - f(c)\|^2 = \sum_{k=1}^{n} |f_k(x) - f_k(c)|^2 \geq |f_k(x) - f_k(c)|^2,$$

for each $k = 1, \ldots, n$. That is, for each k, we have

$$|f_k(x) - f_k(c)| \leq \|f(x) - f(c)\|.$$

Therefore, it follows that

$$d(x, c) < \delta \quad \Rightarrow \quad |f_k(x) - f_k(c)| < \epsilon.$$

Hence, each function f_k is continuous, for $k = 1, \ldots, n$. To prove the converse, suppose each *component function* f_k of a vector-valued function $f : X \to \mathbb{R}^n$ is continuous at some $x = c \in X$, and $\epsilon > 0$ be arbitrary. Then, for some $\delta_k > 0$, we have

$$d(x, c) < \delta_k \quad \Rightarrow \quad |f_k(x) - f_k(c)| < \epsilon/\sqrt{n}.$$

Take $\delta = \min\{\delta_1, \ldots, \delta_n\}$. Then, for all $1 \le k \le n$, we have

$$d(x, c) < \delta \Rightarrow |f_k(x) - f_k(c)| < \epsilon/\sqrt{n}.$$

Therefore, for $d(x, c) < \delta$, we have

$$\|f(x) - f(c)\|^2 = \sum_{k=1}^{n} |f_k(x) - f_k(c)|^2$$

$$< \sum_{k=1}^{n} \epsilon^2/n = \epsilon^2. \text{That is,} \|f(x) - f(c)\| \qquad < \epsilon.$$

Hence, the function f is continuous at $x = c$. This completes the proof.

Corollary 7.4 *Let X be a metric space, and $f, g : X \to \mathbb{R}^n$ be continuous functions. Then, both the sum function $f + g : X \to \mathbb{R}^n$ and the scalar product $\alpha f : X \to \mathbb{R}^n$ are continuous on X, for all $\alpha \in \mathbb{R}$.*

Proof For $x \in X$, the functions $f + g$ and αf are given by

$$(f + g)(x) = f(x) + g(x) \text{ and } (\alpha f)(x) = \alpha f(x).$$

We may write the functions f and g as follows:

$$f = (f_1, \ldots, f_n) \text{ and } g = (g_1, \ldots, g_n).$$

Then, for any $x \in X$, we have

$$(f + g)(x) = (f_1(x) + g_1(x), \ldots, f_n(x) + g_n(x));$$
$$(\alpha f)(x) = (\alpha f_1(x), \ldots, \alpha f_n(x)).$$

As both f and g are continuous, all the component functions f_k, g_k are continuous, for all $1 \le k \le n$. Then, as said in Chap. 4, it follows that $f_k + g_k$ and αf_k are continuous. Hence, by Theorem 7.10, $f + g$ and αf are continuous on X, for all $\alpha \in \mathbb{R}$.

In general, for an open set $E \subset \mathbb{R}^m$, we write $C_n(E)$ for the collection of multi-variable vector-valued continuous functions $f : E \to \mathbb{R}^n$. The corollary above helps to prove that $C_n(E)$ is a real linear space with respect to *pointwise operations*. Since $C_n(E)$ contains all m-variable polynomial functions of any *degree*, it is an infinite dimensional space. We first consider the case when $n = 1$, and write $C_1(E)$ simply as $C(E)$.

We provide here a large *class of continuous functions* contained in the space $C_n(\mathbb{R}^m)$. First, we need to introduce the concept of *linear maps* defined between real linear spaces.

Definition 7.4 Let V, W be two real linear spaces. We say a function $T : V \to W$ is a **linear map** if, for any $u, v \in V$ and $a, b \in \mathbb{R}$, we have

$$T(a\,u + b\,v) = a\,T(u) + b\,T(v).$$

We write $\mathscr{L}(V, W)$ for the collection of linear maps $T : V \to W$. In particular, we write $V^* = \mathscr{L}(V, \mathbb{R})$, which is called the *dual space* of V.

We prove that a every linear map $T \in \mathscr{L}(\mathbb{R}^m, \mathbb{R}^n)$ is continuous.[3] More generally, if V and V' are normed spaces (Definition 6.5), and $T : V \to V'$ is a linear map, we write

$$\|T\| := \sup\left\{\|T(x)\| : \, x \in \mathbb{R}^m\right\} = \sup\left\{\|T(u)\| : \, u \in \mathbb{S}^{m-1}\right\}, \qquad (7.2.2)$$

which is called the *operator norm* of the linear map T, where as usual $\mathbb{S}^{m-1}$ is the *unit sphere* in $\mathbb{R}^m$. We say T is *bounded* if $\|T\| < \infty$. Notice that, if $T \in \mathscr{L}(\mathbb{R}^n, \mathbb{R}^n)$ is given by

$$T(e_i) = \sum_{j=1}^{n} a_{ij} e_j, \quad \text{for} \quad i = 1, \ldots, n,$$

then it follows from the Cauchy-Schwartz inequality that

$$\|T(x)\| \le \|T\|\|x\|, \quad \text{for all} \quad x \in \mathbb{R}^n, \quad \text{with} \quad \|T\| = \sqrt{\sum_{i,j=1}^{n} a_{ij}^2}.$$

To prove our main assertion here, let $V = \mathbb{R}^m$ and $V' = \mathbb{R}^n$. As before, we write $\{e_1, \ldots, e_m\}$ for the standard basis, and let

$$\alpha := \max\left\{\|T(e_i)\| : \, 1 \le i \le m\right\}.$$

Recall that each $x \in \mathbb{R}^m$ can be written uniquely as

$$x = \langle x, e_1 \rangle\, e_1 + \cdots + \langle x, e_m \rangle\, e_m.$$

It thus follows that

$$T(x) = \langle x, e_1 \rangle\, T(e_1) + \cdots + \langle x, e_m \rangle\, T(e_m),$$

---

[3] More generally, a linear map $T : V \to V'$ between normed spaces is continuous if and only if T is continuous at $0 \in V$. In fact, we can replace the term *continuous* by terms such as *bounded*, *Lipschitz continuous*, *uniformly continuous*, and even *continuous at a point*.

and so we have

$$\begin{aligned} \|T(x)\| &= \left\| \langle x, e_1 \rangle\, T(e_1) + \cdots + \langle x, e_m \rangle\, T(e_m) \right\| \\ &\leq |\langle x, e_1 \rangle|\, \|T(e_1)\| + \cdots + |\langle x, e_m \rangle|\, \|T(e_m)\| \\ &\leq \big(|\langle x, e_1 \rangle| + \cdots + |\langle x, e_m \rangle| \big)\alpha \\ &= \alpha\, \|x\|_1, \end{aligned}$$

where $\|x\|_1 = d_1(x, 0) \leq \|x\|$. This proves the next theorem.

Theorem 7.11 *Every linear map $T \in \mathscr{L}(\mathbb{R}^m, \mathbb{R}^n)$ is Lipschitz continuous, and hence uniformly continuous.*

In particular, it follows that the *coordinate function* $x^i : \mathbb{R}^m \to \mathbb{R}$ given by

$$x^i(x) := \langle x, e_i \rangle, \quad \text{for} \ \ x \in \mathbb{R}^m, \tag{7.2.3}$$

are continuous functions, for $i = 1, \ldots, m$. Notice that, if $\{e_1, \ldots, e_m\}$ is an *orthonormal basis* of an m dimension inner product space V, then the *coordinate function* $x^i : V \to \mathbb{R}$ defined as in (7.2.3) is canonical basis of the dual space V^*. Theorem 7.11 has an obvious generalisation for $T \in \mathscr{L}(V, V')$, where V' is any n-dimensional inner product space.

More generally, we see that every *isometry* $T \in \mathscr{L}(\mathbb{R}^n, \mathbb{R}^n)$ is a continuous function. Notice that a linear map $T \in \mathscr{L}(\mathbb{R}^n, \mathbb{R}^n)$ is an isometry (Definition 6.10) if and only if $\|T(x)\| = \|x\|$, for all $x \in \mathbb{R}^n$. Also, since $T(0) = 0$, it follows that T is a *rotation*. Notice that the *translation maps* $x \mapsto x + a : \mathbb{R}^m \to \mathbb{R}^m$ by a nonzero vector $a \in \mathbb{R}^m$ is a non-linear continuous function. In the next two examples, we discuss the continuity of some more (non-linear) functions $\varphi : E \to \mathbb{R}$, with $E \subseteq \mathbb{R}^2$.

Example 7.3 Consider a function $\varphi : \mathbb{R}^2 \to \mathbb{R}$ given by

$$\varphi(x) = \begin{cases} \dfrac{x^6 - 2y^4}{x^2 + y^2}, & \text{for} \ x = (x, y) \neq 0 \\ 0, & \text{for} \ x = 0 \end{cases}$$

Since $\|x\|^2 = x^2 + y^2$, we have $|x|^2 \leq \|x\|^2$, and so $|x| \leq \|x\|$. Similarly, $|y| \leq \|x\|$. For $0 < \|x\| < 1$, we have

$$\begin{aligned} |\varphi(x) - \varphi(0)| &\leq \frac{|x|^6 + 2|y|^4}{\|x\|^2} \\ &\leq \frac{\|x\|^6 + 2\|x\|^4}{\|x\|^2} \\ &= \|x\|^2 (\|x\|^2 + 2) \\ &< 3\|x\|^2 < 3\|x\|. \end{aligned}$$

For any given $\epsilon > 0$, choose $\delta = \min\left\{1, \epsilon/3\right\}$. Then, for all x with $0 < \|x - 0\| < \delta$, we have

$$|\varphi(x) - 0| \leq 3\|x\| < 3\delta \leq \epsilon.$$

Therefore, $\lim\limits_{x \to 0} \varphi(x) = 0 = f(0)$. Hence, the function φ is continuous at $x = 0$. Clearly, φ is continuous for all $x \neq 0$.

Example 7.4 Consider the function $\varphi : \mathbb{R}^2 \to \mathbb{R}$ given by $\varphi(x) = 3x + 2y^2$, where $x = (x, y) \in \mathbb{R}^2$. As in the previous example, since $\|x\|^2 = x^2 + y^2$, we have $|x| \leq \|x\|$ and $|y| \leq \|x\|$. Using these, for $\|x\| < 1$, we have

$$|\varphi(x) - 0| \leq 3|x| + 2|y|^2 \leq 3\|x\| + 2\|x\|^2 \leq 5\|x\|.$$

For given $\epsilon > 0$, choose $\delta = \min\{1, \epsilon/5\}$. Then, for all x with $|x - 0| < \delta$, we have

$$|\varphi(x) - 0| \leq 5|x| < 5\delta \leq \epsilon.$$

So, $\lim\limits_{x \to 0} \varphi(x) = 0 = f(0)$. Hence, the function φ is continuous at $x = 0$. Clearly, φ is continuous for all $x \neq 0$.

Examples 7.1 and 7.2 provide examples of functions $\varphi : \mathbb{R}^2 \to \mathbb{R}$ which are not continuous at $x = 0 \in \mathbb{R}^2$.

Definition 7.5 Let $E \subset \mathbb{R}^m$ be an open set. A function $\varphi \in C(E)$ is called an m-dimensional *Scalar field* defined over the set E.

In practical situations, we use *scalar fields* to represent the physical quantities such as the *temperature, pressure, mass density, electrostatic charge*, etc. Recall that the *graph* of a scalar fields $\varphi : E \to \mathbb{R}$ is the set

$$\Gamma_\varphi := \left\{(x, \varphi(x)) \in \mathbb{R}^{m+1} : x \in E\right\}.$$

In particular, if φ is a linear map, then Γ_φ is a linear subspace of dimension m, with a basis

$$\left(e_1, \varphi(e_1)\right), \ \ldots, \left(e_m, \varphi(e_m)\right),$$

where as usual $\{e_1, \ldots, e_m\}$ is the standard basis of $\mathbb{R}^m$. In simpler situations, scalar fields are studied over some nice *regions* $E \subseteq \mathbb{R}^2$. In such cases, if $\varphi : E \to \mathbb{R}^2$ is a scalar field, then its *graph* as the 2-*dimensional* surface given by

$$S_\varphi(x, y, z) : z - \varphi(x, y) = 0, \quad (x, y) \in E,$$

is usually visualised by considering its *sections* by planes. In actual applications, the most interesting sections are those that include intersections with the coordinate planes $x = 0$ and $y = 0$. In general, a curve on a surface $S \subseteq \mathbb{R}^3$ (or simply a *surface curve*) along which a 2-dimensional scalar field is constant is called a *contour*.

Definition 7.6 Let $E \subseteq \mathbb{R}^m$ be *region*, and consider a field $\varphi \in C(E)$. For $c \in \mathbb{R}$, the set

$$\mathscr{L}\varphi_c := \{u \in E : \varphi(u) = c\}$$

is called the *level set* of the field φ, at the level c. For $m = 2$, level set is called the *level curve* ; and, for $m = 3$, the *level surface*.

Example 7.5 The level curves for the field $\varphi(x, y) = 4x^2 + 9y^2$, $(x, y) \in \mathbb{R}^2$, are *ellipses*. For the field $\varphi(x, y) = xy$, $(x, y) \in \mathbb{R}^2$, level curves are *hyperbolas*. And, for the field $\varphi(x, y) = x^2 + y^2 - 1$, $(x, y) \in \mathbb{R}^2$, level curves are *circles*.

Definition 7.7 Let $E \subset \mathbb{R}^n$ be an open set. A function $f \in C_n(E)$ is called an n-dimensional *vector field* defined over the set E.

The three most important 3-dimensional *vector fields* representing the *gravitational, electric* and *magnetic* fields are extremely useful in physics and engineering. We also come across some other types of 2-dimensional and 3-dimensional vector fields concerning problems related to a study of *fluids*, which usually represent vectors such as *velocity* of the flow, etc. For example, the 2-dimensional vector field $f(x, y) = (-y, x)$ represents a counter-clockwise flow about the origin.

As said before, sometimes, we assume the set $E \subseteq \mathbb{R}^m$ to be an open connected set. In particular, for $m = 1$, we take $E = (a, b)$, for some $a < b$. Notice that, by Theorem 7.10, a *vector function* $\gamma : (a, b) \to \mathbb{R}^n$ is continuous if and only if each of the component function $\gamma_k : (a, b) \to \mathbb{R}$ is continuous, for $k = 1, \ldots, n$.

7.3 Differentiability

Let $E \subseteq \mathbb{R}^m$ be an open set, and $f : E \to \mathbb{R}^n$ be a multivariable vector-valued function, which is assumed to be continuous on $\overline{E} = E \cup E'$. Recall that f is continuous at a point $c \in \overline{E}$ if $f(x) \approx f(c)$ for points x very close to c. As said before, the *local geometry* at point $x = c \in E$ of the *level surfaces* of f is studied mostly in terms of *tangent planes* passing through the point $x = c$.

In practical terms, f represents the *flow field* of some form of energy as a function of position and time, where we usually take the last coordinate of the point x to represent the time variable $t \geq 0$. Therefore, *linearisation principle* plays a significant role in solving some linear partial differential equations,[4] which arise as *models* to solve problems concerning certain important phenomena in science and engineering.

Our main focus here is to discuss that f is *differentiable* at $c \in E$ if, as $\|h\| \to 0$, the vectors $(f(c + h) - f(c))$ approximate the *tangent plane* to the *level surface* of f at the point c given by a linear map $L_c : \mathbb{R}^n \to \mathbb{R}^n$. Said differently, we prove if f

---

[4] A solution of a linear partial differential equation is found by studying the variations of a function f along trajectories on the $(n + 1)$-dimensional (regular) surface defined by f.

is differentiable at $x = c$ then the *vectors* $f(x)$ for points x *very close* to c lie in a tangent plane defined by the partial derivatives of the *component functions* of f at the point $x = c$.

7.3.1 Partial Derivatives

For an open set $E \subseteq \mathbb{R}^2$, the *partial derivatives* of a function $f : E \to \mathbb{R}$ at a point $a = (u, v) \in E$ are given by

$$\partial_x f(a) = \frac{\partial f}{\partial x}(u, v) = \lim_{t \to 0} \frac{f(u + t, v) - f(u, v)}{t} \qquad (7.3.1a)$$

$$\partial_y f(a) = \frac{\partial f}{\partial y}(u, v) = \lim_{t \to 0} \frac{f(u, v + t) - f(u, v)}{t} \qquad (7.3.1b)$$

Recall that, in geometrical terms, the partial derivative $\partial_x f(a)$ is the *ordinary derivative* along the *curve of intersection* of the plane $y = v$ with the *surface* $z = f(x, y)$. We also say that $\partial_x f(a)$ is the derivative of the function f at the point $a = (u, v)$ in the direction of the x-axis or *along the unit vector* $\hat{i} = (1, 0)$. Similarly, the partial derivative $\partial_y f(a)$ is the derivative of the function f at the point $a = (u, v)$ in the direction of the y-axis or *along the unit vector* $\hat{j} = (0, 1)$. Therefore, the two equations in (7.3.1) can also be written as

$$\partial_x f(a) = \lim_{t \to 0} \frac{f(a + t\,u) - f(a)}{t}, \quad \text{with } u = \hat{i}; \qquad (7.3.2a)$$

$$\partial_y f(a) = \lim_{t \to 0} \frac{f(a + t\,u) - f(a)}{t}, \quad \text{with } u = \hat{j}. \qquad (7.3.2b)$$

In general, let $E \subset \mathbb{R}^m$ be an open set, and $f : E \to \mathbb{R}$ be a function. Consider the standard basis of $\mathbb{R}^m$ given by

$$e_i = (0, \dots, 1, \dots, 0), \quad \text{for } i = 1, 2, \dots, m,$$

where e_i has the value 1 at the ith place, and 0 elsewhere.

Definition 7.8 For $a = (a_1, \dots, a_m) \in E$, and $i = 1, \dots, m$, the **ith partial derivative** $\partial_{x_i} f(a) := D_i f(a)$ of f at a along the unit vector e_i (or in x_i-direction) is given by

$$\partial_{x_i} f(a) = \frac{\partial f}{\partial x_i}(a)$$

$$= \lim_{t \to 0} \frac{f(a + t\,u) - f(a)}{t}, \quad \text{with } u = e_i, \qquad (7.3.3)$$

provided the limit exists.

Said differently, the ith partial derivative $\partial_{x_i} f(a)$ is the derivative at $t = 0$ of the function $\hat{f}_i : (-\epsilon, \epsilon) \to \mathbb{R}$ given by

$$\hat{f}_i(t) = f(a + te_i), \quad \text{for} \ t \in (-\epsilon, \epsilon), \tag{7.3.4}$$

for some $\epsilon > 0$. In particular, it follows that all rules of one-variable differentiation apply. We may write

$$E_i = E\big|_{x_i = a_i} = \{(x_1, \ldots, x_n) \in E : x_i = a_i\},$$

so that $\partial_{x_i} f(a)$ is tangent to the m-dimensional surface given by the function $f\big|_{E_i}$. The higher order *mixed* partial derivatives of a function $f : E \to \mathbb{R}$ at a point $a \in E$ are defined in the same way. For $1 \leq i, j \leq m$, we write the *second order* partial derivatives as

$$D_i[D_j f] := \partial_{x_i}(\partial_{x_j} f) = \frac{\partial^2 f}{\partial x_i \partial x_j}; \tag{7.3.5a}$$

$$D_j[D_i f] := \partial_{x_j}(\partial_{x_i} f) = \frac{\partial^2 f}{\partial x_j \partial x_i}, \tag{7.3.5b}$$

and so on. It is also common to use notations such as $f_{ij} = D_{ij} f = D_i[D_j f]$. We left to the reader finding examples of multivariable real-valued functions that illustrate, in general, the second order mixed partial derivatives at some point may not be defined or they are not equal, even when the two exist.

Definition 7.9 For $r \geq 1$, we say $f : E \to \mathbb{R}$ is a **C^r function** (or a **class-r**) if f has partial derivatives of order $\leq r$, and all these partial derivatives are continuous on $\overline{E}$. We write $C^r(E)$ for the space of C^r-*functions* defined on the set E. Also, we say $f : E \to \mathbb{R}$ is a *Smooth function* if $f \in C^\infty(E) := \bigcup_{r \geq 1} C^r(E)$.

A function $f \in C^1(E)$ is also known as a *continuously differentiable* function on E. As before, we write $C(E)$ for the linear space of real-valued continuous functions defined on an open set $E \subset \mathbb{R}^m$.

The next theorem due to *Alexis Clairaut* (1713–1765) and *Hermann Amandus Schwartz* (1843–1921) proves that the equality of mixed partial derivatives of second order holds for functions $f \in C^2(E)$.

Theorem 7.12 (Schwartz) *Let $E \subset \mathbb{R}^m$ be an open set, $a \in E$, and $f \in C^2(E)$. Then, we have*

$$\frac{\partial^2 f}{\partial x_j \partial x_i}\bigg|_a = \frac{\partial^2 f}{\partial x_i \partial x_j}\bigg|_a, \quad \text{for all} \ 1 \leq i, j \leq m.$$

Proof For brevity, we write

$$D_{ij}(f)(a) = \frac{\partial^2 f}{\partial x_j \partial x_i} \quad \text{and} \quad D_{ji} f(a) = \frac{\partial^2 f}{\partial x_i \partial x_j},$$

With standard basis vectors e_i as above, by definition, we have

$$
\begin{aligned}
D_{ji}(f)(a) &= \lim_{s \to 0} \frac{1}{s}[D_i f(a + se_j) - D_i f(a)] \\
&= \lim_{s \to 0} \frac{1}{s}\left(\left[\lim_{t \to 0} \frac{1}{t}((f(a + se_j + te_i) - f(a + se_j)) \right] \right. \\
&\qquad\qquad \left. - \left[\lim_{t \to 0} \frac{1}{t}(f(a + te_i) - f(a)) \right] \right) \\
&= \lim_{s,t \to 0} \frac{1}{st}[g(s, t) - g(0, t)],
\end{aligned}
\tag{7.3.6}
$$

where the function $g(s, t)$ is given by

$$g(s, t) = f(a + se_j + te_i) - f(a + se_j). \tag{7.3.7}$$

As $a \in E$ is an interior point, for some small s and fixed t, the function $g(x, t)$ is well-defined over the interval $[0, s]$. Applying the *mean value theorem* (Corollary 4.2) to the function $x \mapsto g(x, t)$, we find some $s_0 \in (0, s)$ such that

$$g(x, t) - g(0, t) = (x - 0)\partial_x g(s_0, t).$$

However, by (7.3.7) and definition of the partial derivative, we have

$$
\begin{aligned}
\partial_x g(s_0, t) &= D_j f(a + s_0 e_j + te_i) - D_j f(a + s_0 e_j) \\
&= h(t, s_0) - h(0, s_0),
\end{aligned}
$$

with $h(t, s_0) = D_j f(a + te_i + s_0 e_j)$, which is well-defined over the interval $[0, t]$, for some small t. Again, by using Corollary 4.2), we find $t_0 \in (0, t)$ such that

$$
\begin{aligned}
h(t, s_0) - h(0, s_0) &= (t - 0)\partial_t h(t_0, s_0) \\
&= t D_i D_j f(a + s_0 e_j + t_0 e_i).
\end{aligned}
$$

It thus follows from (7.3.6) that

$$
\begin{aligned}
D_{ji}(f)(a) &= \lim_{s,t \to 0} \frac{1}{st}[g(s, t) - g(0, t)] \\
&= \lim_{s,t \to 0} \frac{1}{st}\left(st D_{ij} f(a + s_0 e_j + t_0 e_i)\right) \\
&= D_{ij} f(a).
\end{aligned}
$$

Notice that the last equality follows from the fact that $(s, t) \to (0, 0)$ implies $(s_0, t_0) \to (0, 0)$, and also the assumption that $f \in C^2(E)$.

The assertion in above theorem extends to the *boundary* ∂E, by continuity. Notice that, as before, we define the partial derivatives of a vector-valued function $f : E \to \mathbb{R}^n$ in terms of its *n-component functions*. That is, if $f = (f_1, \ldots, f_n)$, then the partial derivatives of $f : E \to \mathbb{R}^n$ exists at a point $x = a \in E$ if so is the case for every component function $f_i : E \to \mathbb{R}$, $1 \le i \le n$.

7.3.2 Directional Derivatives

The concept of the *directional derivatives* of a function $f : E \to \mathbb{R}$ at a point $a \in E$ is a natural generalisation of the partial derivatives of f at a in the sense that we now take the (partial) derivatives in *any specified direction*, say given by a *nonzero* vector $v \in \mathbb{R}^m$. That is, we extend Definition 7.8 replacing the standard basis vector e_i in (7.3.3) by the *unit vector* $\hat{v} = v / \|v\|$.

Definition 7.10 Let $E \subset \mathbb{R}^m$ be an open set. The **directional derivative** of a function $f : E \to \mathbb{R}$ at a point $a \in E$ in the direction of the unit vector $u \in \mathbb{R}^m$, denoted by $D_u f(a)$, is given by

$$D_u f(a) := \lim_{t \to 0} \frac{f(a + tu) - f(a)}{t}, \tag{7.3.8}$$

provided the limit exists. We also write $D_u f(a)$ as $f'(a; u)$.

Notice that the trace of the continuous vector function $\gamma : \mathbb{R} \to E$ given by $\gamma(t) = a + tu$ is a (parametric) *line passing through a in the direction of the unit vector u*. Let $\epsilon > 0$ be such that the ball $B(a; \epsilon) \subset E$. Clearly then,

$$\gamma(t) = a + tu \in B(a; \epsilon), \quad \text{for all } t \text{ with } |t| < \epsilon.$$

As in (7.3.4), we consider the function $\hat{f}_u : (-\epsilon, \epsilon) \to \mathbb{R}$ given by

$$\hat{f}_u(t) = f(a + tu), \quad \text{for } t \in (-\epsilon, \epsilon), \tag{7.3.9}$$

Then, the *directional derivatives* $D_u f(a)$ is the usual derivative of the function $\hat{f}_u(t)$ at $t = 0$, if the latter exists. In terms of physical applications, $\gamma(t)$ represents the motion of a particle in $E \subseteq \mathbb{R}^m$ moving with the constant velocity u, which is at the point $x = a$ at time $t = 0$. The fundamental relation (7.3.17) derived later explains further the significance of the geometric interpretation given above of the directional derivatives $D_u f(a)$.

Example 7.6 Let $\varphi : \mathbb{R}^2 \to \mathbb{R}$ be the scalar field given by $f(x, y) = xy$. Let $a = (a, b)$ and $u = (u_1, u_2)$ be a unit vector. Then

$$
\begin{aligned}
D_u f(a) &= \lim_{t \to 0} \frac{f(a + tu) - f(a)}{t} \\
&= \lim_{t \to 0} \frac{(a + tu_1)(b + tu_2) - ab}{t} \\
&= au_2 + bu_1.
\end{aligned}
$$

In particular, for $u = (1, 0)$, we have $\partial_x f(a) = f'(a; u) = b$. Similarly, , for $u = (0, 1)$, we have $\partial_y f(a) = f'(a; u) = a$.

Example 7.7 Consider the function $f : \mathbb{R}^2 \to \mathbb{R}$ given by

$$
f(x) = \begin{cases} \dfrac{xy}{x^2 + y^2}, & \text{for } x \neq 0 \\ 0, & \text{for } x = 0 \end{cases}, \quad x = (x, y) \in \mathbb{R}^2.
$$

We first compute the directional derivatives. Let $u = (u, v)$ be a unit vector. Then, we have

$$
\frac{f(0 + tu) - f(0)}{t} = \frac{f(tu)}{t} = \frac{uv}{t(u^2 + v^2)} = \frac{1}{t}(uv),
$$

which implies that $D_u f(0) = 0$, if $uv = 0$. Notice that $D_u f(0)$ does not exist in any other case. Recall that f is not continuous at $x = 0$ (Example 7.1).

7.3.3   The Differential

Let $E \subseteq \mathbb{R}^m$ be an open set, and consider a function $f : E \to \mathbb{R}^n$. We introduce here the fundamental idea of the *derivative* of f at a point $a \in E$, denote by $Df(a)$, in a way so that we have most of the usual properties of the *differentiation*. For example, in every possible scenario, it is desirable to have the fundamental property: *differentiability $\Rightarrow$ continuity*.

The concept of *directional derivatives* introduced above may appear to be a step forward to formulate the definition of the *derivative* $Df(a)$, however, there are functions having the directional derivatives $D_u f(a)$ in every direction, but fail to be continuous at the point a, and hence cannot be *differentiable* at a, in whatever sense one may propose to define the *derivative* $Df(a)$.

Example 7.8 Consider the function $f : \mathbb{R}^2 \to \mathbb{R}$ given by

$$
f(x) = \begin{cases} \dfrac{x^2 y}{x^4 + y^2}, & \text{for } x = (x, y) \neq 0 \\ 0, & \text{for } x = 0 \end{cases}
$$

It follows easily that the directional derivatives of the function f at $\mathbf{0} = (0, 0)$ exists in all direction. Indeed, for $\mathbf{u} = (u, v)$ with $\|\mathbf{u}\| = 1$, since

$$\frac{f(\mathbf{0} + t\mathbf{u}) - f(\mathbf{0})}{t} = \frac{f(t\mathbf{u})}{t} = \frac{t^3 u^2 v}{t^3 (t^2 u^4 + uv^2)},$$

we have

$$f'(\mathbf{0}; \mathbf{u}) = \lim_{t \to 0} \frac{f(\mathbf{0} + t\mathbf{u}) - f(\mathbf{0})}{t} = \begin{cases} u^2/v & v \neq 0; \\ 0 & v = 0. \end{cases}$$

However, as have seen before, this function is not continuous at $\mathbf{0}$ (Example 7.2), and hence cannot have the *derivative* $Df(\mathbf{0})$.

Therefore, mere existence of the directional derivatives $D_{\mathbf{u}} f(\mathbf{a})$ in every direction $\mathbf{u}$ does not *qualify* f to have the *derivative* $Df(\mathbf{a})$. In addition to having the continuity at the point of differentiability, it is also desirable to have the existence of the directional derivatives in every direction at the point where the *derivative* $Df(\mathbf{a})$ exists.

It was Fréchet who discovered that the following reformulation of the derivative of a function $f : [a, b] \to \mathbb{R}$ at a point $x = c$ fits the scheme:

$$f'(c) = \lim_{t \to 0} \frac{f(c + t) - f(c)}{t}$$

$$\Leftrightarrow \quad 0 = \lim_{t \to 0} \frac{f(c + t) - f(c) - f'(c)t}{t},$$

More precisely, the essence of the idea is the fact that the map $\lambda : \mathbb{R} \to \mathbb{R}$ given by $\lambda(t) = f'(c) t$ is a *linear map*. So, taking

$$Df(c) = f'(c) = \lambda(c),$$

it follows that the existence of $f'(c)$ is equivalent to existence of a linear map $\lambda(c) : \mathbb{R} \to \mathbb{R}$ such that

$$\lim_{t \to 0} \frac{f(c + t) - f(c) - \lambda(c) t}{t} = 0.$$

This discussion suggests, to formulate the definition the *derivative* $Df(\mathbf{a})$, we need some preparation with linear maps between real linear spaces (Definition 7.4).

Lemma 7.1 *(Riesz) Let V be the Euclidean space $\mathbb{R}^m$, and $T \in V^*$. Then, for some $v \in V$, we have*

$$T(x) = \langle v, x \rangle, \quad x \in V.$$

Proof Let $\{e_1, \ldots, e_m\}$ be the standard basis of V, and take

$$v = \big(T(e_1), \ldots, T(e_m)\big) \in V.$$

Now, since any $x \in V$ can be written uniquely as

$$x = \langle x, e_1 \rangle e_1 + \cdots + \langle x, e_m \rangle e_m, \tag{7.3.10}$$

it follows that

$$
\begin{aligned}
\langle v, x \rangle &= \langle x, e_1 \rangle T(e_1) + \cdots + \langle x, e_m \rangle T(e_m) \\
&= T\big(\langle x, e_1 \rangle e_1 + \cdots + T(\langle x, e_m \rangle e_m\big) \\
&= T(x), \quad \text{by (7.3.10).} \qquad \qquad \square
\end{aligned}
$$

Notice that, as introduced earlier, the function $x^i : V \to \mathbb{R}$ given by $x^i(x) = \langle x, e_i \rangle$ is the ith *coordinate function*, for $i = 1, \ldots, m$. More generally, if $T = \big(\tau_1, \ldots, \tau_n\big) \in \mathscr{L}(\mathbb{R}^m, \mathbb{R}^n)$, so that the *component functions* $\tau_i : \mathbb{R}^m \to \mathbb{R}$ are linear maps, for all $1 \le i \le n$, then by using Lemma 7.1 we can find vector $v_i \in \mathbb{R}^m$ such that, for all $x \in \mathbb{R}^m$,

$$\tau_i(x) = \langle v_i, x \rangle, \quad \text{for } i = 1, \ldots, n.$$

Let $A = A_T$ be the $n \times m$ matrix with the vector v_i as the ith row, for $i = 1, \ldots, m$. It thus follows that

$$T(x) = Ax, \quad \text{for } x \in \mathbb{R}^m,$$

where both x and $T(x)$ are viewed as the column vectors. Conversely, given any $n \times m$ matrix A over the field $\mathbb{R}$, the function $T = T_A : \mathbb{R}^m \to \mathbb{R}^n$ given by

$$T(x) = Ax, \quad \text{for } x \in \mathbb{R}^m,$$

is a linear map. Notice that, once again, x and $T(x)$ are viewed as the column vectors. It is a theorem in linear algebra that two $n \times m$ matrices over $\mathbb{R}$ define the same linear map : $\mathbb{R}^m \to \mathbb{R}^n$ if and only if the two matrices are the same, up to certain number of *permutations of rows and columns*.

We are now ready to introduce the main concept of the chapter. We remark that, in general, the context should guide when points in some finite dimensional are viewed as the column vectors.

Definition 7.11 (*Fréchet*) Let $E \subseteq \mathbb{R}^m$ be an open set. A function $f : E \to \mathbb{R}^n$ is *Differentiable* at $x = a \in E$ if, for some $L_a = L_a f \in \mathscr{L}(\mathbb{R}^m, \mathbb{R}^n)$, we have

$$\lim_{h \to 0} \frac{f(a + h) - f(a) - L_a(h)}{\|h\|} = 0. \tag{7.3.11}$$

The linear map $L_a f := Df(a)$ is called the **differential** of the function f at the point a. We also say that $Df(a)$ is the (Fréchet) (or *total*) *derivative* of f at a. We say a function $f : E \to \mathbb{R}^n$ is *differentiable function* over E if the derivative $Df(a)$ exists at each point $a \in E$.

In practical situations, we usually assume that f is defined and continuous on the boundary ∂E of the domain E. Sometimes, we also take E to be a connected set. Notice that, in (7.3.11), $\|h\|$ is used as the denominator because it makes no sense to divide by the vector h. Also, since $E - x$ is an open set[5] that contains $\mathbf{0}$, we can have the set $E - x \subset \mathbb{R}^m$ as the domain for taking the limit $h \to \mathbf{0}$.

Further, as E is an open set, we can find some $\delta > 0$ such that $B(a; \delta) \subseteq E$. Therefore, in topological terms, the condition (7.3.11) can be reformulated as follows: For $0 < \|h\| < \delta$,

$$f(a + h) - f(a) - L_a(h) = \epsilon(a, h)\|h\|, \qquad (7.3.12)$$

where the *error function* $\epsilon(a, h)$ satisfies the condition

$$\|\epsilon(a, h)\| \to 0 \quad \text{as} \quad \|h\| \to 0.$$

In Landau's notation introduced earlier, we write

$$f(a + h) - f(a) - L_a(h) = o(\|h\|), \quad \text{as} \quad \|h\| \to 0. \qquad (7.3.13)$$

By what we said above, it follows that $o(\|h\|)$ is defined and continuous at $\mathbf{0}$, provided we take $o(\mathbf{0}) = \mathbf{0}$. This conforms to the following geometric interpretation of the *differential* $L_a f$: *In some small neighborhood of the point* $a \in E$, $L_a f$ *gives the best linear approximation of* $(f(a + h) - f(a))$. Definition 7.11 has the following matrix form formulation.

Definition 7.12 (*Matrix form*) Let $E \subset \mathbb{R}^m$ be an open set. We say $f : E \to \mathbb{R}^n$ is a *differentiable function* at $x = a$ if, for some $n \times m$ matrix $B_a = B_a(f)$, we have

$$\lim_{h \to 0} \frac{f(a + h) - f(a) - B_a h}{\|h\|} = \mathbf{0}. \qquad (7.3.14)$$

We call B_a as the *derivative* of f at a, and write $Df(a) = B_a$.

As we shall see shortly, B_a is the matrix determined by the linear map L_a with respect to the standard basis. In an "*everywhere differentiable*" situation, we may write $B_a(f)$ (or $L_a f$) simply as $B(f)$ (or $L(f)$, respectively).

Example 7.9 Recall that every linear map $T \in \mathscr{L}(\mathbb{R}^m, \mathbb{R}^n)$ is continuous. In fact, every map $T \in \mathscr{L}(\mathbb{R}^m, \mathbb{R}^n)$ is differentiable over the space $\mathbb{R}^m$, with the differential T. Notice that, in this case, the condition (7.3.11) is satisfied trivially for any $a \in \mathbb{R}^m$. In particular, it follows that the identity $I \in \mathscr{L}(\mathbb{R}^p, \mathbb{R}^p)$ given by $I(x) = x$ is differentiable, for all $p \geq 1$, with $DI(x) = I$, for all $x \in \mathbb{R}^p$. More generally, if A is an $n \times m$ matrix and $b \in \mathbb{R}^n$ is some fixed column vector, then the *affine map* $g : \mathbb{R}^m \to \mathbb{R}^n$ given by

---

[5] It is the image of the open set E under a *translation* $\mathbb{R}^m \to \mathbb{R}^m$, which is a linear homeomorphism.

$$g(x) = Ax + b, \quad \text{for} \quad x = (x_1 \cdots x_m)^t \in \mathbb{R}^m,$$

is an *everywhere differentiable* map. For, with $x \in \mathbb{R}^m$ as above, we have

$$\lim_{h \to 0} \frac{f(x + h) - f(x) - Ah}{\|h\|}$$
$$= \lim_{h \to 0} \frac{(A(x + h) + b) - (Ax + b) - Ah}{\|h\|}$$
$$= 0,$$

implying that the affine map g is differentiable over $\mathbb{R}^m$, and $Df = A$.

Example 7.10 The function $f : \mathbb{R}^m \to \mathbb{R}$ given by $f(x) = \langle x, x \rangle = \|x\|^2$ is everywhere differentiable, with the derivative $Df(x) = 2x$, for any $x \in \mathbb{R}^m$. For, in this case, we have

$$\lim_{h \to 0} \frac{f(x + h) - f(x) - 2\langle x, h \rangle}{\|h\|}$$
$$= \lim_{h \to 0} \frac{\langle x + h, x + h \rangle - \langle x, x \rangle - 2\langle x, h \rangle}{\|h\|}$$
$$= \lim_{h \to 0} \frac{\langle x, x \rangle + 2\langle x, h \rangle + \langle h, h \rangle - \langle x, x \rangle - 2\langle x, h \rangle}{\|h\|}$$
$$= \lim_{h \to 0} \frac{\langle h, h \rangle}{\|h\|} = \lim_{h \to 0} \frac{\|h\|^2}{\|h\|}$$
$$= \lim_{h \to 0} \|h\| = 0.$$

Example 7.11 Let $x = (x, y) \in \mathbb{R}^2$. Consider the two functions $f, g : \mathbb{R}^2 \to \mathbb{R}$ given by

$$f(x) = \frac{xy}{x^2 + y^2}, \quad \text{for} \quad x \neq 0; \text{ and, } \quad f(0) = 0;$$

$$g(x) = \frac{x^2 y}{x^4 + y^2}, \quad \text{for} \quad x \neq 0; \text{ and, } \quad f(0) = 0.$$

These functions are not differentiable at $x = 0$, because these are not continuous at $x = 0$ (see Examples 7.1 and 7.2).

We now start establishing some desirable properties of the *differential L_a*. We find it convenient to use Definition 7.12. As a necessity, we first prove *uniqueness* of the *differential $Df = B$*, if it exists.

Theorem 7.13 *Suppose a function $f : E \to \mathbb{R}^n$ is differentiable over an open set $E \subset \mathbb{R}^m$. If the $n \times m$ matrices B and C are differentials of f over E, then $B = C$.*

Proof Let $x = a \in E$ be arbitrary. Then, by definition, we have

$$\lim_{h \to 0} \frac{f(a+h) - f(a) - Bh}{\|h\|} = 0 = \lim_{h \to 0} \frac{f(a+h) - f(a) - Ch}{\|h\|}$$

and so it follows that

$$\lim_{h \to 0} \frac{(B-C)h}{\|h\|} = 0.$$

Take an arbitrary unit vector $u \in E$, and fix it. Let $h = tu$ so that $\|h\| = |t|$, and $h \to 0$ is equivalent to $t \to 0$. Considering two cases when $t \to 0$ through positive or negative values, we have

$$\lim_{t \to 0} \frac{(B-C)tu}{|t|} = 0 \quad \Rightarrow \quad (B-C)u = 0.$$

Thus, for every choice of m linearly independent unit vectors

$$u_1, u_2, \ldots, u_m \in E,$$

the square matrix $U = [u_1 \ \ldots \ u_m]$ is nonsingular, and we have

$$\begin{aligned} (B - C)U &= (B - C)[u_1 \ \ldots \ u_m] \\ &= [(B - C)u_1 \ \ldots \ (B - C)u_m] \\ &= (0 \ldots 0) = O. \end{aligned}$$

Here O denotes the *zero matrix*. So,

$$B - C = OU^{-1} = O \quad \Rightarrow \quad B = C. \qquad \square$$

Next, we prove the most sought after property, namely, *every differentiable function is continuous*.

Theorem 7.14 *Let $E \subset \mathbb{R}^m$ be an open set. If a function $f : E \to \mathbb{R}^n$ is differentiable at $x = a \in E$, with $L_a = B_a$, then f is continuous at a.*

Proof By definition, we have

$$\lim_{h \to 0} \frac{f(a+h) - f(a) - B_a h}{\|h\|} = 0,$$

and so

$$\lim_{h \to 0} (f(a+h) - f(a))$$

$$= \lim_{h \to 0} \left(\|h\| \frac{f(a+h) - f(a) - B_a h}{\|h\|} + B_a h \right)$$

$$= 00 + 0 = 0.$$

Therefore, $\lim_{h \to 0} f(a+h) = f(a)$, which implies that the function f is continuous at a.

It is easy to give a proof of the standard properties of the *differential operator* D stated in the next theorem.

Theorem 7.15 (Algebra of Derivatives) *Let both $E \subset \mathbb{R}^m$ be an open set, $a \in E$, and functions $f, g : E \to \mathbb{R}^n$ be differentiable at a. Then, for any $\alpha \in \mathbb{R}$, the functions $\alpha f(E)$, $f + g$, $f \cdot g$, and f/g $(g(a) \neq 0)$, defined pointwise, are differentiable at a, and we have*

1. $D(\alpha f)(a) = \alpha Df(a)$.
2. $D(f + g)(a) = Df(a) + Dg(a)$.
3. $D(f \cdot g)(a) = g(a)Df(a) + f(a)Dg(a)$.
4. $D(f/g)(a) = \dfrac{g(a)Df(a) - f(a)Dg(a)}{\left(g(a)\right)^2}$.

Also, the chain rule holds for multivariable vector-valued functions.

Theorem 7.16 (Chain Rule) *Let both $E \subset \mathbb{R}^m$ and $F \subset \mathbb{R}^n$ be open connected sets. If the functions $f : E \to \mathbb{R}^n$ and $g : F \to \mathbb{R}^p$ are differentiable on E and F, respectively, and $f(E) \subset F$, then $g \circ f : E \to \mathbb{R}^p$ is also differentiable on E, and*

$$D(g \circ f)(a) = Dg(f(a))Df(a).$$

Further, if $f \in C^r(E)$ and $g \in C^2(F)$, then $g \circ f \in C^r(E)$.

Proof Left to the reader as exercise.

Definition 7.13 Let $U \subset \mathbb{R}^m$ be open set, and $r \geq 0$. We say a function $f = (f_1, \ldots, f_n) : U \to \mathbb{R}^n$ is a C^r-*Function* (or of *Class C^r*) if the each component function $f_i : U \to \mathbb{R}$ belongs to $C^r(U)$. And, we say f is *smooth* if it is a C^∞ *function* in the same sense.

Theorem 7.17 *Let $E \subset \mathbb{R}^m$ be an open set. If a function $f : E \to \mathbb{R}^n$ is differentiable at $a \in E$, then the directional derivatives $D_u f(a)$ exists in every direction u, and $D_u f(a) = L_a(u)$.*

Proof As before, we assume $\|u\| = 1$. Suppose $L_a = B_a$ so that we have

$$\lim_{h \to 0} \frac{f(a+h) - f(a) - B_a h}{\|h\|} = 0,$$

and so, by taking $h = tu$ so that $\|h\| = |t|$, we obtain

$$\lim_{t \to 0} \frac{f(a + tu) - f(a) - B_a t u}{|t|} = 0$$

Since the existence of limit implies the existence of left and right limit, we have

$$\lim_{t \to 0+} \frac{f(a + tu) - f(a)}{t} = B_a u;$$

$$\lim_{t \to 0-} \frac{f(a + tu) - f(a)}{t} = B_a u.$$

It thus follows that

$$D_u f(a) = \lim_{t \to 0} \frac{f(a + tu) - f(a)}{t} = B_a u = L_a(u).$$

$\square$

Definition 7.14 Let $E \subset \mathbb{R}^m$ be open set. Suppose a real-valued function $\varphi : E \to \mathbb{R}$ is differentiable at a point $a \in E$. Then, the map $\tau_a : \mathbb{S}^{m-1} \to \mathbb{R}$ given by $\tau_a(u) = D_u\varphi(u)$ is called the *tangent map* to φ at the point a, where $\mathbb{S}^{m-1} = \{v \in \mathbb{R}^m : \|v\| = 1\}$.

Notice that, if the *direction* u in Definition 7.3.8 is given by an arbitrary nonzero vector in $\mathbb{R}^m$, and $0 \neq \alpha \in \mathbb{R}$, then the relation

$$D_{\alpha u}\varphi(a) = \lim_{t \to 0} \frac{\varphi(a + (t\alpha)u) - \varphi(a)}{t} = \alpha \, D_u\varphi(a)$$

implies that φ has the directional derivative in the direction αu. That is, in geometrical terms, the collection $K \subset \mathbb{R}^m$ of all *direction* with respect to which φ has the directional derivatives forms a *cone*. Therefore, in general, the *tangent map* τ_a is defined over the cone K, and its *graph*

$$\Gamma_{\tau_a} = \{(u, \alpha) \in K \times \mathbb{R} : D_u\varphi(a) = \alpha\}$$

is a *plane* if and only if τ_a is a linear map, which may not be the case, in general (Exercise 7.8). The significance of this remark, for example, becomes clear while solving a first order quasilinear partial differential equation by using the *method of characteristics*.

Also, by Theorem 7.17, we have $\tau_a(u) = L_a(u)$, for all $u \in \mathbb{S}^{m-1}$. The next corollary justifies the terminology.

Corollary 7.5 Let $E \subset \mathbb{R}^m$ be open set. If a real-valued function $\varphi : E \to \mathbb{R}$ is differentiable at a point $a \in E$, then

$$L_a\varphi = \big(D_1\varphi(a),\, D_2\varphi(a),\, \cdots,\, D_m\varphi(a)\big),$$

where $D_i\varphi(a) = \partial_{x_i}\varphi(a)$, *for* $i = 1, 2, \ldots, n$.

Proof It follows from Theorem 7.17 by taking $u = e_i$, for $i = 1, \ldots, m$, viewed as a column vector. A more direct proof can be given as follows. Since the derivative $D\varphi(a)$ is a $1 \times m$ matrix, we may write

$$D\varphi(a) = \big(\lambda_1 \; \lambda_2 \; \cdots \; \lambda_m\big).$$

Also, since $D\varphi(a)u = \varphi'(a; u)$, we have

$$D\varphi(a)e_j = \varphi'(a; e_j) = D_j\varphi(a).$$

However, since

$$D\varphi(a)e_j = \big(\lambda_1 \; \lambda_2 \; \cdots \; \lambda_m\big) [0 \cdots 0 \; 1 \; 0 \cdots 0]^t = \lambda_j.$$

It thus follows that $\lambda_j = D_j\varphi(a)$. Hence,

$$D\varphi(a) = \big(D_1\varphi(a),\, D_2\varphi(a),\, \cdots,\, D_m\varphi(a)\big),$$

as asserted. This completes the proof.

Example 7.12 In view of Theorem 7.17, to show that the function $f : \mathbb{R}^2 \to \mathbb{R}$ given by

$$f(x) = |x| + |y|, \qquad x = (x, y) \in \mathbb{R}^2,$$

is not differentiable at $x = 0$, we only have to show that the directional derivatives of f at $x = 0$ do not exist. For, taking $u = (u_1, u_2)$, we have

$$\frac{f(0 + tu) - f(0)}{t} = \frac{f(tu)}{t} = \frac{|t|}{t}(|u_1| + |u_2|).$$

Clearly, as $t \to 0$, the limit does not exist.

Example 7.13 Consider the function $f : \mathbb{R}^2 \to \mathbb{R}$ given by

$$f(x) = \frac{xy}{x^2 + y^2}, \quad \text{for } x = (x, y) \neq 0; \text{ and, } f(0) = 0.$$

Clearly, the partial derivatives of f exist, and both equal 0. On the other hand, if f is differentiable at $x = 0$, then we must have $B_0 = (0 \; 0)$ (Corollary 7.5). However, with this B_0 and $h = (h_1, h_2)$, we have

$$\frac{f(0 + h) - f(0) - B_0 h}{\|h\|} = \frac{f(h)}{\|h\|} = \frac{h_1 h_2}{(h_1^2 + h_2^2)^{3/2}}.$$

The last expression has no limit when $h \to 0$, and so f cannot be differentiable at $x = 0$.

This example illustrates the fact that the converse of Theorem 7.19 is not true, in general. The next theorem says that, if $f \in C^1(E)$, then the converse holds.

Theorem 7.18 *Let $E \subset \mathbb{R}^m$ be an open set, $a \in E$, and suppose the partial derivatives of a function $f : E \to \mathbb{R}$ exist, and are continuous, in a ball $B(a; \epsilon) \subset E$, for some $\epsilon > 0$. Then, f is differentiable at a.*

Proof For brevity, we take $m = 2$. Let $a = (x_0, y_0)$, and $x = (x, y) \in B(a; \epsilon)$. By our assumption, the function $\gamma_1 : [x_0 - \delta_1, x_0 + \delta_1] \to \mathbb{R}^2$ given by $\gamma_1(t) = f(t, y_0)$ is differentiable, where $x \in [x_0 - \delta_1, x_0 + \delta_1]$ for some $\delta_1 > 0$. So, by Lagrange's theorem, we can write

$$
\begin{aligned}
f(x, y_0) &- f(x_0, y_0) \\
&= \partial_x f(\xi, y_0)(x - x_0) \\
&= \partial_x f(x_0, y_0)(x - x_0) + [\partial_x f(\xi, y_0) - \partial_x f(x_0, y_0)](x - x_0), \quad (7.3.15)
\end{aligned}
$$

where $\xi = \xi(x)$ satisfy the property $0 < |\xi - x_0| < |x - x_0|$. Similarly, for any fixed x, the function $\gamma_2 : [x_0 - \delta_1, x_0 + \delta_1] \to \mathbb{R}^2$ given by $\gamma_2(t) = f(x, t)$ is differentiable, where $x \in [y_0 - \delta_2, y_0 + \delta_2]$ for some $\delta_2 > 0$. Once again, by Lagrange's theorem, we can write

$$
\begin{aligned}
f(x, y) &- f(x, y_0) \\
&= \partial_y f(x, \eta)(y - y_0) \\
&= \partial_y f(x_0, y_0)(y - y_0) + [\partial_y f(x, \eta) - \partial_y f(x_0, y_0)](y - y_0), \quad (7.3.16)
\end{aligned}
$$

where $\eta = \eta(y)$ satisfy the property $0 < |\xi - y_0| < |y - y_0|$. Notice that

$$
\max \left\{ \|(\xi, y_0) - (x_0, y_0)\|, \|(x, \eta) - (x_0, y_0)\| \right\} \leq \|x - a\|,
$$

and so, as $x \to a$, we have

$$
(\xi, y_0) = (\xi(x), y_0) \to a \quad \text{and} \quad (x, \eta) = (x, \eta(x, y)) \to a.
$$

Also, by using the continuity of the partial derivatives at a, we have

$$
|\partial_x f(\xi, y_0) - \partial_x f(a)||x - x_0| + |\partial_y f(x, \eta) - \partial_y f(a)||y - y_0| = o(\|x - a\|),
$$

as $x \to a$. It thus follows from (7.3.15) and (7.3.16) that

$$
\begin{aligned}
f(x) - f(a) &= [f(x) - f(x, y_0)] + [f(x, y_0) - f(a)] \\
&= \partial_x(a)(x - x_0) + \partial_y(a)(y - y_0) + o(\|x - a\|),
\end{aligned}
$$

as $x \to a$. This completes the proof.

Let $\varphi : E \to \mathbb{R}$ be any scalar field that is differentiable at a. By Corollary 7.5, the differential $D\varphi(a) : \mathbb{R}^m \to \mathbb{R}$ can be written uniquely as $D\varphi(a)(x) = \langle x, v \rangle$, for any $x \in \mathbb{R}^m$, where the vector

$$v = \operatorname{grad} \varphi(a) := \left(D_1\varphi(a), \cdots, D_m\varphi(a) \right) = \sum_{i=1}^{m} \frac{\partial \varphi}{\partial x_i}(a)e_i, \qquad (7.3.17)$$

is called the *Gradient* of the field φ at the point a.

Definition 7.15 The vector *differential operator* of the form

$$\nabla := \left(\frac{\partial}{\partial x_1}, \ldots, \frac{\partial}{\partial x_m} \right)$$

is called the **gradient** in m variable $(x_1, \ldots, x_m)$. And, a scalar differential operator of the form

$$\nabla^2 := \frac{\partial^2}{\partial x_1^2} + \cdots + \frac{\partial^2}{\partial x_m^2}$$

is called the *laplace operator* (or simply *laplacian*) in m variable $(x_1, \ldots, x_m)$. We also write $\triangle := \nabla^2$ for the Laplacian.

Notice that grad $\varphi = \nabla\varphi$. It follows from the relation (7.3.17) that *the differential of a scalar field $\varphi : E \to \mathbb{R}$ at a point is uniquely determined by its gradient at that point.* The following two properties of the *Gradient* makes it fundamental to diverse applications in science and engineering:

1. For any scalar field $\varphi \in C^1(E)$, the gradient $\nabla\varphi$ points in the direction of steepest increase of φ, and $(-\nabla\varphi(a))$ points in the direction of steepest decrease. More precisely, we have

$$\left| D_u\varphi(a) \right| = \left| u \cdot \nabla\varphi(a) \right| \leq \|u\|\|\nabla\varphi(u)\|,$$

and, if $\nabla\varphi(a) \neq 0$, the equality holds if and only if $u = a\,\nabla\varphi(a)$, for some $a > 0$. Therefore, by Definition 7.14, the *tangent map* $u \mapsto D_a\varphi(u)$ has the *maximum value* $\|\nabla\varphi(a)\|$; and, if $\nabla\varphi(a) \neq 0$, this maximum is attained at $u = \nabla\varphi(a)/\|\nabla\varphi(a)\|$. The reader may supply the missing details (Exercise 7.19).
2. Let $\varphi : E \to \mathbb{R}$ be scalar field differentiable at $a \in E$. The *graph of the tangent map* τ_a of φ at a given by

$$\Gamma_{(\varphi, \tau_a)} := \left\{ \left(x, D_a\varphi(x) \right) : x \in E \right\}, \qquad (7.3.18)$$

is called the *tangent space* to the graph Γ_φ at the point a. As said before, if $\{e_1, \ldots, e_m\}$ is the standard basis of $\mathbb{R}^m$, then the tangent space $\Gamma_{(\varphi, \tau_a)}$ has the dimension n with a basis given by the column vectors of the $(m + 1) \times m$ matrix

$$
\begin{pmatrix}
1 & 0 & \cdots & 0 \\
0 & 1 & \cdots & 0 \\
\vdots & \vdots & \vdots & \vdots \\
0 & 0 & \cdots & 1 \\
\partial_{x_1}\varphi(a) & \partial_{x_2}\varphi(a) & \cdots & \partial_{x_m}\varphi(a)
\end{pmatrix}
\tag{7.3.19}
$$

Notice that the *tangent plane* to the graph Γ_φ at a is obtained by translating $\Gamma_{(\varphi,\tau_a)}$ to the point $(a, \varphi(a))$, which we write as

$$
\{(x, y) : \; y = \varphi(a) + D_a\varphi(x - a)\}.
$$

Notice that, the intersection of a line $y = mx$ with the graph Γ_φ of the function φ given by $\varphi(x, y) = (x^3 - 3xy^2)/x^2 + y^2$, for $(x, y) \neq (0, 0)$, is line, but Γ_φ has no *tangent plane* at $(0, 0)$. We may take $\varphi(0, 0) = 0$.

More generally, let $E \subset \mathbb{R}^m$ be an open set, $a \in E$, and consider a function $f : E \to \mathbb{R}^n$, written in column vector form as

$$
f = \begin{pmatrix} f_1 & f_2 & \cdots & f_n \end{pmatrix}^t.
$$

Theorem 7.19 *If f is differentiable at $a \in E$, then each partial derivative $D_i f(a)$ exists, for $1 \leq i \leq n$. In this case, the derivative of f at a in matrix form is given by*

$$
Df(a) = \begin{pmatrix} Df_1(a) \\ Df_2(a) \\ \vdots \\ Df_n(a) \end{pmatrix} = \begin{pmatrix} D_1 f_1(a) & D_2 f_1(a) & \cdots & D_m f_1(a) \\ D_1 f_2(a) & D_2 f_2(a) & \cdots & D_m f_2(a) \\ \vdots & \vdots & \vdots & \vdots \\ D_1 f_n(a) & D_2 f_n(a) & \cdots & D_m f_n(a) \end{pmatrix}.
$$

Proof Let $B = \begin{pmatrix} B_1 & B_2 & \cdots & B_n \end{pmatrix}^t$ be any $n \times m$ matrix, with row vectors $B_1, \ldots, B_n$. Then, we have

$$
\frac{f(a + h) - f(a) - Bh}{\|h\|}
$$

$$
= \frac{1}{\|h\|}\left(\begin{pmatrix} f_1(a + h) \\ f_2(a + h) \\ \vdots \\ f_n(a + h) \end{pmatrix} - \begin{pmatrix} f_1(a) \\ f_2(a) \\ \vdots \\ f_n(a) \end{pmatrix} - \begin{pmatrix} B_1 \\ B_2 \\ \vdots \\ B_n \end{pmatrix} h \right)
$$

$$
= \begin{pmatrix} \frac{f_1(a+h) - f_1(a) - B_1 h}{\|h\|} \\ \frac{f_2(a+h) - f_2(a) - B_2 h}{\|h\|} \\ \vdots \\ \frac{f_n(a+h) - f_n(a) - B_n h}{\|h\|} \end{pmatrix}.
$$

It follows from this equation, and (7.3.14), that f is differentiable at a, with $Df(a) = B$, if and only if each component function $f_i : E \to \mathbb{R}$ is differentiable at a and $Df_i(a) = B_i$, for $i = 1, \ldots, n$. So, if f is differentiable, taking limit as $h \to 0$, we obtain

$$Df(a) = \left(Df_1(a)\ Df_2(a)\ \cdots\ Df_n(a) \right)^t,$$

Therefore, by using Corollary 7.5, we have

$$Df(a) = \begin{pmatrix} D_1 f_1(a) & D_2 f_1(a) & \cdots & D_m f_1(a) \\ D_1 f_2(a) & D_2 f_2(a) & \cdots & D_m f_2(a) \\ \vdots & \vdots & \vdots & \vdots \\ D_1 f_n(a) & D_2 f_n(a) & \cdots & D_m f_n(a) \end{pmatrix},$$

as asserted. This completes the proof.

Clearly, the concept of *tangent space* at a point $a \in E$ introduced earlier for a function $\varphi : E \to \mathbb{R}$ (see (7.3.18)), has a natural extension for a function $f : E \to \mathbb{R}^n$ differentiable at a. and its translation given by

$$\left\{ (x, y) \in \mathbb{R}^m \times \mathbb{R}^n :\ y = f(a) + Df(a)(x - a) \right\}$$

is the *tangent plane* at a to the graph Γ_f of the function f. Notice that the *tangent space* at a point $a \in E$ to the graph Γ_f is the image of the injective linear map $x \mapsto (x, Df(x))$, and so has the dimension m. In fact, with respect to standard bases of $\mathbb{R}^m$ and $\mathbb{R}^n$, the *tangent space* at a point $a \in E$ to the graph Γ_f is given by

$$\left\{ (x, y) \in \mathbb{R}^m \times \mathbb{R}^n :\ y = Df(a)(x) \right\},$$

with a basis given by the column vectors of the $(m + n) \times m$ block matrix

$$\mathbf{A} := \left(\mathbf{I}_m\ J_f(a) \right)^t,$$

where $\mathbf{I}_m$ is the $m \times m$ identity matrix, and $J_f(a)$ is the *Jacobian* of f at a, as defined next. Further, we write

$$\mathbf{B} := \left(J_f(a)\ -\mathbf{I}_m \right).$$

Now, assuming that the product linear space $\mathbb{R}^m \times \mathbb{R}^n$ has the *natural* inner product, the n column vectors of the $(m + n) \times m$ *adjoint matrix* $\mathbf{B}^*$ span the *normal space* at $a \in E$ to the graph Γ_f of the function f.

Definition 7.16 Let $E \subset \mathbb{R}^m$ be open, and $a \in E$. Suppose the components functions of a function $f : E \to \mathbb{R}^n$ are given by $f_i : E \to \mathbb{R}$, for $i = 1, \ldots, n$. If f is differentiable at a, with the differential $L_a f : \mathbb{R}^m \to \mathbb{R}^n$, then the matrix of the linear map $L_a f$ given by

$$J_f(a) := \left(D_j f_i(a)\right) = \left(\frac{\partial f_i}{\partial x_j}\Big|_{x=a}\right)_{n\times m}$$

is called the **Jacobian** of f at the point a.

Definition 7.17 Let $E \subset \mathbb{R}^m$ be an open set, and $\varphi : E \to \mathbb{R}$ be any scalar field that is differentiable at a. We say $a \in E$ is a *critical point* (or *stationary point*) of a scalar field φ if $\nabla\varphi(a)$ is undefined or equal to zero. In this case, $\varphi(a)$ is called a *critical value*. We say $a \in E$ is a *regular point* if $\nabla\varphi(a) \neq 0$.

If $D\varphi(a)$ is defined for a the scalar field $\varphi : E \to \mathbb{R}$ at a *critical point* $a \in E$, then a is a point of a *local maxima* or a *local minima* or a *saddle point*. When $\varphi \in C^2(E)$, it is possible to analyse a critical point a of φ by considering the eigenvalues of the *symmetric matrix*[6] of order m given by

$$H_\varphi(a) := \begin{pmatrix} \frac{\partial^2\varphi}{\partial x_1^2}(a) & \frac{\partial^2\varphi}{\partial x_1\partial x_2}(a) & \cdots & \frac{\partial^2\varphi}{\partial x_1\partial x_m}(a) \\ \frac{\partial^2\varphi}{\partial x_2\partial x_1}(a) & \frac{\partial^2\varphi}{\partial x_2^2}(a) & \cdots & \frac{\partial^2\varphi}{\partial x_2\partial x_m}(a) \\ \vdots & \vdots & \vdots & \vdots \\ \frac{\partial^2\varphi}{\partial x_n\partial x_1}(a) & \frac{\partial^2\varphi}{\partial x_n\partial x_2}(a) & \cdots & \frac{\partial^2\varphi}{\partial x_m^2}(a) \end{pmatrix} \qquad (7.3.20)$$

which is called the *hessian matrix* of φ at the point $x = a$ ([2], Chap. 5). Notice that, in terms of the *jacobian* $J_f(a)$, $a \in E$ is a *critical point* of f if the differential $L_a f$ is not defined or the *rank* of the matrix $J_f(a)$ is not *maximal*. In particular, for $m = n$, $a \in E$ is a critical point of a vector field $f \in C^1(E)$ if the determinant $\det\left(J_f(a)\right) = 0$. Or, equivalently, the square matrix $J_f(a)$ of order n is *singular*. In this case, we say $f(a)$ is called a *critical value* of the vector field f. Notice that, if $\varphi \in C^2$ is a scalar field, then

$$H_\varphi(x) = J_{\nabla\varphi}(x), \quad \text{for all } x \in E.$$

Definition 7.18 Let $E \subseteq \mathbb{R}^n$ be an open set, and let $f : E \to \mathbb{R}^n$ be a vector filed with component functions $f_i \in C^1(E)$ for $i = 1, \ldots, n$. The scalar field div f given by

$$\operatorname{div} f := \nabla \cdot f = \frac{\partial f_1}{\partial x_1} + \cdots + \frac{\partial f_n}{\partial x_n}, \qquad (7.3.21)$$

is called the *divergence* of the field f. For $n = 3$, if the component functions $f_i \in C^2(E)$ ($1 \le i \le 3$), then the vector field curl f given by

$$\operatorname{curl} f := \nabla \times f = \left(\frac{\partial f_3}{\partial x_2} - \frac{\partial f_2}{\partial x_3}, \frac{\partial f_1}{\partial x_3} - \frac{\partial f_3}{\partial x_1}, \frac{\partial f_2}{\partial x_1} - \frac{\partial f_1}{\partial x_2}\right), \qquad (7.3.22)$$

is called the *curl* of the field f.

---

[6] Notice that *symmetry* is ensured by Theorem 7.19.

The two operators defined above are fundamental to many important applications in science and engineering. The *divergence* $\nabla \cdot f$ helps to study the amount of *flux diverging from a volume* (a source) or the amount of *flux converging into a volume* (a sink) of some form of energy represented by the vector field f, and the *curl* $\nabla \times f$ is used to study local *rotational behaviour* of a vector field f in $\mathbb{R}^3$. A vector field $f : E \to \mathbb{R}^n$ is called *solenoidal* if $\nabla \cdot f = 0$; and, it is called *irrotational* if $\nabla \times f = 0$.

Further, since $\nabla^2 = \mathrm{div} \circ \mathrm{grad}$, the *Laplacian* $\nabla^2 \varphi$ of a scalar field $\varphi \in C^2(E)$ is fundamental to the study of various types of physical problems concerning phenomena such as *electromagnetism, fluid flow, thermodynamics,* and *elasticity*. For example, the *velocity field* $f = f(x, t)$ of an ideal fluid is a solution of the *Euler's equation*

$$\partial_t f + (f \cdot \nabla) f = g - \frac{1}{\rho} \nabla \varphi,$$

where g is the external force, $\rho = \rho(t, x)$ the density of the fluid, and $\varphi = \varphi(t, x)$ represents the pressure. The transport phenomenon is governed by the following *continuity equation*:

$$\partial_t \rho + \mathrm{div}\left(\rho f\right) = 0.$$

In particular, for *incompressible fluids* (i.e., when $\partial_t \rho = constant$), we obtain the fact that f is *solenoidal*.

Definition 7.19 Let $E \subset \mathbb{R}^n$ be an open set. We say a scalar field $\varphi \in C^2(E)$ is *harmonic* if $\Delta \varphi = 0$ over E. That is, if φ is a solution of the *Laplace equation*

$$\frac{\partial^2 \varphi}{\partial x_1^2} + \cdots + \frac{\partial^2 \varphi}{\partial x_n^2} = 0.$$

For example, if a complex-valued function $w = f(z) = u + iv$ is *analytic* over a domain $\Omega \subseteq \mathbb{C} \equiv \mathbb{R}^2$, then the functions $u, v : D \to \mathbb{R}$ satisfy the *Cauchy-Riemann equations*, and hence are harmonic functions. The next theorem is an important result that we need at a later stage.

Theorem 7.20 (Maximum Principle) *Let* $\Omega \subset \mathbb{R}^n$ *be an open set, and* $\varphi \in C(\overline{\Omega}) \cap C^2(\Omega)$ *be a harmonic function. Then*

$$\sup_{\Omega} |\varphi| = \sup_{\partial \Omega} |\Omega|.$$

Proof (*Outline*) For $a \in \Omega$, and $\epsilon > 0$, take

$$\varphi_\epsilon(x) = \varphi(x) - \epsilon \|x - a\|^2, \quad \text{for } x \in \Omega.$$

It then follows that

$$\nabla^2 \left(\varphi_\epsilon(x)\right) = -2n\epsilon < 0,$$

which implies that a cannot be a point of maximum for φ_ϵ. Notice that, in this case, the Hessian $H_{\varphi_\epsilon}(a) \geq 0$.

A typical application of this theorem is to prove the uniqueness of a solution of *Dirichlet's problems*. We will continue the story in the companion volume on partial differential equations.

7.4 Geometry of Curves and Surfaces

Throughout this section, $E \subseteq \mathbb{R}^n$ is an open connected set. The concepts and terminology introduced here concern the *surface geometry* of a scalar field $\varphi \in C^1(E)$, which is described in terms of the partial derivatives of φ. In physics, and elsewhere, such types of functions are used to describe the *potential energy* associated with a (*conservative*) vector field $f : E \to \mathbb{R}^n$ in the sense that $\nabla\varphi = f$. For example, $\varphi(x) = xy$ is the *potential function* of the vector field $f(x) = (y, x)$, for $x = (x, y) \in \mathbb{R}^2$. Notice that this particular potential function has discontinuous contour curves. Some commonly used potential functions represent *gravitational* or *electric potential*.

We begin by introducing the concept of a smooth curves in the Euclidean space $\mathbb{R}^n$ *parametrised* by some differentiable function $\gamma : I \to \mathbb{R}^n$, where $I \subseteq \mathbb{R}$ is an interval. More generally, as defined in Euclid's *Elements*, a *parametric curve* (or simply a *curve*) in $\mathbb{R}^n$ is the **trace** (or the *trajectory*) of a continuous function $\gamma : I \to \mathbb{R}^n$ given by

$$\mathscr{C}_\gamma := \left\{ x \in \mathbb{R}^n : x = \gamma(t), \text{ for some } t \in I \right\}.$$

We also say that the *path* $\gamma : I \to \mathbb{R}^n$ is a *parametric representation* of the curve $\mathscr{C}_\gamma$. Notice that, however, this definition of $\mathscr{C}_\gamma$ excludes *implicit curves* such as *level curves* and *algebraic curves*, but it does include ghastly curves such as *space-filling curves* and *fractal curves*.

When γ is injective, we say $\mathscr{C}_\gamma$ is *simple curve* (i.e., without self-intersections); and, the curve $\mathscr{C}_\gamma$ is a *closed curve* if $\gamma(a) = \gamma(b)$. For example, if $y = f(t) : I \to \mathbb{R}$ is a continuously differentiable function, then the *graph* Γ_f of f is a *plane curve*, parametrised *naturally* by the parameter t. Notice that, in this case, Γ_f is *simple*, and no vertical line cuts it at more than one point. In general, every parametrised curve $\mathscr{C}_\gamma$ has a natural *orientation*: *positive* for increasing t, and *negative*, otherwise.

Definition 7.20 For an interval $I \subseteq \mathbb{R}$, a function $\gamma : I \to \mathbb{R}^n$ is called an n-dimensional *parametric function* if each *component function* $\gamma_k : I \to \mathbb{R}$ of γ is continuous; and, we say γ is *smooth* if each $\gamma_k \in C^1(I)$, for $k = 1, \ldots, n$.

Recall that $\gamma(t)$ is continuous at a point $c \in I$ if and only if each component function γ_k is continuous at c, for $k = 1, \ldots, n$ (Theorem 7.10). Also, the derivative $D\gamma(c) = \gamma'(c)$ is given by

$$Dy(c) := \lim_{t \to 0} \frac{\gamma(c+t) - \gamma(c)}{t}, \tag{7.4.1}$$

provided the limit exists. Clearly then, γ is differentiable at $c \in I$ if and only if each function $\gamma_i : I \to \mathbb{R}$ is differentiable at c, and we write

$$\gamma'(c) = \big(\gamma_1'(c), \dots, \gamma_n'(c)\big).$$

In physical terms, when the variable $t \in I$ represents time, the vector $D\gamma(t)$ gives the *velocity vector* along the smooth curve $\mathscr{C}_\gamma$. And, in geometrical terms, we use $D\gamma(t)$ to define important concepts such as the *tangent, normal, curvature, torsion*, etc., at a point on the smooth curve $\mathscr{C}_\gamma$. For example, the linear tangent map $t \mapsto \gamma'(c)t$ parametrises the *tangent line* to the curve $\mathscr{C}_\gamma$ at the point $\gamma(c)$, provided $\gamma'(c) \neq 0$. As a special case of the general description given earlier, the graph Γ_γ of γ is a curve in the $(n+1)$-dimensional space $\mathbb{R} \times \mathbb{R}^n$, which has the line

$$t \mapsto \begin{pmatrix} t \\ \gamma'(c)t \end{pmatrix} = \begin{pmatrix} 1 \\ \gamma'(c) \end{pmatrix} t$$

as its *tangent space* at $t = c$, and the *normal plane* to Γ_γ at $(t, \gamma(t))$ is defined by the n row vectors of the matrix

$$\begin{pmatrix} \gamma_1'(t) & -1 & 0 & \cdots & 0 \\ \gamma_2'(t) & 0 & -1 & \cdots & 0 \\ \vdots & \vdots & \vdots & & \vdots \\ \gamma_n'(t) & 0 & 0 & \cdots & -1 \end{pmatrix}$$

We remark that, according to the *fundamental theorem for spaces curves*, every pair of smooth *curvature* and the *torsion* functions satisfying certain fundamental relations determine a unique smooth curve in $\mathbb{R}^n$.

In any particular situation, the domain interval I is chosen considering the validity of functions $\gamma_k : I \to \mathbb{R}$, for all $k = 1, \dots, n$. For example, if the component functions of a smooth curve $\gamma : I \to \mathbb{R}^3$ are given by

$$\gamma_1(t) = a \cos t, \quad \gamma_2(t) = a \sin t, \quad \gamma_3(t) = bt, \quad a, b > 0,$$

we can have $I = \mathbb{R}$. On the other hand, we must have

$$t \in \mathbb{R} \setminus \{(2n+1)\pi/2 : n \in \mathbb{Z}\}$$

for the smooth curve $\mathscr{C}_\gamma \subset \mathbb{R}^3$ to be parametrised by the differentiable function $\gamma(t) = (3 \tan t, 4 \sec t, 5t)$. Our main concern here is to discuss some basic properties of *smooth curves* in $\mathbb{R}^n$.

Definition 7.21 The trace $\mathscr{C} \subset \mathbb{R}^n$ of a differentiable function $\gamma : I \to \mathbb{R}^n$ is called a *smooth curve* (or simply a *curve*), and γ itself is called a (differentiable) *parametrisation* of the curve. For emphasis, we may write $\mathscr{C}_\gamma$. We say $t = t_0 \in I$ is a *regular point* of the curve $\mathscr{C}_\gamma$ if $\gamma'(t_0) \neq 0$. Otherwise, $t = t_0$ is called a *critical point*. We say $\gamma : I \to \mathbb{R}^n$ is a *regular parametrisation* of a curve $\mathscr{C} \subset \mathbb{R}^n$ if every point of I is regular.

For example, the curve $\mathscr{C}_\gamma \subset \mathbb{R}^3$ of the function $\gamma : \mathbb{R} \to \mathbb{R}$ given by

$$\gamma(t) = (5 \cos t, 3 \sin t, 4 \sin t)$$

is a spherical curve because $\|\gamma(t)\| = 5$, for all t. Also, the *Viviani curve* $\mathscr{C}$, obtained as the intersection of the sphere $x^2 + y^2 + z^2 = a^2$ with the cylinder $x^2 + y^2 = ax$, has a parametrisation given by the function

$$\gamma(\theta) := (a \cos^2 \theta, \ a \cos \theta \sin \theta, \ a \sin \theta), \quad \theta \in [0, 2\pi].$$

A simple way to visualise a smooth curve, say in $\mathbb{R}^3$, is by using the geometry of the plane curve defined in terms of the coordinate functions γ_1 and γ_2. For example, the space curve traced by the function

$$\gamma(t) = (4 \cos t, \ 3 \sin t, \ 6t), \quad t \in \mathbb{R},$$

is an *elliptical helix* because $\gamma_1(t) = 4 \cos t$ and $\gamma_2(t) = 3 \sin t$ satisfy the following equation of an ellipse

$$\frac{x^2(t)}{4^2} + \frac{y^2(t)}{3^2} = 1, \quad \text{for all } t \in \mathbb{R}.$$

By the same reasoning, the space curve traced by the function

$$\varphi(t) = (at \cos t, \ at \sin t, \ bt), \quad t \in \mathbb{R},$$

lies on the cone $b^2(x^2 + y^2) = a^2 z^2$, which is called a circular *conic helix*. Also, the space curve traced by the function

$$\varphi(t) = (a \cos t, \ a \sin t, \ bt), \quad t \in \mathbb{R},$$

is a *helix* on a right circular cylinder of base radius a winding around the z-axis at the rate b/a, with rise per revolution i.e., the *pitch* given by $2\pi b$ and the *total height* attained is given by $\ell\sqrt{a^2 + b^2}$, for $t \in [0, \ell]$. Here, the parameter t measures the angle between the x-axis and the line joining the origin to the projection of the point $\varphi(t)$ over xy-plane.

In a typical application involving a first order partial differential equation[7] for a smooth function $u = u(x) \in C^r(E)$, $r \geq 1$, the geometry of the *integral surface*

$$\Gamma_u = \{(x, u(x)) : x \in E\}$$

is described using *trajectories between points* on Γ_u given by some smooth curve $\mathscr{C}_\gamma$, where $\gamma : I \to \mathbb{R}^n$. At any instant $t \in (a, b)$, the point $\gamma(t)$ specifies the position vector of a *flow field*, and the derivative $\gamma'(t)$ gives the velocity, so that the real number $\|\gamma'(t)\|$ represents the speed at time t. The *length* ℓ of the *flow trajectory* of $\gamma(t)$ between the points $\gamma(t_0) = x_0$ and $\gamma(t_1) = x_1$, for $t_0 < t_1 \in I$, is given by the integral

$$\ell(\gamma) := \int_{t_0}^{t_1} \|\gamma'(t)\| dt. \tag{7.4.2}$$

For example, the vector function $\gamma : [0, 2\pi) \to \mathbb{R}^2$ given by

$$\gamma(t) = (a \cos t, a \sin t), \quad t \in [0, 2\pi], \tag{7.4.3}$$

parametrises the (counter-clockwise) circular motion of a particle along a circle of radius a about the origin, with the velocity vector given by the derivative $D\gamma(t) = (-a \sin t, a \cos t)$, at time t. Notice that, in this case, we have $\gamma'(t) \neq 0$, for all $t \in (0, 2\pi)$, i.e., γ is *regular parametrisation* of the circle. Also, the (constant) *speed* of the particle is given by

$$\|\gamma'(t)\| = \sqrt{a^2 \sin^2 t + a^2 \cos^2 t} = a. \tag{7.4.4}$$

So, the (arc) length between the starting point $\gamma(0) = (a, 0)$ to the point $\gamma(\pi) = (-a, 0)$ comes out to be πa, as expected.

In general, there could be infinitely many parametrisations of a smooth curve $\mathscr{C}_\gamma \subset \mathbb{R}^n$, where a *change of parameter* is obtained by using a strictly monotonic continuous function $\theta : I \to \mathbb{R}$ so that, if $J = \theta(I)$, then $\gamma \circ \theta^{-1} : J \to \mathbb{R}^n$ provides another parametrisation of the curve $\mathscr{C}_\gamma$.

Definition 7.22 Let $\mathscr{C}_\gamma \in \mathbb{R}^n$ be a smooth curve traced by a differentiable function $\gamma : [a, b] \to \mathbb{R}^n$. For any surjective continuous increasing function $\alpha : [c, d] \to [a, b]$, the function $\beta = \gamma \circ \alpha : [c, d] \to \mathbb{R}^n$ is called a *re-parametrisation* of the curve $\mathscr{C}_\gamma$.

For example, by using the following functions any finite number of times to change the parameter t:

$$t \mapsto t^2 \ (t > 0); \quad t \mapsto at + b; \quad t \mapsto e^t, \quad \text{etc..},$$

---

[7] Recall that it is an equation involving u, the first order partial derivatives of u, and possibly some other known functions of variables $x \in E$.

any curve $\mathscr{C}_\gamma$ can be re-parametrised. Notice that, however, *re-parametrisation* does not change the *shape* of the curve $\mathscr{C}_\gamma$, it only changes the *speed* of the point $\gamma(t)$ tracing the curve $\mathscr{C}_\gamma$, say between the points $\gamma(a)$ and $\gamma(b)$.

Suppose $\mathscr{C}_\gamma \subset \mathbb{R}^n$ is a smooth curve. If an instantaneous change δt leads to an infinitesimal change in $\gamma(t)$ given by $\delta\gamma$ then the *tangent vector* $\gamma'(t)$ to the curve $\mathscr{C}_\gamma$ at point $P = \gamma(t_0)$ is given by

$$
\begin{aligned}
\gamma'(t) &= \frac{d\gamma(t)}{dt}\bigg|_{t=t_0} \\
&= \lim_{\delta t \to 0} \frac{\gamma(t_0 + \delta t) - \gamma(t_0)}{\delta t} \\
&= \left(\gamma_1'(t_0), \ldots, \gamma_n'(t_0)\right),
\end{aligned}
\tag{7.4.5}
$$

provided limit exists. So, the components of the tangent vector

$$
\gamma'(t) := \left(\gamma_1'(t), \ldots, \gamma_n'(t)\right),
$$

at a point $P(t)$ in the direction of a vector $\boldsymbol{u}$ are given by

$$
\gamma'(t)\big|_P \cdot \frac{\boldsymbol{u}}{|\boldsymbol{u}|}\bigg|_P.
$$

For example, the tangent vector $\gamma'(t)$ of the curve traced by the function

$$
\gamma(t) = (t, t - \sin t, 1 - \cos t), \quad t \in [0, 1],
$$

is a unit vector at the origin. Therefore, for a smooth curve $\mathscr{C}_\gamma$ parametrised by a differentiable function $\gamma(t)$, the *arc length* $s(t)$ of $\mathscr{C}_\gamma$ from a point $\gamma(t_0)$ to $\gamma(t)$ is given by

$$
s(t) = \int_{t_0}^{t} |\gamma'(t)|dt = \int_{t_0}^{t_1} \sqrt{[\gamma_1']^2 + \cdots + [\gamma_n']^2}\, dt.
\tag{7.4.6}
$$

Notice that, by *chain rule*, arc length is independent of the parametrisation γ. Also, by the fundamental theorem, $s'(t) = |\gamma'(t)|$, for all $t \in (a, b)$. In particular, the length s of the *graph* of a continuously differentiable function $y = f(x)$ defined on a closed interval $[a, b]$ is given by

$$
s = \int_{a}^{b} |f'(x)|dx = \int_{a}^{b} \sqrt{1 + [f'(x)]^2}\, dx.
\tag{7.4.7}
$$

Recall that one unit of arc length on a unit circle is called a *radian*. A smooth curve $\mathscr{C}_\gamma$ parametrised by the arc length s is a *unit-speed* curve because point $\gamma(s)$ along the curve travels at the constant speed of one unit per unit of time.

It is in general true that a differentiable curve $\mathscr{C}_\gamma$ can be parametrised by arc length, provided it is **regular**, i.e., $\gamma'(t) \neq 0$ for all $t \in (a, b]$. We need a formula for $t = t(s)$

to obtain the *arc length parametrisation* of a (regular) space curve. For, notice that it is always possible to invert the function $s = s(t)$ *locally* by regularity condition and using the *inverse function theorem*. For example, the *arc length re-parametrisation* of the helix $\gamma(t) = (a \cos t, a \sin t, bt)$ is given by the function

$$\varphi(s) = \left(a \cos \frac{s}{\sqrt{a^2 + b^2}}, a \sin \frac{s}{\sqrt{a^2 + b^2}}, \frac{bs}{\sqrt{a^2 + b^2}} \right),$$

using the formula $t = s/\sqrt{a^2 + b^2}$. Indeed, since $|\varphi'(s)| = 1$ for all s, the curve $\varphi(s)$ is *unit speed*.

In practical applications, we use a re-parametrisation to *adjust the speed* of the flow along the curve $\mathscr{C}_\gamma$. For example, by using the function $\alpha(t) = 2t$ ($t \in [0, \pi]$), we obtain a re-parametrisation $\beta = \gamma \circ \alpha$ of the circle traced by the function γ as in (7.4.3). Since

$$\beta(t) = (a \cos 2t, a \sin 2t), \quad t \in [0, \pi],$$

we see that $\beta(t)$ traverses the circle $\mathscr{C}_\gamma$ with the speed *double* than what $\gamma(t)$ achieves. Notice that, for the function $\gamma(t)$ given by (7.4.3), we have

$$\gamma(2\pi) - \gamma(0) \neq (2\pi - 0)D\gamma(0),$$

which proves the usual *mean value property* as stated in Corollary 4.2 does not hold for an arbitrary vector functions $\gamma : [a, b] \to \mathbb{R}^n$.

The *mean value property* of a smooth function $f : E \to \mathbb{R}^n$ is a fundamental idea which provides an estimate about how much *local growth bounds the global growth*. As growth aspects of such types of functions are significant in many applications, it is important to know what part of the single-variable *mean value theorem* holds in this general case. We first prove a weaker version of Corollary 4.2 for n-variable vector functions defined on some interval $I \subseteq \mathbb{R}$.

Theorem 7.21 *If $\gamma : [a, b] \to \mathbb{R}^n$ is continuous on $[a, b]$, and is differentiable on (a, b), then for some $c \in (a, b)$ we have*

$$\|\gamma(b) - \gamma(a)\| \leq (b - a)\|D\gamma(c)\|.$$

Proof Let $u = \gamma(b) - \gamma(a)$. If $u = 0$, there is nothing to prove, so we may assume that $u \neq 0$. Consider the function $\phi : [a, b] \to \mathbb{R}$ given by

$$\phi(t) := u \cdot \gamma(t) = \sum_{k=1}^{n} (\gamma_k(b) - \gamma_k(a))\gamma_k(t), \quad t \in [a, b].$$

Each function $\gamma_k : [a, b] \to \mathbb{R}$ is continuous on $[a, b]$, and is differentiable on (a, b), and hence so is the scalar function ϕ. By using Corollary 4.2, there is some $c \in (a, b)$ such that

$$\phi(b) - \phi(a) = (b - a)\phi'(c).$$

We also have

$$\phi(b) - \phi(a) = \boldsymbol{u} \cdot \gamma(b) - \boldsymbol{u} \cdot \gamma(a) = \|\boldsymbol{u}\|^2.$$

Also, by Cauchy-Schwarz inequality (6.1.6), we have

$$|\phi'(t)| \leq \sum_{k=1}^{n} |(\gamma_k(b) - \gamma_k(a))\gamma_k'(t)|$$

$$\leq \sqrt{\sum_{k=1}^{n}(\gamma_k(b) - \gamma_k(a))^2 \sum_{k=1}^{n}\gamma_k'(t)^2}$$

$$= \|\boldsymbol{u}\| \, \|\gamma'(t)\|.$$

It thus follows that

$$\|\boldsymbol{u}\|^2 = |\phi(b) - \phi(a)|$$
$$= (b - a)|\phi'(c)|$$
$$\leq (b - a)\|\boldsymbol{u}\| \, \|D\gamma(c)\|.$$

As $\boldsymbol{u} \neq \boldsymbol{0}$, the assertion follows by cancelling out $\|\boldsymbol{u}\|$. This completes the proof.

Next, consider a scalar field $\varphi : E \to \mathbb{R}$, where $E \subseteq \mathbb{R}^m$ is an open connected set. We write $D(U)$ for the space of differentiable functions $\varphi : U \to \mathbb{R}$, where $U \subset \mathbb{R}^m$ is an open set. It is easy to verify the following *Leibniz rule* for $\varphi, \phi \in D(U)$:

$$D\langle \varphi, \phi \rangle = \langle D\varphi, \phi \rangle + \langle \varphi, D\phi \rangle. \tag{7.4.8}$$

In most applications, we usually have that a function $\varphi \in D(U)$ is defined and continuous on the *boundary* ∂U of the open set U. We also need the following simple observation. Notice that, for differentiable functions $\gamma : I \to \mathbb{R}^m$ and $\boldsymbol{f} : U \to \mathbb{R}^n$, with $\gamma(I) \subset U$, the chain rule implies

$$\beta(t) := \boldsymbol{f} \circ \gamma \quad \Rightarrow \quad \beta'(t) = D\boldsymbol{f}(\gamma(t))\gamma'(t),$$

so that, if $\gamma'(t)$ and $\beta'(t)$ are nonzero, then the Jacobian $J_f(\gamma(t_0))$ $(t_0 \in I)$ maps the tangent line at t_0 to the curve $\gamma(t)$ into the tangent lines to the curve image $\beta(t)$ at point t_0.

The next theorem generalises Theorem 7.21 for scalar fields.

Theorem 7.22 (Mean Value Theorem) *Let $E \subset \mathbb{R}^m$ be an open connected set, and suppose two points $\boldsymbol{a}, \boldsymbol{b} \in E$ are such that the line segment given by*

$$L(\boldsymbol{a}; \boldsymbol{b}) := \{(1 - t)\boldsymbol{a} + t\boldsymbol{b} : 0 \leq t \leq 1\},$$

with endpoints a and b, lies inside the region E. Then, for any $\varphi \in C^1(E)$, there is an interior point c of $L(a; b)$, such that

$$\varphi(b) - \varphi(a) = \nabla\varphi(c) \cdot (b - a).$$

Proof Consider the function $f : [0, 1] \to \mathbb{R}$ given by

$$f(t) := \varphi[(1 - t)a + tb], \quad t \in [0, 1].$$

By our assumption, and above remark, we have $f \in C^1([0, 1])$, and so there is some $t_0 \in (0, 1)$ such that
$$f(1) - f(0) = (1 - 0) f'(t_0).$$

By applying the chain rule, we obtain

$$f'(t) = \nabla\varphi\big[(1 - t)a + tb\big] \cdot (b - a).$$

Therefore, by taking $c = (1 - t_0)a + t_0 b \in L$, the assertion follows.

The next theorem proves *mean value inequality* for a multivariable vector-valued function, where.

Theorem 7.23 (Mean Value Inequality) *Let $E \subset \mathbb{R}^m$ be an open set, and $f : E \to \mathbb{R}^n$ be a function differentiable over E. Suppose the line segment $L[a; b]$, with endpoints $a, b \in E$, given by*

$$L(a; b) = \big\{(1 - t)a + tb : 0 \le t \le 1\big\},$$

lie inside the set E. Then, for some interior point x_0 in $L[a; b]$, we have

$$\|f(a) - f(b)\| \le \|Df(x_0)\| \|a - b\|.$$

Proof Consider the function $\gamma : [0, 1] \to \mathbb{R}^n$ given by

$$\gamma(t) := f[(1 - t)a + tb], \quad t \in [0, 1].$$

Since f is differentiable over E, by Theorem 7.21, we can find some $t_0 \in (0, 1)$ such that the function γ satisfies the inequality

$$\|\gamma(1) - \gamma(0)\| \le \|D\gamma(t_0)\| (1 - 0). \tag{7.4.9}$$

Also, as $\gamma(1) = f(b)$ and $\gamma(0) = f(a)$, we have

$$D\gamma(t) = Df[(1 - t)a + tb](b - a).$$

Therefore, by using Theorem 7.16, the proof follows from the inequality (7.4.9). Notice that, in particular, $D\gamma(t_0) = Df(x_0)(b - a)$, where $x_0 = (1 - t_0)a + t_0 b$ lies in the interior of the line ℓ.

A generalisation of this theorem can be obtained by modifying above argument, and using (7.4.2) (see Exercise 7.22).

Example 7.14 Let $U = \{x \in \mathbb{R}^2 : \|x\| > 1\} \setminus \{(x, 0)^T : x \leq 0\}$. If we take $\theta(x)$ to be the unique solution of

$$\cos(\theta(x)) = x, \ \sin(\theta(x)) = y, \ -\pi < \theta(x) < \pi$$

for $x = (x, y)^T \in U$ then $\theta : U \to \mathbb{R}$ is everywhere differentiable with $\|D\theta(x)\| < 1$. However, if $a = (-1, 10^{-1})^T$, $b = (-1, -10^{-1})^T$ then

$$|\theta(a) - \theta(b)| > \|a - b\|.$$

Let $E \subset \mathbb{R}^m$ be an open set, $a \in E$, and suppose $f : E \to \mathbb{R}$ is differentiable at a. By applying a the *translation map* $x \mapsto x - a$, we can assume that $a = 0 \in E$. In general, the mean value inequality in Theorem 7.23 is used to study *local behaviour* of a reasonably well behaved differentiable functions at the origin $0 \in E$. For example, in the case when $m = 2$ and $n = 1$, the mean value inequality implies that every bounded differentiable function f is Lipschitz in the first variable, as shown earlier in Chap. 4.

Definition 7.23 A connected set $S \subset \mathbb{R}^3$ is called a piecewise **smooth surface** if, for each point $P \in S$, there is an open ball $U_P := B(P; \epsilon) \subset \mathbb{R}^3$ and an injective function $f \in C^1(U_P; \mathbb{R}^3)$ that maps the open set $W_P := S \cap U_P$ onto an open set $V \subset \mathbb{R}^2$ (seen as a uv-plane in $\mathbb{R}^3$). If $F = \left(f|_{W_P}\right)^{-1}$ then $F(V) = W_P$, and $F : V \to \mathbb{R}^3$ defines a function

$$\mathbf{r}(u, v) = (x(u, v), y(u, v), z(u, v)) \in C^1(V),$$

which is called a *regular local representation* of the surface S. We also say S is a *parametrised surface*, with *parametrisation* given by $\mathbf{r}$. A surface S is said to be of class C^r if the following coordinate functions of a *parametrisation* $\mathbf{r}$ are so:

$$x = x(u, v), \quad y = y(u, v), \quad z = z(u, v), \quad (u, v) \in V,$$

We say S is **regular surface**, if $\mathbf{r}_u \times \mathbf{r}_v \neq 0$.

Informally, with $E \in \mathbb{R}^n$ an open connected set, the *graph* of a C^2-smooth scalar field $\varphi : E \to \mathbb{R}$ is an $(n + 1)$-dimensional *regular surface*. Here, we consider a *regular surface* in $\mathbb{R}^3$ obtained as a 2-dimensional analogue of a parametrised curve. Let $U \subseteq \mathbb{R}^2$ is an open connected set. A function $\mathbf{r} : U \to \mathbb{R}^3$ of the form

$$\mathbf{r}(u, v) = \big(x(u, v), y(u, v), z(u, v)\big), \quad \text{for } (u, v) \in U,$$

is called a *parametrisation* of the 2 -dimensional surface $\mathbf{r}(U) \subset \mathbb{R}^3$. For example, if $\Pi \subset \mathbb{R}^3$ is the plane passing through a point $a = (a_1, a_2, a_3)$, and parallel to vectors $u = (u_1, u_2, u_3)$ and $v = (v_1, v_2, v_3)$, then the function $\mathbf{r} : \mathbb{R}^2 \to \Pi$ given by

$$
\begin{aligned}
\mathbf{r}(s, t) &= a + su + tv \\
&= \big(a_1 + su_1 + tv_1, a_3 + su_2 + tv_2, a_3 + su_3 + tv_3\big),
\end{aligned}
$$

gives a parametrisation of the plane Π, where $(s, t) \in U = \mathbb{R}^2$. Also, we can parametrise the sphere

$$S(\mathbf{0}; a) = \big\{(x, y, z) \in \mathbb{R}^3 : x^2 + y^2 + z^2 = a^2\big\},$$

by using the *spherical coordinates* (θ, ϕ). In this case, the function $\mathbf{r} : [0, 2\pi) \times [0, \pi] \to S(\mathbf{0}; a)$ given by

$$\mathbf{r}(\theta, \phi) = \big(a \cos \theta \sin \phi, a \sin \theta \cos \phi, a \cos \phi\big),$$

gives a parametrisation of $S(\mathbf{0}; a)$, where $\theta \in [0, 2\pi)$ is the *longitudinal angle* and $\phi \in [0, \pi]$ is the *latitudinal angle*.

In general, let $I = (a, b)$ and $J = (c, d)$ be two intervals, and $U = I \times J$ be a set of *parameters* (u, v). Then, any continuous function $\mathbf{r} : I \times J \to \mathbb{R}^3$ defines a *parametrised surface* $S_{\mathbf{r}} \subset \mathbb{R}^3$ given by

$$S_{\mathbf{r}} := \big\{x \in \mathbb{R}^3 : x = \mathbf{r}(u, v), \text{ for some } (u, v) \in I \times J\big\}.$$

Here, we may write the *coordinate functions* $x_i : I \times J \to \mathbb{R}$ of a point $x \in S(\mathbf{r})$ as

$$x_1 = x_1(u, v), \quad x_2 = x_2(u, v), \quad x_3 = x_3(u, v)), \quad (u, v) \in I \times J.$$

A *helicoid* is (ruled) surface like a spiral staircase in a building, described by a straight line that rotates at a constant angular rate around a fixed axis, intersects the axis at a constant angle θ. For $(u, v) \in U = \mathbb{R}^2$, the following functions give a parametrisation of the helicoid:

$$
\begin{aligned}
x_1(u, v) &= u \cos(v\,\theta), \\
x_2(u, v) &= u \sin(v\,\theta), \\
x_3(u, v) &= v.
\end{aligned}
$$

Notice that the graph $\Gamma_\varphi \subset \mathbb{R}^3$ of a scalar field $\varphi : U \to \mathbb{R}$ defines surface parametrised naturally by *parameters* $(x, y) \in U$. We write

$$\Gamma_\varphi = \big\{(x, y, z) \in \mathbb{R}^3 : \varphi(x, y) - z = 0\big\}.$$

Similarly, the graph of a 3-variable scalar field $w = \varphi(x, y, z)$ defines a three-parameters 3-dimensional surface $\Gamma_\varphi \subset \mathbb{R}^4$ parametrised naturally by *parameters* (x, y, z) in some open connected set in $\mathbb{R}^3$, and so on. So, in general, we can define an n-parameter *hypersurface* in $\mathbb{R}^{n+2}$ as the graph of an n-variable scalar field $x_{n+1} = \varphi(x)$, where x varies over some open connected set in $\mathbb{R}^n$. Notice that *connectedness* ensures that the surface is in one piece.

In general, let a differentiable function $\mathbf{r} : U \to \mathbb{R}^3$ given by

$$\mathbf{r}(u, v) = \big(x(u, v), y(u, v), z(u, v)\big), \quad \text{for } (u, v) \in U,$$

be a parametrisation of a surface $S \subset \mathbb{R}^3$ so that $S = \mathbf{r}(U)$. In this case, for $\mathbf{a} = (u_0, v_0) \in U$, we usually write the Jacobian matrix $J_{\mathbf{r}}(\mathbf{a})$ as

$$J_{\mathbf{r}}(\mathbf{a}) = \begin{pmatrix} x_u & x_v \\ y_u & y_v \\ z_u & z_v \end{pmatrix}, \quad \text{where } x_u = \partial_u x(\mathbf{a}), \ x_v = \partial_v x(\mathbf{a}), \text{ etc.}$$

Then, the two column vectors

$$\mathbf{r}_u := \begin{pmatrix} x_u \\ y_u \\ z_u \end{pmatrix} \quad \text{and} \quad \mathbf{r}_v := \begin{pmatrix} x_v \\ y_v \\ z_v \end{pmatrix}$$

are the *velocity vectors* respectively of the u-curve $u \mapsto \mathbf{r}(u, v_0)$ and the v-curve $v \mapsto \mathbf{r}(u_0, v)$, at the point $\mathbf{a} = (u_0, v_0)$. The *tangent space* to graph $\Gamma_{\mathbf{r}}$ at $\mathbf{a}$ is the 2-dimensional subspace of $\mathbb{R}^2 \times \mathbb{R}^3$ given by

$$\left\{ (u, v; x, y, z) \in \mathbb{R}^2 \times \mathbb{R}^3 : \begin{pmatrix} x \\ y \\ z \end{pmatrix} = \begin{pmatrix} x_u & x_v \\ y_u & y_v \\ z_u & z_v \end{pmatrix} \begin{pmatrix} u \\ v \end{pmatrix} \right\},$$

with a basis given by column vectors

$$\begin{pmatrix} 1 & 0 & x_u & y_u & z_u \end{pmatrix}^t \quad \text{and} \quad \begin{pmatrix} 0 & 1 & x_v & y_v & z_v \end{pmatrix}^t$$

And, a basis of the 3-dimensional *normal space* is given by the row vectors of the matrix

$$\begin{pmatrix} x_u & x_v & -1 & 0 & 0 \\ y_u & y_v & 0 & -1 & 0 \\ z_u & z_v & 0 & 0 & -1 \end{pmatrix}.$$

Next, as said before, the geometry of a 2-dimensional parametrised surface Γ_φ is described by using the *level curves*

$$L_\varphi(c) := \big\{ (x, y) \in U : \varphi(x, y) = c \big\}, \quad c \in \mathbb{R},$$

which are obtained by slicing $\Gamma_\varphi \subset \mathbb{R}^3$ horizontally by the plane $z = c$. In particular, the surface $z = f(x, y)$ in $\mathbb{R}^3$ is the level surface $w = 0$ of the scalar field $\varphi(x, y, z) = f(x, y) - z$.

There are many different ways to view a level surface $w = c$ of a scalar field $\varphi = \varphi(x, y, z)$. For example, a visually intuitive method is when one of the three variables is kept fixed to define a *contour map* in the remaining two dimensions of the level surface. Geometrically, if $x = a$ is kept fixed, the contour map is the intersection of the level surface $w = c$ with the plane $x = a$. The contour maps on a level surface obtained in this manner are called the *level curves* of the scalar fields $\varphi = \varphi(x, y, z)$.

Example 7.15 If a scalar field is given by $w = \varphi(x, y, z) = 4x^2 + y^2 - z$, then for $w = 0$

$$z = f(x, y) = 4x^2 + y^2,$$

defines the level surface $w = 0$ of the scalar field. Notice that the level surface $z = f(x, y) = 4x^2 + y^2$ is the bowl-shaped *elliptical paraboloid*, with bottom at the origin, which is its *level curve* for $z = 0$. In general, the *level curves* of the level surface $w = 0$ are the *ellipses* $4x^2 + y^2 = a$. The scalar fields

$$\varphi(x, y, z) = \sqrt{x^2 + y^2 + z^2}$$

can be viewed as the level suface $w = r > 0$, which is the sphere in $\mathbb{R}^3$ of radius r. In this case, the *level curves* of the level surface $w = r$ are the *equators* (circles of lattitude). Also, for the scalar field defined by

$$w = \varphi(x, y, z) = x^2 + y^2 - z^2, \quad (x, y, z) \in \Omega,$$

we obtain three particularly important surfaces as level surfaces: Two-sheeted hyperboloid for $w = -1$; a double cone for $w = 0$; and, a single-sheeted hyperboloid for $w = 1$.

We use level curves for different values of z to study properties of a scalar field. In what follows, we reserve the term *level curves* for the curves obtained by *slicing the surface* along the z-axis. In practical terms, if $w = \varphi(x, y, z)$ represents the potential energy of a field then the level surfaces represents layers of constant potential i.e., the *equipotential surfaces*. Similarly, if $w = \varphi(x, y, z)$ represents the temperature distribution on a surface then layers of constant temperature $w = c$ are called *isothermal surfaces*. This concept plays an important role in analysing an analytical solution of a physical problem modelled by some partial differential equation.

Notice that if $\gamma : I \to \mathbb{R}^n$ is a differentiable curve passing through the point $a = \gamma(t_0) \in E$ and $\gamma'(t_0) = v$. It then follows easily that, if φ is differentiable at a, then the map $t \mapsto \varphi(\gamma(t))$ is differentiable at $t = t_0$, and

$$\frac{d(\varphi \circ \gamma)}{dt}(t_0) = \nabla\varphi(a) \cdot \gamma'(t_0). \tag{7.4.10}$$

So, in particular, the growth of the scalar field φ at a along every regular curve γ passing through a depends only on the velocity $\gamma'(t_0) = v$. It thus makes sense to introduce the number $D_v\varphi(a)$ as the *slope of φ at a in the direction v*.

The next theorem proves that, if $a \in E$ is not a critical point of a differentiable scalar field $\varphi : E \to \mathbb{R}$ (Definition 7.17), the gradient $\nabla\varphi(a)$ defines the *normal* to the level sets of the *graph Γ_φ of φ* at $x = a$.

Theorem 7.24 *Let $E \subseteq \mathbb{R}^n$ be a connected open set, $a \in E$, and $\varphi : E \to \mathbb{R}$ be differentiable at a regular point a. Then, the gradient $\nabla\varphi(a)$ is normal to the surface Γ_φ at the point a.*

Proof Suppose $\gamma : I \to \mathbb{R}^n$ is a differentiable curve passing through the point a that lies on the level surface $\varphi \equiv \alpha$, where α is some constant. Then, as an arbitrary point on $\mathscr{C}_\gamma$ is given by

$$\gamma(t) = (x_1(t), x_2(t), \ldots, x_n(t)), \quad t \in I,$$

we have $\varphi(\gamma(t)) = \alpha$, so that

$$\frac{d}{dt}\varphi(\gamma(t)) = 0 \quad \Rightarrow \quad \nabla\varphi \cdot \gamma'(t) = 0,$$

by (7.4.10), which implies that the gradient $\nabla\varphi$ is *orthogonal* to the tangent vector $\gamma'(t)$ for each t, with $\gamma'(t) \neq 0$. Hence, the vector $\nabla\varphi(a)$ is normal to surface Γ_φ at each point, provided γ is a *regular curve*. Notice that, the latter condition is no restriction because, by differentiability of φ, it is always possible to choose an *arc-length parameterisation* for C_γ so that $\gamma' \not\equiv 0$.

Moreover, if θ is the angle between the gradient vector $\nabla\varphi(x)$ at a *regular point* $x \in E$ and a vector v then

$$D_v\varphi = \nabla\varphi \cdot \frac{v}{|v|} = |\nabla\varphi| \cos\theta \qquad (7.4.11)$$

implies that $D_v\varphi$ attains its *maximum value* $|\nabla\varphi|$ when $\cos\theta = 1$ i.e., when $\theta = 0$. Therefore, the directional derivative $D_v\varphi$ of φ is *maximum* when it is evaluated in the direction of the vector $\nabla\varphi(x)$. This simple remark, together with Theorem 7.24, explains the importance of the following concept.

Definition 7.24 The directional derivative $D_n\varphi(x)$ taken in the direction of the (outward) normal n to a surface is called the *normal derivative*, which is denoted by $\partial_n\varphi = \partial\varphi/\partial n$.

The *normal derivative* at any point on the graph Γ_φ of a differentiable function $\varphi = \varphi(x)$ ($x \in \Omega$) is given by

$$\partial_n\varphi = D_{\nabla\varphi}\varphi = \nabla\varphi \cdot n = |\nabla\varphi|, \quad \text{where} \quad n = \frac{\nabla\varphi}{|\nabla\varphi|},$$

which is *maximum value* of the gradient $\nabla\varphi$. So, in physical terms, $\partial_n\varphi$ at any point is the rate of change in values of the quantity $\varphi(x)$ as the point x move *orthogonally* across the *boundary* $\partial\Gamma_\varphi$ of the surface Γ_φ.

7.5   Two Fundamental Theorems

We discuss here two fundamental theorems of multivariable analysis, known as the *Inverse Function Theorem* (or simply IFT) and the *Picard-Lindelöf Theorem*.[8] In each case, we use an important property of any complete metric space to give a proof, known as the *contraction principle* (or Banach *fixed-point theorem*).

Let (X, d) be a metric space. Recall that X is said to be *complete* if every Cauchy sequence in X converges to a point in X. We say a function $T : X \to X$ is a *contraction* if, for some $\lambda \in (0, 1)$, we have

$$d(Tx, Ty) \le \lambda\, d(x, y), \quad \text{for all} \quad x, y \in X. \tag{7.5.1}$$

The constant λ is called the *contraction factor* of T. Notice that, a contraction is Lipschitz map, and so uniformly continuous.

Theorem 7.25 (Contraction Principle) *Let (X, d) be a complete metric space, and $T : X \to X$ be a contraction, with contraction factor $\lambda \in (0, 1)$. Then, there is a unique $x_0 \in X$ such that $T(x_0) = x_0$. More precisely, for any $x \in X$, $T^n(x) \to T(x_0)$ as $n \to \infty$.*

Proof A proof of uniqueness of x_0 is a simpler part of the story. For, if $x_0' \in X$ is such that $Tx_0' = x_0'$, then

$$
\begin{aligned}
d(x_0, x_0') &\le \lambda\, d(Tx_0, Tx_0') \\
&= d(x_0, x_0') \\
\Rightarrow \quad x_0 &= x_0', \quad \text{because} \quad \lambda < 1.
\end{aligned}
$$

Finally, to prove the existence, let $x \in X$ be arbitrary, but otherwise fixed. Consider the sequence $\langle T^n x \rangle$. Applying the iterates $T^n = T^{n-1} \circ T$, $n \ge 1$, we have

$$d(T^n x, T^n y) \le \lambda\, d(T^{n-1}x, T^{n-1}y)$$

$$\vdots$$

$$\le \lambda^n d(x, y), \quad \text{for all} \quad x, y \in X,$$

---

[8] The theorem is named after the French mathematician *Émile Picard* (1856–1941) and the Finnish mathematician *Ernst Lindelöf* (1870–1946). It is also known as the *Cauchy-Lipschitz theorem*.

by using inequality (7.5.1) n number of times. That is, for each $n \geq 1$, the nth iterate of T is a contraction, with *contraction factor* λ^n. So, for any $n \in \mathbb{N}$,

$$d(T^n x, T^{n+1} x) \leq \lambda^n d(x, Tx).$$

Then, by the triangle inequality and repeated use of the previous inequality, it follows that, for any $n, k \in \mathbb{N}$,

$$
\begin{aligned}
d(T^n x, T^{n+k} x) &\leq d(T^n x, T^{n+1} x) + \cdots + d(T^{n+k-1} x, T^{n+k} x) \\
&\leq \lambda^n d(x, Tx) + \lambda^{n+1} d(x, Tx) + \cdots + \lambda^{n+k-1} d(x, Tx) \\
&\leq (\lambda^n + \lambda^{n+1} + \cdots + \lambda^{n+k-1}) d(x, Tx) \\
&= \lambda^n \frac{1 - \lambda^k}{1 - \lambda} d(x, Tx) \\
&\leq \frac{\lambda^n}{1 - \lambda} d(x, Tx) \longrightarrow 0 \text{ as } n \to \infty,
\end{aligned}
$$

by using the assumption that $\lambda < 1$. We thus conclude that $\langle T^n x \rangle$ is a Cauchy sequence. Now, as (X, d) is complete, we can find $x_0 \in X$ such that $\lim_{n \to \infty} T^n x = x_0$. Notice that the continuity of T implies that

$$T x_0 = T \left(\lim_{n \to \infty} T^n x \right) = \lim_{n \to \infty} T^{n+1} x = x_0.$$

This completes the proof.

The conclusion of this theorem fails for $\lambda = 1$. For example, consider the closed set $X = [1, \infty) \subset \mathbb{R}$ viewed as a metric space under the induced *modulus metric*. By Theorem 6.19, X is complete. Notice that the function $T : X \to X$ defined by $T(x) = x + 1/x$ satisfies

$$|T(x) - T(y)| < |x - y|, \quad \text{for all} \quad x, y \in X,$$

but T has *no fixed point*.

Let us now start with a description about the first main theorem of the section. Let $E \subset \mathbb{R}^n$ is an open set, and $f : E \to \mathbb{R}^n$ be a C^r-function over E, for some $r \geq 1$. Recall that if the function f is a *linear map*, say given by a square matrix $A := A_f$ of order n, i.e., $f(x) = Ax$, for a (column) vector $x \in \mathbb{R}^n$, then the system of *linear equations* $Ax = b$ has a unique solution, for any vector $b \in \mathbb{R}^n$, if and only if A is *invertible* or, equivalently, $\det(f) \neq 0$. More generally, the next theorem proves that an arbitrary non-linear system $f(x) = b$ admits a *local solution* near any point $a \in E$ if the Jacobian matrix $J_f(a)$ is *invertible*, which is called the *non-degeneracy condition*.

Theorem 7.26 (Inverse Function Theorem) *Let $U \subset \mathbb{R}^n$ be an open set, $a \in U$, and $f : U \to \mathbb{R}^n$ be a C^r-function, for some $r \geq 1$, such that the Jacobian matrix $J_f(a)$ is nonsingular. Then, there exists an open neighbourhood $a \in B_a \subset U$, such that*

1. *$g := f|_{B_a}$ is an injective function;*
2. *$V := f(B_a)$ is open, and $g^{-1} : V \to B_a$ is continuous;*
3. *g^{-1} is also a C^r-function, and*

$$Dg(b) = [Df(a)]^{-1}, \quad \text{with} \quad b = f(a).$$

In short, f is locally invertible, *open, with local inverse a C^r-function.*

Said differently, this theorem says, if $\det \left(J_f(a) \right) \neq 0$, for some $a \in E$, then in a small neighbourhood around the point a the function f is nicely approximated by a linear map $L_a f : \mathbb{R}^n \to \mathbb{R}^n$, up to the constant $f(a)$. That is, the local information carried by the function f near a *regular point* a can be obtained from the invertible linear map $L_a f$. This is known as the *principle of linearisation*. In general, this fundamental principle is used extensively in science and engineering to obtain a *local solution* of some intricate non-linear problems. However, we shall use it mainly to study the geometry of the integral surface of some important partial differential equations (see the companion volume *Fundamentals of Partial Differential Equations* by the same authors).

To give a proof of Theorem 7.26, we need some preparation. First, we introduce the concept of *diffeomorphism*. Let $U, V \subset \mathbb{R}^n$ be two open sets. We say a bijective function $f : U \to V$ is a **diffeomorphism** if both f and f^{-1} are smooth functions. It should be clear that, if $f : U \to V$ is a diffeomorphism, then the determinant $|J_f(a)| \neq 0$, for all $a \in U$, and so the Jacobian matrix $J_f(a)$ is nonsingular, for all $a \in U$. That is, $Df(a) : \mathbb{R}^n \to \mathbb{R}^n$ is an isomorphism, for all $a \in U$. Moreover, by the next theorem, we have $[Df]^{-1} = Df^{-1}$ over $\mathbb{R}^n$.

Theorem 7.27 *Let $E \subset \mathbb{R}^m$ be an open connected set, and $f : E \to \mathbb{R}^m$ be a function differentiable at a point $x = a \in E$. Suppose $V \subset \mathbb{R}^m$ is a neighborhood of the point $b = f(a)$, and $g : V \to \mathbb{R}^m$ be a function differentiable at the point b, $g(b) = a$, and*

$$g(f(x)) = x, \quad \text{for all} \quad x \in U,$$

where U is some neighborhood of the point a. Then $Dg(b) = [Df(a)]^{-1}$.

Proof As the derivative of the identity function is the identity matrix, and $g(f(x)) = x$, it follows by Theorem 7.16 that $Dg(b)Df(a) = I$ and hence $Dg(b) = [Df(a)]^{-1}$. $\blacksquare$

Therefore, if a bijective function and its inverse are known to be differentiable, then the chain rule proves that the derivative of the inverse function is the reciprocal of the derivative of the function. Theorem 7.26 says that the converse holds, but only *locally*. That is,

$$\text{local isomorphism} \quad \Rightarrow \quad \text{local diffeomorphism.}$$

In particular, it is hoped to find a *local solution* to some non-linear system of the form $f(a) = b$, provided $J_f(a) \neq 0$.

More generally, let $U \subseteq \mathbb{R}^m$ and $V \subseteq \mathbb{R}^n$ be two open sets, and suppose $f : U \to V$ is a smooth map. Let $\Phi : U \to U_1$ and $\Psi : V \to V_1$ be two diffeomorphisms, where $U_1 \subseteq \mathbb{R}^m$ and $V_1 \subseteq \mathbb{R}^n$ are open sets. The central idea of the proof is to find a suitable *change of coordinates* function

$$f_1 = \Psi \circ f \circ \Phi^{-1} : U_1 \to V_1$$

such that f_1 has a simpler description. For example, suppose $f : U \to V$ is smooth, and $a \in U$. By applying the translations $\tau_1 : x \mapsto x - a :: \mathbb{R}^m \to \mathbb{R}^m$ and $\tau_2 : x \mapsto x - f(a) : \mathbb{R}^n \to \mathbb{R}^n$, we may write $U_1 = \tau_1(U)$ and $V_1 = \tau_2(V)$. Then, in the new coordinates, the point a is replaced by the origin $0 \in \mathbb{R}^m$ and f_1 satisfies the condition $f_1(0) = 0$.

Proof (*Proof of Theorem* 7.26) In view of above remarks, we assume that $a = 0 = f(0)$, and that $J_f(0) = I_n$, the $n \times n$ identity matrix. Further, considering the function $g : U \to \mathbb{R}^n$ given by $g(x) = x - f(x)$, we see that $g(0) = 0$ and $J_g(0) = 0$. For $r > 0$, write $D_r = \overline{B}(0; r)$. Notice that each D_r is a complete metric space under the induced Euclidean norm. By our assumption, there is some small $r' > 0$ such that $J_f(x)$ is nonsingular, for all $x \in D_{r'}$. We fix this $r' > 0$ for the rest of the proof.

Now, by using the fact that g is a C^1-function with $J_g(0) = 0$, and applying the mean value theorem as given in Exercise 7.22, it follows that, for some small $r > 0$, we have

$$x_1, x_2 \in D_r \quad \Rightarrow \quad \|g(x_1) - g(x_2)\| \leq \frac{1}{2}\|x_1 - x_2\|. \tag{7.5.2}$$

And, since $g(x) = x - f(x)$, we also have

$$x_1, x_2 \in D_r \quad \Rightarrow \quad \|f(x_1) - f(x_2)\| \geq \frac{1}{2}\|x_1 - x_2\|. \tag{7.5.3}$$

We assert that, for each $y \in D_{r/2}$, there is a unique $x \in D_r$ such that $f(x) = y$. For, consider the function T_y defined by $T_z = y + g(z)$, for $z \in D_r$. Then, by (7.5.2), we have $T_y(D_r) \subseteq D_r$, and T_y is a contraction (with contraction factor $1/2$), and so Theorem 7.25 applies to the function $T_y : D_r \to D_r$. Thus, we may take $x \in D_r$ to be the unique fixed point of the contraction T_y. Clearly then,

$$x = T_y(x) = y + g(x) = y + x - f(x) \quad \Leftrightarrow \quad f(x) = y.$$

Let $V_1 = B_{r/2} \subset \mathbb{R}^n$, and $U_1 = f^{-1}(V_1)$ be the neighbourhood of $\mathbf{0}$ contained in U. By the previous assertion, we have $f|_{U_1} = f_1 : U_1 \to V_1$ is a bijection. Also, by (7.5.3), we have

$$\| f_1^{-1}(y_1) - f_1^{-1}(y_2) \| = \| x_1 - x_1 \| \leq 2\| y_1 - y_2 \|,$$

which proves that f_1^{-1} is continuous, and so f_1 is a homeomorphism. In fact, f_1^{-1} is differentiable at each point of V_1, with $J_{f_1^{-1}} = (J_{f_1})^{-1} \circ f_1^{-1}$. To prove this, let $v = f_1(u) \in V_1$, with $u \in U_1$. Since f_1 is known to be differentiable at u, we may write

$$f_1(x) - f_1(u) = J_{f_1}(u) \cdot (x - u) + \| x - u \| \epsilon(x, u),$$

where the *error function* $\epsilon(x, u) \to \mathbf{0}$ as $x \to u$. By our initial assumption, we know that $J_f(x)$ is nonsingular, for all $x \in D_{r'}$. So, we obtain

$$x - u = [J_{f_1}(u)]^{-1} \cdot (f_1(x) - f_1(u)) - \| x - u \| [J_{f_1}(u)]^{-1} \epsilon(x, u).$$

Thus, with $x = f_1(y)$, the equation gives

$$f_1(y) - f_1(v) = [J_{f_1}(u)]^{-1} \cdot (y - v) - \| x - u \| [J_{f_1}(u)]^{-1} \epsilon(x, u)$$
$$= [J_{f_1}(u)]^{-1} \cdot (y - v) - \| y - v \| \delta(y, v)$$

where

$$\delta(y, v) = \frac{\| x - u \|}{\| y - v \|} [J_{f_1}(u)]^{-1} \epsilon(x, u).$$

Since $y \to v \Leftrightarrow x \to u$, by (7.5.3), we have $\| x - u \| / \| y - v \| \leq 2$. That is, $\lim_{y \to v} \delta(y, v) = 0$. Hence, we conclude that f_1^{-1} is differentiable, with

$$J_{f_1^{-1}}(v) = [j_{f_1}(u)] = J_{f_1}[f_1^{-1}(v)].$$

Finally, note that since f is a C^r-function, the entries of the matrix $[J_{f_1}]^{-1}$ are C^{r-1}-functions. So, by the previous details, and the fact that f_1 is a homeomorphism, it follows that $J_{f_1^{-1}}$ is continuous. That is, $f_1 \in C^1$. Thus, we are done for $r = 1$. Otherwise, we may keep repeating the differentiability argument for f_1^{-1}. This completes the proof.

The *implicit function theorem*, as stated below, is equivalent to Theorem 7.26. To describe it, let an open set $E \subset \mathbb{R}^{k+n}$ be the domain of a continuously differentiable function $f : E \to \mathbb{R}^n$. Suppose for some $a \in \mathbb{R}^k$ and $b \in \mathbb{R}^n$, with $(a, b) \in E$, we have $f(a, b) = 0$. If $Df(a, b)$ is invertible, i.e., $\det\left(J_f(a, b)\right) \neq 0$, then there exists some ball $B(a; \epsilon_1) = U \subset \mathbb{R}^k$ and a continuous function $g : U \to \mathbb{R}^n$ such that $g(a) = b$ and, for every $u \in U$, $f(u, g(u)) = 0$.

Theorem 7.28 (Implicit Function Theorem) *For $m = k + n$, let $E \subset \mathbb{R}^m$ be an open sets, and $a \in E$. Let $f : E \to \mathbb{R}$ be a smooth map with $f(a) = b$. If $D_k f(a) \neq \mathbf{0}$,* then on some neighbourhood $U \ni a$ in E, the set of solutions to the equation $f(x) = b$ is the graph of a smooth function $x_k = g(x_1, \ldots, x_{k-1}, x_{k+1}, \ldots, x_m)$.

To conclude the chapter, we describe the *Picard-Lindelöff theorem*. For, let $V \subset \mathbb{R}^m$ and $U \subset \mathbb{R}^n$ be open sets, and let functions $f_i \in C^\infty[(t_0 - \epsilon, t_0 + \epsilon) \times V \times U]$, for some $\epsilon > 0$. In general, the theorem ensures the *existence* and *uniqueness* of a *solution* of a system of first order *initial value problem* of the form

$$\frac{dx_i}{dt} = f_i(t, x_1(t, b), \ldots, x_n(t, b); b), \quad x_i(t_0, b) = a_i, \tag{7.5.4}$$

for $i = 1, \ldots, n$. In actual applications, the vector $b = (b_1, \ldots, b_m) \in V$ represents the *parameters* of the system governed by such type of differential equation. Notice that, by attaching to (7.5.4) the following $(m + 1)$ first order differential equations

$$\frac{dt}{dt} = 1, \quad \frac{db_i}{dt} = 0, \quad 1 \le i \le m,$$

we can treat the equations given above as an *autonomous* system, without parameters. We are thus led to consider an autonomous system of the form

$$\frac{dx_i}{dt} = f_i(t, x_1(t), \ldots, x_n(t)), \quad \text{with } x(t_0) = a, \tag{7.5.5}$$

for $i = 1, \ldots, n$. Further, by shifting the *origin*, it can be assumed that $t_0 = 0$. With these simplifications, the Picard-Lindelöff theorem states as follows.

Theorem 7.29 (Picard–Lindelöff Theorem) *Let $E \subset \mathbb{R}^m$ and $D \subset \mathbb{R}^n$ be open sets, $c > 0$, and*

$$f_i \in C^\infty[(-c, c) \times E \times D], \quad 1 \le i \le n.$$

Consider the following system of ordinary differential equations:

$$\frac{dx_i}{dt} = f_i(t, x_1(t), \ldots, x_n(t)), \quad 1 \le i \le n. \tag{7.5.6}$$

Then, for any $a = (a_1, \ldots, a_n) \in U$, there exists n smooth functions $x_i = x_i(t)$ defined on a non-degenerate interval I around 0 satisfying the system (7.5.6), and the initial conditions

$$x_i(0) = a_i, \quad 1 \le i \le n. \tag{7.5.7}$$

In fact, the *uniqueness*, and *smooth dependence* on initial conditions, of a solution hold too. That is, if there are some other n smooth functions $\widehat{x}_i = \widehat{x}_i(t)$ defined on some non-degenerate interval J around 0, satisfying the system (7.5.6), and also the initial conditions (7.5.7), then $x_i = \widehat{x}_i$ on $I \cap J$, for all $1 \le i \le n$. Finally, considering the dependence of the solutions on the point a, we may write $x_i = x_i(t, a)$. and view these being defined on the set $I \times E \times D$. Then, there are neighbourhoods $U_a \subset D$, and a choice of number $\epsilon > 0$, such that the solutions $x_i = x_i(t, x)$ are defined and smooth on the open set $(-\epsilon, \epsilon) \times U_a \subset \mathbb{R}^{m+1}$. Further, as said above, it is possible to view the system (7.5.6) as an *autonomous system* without parameters.

To gain an insight into a proof of the general theorem, we prove here a simpler version. The argument used here can be extended easily to construct a proof of Theorem 7.29. Let us start with a simple illustration.

Example 7.16 For $I = [0, 1]$, the space $C(I)$ of real-valued continuous functions defined on the interval I is a complete metric space with respect to the *sup metric* d_∞ given by

$$d_\infty(f, g) := \sup \{|f(t) - g(t)| : t \in I\}.$$

Consider the function $T : C(I) \to C(I)$ given by

$$Tf(t) = 1 - \int_0^t u f(u)\, du, \quad \text{for} \ \ f \in C(I) \ \text{and} \ t \in [0, 1].$$

It then follows that

$$d_\infty(Tf, Tg) \leq (1/2) d_\infty(f, g).$$

So, by Theorem 7.25, the equation $Tf = f$ has a *unique* solution. For such a function $f \in C(I)$, since the mapping

$$t \mapsto (Tf)(t) : I \to \mathbb{R}$$

is differentiable, we conclude that the mapping $t \mapsto f(t) : I \to \mathbb{R}$ is differentiable, with

$$f'(t) = \frac{d}{dt}(Tf)(t) = -tf(t).$$

Therefore, as $f(0) = 1$, we obtain

$$f(t) = e^{-t^2/2}, \quad \text{for} \ \ t \in I.$$

In fact, it can be shown that this solution extends to $\mathbb{R}$.

We discuss here the case $m = n = 1$ of the system (7.5.5). That is, we consider the following 1-dimensional first order initial value problem[9]:

$$\frac{dx}{dt} = f(t, x), \quad \text{with} \ \ x(t_0) = x_0. \tag{7.5.8}$$

Let $D \subset \mathbb{R} \times \mathbb{R}$ be a region such as an open rectangle, containing the point (t_0, x_0), and suppose $f : D \to \mathbb{R}$ is a continuous function with *Lipschitz constant* $M > 0$. That is,

---

[9] Observe that all assertions proved in the subsequent formulation remain valid even when we take $D \subset \mathbb{R} \times \mathbb{R}^n$ by replacing $x \in \mathbb{R}$ by a vector $x \in \mathbb{R}^n$ so that assertions extends *mutatis mutandis* to a system of n - *initial value problems*. Of course, in latter case, y is a *parametrisation* of a smooth solution seen as a curve $C_y \subset \mathbb{R}^n$.

$$|f(t, x_1) - f(t, x_1)| \leq M|x_1 - x_2|, \quad \text{for all} \quad (t, x_i) \in D.$$

Notice that, by using Theorem 4.22, a differentiable function $\gamma : I \to \mathbb{R}$ provides a *local solution* to the above initial value problem if $(t, \gamma(t)) \in D$, for all $t \in I$, and γ satisfies the Eq. (7.5.8). To show the uniqueness of a *local solution* passing through the point (t_0, x_0), consider a sub-region $D' \subset D$ containing the point (t_0, x_0) such that $|f(t, x)| < B$, for some $B > 0$, and for all $(t, x) \in D'$. We may choose a number $b > 0$ such that $bB < 1$, and

$$[t_0 - b, t_0 + b] \times [x_0 - bB, x_0 + bB] \subset D'.$$

It then remains to prove the next theorem.

Theorem 7.30 *For the interval $J = [t_0 - b, t_0 + b]$, let Y be the subspace of functions $\gamma(t)$ in the complete metric space $(C(J), d_\infty)$ satisfying $d_\infty(\gamma(t), x_0) < bB$. Then, the function $F : Y \to Y$ defined by*

$$F(\gamma)(t) = x_0 + \int_{t_0}^{t} f(s, \gamma(s)) \, ds, \quad t \in J, \tag{7.5.9}$$

is a contraction.

Proof Notice that $\gamma(t)$, $t \in I$, is a *local solution* of the Eq. (7.5.8) if and only if

$$\gamma(t) = x_0 + \int_{t_0}^{t} f(s, \gamma(s)) \, ds, \quad \text{for every} \quad t \in I.$$

Moreover, $F(\gamma(t_0)) = x_0$, $F(\gamma)$ is continuous, and for every $t \in J$,

$$|F(\gamma(t)) - x_0| = \left| \int_{t_0}^{t} f(s, \gamma(s)) \, ds \right|$$
$$\leq \int_{t_0}^{t} |f(s, \gamma(s))| \, ds$$
$$\leq bB.$$

Hence, $F(\gamma) \in Y$. Finally, for $\gamma_1, \gamma_2 \in Y$, we have

$$|F(\gamma_1(t)) - F(\gamma_1(t))| = \left| \int_{t_0}^{t} (f(s, \gamma_1(s)) - f(s, \gamma_2(s))) \, ds \right|$$
$$\leq \int_{t_0}^{t} |f(s, \gamma_1(s)) - f(s, \gamma_2(s))| \, ds$$
$$\leq M \int_{t_0}^{t} |\gamma_1(s) - \gamma_2(s)| \, ds$$
$$\leq M b \, d_\infty(\gamma_1, \gamma_2)$$

That is,

$$d_\infty(F(\gamma_1(t)), F(\gamma_1(t))) \le M \, b \, d_\infty(\gamma_1, \gamma_2).$$

Therefore, F is a contraction of the complete metric space Y.

Hence, by Theorem 7.25, F has a unique fixed point, which is desired *unique solution* of the Eq. (7.5.8). Finally, to extend the local solution across D', we may write

$$J_1 = J, \quad b_1 = b, \quad t_1 = t_0 + b, \quad \text{and} \quad x_1 = \gamma(t_1),$$

and then apply Theorem 7.30 to the pair (t_1, x_1) to obtain new J_2, b_2, and the point (t_2, x_2). The solutions $\gamma_1 = \gamma$ on J_1, and γ' on J_2, both agree on an interval so that γ' provides a unique solution on the set $J_1 \cup J_2$, by a typical connectedness argument. Proceeding inductively, we find a sequence $\langle t_n \rangle$, with $t_{n+1} > t_n$, and points $(t_n, x_n) \in D'$ having all the properties as stated above. So, if D' is bounded, then

$$a_n = d_2((t_n, x_n), \partial D') \longrightarrow 0, \quad \text{as} \quad n \to \infty,$$

where $\partial D'$ denotes the boundary of the set D'. Thus, if the sequence $\langle b_n \rangle$ is chosen to satisfy the condition that

$$b_n = \min\left\{ \frac{a_n}{B^2 + 1}, \frac{1}{2M} \right\},$$

it then follows that

$$\sum_{n=1}^{\infty} a_n < \infty \quad \text{and} \quad \lim_{n \to \infty} a_n = 0,$$

because the series $\sum_{n=1}^{\infty} b_n$ is convergent. Hence, by using the fact that D is the union of an increasing sequence of sets each having the properties of D', we obtain a proof of the next theorem.

Theorem 7.31 (Picard–Lindelöf) *Let $D \subset \mathbb{R} \times \mathbb{R}$ be a region, and $f : D \to \mathbb{R}$ be a continuous function such that, for some $M > 0$,*

$$|f(t, x_1) - f(t, x_1)| \le M|x_1 - x_2|,$$

for all $(t, x_1), (t, x_2) \in D$. Then, for every $(t_0, x_0) \in D$, the Eq. (7.5.8) has a unique solution $y = \gamma(t)$, with $\gamma(t_0) = x_0$, such that the solution curve C_γ passes across D from the boundary to boundary.

Notice that the iterated solutions to (7.5.8) as obtained by above procedure form a *simple covering* of the domain D.

Exercises

7.1 Let X be a metric space. Show that a function $f : X \to \mathbb{R}^n$ has limit $(l_1, \ldots, l_n)$ as $x \to c$ if and only if each of the component function $f_k : X \to \mathbb{R}, 1 \leq k \leq n$, has limit l_k as $x \to c$.

7.2 Let X and Y be metric spaces. Show that a function $f : X \to Y$ is continuous on X if and only if for each $c \in X$ and each neighbourhood V of $f(c)$, there is a neighbourhood U of c such that $f(U) \subset V$.

7.3 Let X, Y, and Z be metric spaces. If $f : X \to Y$ and $g : Y \to Z$ are continuous functions on X and Y, respectively, then show that $g \circ f : X \to Z$, defined by $(g \circ f)(x) = g(f(x))$, is also continuous on X.

7.4 Show that L' Hospital's rule fails for some function $f : [0, 2\pi] \to \mathbb{R}^2$.

7.5 Let $f(x, y) = 5x^2 y/(x^2 + y^2)$ for $x = (x, y) \neq (0, 0)$. Find the limit of $f(x)$ as $x \to (0, 0)$, if it exists.

7.6 Test the continuity of $f : \mathbb{R}^2 \to \mathbb{R}$ defined by $f(x) = x^4 y/(x^8 + y^2)$ where $x = (x, y)$ and $f(0, 0) = 0$.

7.7 (Peano) Let $\varphi(x, y) = (x^2 - y^2)/(x^2 + y^2)$, and $f(x, y) = xt\varphi(y, x)$. Show that $D_1 D_2 f(0, 0) \neq D_2 D_1 f(0, 0)$.

7.8 Let $\varphi(x, y) = x^2 y/(x^2 + y^2)$ for $x = (x, y) \neq (0, 0)$, with $\varphi(0, 0) = 0$. Show that, at $(0, 0)$, φ is continuous and all its directional derivatives vanish, but the tangent map $\tau_{(0,0)}$ is not a linear map.

7.9 Let X and Y be metric spaces, and $f : X \to Y$ be a uniformly continuous function. Show that, if $\langle x_n \rangle$ is a Cauchy sequence in X, then $\langle f(x_n) \rangle$ is a Cauchy sequence in Y. Does the statement holds for continuous functions?

7.10 Let $1 \leq k \leq n$ be fixed. Is the kth projection function $\pi_k : \mathbb{R}^n \to \mathbb{R}$ given by $\pi_k(x_1, \ldots, x_m) = x_k$, differentiable? If yes, what is the derivative?

7.11 Find the derivative of the function $f : \mathbb{R}^n \to \mathbb{R}$ given by $f(x) = x^t A x$, where $x \in \mathbb{R}^n$ is viewed as a column vector.

7.12 Let $y \in \mathbb{R}^n$ be a fixed vector. Is the function $f : \mathbb{R}^n \to \mathbb{R}$ given by $f(x) = \langle x, y \rangle$ differentiable? Justify your answer.

7.13 Show that the derivative of $f : \mathbb{R}^n \to \mathbb{R}$ given by $f(x) = \|x\|$ is $x/\|x\|$, for any $0 \neq x \in \mathbb{R}^n$.

7.14 Is the function $f : \mathbb{R}^n \to \mathbb{R}^n$ given by $f(x) = x\|x\|$ differentiable?

7.15 If the function $f : \mathbb{R}^m \to \mathbb{R}$ satisfy $|f(x)| \leq |x|^2$, show that it is differentiable at $\mathbf{0}$.

7.16 Let $f : \mathbb{R} \to \mathbb{R}$ be an even functions, and let $f : \mathbb{R}^n \to \mathbb{R}$ be defined by $f(x) = f(\|x\|)$. If f is differentiable at 0, show that the function f is differentiable at $\mathbf{0}$.

7.17 Let $E \subseteq \mathbb{R}^m$ be an open connected set, and $\varphi : E \to \mathbb{R}^n$ be a differentiable function such that $Df(x) = \mathbf{0}$, for all $x \in E$. Prove that f is constant.

7.18 Let $\varphi(x, y) = e^{xy}$, for $(x, y) \in \mathbb{R}^2$. Write the equation of the tangent plane to the graph Γ_φ at the origin.

7.19 Let $E \subseteq \mathbb{R}^m$ be an open connected set, $\varphi : E \to \mathbb{R}$ be a differentiable function such that grad $\varphi(x) \neq 0$, and $u \in E$ be a unit vector. Show that $D_u \varphi(x)$ is *maximum* when u is in the direction of the vector $\nabla \varphi(x)$, and $D_u \varphi(x)$ is *minimum* when u is in the opposite direction, i.e., along the vector $(-\nabla \varphi(x))$.

7.20 Let $E \subseteq \mathbb{R}^m$ be an open connected set, and $\varphi : E \to \mathbb{R}$ be a differentiable function, with $\varphi(a) = c$. If $\gamma : \mathbb{R} \to \mathbb{R}^m$ is a curve lying entirely in the level set $\varphi(x) = c$, with $\gamma(t_0) = a$, show that grad $\varphi(a)$ is orthogonal to the tangent vector $\gamma'(t)$ at $t = t_0$.

7.21 Let $E \subseteq \mathbb{R}^m$ be an open connected set, and $\varphi : E \to \mathbb{R}$ be a differentiable function. Show that, for any unit vector $u \in E$, $D_u \varphi(x) = \nabla \varphi(x) \cdot u$, for all $x \in E$. [**Hint:** Apply the chain rule to the function $\gamma(t) = a + tu$, $t \in \mathbb{R}$.]

7.22 Suppose that $E \subset \mathbb{R}^m$ is an open set, and let $f : E \to \mathbb{R}^n$ be a differentiable function. Let $a, b \in E$ be such that the line segment $L[a; b]$ with end points a and b lies inside E. Show that

$$\|f(a) - f(b)\| = (b - a) \int_0^1 J_f[(1 - t)a + tb] dt.$$

7.23 If $f : \mathbb{R}^m \to \mathbb{R}^n$ is differentiable at a, prove that there exists $\delta > 0$ and $M > 0$ such that

$$x \in B(a; \delta), \ x \neq a, \ \Rightarrow \ \frac{\|f(x) - f(a)\|}{\|x - a\|} < M.$$

Hence, prove Theorem 7.16.

7.24 Let $I \subseteq \mathbb{R}$ be an interval, and $f : I \to \mathbb{R}$ be a differentiable function such that $f(t, f(t)) = 0$, for all $t \in I$ and some differentiable function $f = f(x, y)$. Show that $f'(t) = -(\partial_x f(t, f(t))/\partial_y f(t, f(t)))$, wherever the denominator is nonzero. How about the formula if $f, g : I \to \mathbb{R}$ are two differentiable functions such that

$$\begin{cases} f_1(t, f(t), g(t)) = 0 \\ f_2(t, f(t), g(t)) = 0 \end{cases}, \ \text{for} \ t \in I,$$

and $f_1 = f_1(x, y, z)$, $f_2 = f_2(x, y, z)$ are some differentiable functions.

7.25 (*Euler Theorem*) Let $E \subset \mathbb{R}^m \setminus \{0\}$ be an open set, and $\varphi \in C^1(E)$. For $a \in \mathbb{R}$, we say φ is a *homogeneous function* of degree a (or simply a-*homogeneous*) over E if $\varphi(tx) = t^a \varphi(x)$, for all $x \in E$ and $t > 0$. Show that φ is a-homogeneous if and only if $\nabla \varphi(x) \cdot x = a\varphi(x)$, for all $x \in E$.

7.26 Let $E \subset \mathbb{R}^m$ be an open set, $a \in E$, and $f : E \to \mathbb{R}^n$ be a function differentiable at a. Prove that

a. For any $\epsilon > 0$, there exists a ball $B(a; \epsilon)$ such that

$$\|f(x) - f(a)\| \leq (\|Df(a)\| + \epsilon) \|x - a\|, \quad \text{for all} \ x \in B(a; \epsilon).$$

b. For any $\epsilon > 0$, there exists a ball $B(a; \epsilon)$ such that

$$\|f(x) - f(y)\| \leq \left(\|Df(a)\| + \epsilon\right)\|x - y\|, \quad \text{for all} \quad x, y \in B(a; \epsilon).$$

References

1. J.R. Munkres, *Analysis on Manifolds* Advanced Book Program. (Addison-Wesley Publishing Company, Redwood City, CA, 1991)
2. T. Shifrin, *Multivariable Mathematics—Linear Algebra, Multivariable Calculus, and Manifolds* (Wiley, Inc., 2005)

Chapter 8
Sequences and Series of Functions

> *When I wrote this, only God and I understood what I was doing.*
> *Now, God only knows.*

Karl W. Weierstrass (1815–1897)

In Chapter 3, we discussed the convergence (or *non-convergence*) of sequences and series over the linear space $\mathbb{R}$ with respect to the *modulus metric*. More generally, in Chapter 6, we introduced certain *types of convergence* in interesting linear spaces such as $X = \mathbb{R}^n$, ℓ_p, $C([a, b])$, etc., using many different types of metrics on these spaces. We introduce here an important type of convergence for sequences (and hence of series) of bounded function $f : I \subset \mathbb{R}$, where $I \subseteq \mathbb{R}$ is some interval. In this case, we say a *convergence type* is *good* if the *pointwise limit* (or the *limit function*) of a sequence $\langle f_n \rangle$ given by

$$f(x) := \lim_{n \to \infty} f_n(x), \quad x \in I,$$

is defined, and it enjoys all the analytical properties that the functions f_n have. Our main concern in this chapter is to investigate the function f for properties such as continuity, differentiability, and integrability, where the convergence is defined with respect to the *uniform metric* d_∞ (Definition 6.13). Recall that a sequence $\langle f_n \rangle$ of bounded functions defined on an interval $I \subseteq \mathbb{R}$ is said to *converge uniformly* if it converges with respect to the uniform metric d_∞. Therefore, a series $\sum_{k=1}^{\infty} f_k$ of functions converges uniformly on the interval $I \subseteq \mathbb{R}$ if and only if the sequence $\langle s_n \rangle$ of *partial sums* given by

© The Author(s), under exclusive license to Springer Nature Singapore Pte Ltd. 2022   333
A. K. Razdan and V. Ravichandran, *Fundamentals of Analysis with Applications*,
https://doi.org/10.1007/978-981-16-8383-1_8

$$s_n(x) = \sum_{k=1}^{n} f_k(x), \qquad x \in I,$$

converges uniformly with respect to the metric d_∞. We prove here some useful *convergence theorems* for series of functions such as *Cauchy criterion, dominated convergence, Weierstrass M-test*, etc.

The ideas presented in the early part of the chapter, and the theorems proved, will be very helpful in the later part to study the uniform convergence of some *special types* of power series about a point of sufficiently smooth real-valued functions, known as the *Taylor series*, and also to deal with some pathological series of functions as in Example 8.7. As an application of the concepts developed here, in Chapter 10, we shall discuss the convergence issues related to the Fourier series representation of a sufficiently nice periodic function over $\mathbb{R}$.

8.1 Pointwise Convergence

To begin with, we remark that, in general, the equality of the iterated limits does not imply the existence of the limit of the function. Observe that for the function $f : \mathbb{R}^2 \setminus \{(0, 0)\} \to \mathbb{R}$ given by

$$f(x, y) = \frac{x^2 - y^2}{x^2 + y^2}, \quad x = (x, y) \in \mathbb{R}^2 \setminus \{(0, 0)\},$$

we have

$$\lim_{x \to 0} \left(\lim_{y \to 0} f(x, y) \right) = 1 \neq -1 = \lim_{y \to 0} \left(\lim_{x \to 0} f(x, y) \right).$$

The function has no limit when $(x, y) \to (0, 0)$, and hence the iterated limits are not equal. The situation is no different for the limit of a sequence of functions.

Let $E \subset \mathbb{R}$. We say a *sequence of functions* $f_n : E \to \mathbb{R}$ converges pointwise to a function $f : E \to \mathbb{R}$ if the sequence $\langle f_n(x) \rangle$ of real numbers converges to the real number $f(x)$, for each $x \in E$. So, the concept of pointwise convergence is a *local property* of a sequence of functions. Notice that, *no metric on a class of bounded functions defines pointwise convergence*. More generally, we have the following definition.

Definition 8.1 (Pointwise Convergence) Let $E \subset \mathbb{R}^m$. A sequence of functions $f_n : E \to \mathbb{R}$ *converges pointwise* to a function $f : E \to \mathbb{R}$ if, for each $\epsilon > 0$ and for each $x \in E$, there is a natural number $N = N(\epsilon, x)$ such that

$$n \geq N \quad \Rightarrow \quad |f_n(x) - f(x)| < \epsilon.$$

In this case, we write $\lim_{n \to \infty} f_n = f$ or $f_n \to f$ as $n \to \infty$, pointwise over E, and call the function f a *pointwise limit* of the sequence $\langle f_n \rangle$ over the set E.

Fig. 8.1 Graphs of
$f_n(x) = x^n$, for $n = 1, 5, 15$

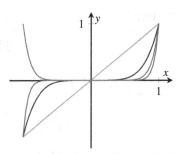

The next few examples will illustrate why pointwise convergence is not good for analytic properties such as continuity, differentiability, and integrability. Notice that, if a sequence $f_n : E \to \mathbb{R}$ of continuous function converges pointwise to a function $f : E \to \mathbb{R}$, then f is continuous at a point $x = x_0 \in E$ if and only if $\lim_{x \to x_0} f(x) = f(x_0)$. We have

$$\lim_{x \to x_0} f(x) = \lim_{x \to x_0} \lim_{n \to \infty} f_n(x) \quad \text{and} \quad \lim_{n \to \infty} \lim_{x \to x_0} f_n(x) = f(x_0).$$

So, the limit function f is continuous at $x = x_0$, provided

$$\lim_{x \to x_0} \lim_{n \to \infty} f_n(x) = \lim_{n \to \infty} \lim_{x \to x_0} f_n(x).$$

Therefore, continuity fails if interchanging the order of limits does not give the same value. For convenience, we consider the case when $E \subseteq \mathbb{R}$ is an interval.

Example 8.1 For the sequence of functions $f_n : (-1, 1] \to \mathbb{R}$ given by

$$f_n(x) = x^n, \quad x \in (-1, 1],$$

we have

$$\lim_{n \to \infty} f_n(x) = \begin{cases} 0 & (x \neq 1) \\ 1 & (x = 1) \end{cases} \quad \text{and} \quad \lim_{x \to 1-} f_n(x) = 1.$$

(See Fig. 8.1) Therefore, it follows that

$$\lim_{x \to 1-} \left(\lim_{n \to \infty} f_n(x) \right) = 0 \neq 1 = \lim_{n \to \infty} \left(\lim_{x \to 1-} f_n(x) \right).$$

Example 8.2 We show here the limit function of a sequence of continuous functions is not necessarily continuous. Consider the sequence of functions $f_n \in C(\mathbb{R})$ given by

$$f_n(x) = \frac{nx^2 - 1}{nx^2 + 1}, \quad x \in \mathbb{R}.$$

We then have

$$f(x) := \lim_{n \to \infty} f_n(x) = \begin{cases} 1 & (x \neq 0) \\ -1 & (x = 0) \end{cases},$$

and $\lim_{x \to 0} f_n(x) = -1$. It thus follows that

$$\lim_{x \to 0} \lim_{n \to \infty} f_n(x) = 1 \neq -1 = \lim_{n \to \infty} \lim_{x \to 0} f_n(x);$$

Notice the limit function f is not a continuous function.

The next example shows that the pointwise limit of a sequence of continuous functions could be an everywhere discontinuous function.

Example 8.3 Let $n \geq 0$, and $f_n : \mathbb{R} \to \mathbb{R}$ be the sequence of functions given by

$$f_n(x) = \lim_{m \to \infty} (\cos(n!\pi x))^{2m}, \quad x \in \mathbb{R}.$$

As the limits over n and m commute, each f_n is a continuous function. Let $f : \mathbb{R} \to \mathbb{R}$ be the limit function. That is,

$$f(x) = \lim_{n \to \infty} f_n(x), \quad x \in \mathbb{R}.$$

Recall that $\cos \pi x = \pm 1$ if and only if $x \in \mathbb{Z}$. However, if $x \in \mathbb{Q}$ then $n!x$ can never an integer, and hence

$$0 \leq \alpha := |\cos(n!\pi x)| < 1.$$

This shows that

$$f_n(x) = \lim_{m \to \infty} \alpha^{2m} = 0 \quad \text{and} \quad f(x) = 0.$$

If $x = p/q \in \mathbb{Q}$ then $n!\pi p/q \in \mathbb{Z}$ when $n \geq q$, and so

$$f_n(x) = \lim_{n \to \infty} (\pm 1)^{2m} = 1, \quad \text{for } n \geq q.$$

This shows that

$$f(x) = \lim_{n \to \infty} f_n(x) = 1, \quad \text{for all } x \in \mathbb{Q}.$$

It, thus, follows that

$$f(x) = \begin{cases} 1, & \text{for } (x \in \mathbb{Q}) \\ 0, & \text{for } (x \in \mathbb{R} \setminus \mathbb{Q}). \end{cases}$$

It follows easily that $f(x)$ is *everywhere discontinuous* function (Example 4.5).

Fig. 8.2 Graphs of
$f_n(x) = \sin(nx)/n$, for
$n = 1, 10, 20$

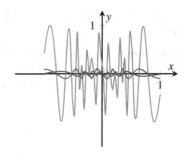

Example 8.4 We show here a sequence of first derivatives of differentiable functions may not converge to the derivative of the limit function. Let $n \geq 1$, and $f_n : \mathbb{R} \to \mathbb{R}$ be the sequence of functions given by

$$f_n(x) = \frac{\sin(nx)}{n}, \quad x \in \mathbb{R}.$$

Let $f : \mathbb{R} \to \mathbb{R}$ be the limit function. That is,

$$f(x) := \lim_{n \to \infty} f_n(x), \quad x \in \mathbb{R}.$$

Then, we have

$$0 \leq |f_n(x)| \leq 1/n, \quad \text{for all} \quad x \in \mathbb{R},$$

and hence $f_n(x) \to 0$ as $n \to \infty$. So, $f(x) = 0$ for all $x \in \mathbb{R}$ (see Fig. 8.2). Clearly $f'(x) = 0$ and $f'_n(x) = \cos(nx)$. Therefore,

$$f'_n(0) = 1 \to 1 \neq 0 = f'(0).$$

Example 8.5 We show here that the integral of the limit function may not be same as the limit of the integral of a sequence of integrable functions. Let $n \in \mathbb{N}$, and $f_n : [0, 1] \to \mathbb{R}$ be the sequence of functions given by

$$f_n(x) = \begin{cases} n & (0 \leq x \leq 1/n) \\ 0 & (1/n < x \leq 1). \end{cases}$$

Clearly then, $f(x) := \lim_{n \to \infty} f_n(x) = 0$. Also,

$$\int_0^1 f_n(x)dx = \int_0^{1/n} n \, dx = 1 \neq 0 = \int_0^1 f(x)dx.$$

Theorem 8.1 *The limit function of a sequence in the space $\mathcal{R}([a, b])$ may not be Riemann integrable.*

Proof Left for the reader as Exercise 8.3. □

Considering Definition 8.1, we define the pointwise convergence of a series of functions as follows.

Definition 8.2 (Pointwise Convergence of Series) Let $E \subseteq \mathbb{R}$, and $f_k : E \to \mathbb{R}$ be a sequence of functions. We say the series $\sum_{k=0}^{\infty} f_k$ *converges pointwise* to a function $f : E \to \mathbb{R}$ if for each $x \in E$ the sequence of partial sums given by

$$s_n(x) := \sum_{k=0}^{n} f_k(x), \quad n \geq 0,$$

converges pointwise to the function f. In this case, we write

$$f(x) = \sum_{k=0}^{\infty} f_k(x), \quad \text{for} \quad x \in E.$$

The next example shows that a series of continuous functions may not converge to a continuous function.

Example 8.6 Let $n \geq 0$, and $f_n : \mathbb{R} \to \mathbb{R}$ be the sequence of continuous functions given by

$$f_n(x) = \frac{x}{(1 + |x|)^n}, \quad x \in \mathbb{R}.$$

Then, we have

$$f(x) = \sum_{k=0}^{\infty} f_k(x) = \begin{cases} x + 1 & (x > 0) \\ 0 & (x = 0) \\ x - 1 & (x < 0) \end{cases},$$

which is clearly discontinuous at $x = 0$.

All the illustrations discussed above point to the fact that, in order to do analysis with sequences and series of functions, we need to replace the notion of pointwise convergence with a concept that provides a stronger form of convergence so that the limit function has every analytical property like continuity, differentiability, or integrability, if so have the terms of the sequence. Further, in case of series of functions, we should be able to perform *termwise differentiation* and *termwise integration*.

8.2 Uniform Convergence

The central idea here is that we are now considering a *global convergence* for a sequence of functions in the sense that the number $N \in \mathbb{N}$ as in Definition 8.1 is independent of the point $x \in E$. Notice that, as we are dealing all through with *sup metric*, every sequence of functions considered here is from a class of bounded functions defined on a common domain $E \subseteq \mathbb{R}$.

Definition 8.3 (Uniform Convergence) A sequence of functions $f_n : E \to \mathbb{R}$ is called *uniformly convergent* if there is some function $f : E \to \mathbb{R}$ such that, for any $\epsilon > 0$ and some $N = N(\epsilon) \in \mathbb{N}$, we have

$$n \geq N \quad \Rightarrow \quad |f_n(x) - f(x)| < \epsilon, \quad \text{for all} \quad x \in E.$$

In this situation, we write u-$\lim_{n \to \infty} f_n = f$ or $f_n \rightrightarrows f$ as $n \to \infty$.

We know that every convergent sequence is Cauchy, in general. In particular, for a sequence of functions $\langle f_n \rangle$, the concept of *uniformly Cauchy* is a useful way to deal with uniform convergence of $\langle f_n \rangle$.

Definition 8.4 Let $E \subseteq \mathbb{R}$. A sequence of functions $f_n : E \to \mathbb{R}$ is said to be **uniformly Cauchy** if for a given $\epsilon > 0$ there is some $N = N(\epsilon) \in \mathbb{N}$ such that, for all $x \in E$,

$$n, m \geq N \quad \Rightarrow \quad |f_n(x) - f_m(x)| < \epsilon.$$

Theorem 8.2 (Cauchy Criterion) *Let $E \subseteq \mathbb{R}$. A sequence of functions $f_n : E \to \mathbb{R}$ converges uniformly to a function $f : E \to \mathbb{R}$ if and only if it is uniformly Cauchy.*

Proof Let $f_n \rightrightarrows f$. Let $\epsilon > 0$ be given. Choose a positive integer N such that

$$|f_n(x) - f(x)| < \frac{\epsilon}{2} \quad \text{for all } n \geq N \text{ and for all } x \in E.$$

Then, for $n, m \geq N$, we have

$$\begin{aligned}
|f_n(x) - f_m(x)| &\leq |(f_n(x) - f(x)) - (f_m(x) - f(x))| \\
&\leq |f_n(x) - f(x)| + |f_m(x) - f(x)| \\
&< \frac{\epsilon}{2} + \frac{\epsilon}{2} = \epsilon.
\end{aligned}$$

This proves that $\langle f_n \rangle$ is uniformly Cauchy. Conversely, assume that $\langle f_n \rangle$ is uniformly Cauchy. For a given $\epsilon > 0$, choose a positive integer N such that

$$|f_n(x) - f_m(x)| < \frac{\epsilon}{2} \quad \text{for all } n, m \geq N \text{ and for all } x \in E. \tag{8.2.1}$$

For each fixed $x \in E$, the sequence $\langle f_n(x) \rangle$ is Cauchy, and hence, there is a real number $f(x)$ such that $f_n(x) \to f(x)$. This defines a function $f : E \to \mathbb{R}$. We show that $f_n \rightrightarrows f$. Let $m \to \infty$ in (8.2.1) to get, for $n \geq N$ and for all $x \in E$,

$$|f_n(x) - f(x)| \leq \frac{\epsilon}{2} < \epsilon.$$

This proves that $f_n \rightrightarrows f$. $\qquad \square$

The next theorem proves uniform convergence preserves continuity.

Theorem 8.3 *If a sequence* $f_n \in C(E)$ *converges uniformly to the limit function* $f : E \to \mathbb{R}$ *then* $f \in C(E)$.

Proof Let $\epsilon > 0$ be given, and $x_0 \in E$ be fixed. By the uniform convergence of $\langle f_n \rangle$, choose a positive integer N such that

$$|f_n(x) - f(x)| < \frac{\epsilon}{3} \quad \text{for all } n \geq N \text{ and for all } x \in E.$$

By using continuity of f_N, choose a $\delta > 0$ such that

$$|x - x_0| < \delta, x \in E \Rightarrow |f_N(x) - f_N(x_0)| < \frac{\epsilon}{3}.$$

With these inequalities, we see that, for $x \in E$ with $|x - x_0| < \delta$,

$$
\begin{aligned}
|f(x) - f(x_0)| \\
&= |(f(x) - f_N(x)) + (f_N(x) - f_N(x_0)) + (f_N(x_0) - f(x_0))| \\
&\leq |f_N(x) - f(x)| + |f_N(x) - f_N(x_0)| + |f_N(x_0) - f(x_0)| \\
&< \frac{\epsilon}{3} + \frac{\epsilon}{3} + \frac{\epsilon}{3} \\
&= \epsilon.
\end{aligned}
$$

Therefore, the function f is continuous at $x = x_0$. Since $x_0 \in E$ is arbitrary, the function f is continuous on E. $\qquad\square$

Example 8.7 Let the function $\phi : [-1, 1] \to \mathbb{R}$ be given by $\phi(x) = |x|$, and extend it to $\mathbb{R}$ by taking $\phi(x + 2) = \phi(x)$, for all $x \in \mathbb{R}$. Clearly, the extended function is periodic of period 2. It is called a *sawtooth function* (see Fig. 8.3). Then the series

$$f(x) := \sum_{k=0}^{\infty} \left(\frac{3}{4}\right)^n \phi(4^n x)$$

converges uniformly as each of the terms of the series satisfies

$$\left|\left(\frac{3}{4}\right)^n \phi(4^n x)\right| \leq \left(\frac{3}{4}\right)^n \quad \text{and} \quad \sum_{k=0}^{\infty} \left(\frac{3}{4}\right)^n \text{ converges.}$$

Since each term is continuous and the convergence is uniform, the function f is continuous on $\mathbb{R}$. It can be shown that this function is nowhere differentiable. The reader can fill the details by referring to any standard text on real analysis.

As a simple application of Theorem 8.3, we prove the following important theorem about the completeness of the space $C(E)$.

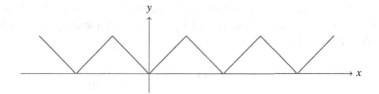

Fig. 8.3 The graph of the $\phi(x)$ as in Example 8.7

Theorem 8.4 *For any compact set $E \subset \mathbb{R}$, the space $C(E)$ of continuous functions $f : E \to \mathbb{R}$ is complete with respect to the sup metric.*

Proof Suppose $\langle f_n \rangle$ is a Cauchy sequence in the space $C(E)$. Then, for any $\epsilon > 0$, there is some $N \in \mathbb{N}$ such that

$$k, \ell \geq N \quad \Rightarrow \quad \sup_{y \in E} |f_k(y) - f_\ell(y)| < \epsilon. \tag{8.2.2}$$

Also, for any fixed $x \in E$, we have

$$k, \ell \geq N \quad \Rightarrow \quad \|f_k(x) - f_\ell(x)\| \leq \sup_{y \in E} \|f_k(y) - f_\ell(y)\| < \epsilon. \tag{8.2.3}$$

That is, the sequence of real numbers $\langle f_n(x) \rangle$ is a Cauchy sequence. By completeness of $\mathbb{R}$, we obtain the limit function $f : E \to \mathbb{R}$ given by

$$f(x) := \lim_{n \to \infty} f_n(x), \quad x \in E.$$

Notice that, with ϵ and N as above, we have

$$m \geq N \quad \Rightarrow \quad |f(x) - f_m(x)| = \lim_{n \to \infty} |f_n(x) - f_m(x)| < \epsilon,$$

for all $x \in E$. That is,

$$\sup_{x \in E} \|f(x) - f_n(x)\| < \epsilon.$$

Therefore, $f_n \rightrightarrows f$ as $n \to \infty$. As $f \in C(E)$ by Theorem 8.3, proof is complete. $\square$

Notice that the same proof works to show that the space $B(E)$ of bounded functions $f : E \to \mathbb{R}$ is complete with respect to the *sup metric*. The next theorem proves uniform convergence preserves integrability.

Theorem 8.5 *Suppose $\langle f_n \rangle$ is sequence in $\mathscr{R}([a, b])$ converges uniformly to the limit function $f : [a, b] \to \mathbb{R}$. Then $f \in \mathscr{R}([a, b])$, and*

$$\int_a^b f(x)dx = \lim_{n \to \infty} \int_a^b f_n(x)dx. \tag{8.2.4}$$

Proof A proof of the first assertion is left for the reader. So, we assume that $f \in \mathcal{R}([a, b])$. To prove the second assertion, let $f_n \rightrightarrows f$. Then, for any given $\epsilon > 0$, we choose $N \in \mathbb{N}$ such that, for all $x \in E$,

$$n \geq N \quad \Rightarrow \quad |f_n(x) - f(x)| < \frac{\epsilon}{2(b-a)}.$$

That is, for $n \geq N$, we have

$$\left| \int_a^b f_n(x) dx - \int_a^b f(x) dx \right| \leq \left| \int_a^b (f_n(x) - f(x)) dx \right|$$

$$\leq \int_a^b |f_n(x) - f(x)| \, dx$$

$$\leq \frac{\epsilon}{2(b-a)} (b-a)$$

$$< \epsilon,$$

which implies $\int_a^b f_n(x) dx$ converges to $\int_a^b f x) dx$. This completes the proof. $\qquad\square$

Notice that the Eq. (8.2.4) is equivalent to

$$\lim_{n \to \infty} \int_a^b f_n(x) dx = \int_a^b \lim_{n \to \infty} f_n(x) dx,$$

so the limit and integration can be interchanged for a sequence of Riemann integrable functions, provided the convergence is uniform.

As before, for the series of functions $\sum_{k=1}^{\infty} f_k$, uniform convergence can be defined in terms of its partial sums. That is, a series $\sum_{k=1}^{\infty} f_k$ converges uniformly to function f if $\sum_{k=1}^{n} f_k \rightrightarrows f$.

Theorem 8.6 *Let $f_k : E \to \mathbb{R}$ be a sequence of functions. The series $\sum_{k=0}^{\infty} f_k(x)$ converges uniformly if for every $\epsilon > 0$ there is some $N \in \mathbb{N}$ such that*

$$n > m \geq N \quad \Rightarrow \quad \sum_{k=m+1}^{n} |f_k(x)| < \epsilon, \quad \text{for all} \ \ x \in E.$$

Proof Consider the sequence partial sums $s_n(x) = \sum_{k=0}^{n} f_k(x)$ of the series $\sum_{k=0}^{\infty} f_k(x)$. Clearly then, for $n > m$,

$$|s_n(x) - s_m(x)| = \left| \sum_{k=m+1}^{n} f_k(x) \right| \leq \sum_{k=m+1}^{n} |f_k(x)|.$$

So, by the given hypothesis, for any $\epsilon > 0$ there is some $N \in \mathbb{N}$ such that

$$n > m \geq N \quad \Rightarrow \quad \left|s_n(x) - s_m(x)\right| < \epsilon, \quad \text{for all } x \in E,$$

and hence for all $n, m \geq N$. This shows that the sequence $\langle s_n \rangle$ of functions converges uniformly. Therefore, the series $\sum_{k=0}^{\infty} f_k(x)$ converges uniformly. $\square$

Theorem 8.7 (Dominated Convergence) *Let $f_n, g_n : E \to \mathbb{R}$ be functions with $g_n(x) \geq 0$ and $|f_n(x)| \leq g_n(x)$. If the series $\sum_{k=0}^{\infty} g_k(x)$ converges uniformly, then the series $\sum_{k=0}^{\infty} f_k(x)$ converges uniformly.*

Proof Let $s_n(x)$ and $t_n(x)$, respectively, be the sequences of partial sums of the series $\sum_{k=0}^{\infty} f_k(x)$ and $\sum_{k=0}^{\infty} g_k(x)$. Then, for $n > m$, we have

$$\left|s_n(x) - s_m(x)\right| = \left| \sum_{k=m+1}^{n} f_k(x) \right| \leq \sum_{k=m+1}^{n} \left| f_k(x) \right|$$

$$\leq \sum_{k=m+1}^{n} g_k(x)$$

$$= t_n(x) - t_m(x)$$

$$= \left| t_n(x) - t_m(x) \right|,$$

which proves the assertion. $\square$

Corollary 8.1 (Weierstrass M-test) *Let $f_n : E \to \mathbb{R}$ be functions with $|f_n(x)| \leq M_n$. If the series $\sum_{k=0}^{\infty} M_k$ converges, then the series $\sum_{k=0}^{\infty} f_k(x)$ converges uniformly.*

Proof Take $g_n(x) = M_n$ in the previous theorem. Notice that, in this case, uniform convergence is the same as the convergence of the series (of numbers). $\square$

8.3 Power Series

A series of the form

$$\sum_{k=0}^{\infty} a_k(x - a)^k = a_0 + a_1(x - a) + a_2(x - a)^2 + \cdots \qquad (8.3.1)$$

is called a *power series* about the point $x = a$, and a_k is called the kth coefficient of the series. Notice that the series converges at $x = a$, trivially, and it may be the only point where it may converge. On the other hand, the series like

$$\sum_{k=0}^{\infty} \frac{(x - a)^k}{k!}$$

converges for all $a \in \mathbb{R}$, and the geometric series $\sum_{k=0}^{\infty} (x - a)^k$ converges in the interval $|x - a| < 1$. In general, by applying Theorem 3.29 with $x_n = a_n(x - a)^n$, we conclude that the series converges (absolutely) for all x in the interval

$$|x - a| < R := \lim_{n \to \infty} \left| \frac{a_n}{a_{n+1}} \right|$$

, and it diverges for x in the region $|x - a| > R$. Also, by applying Theorem 3.30 with $x_n = a_n(x - a)^n$, it follows that for

$$\frac{1}{R} := \lim_{n \to \infty} |a_n|^{1/n},$$

the series (8.3.1) converges (absolutely) for all x in the interval $|x - a| < R$ and it diverges if $|x - a| > R$. Therefore, for some $R \geq 0$, a power series about a point $x = a$ converges in an interval $|x - a| < R$ and diverges at every point outside it.

Definition 8.5 The number $R \geq 0$ with the above-stated property is called the *radius of convergence* (or simply ROC) of the power series.

To derive a general formula for R, we shall use the concept of *limit superior* of a sequence, as introduced in Chapter 3.

Theorem 8.8 (Hadamard) *The radius of convergence R of a power series about a point a, with coefficients a_n, is given by*

$$\frac{1}{R} = \limsup_{n \to \infty} |a_n|^{1/n}.$$

Proof Left for the reader as an exercise.   □

As mentioned above, the ratio and root tests are two convenient ways to find the radius of convergence of a power series, provided the limit in Theorem 8.8 exists.

Example 8.8 The series $\sum_{n=0}^{\infty} n^n x^n$ converges only at $x = 0$ as

$$R = \frac{1}{\lim_{n \to \infty} |n^n|^{1/n}} = 0.$$

The series $\sum_{n=0}^{\infty} x^n$ converges for $|x| < 1$ as $R = 1$. The series $\sum_{n=0}^{\infty} x^n/n!$ converges for all $x \in \mathbb{R}$ as

$$R = \lim_{n \to \infty} (n!)^{1/n} = \infty.$$

To see the later limit is ∞, notice that there are at least $n/2$ integers in $\{1, \ldots, n\}$ that are greater than or equal to $n/2$. This shows that

$$n! \geq \left(\frac{n}{2} \right)^{n/2}$$

and this immediately implies that

$$(n!)^{1/n} \geq \sqrt{\frac{n}{2}} \to \infty.$$

Hence, $\lim_{n \to \infty} (n!)^{1/n} = \infty$.

Theorem 8.9 (Taylor's Theorem with Integral Remainder) *Let $f : [a, b] \to \mathbb{R}$ be a function such that $f \in C^n([a, b])$ and $f^{(n+1)}(x) \in \mathcal{R}([a, b])$. Then, for any $x_0 \in [a, b]$,*

$$f(x) = \sum_{k=0}^{n} \frac{f^{(k)}(x_0)}{k!} (x - x_0)^k + \int_{x_0}^{x} \frac{f^{(n+1)}(t)}{n!} (x - t)^n \, dt.$$

Proof Let $x_0 \in [a, b]$. Then, we can write

$$f(x) = f(x_0) + \int_{x_0}^{x} f'(t) \, dt = f(x_0) + \int_{x_0}^{x} f'(t) \, d(t - x).$$

Integrating by parts, we get

$$f(x) = f(x_0) + f'(x_0)(x - x_0) + \int_{x_0}^{x} (t - x) f''(t) dt.$$

Let R_m be defined by

$$R_m(x) = \int_{x_0}^{x} \frac{f^{(m+1)}(t)}{m!} (x - t)^m \, dt.$$

Then

$$\begin{aligned}
R_m(x) &= -\int_{x_0}^{x} \frac{f^{(m+1)}(t)}{m!} d \frac{(x - t)^{m+1}}{m+1} \\
&= -\frac{f^{(m+1)}(t)}{m!} \frac{(x - t)^{m+1}}{m+1} \Big|_{t=x_0}^{x} + \int_{x_0}^{x} \frac{f^{(m+2)}(t)}{m!} \frac{(x - t)^{m+1}}{m+1} \, dt \\
&= \frac{f^{(m+1)}(t)}{m!} \frac{(x - x_0)^{m+1}}{m+1} + \int_{x_0}^{x} \frac{f^{(m+2)}(t)}{(m+1)!} (x - t)^{m+1} \, dt \\
&= \frac{f^{(m+1)}(t)}{m!} \frac{(x - x_0)^{m+1}}{m+1} + R_{m+1}.
\end{aligned}$$

Using this, the result follows by an application of the principle of mathematical induction. $\qquad\square$

Theorem 8.10 (Taylor's Theorem with Lagrange's Form of Remainder) *Let $f : [a, b] \to \mathbb{R}$ be a function and $x_0 \in [a, b]$. If $f^{(n)}(x)$ is continuous on $[a, b]$ and*

$f^{(n+1)}$ *is continuous on* $[a, b]$, *then there is a c* $= c(x) \in [a, b]$ *such that*

$$f(x) = \sum_{k=0}^{n} \frac{f^{(k)}(x_0)}{k!}(x - x_0)^k + \frac{f^{(n+1)}(c)}{(n+1)!}(x - x_0)^{n+1}.$$

Proof The weighted mean value theorem (Theorem 5.24) states that there exists $c \in (a, b)$ such that $\int_a^b f(x)g(x)dx = f(c)\int_a^b g(x)dx$ for continuous functions $f, g : [a, b] \to \mathbb{R}$. By using this, we see that

$$R_n(x) = \int_{x_0}^{x} \frac{f^{(n+1)}(t)}{n!}(x - t)^n \, dt = \frac{f^{(n+1)}(c)}{n!} \frac{(x - x_0)^{n+1}}{n+1}.$$

$\square$

Theorem 8.11 (Taylor's series) *Let* $f : [a, b] \to \mathbb{R}$ *be a function and* $x_0 \in [a, b]$. *If f has derivative of all orders on* $[a, b]$ *and* $|f^{(n)}(x)| \le M^n$ *for some* $M \ge 0$ *and for all* $x \in [a, b]$, *then*

$$f(x) = \sum_{k=0}^{\infty} \frac{f^{(k)}(x_0)}{k!}(x - x_0)^k.$$

Proof The remainder term in Theorem 8.10 given by

$$R_n(x) = \frac{f^{(n+1)}(c)}{(n+1)!}(x - x_0)^{n+1}$$

satisfies

$$|R_n(x)| \le \frac{M^{n+1}|x - x_0|^{n+1}}{(n+1)!} \to 0,$$

as $n \to \infty$. The last expression is the general term in the Taylor series of $e^{M|x-x_0|}$ and so it goes to 0 as $n \to \infty$. $\square$

The series given in Theorem 8.11 is known as Taylor's series of f around $x = x_0$. It should be noted that Taylor's series of functions having derivatives of all orders need not be equal to the function. For example, consider the function $f : [0, \infty) \to \mathbb{R}$ defined by $f(0) = 0$ and $f(x) = e^{-1/x^2}$ for $x > 0$.

In actual applications, we mostly deal with unknown functions $y = f(x)$, with values given only at finitely many values of the argument x. The first approach in all such situations is to approximate the function f using a class of functions that are known to have certain *nice properties*. Of course, we start with *simple polynomials* involving only powers. In doing so, Taylor's theorem proves to be very helpful. As said above, it states that if a function $y = f(x)$ is *sufficiently smooth* over a small interval (a, b) around a point x_0 then the polynomials defined by a sequence of partial sums of its Taylor series provide an approximation of the function $y = f(x)$ at each point $x \in (a, b)$.

Clearly, this *approximation method* is very limited in its scope, and the approximation *error* is too high. Instead, for a real-valued function $f \in C([a, b])$, the next theorem guarantees that the function f can be approximated by a sequence of polynomials. It proves the possibility of *uniform approximation*, and uses the following sequence of *Bernstein polynomials*

$$B_n(f, x) = \sum_{k=0}^{n} f\left(\frac{k}{n}\right) \binom{n}{k} x^k (1 - x)^{n-k}, \quad x \in [0, 1].$$

However, the *convergence* with Bernstein polynomials is very slow.

Theorem 8.12 (Weierstrass) *Given $f \in C([a, b])$, and $\epsilon > 0$, there exists some $n = n(\epsilon) \in \mathbb{N}$ and a polynomial $p(x)$ of degree n such that $|f(x) - p(x)| < \epsilon$, $\forall x \in [a, b]$.*

Proof Left to the reader as an exercise. □

In general, *Sturm-Liouville theory* provides a general framework to study some very effective methods of approximating piecewise smooth functions. The concept of the Fourier series discussed in Chapter 10 is a special case of this profound theory.

Exercises

8.1 Suppose $f_n : [a, b] \to \mathbb{R}$ is a monotonically increasing sequence converging poitwise to $f \in C([a, b])$. Show that the convergence is uniform.

8.2 Suppose $f \in C([0, \infty))$ is such that $f(nx) \to 0$ as $n \to \infty$, for any $x \geq 0$. Prove that $f(x) \to 0$ as $x \to \infty$.

8.3 Verify whether the double sequence $a_{m,n} = m/(m + n)$ satisfies

$$\lim_{n \to \infty} \lim_{m \to \infty} a_{m,n} = \lim_{m \to \infty} \lim_{n \to \infty} a_{m,n}.$$

8.4 Find the sum of the series of continuous functions $\sum_{n=0}^{\infty} x^2/(1 + x^2)^n$. Is the sum continuous in $\mathbb{R}$?

8.5 Supply an example to prove the assertion in Theorem 8.1.

8.6 For $n \geq 1$, let $f_n : [0, 1] \to \mathbb{R}$ be given by $f_n(x) = n^2 x(1 - x^2)^n$. Verify whether $\lim_{n \to \infty} \int_0^1 f_n(x)dx = \int_0^1 \lim_{n \to \infty} f_n(x)dx$.

8.7 For $n \geq 1$, let $f_n : [0, 1] \to \mathbb{R}$ be given by $f_n(x) = 1/(1 + nx^2)$ and let $f(x) = \lim_{n \to \infty} f_n(x)$. Verify whether $\lim_{n \to \infty} f_n'(x) = f'(x)$.

8.8 For $n \geq 1$, let $f_n : [0, 1] \to \mathbb{R}$ be given by $f_n(x) = nx/(n + x)$. Verify whether $f_n(x) \rightrightarrows x$.

8.9 For $n \geq 1$, let $f_n : [0, 1] \to \mathbb{R}$ be given by $f_n(x) = \sqrt{x^2 + 1/n^2}$. Verify whether $f_n(x) \rightrightarrows x$.

8.10 Does the sequence of functions $f_n : [1, \infty) \to \mathbb{R}$, defined by $f_n(x) = (x/n)^n e^{-x}$, converge uniformly?

8.11 Prove that $f_n : \mathbb{R} \to \mathbb{R}$ defined by $f_n(x) = \sin(nx)/nx$ converges uniformly on $[1, \infty)$ but not on $(0, 1)$.

8.12 For $n \geq 1$, let $f_n : [0, 1] \to \mathbb{R}$ be given by $f_n(x) = x^n(1 - x^n)$. Show that $f_n(x) \rightrightarrows 0$.

8.13 Suppose that $f : \mathbb{R} \to \mathbb{R}$ is uniformly continuous and let $f_n(x) = f(x + 1/n)$. Prove that $f_n \rightrightarrows f$.

8.14 Prove that the function $f(x)$ defined in Example 8.7 is nowhere differentiable.

8.15 Give a proof of Theorem 8.8.

8.16 Let the function $f : [0, \infty) \to \mathbb{R}$ be defined by $f(0) = 0$ and $f(x) = e^{-1/x^2}$ for $x > 0$. Show that $f^{(n)}(0) = 0$ for all n. Does the Taylor's series around $x = 0$ equals the given function f?

8.17 Give a proof of Theorem 8.12.

8.18 Consider the piecewise linear function

$$f(x) := \begin{cases} x, & \text{for } 0 \leq x \leq 1/2 \\ 1 - x, & \text{for } 1/2 \leq x \leq 1 \end{cases},$$

and extend it periodically to $\mathbb{R}$ by setting $f(x + 1) = f(x)$ for all $x \in \mathbb{R}$. Show that the function

$$g(x) := \sum_{n=0}^{\infty} \frac{1}{2^n} g(2^n x), \quad x \in \mathbb{R},$$

is nowhere differentiable continuous function.

8.19 Find a sequence $\langle f_n \rangle$ in $C([0, 1])$ such that $0 \leq f_n(x) \leq 1$, and

$$\int_0^1 f_n(x)dx \to 0, \quad \text{as } n \to \infty,$$

but the sequence $\langle f_n(x) \rangle$ in $\mathbb{R}$ doesn't converge for any $x \in [0, 1]$.

8.20 Show that the series

$$\sum_{n=0}^{\infty} \frac{(-1)^n}{n!} x^n [\log x]^n, \quad x \in (0, 1],$$

converges uniformly, and hence deduce that

$$\int_0^1 x^{-x}dx = \frac{1}{1^1} + \frac{1}{2^2} + \frac{1}{3^3} + \cdots + \frac{1}{n^n} + \cdots.$$

Chapter 9
Measure and Integration

The only teaching that a professor can give, in my opinion, is that of thinking in front of his students.

Henri Léon Lebesgue (1875–1941)

We introduce here a *set function m* defined over a class $\mathscr{M}$ of subsets $A \subseteq \mathbb{R}$, called *Lebesgue measurable set.* We prove m is an extension of the usual *length function* ℓ defined on the class of intervals (Theorem 9.9), which is ultimately used to define the concept of the integral $\int_A f \, dm$ of a real-valued *"measurable function"* f over a set $A \in \mathscr{M}$. A set $M \in \mathscr{M}$ is called a *measurable set*[1] in $\mathbb{R}$, and the extended real-valued set function $m : \mathscr{M} \to [0, \infty]$ is the *measure* on $\mathbb{R}$. As seen in Chap. 4, the Riemann integral $\int_I f(x) \, dx$ of a real-valued *bounded* function f defined on a bounded interval $I \subset \mathbb{R}$ uses the *lengths* $\ell(I_k)$ of intervals in (ordered) *tagged partitions* $\{(I_k, t_k)\}$ of the interval I to *approximate* the *area under the curve*, if it exists, by using the sums of *rectangular areas* of the form

$$\sum_{t_k \in I_k} f(t_k)\ell(I_k), \quad \text{taking the limit over all such partitions as} \quad \ell(I_k) \to 0.$$

A far-reaching theory of integration, introduced by the French mathematician *Henri Lebesgue* (1875–1941), differs fundamentally in many ways from the Riemann's theory. Unlike Riemann, Lebesgue considered a subdivision

$$\min(f) \le y_0 = f(a) < y_1 < \cdots < y_{n-1} < y_n = f(b) \le \max(f),$$

---

[1] Very roughly, we say $A \subseteq \mathbb{R}$ is *measurable* if it can be covered by a countable collection of intervals of arbitrarily small length.

© The Author(s), under exclusive license to Springer Nature Singapore Pte Ltd. 2022
A. K. Razdan and V. Ravichandran, *Fundamentals of Analysis with Applications*,
https://doi.org/10.1007/978-981-16-8383-1_9

of the *range* of f, and used the sums of the form

$$\sum_{k=1}^{n} y_{k-1} m_k, \quad \text{where} \quad m_k = m\big(f^{-1}[y_{k-1}, y_k]\big), \text{ for } 1 \le k \le n,$$

to approximate the integral $\int_A f \, dm(A)$, with $A = \bigcup_{k=1}^{n} f^{-1}[y_{k-1}, y_k]$. This simple flip of approach led Lebesgue *a priori* to define the *measure* $m(A)$ of a measurable set $A \subset \mathbb{R}$ in a way such that m_k turns out to the sum of the lengths of the subintervals of the interval $[a, b]$ over which $y_{k-1} \le f(x) < y_k$. An interesting outcome of Lebesgue's theory is the concept of a *null set*, i.e., a set $A \subseteq \mathbb{R}$ with $m(A) = 0$. It leads to the fundamental concept of *almost everywhere*, which is the key argument to prove that a bounded function is Riemann integrable if and only if it is not *too badly discontinuous* (Theorem 9.24).

In abstract terms, a similar scheme helps to develop a theory of measure over an arbitrary topological space X, and to introduce the concept of the integral $\int_A f \, d\mu$ of a real-valued *measurable* function f defined over a set $A \in \mathscr{A}$, where $\mu : \mathscr{A} \to [0, \infty]$ is a *measure* on X defined on a special type of class $\mathscr{A}$ of sets $A \subseteq X$. Lebesgue developed the abstract theory as part of his PhD dissertation, partly to fill the *gaps* in the definition of the Riemann integral of a bounded function, and also to solve some previously existing problems such as *representation of a periodic function by a Fourier series* (see Chap. 10).

Since abstract measure theory has numerous applications in science and engineering, we will start our discussion with general measure spaces. Notice that it also provides mathematical foundation to probability theory.[2] We refer the reader to [1] or [2] or [3], for further details.

9.1 Measure Space

In this section, we give a *crisp introduction* to the abstract measure theory, and review some important theorems inasmuch as we need to introduce some basic concepts of Lebesgue measure on $\mathbb{R}$, as in the next section, and the concept of the Lebesgue integral of a real-valued *"measurable function"* f over a set $A \subseteq \mathbb{R}$, in Sect. 9.3. Once the *Lebesgue measure* on $\mathbb{R}$ is introduced, and some related theorems are proved, the first part of the story is complete. Notice that since the *functions* are now allowed to attain $\pm\infty$, we are working here with the *extended real line* given by

$$\overline{\mathbb{R}} := \mathbb{R} \cup \{\pm\infty\} = [-\infty, \infty].$$

It is assumed that a topology on $\overline{\mathbb{R}}$ is given by the open sets of the form

---

[2] It is a branch of mathematics that provides tools to model *random phenomena* of the real world.

$$[-\infty, b), \quad (a, b), \quad (a, +\infty],$$

and all their countable unions. We may also use the notation $[0, \infty] := \mathbb{R}^+ \cup \{+\infty\}$.

Here, we are dealing mainly with sets, and certain special type of nonnegative real-valued *set function* ν defined on a class $\mathscr{C}$ of sets, where $\nu(A)$ in concrete situations is supposed to give a sort of *measure* of a set $A \in \mathscr{C}$. As said above, we may have $\nu(A) = \pm\infty$, for some sets A. Let us start with some set theoretic definitions.

Let X be a nonempty set, and $\mathscr{P}(X)$ denote the *power set* of the set X, i.e., the class of all subsets of X. Recall that, for $A, B \in \mathscr{P}(X)$, we write the (absolute) *complement* $X \setminus A$ as A^c,

$$A \triangle B = (A \setminus B) \cup (B \setminus A)$$

is the *symmetric difference* of the sets A and B, and we have $A \setminus B = A \cap B^c$. Let $\mathscr{C}$ be a class of sets in $\mathscr{P}(X)$. We say $\{A_n\} \subset \mathscr{C}$ is a collection (or a sequence $\langle A_n \rangle$) of *pairwise disjoint* sets if $A_k \cap A_j = \emptyset$, for all $k \neq j$. The next lemma is used frequently in the sequel.

Lemma 9.1 *Let $\langle A_n \rangle$ be a sequence of sets in $\mathscr{C}$. Take $B_n = A_n \setminus \left(\bigcup_{k=1}^{n-1} A_k \right)$, for $n \geq 1$, with $A_0 = \emptyset$. Then, we have*

1. *$B_n \subseteq A_n$, for all $n \geq 1$.*
2. *$\bigcup_{n=1}^{m} B_n = \bigcup_{n=1}^{m} A_n$, for all $m \geq 1$, and hence $\bigcup_{n=1}^{\infty} B_n = \bigcup_{n=1}^{\infty} A_n$.*
3. *The sets in the collection $\{B_n\}$ are pairwise disjoint.*

Proof Notice that (1) holds, by definition. For (2) and (3), notice that

$$B_n = A_n \setminus \left(\bigcup_{k=1}^{n-1} A_k \right) = A_n \cap \left(\bigcup_{k=1}^{n-1} A_k \right)^c$$
$$= A_1^c \cap A_2^c \cap \cdots \cap A_{n-1}^c \cap A_n.$$

Clearly then, for $i < j$, we have $B_i \cap B_j \subset A_i \cap B_j = \emptyset$, and so $\langle B_n \rangle$ is a sequence of pairwise disjoint sets in $\mathscr{C}$, with $\bigcup_{n=1}^{\infty} B_n \subseteq \bigcup_{n=1}^{\infty} A_n$. To prove the reverse inclusion, suppose $a \in \bigcup_{n=1}^{\infty} A_n$, and let $k \in \mathbb{N}$ be the least such that $a \in A_k$. If $k = 1$ then $a \in A_1 = B_1$. And, for $k > 1$, we have

$$a \in A_k \setminus \bigcup_{k=1}^{n-1} A_k = B_k.$$

Hence, (2) and (3) hold. This completes the proof. $\square$

Definition 9.1 Let $A_n \in \mathscr{C}$, for $n \in \mathbb{N}$. We say $\langle A_n \rangle$ is an **expanding sequence** (or an *increasing sequence*) if $A_n \subseteq A_{n+1}$, for all $n \geq 1$. In this case, we write $A_n \uparrow$. Similarly, we say a sequence $\langle A_n \rangle$ is a **contracting sequence** (or a *decreasing sequence*) if $A_n \supseteq A_{n+1}$, for all $n \geq 1$. In this case, we write $A_n \downarrow$. We say a sequence $\langle A_n \rangle$ in $\mathscr{C}$ is **monotone** if $A_n \uparrow$ or $A_n \downarrow$, and we define

$$A_n \uparrow \quad \Rightarrow \quad \lim_{n \to \infty} A_n = \bigcup_{n=1}^{\infty} A_n \tag{9.1.1a}$$

$$A_n \downarrow \quad \Rightarrow \quad \lim_{n \to \infty} A_n = \bigcap_{n=1}^{\infty} A_n. \tag{9.1.1b}$$

We introduce some useful notations. The class of sets formed by taking countable unions of the members of a class $\mathscr{C} \subseteq \mathscr{P}(X)$ is denoted by $\mathscr{C}_\sigma$. And, we write $\mathscr{C}_\delta$ for the class of sets formed by taking countable intersections of the members of $\mathscr{C}$. A similar meaning stands for the extended notions such as $\mathscr{C}_{\delta\sigma}$ and $\mathscr{C}_{\sigma\delta}$. In particular, when X is a topological space, we write $\mathscr{G}_\delta$ for the family of countable intersection of open sets in X, and $\mathscr{F}_\sigma$ for the family of countable union of closed sets in X.

The concept introduced in the next definition is analogous to the concept we introduced earlier for sequences of real numbers (Chap. 3).

Definition 9.2 The *limit superior* of a sequence $\langle A_n \rangle$ in a class $\mathscr{C} \mathscr{P}(X)$ is given by

$$\limsup_n A_n = \bigcap_{n=1}^{\infty} \bigcup_{k=n}^{\infty} A_k \in \mathscr{C}_{\sigma\delta},$$

And, the *limit inferior* of $\langle A_n \rangle$ is given by

$$\liminf_n A_n = \bigcup_{n=1}^{\infty} \bigcap_{k=n}^{\infty} A_k \in \mathscr{C}_{\delta\sigma}.$$

Notice that since $\langle \bigcup_{k=n}^{\infty} A_k \rangle$ is a *contracting sequence* and $\langle \bigcap_{k=n}^{\infty} A_k \rangle$ is an *expanding sequence*, so both $\limsup_n A_n$ and $\liminf_n A_n$ exist. Also, since

$$x \in \limsup_n A_n \quad \Leftrightarrow \quad x \in A_n \text{ for infinitely many } n; \text{ and,}$$

$$x \in \liminf_n A_n \quad \Leftrightarrow \quad x \in A_n \text{ for all but finitely many } n,$$

it follows that we always have

$$\liminf_n A_n \subset \limsup_n A_n, \quad \text{for} \quad A_n \in \mathscr{C}.$$

We say a sequence $\langle A_n \rangle$ in $\mathscr{C}$ **converges** if $\liminf_n A_n = \limsup_n A_n$. In this case, we write

$$\lim_{n \to \infty} A_n = \liminf_n A_n = \limsup_n A_n.$$

In particular, we have that *every monotone sequence converges*, and so we write $A = \lim_{n \to \infty} A_n$. Also, it follows that the complementation operation interchanges the two types of limits, i.e.,

$$\left(\limsup_n A_n \right)^c = \liminf_n A_n^c \quad \text{and} \quad \left(\liminf_n A_n \right)^c = \limsup_n A_n^c.$$

Let $A \in \mathscr{C}$, and $\langle A_n \rangle$ be a sequence in $\mathscr{C}$. If $\langle A_n \rangle$ is *expanding* so that

$$A = \lim_{n \to \infty} A_n = \bigcup_{n=1}^{\infty} A_n,$$

we write $A_n \uparrow A$. Also, if $\langle A_n \rangle$ is a *contracting sequence* so that

$$A = \lim_{n \to \infty} A_n = \bigcap_{n=1}^{\infty} A_n,$$

we write $A_n \downarrow A$. For example, in a typical application of these notions, we would use these to express intervals of $\mathbb{R}$ in the form

$$[a, b) = \bigcap_{n=k}^{\infty} \left(a + \frac{1}{n}, b + \frac{1}{n} \right]; \quad (a, b) = \bigcup_{n=k}^{\infty} \left(a, b + \frac{1}{n} \right], \quad \text{etc.,} \qquad (9.1.2)$$

for $k > (b - a)^{-1}$.

Definition 9.3 A collection $\mathscr{R} \subseteq \mathscr{P}(X)$ is called a **ring** if $\emptyset \in \mathscr{R}$, and $\mathscr{R}$ is closed with respect to taking finite unions, finite intersections, and relative complementation, i.e.,

$$A, B \in \mathscr{R} \quad \Rightarrow \quad A \cup B, \quad A \cap B \in \mathscr{R}, \quad \text{and} \quad A \setminus B \in \mathscr{R}.$$

A *ring* $\mathscr{R}$ is called an **algebra** if $X \in \mathscr{R}$. We usually write an algebra as $\mathscr{A}$. Further, we say an *algebra* $\mathscr{A}$ is a σ -**algebra** (or a σ-**field**) if $\mathscr{A}$ is closed with respect to taking countable unions. That is, for every countable class $\{ A_n \in \mathscr{A} : n \geq 0 \}$, we have $\bigcup_{n=0}^{\infty} A_n \in \mathscr{A}$.

Notice that, by the de Morgan's Laws, we have every σ -*algebra* $\mathscr{A}$ is also closed with respect to taking countable intersections, and hence also with respect to taking the *limit superior* and the *limit inferior* of sequences of sets in $\mathscr{A}$.

Example 9.1 Trivially, $\mathscr{P}(X)$ is a σ -algebra. In fact, it is the largest σ - algebra $\mathscr{A}$ of subsets of the set X. On the other hand, for any $A \subset X$, the collection

$$\Omega_A = \{ \emptyset, X, A, A^c \}$$

is a σ -algebra, which is actually the *smallest* σ -algebra containing the set A.

More generally, given a nonempty class $\mathscr{C} \subseteq \mathscr{P}(X)$, the intersection of all algebras containing $\mathscr{C}$ is the smallest algebra that contains $\mathscr{C}$. It is called the *algebra generated by* the class $\mathscr{C}$, which is given by $\bigcup_{k \geq 0} \mathscr{C}^{(k)}$, where $\mathscr{C}^{(k)}$ is defined inductively as follows:

$$\mathscr{C}^{(0)} = \mathscr{C} \cup \{\emptyset\};$$
$$\mathscr{C}^{(k+1)} = \bigcup \{A \cup B, \ A^c : \ A, B \in \mathscr{C}^{(k)}\}, \quad \text{for } k \geq 0.$$

Similarly, the σ-*algebra generated by* a nonempty class $\mathscr{C} \subseteq \mathscr{P}(X)$ is the intersection of all σ-algebras containing the class $\mathscr{C}$, which we write as $\sigma(\mathscr{C})$. Notice that, for any set $A \in \mathscr{P}(X)$, we have

$$\sigma(\mathscr{C} \cap A) = \sigma(\mathscr{C}) \cap A = \{B \cap A : \ B \in \sigma(\mathscr{C})\}. \tag{9.1.3}$$

In some interesting situations, it suffices to apply countable unions and complements, to construct the σ-algebra $\sigma(\mathscr{C})$. The concept introduced in the next definition is fundament to the Lebesgue's theory. It is named after *Émile Borel* (1871–1956).

Definition 9.4 If X is a topological space, with a topology given by $\mathscr{T} \subseteq \mathscr{P}(X)$, then the σ-algebra $\mathscr{B}_X$ generated by the collection $\mathscr{T}$ of open sets in X is called the **Borel σ-algebra**. A set $B \in \mathscr{B}_X$ is called a **Borel set** of the topological space X.

Our main concern in this chapter is the *Borel σ-algebra* $\mathscr{B}_{\mathbb{R}}$, generated by the collection $\mathscr{C}$ of *right-closed* intervals $(a, b]$. Notice that, by using (9.1.2), it follows that the Borel σ-algebra $\mathscr{B}_{\mathbb{R}}$ contains all types of intervals in the real line $\mathbb{R}$. Therefore, by Theorem 5.3, every open set (and hence every closed set) in $\mathbb{R}$ is a *Borel set*. In general, once again, for any $B \in \mathscr{B}_{\mathbb{R}}$, we have

$$\mathscr{B}(B) = B \cap \mathscr{B}_{\mathbb{R}} = \{B' \in \mathscr{B}(\mathbb{R}) : \ B' \subset B\}. \tag{9.1.4}$$

Recall that by an *extended nonnegative real-valued set function* on a class $\mathscr{C} \subseteq \mathscr{P}(X)$ we mean a function $\mu : \mathscr{C} \to [0, \infty]$, where $[0, \infty] = \mathbb{R}^+ \bigcup \{\infty\}$. In all that follows, the term *set function* is used in this sense only, and we will write μ *is a function on a class* $\mathscr{C}$.

Definition 9.5 Let a class $\mathscr{C} \subseteq \mathscr{P}(X)$ be closed with respect to taking countable unions. A function μ on a class $\mathscr{C}$ is called **countably additive** if, for any sequence $\langle A_n \rangle$ of *pairwise disjoint* sets in $\mathscr{C}$, we have

$$\mu\left(\bigcup_{n=1}^{\infty} A_n\right) = \sum_{n=1}^{\infty} \mu(A_n). \tag{9.1.5}$$

A countably additive function μ on a class $\mathscr{C}$ is called σ-*additive* if $\mu(\emptyset) = 0$. Also, we say a function μ on a class $\mathscr{C}$ is **countably subadditive** if, for any sequence $\langle A_n \rangle$ in $\mathscr{C}$, we have

$$\mu \left(\bigcup_{n=1}^{\infty} A_n \right) \le \sum_{n=1}^{\infty} \mu(A_n). \tag{9.1.6}$$

A countably subadditive function μ on a class $\mathscr{C}$ is called σ - *subadditive* if $\mu(\emptyset) = 0$.

We say a function μ on a class $\mathscr{C}$ is *finitely additive* if it satisfies the condition (9.1.5) for every finite collection of *pairwise disjoint* sets in $\mathscr{C}$. And, a function μ on a class $\mathscr{C}$ is called *finitely subadditive* if it satisfies the condition (9.1.6) for every finite collection of sets in $\mathscr{C}$. Further, by adding the condition that $\mu(\emptyset) = 0$, we have for a function ν on a class $\mathscr{C}$ the concepts of μ being **additive** or **subadditive**. The next theorem proves some of the properties that every *additive* function ν on an algebra $\mathscr{A}$ satisfy.

Theorem 9.1 *Let $\mathscr{A}$ be an algebra of subsets of a set X, and ν be an extended real-valued additive set function on $\mathscr{A}$. Then, we have*

1. *If there is some $A \in \mathscr{A}$, with $\nu(A) < \infty$, then $\nu(\emptyset) = 0$;*
2. *For all $A, B \in \mathscr{A}$, we have*

$$\nu(A \cup B) + \nu(A \cap B) = \nu(A) + \nu(B);$$

3. *For all $B \subset A \in \mathscr{A}$, $\nu(A) = \nu(B) + \nu(A \setminus B)$. In particular,*

$$\nu(B) < \infty \quad \Rightarrow \quad \nu(A \setminus B) = \nu(A) - \nu(B); \quad and$$
$$\nu(A \setminus B) \ge 0 \quad \Rightarrow \quad \nu(B) \le \nu(A).$$

4. *If ν is nonnegative, then it is finitely subadditive.*

Proof Applying additivity of ν to the identity $A = A \bigcup \emptyset$, we have (1). Next, to prove (2), we use additivity of ν on $\mathscr{A}$ to write

$$\nu(A) = \nu(A \cap B) + \nu(A \setminus B) \quad \text{and} \quad \nu(B) = \nu(A \cap B) + \nu(B \setminus A),$$

then the assertion follows from the relation

$$\nu(A \cup B) = \nu(A \setminus B) + \nu(B \setminus A) + \nu(A \cap B).$$

Also, since $B \setminus A \in \mathscr{A}$ for every $A, B \in \mathscr{A}$, we have

$$B \subset A \quad \Rightarrow \quad \nu(A) = \nu(B) + \nu(A \setminus B).$$

This proves (3). Finally, (4) follows by using Lemma 9.1 and (3). $\qquad\square$

Definition 9.6 Let X be a nonempty set, $\Omega \subseteq \mathscr{P}(X)$ be a σ - algebra, and ν be σ -additive function on Ω. Then, the triplet (X, Ω, ν) (or simply the set X) is called a *measure space*; the functionν is called a *measure* on the set X, and each set $A \in \Omega$ is called a ν - *measurable set* (or simply a *measurable set*).

A *measurable set* A in a measure space (X, Ω, v) is said to have the *finite measure* if $v(A) < \infty$. Also, we say A has the σ-*finite measure* if there exists a sequence $\langle A_n \rangle$ in Ω such that

$$A = \bigcup_{n=1}^{\infty} A_n, \quad \text{with} \quad v(A_n) < \infty, \quad \text{for all} \ n \in \mathbb{N}.$$

Notice that, by using Lemma 9.1, it can always be arranged to assume that the sets in the sequence $\langle A_n \rangle$ are *pairwise disjoint*.

Definition 9.7 Let (X, Ω, v) be a measure space. We say v is a **finite measure** on X if the set X has the finite measure. And, we say v is a σ-**finite measure** on X if the set X has the σ-finite measure.

Example 9.2 Let X be a nonempty set, $\Omega = \mathscr{P}(X)$, and define

$$c(E) = \begin{cases} \#E, & \text{if } E \in \Omega \text{ is finite} \\ \infty, & \text{if } E \text{ is infinite} \end{cases}$$

Then, $c : \Omega \to [0, \infty]$ is a measure, which is called the *counting measure* on the set X. Notice that the counting measure c is σ-*finite* if X is a countably infinite set. More generally, suppose $Y \subset X$ be a *denumerable set*, and $c : Y \to [0, \infty]$ is an arbitrary function. Then, the set function on $\mathscr{P}(X)$ given by

$$v(A) := \sum_{x \in A \cap Y} c(x), \quad \text{for} \ A \subseteq X,$$

defines a measure on X, which is called the *discrete measure*. Notice that, in this case, v is a *finite measure* if and only if $\sum_{y \in Y} c(y) < \infty$. And, v is a σ-*finite measure* if and only if $c(y) \in [0, \infty]$, for all $y \in Y$. The construction given above generalises easily to the case when Y is an uncountable set.

Notice that, by using the part (3) of Theorem 9.1, it follows that if (X, Ω, v) is a measure space, with $v(X) < \infty$, then v is *monotone*. Further, for any sequence $\langle A_n \rangle$ in Ω, it follows from the construction as in Lemma 9.1 that $v(B_n) \le v(A_n)$, for all $n \ge 1$, by using the monotone property of v. Therefore, we have

$$v \left(\bigcup A_n \right) = v \left(\bigcup B_n \right) = \sum v(B_n) \le \sum v(A_n).$$

Hence, v is *countably subadditive*.

Definition 9.8 Let (X, Ω, v) be a measure space. We say v is *continuous from below* at A if, for any expanding sequence $\langle A_n \rangle$ in Ω, we have

$$A_n \uparrow A \quad \Rightarrow \quad \lim_{n \to \infty} v(A_n) = v(A). \tag{9.1.7}$$

Also, we say v is *continuous from above* at A if, for any contracting sequence $\langle A_n \rangle$ in Ω, we have

$$A_n \downarrow A \quad \Rightarrow \quad \lim_{n \to \infty} v(A_n) = v(A). \tag{9.1.8}$$

Theorem 9.2 *Let (X, Ω, v) be a measure space. Then, v is continuous from below. Further, if $v(X) < \infty$, then v is also continuous from above.*

Proof To prove the continuity of a measure v from below at A, let $\langle A_n \rangle$ be an expanding sequence in Ω with $A_n \uparrow A$. Choose $B_1 = A_1$ and $B_k = A_k \setminus A_{k-1}$, $k > 1$, so that $\langle B_k \rangle$ is a pairwise disjoint sequence in the σ - algebra Ω, with

$$\bigcup_{k=1}^{\infty} B_k = \bigcup_{n=1}^{\infty} A_n = A \quad \text{and} \quad A_n = \bigcup_{k=1}^{n} B_k.$$

Then, by the countability and finite additivity of v, we have

$$v(A) = \sum_{n=1}^{\infty} v(B_n) = \lim_{n \to \infty} \sum_{k=1}^{n} v(B_k) = \lim_{n \to \infty} v(A_n).$$

Next, to prove the continuity of v from above at A, let $\langle A_n \rangle$ be a contracting sequence in Ω with $A_n \downarrow A$. Notice that, as $v(X) < \infty$, we have $v(A_1) < \infty$. Since $(A_1 \setminus A_n) \uparrow (A_1 \setminus A)$, the above discussion implies that

$$v(A_1) - v(A_n) = v(A_1 \setminus A_n) \uparrow v(A_1 \setminus A) = v(A_1) - v(A).$$

So, the assumption $v(A_1) < \infty$ completes the proof. $\qquad \square$

The *continuity from above* fails for a measure v when $v(A_k) = \infty$, for some k. For example, the *counting measure* c on the set $X = \mathbb{N}$ is not *continuous from above* at $\emptyset$. For, take $A_n = \{k \in \mathbb{N} : k \geq n\}$ so that $c(A_n) = \infty$, for each n, but

$$c \left(\bigcap_{n=1}^{\infty} A_n \right) = c(\emptyset) = 0 \neq \lim_{n \to \infty} c(A_n).$$

We can use the continuity properties given in Theorem 9.2 to prove the σ-additivity of an additive set function defined on a σ-algebra.

Theorem 9.3 *Let Ω be a σ-algebra, and suppose an additive function $v : \Omega \to [0, \infty]$ is continuous from above at $\emptyset$, i.e., for any contracting sequence $\langle A_n \rangle$ in Ω with $A_n \downarrow \emptyset$, we have $\lim_{n \to \infty} v(A_n) = 0$. Then, v is σ-additive on Ω.*

Proof In general, let $v : \Omega \to \mathbb{R}^+$ be a finitely additive set function, with $v(\emptyset) = 0$ and $v(X) < \infty$, which is continuous from above. To prove the σ- additivity of v, let

$\langle A_n \rangle$ be a sequence of *pairwise disjoint* sets in Ω, and take $B_n = \bigcup_{k=1}^{n} A_k$. Clearly then, $\langle B_n \rangle$ is an expanding sequence. So, we have

$$\nu\left(\bigcap_{n=1}^{\infty} B_n^c\right) = \lim_{n\to\infty} \nu\left(B_n^c\right),$$

which by finite additivity of ν implies that

$$\nu(X) - \nu\left(\bigcup_{n=1}^{\infty} B_n\right) = \lim_{n\to\infty} \left[\nu(X) - \nu(B_n)\right].$$

It thus follows that

$$\nu\left(\bigcup_{n=1}^{\infty} B_n\right) = \lim_{n\to\infty} \nu\left(B_n\right).$$

Hence, by using the fact that $\bigcup_{n=1}^{\infty} A_n = \bigcup_{n=1}^{\infty} B_n$ and

$$\nu(B_n) = \sum_{k=1}^{n} \nu(A_k), \quad \text{(by finite additivity of } \nu\text{)}$$

we conclude that

$$\nu\left(\bigcup_{n=1}^{\infty} A_n\right) = \lim_{n\to\infty} \sum_{k=1}^{n} \nu(A_k) = \sum_{n=1}^{\infty} \nu(A_n).$$

$\square$

Notice that it follows from the part (3) of Theorem 9.1 that a nonnegative finitely additive set function ν defined on an algebra $\mathscr{A}$ is finite if and only if it is *bounded*, i.e.,

$$\sup\left\{\nu(A) : A \in \mathscr{A}\right\} < \infty.$$

An illuminating application of the Lebesgue's theory is provided by the mathematical foundation of the probability theory, first published in 1932 by the Russian mathematician *Andrey Kolmogorov* (1903–1987). In this case, a set X consists of all possible outcomes of a random experiment, usually called the *sample space*; a class Ω of subsets of X representing the *events*; and there is a *probability function* $P : \Omega \to [0, 1]$ so that the (cardinal) *measure* $P(E)$ gives the *probability* of an event $E \in \Omega$.

Definition 9.9 Let (X, Ω, ν) be a measure space. If ν satisfies the condition $\nu(X) = 1$ so that it takes values in the interval $[0, 1]$, we say ν is a **probability measure**. In this case, the triple (X, Ω, ν) is called a *probability space*; the members of the

σ-algebra Ω are called the *events*; and, for any $E \in \Omega$, the real number $v(E)$ is called the **probability** of the event E.

Notice that, since $P(\emptyset) = 0$, countable additivity of a *probability measure* v implies finite additivity.

Example 9.3 For any fixed $x \in X$, the simplest example of a *probability measure* is provided by the function given by

$$v_x(A) = \chi_A(x) = \begin{cases} 1, & \text{if } x \in A \\ 0, & \text{if } x \notin A \end{cases}$$

It is called the *Dirac unit mass*. Further, if X is a denumerable *sample space* of a random experiment so that

$$X = \{x_1, x_2, x_3, \ldots\}, \quad \text{with} \quad P[X = x_i] = p_i,$$

and Ω is a σ-algebra of *events*, then the function $v : \Omega \to [0, \infty)$ given by

$$v(A) = \sum_{x_i \in A} p_i, \quad A \in \Omega,$$

is a probability measure. Of course, we have $v(X) = 1$, because $\sum p_i = 1$.

Notice that the concept of *conditional probability* of an event with respect to a previously occurred event $A \in \Omega$ is based on the construction as given in (9.1.4). In general, it helps to study the *probability context* of a concept or a theorem that are being stated here in abstract terms.

In his original paper, Kolmogorov used finitely additivity and the continuity from above at $\emptyset$ to define the probability measure. In the next theorem, we use the continuity from below.

Theorem 9.4 *Let X be a nonempty set, and $\mathcal{R} \subset \mathcal{P}(X)$ be a ring. Suppose there is a nonnegative real-valued set function v on the ring $\mathcal{R}$, with $v(\emptyset) = 0$, such that*

$$v(C \cup D) + v(C \cap D) = v(C) + v(D), \quad C, D \in \mathcal{R}, \tag{9.1.9}$$

and v is continuous from below at each $A \in \mathcal{R}$. For $A \subseteq X$, define

$$v^*(A) = \inf \{v(C) : C \in \mathcal{R}, C \supset A\}. \tag{9.1.10}$$

Then, we have

1. *$v^*(A) \geq 0$, and $v^*(A)\big|_{\mathcal{R}} = v$;*
2. *$v^*(A \cup B) + v^*(A \cap B) \leq v^*(A) + v^*(B)$ so that $v^*(A) + v^*(A^c) \geq 1$;*
3. *v^* is monotone;*

4. $A_n \uparrow A \implies v^*(A_n) \to v^*(A)$.

Proof Observe that the first part of (1) follows from the monotonicity of function v because $v(\emptyset) = 0$; and the second part follows from the definition of v^*. For (2), let $\epsilon > 0$. Then, the definition of infimum gives $C, D \in \mathcal{R}$ such that

$$C \supset A \quad \text{and} \quad v(C) \leq v^*(A) + \frac{\epsilon}{2};$$
$$D \supset B \quad \text{and} \quad v(D) \leq v^*(B) + \frac{\epsilon}{2}.$$

So, we have

$$
\begin{aligned}
v^*(A) + v^*(B) + \epsilon &\geq v(C) + v(D) \\
&= v(C \cup D) + v(C \cap D) \quad \text{(by (9.1.9))} \\
&\geq v^*(A \cup B) + v^*(A \cap B),
\end{aligned}
$$

which proves (2). Next, (3) follows directly from the definition of v^*. Finally, by (3), we have

$$v^*(A) \geq \lim_{n \to \infty} v^*(A_n).$$

So, given $\epsilon > 0$ and $n \in \mathbb{N}$, we may choose $C_n \in \mathcal{R}$, with $C_n \supset A_n$, such that

$$v(C_n) \leq v^*(A) + \frac{\epsilon}{2^n}.$$

So, $A = \bigcup A_n \subset \bigcup C_n \in \mathcal{R}$ implies that

$$v^*(A) \leq v^* \left(\bigcup C_n \right) = v \left(\bigcup C_n \right) = \lim_{n \to \infty} v \left(\bigcup_{j=1}^{n} C_j \right).$$

To complete the proof, we assert that

$$v \left(\bigcup_{j=1}^{n} C_j \right) \leq v^*(A_n) + \epsilon \sum_{j=1}^{n} \frac{1}{2^j}, \quad n = 1, 2, \ldots .$$

The assertion holds for $n = 1$, by our choice of C_1. Assuming it holds for any $n \geq 1$, we apply (9.1.9) to sets $\bigcup_{j=1}^{n} C_j$ and C_{n+1} to conclude that

$$v \left(\bigcup_{j=1}^{n+1} C_j \right) = v \left(\bigcup_{j=1}^{n} C_j \right) + v(C_{n+1}) - v \left[\left(\bigcup_{j=1}^{n} C_j \right) \cap C_{n+} \right].$$

However, since we have

$$\left(\bigcup_{j=1}^{n} C_j\right) \cap C_{n+1} \supset C_n \cap C_{n+1} \supset A_n \cap A_{n+1} = A_n,$$

a simple inductive argument gives

$$\nu\left(\bigcup_{j=1}^{n+1} C_j\right) \leq \nu^*(A_n) + \epsilon \sum_{j=1}^{n} \frac{1}{2^j} + \nu^*(A_{n+1}) + \frac{\epsilon}{2^{n+1}} - \nu^*(A_n)$$

$$\leq \nu^*(A_{n+1}) + \epsilon \sum_{j=1}^{n+1} \frac{1}{2^j}.\square$$

$\square$

9.1.1 Outer Measure

Notice that, if $\langle A_n \rangle$ is a sequence in $\mathscr{P}(X)$, then by parts (4) and (2) of Theorem 9.4 we have

$$\nu^*\left(\bigcup_{n=1}^{\infty} A_n\right) = \lim_{n\to\infty} \nu^*\left(\bigcup_{j=1}^{n} A_j\right) \leq \lim_{n\to\infty} \sum_{j=1}^{n} \nu^*(A_j),$$

and so ν^* is a countably subadditive set function. Also, it follows from parts (1) and (3) of Theorem 9.4 that $\nu^*(\emptyset) = 0$, and the function ν^* is monotone. Hence, the *extended* set function ν^* as defined in (9.1.10) is an *outer measure* on X.

Definition 9.10 Let X be a nonempty set. A σ-subadditive monotone extended real-valued set function $\mu^* : \mathscr{P}(X) \to [0, \infty]$ is called an **outer measure** on the set X.

Example 9.4 Any measure given on a set X is an outer measure. However, the converse may not hold. For example, let X be a countably infinite set such as $\mathbb{N}$, and define

$$\nu^*(A) = \begin{cases} 1, & \text{for } A \neq \emptyset \\ 0, & \text{for } A = \emptyset \end{cases}, \quad A \subseteq X.$$

By definition, $\nu^*(\emptyset) = 0$. Also, for $A \subseteq B$,

$$B = \emptyset \implies A = \emptyset \implies \nu^*(A) = 0 = \nu^*(B);$$
$$B \neq \emptyset \implies A = \emptyset \text{ or } A \neq \emptyset \implies \nu^*(A) \leq \nu^*(B), \text{ in both cases.}$$

Similarly, we may consider other cases to prove that ν^* is countably subadditive. However, since we can write $X = \cup_{n=1}^{\infty}\{x_n\}$, so ν^* cannot be a measure.

The concept of *measurable set* as introduced in the next definition is due to the Greek mathematician *Constantin Carathéodory* (1873–1950). In the original paper, Carathéodory called such types of sets as *additive*.

Definition 9.11 (*Carathéodory*) Let X be a nonempty set, and $\mu^* : \mathscr{P}(X) \to [0, \infty]$ is an outer measure. We say $E \subset X$ is a μ^*-**measurable** (or simply *measurable*) if, for all $A \subset X$, we have

$$\mu^*(A) = \mu^*(A \cap E) + \mu^*(A \cap E^c).$$

The set A is called a *test set*.

Let $E \subseteq X$ be arbitrary, and $\mu^* : \mathscr{P}(X) \to [0, \infty]$ is an outer measure. Since the following identity holds, for any set $A \subseteq X$,

$$A = (A \cap E) \cup (A \cap E^c)$$

it follows from the subadditivity of μ^* that

$$\mu^*(A) \leq \mu^*(A \cap E) + \mu^*(A \cap E^c).$$

Therefore, equivalently, $E \subseteq X$ is μ^*-*measurable* μ^*-*measurable* if

$$\mu^*(A) \geq \mu^*(A \cap E) + \mu^*(A \cap E^c). \tag{9.1.11}$$

Notice that there is nothing to prove if $\mu^*(A) = \infty$. So, to verify some $E \subseteq X$ is μ^*-*measurable*, it suffices to use $A \subseteq X$ as a *test set*, with $\mu^*(A) < \infty$. The next theorem proves that every outer measure on a set induces a measure.

Theorem 9.5 *Let $\mu^* : \mathscr{P}(X) \to [0, \infty]$ be an outer measure on a set X. Then the class $\mathscr{M}$ of μ^*-measurable subsets of the set X is a σ-algebra, and the restriction*

$$\mu = \mu^*\Big|_{\mathscr{M}} : \mathscr{M} \to [0, \infty], \tag{9.1.12}$$

is a measure on X.

Proof It follows from the condition (9.1.11) that $\emptyset, X \in \mathscr{M}$. And, since the condition (9.1.11) is symmetrical with respect to the operation of *complementation*, it follows that $E \in \mathscr{M} \Rightarrow E^c \in \mathscr{M}$. Now, let $E, F \in \mathscr{M}$. Then, for any $A \subset X$, we have

$$\mu^*(A) = \mu^*(A \cap E) + \mu^*(A \cap E^c).$$

So, by taking $A \cap E^c$ the a test set for $F \in \mathscr{M}$, we obtain

$$\mu^*(A \cap E^c) = \mu^*((A \cap E^c) \cap F) + \mu^*((A \cap E^c) \cap F^c)$$
$$= \mu^*(A \cap (E^c \cap F)) + \mu^*(A \cap (E \cup F)^c)$$
$$\Rightarrow \quad \mu^*(A) = \mu^*(A \cap E) + \mu^*(A \cap (E^c \cap F)) + \mu^*(A \cap (E \cup F)^c)$$
$$\geq \mu^*((A \cap E) \cup (A \cap (E^c \cap F)) + \mu^*(A \cap (E \cup F)^c)$$
$$= \mu^*((A \cap (E \cup F)) + \mu^*(A \cap (E \cup F)^c), \tag{9.1.13}$$

by using the identity

$$(A \cap E) \cup (A \cap (E^c \cap F)) = A \cap (\emptyset \cap (E \cup F)) = A \cap (E \cup F).$$

It thus follows that $E \bigcup F \in \mathcal{M}$. Next, let $\langle E_n \rangle$ be a sequence of pairwise disjoint sets in the class $\mathcal{M}$, and write $F_n = \bigcup_{k=1}^{n} E_k$. Notice that, by what is shown above, $F_n \in \mathcal{M}$ for all $n \geq 1$. We now proceed to show that, for any $A \subseteq X$,

$$\mu^*(A \cap F_n) = \sum_{k=1}^{n} \mu^*(A \cap E_k), \quad \text{for all } n \geq 1. \tag{9.1.14}$$

Notice that it holds trivially for $n = 1$. So, to apply the mathematical induction, we may assume that (9.1.14) holds for some $n \geq 1$. Then, using $A \cap F_{n+1}$ as a test set for $F_n \in \mathcal{M}$, we have

$$\mu^*(A \cap F_{n+1}) = \mu^*(A \cap F_{n+1} \cap F_n) + \mu^*(A \cap F_{n+1} \cap F_n^c)$$
$$= \mu^*(A \cap F_n) + \mu^*(A \cap E_{n+1})$$
$$= \sum_{k=1}^{n} \mu^*(A \cap E_k) + \mu^*(A \cap E_{n+1})$$
$$= \sum_{k=1}^{n+1} \mu^*(A \cap E_k).$$

Therefore, (9.1.14) holds for all $n \geq 1$. Finally, let $F = \bigcup_{n=1}^{\infty} E_n$ so that

$$F_n \subset F \quad \Rightarrow \quad A \cap F_n \subset A \cap F$$

implies that

$$\mu^*(A \cap F) \geq \mu^*(A \cap F_n) = \sum_{k=1}^{n} \mu^*(A \cap E_k).$$

Then, by taking limit as $n \to \infty$, it follows that

$$\mu^*(A \cap F) \geq \sum_{k=1}^{\infty} \mu^*(A \cap E_k).$$

We also have

$$\mu^*(A \cap F) = \mu^* \left(\bigcup_{n=1}^{\infty} (A \cap F_n) \right) \leq \sum_{n=1}^{\infty} \mu^*(A \cap E_n),$$

by the subadditivity of μ^*. Hence,

$$\mu^*(A \cap F) = \sum_{n=1}^{\infty} \mu^*(A \cap E_n), \quad \text{with } F = \bigcup_{n=1}^{\infty} E_n. \tag{9.1.15}$$

Notice that, as $F_n \in \mathscr{M}$, for any $A \subseteq X$ we have

$$\begin{aligned}
\mu^*(A) &= \mu^*(A \cap F_n) + \mu^*(A \cap F_n^c) \\
&\geq \mu^*(A \cap F_n) + \mu^*(A \cap F^c) \quad (F^c \subset F_n^c \Rightarrow A \cap F^c \subset A \cap F_n^c) \\
&= \sum_{k=1}^{n} \mu^*(A \cap E_k) + \mu^*(A \cap F^c) \\
&= \sum_{n=1}^{\infty} \mu^*(A \cap E_n) + \mu^*(A \cap F^c) \quad (\text{as } n \to \infty) \\
&= \mu^*(A \cap F) + \mu^*(A \cap F^c).
\end{aligned}$$

Therefore, for any sequence $\langle E_n \rangle$ of pairwise disjoint sets in $\mathscr{M}$, $\bigcup_{n=1}^{\infty} E_n \in \mathscr{M}$. To complete the proof, let $\langle A_n \rangle$ be an arbitrary sequence in $\mathscr{M}$. Then, by Lemma 9.1, we can find a sequence $\langle B_n \rangle$ of pairwise disjoint sets in $\mathscr{M}$, with $\bigcup_{n=1}^{\infty} B_n = \bigcup_{n=1}^{\infty} A_n$, and so by the previous argument we have

$$\bigcup_{n=1}^{\infty} A_n = \bigcup_{n=1}^{\infty} B_n \in \mathscr{M},$$

which proves that $\mathscr{M}$ is a σ-algebra. Let $\mu = \mu^*|_{\mathscr{M}}$. Clearly, $\mu(\emptyset) = \mu^*(\emptyset) = 0$. Also, putting $A = X$ in (9.1.15), it follows that μ is countably additive. Hence, μ defines a *measure* on X. $\qquad\square$

Therefore, $(X, \mathscr{M}, \mu)$ is a *measure space*, where μ as in (9.1.12) is called the **measure induced by** the outer measure μ^*. It thus makes sense to discuss some methods of constructing an outer measure on a given (nonempty) set. In the sequel, the *construction* given in the next theorem is referred to as the **Method−I**.

Definition 9.12 We say a class $\mathscr{C}$ of subsets of a set X has the **sequential covering property** if, for each $A \subset X$, there is a sequence $\langle E_n \rangle$ in $\mathscr{C}$ such that $A \subset \bigcup_{n=1}^{\infty} E_n$. We also say that $\mathscr{C}$ *sequential covering class*.

For example, by using (9.1.2), it follows that the class $\mathscr{C}$ of all intervals of the form $(a, b]$ has the *sequential covering property* for all open sets in $\mathbb{R}$ (Theorem 5.3).

Theorem 9.6 (Carathéodory) *Let X be a nonempty set, $\mathscr{C} \subseteq \mathscr{P}(X)$ be a sequential covering class containing $\emptyset$, and $\tau : \mathscr{C} \to [0, \infty]$ be such that $\tau(\emptyset) = 0$. Then, the set function $\mu^* : \mathscr{P}(X) \to [0, \infty]$ given by*

$$\mu^*(A) = \inf \left\{ \sum_{n=1}^{\infty} \tau(E_n) : E_n \in \mathscr{C}, \ A \subseteq \bigcup_{n=1}^{\infty} E_n \right\}, \quad A \subseteq X, \tag{9.1.16}$$

is an outer measure on X. Further, it is unique with the given conditions.

Proof Notice that it is assumed that $\inf \emptyset = \infty$. Observe that for any $E \in \mathscr{C}$ we have $\mu^*(E) \leq \tau(E)$. Also, by definition, we have that μ^* is nonnegative, and $\mu^*(\emptyset) = 0$. Now, let $A \subseteq B$ be two sets in $\mathscr{P}(X)$, and $\epsilon > 0$. Then, by definition of infimum, there exists a sequence $\langle E_n \rangle$ in $\mathscr{C}$ such that

$$B \subseteq \bigcup_{n=1}^{\infty} E_n \quad \Rightarrow \quad \sum_{n=1}^{\infty} \tau(E_n) \leq \mu^*(B) + \epsilon.$$

But that, $A \subseteq B$ implies that

$$\mu^*(A) \leq \sum_{n=1}^{\infty} \tau(E_n) \leq \mu^*(B) + \frac{\epsilon}{2}.$$

Thus, μ^* is monotone. Next, let $\langle C_n \rangle$ be an arbitrary sequence in $\mathscr{C}$, and $\epsilon > 0$. Then, for each $n \geq 1$, there exists a sequence $\langle E_{n,k} \rangle$ in $\mathscr{C}$ such that

$$C_n \subseteq \bigcup_{k=1}^{\infty} E_{n,k} \quad \Rightarrow \quad \sum_{k=1}^{\infty} \tau(E_{n,k}) \leq \mu^*(C_n) + \frac{\epsilon}{2^n}. \tag{9.1.17}$$

But then, since

$$\bigcup_{n=1}^{\infty} C_n \subset \bigcup_{n=1}^{\infty} \bigcup_{k=1}^{\infty} E_{n,k},$$

it follows that

$$\begin{aligned}
\mu^* \left(\bigcup_{n=1}^{\infty} C_n \right) &\leq \sum_{n=1}^{\infty} \sum_{k=1}^{\infty} \tau(E_{n,k}) \\
&\leq \sum_{n=1}^{\infty} \left[\mu^*(C_n) + \frac{\epsilon}{2^n} \right] \quad \text{(by (9.1.17))} \\
&\leq \sum_{n=1}^{\infty} \mu^*(C_n) + \epsilon.
\end{aligned}$$

Since $\epsilon > 0$ is arbitrary, it follows that μ^* is countably subadditive. To prove the uniqueness, let $\tau_1, \tau_2 : \mathscr{C} \to \mathbb{R}^+ \cup \{+\infty\}$ be two set functions such that $\tau_1(\emptyset) = \tau_2(\emptyset) = 0$. Let μ_1^* and μ_2^* be outer measures defined respectively by τ_1 and τ_2 such that $\mu_1^*(E) = \mu_2^*(E)$ for all $E \in \mathscr{C}$. Then, for any $A \subset X$ and $\epsilon > 0$, there exists a sequence $\langle E_n \rangle$ in $\mathscr{C}$ such that $A \subseteq \cup_{n=1}^\infty E_n$ implies that

$$\mu_1^*(A) + \epsilon \geq \sum_{n=1}^\infty \tau(E_n)$$
$$\geq \sum_{n=1}^\infty \mu_1^*(E_n) = \sum_{n=1}^\infty \mu_1^*(E_n)$$
$$\geq \mu_2^*(A),$$

which proves the uniqueness of μ^*. $\square$

Notice that, in general, the outer measure μ^* constructed in Theorem 9.6 may not be σ-additive, i.e., it may not give a measure on X.

9.1.2 Null Sets and Completion

Definition 9.13 Let (X, Ω, ν) be a measure space. A set $A \in \Omega$ is called ν - **null** (or simply null) if $\nu(A) = 0$.

Notice that, by the monotone property, a subset of a *null set* is null; and, by countable subadditivity, a countable union of null sets is again a null set.

Definition 9.14 A measure space (X, Ω, ν) is said to be **complete** if for every $A \in \Omega$, $\nu(A) = 0$, and any $N \subset A$, we have $N \in \Omega$ (and so $\nu(N) = 0$).

Let (X, Ω, ν) be a measure space. We say a measure space $(X, \overline{\Omega}, \overline{\nu})$ is a *completion* of the measure space (X, Ω, ν) if $\overline{\Omega}$ is the smallest σ-algebra containing Ω, $\overline{\nu}\big|_\Omega = \nu$, and for every $A \in \overline{\Omega}$, we have

$$\overline{\nu}(A) = 0 \quad \text{and} \quad N \subset A \quad \Rightarrow \quad N \in \overline{\Omega}.$$

As we shall see shortly, a completion of a measure space is always unique. Said differently, the measure space $(X, \overline{\Omega}, \overline{\nu})$ is *complete*, and it is the *smallest extension* of the measure space (X, Ω, ν).

Now, since a measure ν on a set X is complete if the σ-algebra Ω contains all ν-null sets, there is a "*natural procedure*" to find the *complete extension* of a measure space (X, Ω, ν).

Theorem 9.7 *Let (X, Ω, ν) be a measure space, and $(X, \overline{\Omega}, \overline{\nu})$ be a completion of (X, Ω, ν). Then*

$$\overline{\Omega} = \{A \cup N : A \in \Omega \text{ and } N \subset B \in \Omega \text{ with } v(B) = 0\}.$$

Hence, the completion $(X, \overline{\Omega}, \overline{v})$ *is unique.*

Proof First, we prove $\overline{\Omega}$ is a σ-algebra. Taking $A = N = B = \emptyset$, we have $\emptyset \in \overline{\Omega}$. Next, if $E = A \cup N \in \overline{\Omega}$, then

$$E^c = \left(A \cup B\right)^c \cup \left(B \setminus (A \cup N)\right)$$

is such that $(A \cup B)^c \in \Omega$, $B \setminus (A \cup N) \subset B \in \Omega$, with $v(B) = 0$. So, $E^c \in \overline{\Omega}$. Now, if $E_n = A_n \bigcup N_n \in \overline{\Omega}$ are pairwise disjoint, with

$$A_n \in \Omega \quad \text{and} \quad N_n \subset B_n \in \Omega, \quad \text{with } v(B_n) = 0,$$

then A_n's are pairwise disjoint such that

$$\bigcup_{n=1}^{\infty} A_n \in \Omega \text{ and } \bigcup_{n=1}^{\infty} N_n \subset \bigcup_{n=1}^{\infty} B_n \in \Omega, \quad \text{with } v\left(\bigcup_{n=1}^{\infty} B_n\right) = 0.$$

Thus, $\bigcup_{n=1}^{\infty} E_n \in \overline{\Omega}$. It is now possible to extend v uniquely to a measure $\overline{v}$ on $\overline{\Omega}$ by defining

$$\overline{v}(A \cup N) = v(A), \quad \text{for any } A \in \Omega.$$

Of course, $\overline{v}$ is well-defined. For, if $A \cup N = A' \cup N'$, where $N' \subset B' \in \Omega$ with $v(B) = 0$, then

$$A \subset A \cup N = A' \cup N' \implies v(A) \le v(A') + v(N') = v(A').$$

Similarly, we obtain $v(A') \le v(A)$. It is easy to see that $\overline{v}$ is a measure. To prove that $\overline{v}$ is complete, let $E \in \overline{\Omega}$ with $\overline{v}(E) = 0$, and $F \subset E = A \cup N$. Then, for some $B \in \Omega$, we have $N \subset B$ and $v(B) = 0$. Notice that $0 = \overline{v}(E) = v(A)$. Take $F = \emptyset \cup B'$, where $B' = A \cup B \in \Omega$, with $v(B') \le v(A) + v(B) = 0$. Thus, $F \in \overline{\Omega}$. Clearly, $(X, \overline{\Omega}, \overline{v})$ is an extension of (X, Ω, v), because every $A \in \Omega$ belongs to $\overline{\Omega}$ as $A = A \cup \emptyset$. The minimality by construction proves the uniqueness. $\qquad \square$

We conclude our discussion here with the following important result.

Theorem 9.8 *With notations as in Theorem 9.5, we have* $(X, \mathcal{M}, \mu)$ *is a complete measure space.*

Proof Let $M \in \mathcal{M}$ be such that $\mu(M) = 0$, and $A \subseteq M$. Then, $\mu^*(A) \le \mu(M) = 0$. Also, for any $E \subseteq X$, we have

$$\mu^*(E \cap A) + \mu^*(E \cap A^c) \le \mu^*(A) + \mu^*(E) = \mu^*(E).$$

Therefore, it follows that $A \in \mathcal{M}$. $\qquad \square$

In particular, for any $A \subseteq X$, $\mu^*(A) = 0 \Rightarrow A \in \mathcal{M}$, and for every collection $\{M_n\}$ of pairwise disjoint sets in $\mathcal{M}$, we have

$$\sum_{n=1}^{\infty} \mu^*(E \cap M_n) = \mu^*\left(E \cap \left(\cup_{n=1}^{\infty} M_n\right)\right).$$

9.2 Lebesgue Measure

We introduce here the concept of *Lebesgue measure* on $\mathbb{R}$. Let $\mathscr{C}$ denote the class of right-closed intervals of the form $(a, b]$, together with the *degenerate interval* $\emptyset := (a, a)$. Recall that the class $\mathscr{C}$ has the *sequential covering property* for all open sets in $\mathbb{R}$. We have to work ultimately with the countable unions of intervals of *any type*. And, in view of (9.1.2), it is immaterial which particular types of intervals we take in $\mathscr{C}$ to cover a set $A \subset \mathbb{R}$. Further, as usual, we write $\ell(I)$ for the *length* of an interval $I \subseteq \mathbb{R}$. Let m^* denote the outer measure on $\mathbb{R}$ constructed by the **Method $-$ I** so that, by Theorem 9.6, we have

$$m^*(A) = \inf \left\{ \sum_{n=0}^{\infty} \ell(I_n) : \bigcup_{n=0}^{\infty} I_n \supseteq A \right\}, \quad \text{for} \ \ A \subseteq \mathbb{R}, \tag{9.2.1}$$

where the infimum is taken over all sequences $\langle I_n \rangle$ in $\mathscr{C}$ covering the set A. As shown in the previous section, $m^* : \mathscr{P}(\mathbb{R}) \to [0, \infty]$ is an extended nonnegative monotone countably subadditive set function.

Definition 9.15 The function m^* as given in (9.2.1) is called the **Lebesgue outer measure** on $\mathbb{R}$.

The next theorem proves that m^* is an extension of the length function ℓ.

Theorem 9.9 *For any interval $I \subset \mathbb{R}$, we have $m^*(I) = \ell(I)$.*

Proof Clearly, for each $I \in \mathscr{C}$, we have $m^*(I) \leq \ell(I)$. However, it requires some effort to prove that $\ell(I)$ cannot be greater than $m^*(I)$, for *any type of interval I*. If I is unbounded, then it cannot be covered by any sequence of intervals $\langle I_n \rangle$ in $\mathscr{C}$, with finite total length, which implies that $m^*(I) = \infty$. So, we assume that I is a bounded interval. First, let $I = [a, b]$, and $\epsilon > 0$. Then, by the definition of infimum, there exists a sequence $\langle I_n \rangle$ in $\mathscr{C}$ such that

$$[a, b] \subset \bigcup_{n=1}^{\infty} I_n \quad \text{and} \quad \sum_{n=1}^{\infty} \ell(I_n) < m^*([a, b]) + \frac{\epsilon}{2}. \tag{9.2.2}$$

If required, we may *enlarge infinitesimally* the two ends a_n, b_n of intervals I_n so that

$$I_n \subset J_n = \left(a_n - \frac{\epsilon}{2^{n+2}}, b_n + \frac{\epsilon}{2^{n+2}} \right).$$

Since $\ell(I_n) = \ell(J_n) - \frac{\epsilon}{2^{n+1}}$, we have

$$\sum_{n=1}^{\infty} \ell(I_n) = \sum_{n=1}^{\infty} \ell(J_n) - \frac{\epsilon}{2}$$

$$\Rightarrow \quad \sum_{n=1}^{\infty} \ell(J_n) \leq \sum_{n=1}^{\infty} \ell(I_n) + \frac{\epsilon}{2} \leq m^*([a, b]) + \epsilon, \qquad (9.2.3)$$

where the last inequality follows using (9.2.2). Now, since $\{J_n : n \geq 1\}$ is an open cover of the interval $[a, b]$, by Theorem 6.27, only finitely many of the open intervals J_n are enough to cover $[a, b]$. Re-indexing, if necessary, we assume

$$[a, b] \subset \bigcup_{k=1}^{m} J_k = \bigcup_{k=1}^{m} (c_k, d_k), \quad \text{say.}$$

So, taking $c = \min\{c_1, \ldots, c_m\}$ and $d = \max\{d_1, \ldots, d_m\}$, it follows that

$$\ell([a, b]) < \ell([c, d]) < \sum_{n=1}^{\infty} \ell(J_n) \leq m^*([a, b]) + \epsilon,$$

where the last inequality follows using (9.2.3). Hence, $m^*([a, b]) = \ell([a, b])$. Next, if $I = (a, b)$ and $\epsilon > 0$ is fixed, then

$$\ell((a, b)) = \ell\left([a + \frac{\epsilon}{2}, b - \frac{\epsilon}{2}] \right) + \epsilon$$

$$\leq m^*\left([a + \frac{\epsilon}{2}, b - \frac{\epsilon}{2}] \right) + \epsilon \quad \text{(by previous case)}$$

$$\leq m^*((a, b)) + \epsilon. \quad \text{(by monotonicity of } m^*)$$

Finally, if I is a semi-open interval of the type $[a, b)$, we have

$$\ell(I) = \ell((a, b)) \leq m^*((a, b)) \leq m^*(I),$$

where the last inequality follows by using the monotone property of m^*. This completes the proof. $\qquad \square$

Recall that a set $A \subset \mathbb{R}$ is m^*-null (or simple a *null set*) if $m^*(A) = 0$. For example, clearly, $\emptyset$ is a null set. Notice that, in general, a set $A \subset \mathbb{R}$ is null with respect to the Lebesgue outer measure m^* if and only if for each $\epsilon > 0$ there exists a sequence of intervals $\langle I_n \rangle$ such that

$$A \subset \bigcup_{n=1}^{\infty} I_n \quad \text{and} \quad \sum_{n=1}^{\infty} \ell(I_n) < \epsilon.$$

We use this equivalent condition to find other *null sets* in $\mathbb{R}$. To start with, let $A = \{a\}$ and $\epsilon > 0$. Then, we may take $I_1 = [a - (\epsilon/4), a + (\epsilon/4)]$ and $I_n = \emptyset$, for $n \geq 2$, so that

$$\sum_{n=1}^{\infty} \ell(I_n) = \frac{\epsilon}{2} < \epsilon \quad \Rightarrow \quad m(\{x\}) = 0.$$

By a similar argument, it follows that every finite set in $\mathbb{R}$ is a null set. The next theorem proves that each countable union of null sets is a null set. In particular, it follows that every countable set such as the set $\mathbb{Q}$ of rational numbers is a null set.

Theorem 9.10 *Countable union of null sets in $\mathbb{R}$ is a null set.*

Proof Let $\langle A_n \rangle$ be a sequence of null sets in $\mathbb{R}$ and $\epsilon > 0$. Then, for each $n \geq 1$, there exists a sequence of intervals $\langle I_k^n \rangle$ such that

$$A_n \subset \bigcup_{k=1}^{\infty} I_k^n \quad \text{and} \quad \sum_{k=1}^{\infty} \ell(I_k^n) < \frac{\epsilon}{2^n}.$$

It is then possible to define a sequence of intervals $\langle J_n \rangle$ picking all the intervals in $\{I_k^n : n, k \geq 1\}$, say by Cantor's *diagonal process* as in Example 1.2, such that

$$\bigcup_{n=1}^{\infty} A_n \subset \bigcup_{n=1}^{\infty} \bigcup_{k=1}^{\infty} I_k^n = \bigcup_{s=1}^{\infty} J_s,$$

and we also have

$$\sum_{s=1}^{\infty} \ell(J_s) = \sum_{n=1}^{\infty} \left(\sum_{k=1}^{\infty} \ell(I_k^n) \right) < \sum_{n=1}^{\infty} \frac{\epsilon}{2^n} = \epsilon.$$

Hence, $\bigcup_{n=1}^{\infty} A_n$ is a null set. $\qquad\square$

Corollary 9.1 *Every set $A \subseteq \mathbb{R}$ that is not a null set is uncountable.*

In particular, as shown earlier in Chap. 2, every interval of positive length is an uncountable set. Notice that the set $\mathbb{R}$ of real numbers also contains some *uncountable* null sets.

Example 9.5 Let $C \subset [0, 1]$ be the *Cantor ternary set* as constructed in Example 2.6. We know that it is also given by

$$C = \bigcap_{n=1}^{\infty} C_n = \left\{ \sum_{k=1}^{\infty} \frac{a_k}{3^k} : a_k \in \{0, 2\} \right\},$$

where the set $C_n \subset [0, 1]$ is a union of 2^n disjoint closed intervals, with each interval being of the length $1/3^n$. So, $\ell(C_n) = (2/3)^n$. Since

$$\lim_{n \to \infty} \left(\frac{1}{3} + \frac{1}{3^2} + \dots \right) = \lim_{n \to \infty} \frac{2^n}{3^n} = 0,$$

for any $\epsilon > 0$, we choose $n \in \mathbb{N}$ such that $(2/3)^n < \epsilon$. It then follows that the set C is a *null set*. Notice that $C \neq \emptyset$ because it always contains end points of all the intervals in C_n. Recall that C is an uncountable compact set.

Let $\mathcal{M}_\mathbb{R}$ denote the m^*-measurable sets in $\mathbb{R}$. Recall that, by Theorem 9.5, $\mathcal{M}_\mathbb{R}$ is a σ-algebra.

Definition 9.16 A set $A \in \mathcal{M}_\mathbb{R}$ is called a **Lebesgue measurable set** (or simply a **measurable set**) in $\mathbb{R}$.

As before, $\emptyset, \mathbb{R} \in \mathcal{M}_\mathbb{R}$. Also, every null set in $\mathbb{R}$ is a *Lebesgue measurable* set (Exercise 9.7). In particular, every countable set in $\mathbb{R}$ is measurable. So, the set $\mathbb{Q}$, and hence the set $\mathbb{R} \setminus \mathbb{Q}$ of irrational numbers, are measurable sets. The next fundamental theorem proves that the class $\mathcal{M}_\mathbb{R}$ of Lebesgue measurable sets contains the Borel σ-algebra $\mathcal{B}_\mathbb{R}$.

Theorem 9.11 *Every Borel set in $\mathbb{R}$ is a measurable set. Further, every subset of a null set in $\mathcal{B}_\mathbb{R}$ is measurable.*

Proof Since $\mathcal{B}_\mathbb{R}$ is the σ-algebra generated by all open sets in $\mathbb{R}$, it suffices to prove that every interval I (of any type) is a measurable set. First, suppose $I = [a, b]$. Let $A \subseteq \mathbb{R}$ and $\epsilon > 0$ be arbitrary. Then, for some covering $\{I_n\}$ of the set A, we have

$$m^*(A) \leq \sum_{n=1}^{\infty} \ell(I_n) < m^*(A) + \epsilon.$$

Now, as $\{I_n \cap [a, b]\}$ forms a cover of the set $A \cap [a, b]$, and the collection

$$\{I_n \cap (-\infty, b)\} \bigcup \{I_n \cap (a, +\infty)\}$$

covers the set $A \cap [a, b]^c$, we have

$$m^*(A \cap [a, b]) \leq \sum_{n=1}^{\infty} \ell(I_n \cap [a, b]);$$

$$m^*(A \cap [a, b]^c) \leq \sum_{n=1}^{\infty} \ell(I_n \cap (-\infty, b)) + \sum_{n=1}^{\infty} \ell(I_n \cap (a, +\infty)).$$

Notice that the desired conclusion follows by using the relation

$$\ell(I_n) = \ell\big(I_n \cap [a, b]\big) + \ell\big(I_n \cap (-\infty, b)\big) + \ell\big(I_n \cap (a, +\infty)\big),$$

and putting together the three inequalities as obtained above. Also, a proof for the case when I is an unbounded interval is much simpler (why?). Next, to prove the second assertion, let $N \subset B \in \mathscr{B}_{\mathbb{R}}$, where $m^*(B) = 0$. Then, for $A \subset \mathbb{R}$, since

$$m^*(A \cap N) \le m^*(N) \le m^*(B) = 0,$$

it suffices to note that $m^*(A) \ge m^*(A \cap N')$, by the monotone property of m^*. Hence, $N \in \mathscr{M}_{\mathbb{R}}$.   □

It follows from Theorem 9.5 that the Lebesgue outer measure m^* restricted to the class $\mathscr{M}_{\mathbb{R}}$ of Lebesgue measurable sets gives a measure on $\mathbb{R}$. We write

$$m = m^*\Big|_{\mathscr{M}_{\mathbb{R}}} : \mathscr{M}_{\mathbb{R}} \to [0, \infty]. \tag{9.2.4}$$

We also know that m is a *complete measure* on $\mathbb{R}$ (Theorem 9.8).

Definition 9.17 The measure m as in (9.2.4) is called the **Lebesgue measure** (or simple *measure*) on $\mathbb{R}$.

Notice that the *Lebesgue measure* m on $\mathbb{R}$ has all the general properties of a measure proved in the previous section. It also has some additional (topological) properties. First, we introduce some terminology. For, let $\tau_a : \mathbb{R} \to \mathbb{R}$ denote the *translation mapping* by $a \in \mathbb{R}$; $\delta_t : \mathbb{R} \to \mathbb{R}$ denote the *dilation* by a factor $0 < t \in \mathbb{R}$; and $\rho : \mathbb{R} \to \mathbb{R}$ denote the *reflection* about the point $0 \in \mathbb{R}$. That is,

$$\tau_a(x) = x + a, \quad \text{for } x \in \mathbb{R};$$
$$\delta_t(x) = t\,x, \quad \text{for } x \in \mathbb{R};$$
$$\rho(x) = -x, \quad \text{for } x \in \mathbb{R}.$$

Notice that the length $\ell(I)$ of a finite interval $I \subset \mathbb{R}$ does not change by applying a translation τ_a or the reflection ρ about the *origin*. However, the dilation δ_t increases $\ell(I)$ by a factor of t. Therefore, the class $\mathscr{M}_{\mathbb{R}}$ of Lebesgue measurable sets in $\mathbb{R}$ remains invariant under the three transformations as defined above.

Theorem 9.12 *The Lebesgue measure m on R is* translation invariant, *i.e., for any measurable set $A \subset \mathbb{R}$, $m(A + a) = m(A)$, for every $a \in \mathbb{R}$. Also, it is invariant under the transformations ρ, and positive-homogeneous of degree 1 under the dilation δ_t.*

Proof By the remark above, it follows easily that the Lebesgue outer measure m^* is invariant with respect to transformations τ_a and ρ. So, in particular, m is invariant with respect to these two transformations on measurable sets. Also, since

$$A \subseteq \bigcup_{n=1}^{\infty} I_n \quad \Rightarrow \quad tA \subseteq \bigcup_{n=1}^{\infty} (tI_n),$$

it follows that

$$m^*(tA) \le t \sum_{n=1}^{\infty} \ell(I_n) \le tm^*(A).$$

Following the same procedure applying $1/t$ to the set tA, we obtain the equality. $\square$

We know that, for a general measure, the *continuity from above* at $\emptyset$ fails if for no member of a contracting sequence $\langle A_n \rangle$ we have $\nu(A_n) < \infty$. For example, consider the contracting sequence given by sets

$$E_n = (n, \infty), \quad n \ge 1,$$

then we have $E_n \downarrow \emptyset$, but $m(E_n) = +\infty$ for each n. So, the Lebesgue measure m in general is not continuous from above at $\emptyset$.

We next discuss a situation that is a sort of perfect (probability based) prototype to illustrate why it is important that the Lebesgue measure m must be defined as the extension of the *length function* ℓ.

Example 9.6 (*A Probability Connection*) The problem we are presenting here is known as the *law of large numbers*, which was first formulated by *James Bernoulli* (1654–1705). Suppose each trial of a random experiment such as tossing of a fair coin has only two possible outcomes, say a HEAD (H) or a TAIL (T). Then, a sequence $s : \mathbb{N} \to \{H, T\}$ of the form

$$\langle H, T, T, H, H, T, T, H, T, \ldots \rangle.$$

is called a *Bernoulli sequence*. If h_n denote the number of H's occurring in first n tosses, then the *Bernoulli's law* states (very roughly) that

$$\frac{h_n}{n} \to \frac{1}{2}, \quad \text{as} \quad n \to \infty.$$

We need to introduce some measure theoretic terminology to give a precise statement of the Bernoulli's law. Recall that the *sample space* B of all Bernoulli sequences is an *uncountable set* (Example 2.13). We associate with $s = \langle s_1, s_2, s_2, \ldots \rangle \in B$ a (non-terminating) binary representation of the form

$$\alpha_s = 0.a_1 a_2 a_3 \cdots = \sum_{i=1}^{\infty} \frac{a_i}{2^i}, \quad \text{with} \quad a_i = \begin{cases} 1, & \text{if } s_i = H \\ 0, & \text{if } s_i = T \end{cases}.$$

Then, each *event* $E \subset B$ is identified with the subset B_E of $(0, 1]$. For example, if $E_1 \subset B$ contains only those Bernoulli sequences that start with H then $\alpha \in B_{E_1}$ if and only if

$$\alpha \geq 0.1000 \cdots \quad \text{and} \quad \alpha \leq 0.111 \cdots .$$

That is, $B_{E_1} = [1/2, 1]$. Likewise, if $E_2 \subset B$ is the event wherein the first n outcomes are prescribed as $s_1, \ldots, s_n$, then $\alpha \in B_{E_2}$ if and only if

$$\alpha \geq \alpha_1 = 0.s_1 s_2 \cdots s_n 000 \cdots \quad \text{and} \quad \alpha \leq 0.s_1 s_2 \cdots s_n 111 \cdots .$$

Therefore, we have

$$0.s_1 s_2 \cdots s_n 111 \cdots = \alpha_1 + \sum_{i=n+1}^{\infty} \frac{a_i}{2^i} = \alpha_1 + \frac{1}{2^n},$$

and so it follows that

$$B_E = \left[\alpha_1, \alpha_1 + (1/2)^n \right].$$

In measure theoretic terms, we say that the probability $P(E)$ is given by the *measure* of the set $B_E \subset (0, 1]$. Hence, as long as the set B_E is an interval, we can the length $\ell(B_E)$ of B_E to obtain the *probability* $P(E)$. For example,

$$P(E_1) = \ell([1/2, 1]) = \frac{1}{2}, \quad \text{and},$$

$$P(E_2) = \ell([\alpha_1, \alpha_1 + (1/2)^n]) = \frac{1}{2^n}.$$

However, since the (measurable) sets $B_E \subset (0, 1]$ in general are very complicated, we need the *Lebesgue measure* $m(B_E)$ to compute the probability of the event E. Now, to give a formal statement of Bernoulli's *law of large number*, we write

$$s_n(\alpha) = a_1 + \cdots + a_n, \quad \text{for } \alpha \in (0, 1],$$

where $0.a_1 a_2 \cdots$ is a *binary representation* of α. For a fixed $\epsilon > 0$, let B_n denotes the event : *after first n tosses, the number of H is not close to* $1/2$, i.e.,

$$B_n = \left\{ \alpha \in (0, 1] : \left| \frac{s_n(\alpha)}{n} - \frac{1}{2} \right| > \epsilon \right\}, \quad n \in \mathbb{N}.$$

Then, the *weak law of large number* states that $m(B_n) \to 0$ as $n \to \infty$. And, for

$$B = \left\{ \alpha \in (0, 1] : \lim_{n \to \infty} \frac{s_n(\alpha)}{n} = \frac{1}{2} \right\},$$

the *strong law of large number* states that $m((0, 1] \setminus B) = 0$. So, $(0, 1] \setminus B$ is also an example of an *uncountable null set* in $\mathbb{R}$.

The illustration given in the above example is an adapted version of ([4], Sect. 3.1). The motivated reader may also like to read the related material from [5].

9.2.1 Regularity Property

The next theorem proves the *regularity property* of the Lebesgue outer measure m^* which, in particular, implies that every Lebesgue measurable set in $\mathbb{R}$ can be *approximated* by open sets that contain it. Notice that it also explains why the term *outer* is used in the first place.

Theorem 9.13 *Let $\epsilon > 0$ be arbitrary. Then, for each $A \subset \mathbb{R}$, there exists an open set $O \supset A$ such that*

$$m(O) \leq m^*(A) + \epsilon.$$

Consequently, for each measurable set E, there exists an open set $O \supset E$ such that $m(O \setminus E) < \epsilon$.

Proof Let $\langle I_n \rangle$ be a sequence of intervals such that

$$A \subset \bigcup_{n=1}^{\infty} I_n \quad \text{and} \quad \sum_{n=1}^{\infty} \ell(I_n) < m^*(A) + \frac{\epsilon}{2}. \tag{9.2.5}$$

As before, we may *enlarge infinitesimally* the two ends a_n, b_n of intervals I_n so that

$$I_n \subset J_n = \left(a_n - \frac{\epsilon}{2^{n+2}}, b_n + \frac{\epsilon}{2^{n+2}} \right).$$

Thus, $\ell(I_n) = \ell(J_n) - \frac{\epsilon}{2^{n+1}}$ implies that

$$\sum_{n=1}^{\infty} \ell(I_n) = \sum_{n=1}^{\infty} \ell(J_n) - \frac{\epsilon}{2}$$

$$\Rightarrow \quad \sum_{n=1}^{\infty} \ell(J_n) \leq \sum_{n=1}^{\infty} \ell(I_n) + \frac{\epsilon}{2} \leq m^*([a, b]) + \epsilon,$$

where the last inequality follows from (9.2.5). So, we may take $O = \cup_{n=1}^{\infty} J_n$, which contains A. Next, when $A = E$ is a Lebesgue measurable set, we have

$$m(E) < \infty \quad \Rightarrow \quad m(O \setminus E) = m(O) - m(E) \leq \epsilon.$$

So, to complete the proof, let us assume that $m(E) = \infty$. Notice that, for each $n \in \mathbb{N}$, the Lebesgue measurable sets $E_n = E \cap (-n, n)$ have finite measure. So, it is possible to find open sets $O_n \supset E_n$ with

$$m(O_n \setminus E_n) \leq \frac{\epsilon}{2^n}.$$

But then, $O = \bigcup_{n=1}^{\infty} O_n \supset \bigcup_{n=1}^{\infty} E_n = E \cap (\bigcup_{n=1}^{\infty}(-n, n)) = E$, and

$$O \setminus E = \left(\bigcup_{n=1}^{\infty} O_n \right) \setminus \left(\bigcup_{n=1}^{\infty} E_n \right) \subset \bigcup_{n=1}^{\infty} (O_n \setminus E_n)$$

implies that

$$m(O \setminus E) \leq \sum_{n=1}^{\infty} m(O_n \setminus E_n) = \sum_{n=1}^{\infty} \frac{\epsilon}{2^n} = \epsilon.$$

This completes the proof.   □

Equivalently, we prove that every measurable set can be *approximated* by closed sets it contains.

Theorem 9.14 *For any $\epsilon > 0$, and $E \in \mathcal{M}_{\mathbb{R}}$, there exists a closed set $F \subset E$ such that $m(E \setminus F) < \epsilon$. Consequently, there exists closed sets F_n such that the $\mathcal{F}_\sigma$ set*

$$B = \bigcup_{n=1}^{\infty} F_n \subset E, \quad with \ m(E \setminus B) = 0.$$

Proof Since $E' \in \mathcal{M}$, by Theorem 9.13, we find an open set $O \supset E'$ such that $m(O \setminus E') \leq \epsilon$. Then, taking $F = O' \subset E$, we have

$$O \setminus E' = O \cap E = E \setminus O'.$$

The rest of the details are similar as in the proof of Theorem 9.13.   □

It follows, in particular, that for every measurable set E there exists a Borel set $B \,(= \bigcap_{n=1}^{\infty} O_n)$ containing E such that $m(E) = m(B)$. So, we have

$$m(B \triangle E) = m(B \setminus E) = 0.$$

Theorem 9.15 *The σ-algebra $\mathcal{M}_{\mathbb{R}}$ of measurable sets is a completion of $\mathcal{B}(\mathbb{R})$.*

Proof By Theorem 9.11, we know that $\mathcal{B}(\mathbb{R}) \subset \mathcal{M}_{\mathbb{R}}$, and every subset of a null set in $\mathcal{B}(\mathbb{R})$ is measurable. By applying the first assertion of Theorem 9.13, with $\epsilon = \frac{1}{n}$, we obtain a sequence of open sets $\langle O_n \rangle$, each containing A, such that

$$m(O_n) \leq m^*(A) + \frac{1}{n}, \quad n \geq 1.$$

It thus follows that

$$E = \bigcap_{n=1}^{\infty} O_n \in \mathscr{G}_\delta$$

is a measurable set containing A, with

$$m(E) < m(O_n) \le m^*(A) + \frac{1}{n}, \quad \text{for all } n \ge 1.$$

So, $m(E) \le m^*(A)$. And, since the monotone property of m^* gives the reverse inequality, we obtain

$$m\left(\bigcap_{n=1}^{\infty} O_n\right) = m^*(A).$$

This proves that, for any $A \subset \mathbb{R}$, $m^*(A)$ can be *approximated from above* by $m(O_n)$, where $\{O_n\}$ is a sequence of open sets in $\mathbb{R}$. Likewise, by Theorem 9.14, we have that for any $A \subset \mathbb{R}$, $m^*(A)$ can be *approximated from below* by $m(F_n)$, where $\{F_n\}$ is a sequence of closed sets in $\mathbb{R}$. $\qquad\square$

9.3 Lebesgue Integration

In abstract terms, to compute the integral of a *nice function* $f : X \to Y$ over a *nice set* $E \subseteq X$, say representing some sort of (physical) *content* of f over E, a common approach is to choose some specific method (based on some *sufficiently evolved* theory) to form the smaller pieces of the *whole*, and subsequently make convenient the *"measurements with infinitesimals"*, by using a suitable *limit process*. A more significant aspect of the procedure is to understand what limiting process one may apply in a particular situation to *put together* the *"infinitely many measurements"*.

For example, the *length* $L(\gamma)$ of the curve traced by a *rectifiable path* $\gamma : [0, 1]) \to \mathbb{R}^n$ is found by *summing together* the lengths of all the *infinitesimal line segments* between *nearby points* on the curve, which are usually obtained by considering a fine enough partitions of the interval $[0, 1]$ in the Riemann sense. If γ is Lipschitz, then we have

$$L(\gamma) = \int_0^1 \left\| x'(t) \right\| dt, \quad \text{where } \gamma(t) = x(t) \in \mathbb{R}^n.$$

In general, the Lebesgue integration is about finding the geometric quantities such as *length, area, volume*, etc., in several different ways. The importance of Lebesgue's theory can be gauged from the fact that the story can be started from *almost anywhere*. In modern mathematics, the *Lebesgue integral* of a function is used to define an association between the objects of two spaces by putting together *coherently obtained data* for certain suitably chosen (measurable) infinitesimals of the whole.

However, as we started looking only for an improved theory of integration (at least in comparison to the one presented earlier in Chap. 4), it makes sense to recall that the (Riemann) *integral operator*

$$\int_{[a,b]} : B[a, b] \to C[a, b],$$

has many interesting properties of practical importance. For example, it behaves nicely with infinite sums, differentiation, certain improper functions, etc. Yet it lacks many other properties that an integral operator in general is expected to have. The following three limitations of Riemann's theory are more prominent:

1. The class $\mathcal{R}$ of Riemann integrable functions is too small to serve practical needs of many interesting applications.
2. There is no *easy-to-use* characterisation to decide the integrability of a function.
3. It lacks *completeness* in the sense that if $\langle f_n \rangle$ is a sequence of Riemann integrable functions, with $f_n \to f$ as $n \to \infty$, then it may not be true that

$$\int f_n \to \int f \quad \text{as} \quad n \to \infty. \quad \text{(Example 8.5)}$$

   The limit function f may not be Riemann integrable if convergence is *pointwise*.
4. None of the *fundamental convergence theorems*, as discussed at the end of the chapter, is valid in Riemann's theory of integration, without assuming the *uniform convergence*.

9.3.1 Measurable Functions

The Lebesgue's theory of integration is facilitated by the fact that countable sets in $\mathbb{R}$ are *null sets* (Theorem 9.10). In some cases, this fact helps to ignore countably many points of E where the function f is not *too badly discontinuous*. Therefore, it is possible to integrate a much wider class of functions such as *Dirichlet function*,[3] which are *not integrable* in the Riemann's theory (Theorem 9.24).

So, we start here with the fundamental concept of *almost everywhere* (or **a.e.**, in short). For example, we say a function $f : \mathbb{R} \to [-\infty, \infty]$ satisfies a property P **a.e.** if the set A of points $x \in \mathbb{R}$ where f fails the property P, is contained in some null set of $\mathbb{R}$. In this case, we write f **a.e.** P. So, if P refers to the *continuity*, then in Lebesgue sense the *Thomae function* is continuous **a.e.**, because it is continuous precisely at irrational numbers (Example 4.6), and the set $\mathbb{Q}$ is a null set.

---

[3] A yet another nasty function was constructed by Lebesgue himself, called the *Lebesgue's function* (or *devil's staircase*), which is a continuous function; it takes a measurable set to a non-measurable set; and it maps null set of the Cantor ternary set to a set of positive Lebesgue measure. The reader may refer to [1], for further details.

The next definition is given for an abstract measure space.

Definition 9.18 Let $(X, \mathscr{A}, \nu)$ be a measure space. We say a property P holds **almost everywhere** (in short **a. e.**) on X if there exists a set $A \in \mathscr{A}$ such that the property P holds on the set $A^c = X \setminus A$, and $A \subseteq B \in \mathscr{A}$, with $\nu(B) = 0$.

In all that follows, we are dealing with the measure space $(\mathbb{R}, \mathscr{M}, m)$, where m is the Lebesgue measure and $E \in \mathscr{M}$ is a Lebesgue measurable subset of $\mathbb{R}$. By above definition, if $f, g : \mathbb{R} \to [-\infty, \infty]$ are functions, we say $f = g$ **a. e.** if the set

$$E = \big\{ x \in \mathbb{R} : f(x) \neq g(x) \big\} \in \mathscr{M} \quad \text{with} \quad m(E) = 0.$$

In this case, we write $f = g$ **a. e.**. In the sequel, the equality of two functions defined on $\mathbb{R}$ is taken in this sense only. It is in general true that, in order to study a function $f : \mathbb{R} \to [-\infty, \infty]$ for any type of *measure theoretic* property P, the set E of points $x \in \mathbb{R}$ where f fails to satisfy P is a Lebesgue measurable set.

Definition 9.19 We say $f : \mathbb{R} \to [-\infty, \infty]$ is a **measurable function** if the set

$$f^{-1}\big((a, \infty]\big) = \big\{ x \in \mathbb{R} : f(x) > a \big\}$$

is Lebesgue measurable, for all $a \in \mathbb{R}$.

Notice that we can replace the condition $f(x) > a$ in Definition 9.19 by any one of the following three inequalities:

$$f(x) \geq a, \qquad f(x) < a, \qquad f(x) \leq a.$$

Also, since m is complete, for $f = g$ **a. e.** to hold the sets

$$E = h^{-1}\big((0, \infty]\big) \quad \text{and} \quad F = h^{-1}\big([-\infty, 0)\big), \quad \text{with} \quad h = f - g,$$

must be *Lebesgue measurable*, i.e., $E, F \in \mathscr{M}$. Therefore, if f is measurable and $f = g$ **a. e.**, then we have g is measurable.

As said earlier, the interval $(a, \infty]$ is open in the extended line $[-\infty, \infty]$. Also, since Borel sets are measurable, it follows that every continuous function $f : \mathbb{R} \to [-\infty, \infty]$ is measurable. In fact, it can be shown that monotone functions are also Borel measurable. Further, the composition of a measurable function $f : \mathbb{R} \to [-\infty, \infty]$ with a translation or dilation is again a measurable function. Further, given two measurable functions $f, g : \mathbb{R} \to [-\infty, \infty]$, it follows easily that the functions

$$f \pm g, \quad f \cdot g, \quad f^+ := \max\{f, 0\}, \quad f^- := \min(f, 0) = \max\{-f, 0\},$$

are measurable. For example, assuming the measurability of $f \pm g$, the familiar *polarisation identity*

$$f \cdot g = \frac{1}{4}(f+g)^2 - \frac{1}{4}(f-g)^2, \quad f \neq g,$$

proves a part of the second assertion; assertion for the case when $f = g$ follows by using

$$\{x \in \mathbb{R} : f^2(x) > a\} = \{x \in \mathbb{R} : f(x) > \sqrt{a}\} \cup \{x \in \mathbb{R} : f(x) < -\sqrt{a}\}.$$

Notice that we always have

$$f = f^+ - f^- \quad \text{and} \quad |f| = f^+ + f^-.$$

In particular, it follows that f is measurable $\Rightarrow |f|^p$ is measurable, for all $p \geq 1$. However, notice that there do exists non-measurable functions f such that $|f|$ is a measurable function. Further, for any sequence of measurable functions $\langle f_n \rangle$, the functions

$$\inf_n f_n, \quad \sup_n f_n, \quad \liminf f_k, \quad \limsup f_k$$

are measurable. For example,

$$\{x \in \mathbb{R} : (\sup_n f_n)(x) > a\} = \cup_{n=1}^{\infty} \{x \in \mathbb{R} : f_n(x) > a\}$$

proves that $\sup_n f_n$ is a measurable function.

In general, the best way to deal with an *arbitrary* measurable function on $\mathbb{R}$ is to use certain *simpler* types of functions as *approximates* (Theorem 9.16). In general, we say $s : \mathbb{R} \to \mathbb{R}$ is a *simple function* if the range $\text{Ran}(f) \subset \mathbb{R}$ is a finite set. For example, the *characteristic function* χ_E of a set $\emptyset \neq E \subset \mathbb{R}$ given by

$$\chi_E(x) = \begin{cases} 1, & x \in E \\ 0, & x \notin E \end{cases}, \quad \text{for } x \in \mathbb{R},$$

is a simple function. More generally, if $s : \mathbb{R} \to \mathbb{R}$ is a simple function, with

$$\text{Ran}(s) = \{a_1, \ldots, a_n\} \subset \mathbb{R},$$

then the sets $E_k = s^{-1}(a_k) \subset \mathbb{R}$, $1 \leq k \leq n$, form a *partition* of $\mathbb{R}$, and we have

$$s := a_1 \chi_{E_1} + \cdots + a_n \chi_{E_n}, \tag{9.3.1}$$

where the finite linear combination of characteristic functions on the right side of (9.3.1) is defined using the properties as given in Theorem 1.6. It is called the **standard representation** of the simple function s. For example, the *standard representation* of the simple function $s = \chi_{[0,2]} + \chi_{[1,3]}$ is given by

$$s = 0 \cdot \chi_{E_1} + 1 \cdot \chi_{E_2} + 2 \cdot \chi_{E_3},$$

where $E_1 = (-\infty, 0) \cup (3, \infty)$, $E_2 = [0, 1) \cup (2, 3]$, and $E_3 = [1, 2]$. Notice that *standard representation* of a simple function is unique only up to a rearrangement of terms on the right side of (9.3.1).

Every finite linear combination of simple functions, with real coefficients, is again a simple function, and so every finite linear combination of the form as in (9.3.1) is also a simple function. Also, it follows from Theorem 1.6 that the finite product of simple functions is a simple function. Further, for any two simple functions s, t, we have

$$\max(s, t) \quad \text{and} \quad \min(s, t)$$

are simple functions. In particular, so are the functions

$$s^+ = \max\{f, 0\}, \quad s^- = \min\{f, 0\}, \quad \text{and} \quad |s| = s^+ + s^-.$$

Notice that, since

$$E = \chi_E^{-1}(\{1\}), \quad \text{for any nonempty set} \ \ E \subset \mathbb{R},$$

we have that E is *measurable* if and only if so is the function χ_E. In general, by our comments given above, a simple function s of the form as in (9.3.1) is measurable if and only if $E_k \subset \mathbb{R}$ is a measurable set, for all $k = 1, \ldots, n$. In all that follows, we shall be working only with measurable simple functions.

Definition 9.20 A function $s : \mathbb{R} \to \mathbb{R}$ is called a **simple function** if it is a measurable function, with a finite range set $\mathrm{Ran}(f) \subset \mathbb{R}$.

The next theorem is the foundation of the concept of the Lebesgue integral of a measurable function defined over a measurable set in $\mathbb{R}$. Notice that since $f = f^+ - f^-$, where both

$$f^+ = \max\{f, 0\} \quad \text{and} \quad f^- = \max\{-f, 0\}$$

are positive functions, it suffices to prove the next theorem for the case when $f \geq 0$.

Theorem 9.16 (Approximation Theorem) *Any function $f : \mathbb{R} \to [0, \infty]$ can be approximated by an increasing sequence $\langle s_n \rangle$ of nonnegative simple functions on $\mathbb{R}$, i.e.,*

$$\lim_{n \to \infty} s_n(x) = f(x), \quad \text{for every} \ x \in \mathbb{R}.$$

The convergence is uniform on every subset of $\mathbb{R}$ where f is bounded. Moreover, if f is measurable, then we can chose the functions s_n to be measurable.

Proof The proof is based on a typical Lebesgue type of argument, namely, by dividing the range of the function f so that *horizontal lines* can be used as approximations.

Clearly, we need to partition the unbounded interval $[0, \infty]$. Let $G = f^{-1}(\{0\})$. In general, for $n, k \in \mathbb{N}$, with $1 \leq k \leq 2^{2n}$, we consider the sets

$$F_n = f^{-1}([2^n, \infty]) \quad \text{and} \quad E_n^{(k)} = f^{-1}\left(\left(\frac{k-1}{2^n}, \frac{k}{2^n}\right)\right).$$

Notice that, for example, at the first stage we are using the partition

$$[0, \infty] = \{0\} \cup (0, 1/2) \cup [1/2, 1) \cup [1, 3/2) \cup [3/2, 2) \cup [2, \infty]$$

so that $F_1 = f^{-1}([2, \infty])$, $E_1^{(1)} = f^{-1}((0, 1/2))$, $E_1^{(2)} = f^{-1}([1/2, 1))$, and so on. So, for each $n \in \mathbb{N}$, such type of $(2^{2n} + 1)$ sets form a partition of the set $f^{-1}((0, \infty])$, and we have

$$E_{n+1}^{(2k-1)} \bigcup E_{n+1}^{(2k)} = E_n^{(k)}, \quad \text{for } 1 \leq k \leq 2^{2n}; \tag{9.3.2}$$

$$\left(\bigcup_{m=2^{2n+1}+1}^{2^{2n+2}} E_{n+1}^{(m)} \right) \bigcup F_{n+1} = F_n. \tag{9.3.3}$$

Let $\langle s_n \rangle$ be the sequence of simple functions given by

$$s_n = 2^n \, \chi_{F_n} + \sum_{k=1}^{2^{2n}} \frac{k-1}{2^n} \chi_{E_n^{(k)}}, \quad x \in \mathbb{R}.$$

For example, we have

$$s_1 = 2 \cdot \chi_{F_1} + 0 \cdot \chi_{E_1^{(1)}} + \frac{1}{2} \cdot \chi_{E_1^{(2)}} + 1 \cdot \chi_{E_1^{(3)}} + \frac{3}{2} \cdot \chi_{E_1^{(4)}}.$$

Clearly, by construction, if f is measurable, then so are the functions s_n. By using (9.3.3), we also have

$$s_n = f = 0, \quad \text{(on the set } G\text{)};$$

$$s_n = \frac{k-1}{2^n} < f \leq \frac{k}{2^n} = s_n + \frac{1}{2^n}, \quad \text{(on sets } E_n^{(k)}\text{)};$$

$$s_n = 2^n < f, \quad \text{(on sets } F_n\text{)}.$$

Now, for $f(x) = \infty$, we have $x \in F_n$ for every n, and so

$$s_n(x) = 2^n \quad \rightarrow \quad \infty = f(x).$$

Next, for $f(x) < \infty$, we have $f(x) \leq 2^n$, for sufficiently large n, and so

$$0 \le f(x) - s_n(x) \le \frac{1}{2^n} \quad \Rightarrow \quad s_n(x) \to f(x).$$

If $K \subset \mathbb{R}$ is such that f is bounded on K then there is some N such that $f(x) \le 2^N$, for all $x \in K$. So, for all $n \ge N$, we have

$$0 \le f(x) - s_n(x) \le \frac{1}{2^n} \text{ for all } x \in K \quad \Rightarrow \quad s_n \to f,$$

uniformly over K. Finally, to prove that $\langle s_n \rangle$ is increasing, notice that $s_n = s_{n+1} = f$ on the set G. Next, by (9.3.2), we have

$$x \in E_n^{(k)} \quad \Rightarrow \quad s_n(x) = \frac{k-1}{2^n} \text{ and } s_{n+1}(x) = \frac{2k-2}{2^{n+1}} \text{ or } \frac{2k-1}{2^{n+1}},$$

which implies that $s_n \le s_{n+1}$ over the set $E_n^{(k)}$. Similarly, by using (9.3.3), it follows that

$$x \in F_n \quad \Rightarrow \quad s_n(x) \le s_{n+1}(x). \square$$

Therefore, the proof is complete. $\qquad\qquad\qquad\qquad\qquad\qquad\qquad\qquad$ $\square$

We say a function $s : \mathbb{R} \to \mathbb{R}$ is *elementary* if $\mathrm{Ran}(s)$ is at most countable. Notice that Theorem 9.16 also holds for elementary functions and, in this case as well, the convergence is uniform on sets over which f is bounded.

In 1912, the Russian mathematician *Nikolai Lusin* (1883–1950) proved that measurable functions are *almost* continuous (Theorem 9.18). To prove it, we apply the above *approximation theorem* and the following fundamental theorem of *Dmitri Egoroff* (1869–1931), which gives conditions so the *pointwise convergence* of a sequence $\langle f_n \rangle$ of real-valued measurable functions *almost everywhere* on a measurable set of finite measure imply the *almost uniform convergence*. Of course, the converse holds in general (Exercise 9.17).

Theorem 9.17 (Egoroff Theorem) *Let $E \subset \mathbb{R}$ be a measurable set of finite measure, and $\langle f_n \rangle$ be a sequence of measurable functions such that $f_n \to f$ almost everywhere on E, where $f : E \to \mathbb{R}$. Then, for any $\epsilon > 0$, there exists a measurable set $E_\epsilon \subset E$ such that $m(E \setminus E_\epsilon) < \epsilon$, and $f_n \to f$ uniformly over E_ϵ.*

Proof The proof is completed by constructing a sequence $\langle g_k \rangle$ of monotonically decreasing nonnegative measurable functions such that $g_k \to 0$ uniformly on E_ϵ. First, let

$$N = \left\{ x \in E : f_n(x) \not\to f(x) \right\}.$$

Then, by our assumption, we have N is a null set, and so $m(E) = m(E \setminus N)$, and $f_n \to f$ over $E \setminus N$. By an earlier comment, we have that f is measurable on $E \setminus N$. We take

$$g_k := \sup \left\{ |f_k - f|, |f_{k+1} - f|, \dots \right\}, \qquad \text{for } k \in \mathbb{N}.$$

Notice that we have each g_k is a measurable function on $E \setminus N, 0 \leq g_{k+1} \leq g_k$, and $g_k \to 0$ over the set $E \setminus N$, as $n \to \infty$. We claim that this convergence is uniform on whole of the set $E \setminus N$, except possibly a *negligible* part of it.

Let $\epsilon > 0$ be arbitrary. For $p \geq 1$, consider the measurable sets

$$E_k^{(p)} := \left\{ x \in E \setminus N : g_k(x) < \frac{1}{p} \right\}.$$

Notice that, for each $p \geq 1$, we have

$$E_1^{(p)} \subset E_2^{(p)} \subset \cdots \subset \lim_{k \to \infty} E_k^{(p)} = E \setminus N,$$

and $m(E_k^{(p)}) = m(E \setminus N)$, by measure properties of m. So, for each $p \geq 1$, we can choose some sufficiently large K_p such that

$$0 \leq m(E \setminus N) - m(E_{K_p}^{(p)}) < \frac{\epsilon}{2^p}.$$

The sets $E_{K_p}^{(p)}$ are *almost* the same as the set $E \setminus N$, and on such sets we have

$$0 \leq g_k < \frac{1}{2^p}, \qquad \text{for all } k \geq K_p.$$

Finally, we take $E_\epsilon = \bigcap_{p=1}^{\infty} E_{K_p}^{(p)}$. Then,

$$m\left((E \setminus N) \setminus \bigcap_{p=1}^{\infty} E_{K_p}^{(p)} \right) = m\left(\bigcup_{p=1}^{\infty} \left((E \setminus N) \setminus E_{K_p}^{(p)} \right) \right)$$

$$\leq \sum_{p=1}^{\infty} m\left((E \setminus N) \setminus E_{K_p}^{(p)} \right)$$

$$= \sum_{p=1}^{\infty} \left(m(E \setminus N) - m(E_{K_p}^{(p)}) \right)$$

$$< \sum_{p=1}^{\infty} \frac{\epsilon}{2^p} = \epsilon.$$

To complete the proof, we show that $g_k \to 0$ on E_ϵ. For, let $\delta > 0$ be arbitrary. By using the archimedean property, we choose $N \in \mathbb{N}$ such that $\frac{1}{N} < \delta$. Notice that

$$E_\epsilon = \bigcap_{p=1}^{\infty} E_{K_p}^{(p)} \subset E_{K_N}^{(N)} = \left\{ x \in E \setminus N : g_{K_N} < \frac{1}{N} \right\},$$

and $g_k \le g_{K_N}$, for all $k \ge K_N$. That is, $0 \le g_k < \delta$, for all $k \ge K_N$, and all $x \in E_\epsilon$.

$\square$

Notice that, by considering the sequence of measurable function $f_n = \chi_{[n,n+1]}$, it follows that the condition of finiteness of measure of E in the above cannot be relaxed.

Theorem 9.18 (Lusin) *Let $E \subset \mathbb{R}$ be a measurable set, with $m(E) < \infty$, and $f : E \to \mathbb{R}$ be a measurable function. Then, for any $\epsilon > 0$, there is some closed set $F_\epsilon \subset E$ such that $m(E \setminus F_\epsilon) < \epsilon$, and $f\big|_{F_\epsilon}$ is continuous.*

Proof By Theorem 9.16, we write f as the limit of a sequence $\langle s_k \rangle$ of simple functions on E, where each s_k is continuous except possibly at a finite set of points. So, the set of discontinuities D of all s_k is a countable set, which is a null set. Thus, there exists a sequence $\langle I_n \rangle$ of intervals such that

$$D \subseteq \bigcup_{n=1}^{\infty} I_n, \quad \text{and} \quad \sum_{n=1}^{\infty} \ell(I_n) < \frac{\epsilon}{2}.$$

Clearly then, on the closed set $E \setminus \bigcup_{n=1}^{\infty} I_n$, the simple functions s_k are continuous, and converge to f. Therefore, by using Theorem 9.17, we obtain a closed set $F_\epsilon \subset E \setminus \bigcup_{n=1}^{\infty} I_n$ such that

$$m\left(\left(E \setminus \cup_{n=1}^{\infty} I_n \right) \setminus F_\epsilon \right) < \frac{\epsilon}{2},$$

and $s_k \to f$ uniformly over F_ϵ. Hence, f is continuous. This completes the proof.

$\square$

9.3.2 Lebesgue Integral

Let $(\mathbb{R}, \mathcal{M}, m)$ be the Lebesgue measure space, $E \in \mathcal{M}$ be a (Lebesgue) *measurable set*, and $f : \mathbb{R} \to \mathbb{R}$ be a *measurable* function. Recall that the set $f^{-1}\big((a, \infty)\big) \in \mathcal{M}$, for all $a \in \mathbb{R}$. We prove here the *Lebesgue integral* of f over E is given as the limiting value of the integrals of certain *simple functions* approximating the function $\chi_E f$, which we will write as

$$\int_E f \, dm = \int_{\mathbb{R}} (\chi_E f) \, dm.$$

Our final aim is to formulate a definition of the integral $\int_E f \, dm$. Let us start with the case when f is simple function.

Definition 9.21 (*For Simple Functions*) Let $s : \mathbb{R} \to [0, \infty)$ be a simple measurable function, with the *standard representation* given by

$$s = \alpha_1 \chi_{E_1} + \cdots + \alpha_n \chi_{E_n}, \quad \alpha_i \in \mathbb{R}, \tag{9.3.4}$$

so that each $E_i \subset \mathbb{R}$ is measurable. Then, the **Lebesgue integral** of s over $\mathbb{R}$ is defined as follows:

$$\int_{\mathbb{R}} s \, dm = \sum_{k=1}^{n} \alpha_k \, m(E_k), \tag{9.3.5}$$

where m is the Lebesgue measure.

It is straightforward to see that equality (9.3.5) holds for any simple function of the form (9.3.4) even when it is not given in terms of a standard representation, i.e., when not all α_k are distinct or some E_k may be empty or the collection $\{E_k\}$ does not cover the line $\mathbb{R}$. Further, when some $\alpha_k = 0$ and $m(E_k) = \infty$, we shall follow the convention that $\alpha_k \, m(E_k) = 0$. Said differently, it is assumed that the value of integral is independent of points $x \in \mathbb{R}$, where $s(x) = 0$. We also have

$$\int_{\mathbb{R}} s \, dm < \infty \quad \Leftrightarrow \quad m(E_k) < \infty, \text{ for all other } k. \tag{9.3.6}$$

Therefore, the Lebesgue integral of a nonnegative simple (measurable) function $s : \mathbb{R} \to [0, \infty)$ is finite if and only if $s = 0$ outside a set of finite measure. Also,

$$\int_{\mathbb{R}} s \, dm = 0 \quad \Leftrightarrow \quad m(\{x \in \mathbb{R} : s(x) > 0\}) < \infty. \tag{9.3.7}$$

The next theorem proves some basic properties of the integral given by (9.3.5).

Theorem 9.19 *Let $s_1, s_2 : \mathbb{R} \to [0, \infty]$ be (measurable) simple functions, and α be a positive real number. Then, we have*

1. $\displaystyle\int_{\mathbb{R}} (s_1 + s_2) \, dm = \int_{\mathbb{R}} s_1 \, dm + \int_{\mathbb{R}} s_2 \, dm.$
2. $\displaystyle\int_{\mathbb{R}} (\alpha \, s_1) \, dm = \alpha \int_{\mathbb{R}} s_1 \, dm.$
3. $s_1 \leq s_2 \implies \displaystyle\int_{\mathbb{R}} s_1 \, dm \leq \int_{\mathbb{R}} s_2 \, dm.$
4. *If $E_n \subset \mathbb{R}$ are such that $E_n \uparrow \mathbb{R}$, then*

$$\lim_{n \to \infty} \int_{\mathbb{R}} (s_1 \chi_{E_n}) \, dm = \int_{\mathbb{R}} s_1 \, dm.$$

Proof Notice that if

$$s_1 = \sum_{i=1}^{n} \alpha_i \, \chi_{E_i} \quad \text{and} \quad s_2 = \sum_{j=1}^{m} \beta_j \, \chi_{F_j}$$

are standard representations, then

$$\sum_{i=1}^{n} \sum_{j=1}^{m} (\alpha_i + \beta_j) \chi_{E_i \cap F_j}, \tag{9.3.8}$$

is the standard representation of the sum function $s_1 + s_2$. For (4), apply the *continuity from below* property of m. The rest of the details are left as an easy exercise for the reader. $\qquad\square$

Theorem 9.20 *For $n \geq 1$, let $s, s_n : \mathbb{R} \to [0, \infty)$ be simple measurable functions so that $s_n \leq s_{n+1}$ on $\mathbb{R}$, for all n. Then*

1. $\lim\limits_{n \to \infty} s_n \leq s$ *on $\mathbb{R}$ implies that*

$$\lim_{n \to \infty} \int_{\mathbb{R}} s_n dm \leq \int_{\mathbb{R}} s\, dm;$$

2. $s \leq \lim\limits_{n \to \infty} s_n$ *on $\mathbb{R}$ implies that* $\int_{\mathbb{R}} s\, dm \leq \lim\limits_{n \to \infty} \int_{\mathbb{R}} s_n\, dm.$

Proof Notice that, by (3) of Theorem 9.19, the sequence of positive real numbers $\left\langle \int_{\mathbb{R}} s_n\, dm \right\rangle$ is monotone, and so $\lim\limits_{n \to \infty} \int_{\mathbb{R}} s_n\, dm$ exists in $[0, \infty]$. Also, by the same condition, we have

$$\int_{\mathbb{R}} s_n\, dm \leq \int_{\mathbb{R}} s\, dm.$$

This proves (1). Next, for (2), take $a \in [0, 1)$ and define

$$E_n := \{x \in \mathbb{R} : (as)(x) \leq s_n(x)\} \in \mathcal{M}.$$

Then, we have $E_n \uparrow \mathbb{R}$, and so by (4) of Theorem 9.19, it follows that

$$a \int_{\mathbb{R}} s\, dm = \int_{\mathbb{R}} (as)\, dm$$
$$= \lim_{n \to \infty} \int_{\mathbb{R}} (as\, \chi_{E_n})\, dm$$
$$\leq \lim_{n \to \infty} \int_{\mathbb{R}} s_n\, dm.$$

Finally, taking limit $a \to 1^-$, the assertion (2) follows. $\qquad\square$

In particular, it follows from (1) above that if $s_n, \sigma_n : \mathbb{R} \to [0, \infty]$ are two increasing sequences of simple measurable functions such that $\lim\limits_{n \to \infty} s_n = \lim\limits_{n \to \infty} \sigma_n$ holds over $\mathbb{R}$ then

$$\lim_{n \to \infty} \int_{\mathbb{R}} s_n\, dm = \lim_{n \to \infty} \int_{\mathbb{R}} \sigma_n\, dm. \tag{9.3.9}$$

Definition 9.22 (*For Nonnegative Functions*) Let $f : \mathbb{R} \to [0, \infty]$ be a measurable function, and $\langle s_n \rangle$ be any increasing sequence of nonnegative simple measurable functions defined over $\mathbb{R}$ such that $\lim_{n \to \infty} s_n = f$ over $\mathbb{R}$. We define the **Lebesgue integral** of f over $\mathbb{R}$ as follows:

$$\int_{\mathbb{R}} f \, dm := \lim_{n \to \infty} \int_{\mathbb{R}} s_n \, dm. \tag{9.3.10}$$

Notice that, by Theorem 9.16, there always exists an increasing sequence $\langle s_n \rangle$ of nonnegative measurable simple functions defined on $\mathbb{R}$ such that $\lim_{n \to \infty} s_n = f$. Also, by (9.3.9), the integral $\int_{\mathbb{R}} f \, dm$ is well-defined, which in general could be infinite. The next theorem proves some basic properties of the integral (9.3.10).

Theorem 9.21 *Let* $f, g : \mathbb{R} \to [0, \infty]$ *be measurable functions, and* $\alpha, \beta \in \mathbb{R}$. *Then*

1. *Lebesgue integral is linear:*

$$\int_{\mathbb{R}} (\alpha f + \beta g) dm = \alpha \int_{\mathbb{R}} f \, dm + \beta \int_{\mathbb{R}} g \, dm;$$

2. *Lebesgue integral is monotone:*

$$f \le g, \;\; over \;\; \mathbb{R} \;\; \Rightarrow \;\; \int_{\mathbb{R}} f \, dm \le \int_{\mathbb{R}} g \, dm;$$

3. $\int_{\mathbb{R}} f \, dm = 0 \Leftrightarrow f = 0 \, \boldsymbol{a.e.}$; *in particular,*

$$\int_{\mathbb{R}} f \, dm = \int_{\mathbb{R}} g \, dm \;\; \Leftrightarrow \;\; f = g \, \boldsymbol{a.e.}.$$

Proof The first two assertions follows directly from Theorem 9.19, using Definition 9.22. For (3), suppose $\int_{\mathbb{R}} f \, dm = 0$, and take $E_n = f^{-1}([1/n, \infty])$, for $n \in \mathbb{N}$. Then,

$$\frac{1}{n} \chi_{E_n} \le f \;\; \Rightarrow \;\; \frac{1}{n} m(E_n) = \int_{\mathbb{R}} \frac{1}{n} \chi_{E_n} dm \le \int_{\mathbb{R}} f \, dm = 0,$$

by (2). So, $m(E_n) = 0$ for all $n \in \mathbb{N}$. Thus, as

$$\{x \in \mathbb{R} : f(x) \ne 0\} = \bigcup_{n=1}^{\infty} E_n,$$

it follows that $m(\{x \in \mathbb{R} : f(x) \ne 0\}) = 0$. Conversely, suppose $f = 0$ **a.e.** on $\mathbb{R}$, and $\langle s_n \rangle$ be an increasing sequence of nonnegative measurable simple functions defined on $\mathbb{R}$ such that $\lim_{n\to\infty} s_n = f$ on $\mathbb{R}$. Clearly then, each $s_n = 0$ **a.e.** on $\mathbb{R}$, and so by our remark given after (9.3.5), we have $\int_{\mathbb{R}} s_n dm = 0$ for all $n \in \mathbb{N}$. Hence,

$$\int_{\mathbb{R}} f \, dm = \lim_{n\to\infty} \int_{\mathbb{R}} s_n dm = 0.$$

$\square$

Definition 9.23 (*For Arbitrary Function*) Let $f : \mathbb{R} \to [-\infty, \infty]$ be an arbitrary measurable function, and consider the functions f^+, $f^- : \mathbb{R} \to [0, \infty]$. If

$$\int_{\mathbb{R}} f^+ \, dm < \infty \quad \text{or} \quad \int_{\mathbb{R}} f^- \, dm < \infty,$$

then we define the **Lebesgue integral** of f over $\mathbb{R}$ by

$$\int_{\mathbb{R}} f \, dm := \int_{\mathbb{R}} f^+ \, dm - \int_{\mathbb{R}} f^- \, dm. \tag{9.3.11}$$

We say f is *Lebesgue integrable* over $\mathbb{R}$ if $\int_{\mathbb{R}} f \, dm < \infty$.

Notice that, in general, $\int_{\mathbb{R}} f \, dm$ could be infinite. Also, as $|f| = f^+ + f^-$ so that f^+, $f^- \le |f|$ over $\mathbb{R}$, we have respectively

$$\int_{\mathbb{R}} |f| \, dm := \int_{\mathbb{R}} f^+ \, dm + \int_{\mathbb{R}} f^- \, dm; \tag{9.3.12a}$$

$$\int_{\mathbb{R}} f^+ \, dm, \le \int_{\mathbb{R}} |f| \, dm; \tag{9.3.12b}$$

$$\int_{\mathbb{R}} f^- \, dm \le \int_{\mathbb{R}} |f| \, dm. \tag{9.3.12c}$$

So, equivalently, $f : \mathbb{R} \to [-\infty, \infty]$ is integrable if any one of the following two conditions hold:

1. $\int_{\mathbb{R}} f^+ \, dm < \infty$ and $\int_{\mathbb{R}} f^- \, dm < \infty$.

2. $\int_{\mathbb{R}} |f| \, dm < \infty$.

Further, if $E = \{x \in \mathbb{R} : |f(x)| = \infty\}$, then $\alpha \chi_E \le |f|$, for every $\alpha \in (0, \infty)$, implies that

$$\alpha \, m(E) = \int_{\mathbb{R}} (\alpha \, \chi_E) dm \le \int_{\mathbb{R}} |f| dm < \infty. \tag{9.3.13}$$

It thus follows that

$$m(E) \le \frac{1}{\alpha} \int_{\mathbb{R}} |f| dm.$$

Taking limit as $\alpha \to \infty$, we obtain $m(E) = 0$.

Therefore, if $f : \mathbb{R} \to [-\infty, \infty]$ is integrable, then $f(x) \in \mathbb{R}$ **a. e.** over $\mathbb{R}$. That is, $f : \mathbb{R} \to \mathbb{R}$ is integrable if and only if it is *absolutely integrable*. So, we see that a statement about Lebesgue integrability is actually a statement about *absolute integrability*. We write $L^1(\mathbb{R})$ for the set of all Lebesgue integrable functions $f : \mathbb{R} \to \mathbb{R}$ in the sense as described above. Also, it follows easily that, for $f \in L^1(\mathbb{R})$, the set of points $x \in \mathbb{R}$ such that $f(x) \neq 0$ is of σ-finite measure (Exercise 9.19).

The next theorem proves some basic properties of the integral (9.3.11).

Theorem 9.22 *Let $f, g : \mathbb{R} \to [-\infty, \infty]$ be measurable functions.*

1. *For any $\alpha \in \mathbb{R}$, if $\int_\mathbb{R} f \, dm$ is defined, then so is $\int_\mathbb{R} \alpha f \, dm$, and*

$$\int_\mathbb{R} \alpha f \, dm = \alpha \int_\mathbb{R} f \, dm.$$

   So, if f is integrable then so is αf.
2. *If both $\int_\mathbb{R} f \, dm$ and $\int_\mathbb{R} g \, dm$ are defined, and they are not opposite infinities, then $\int_\mathbb{R} (f + g) \, dm$ is defined and*

$$\int_\mathbb{R} (f + g) dm = \int_\mathbb{R} f \, dm + \int_E g \, dm.$$

   So, if f, g are integrable then so is $f + g$.
3. *If both $\int_\mathbb{R} f \, dm$ and $\int_\mathbb{R} g \, dm$ are defined, then*

$$f \leq g \ \text{ over } \ \mathbb{R} \ \Rightarrow \ \int_\mathbb{R} f \, dm \leq \int_\mathbb{R} g \, dm.$$

4. *If $\int_\mathbb{R} f \, dm$ is defined, then*

$$\left| \int_\mathbb{R} f \, dm \right| \leq \int_\mathbb{R} |f| \, dm.$$

Proof For (1), the proof follows easily by using

$$0 < \alpha < \infty \ \Rightarrow \ (\alpha f)^+ = \alpha f^+ \text{ and } (\alpha f)^- = \alpha f^-;$$
$$-\infty < \alpha < 0 \ \Rightarrow \ (\alpha f)^+ = -\alpha f^- \text{ and } (\alpha f)^- = -\alpha f^+.$$

For (2), suppose $\int_\mathbb{R} f^- dm < \infty$ and $\int_\mathbb{R} g^- dm < \infty$. Then, by Exercise 9.20, it follows that if

$$E = \{x \in \mathbb{R} : \ f(x) \neq -\infty \text{ and } g(x) \neq -\infty\},$$

we have $m(\mathbb{R} \setminus E) = 0$. The rest of the argument follows by applying Theorem 9.21 to the *positive* and *negative part* of the measurable functions $F = f \chi_E$ and $G = g \chi_E$, and by using Definition 9.23. Other cases can be dealt similarly. For (3), observe that

$$f \leq g = g^+ - g^- \leq g^+ \quad \Rightarrow \quad f^+ \leq g^+.$$

And, similarly we have $g^- \leq f^-$. Therefore, if $\int_{\mathbb{R}} g^+ dm < \infty$, then $\int_{\mathbb{R}} f^+ dm < \infty$. Likewise, if $\int_{\mathbb{R}} f^- dm < \infty$, then $\int_{\mathbb{R}} g^- dm < \infty$. So, the assertion follows by subtracting the inequalities

$$\int_{\mathbb{R}} f^+ dm \leq \int_{\mathbb{R}} g^+ dm, \quad \int_{\mathbb{R}} g^- dm \leq \int_{\mathbb{R}} f^- dm.$$

For (4), notice that

$$\left| \int_{\mathbb{R}} f\, dm \right| = \left| \int_{\mathbb{R}} f^+ dm - \int_{\mathbb{R}} f^- dm \right|$$
$$\leq \int_{\mathbb{R}} f^+ dm + \int_{\mathbb{R}} f^- dm$$
$$= \int_{\mathbb{R}} |f| dm.$$

This completes the proof.   □

More generally, as mentioned earlier, we say a measurable function $f : \mathbb{R} \to [-\infty, \infty]$ is **Lebesgue integrable** over a measurable set $E \subset \mathbb{R}$ if the function $f \chi_E : \mathbb{R} \to [-\infty, \infty]$ is integrable over $\mathbb{R}$. That is, if

$$\int_{\mathbb{R}} f^+ \chi_E\, dm < \infty \quad \text{or} \quad \int_{\mathbb{R}} f^- \chi_E\, dm < \infty.$$

So, in this case, the **Lebesgue integral** of f over E is given by

$$\int_E f\, dm := \int_E f^+\, dm - \int_E f^-\, dm$$
$$= \int_{\mathbb{R}} f^+ \chi_E\, dm - \int_{\mathbb{R}} f^- \chi_E\, dm.$$

(9.3.14)

Finally, we say f is *Lebesgue integrable over E* if $\int_E f\, dm < \infty$. Notice that, over $\mathbb{R}$, we have

$$(f \chi_E)^+ = f^+ \chi_E \leq f^+ \quad \text{and} \quad (f \chi_E)^- = f^- \chi_E \leq f^-.$$

We also have, for any measurable function $f : \mathbb{R} \to [-\infty, \infty]$ with $\int_{\mathbb{R}} f\, dm < \infty$,

$$\int_E f\, dm \in [-\infty, \infty) \quad \text{or} \quad \int_E f\, dm \in (-\infty, \infty].$$

All the theorem that are valid for the operator $\int_{\mathbb{R}}$ remain so for $\int_E$, for any measurable set $E \subseteq \mathbb{R}$. We only need to make some minor changes, such as replacing each appearance of f by $f\chi_E$, to write a proof for any one of these results. The next theorem states some basic properties of the integral as defined in (9.3.14).

Theorem 9.23 *For $E \in \mathcal{M}$, $f, g \in L^1(\mathbb{R})$, and $\alpha \in \mathbb{R}$,*

1. *Lebesgue integral is linear:*

$$\int_E \alpha f \, dm = \alpha \int_E f \, dm \quad and \quad \int_E (f+g)dm = \int_E f \, dm + \int_E g \, dm.$$

2. *Lebesgue integral is monotone:*

$$f \leq g \text{ for all } x \in R \quad \Rightarrow \quad \int_E f \, dm \leq \int_E g \, dm.$$

3. $|f| \in L^1(\mathbb{R})$ *and* $\left| \int_E f \, dm \right| \leq \int_E |f| \, dm.$
4. $m(E) = 0 \Rightarrow \int_E f \, dm = 0$. *In particular, for $F \subset E \in \mathcal{M}$ with $m(E \setminus F) = 0$,*

$$\int_E f \, dm = \int_F f \, dm.$$

Proof Left to the reader as an exercise.   □

Notice that, for any bounded interval $I \subset \mathbb{R}$, we have $m(I) = \ell(I)$ so that $dm(I) = dx$. It therefore follows that, in this situation, the Lebesgue integral of a bounded function $f : I \to \mathbb{R}$ is the same as the Riemann integral of $f(x)$. In fact, the idea extends easily so that we can talk about the integrability of f over any set $A \subset \mathbb{R}$ of finite measure. The main argument is that the integral $\int_A f \, dm$ does not change if countable partitions are used. We can show that lower sums $L(P; f)$ and upper sums $U(P; f)$ associated with any *countable partitions* P of A give an equivalent definition of integrability. Also, in this case, since

$$\sup_P L(P; f) \geq \sup_{P_{finite}} L(P; f);$$

$$\inf_P U(P; f) \leq \inf_{P_{finite}} U(P; f),$$

it is *a priori* more convenient for f to be integrable if countably infinite partitions are allowed.

Theorem 9.24 (Riemann-Lebesgue Theorem) *Let $I = [a, b]$, and $f : I \to \mathbb{R}$ be a bounded function. Then $f \in \mathcal{R}([a, b])$ if and only if the set of discontinuities of f is a null set.*

A proof given here uses the standard argument involving the *oscillations* of a function, as introduced earlier in Exercise 4.3.3 [6]. However, we need to introduce a generalised version of the concept of *oscillations*. For, let $I = [a, b]$, and $f : I \to \mathbb{R}$ be a bounded function. For any $A \subset \mathbb{R}$, let $a \in A \cap I$. We write

$$I_\delta(a) := \{x \in A \cap I : |x - a| < \delta\},$$

and take

$$M(f; a) := \sup\{|f(x) - f(y)| : x, y \in I_\delta(a)\}.$$

Then the *oscillation* of f at point a is given by

$$o(f; a) := \lim_{\delta \to 0^+} M(f; a).$$

Recall that f is continuous at a if and only if $o(f; a) = 0$. The reader is referred to Chap. 4 for any unexplained terminology or notations used in the proof given below.

Proof of Theorem 9.24 First, suppose $f \in \mathcal{R}([a, b])$. Let D be the set of discontinuities of f so that $A := D \cap I$ consists of points a with $o(f; a) > 0$. Therefore, taking

$$A_n := \left\{a \in I : o(f; a) \geq \frac{1}{n}\right\}, \quad n \in \mathbb{N}, \tag{9.3.15}$$

we can write $A = \bigcup_{n=1}^{\infty} A_n$. In order to prove that A is a *null set*, let $\epsilon > 0$ be arbitrary. Then, as $f \in \mathcal{R}([a, b])$, for each $k \in \mathbb{N}$ there is a partition

$$P_k : \quad a = x_0 \leq x_1 \leq \cdots \leq x_{n-1} \leq x_n = b$$

such that we have

$$U(f; P_k) - L(f; P_k) < \frac{\epsilon}{k}.$$

For any fixed i, if $a \in A_k \cap (x_{i-1}, x_i)$, then

$$\begin{aligned}
\frac{\epsilon}{k} &> U(f; P_k) - L(f; P_k) \\
&= \sum_{i=1}^{n} \left(M_i(f) - m_i(f)\right) \Delta x_i \\
&\geq \sum_{x_i \in D_k} \left(M_i(f) - m_i(f)\right) \Delta x_i,
\end{aligned} \tag{9.3.16}$$

where $D_k := A_k \cap (x_{i-1}, x_i) \neq \emptyset$ so that

$$\sum_{x_i \in D_k} \left(M_i(f) - m_i(f)\right) \Delta x_i \geq \frac{1}{k} \sum_{x_i \in D_k} \Delta x_i. \tag{9.3.17}$$

It thus follows from previous two equations that $\epsilon > \sum_{x_i \in D_k} \Delta x_i$. Now, taking J as the indices $0 \leq i \leq n$ appearing in the sum $\sum_{x_i \in D_k} \Delta x_i$, we have that the collection $\bigcup_{j \in J} (x_{j-1}, x_j)$ forms an open cover of the set A_k (except possibly the end points). Further, as

$$\sum_{j \in J} m\big((x_{j-1}, x_j)\big) = \sum_{x_i \in D_k} \Delta x_i < \epsilon,$$

we conclude that

$$m\big(A_k \setminus \{x_0, x_1, \ldots, x_n\}\big) = 0 \quad \text{and} \quad m(A_k) = 0, \quad \text{for all } k.$$

Hence, $m(A) = 0$, by Theorem 9.10.

To prove the converse, suppose the set of discontinuities A of a bounded function $f : [a, b] \to \mathbb{R}$ is a null set in $\mathbb{R}$. By Exercise 5.3, we know that the set

$$A_s := \big\{x \in [a, b] : \ \mathrm{o}(f; x) \geq s\big\}, \quad \text{for any } \ s > 0, \tag{9.3.18}$$

is compact in $[a, b]$. Notice that, if f is discontinuous at $x_0 \in [a, b]$, then $x_0 \in A_s$, for some $s > 0$. It thus follows that $A = \bigcup_{n=1}^{\infty} A_{1/n}$. Recall that we have $m(A) = 0$, and so $m(A_s) = 0$, for all $s > 0$. Let $\epsilon > 0$ be arbitrary. We can find a sequence $\langle I_n \rangle$ of open intervals such that

$$A_{\epsilon/2(b-a)} \subset \bigcup_{n=1}^{\infty} I_n \quad \text{and} \quad \sum_{n=1}^{\infty} \ell(I_n) < \frac{\epsilon}{2(M - m)},$$

where $M := \sup_{x \in [a,b]} f(x)$ and $m := \inf_{x \in [a,b]} f(x)$. Of course, it can be assumed that $M > m$. Since A_s is compact, for every $s > 0$, there exists a finite collection $\big\{I_j : 1 \leq j \leq K\big\}$ of intervals that covers the set $A_{\epsilon/2(b-a)}$. Taking

$$x \in [a, b] \setminus \left(\bigcup_{j=1}^{K} I_j\right) \subset [a, b] \setminus A_{\epsilon/2(b-a)},$$

we have $\mathrm{o}(f; x) < \epsilon/2(b - a)$. So, for each such x there exists some δ_x so that

$$y, z \in (x - \delta_x, x + \delta_x) \quad \Rightarrow \quad |f(y) - f(z)| < \frac{\epsilon}{2(b - a)}. \tag{9.3.19}$$

Clearly, the set $[a, b] \setminus \left(\bigcup_{j=1}^{K} I_j\right)$ is compact, with the collection

$$\big\{(x - \delta_x, x + \delta_x) : \ x \in [a, b] \setminus \big(\cup_{j=1}^{K} I_j\big)\big\}$$

as an open cover, so we can find finitely many points $\overline{x}_1, \ldots, \overline{x}_k$ and numbers $\delta_1, \ldots, \delta_k$ such that

$$[a, b] \setminus \left(\bigcup_{j=1}^{K} I_j \right) \subset \bigcup_{t=1}^{k} (\bar{x}_t - \delta_t, \bar{x}_t + \delta_t).$$

Notice that the (finite) collection C of intervals given by

$$(\bar{x}_t - \delta_t, \bar{x}_t + \delta_t); \quad I_j, \quad \text{for } 1 \le t \le k \text{ and } 1 \le j \le K,$$

forms an open cover of the interval $[a, b]$. Choose a partition

$$P := \{a = x_0, x_1, \ldots, x_{p-1}, x_p = b\}$$

such that each subinterval lies completely in exactly one open set of above cover of $[a, b]$. Suppose the set C_1 refers to those subintervals that are contained in $I_j \in C$, and C_2 refers to those subintervals that are contained in the intervals $(\bar{x}_t - \delta_t, \bar{x}_t + \delta_t) \in C$. We then have

$$U(f; P) - L(f; P) = \sum_{C_1} (M_i - m_i) \Delta x_i + \sum_{C_2} (M_i - m_i) \Delta x_i. \tag{9.3.20}$$

Moreover, we find that

$$\sum_{C_1} (M_i - m_i) \Delta x_i \le (M - m) \sum_{C_1} \Delta x_i$$

$$(M - m) \sum j = 1^K \ell(I_j)$$

$$< (M - m) \frac{\epsilon}{2(M - m)} = \frac{\epsilon}{2}; \tag{9.3.21}$$

and also

$$\sum_{C_2} (M_i - m_i) \Delta x_i \le \frac{\epsilon}{2(b - a)} \sum_{C_2} \Delta x_i \quad \text{(by (9.3.19))}$$

$$< \frac{\epsilon}{2(b - a)} (b - a) = \frac{\epsilon}{2}. \tag{9.3.22}$$

Therefore, by using equations (9.3.21) and (9.3.22) in (9.3.20), we obtain

$$U(f; P) - L(f; P) < \epsilon.$$

Hence, $f \in \mathscr{R}([a, b])$. This completes the proof. □

Notice that the above theorem gives a criterion of Riemann integrability that does not involve partitions of the interval $[a, b]$. It is based only on the analytical properties of the bounded function $f : [a, b] \to \mathbb{R}$. Also, it took care of two major limitations of Riemann's theory, as mentioned in the beginning of the chapter. The convergence theorems given in the next section would take care of the third.

9.4   Fundamental Convergence Theorems

We prove here some important *convergence theorems* for Lebesgue integral. Recall that similar convergence theorems for Riemann integral are based on as stronger condition as the *uniform convergence* (see Theorem 8.5 and Theorem 8.7). We start with the *monotone convergence theorem* due to Lebesgue and Levi.

Theorem 9.25 (Monotone Convergence Theorem) *For $n \geq 1$, let $f, f_n : \mathbb{R} \to [0, \infty]$ be measurable functions, with $f_n \leq f_{n+1}$ **a.e.** over $\mathbb{R}$, such that $\lim_{n \to \infty} f_n = f$ **a.e.** over $\mathbb{R}$. Then*

$$\lim_{n \to \infty} \int_{\mathbb{R}} f_n dm = \int_{\mathbb{R}} f \, dm.$$

Proof First, assume that $f_n \leq f_{n+1}$ and $\lim_{n \to \infty} f_n = f$ over $\mathbb{R}$. By using the monotone property of Lebesgue integrals, we obtain

$$\lim_{n \to \infty} \int_{\mathbb{R}} f_n dm = \int_{\mathbb{R}} f \, dm. \tag{9.4.1}$$

Next, let $\langle s_n \rangle$ be an increasing sequence of nonnegative measurable simple functions defined on $\mathbb{R}$ such that $\lim_{n \to \infty} s_n = f$. Then, $s_k \leq f$ for every $k \in \mathbb{N}$. As before, for $a \in [0, 1]$, take

$$E_n := \big\{ x \in \mathbb{R} : (a\, s_k)(x) \leq f_n(x) \big\} \in \mathcal{M}.$$

Clearly, we have $E_n \uparrow \mathbb{R}$, and $a\, s_k \chi_{E_n} \leq f_n$ over $\mathbb{R}$. So, by Theorem 9.20,

$$a \int_{\mathbb{R}} s_k dm = \int_{\mathbb{R}} (a\, s_k) dm$$

$$= \lim_{n \to \infty} \int_{\mathbb{R}} (a\, s_k \chi_{E_n}) dm$$

$$\leq \lim_{n \to \infty} \int_{\mathbb{R}} f_n \, dm.$$

Thus, by taking the limit $a \to 1^-$, it follows that

$$\int_{\mathbb{R}} s_k dm \leq \lim_{n \to \infty} \int_{\mathbb{R}} f_n dm.$$

Finally, by taking the limit as $k \to \infty$, we obtain the reverse inequality in (9.4.1). The general case follows easily from this. For, let $E \subset \mathbb{R}$ be a measurable set such that $m(\mathbb{R} \setminus E) = 0$, and consider the conditions $f_n \leq f_{n+1}$ and $\lim_{n\to\infty} f_n = f$ over E. Then, by applying the previous argument to the sequence $g_n = f_n \chi_E$, with $g = f \chi_E$, the proof follows by (3) of Theorem 9.21. $\square$

Corollary 9.2 *Let* $f, f_n : \mathbb{R} \to [0, \infty]$ *be measurable functions such that* $\sum_{n=1}^{\infty} f_n = f$ *a. e. over* $\mathbb{R}$. *Then*

$$\sum_{n=1}^{\infty} \int_{\mathbb{R}} f_n dm = \int_{\mathbb{R}} f \, dm.$$

Proof Clearly, the sequence given by $g_n = f_1 + \cdots + f_n$ satisfies the hypotheses of Theorem 9.25, with $\lim_{n\to\infty} g_n = f$ **a.e.** over $\mathbb{R}$. So, the proof follows, by using (1) of Theorem 9.21. $\square$

The following is known as *Fatou's Lemma*.

Theorem 9.26 (Fatou) *Let* $f, f_n : \mathbb{R} \to [0, \infty]$ *be measurable functions. If* $f = \liminf_{n\to\infty} f_n$ *a. e. over* $\mathbb{R}$, *then*

$$\int_{\mathbb{R}} f \, dm \leq \liminf_{n\to\infty} \int_{\mathbb{R}} f_n dm.$$

Proof Apply the monotone property of Lebesgue integral and the Monotone Convergence Theorem to the sequence $g_n = \inf_{k \geq n} f_k$, with $g_n \leq f_n$ on $\mathbb{R}$, for all $n \in \mathbb{N}$. Notice that, by assumption, we have $f = \lim_{n\to\infty} g_n$ **a.e.** on $\mathbb{R}$. $\square$

Theorem 9.27 (Dominated Convergence Theorem) *Let* $f, f_n : \mathbb{R} \to [-\infty, \infty]$ *be measurable functions, and* $g : \mathbb{R} \to [0, \infty]$ *be such that* $\int_{\mathbb{R}} g \, dm < \infty$. *Suppose, on* $\mathbb{R}$, $f = \lim_{n\to\infty} f_n$ *a. e.*, $|f_n| \leq g$ *a. e. for all* n, *and that* $\int_{\mathbb{R}} g \, dm < \infty$. *Then, all* f_n *and* f *are integrable, and*

$$\int_{\mathbb{R}} f \, dm = \lim_{n\to\infty} \int_{\mathbb{R}} f_n dm.$$

Proof Notice that $|f_n| \leq g$ **a.e.** for all n implies that

$$\int_{\mathbb{R}} f_n dm \leq \int_{\mathbb{R}} g \, dm < \infty,$$

and so each f_n is integrable. Further, $f = \lim_{n\to\infty} f_n$ **a.e.** on $\mathbb{R}$ implies that $|f| \leq g$ **a.e.** on $\mathbb{R}$, which implies that f is also integrable. In view of the remark made after

Definition 9.23, we may consider the functions $F, F_n : \mathbb{R} \to \mathbb{R}$ such that $F = f$ and $F_n = f_n$ **a.e.** for all n, on $\mathbb{R}$. It thus suffices to prove that

$$\int_{\mathbb{R}} F \, dm = \lim_{n \to \infty} \int_{\mathbb{R}} F_n dm.$$

For, since $g + F_n, g - F_n \geq 0$ on $\mathbb{R}$, it follows from Theorem 9.26 that

$$\int_{\mathbb{R}} g \, dm \pm \int_{\mathbb{R}} F \, dm \leq \liminf_{n \to \infty} \int_{\mathbb{R}} (g \pm F_n) dm,$$

and so

$$\int_{\mathbb{R}} g \, dm \pm \int_{\mathbb{R}} F \, dm \leq \int_{\mathbb{R}} g \, dm + \liminf_{n \to \infty} \pm \int_{\mathbb{R}} F_n dm.$$

We thus obtain

$$\pm \int_{\mathbb{R}} F \, dm \leq \liminf_{n \to \infty} \int_{\mathbb{R}} (g \pm F_n) dm,$$

and, therefore,

$$\limsup_{n \to \infty} \int_{\mathbb{R}} (g \pm F_n) dm \leq \int_{\mathbb{R}} F \, dm \leq \liminf_{n \to \infty} \int_{\mathbb{R}} (g \pm F_n) dm,$$

which completes the proof. $\qquad \square$

Corollary 9.3 *Let $f, f_n : \mathbb{R} \to [-\infty, \infty]$ be measurable functions such that $\sum_{n=1}^{\infty} \int_{\mathbb{R}} |f_n| \, dm < \infty$. Then, $\sum_{n=1}^{\infty} f_n(x) < \infty$ a.e. for all $x \in \mathbb{R}$. And, if $f = \sum_{n=1}^{\infty} f_n$ a.e. on $\mathbb{R}$, then $\int_{\mathbb{R}} f \, dm = \sum_{n=1}^{\infty} \int_{\mathbb{R}} f_n dm$.*

Proof By applying Corollary 9.2 to the function $g : \mathbb{R} \to [0, \infty]$ given by

$$g(x) := \sum_{n=1}^{\infty} |f_n(x)|, \quad x \in \mathbb{R},$$

it follows that

$$\int_{\mathbb{R}} g \, dm = \sum_{n=1}^{\infty} \int_{\mathbb{R}} |f_n| \, dm < \infty.$$

So, $g < \infty$ **a.e.** on $\mathbb{R}$. That is, the series $\sum_{n=1}^{\infty} f_n(x)$ converges absolutely, and so converges **a.e.** on $\mathbb{R}$. Now, by taking $s_n = \sum_{k=1}^{n} f_k$, we have

$$\lim_{n \to \infty} s_n = f \text{ a. e. } \quad \text{and} \quad |s_n| \le g, \quad \text{on } \mathbb{R},$$

which completes the proof, by using Theorem 9.27.   □

9.5 L^p Spaces

As mentioned earlier in Example 6.4, the notation ℓ_p is a shorthand for $\ell_p(\mathbb{N})$, which is a *discrete case* version of the *Lebesgue space* $L^p(\mathbb{R})$ of real-valued Lebesgue p-integrable functions over $\mathbb{R}$. We give here a brief introduction to such types of spaces. More generally, let $(X, \mathscr{A}, \mu)$ be a measure space, $E \subseteq X$ be a measurable set, and $\mathscr{L}^p(E)$ denote the collection of μ-measurable functions $f : E \to \mathbb{R}$ such that

$$\int_E |f|^p \, d\mu < \infty,$$

where p-integrability of f over E is taken in the sense of definition given in (9.3.14). We then say f is p-*integrable function* over the set $E \in \mathscr{A}$. For $f, g \in \mathscr{L}^p(E)$, we define $f \sim_\mu g$ if and only if $f = g$ **a. e.**, i.e.,

$$\mu(\{x \in E : f(x) \ne g(x)\}) = 0.$$

Clearly then, $\sim_\mu$ is an equivalence relation on the set $\mathscr{L}^p(E)$. We write $L^p(E)$ for the set of all equivalence classes, which is a vector space with respect to pointwise operations. Also, with respect to metric given by

$$d_p(f, g) = \left(\int_E |f - g|^p \, d\mu \right)^{1/p}, \quad \text{for } 1 \le p < \infty; \tag{9.5.1a}$$

$$d_\infty(f, g) = \inf \{\alpha > 0 : \mu(\{x \in E : |f(x) - g(x)| > \alpha\}) = 0\}, \tag{9.5.1b}$$

the set $L^p(E)$ is a metric space, for all $1 \le p \le \infty$. For $1 \le p < \infty$, the triangle inequality of d_p follows from the following famous inequalities:

1. (Hölder's Inequality) For q being the *conjugate exponent* of $1 < p < \infty$, i.e., $(1/p) + (1/q) = 1$, if f and g are measurable functions on E, then

$$\int_E |fg| \, d\mu \le \left(\int_E |f|^p d\mu \right)^{1/p} \left(\int_E |g|^q d\mu \right)^{1/q}. \tag{9.5.2}$$

2. (Minkowski's Inequality) Assuming that the sets

$$f^{-1}(\infty) \bigcap g^{-1}(-\infty) \quad \text{and} \quad f^{-1}(-\infty) \bigcap g^{-1}(\infty)$$

are μ-null sets, then we have

$$\left(\int_E |f + g|^p d\mu\right)^{1/p} \le \left(\int_E |f|^p d\mu\right)^{1/p} + \left(\int_E |g|^p d\mu\right)^{1/p}. \quad (9.5.3)$$

The proof of these inequalities can be written by using arguments as given in Chap. 6. Next, for $p = \infty$, we say a measurable function $f : E \to [-\infty, \infty]$ is **essentially bounded** if there exists $\alpha > 0$ such that the set $\{x \in E : |f(x)| > \alpha\}$ is a μ-null set. Any such α is called an *essential bound* of the function $|f|$, which we may write as esb($|f|$), and the infimum of all esb($|f|$) is called the **essential bound** of the function $|f|$ over E. We may write

$$\text{essup}(|f|) := \inf \left\{ \text{esb}(|f|) \right\}.$$

For example, every simple measurable function on E is essentially bounded. In particular, if any such type of function is zero outside a set of finite measure, then it is p-integrable function. Notice that an essentially bounded measurable function on a measurable set E may not be bounded. A simple example is given by the function

$$f(x) = \begin{cases} n, & \text{for } x = (1/n), \quad n \in \mathbb{N} \\ x, & \text{otherwise} \end{cases}$$

It follows easily that the sum of two essentially bounded measurable functions f and g on a measurable set E is essentially bounded, and

$$\text{essup}(|f + g|) \le \text{essup}(|f|) + \text{essup}(|g|).$$

Therefore, the linear space $L^\infty(E)$ of essentially bounded measurable function on a measurable set E is a metric space with respect to the metric d_∞ as given in (9.5.1).

Theorem 9.28 *For $1 \le p \le \infty$, each the linear space $L^p(E)$ is a complete metric space with respect to the metric d_p as given in (9.5.1).*

Proof Left to the reader as an exercise. $\square$

Definition 9.24 For $1 \le p < \infty$, we call $L^p(E)$ is called the **Lebesgue space** of real-valued p-integrable functions defined over the set E (or simply the **L^p-spaces** over E).

Remark 9.1 In some interesting cases, such as the probability theory, we take $E = [0, 1]$ so that $\mathscr{A}$ is the σ-algebra of Lebesgue measurable sets in the interval $[0, 1]$, and μ is the induced Lebesgue measure on $[0, 1]$. In this case, we write $L^p(E)$ simply as $L^p[0, 1]$. Other terminology such as $L^p[a, b]$ or $L^p[0, \infty)$ has a similar meaning.

Our main concern in the next chapter is to apply the tools developed here in the study of the Fourier series of real-valued periodic functions f in the space $L^1[a, a + 2\pi]$ or $L^2[a, a + 2\pi]$ of period 2π, where in most interesting cases we have $a \in \{0, -\pi\}$.

Definition 9.25 A function $f : \mathbb{R} \to \mathbb{R}$ is called **absolutely integrable** if $f \in L^1(\mathbb{R})$, and **square integrable** if $f \in L^2(\mathbb{R})$.

The results stated in the next theorem are used in the sequel.

Theorem 9.29 *Let $E \subseteq \mathbb{R}$ be a Lebesgue measurable set, and $1 \le p < \infty$. Then, we have*

1. *The set of simple measurable functions on E, which are zero outside a set in $\mathbb{R}$ of finite measure, is dense in $(L^p(E), d_p)$.*
2. *The set of simple measurable functions on E is dense in $(L^\infty(E), d_\infty)$.*
3. *If $m(E) < \infty$, then $B(E) \bigcap C(E)$ is dense in $(L^p(E), d_p)$.*
4. *The set of step functions over E is dense in $L^p[a, b]$.*

Proof Left to the reader as an exercise. □

Exercises

9.1 Let X be a nonempty set. A class $\mathcal{M} \subset \mathcal{P}(X)$ is called a **monotone class** on X if, for every monotone sequence $\langle A_n \rangle$ in $\mathcal{M}$, we have $\lim_{n \to \infty} A_n \in \mathcal{M}$. Show that

   a. For every class $\mathcal{C} \subset \mathcal{P}(X)$, there is a smallest *monotone class* $\mathcal{M}_\mathcal{C}$ containing $\mathcal{C}$.
   b. If $\mathcal{M}_\mathcal{C}$ is closed with respect to the operations of complementation and finite unions, then $\mathcal{M}_\mathcal{C}$ is a σ-algebra.
   c. If $\mathcal{A}$ is an algebra on the set X, then $\mathcal{M}_\mathcal{C} = \sigma(\mathcal{A})$.
   d. (Halmos) If an algebra $\mathcal{A}$ on the set X is contained in a monotone class $\mathcal{M}$, then $\sigma(\mathcal{A}) \subset \mathcal{M}$.

9.2 Let μ be an additive function defined on a ring $\mathcal{R}$ that is closed under countable unions. Show that μ is σ-additive if and only if it is σ-subadditive.

9.3 Let μ be an additive function defined on a ring $\mathcal{R}$ that is closed under countable unions. Show that μ is σ-additive if and only if μ is continuous from below.

9.4 Let μ be a σ-additive function defined on a σ-algebra $\mathcal{A}$, and $\langle A_n \rangle \subset \mathcal{A}$. Show that

$$\mu\left(\liminf_{n \to \infty} A_n\right) \le \liminf_{n \to \infty} \mu(A_n).$$

9.5 Let μ be a σ-additive function defined on a σ-algebra $\mathcal{A}$, and $\langle A_n \rangle \subset \mathcal{A}$ such that $\sum_{n=1}^\infty \mu(A_n) < \infty$. Show that

$$\mu\left(\limsup_{n \to \infty} A_n\right) = 0.$$

9.6 Let (X, Ω, ν) be a measure space. Show that if ν is σ-finite then it is *semifinite*, i.e., for every $A \in \Omega$ with $\nu(A) = \infty$, there is a set $B \in \Omega$ such that $B \subset A$, and $0 < \nu(B) < \infty$.

9.7 Let X be a nonempty set, and $\mu^* : \mathscr{P}(X) \to [0, \infty]$ be an outer measure. Show that if $\mu^*(E) = 0$ then E is a μ^*-measurable set, and

$$\mu^*(A) = \mu^*(A \setminus E), \quad \text{for all} \quad A \subseteq X.$$

9.8 Prove that every outer measure $m^* : \mathscr{P}(\mathbb{R}) \to [0, \infty]$ is a monotone, countably subadditive, translation invariant set function. Further, show that there exists a collection of pairwise disjoint sets $A_n \subset \mathbb{R}$ such that

$$m^*\left(\bigcup_{n=1}^{\infty} A_n\right) \neq \sum_{n=1}^{\infty} m^*(A_n).$$

Do we really need to consider a countable collection to prove the second assertion?

9.9 Let $m^* : \mathscr{P}(\mathbb{R}) \to [0, \infty]$ be an outer measure on $\mathbb{R}$, and $A, B \in \mathscr{P}(\mathbb{R})$ be arbitrary, with $m^*(A) = 0$. Show that $m^*(A \cup B) = m^*(B)$, and hence deduce that

$$m^*(A \triangle B) = 0 \quad \Rightarrow \quad m^*(A) = m^*(B).$$

9.10 Show that the countably additive set function

$$m = m^*\big|_{\mathscr{M}_\mathbb{R}} : \mathscr{M}_\mathbb{R} \to \mathbb{R}^+ \cup \{+\infty\}$$

satisfies all the properties that a general measure has.

9.11 Let $E \subseteq \mathbb{R}$. Prove that the following are equivalent:

   a. $E \in \mathscr{M}_\mathbb{R}$.
   b. For every $\epsilon > 0$, there exists an open set $O \supset E$ such that $m^*(O \setminus E) \leq \epsilon$.
   c. There exists a $\mathscr{G}_\delta$ set $G \supset E$ such that $m^*(G \setminus E) = 0$.
   d. For every $\epsilon > 0$, there exists a closed set $F \subset E$ such that $m^*(E \setminus F) \leq \epsilon$.
   e. There exists a $\mathscr{F}_\sigma$ set $F \subset E$ such that $m^*(E \setminus F) = 0$.

9.12 Show that every monotone function $f : \mathbb{R} \to [-\infty, \infty]$ is Borel measurable.

9.13 Show that composition of a measurable function $f : \mathbb{R} \to [-\infty, \infty]$ with translation or dilation is again a measurable function.

9.14 Let $f, g : \mathbb{R} \to [-\infty, \infty]$ be two measurable functions. Show that all the following three functions are measurable:

$$f \cdot g; \quad f^+ = \max\{f, 0\}; \quad f^- = \max\{-f, 0\}.$$

9.15 Show that f is measurable implies $|f|^p$ is measurable, for all $p \geq 1$. Give an example of a non-measurable functions f such that $|f|$ is measurable.

9.16 Show that, for any sequence $\langle f_n \rangle$ of measurable functions, all the following three functions are measurable:

$$\inf_n f_n, \quad \liminf f_k, \quad \text{and} \quad \limsup f_k.$$

9.17 Let $(X, \mathscr{A}, \mu)$ be a measure space, and $\langle f_n \rangle$ be an almost uniformly Cauchy sequence of almost everywhere finite real-valued measurable functions defined on X. Prove that there exists an almost everywhere finite real-valued measurable function f on X such that $\langle f_n \rangle$ converges almost uniformly and almost everywhere to f over X.

9.18 Show that a simple function $s = \sum_{k=1}^n a_k \chi_{E_k}$ is integrable on a measurable set $E \subset \mathbb{R}$ if and only if $m(\{x \in E : s(x) \neq 0\}) < \infty$.

9.19 Show that, for any Lebesgue integrable function $f : \mathbb{R} \to [-\infty, \infty]$, the set

$$E = \{x \in \mathbb{R} : f(x) \neq 0\}$$

has σ-finite measure.

9.20 Let $f, g : \mathbb{R} \to [-\infty, \infty]$ be measurable functions such that $f = g$ **a. e.** on $\mathbb{R}$. Show that, if f is Lebesgue integrable, then so is g, and $\int_{\mathbb{R}} f \, dm = \int_{\mathbb{R}} g \, dm$.

9.21 Let $f : \mathbb{R} \to [-\infty, \infty]$ be a measurable function. Show that

$$f = 0 \text{ \textbf{a. e.} on } \mathbb{R} \quad \Leftrightarrow \quad \int_{\mathbb{R}} |f| dm = 0 \quad \Leftrightarrow \quad \int_{\mathbb{R}} f \chi_E dm = 0,$$

for any $E \in \mathscr{M}_{\mathbb{R}}$.

9.22 Supply a complete proof of Theorem 9.19.

9.23 Let $f : \mathbb{R} \to [-\infty, \infty]$ be an integrable function. Show that, for every $\epsilon > 0$, there is some measurable simple function $\varphi : \mathbb{R} \to \mathbb{R}$ such that $\int_{\mathbb{R}} |f - \varphi| dm < \epsilon$.

9.24 Supply a proof of Theorem 9.23.

9.25 Show that if $f : \mathbb{R} \to \mathbb{R}$ is a bounded function then the set of discontinuities of f is a countable union of closed sets.

9.26 Supply a proof of Theorem 9.28.

9.27 Supply a proof of Theorem 9.29.

References

1. J. Yeh, *Real Analysis: Theory of Measure and Integration* (World Scientific, 2006)
2. M.E. Munroe, *Introduction to Measure and Integration* (Addison-Wesley, Cambridge, 1953)
3. H.L. Royden, *Real Analysis (3/e)* (Macmillan, New York, 1988)
4. K. Saxe, *Begining Functional Analysis* (Springer, New York, 2002)
5. F. Burk, *Lebesgue Measure and Integration* (John Wiley, 1998)
6. J.R. Kirkwood, *An Introduction to Analysis (2/e)* (PWS Pbulishing, Boston, 1995)

Chapter 10
Fourier Series

Profound study of nature is the most fertile source of mathematical discoveries.

Joseph Fourier (1768–1830)

The chapter is an introduction to the theory of Fourier series. We present here some aspects of Fourier series as applicable to practical problems in physics and engineering concerning a wide range of phenomena such as mechanical vibrations, heat conduction, acoustic, and electromagnetic waves, all governed by *waves*. In a typical practical situation, a *continuous-time* periodic signal $f : [a, b] \to \mathbb{R}$ is used to represent a wave of period $b - a$, where it can be assumed that the associated *periodic extension* $f : \mathbb{R} \to \mathbb{R}$ is an *absolutely integrable* function, i.e., $f \in L^1(\mathbb{R})$. We remark that most periodic signals, such as $\sin(t)$ and $\cos(t)$, we come across in various applications do satisfy this condition.

In general, a *mechanical wave* travels through an elastic medium (with inertia) as a periodic disturbance in energy levels, whereas the *electromagnetic waves* are produced by the vibrations of an electric charge in an electric and/or magnetic fields. It is always assumed that waves in any physical system have a measurable *speed of propagation*, and the direction of energy transfer is called a *ray*.

A mechanical wave is called *longitudinal* (or compressional) if the particles of the medium are displaced along a ray. Such types of waves are formed by a train of periodically alternating regions of compression and rarefaction. On the other hand, a wave such as *ocean tide*, is called *transversal* because the displacement at each point of the medium is perpendicular to the ray. Both types of waves exist in solids—a longitudinal wave may be caused by a compressional or a dilatation strains, whereas a transversal wave is produced by a shear strain.

Some of the commonly used terms that we shall keep referring to all through our discussion are as given below.

1. The number of wave cycles formed per unit time is called the *frequency* of the wave, which we usually write as ν.
2. The maximum vertical distance by which a wave is displaced from the equilibrium state is called the *amplitude* of the wave, which we usually write as α.
3. The length over which a single *peak* and *trough* are formed is called the *wavelength* of the wave, which we usually write as λ.

High amplitude mechanical waves are called the *shock-waves*, and low amplitude waves are called the *sound waves*.[1] A *standing wave* is created when waves interfere with each other.

For a fixed real number $\ell > 0$, and an integer $n \geq 1$, we write

$$\omega_n = n\,\omega_0, \quad \text{where} \quad \omega_0 = \frac{\pi}{\ell}.$$

For an arbitrary $a \in \mathbb{R}$, consider the sequence $\langle h_n \rangle$ of periodic functions $h_n \in L^1([a, a + 2\ell])$ of period 2ℓ given by

$$h_n(t) := a_n \cos(\omega_n t) + b_n \sin(\omega_n t)$$
$$:= c_{-n}\, e^{-i\omega_n t} + c_n\, e^{i\omega_n t}, \qquad t \in [a, a + 2\ell],$$

where $a_n, b_n \in \mathbb{R}$, and we have

$$c_n = \frac{a_n - ib_n}{2}, \qquad c_{-n} = \frac{a_n + ib_n}{2},$$

are complex numbers, with $b_0 = 0$. Notice that the exponential form of h_n is obtained by using the Euler identity:

$$e^{i\,\omega_n t} = \cos(\omega_n t) + i \sin(\omega_n t).$$

A *Fourier series* is a trigonometric or an exponential series of the form

$$\frac{a_0}{2} + \sum_{n=1}^{\infty} h_n(t) \quad \text{or} \quad \sum_{n=-\infty}^{\infty} c_n e^{i\omega_n t}, \quad \text{for } t \in [a, a + 2\ell],$$

where $a_n, b_n \in \mathbb{R}$, and $c_n \in \mathbb{C}$, are called the *Fourier coefficients*. Notice that, if a trigonometric series of the form given above converges uniformly for some suitable choice of values $a_n, b_n \in \mathbb{R}$, then it defines a periodic function $s : [a, a + 2\ell] \to \mathbb{R}$, and we say that the function s is represented by a *term-wise differentiable* and *term-wise integrable* Fourier series, with Fourier coefficients $a_n, b_n \in \mathbb{R}$ given by the integrals

---

[1] The knowledge about *sound* as disturbance in air pressure goes back to antiquity.

$$a_n = \frac{1}{\ell} \int_a^{a+2\ell} f(t) \cos(\omega_n t) dt;$$

$$b_n = \frac{1}{\ell} \int_a^{a+2\ell} f(t) \sin(\omega_n t) dt,$$

In general, if a periodic signal $f \in L^1([a, a + 2\ell])$ is represented by a Fourier series, then ω_n is the *frequency* of its nth mode, and we say f is approximated by a *superposition* of functions h_n, where the Fourier coefficients a_n, $b_n \in \mathbb{R}$ represent the *amplitudes of the moving waves*. The main theorem of the chapter proves that the Fourier series of a *sufficiently nice* periodic signal provides all the information that the signal carries. Therefore, as a *discrete system*, Fourier series takes a nice periodic signal as input and returns its *frequency spectrum* as the output. Hence, the process involved is known as the *spectral analysis*.[2]

10.1 Evolution of Modern Mathematics

Greeks are believed to have *known* a primitive version of a Fourier series as *epicycles*,[3] which they used to study the variations in position and brightness of the apparent motion of the celestial objects. The pre-Galilean astronomers who used the Ptolemy world view had to add a large number of epicycles in order to make accurate predictions about the celestial events.

However, the actual motivation to study trigonometric series arose when some scientists and mathematicians of eighteenth century started investigating problems related to the vibration of strings such as in a *violin*. We start with an early observation of *Brook Taylor* (1685–1731) about simple modes, namely, a single vibration of a string of length ℓ is furnished by the function

$$y(t) = \alpha \sin(n\pi t/\ell).$$

However, it was *Daniel Bernoulli* (1700–1782) who first realised that the sounds of two different *modes* are heard simultaneously. In 1742, based on physical evidences and experimental observations, he *conjectured* that any acoustic vibration (such as sinusoidal waves) could be expressed as a superposition of such elementary vibrations. We now know that, in general, the *superposition* of waves is a powerful concept that is very effective to solve a number of important problems related to *linear systems* in physics and engineering. Notice that addition of waves at different frequencies results into an *intensity modulation* as a function of time.

---

[2] Daniel Bernoulli's work on problems related to vibrating strings is considered to be the beginning of the theory of *spectral analysis*. In the general case, we use integral transforms such as due to Fourier and Laplace to do the spectral analysis of the *aperiodic signals*.

[3] Epicycles are finite combination of circles $z_n = r_n e^{i \alpha_n t}$, $n \in \mathbb{Z}$, used in ancient astronomy as a geometric model. In the present time, it is an important tool to draw computer graphics.

In 1747, the French mathematician *Rond d'Alembert* (1717–1783) was first to publish a *partial differential equation* model for the *vibrating strings problem*. It is now called the fundamental 1-*dimensional wave equation*. d'Alembert derived the wave equation by applying Taylor's expansion of the restoring force, and the momentum law. The story of Fourier series starts from the point in time when d'Alembert published his *travelling wave solution* to the wave equation, which appeared in the paper *Recherches sur la courbe que forme une corde tendue mise en vibration*.

To continue with the story, we first give a modern description of d'Alembert's *wave equation*. Suppose a perfectly elastic thin string of length ℓ is sufficiently stretched before fixing its ends at the points $x = 0$ and $x = \ell$. When the string is set in motion, we write $y(x, t)$ for the *transverse displacement* of the string at a position $0 < x < \ell$ and at time $t \geq 0$. Suppose the *initial deflection* and the *initial velocity* are given by some functions $d(x)$, $v(x)$, for $x \in [0, \ell]$. Then, the function $y = y(x, t)$ satisfies the 1-*dimensional wave equation*

$$\frac{\partial^2}{\partial t^2} y(x, t) = c^2 \frac{\partial^2}{\partial x^2} y(x, t); \tag{10.1.1}$$

and also the following *initial* and *boundary conditions*:

$$y(x, 0) = d(x), \quad \frac{\partial y(x, t)}{\partial t}\bigg|_{t=0} = v(x), \quad 0 < x < \ell; \tag{10.1.2}$$

$$y(0, t) = 0, \quad y(\ell, t) = 0, \quad t > 0. \tag{10.1.3}$$

The parameter $c > 0$ represents the *wave speed* of the string. In his original work, d'Alembert assumed $c = 1$, and that the string is at rest initially, i.e., $v \equiv 0$. Notice that the (boundary) conditions in (10.1.3) hold because the ends of the string are fixed. d'Alembert proved that the general form of displacement function $y(x, t)$ is implicitly given by

$$y(x, t) = \frac{1}{2}\Big[d(x + ct) + d(x - ct)\Big]. \tag{10.1.4}$$

Notice that, by using conditions in (10.1.3), it follows that the function $y = y(x, t)$ is a periodic function of period 2ℓ, with respect to the variable $x + ct$. Further, as $\partial y/\partial t = 0$, for all $x \in [0, \ell]$, $y = y(x, t)$ is an odd function over the interval $[-\ell, \ell]$. Therefore, according to d'Alembert *travelling wave solution* as in (10.1.4), the function $y = y(x, t)$ must be an odd, periodic function of period 2ℓ, having second order derivatives. d'Alembert was aware of the fact that his methods do not allow for a *triangular shape* function $d(x)$ such as a *kink* in a plucked string situation, as in the (initial) conditions (10.1.2).

In 1749, *Leonhard Euler* published "*De vibratione chordarum exercitatio*", in which he developed a theory for string motions under slightly different circumstances. He proposed the following solution for (continuous) *vibrating string problem*:

$$y(x, t) = \sum_{n=1}^{\infty} b_n \sin(\omega_n x) \cos(c\omega_n t). \qquad (10.1.5)$$

which he obtained in response to d'Alembert's derivation of the travelling wave picture of vibrating string. Euler accepted as $d(x)$ any continuous curve with piecewise continuous slope and curvature, and *never bothered* to ensure the existence of the partial derivatives entering the wave equation. In modern language, we need the *distribution theory* to conclude that *the general solution for the motion of a vibrating string is a superposition of an infinite number of simple harmonics.* In 1763, d'Alembert introduced the *method of separation of variables* to deal with the problem of vibrating string of variable thickness. It was one of his greatest contribution to the *spectral problems*.

In 1751, Daniel Bernoulli published two papers dealing with the problem of finding an *expression* for the transversal motions of a vibrating elastic strings. In *"De vibrationibus et sono laminarum elasticarum"*, he discussed the case of a horizontal rod of length ℓ, with one end fixed. In that paper, he used infinite series, and also exponential and trigonometric functions to write a closed form expression. In *"De sonis multifariis quos laminae elasticae diversimodeedunt disquisitiones mechanico-geometricae experimentis acusticis illustratae et confirmatae"*, Bernoulli used a relation between curvature and moment to describe the partial differential equation for the case when ends are free.

In 1753, commenting on the work of d'Alembert and Euler towards the *vibrating strings problem*, Daniel Bernoulli published a lengthy note blaming the two for obscuring the subject by incorporating unverified sophisticated mathematics. He argued that any vibrating body could produce waves as a series of simple modes with a well-defined frequency of oscillation. As said earlier, superposition of these modes could be used to produce more complex vibrations. He rested his argument on the two assertions:

1. d'Alembert's and Euler's supposedly new solutions are nothing but mixtures of simple modes, since in an infinite series of the form

$$d(x) = \sum_{n=1}^{\infty} b_n \sin(2\omega_n x), \qquad (10.1.6)$$

there is a sufficiently large number of constants b_n so that the *series can adapt to any curve*;
2. The aggregation of these partial modes in a single formula is incompatible with the physical character of the decomposition.

Bernoulli had no mathematical proof of (1). Without any rigorous proof, he also stated that any possible movement of the vibrating string takes the form (10.1.5). He only proved that the basic periodicity properties of the solutions of d'Alembert and Euler can be reproduced by principle of superposition.

In his response to Bernoulli's memoirs, Euler acknowledged his contribution as the *best physical model*, but rejected Bernoulli's claim that the superposition of simple modes always preserved the structure of the partial modes. By taking $b_n = \alpha^n$ ($\alpha \in \mathbb{R}$) in (10.1.6), he illustrated the fact that, when the number of terms becomes infinite, it is not possible to have a solution (10.1.5) (as function of x) to be composed of an infinite number of sine curves. Moreover, he observed that any solution of the form (10.1.5) must be periodic and odd with respect to the spatial variable x, whereas d'Alembert's solution is valid for any *displacement function d*. So, he concluded that Bernoulli's solutions are a subclass of his solutions. Euler also analysed the (inhomogeneous) equation of the *heavy vibrating string*

$$\frac{\partial^2}{\partial t^2} y(x,t) - c \frac{\partial^2}{\partial x^2} y(x,t) = g(x,t),$$

and obtained a solution of the form

$$y(x,t) = \varphi(x+ct) + \psi(x-ct) - \frac{x(x-\ell)g}{2c^2}.$$

Euler also attempted the problem of vibrations of rectangular and circular membranes, but the complete theory (in arbitrary domains) was developed much later by *Denis Poisson* (1781–1840) in 1829.

While trying to obtain a solution similar to Euler's for the wave equation, *Joseph-Louis Lagrange* (1736–1813) used for the first time the idea of *the limit of partial sums* to obtain inadvertently the following formulas for Fourier coefficients:

$$b_n = \frac{2}{\ell} \int_0^\ell d(x) \sin(\omega_n x) dx, \quad (n \ge 1).$$

It follows easily from the *orthogonality* of sine and cosine functions (Theorem 10.1). In general, the success of the method of *separation of variables* in getting a unique solution of (10.1.1), subject to given initial and boundary conditions, depends on how well the Fourier series represents the function $f(x)$. Such types of procedures are used, for example, in solving some initial-boundary value problems related to certain important linear partial differential equations, as described in the companion volume [1].

We find that a part of solution of the 1-dimensional wave equation can be expressed as the series (10.1.5). In tune with Bernoulli's theory, we interpret the nth harmonic $\sin(\omega_n x) \cos(\omega_n t)$ physically to represent a *standing wave* of the string, having *nodes* at

$$x = \frac{\ell}{n}, \frac{2\ell}{n}, \ldots, \frac{(n-1)\ell}{n}.$$

The question about possibility of representing an *arbitrary* (periodic) function by a trigonometric series gained prominence only after the work of *Joseph Fourier* (1768 - 1830) on heat conduction in solids appeared in 1806. It is instructive to know how.

A simple case of Fourier problem is about finding the temperature distribution $u = u(x, t)$ in a perfectly insulated thin rod of length ℓ governed by the *heat equation*

$$\frac{\partial}{\partial t}u(x, t) = c^2\,\frac{\partial^2}{\partial x^2}u(x, t), \tag{10.1.7}$$

subject to the following initial-boundary conditions,

$$u(x, 0) = f(x), \quad \text{for all} \quad 0 < x < \ell; \tag{10.1.8}$$
$$u(0, t) = 0 = u(\ell, t), \quad \text{for all} \quad t > 0. \tag{10.1.9}$$

The constant c in this case refers to the *thermal diffusivity* of the material of the rod. In 1811, Fourier presented his four years of work on *thermal conduction* in solids to *Académie des Sciences*, in which he developed for most of such initial-boundary value problems a series solution of the following form

$$u(x, t) = \sum_{n=1}^{\infty} b_n e^{\omega_n^2 ct} \sin(\omega_n x), \quad \text{with} \tag{10.1.10}$$

$$b_n = \frac{2}{\ell}\int_0^{\ell} f(y)\sin(\omega_n x)dy. \tag{10.1.11}$$

Fourier worked with $\ell = 1$ in the original text. Finally, in 1822, he published this work as part of his celebrated book *"Théorie analytíque de la chaleur"* (The Analytical Theory of Heat), in which he based his investigations on Newton's law of cooling. Fourier was the first to have come up with *orthogonality method* of computing the coefficients a_n, b_n, and asserted[4] that every integrable periodic functions admit a *trigonometric series* representation of the form

$$S(t) := \frac{a_0}{2} + \sum_{n=1}^{\infty}[a_n\cos(n\,t) + b_n\sin(n\,t)]. \tag{10.1.12}$$

In his original text, Fourier assumed $a_n = 0$ for all n (see Eq. (10.1.10)). Fourier also faced the problem of writing a constant function as a *cosine series*, and in the process obtained the first proof of convergence of a trigonometric series. In general, the issues concerning convergence of Fourier series (in whatever permissible sense) are very significant when it comes to applying such type of series in actual applications. Technically, convergence of Fourier series of a *sufficiently nice* periodic signal amounts to ensuring that it can be analysed by *harmonically related* sinusoids.

---

[4] This not-so-wrong assertion of Fourier prompted best mathematician of the time to make discoveries that shaped the modern mathematics, the way we understand it today. In the history of mathematics, it has never been the case that an *innocent mistake* led to such an enormous flurry of activities.

The first major breakthrough appeared in 1829 paper of the German mathematician *Gustav Dirichlet* (1805–1859) who proved that the Fourier series of a periodic function *converges pointwise* to the function if it is piecewise smooth (Theorem 10.2). The work of Dirichlet also led him to a formal definition of a function, the way we understand it today, which helped him to study Fourier series of a functions, *with discontinuities*. This paper of Dirichlet is largely considered to be the beginning of the subject, now called *mathematical analysis*. However, it was due to seminal work of Riesz-Fischer that concluded the story of Fourier series in 1907 by proving that the Fourier was *almost* right. Surely, the requisite mathematical foundation for their work was provided by the fundamental ideas of *Henry Lebesgue* about measure and integration, published in 1904 as his doctoral work.

10.2   Definitions and Examples

Notice that, for the differential operator $D \equiv d/dt$, we have

$$D^2[\sin(n\,t)] = -n^2 \sin(n\,t);$$
$$D^2[\cos(n\,t)] = -n^2 \cos(n\,t).$$

We then say that $\lambda = n^2$ is an *eigenvalue* of (positive definite) differential operator $-D^2$, with sine and cosine being the associated *eigenfunctions*. The *separation of variables* and *eigenfunctions expansion* are two powerful methods to solve some important *initial-boundary value problems* related to linear partial differential equations such as mentioned in the previous section. Both these methods use the idea of (generalised) Fourier series representation of a periodic function.

Partly due to such types of considerations, there has always been a quest to understand how to express an arbitrary *periodic signal* as a trigonometric series of the form (10.1.12), where the coefficients a_n and b_n represent the *amplitudes*, and the associated sequence of *partial sums* $\langle s_k(f) \rangle$ given by

$$s_k(f)(t) = \frac{a_0}{2} + \sum_{n=1}^{k} [a_n \cos(\omega_n t) + b_n \sin(\omega_n t)], \qquad (10.2.1)$$

is used to find the *appropriate frequencies* to represent the signal accurately. Recall that $f \in L^1(\mathbb{R})$ means that $\int_{-\infty}^{\infty} |f(t)|dt < \infty$, i.e., f is an *absolutely integrable* function. Also, for any $f \in L^1([a, a + 2\ell])$, with $f(t + 2\ell) = f(t)$ for all $t \in \mathbb{R}$, the integral

$$\int_a^{a+2\ell} f(t)\,dt$$

is independent of the point $a \in \mathbb{R}$. We are thus free to work in any specified period of a given periodic signal. Unless specified otherwise, by *integrable periodic function* we mean a 2ℓ-periodic L^1-function defined over an interval of the form $[a, a + 2\ell]$, for some $a \in \mathbb{R}$. In most situations, we take $a = 0$ or $a = -\ell$.

Definition 10.1 For any fixed $a \in \mathbb{R}$ and $t \in [a, a + 2\ell]$, a trigonometric series of the form

$$\frac{a_0}{2} + \sum_{n=1}^{\infty} [a_n \cos(\omega_n t) + b_n \sin(\omega_n t)], \tag{10.2.2}$$

is called a *Fourier series* of a 2ℓ-periodic function $f(t) \in L^1([a, a + 2\ell])$ if the *Fourier coefficients* a_n, $b_n \in \mathbb{R}$ are given by the relations

$$a_n = \frac{1}{\ell} \int_a^{a+2\ell} f(x) \cos(\omega_n x) dx, \quad (n \ge 0); \tag{10.2.3}$$

$$b_n = \frac{1}{\ell} \int_a^{a+2\ell} f(x) \sin(\omega_n x) dx, \quad (n \ge 1). \tag{10.2.4}$$

We write $s(f)(t)$ for the series (10.2.2). Notice that the term $a_0/2$ is the *average value* of the function $f(t)$ over the interval $[a, a + 2\ell]$. As said above, we are mainly focussing on the cases when $t \in [-\ell, \ell]$ or $t \in [0, 2\ell]$.

Clearly, since a_n and b_n exist for any integrable function $f(t)$, it is always possible to write the series (10.2.2) for such types of functions. However, notice that we are yet to settle the following fundamental problem related to the convergence of the series (10.2.2):

For a given integrable periodic function $f : \mathbb{R} \to \mathbb{R}$, when does the sequence $\langle s_k(f) \rangle$ of partial sums of the series (10.2.2) converge to f as $k \to \infty$?

Therefore, we may write

$$f(t) \sim \frac{a_0}{2} + \sum_{n=1}^{\infty} [a_n \cos(\omega_n t) + b_n \sin(\omega_n t)].$$

We say the Fourier series (10.2.2) *represents the function* $f(t)$ over the interval $[a, a + 2\ell]$ if the equality holds. In general, when it comes to applying these ideas in practical situations, the problems related to convergence of Fourier series (in any manner) are very significant. For example, in the case of initial-boundary value problems related to the heat Eq. (10.1.7), a series solution of the form (10.1.10) makes sense only when the series converge uniformly on the interval $[0, \ell]$ and also its term-wise derivatives converge uniformly to the derivative $u_x(x, t)$.

More generally, we strive to find classes of integrable periodic functions $f : [a, a + 2\ell] \to \mathbb{R}$ that have their Fourier series to be term-wise differentiable and integrable. We know that it is always possible to do so if the convergence is uniform. So, if the series (10.2.2) *converges uniformly* for a periodic function $f \in L^1([a, a + 2\ell])$, we write for $t \in [a, a + 2\ell]$

$$f(t) = \frac{a_0}{2} + \sum_{n=1}^{\infty} [a_n \cos(\omega_n t) + b_n \sin(\omega_n t)],$$

where the series on the right side is *termwise differentiable* and *termwise integrable*, with derivative $f'(t)$ and integral $\int_a^{a+2\ell} f(t)dt$. Clearly, $f(t)$ satisfies the periodicity condition

$$f(t + 2\ell) = f(t), \quad \forall \ t \in \mathbb{R}.$$

The next theorem proves that, in this situation, the Fourier coefficients a_n and b_n are respectively given by (10.2.3) and (10.2.4).

Theorem 10.1 (Euler, 1777) *If the series* (10.2.2) *converges uniformly to a function* $f(t)$ *over the interval* $[a, a + 2\ell]$, *then the coefficients* a_n, b_n *are given by Eqs.* (10.2.3) *and* (10.2.4).

Proof Since we have

$$\frac{a_0}{2} + \sum_{n=1}^{\infty} [a_n \cos(\omega_n t) + b_n \sin(\omega_n t)] \rightrightarrows f(t),$$

the function

$$\frac{a_0}{2} \cos(\omega_m t) + \sum_{n=1}^{\infty} [a_n \cos(\omega_n t) \cos(\omega_m t) + b_n \sin(\omega_n t) \cos(\omega_m t)]$$

$$\rightrightarrows f(x) \cos(\omega_m t)$$

and hence

$$\int_a^{a+2\ell} f(x) \cos(\omega_m x)dx = \frac{a_0}{2} \int_a^{a+2\ell} \cos(\omega_m x)dx$$

$$+ \sum_{n=1}^{\infty} \left(a_n \int_a^{a+2\ell} \cos(\omega_n x) \cos(\omega_m x)dx \right.$$

$$\left. + b_n \int_a^{a+2\ell} \sin(\omega_n x) \cos(\omega_m x)dx \right) \qquad (10.2.5)$$

It is clear that, for $m \neq 0$,

$$\int_a^{a+2\ell} \cos(\omega_m x)dx = \frac{\sin(\omega_m x)}{\omega_m} \bigg|_a^{a+2\ell} = 0$$

and

$$\int_a^{a+2\ell} \sin(\omega_m x)dx = - \frac{\cos(\omega_m x)}{\omega_m} \bigg|_a^{a+2\ell} = 0.$$

By using the last equation and the identity

$$\cos(\omega_n x)\cos(\omega_m x) = \frac{1}{2}\left[\cos(\omega_{m+n} x) + \cos(\omega_{m-n} x)\right],$$

we see that, for $m \neq n$,

$$\int_a^{a+2\ell} \cos(\omega_n x)\cos(\omega_m x)dx$$
$$= \frac{1}{2}\int_a^{a+2\ell} \left[\cos(\omega_{m+n} x) + \cos(\omega_{m-n} x)\right]dx$$
$$= 0$$

For $n = m$, we see that

$$\int_a^{a+2\ell} \cos^2(\omega_m x)dx = \frac{1}{2}\int_a^{a+2\ell} \left[\cos(2\omega_m x) + 1\right]dx = \ell.$$

Similarly, by using

$$\sin(\omega_n x)\cos(\omega_m x) = \frac{1}{2}\left[\sin(\omega_{m+n} x) + \sin(\omega_{m-n} x)\right],$$

we get

$$\int_a^{a+2\ell} \sin(\omega_n x)\cos(\omega_m x)dx$$
$$= \frac{1}{2}\int_a^{a+2\ell} \left[\sin(\omega_{m+n} x) + \sin(\omega_{m-n} x)\right]dx$$
$$= 0.$$

With all these identities, we see that (10.2.5) gives

$$a_m = \frac{1}{\ell}\int_a^{a+2\ell} f(x)\cos(\omega_m x)dx.$$

This formula works for $m = 0$ also. In a similar fashion, we can show that

$$b_m = \frac{1}{\ell}\int_a^{a+2\ell} f(x)\sin(\omega_m x)dx.$$

Therefore, the series (10.2.2) is the Fourier series of $f(t)$.

Remark 10.1 If a periodic function $f \in L^1$ is the *uniform limit* of its Fourier series then it is indeed *represented* by the series (10.2.2), where the Fourier coefficients

a_n, b_n are given by integrals (10.2.3) and (10.2.4). The related convergence issues are discussed in Sect. 10.3.

Now, to illustrate how to write a Fourier series, we consider some interesting integrable periodic function defined over an interval $[-\ell, \ell]$ or $[0, 2\ell]$. Notice that it follows from the proof of Theorem 10.1 that

$$\int_a^{a+2\ell} \sin(\omega_n x)dx = \int_a^{a+2\ell} \cos(\omega_n x)dx = 0,$$

for all $n \geq 1$; and, for $m \neq n$,

$$\int_a^{a+2\ell} \sin(\omega_m x)\cos(\omega_n x)dx = 0;$$
$$\int_a^{a+2\ell} \cos(\omega_n x)\cos(\omega_m x)dx = \int_a^{a+2\ell} \sin(\omega_n x)\sin(\omega_m x)dx = 0.$$

Clearly, the two integrals in the last equation equal ℓ when $m = n$. These simple facts are known as the *orthogonality relations*, which are used frequently in the sequel while computing the values a_n and b_n. Also, the following simple observations may prove useful in certain situations:

$$\sin(n\pi) = 0 \quad \text{and} \quad \cos(n\pi) = (-1)^n, \quad \text{for } n \geq 1;$$
$$\int e^{ax} \cos(bx)dx = \frac{e^{ax}}{a^2 + b^2}\left[a\cos(bx) + b\sin(bx)\right];$$
$$\int e^{ax} \sin(bx)dx = \frac{e^{ax}}{a^2 + b^2}\left[a\sin(bx) - b\cos(bx)\right];$$
$$\int u(x)v(x)\,dx = uv_1 - u'v_2 + u''v_3 + \cdots,$$

where $'$ denotes the derivative,

$$v_1 = \int v\,dx \quad \text{and} \quad v_{i+1} = \int v_i dx, \quad \text{for } i \geq 1.$$

We start with the case when $f \in L^1([-\ell, \ell])$, where $f(t + 2k\ell) = f(t)$, for all $k \in \mathbb{Z}$ and $t \in [-\ell, \ell]$. Then, to write the series (10.2.2) for f, we compute the coefficients a_n and b_n using the following formulas:

$$a_n = \frac{1}{\ell}\int_{-\ell}^{\ell} f(x)\cos(\omega_n x)dx, \quad (n \geq 0); \tag{10.2.6}$$

$$b_n = \frac{1}{\ell}\int_{-\ell}^{\ell} f(x)\sin(\omega_n x)dx, \quad (n \geq 1). \tag{10.2.7}$$

Example 10.1 Consider the *square wave* defined by

$$f(x) = \begin{cases} 0, & \text{for } -\ell < x < -1 \\ 1, & \text{for } -1 < x < 1 \quad , \quad \ell > 1. \\ 0, & \text{for } 1 < x < \ell \end{cases}$$

By Eq. (10.2.6), we have

$$a_0 = \frac{1}{\ell} \int_{-\ell}^{\ell} f(x)dx = \frac{1}{\ell} \int_{-1}^{1} (1)dx = \frac{2}{\ell};$$

and, for $n \geq 1$,

$$a_n = \frac{1}{\ell} \int_{-\ell}^{\ell} f(x) \cos(\omega_n x)dx = \frac{1}{\ell} \int_{-1}^{1} \cos(\omega_n x)dx$$

$$= \frac{2}{\ell} \int_{0}^{1} \cos(\omega_n x)dx = \frac{2}{\ell} \left(\frac{\ell}{2n\pi} \sin(\omega_n x) \right)_{0}^{1}$$

$$= \frac{1}{n\pi} \sin \left(\frac{n\pi}{\ell} \right).$$

By Eq. (10.2.7), we have

$$b_n = \frac{1}{\ell} \int_{-\ell}^{\ell} f(x) \sin(\omega_n x)dx = \frac{1}{\ell} \int_{-1}^{1} \sin(\omega_n x)dx = 0.$$

Hence, we have

$$f(x) \sim \frac{1}{\ell} + \frac{1}{\pi} \sum_{n=1}^{\infty} \frac{1}{n} \sin(\omega_n) \cos(\omega_n x).$$

The procedure used in above example also applies to the case when $f \in L^1([0, 2\ell])$. Notice that, in this case, the Fourier series of f is still the same as in (10.2.2), but the coefficients a_n and b_n have to be computed using the following formulas:

$$a_n = \frac{1}{\ell} \int_{0}^{2\ell} f(x) \cos(\omega_n x)dx; \tag{10.2.8}$$

$$b_n = \frac{1}{\ell} \int_{0}^{2\ell} f(x) \sin(\omega_n x)dx. \tag{10.2.9}$$

Example 10.2 Consider the function $f(x) = x \sin x$, $x \in [0, 2\pi]$. Then, by using Eq. (10.2.8), we have

$$a_n = \frac{1}{\pi} \int_0^{2\pi} x \sin x \cos nx \, dx = \frac{1}{2\pi} \int_0^{2\pi} x(2 \sin x \cos nx) \, dx$$

$$= \frac{1}{2\pi} \left[\int_0^{2\pi} x \sin(1 + n)x \, dx + \int_0^{2\pi} x \sin(1 - n)x \, dx \right]$$

$$= \frac{1}{2\pi} \left[-\frac{2\pi}{(1 + n)} - \frac{2\pi}{(1 - n)} \right] = \frac{2}{n^2 - 1},$$

for $n \neq 1$, and

$$a_1 = \frac{1}{\pi} \int_0^{2\pi} x \sin x \cos x \, dx = \frac{1}{2\pi} \int_0^{2\pi} x \sin 2x \, dx = -\frac{1}{2}.$$

Also, by using equation (10.2.9), we have

$$b_1 = \frac{1}{\pi} \int_0^{2\pi} x \sin^2 x \, dx = \frac{1}{2\pi} \int_0^{2\pi} x[1 - \cos 2x] \, dx = \pi.$$

Observe that $b_n = 0$, for all $n \geq 2$. Hence, we have

$$f(x) \sim -1 - \frac{1}{2} \cos x + \pi \sin x + 2 \sum_{n=2}^{\infty} \frac{1}{n^2 - 1} \cos nx.$$

In certain applications, it is convenient to view a 2π- periodic function $f : (-\pi, \pi] \to \mathbb{R}$, with $f(-\pi) = f(\pi)$, as a function $f : \mathbb{T} \to \mathbb{R}$, where

$$\mathbb{T} := \frac{\mathbb{R}}{2\pi \mathbb{Z}} = \left\{ (x, y) \in \mathbb{R}^2 \mid x^2 + y^2 = 1 \right\}$$

is the *unit circle*. Conversely, each function $f : \mathbb{T} \to \mathbb{R}$ defines a 2π- periodic function $f : (-\pi, \pi] \to \mathbb{R}$. More generally, a *complex Fourier series* of a periodic function $f \in L^1[-\ell, \ell]$, with $f(-\ell) = f(\ell)$, is defined as follows.

Definition 10.2 (*Complex Form*) For any $0 < \ell \in \mathbb{R}$ and $t \in [-\ell, \ell]$, an exponential series of the form

$$\sum_{-\infty}^{\infty} c_n e^{i \omega_n t}, \qquad (10.2.10)$$

is called a (complex) *Fourier series* of a periodic functions $f \in L^1[-\ell, \ell]$, with $f(-\ell) = f(\ell)$, and the complex numbers c_n given by

$$c_n = \frac{1}{2\ell} \int_{-\ell}^{\ell} e^{-i \omega_n x} f(x) dx, \qquad (10.2.11)$$

are called the (complex) *Fourier coefficients* of f.

Example 10.3 Consider the *square wave*

$$f_\ell(t) = \begin{cases} 0, & \text{for } -\ell < t < -1 \\ 1, & \text{for } -1 < t < 1 \\ 0, & \text{for } 1 < t < \ell \end{cases},$$

with $f_\ell(x + 2\ell) = f_\ell(x)$, for all $x \in \mathbb{R}$. Then, by Eq. (10.2.11),

$$c_n = \frac{1}{2\ell} \int_{-1}^{1} e^{-\omega_n ix} dx = \frac{1}{n\pi} \sin(\omega_n).$$

Therefore, we obtain

$$f(t) \sim \frac{1}{\pi} \sum_{n=-\infty}^{\infty} \frac{\sin(\omega_n)}{n} e^{\omega_n it}.$$

Example 10.4 Consider the function $f(x) = e^{-|x|}$, $x \in [-\ell, \ell]$, with $f(x + 2\ell) = f(x)$ for $x \in \mathbb{R}$. Then, by Eq. (10.2.11), we have

$$\begin{aligned} c_n &= \frac{1}{2\ell} \int_{-\ell}^{\ell} e^{-\omega_n ix} e^{-|x|} dx \\ &= \frac{1}{2\ell} \left[\int_{-\ell}^{0} e^{-(\omega_n i-1)x} dx + \int_{0}^{\ell} e^{-(\omega_n i+1)x} dx \right] \\ &= \frac{e^{-\ell}}{2\ell} \left[\frac{e^{n\pi i} - 1}{\omega_n i - 1} + \frac{1 - e^{-n\pi i}}{\omega_n i + 1} \right] \\ &= \frac{[1 - (-1)^n]e^{-\ell}}{2(\ell^2 + n^2\pi^2)}. \end{aligned}$$

Therefore, we obtain

$$f(x) \sim \frac{e^{-\ell}}{2} \sum_{n=-\infty}^{\infty} \frac{1}{\ell^2 + (2n-1)^2\pi^2} e^{i(2n-1)\omega_n x}.$$

It is more common to work with periodic functions $f \in L^1([-\pi, \pi])$ such that $f(t + 2k\pi) = f(t)$, for all $k \in \mathbb{Z}$ and $t \in [-\pi, \pi]$. Notice that, in this case, we have $\omega_n = n$, and the Fourier series is given by the series (10.1.12), where coefficients $a_n, b_n \in \mathbb{R}$ are obtained using the following integrals:

$$a_n = \frac{1}{\pi} \int_{-\pi}^{\pi} f(x) \cos nx \, dx, \tag{10.2.12}$$

$$b_n = \frac{1}{\pi} \int_{-\pi}^{\pi} f(x) \sin nx \, dx. \tag{10.2.13}$$

Example 10.5 Consider the *unit step function*

$$h(t) = \begin{cases} 0, & \text{for } -\pi < t < 0 \\ 1, & \text{for } 0 < t < \pi \end{cases},$$

with $h(t + 2\pi) = h(t)$, for all $t \in \mathbb{R}$. It then follows that

$$a_0 = \frac{1}{\pi} \int_0^\pi dx = 1,$$

and, for all $n > 0$,

$$a_n = \frac{1}{\pi} \int_0^\pi \cos nx \, dx = \frac{1}{n\pi} \left[\sin nx \right]_0^\pi = 0.$$

Next, for $n \geq 1$, we have

$$b_n = \frac{1}{\pi} \int_0^\pi \sin nx \, dx = -\frac{1}{n\pi} \left[\cos nx \right]_0^\pi = \frac{1 - (-1)^n}{n\pi},$$

so that $b_n = 0$, for n even; and, $b_n = 2/n\pi$, for n odd. Therefore, we can write the
Fourier series of the function $h(t)$ as follows:

$$h(t) \sim \frac{1}{2} + \frac{2}{\pi} \sum_{n=1}^\infty \frac{\sin(2n-1)t}{(2n-1)} = \frac{1}{2} + \frac{2}{\pi} \left[\sin t + \frac{\sin 3t}{3} + \cdots \right].$$

Example 10.6 Consider the function $f : [-\pi, \pi] \to \mathbb{R}$ defined by $f(t) = t + t^2$,
with $f(t + 2\pi) = f(t)$ for all $t \in \mathbb{R}$. In this case, we have

$$a_0 = \frac{1}{\pi} \int_{-\pi}^\pi (x + x^2) dx = \frac{2}{\pi} \int_0^\pi x^2 dx = \frac{2}{\pi} \left(\frac{x^3}{3} \right) \Big|_0^\pi = \frac{2\pi^2}{3},$$

and, for $n \geq 1$, we have

$$a_n = \frac{1}{\pi} \int_{-\pi}^\pi (x + x^2) \cos nx \, dx = \frac{2}{\pi} \int_0^\pi x^2 \cos nx \, dx$$

$$= \frac{2}{\pi} \left(x^2 \frac{\sin nx}{n} + 2x \frac{\cos nx}{n^2} - 2 \frac{\sin nx}{n^3} \right) \Big|_0^\pi$$

$$= \frac{2}{\pi n^2} (2\pi(-1)^n) = \frac{4(-1)^n}{n^2}.$$

Also, for $n \geq 1$, we have

$$b_n = \frac{1}{\pi} \int_{-\pi}^{\pi} (x + x^2) \sin nx \, dx = \frac{2}{\pi} \int_0^{\pi} x \sin nx \, dx$$

$$= \frac{2}{\pi} \left(-x \frac{\cos nx}{n} + \frac{\sin nx}{n^2} \right) \Big|_0^{\pi}$$

$$= \frac{2}{\pi n} \left(\pi(-1)^n \right) = \frac{2(-1)^{n+1}}{n}.$$

Therefore, we obtain

$$f(t) \sim \frac{\pi^2}{6} + \sum_1^{\infty} (-1)^n \left[\frac{4}{n^2} \cos nt - \frac{2}{n} \sin nt \right].$$

Remark 10.2 If $f(t) : [-\ell, \ell] \to \mathbb{R}$ is an integrable periodic function of period 2ℓ, then the *transformed function* given by

$$g(x) := f\left(\frac{\ell x}{\pi} \right), \quad x \in [-\pi, \pi],$$

defines an integrable periodic function $g : [-\pi, \pi] \to \mathbb{R}$, of period 2π. So, if $f(t)$ is represented by the Fourier series (10.2.2), then the function $g(x)$ is represented by the series

$$g(x) = \frac{c_0}{2} + \sum_{n=1}^{\infty} \left[c_n \cos(nx) + d_n \sin(nx) \right], \tag{10.2.14}$$

where the Fourier coefficients of $g(x)$ are given by the relations

$$c_n = \frac{1}{\pi} \int_{-\pi}^{\pi} g(u) \cos(nu) \, du \quad \text{and} \quad d_n = \frac{1}{\pi} \int_{-\pi}^{\pi} g(u) \sin(nu) du.$$

Conversely, by using the substitution $x = \pi t / \ell$, one can recover the series (10.2.2) from the series (10.2.14), and the above formulas for coefficients c_n and d_n transform respectively to the relations (10.2.3) and (10.2.4) (with $a = -\ell$).

As shown later, all the following series converge with respect to L^2-norm:

$$\sum_{n=1}^{\infty} |a_n|^2, \quad \sum_{n=1}^{\infty} |b_n|^2, \quad \text{and} \quad \sum_{n=-\infty}^{\infty} |c_n|^2,$$

In particular, we have $a_n, b_n, c_n \to 0$, as $n \to \infty$. That is, absolute values of the Fourier coefficients of a 2π-periodic function $f \in L^2[-\pi, \pi]$, are always arbitrarily small. The next lemma (independently) due to Riemann and Lebesgue says that this assertion holds even for 2π-periodic functions $f \in L^1[-\pi, \pi]$. In general, for any such f and $\alpha, \beta \in \mathbb{R}$, we have

$$\lim_{|\alpha| \to \infty} \int_a^{a+2\ell} f(t) \sin(\alpha t + \beta) dt = 0. \tag{10.2.15}$$

However, it serves the purpose here to prove the following simpler version of it.

Lemma 10.1 (Riemann-Lebesgue Lemma) *If* $f : [-\pi, \pi] \to \mathbb{R}$, *with* $f(t + 2\pi) = f(t)$, $t \in \mathbb{R}$, *is any integrable function, then for any* $\lambda \in \mathbb{R}$

$$\lim_{\lambda \to \infty} \int_{-\pi}^{\pi} f(t) \sin(\lambda t) dt = 0 = \lim_{\lambda \to \infty} \int_{-\pi}^{\pi} f(t) \cos(\lambda t) dt. \tag{10.2.16}$$

Proof We assume that the result of Exercise 10.4 is known. The only nontrivial part is to show that the assertion holds for the function $f = \chi_M$, where $M \subset (-\pi, \pi)$ is a measurable set. Let $\epsilon > 0$ be arbitrary. Then, there exists a *simple function* s such that $\int_{-\pi}^{\pi} |f(t) - s(t)| dt < \epsilon/2$. So,

$$\left| \int_{-\pi}^{\pi} f(t) \sin(\lambda t) dt \right|$$
$$= \left| \int_{-\pi}^{\pi} [f(t) - s(t) + s(t)] \sin(\lambda t) dt \right|$$
$$\leq \left| \int_{-\pi}^{\pi} [f(t) - s(t)] \sin(\lambda t) dt \right| + \left| \int_{-\pi}^{\pi} s(t) \sin(\lambda t) dt \right|.$$

Since the second term on the right side is bounded by part (iii) of Exercise 10.10, the assertion for the left side integral in (10.2.16) holds. A proof for the other integral follows from by taking $\beta = \pi/2$ in Eq. (10.2.15).

10.2.1 Sine and Cosine Fouries Series

We say $f : [-\ell, \ell] \to \mathbb{R}$ is an *even function* if $f(-x) = f(x)$, for all $x \in [-\ell, \ell]$; and, f is called an *odd function* if $f(-x) = -f(x)$, for all $x \in [-\ell, \ell]$. For example, the functions x^{2n} and $\cos nx$ are even functions, while x^{2n+1} and $\sin nx$ are odd functions, for any $n \in \mathbb{N}$. Clearly, the product of two even functions is even, and the product of two odd functions is also even. Also, the product of an odd function with an even function is always odd. Notice that, for any function $f : [-\ell, \ell] \to \mathbb{R}$, the functions $g, h : [-\ell, \ell] \to \mathbb{R}$ given by

$$g(x) = \frac{f(x) + f(-x)}{2} \quad \text{and} \quad h(x) = \frac{f(x) - f(-x)}{2},$$

are *even* and *odd* functions, respectively, and we have

$$f(x) = g(x) + h(x), \quad \text{for all} \quad x \in [-\ell, \ell].$$

Therefore, it follows that every function $f : [-\ell, \ell] \to \mathbb{R}$ can be expressed as a sum of an even and an odd function. We also need these simple observations in the later part of the section. Further, the following simple fact comes handy when dealing with an *even function* or an *odd function*:

$$\int_{-\ell}^{\ell} f(x)dx = \begin{cases} 0 & , \quad \text{when } f \text{ is odd} \\ 2\int_{0}^{\ell} f(x)dx, & \text{when } f \text{ is even} \end{cases} \tag{10.2.17}$$

As will be shown in the next section, for any integrable 2ℓ- periodic function $f :$ $[-\ell, \ell] \to \mathbb{R}$ satisfying the *Dirichlet's conditions* (Theorem 10.2), we can write

$$\frac{f(x^+) + f(x^-)}{2} = \frac{a_0}{2} + \sum_{n=1}^{\infty} [a_n \cos(\omega_n x) + b_n \sin(\omega_n x)], \ x \in (-\ell, \ell),$$
$$\tag{10.2.18}$$

where the coefficients a_n, b_n are given by Eqs. (10.2.6) and (10.2.7). In particular, when $f(x)$ is an even function, we have that the function $f(t)\sin(\omega_n x)$ is odd and the function $f(t)\cos(\omega_n x)$ is even. It then follows that $b_n = 0$, for $n \geq 1$, and

$$a_n = \frac{2}{\ell} \int_{0}^{\ell} f(t)\cos(\omega_n t)dt, \quad \text{for} \ n \geq 0. \tag{10.2.19}$$

Consequently, in this case, we obtain a Fourier *cosine series*:

$$\frac{f(x^+) + f(x^-)}{2} = \frac{a_0}{2} + \sum_{n=1}^{\infty} a_n \cos(\omega_n x), \quad \text{for} \ x \in (-\ell, \ell). \tag{10.2.20}$$

On the other hand, when $f(x)$ is an odd function, we have that the function $f(t)\cos(\omega_n t)$ is odd and the function $f(t)\sin(\omega_n t)$ is even. Thus, $a_n = 0$, for $n \geq 0$, and

$$b_n = \frac{2}{\ell} \int_{0}^{\ell} f(t)\sin(\omega_n t)dt, \quad \text{for} \ n \geq 1. \tag{10.2.21}$$

Consequently, in this case, we obtain a Fourier *sine series*:

$$\frac{f(x^+) + f(x^-)}{2} = \sum_{n=1}^{\infty} b_n \sin(\omega_n x), \quad \text{for} \ x \in (-\ell, \ell). \tag{10.2.22}$$

Observe that, at each point of continuity $x \in (-\ell, \ell)$, we have

$$\frac{f(x^+) + f(x^-)}{2} = f(x).$$

Fig. 10.1 Approximation of $f(x) = x$ by partial sequence $s_n(x)$ for $n = 1, 3$

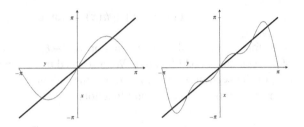

Example 10.7 The *sawtooth function* $f : \mathbb{R} \to \mathbb{R}$ defined by $f(x) = x$, for $-\pi < x < \pi$, and $f(x + 2k\pi) = f(x)$, for $k \in \mathbb{Z}$, is an odd function and hence its Fourier sine series is $\sum_{n=1}^{\infty} b_n \sin nx$, where

$$
b_n = \frac{2}{\pi} \int_0^{\pi} x \sin nx \, dx
$$
$$
= \frac{2}{\pi} \left(-x \frac{\cos nx}{n} + \frac{\sin nx}{n^2} \right) \Big|_0^{\pi} = \frac{2}{n}(-1)^{n+1}.
$$

Therefore, by the above observation, we obtain

$$
x = 2 \sum_{n=1}^{\infty} \frac{(-1)^{n+1}}{n} \sin nx \quad (-\pi < x < \pi).
$$

Notice that, at the end points $x = \pm\pi$, we have

$$
f(x) \sim 2 \sum_{n=1}^{\infty} \frac{(-1)^{n+1}}{n} \sin nx.
$$

Example 10.8 Since the *rectangle wave* function

$$
f(x) = \begin{cases} -a, & \text{for } -\pi < x < 0 \\ a, & \text{for } 0 < x < \pi \end{cases}, \quad \text{with } f(x + 2\pi) = f(x), \ \forall \, x \in \mathbb{R},
$$

is odd in the interval $(-\pi, \pi)$, it follows that $a_n = 0$, for all $n \geq 0$, and

$$
b_n = \frac{2}{\pi} \int_0^{\pi} f(x) \sin(n x) dx = \frac{2a}{\pi} \int_0^{\pi} \sin(n x) dx
$$
$$
= -\frac{2a}{n\pi}[(-1)^n - 1] = \begin{cases} \frac{4a}{n\pi}, & \text{when } n \text{ is odd} \\ 0, & \text{when } n \text{ is even} \end{cases}.
$$

Therefore, for $0 \neq x \in (-\pi, \pi)$, we have

Fig. 10.2 Approximation of $f(x) = |x|$ by partial sequence $s_n(x)$ for $n = 1, 3$

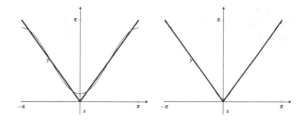

$$f(x) = \sum_{n=1}^{\infty} \frac{4a}{\pi(2n-1)} \sin\left((2n-1)x\right).$$

In particular, for $a = \frac{\pi}{4}$, we obtain

$$\frac{\pi}{4} = \sin x + \frac{1}{3}\sin 3x + \frac{1}{5}\sin 5x + \dots .$$

Example 10.9 The function $f : [-\pi, \pi] \to \mathbb{R}$ given by $f(x) = |x|$ is even, and so it has the following Fourier cosine series

$$\frac{a_0}{2} + \sum_{n=1}^{\infty} a_n \cos nx,$$

where

$$a_0 = \frac{2}{\pi}\int_0^{\pi} |x|dx = \frac{2}{\pi}\int_0^{\pi} xdx = \frac{2}{\pi}\left[\frac{x^2}{2}\right]_0^{\pi} = \pi$$

and, for $n \geq 1$,

$$\begin{aligned}
a_n &= \frac{2}{\pi}\int_0^{\pi} |x|\cos nx dx = \frac{2}{\pi}\int_0^{\pi} x\cos nx dx \\
&= \frac{2}{\pi}\left(x\frac{\sin nx}{n} + \frac{\cos nx}{n^2}\right)\Big|_0^{\pi} \\
&= \frac{2}{\pi n^2}\left((-1)^n - 1\right) = \begin{cases} -\frac{4}{\pi(2k-1)^2} & (n = 2k-1) \\ 0 & (n = 2k). \end{cases}
\end{aligned}$$

Therefore, by the continuity of the function f, we can write (Fig. 10.2)

$$|x| = \frac{\pi}{2} - \frac{4}{\pi}\sum_{k=1}^{\infty} \frac{\cos(2k-1)x}{(2k-1)^2}.$$

Fourier series is an important tool to find the sum of an infinite series. For the 2π-periodic function $f : [-\pi, \pi] \to \mathbb{R}$ as in Example 10.6, the Fourier series of

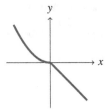

$f(x)$ at $x = \pi$ converges to

$$\frac{f(\pi^+) + f(\pi^-)}{2} = \frac{(-\pi + \pi^2) + (\pi + \pi^2)}{2} = \pi^2,$$

by Theorem 10.2. In particular, we obtain

$$\sum_{1}^{\infty} \frac{1}{n^2} = \frac{\pi^2}{6},$$

which is known as the *Basel theorem*. Further, by taking $x = 0$, we get

$$\sum_{1}^{\infty} \frac{(-1)^{n-1}}{n^2} = \frac{\pi^2}{12}.$$

10.2.2   Half-Range Fourier Series

Notice that there is natural way to extend a real-valued function f defined on some finite interval $[a, b]$ to the whole of $\mathbb{R}$ as a periodic function of period $b - a$, which is called the **periodic extension** of f. So, it is possible to introduce the concept of Fourier series of a function $f : [a, b] \to \mathbb{R}$, provided the *periodic extension* it defines is an absolutely integrable over $\mathbb{R}$. However, it is most unlikely that the periodic extensions of a continuous function f so obtained would be continuous. For example, the *periodic extension* of the function

$$f(t) = \begin{cases} t^2, & \text{for } -1 \leq t \leq 0 \\ -t, & \text{for } 0 < t \leq 1 \end{cases} \tag{10.2.23}$$

obtained by taking $f(t + 2) = f(t)$, $t \in \mathbb{R}$, is discontinuous at each integer point (see Fig. 10.3).

   In particular, for $a = 0$ and $b = \ell$, we have the concept of *half-range* Fourier series. That is, to write Fourier series of a function $f : [0, \ell] \to \mathbb{R}$, we apply the

Fig. 10.4 Even and odd extensions of $f(x) = x^2$, $x \in [0, 1]$

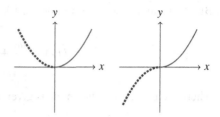

procedure described in the previous part of the section to the periodic extension of the function f, subject to that it has the desirable properties.

Notice that we have two ways to obtain an extension of a function $f : [0, \ell] \to \mathbb{R}$. For, we can use *the reflection in y-axis* of f to obtain an *even extension* $f_e : [-\ell, \ell] \to \mathbb{R}$ given by

$$f_e(x) = \begin{cases} f(x), & \text{if } 0 \le x \le \ell \\ f(-x), & \text{if } -\ell \le x \le 0 \end{cases}$$

Alternatively, we can use *the reflection about the origin* of f to obtain an *odd extension* $f_o : [-\ell, \ell] \to \mathbb{R}$ given by

$$f_o(x) = \begin{cases} f(x), & \text{if } 0 \le x \le \ell \\ -f(-x), & \text{if } -\ell \le x \le 0 \end{cases}.$$

Subsequently, we extend the function $f_e : [-\ell, \ell] \to \mathbb{R}$ (or $f_o : [-\ell, \ell] \to \mathbb{R}$) to the whole of $\mathbb{R}$ by taking $f_e(x + 2\ell) = f(x)$ (or, respectively, $f_o(x + 2\ell) = f(x)$), for all $x \in \mathbb{R}$ (see Fig. 10.4). In each case, we obtain a periodic function $g : \mathbb{R} \to \mathbb{R}$ of period 2ℓ. Now, if the extension (f_e or f_o) meets the Dirichlet conditions (Theorem 10.2) then we may be able to obtain a Fourier cosines or a sine series for the *half-range* function $f : [0, \ell] \to \mathbb{R}$. A choice for a particular type of a series depends on the nature of the problem in hand.

To illustrate the ideas, first consider a function $f : [0, \pi] \to \mathbb{R}$, and let functions $g, h : [-\pi, \pi] \to \mathbb{R}$ be respectively the *even* and *odd extensions* of f. If the periodic extension $\widehat{g} : \mathbb{R} \to \mathbb{R}$ of g is absolutely integrable, then the Fourier cosine series of g is given given by

$$g(x) \sim \frac{a_0}{2} + \sum_{n=1}^{\infty} a_n \cos nx,$$

where the Fourier coefficient a_n is given by

$$a_n = \frac{2}{\pi} \int_0^{\pi} g(x) \cos nx\, dx.$$

But then, since $f(x) = g(x)$, for $x \in [0, \pi]$, we obtain

$$f(x) \sim \frac{a_0}{2} + \sum_{n=1}^{\infty} a_n \cos nx,$$

where the Fourier coefficient a_n is given by

$$a_n = \frac{2}{\pi} \int_0^\pi f(x) \cos nx dx.$$

Therefore, in this case, we get a *Fourier cosine series* of the half-range function $f(x)$, $x \in [0, \pi]$. Similarly, if the periodic extension $\widehat{h} : \mathbb{R} \to \mathbb{R}$ of the *odd extension* h is absolutely integrable, then the Fourier sine series of h is given given by

$$h(x) \sim \sum_{n=1}^{\infty} b_n \sin nx, \quad \text{with} \quad b_n = \frac{2}{\pi} \int_0^\pi h(x) \sin nx \, dx.$$

Therefore, in this case, a *Fourier sine series* of the half-range function f is given by

$$f(x) \sim \sum_{n=1}^{\infty} b_n \sin nx, \quad \text{where } b_n = \frac{2}{\pi} \int_0^\pi f(x) \sin nx \, dx.$$

In a practical situation, one may take a suitable extension (even or odd) of f depending on which type of Fourier series (cosine or sine) one is interested to work with.

Example 10.10   First, we represent the function $f(x) = x^2$, $x \in (0, \pi)$, as a *Fourier cosine series*. For, since

$$a_0 = \frac{2}{\pi} \int_0^\pi x^2 dx = \frac{2\pi^2}{3},$$

and, for $n \geq 1$,

$$a_n = \frac{2}{\pi} \int_0^\pi x^2 \cos nx \, dx = \frac{4(-1)^n}{n^2},$$

it follows that, for $x \in (0, \pi)$, we can write

$$x^2 = \frac{\pi^2}{3} + 4 \sum_{n=1}^{\infty} \frac{(-1)^n}{n^2} \cos nx.$$

Next, we represent f as a *Fourier sine series*. For, in this case, we have

$$b_n = \frac{2}{\pi} \int_0^\pi x^2 \sin nx \, dx$$

$$= \frac{2}{\pi} \left[x^2 \frac{\cos nx}{-n} + \frac{2x \sin nx}{n^2} + \frac{2 \cos nx}{n^3} \right]_0^\pi$$

$$= \frac{2}{\pi} \left[\frac{\pi^2 (-1)^{n+1}}{n} + \frac{2}{n^3} ((-1)^n - 1) \right]$$

$$= \begin{cases} -\frac{2\pi}{n}, & \text{for } n \text{ even} \\ \frac{2\pi}{n} - \frac{8}{n^3 \pi}, & \text{for } n \text{ odd} \end{cases}.$$

So, for $x \in (0, \pi)$, we can also write

$$x^2 = \sum_{n=1}^\infty \left(\frac{2\pi}{(2n-1)} - \frac{8}{(2n-1)^3 \pi} \right) \sin(2n-1)x - \sum_{n=1}^\infty \frac{\pi}{n} \sin 2nx.$$

In general, with suitable modifications in the above discussion, a *half-range* function f defined over an interval $(0, \ell)$ can be represented by a Fourier cosine series or Fourier sine series. A *Fourier cosine series* of the half-range function $f(x)$ ($x \in (0, \ell)$) is given by

$$f(x) \sim \frac{a_0}{2} + \sum_{n=1}^\infty a_n \cos(\omega_n x),$$

where the Fourier coefficient a_n is given by

$$a_n = \frac{2}{\ell} \int_0^\ell f(x) \cos(\omega_n x) dx, \quad n \geq 0;$$

and, a *Fourier sine series* of the half-range function $f(x)$ ($x \in (0, \ell)$) is given by

$$f(x) \sim \sum_{n=1}^\infty b_n \sin(\omega_n x),$$

where the Fourier coefficient b_n is given by

$$b_n = \frac{2}{\ell} \int_0^\ell f(x) \sin(\omega_n x) dx, \quad n \geq 1.$$

Example 10.11 To write a *Fourier cosine series* for the function $f(x) = x$, $x \in (0, 2)$, we have

$$a_0 = \int_0^2 x \, dx = 2,$$

and, for $n \geq 1$,

$$a_n = \int_0^2 x \cos\left(\frac{n\pi x}{2}\right) dx = \frac{-4}{n^2\pi^2}(\cos n\pi - 1) = \frac{4}{n^2\pi^2}((-1)^{n+1} + 1).$$

So, for $x \in (0, 2)$, we can write

$$x = 1 + \frac{8}{\pi^2} \sum_{n=1}^{\infty} \frac{1}{(2n-1)^2} \cos\left(\frac{(2n-1)\pi x}{2}\right).$$

10.3 Convergence Issues

Let $f \in L^1([a, a + 2\ell])$ be a periodic function of period 2ℓ. By using Eqs. (10.2.6) and (10.2.7), we compute the Fourier coefficients a_n and b_n, and write the Fourier series of f as

$$s(f)(t) := \frac{a_0}{2} + \sum_{n=1}^{\infty} \left[a_n \cos(\omega_n t) + b_n \sin(\omega_n t)\right]. \tag{10.3.1}$$

Recall that, as before, we have $\omega_n = n\pi/\ell$, for $n \geq 1$. In general, it is known that there exist periodic functions in $L^1[-\pi, \pi]$ such that the series $s(f)$ diverges at every point $t \in [-\pi, \pi]$. Also, there are continuous functions in $L^1[-\pi, \pi]$ such that even *pointwise convergence* may not be possible. The reader is referred to the excellent text [2] for further details.

Therefore, as has been said before, the fundamental problem of the Fourier analysis is to find the class of functions f such that the series $s(f)$ represents f. Said differently, we need to find the answer to the following two questions:

1. Does the series $s(f)$ converges in some sense?
2. If the answer to (1) is yes, is $s(f) = f$?

Equivalently, if $\langle s_k(f) \rangle$ is the sequence of partial sum of the series (10.3.1), i.e.,

$$s_k(f)(t) = \frac{a_0}{2} + \sum_{n=1}^{k} [a_n \cos(\omega_n t) + b_n \sin(\omega_n t)], \quad k \geq 1, \tag{10.3.2}$$

we need to investigate the convergence $s_k(f) \to s(f)$, as $k \to \infty$, in some appropriate metric linear space. For example, Fig. 10.5 demonstrates approximation of the function $f(t) = t + t^2$ by the second and third partial sums of its Fourier series, as obtained in Example 10.6. Our focus here is to discuss the following three types of convergence:

1. $s_k(f)(t) \to s(f)(t)$, as $k \to \infty$, for all $t \in [a, a + 2\ell]$;

Fig. 10.5 Approximation of $f(t) = t + t^2$ by $s_2(f)$ and $s_3(f)$ of its Fourier series

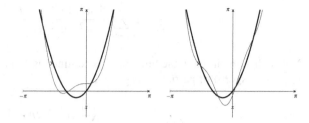

2. $\displaystyle\sup_{t\in[a,a+2\ell]} |s(f)(t) - s_k(f)(t)| \to 0$, as $k \to \infty$

3. $\displaystyle\lim_{k\to\infty} \int_a^{a+2\ell} [s(f)(t) - s_k(f)(t)]^2 dt = 0.$

Recall that the condition (1) says $s(f) \to f$ *pointwise* over the interval $[a, a + 2\ell]$; the condition (2) says f is the *uniform limit* of the series $s(f)$; and, the condition (3) gives the L^2-*convergence* of $s(f)$ to the function f. We are in the best position to apply Fourier theory in practical situations if f is the uniform limit of the series $s(f)$. Recall that, in every suitable function space, we have

$$uniform\ convergence \quad \Rightarrow \quad pointwise\ convergence;$$
$$uniform\ convergence \quad \Rightarrow \quad L^2\ convergence,$$

but none of the reverse implication holds. Also, the sequence of functions $f_n(t) = \sqrt{n}\, t^n$, $0 \le t \le 1$, does converge pointwise to the *zero function* on $[0, 1]$, but not with respect to the L^2-metric. The reader may find an example to show that even the converse is not true. The next example illustrates some types of complexities involved with the issue of convergence of the Fourier series.

Example 10.12 The Fourier series of the *triangular wave* function (see Fig. 10.6)

$$f(t) = \begin{cases} t, & 0 \le t \le 1/2 \\ 1 - t, & 1/2 \le t \le 1 \end{cases}, \quad \text{with } f(t + 1) = f(t),\ \forall\, t \in \mathbb{R},$$

is given by

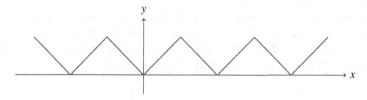

Fig. 10.6 A triangular wave function

$$\frac{1}{4} - 2 \sum_{n=0}^{\infty} \frac{\cos(2(2n+1)\pi t)}{\pi^2 (2n+1)^2}.$$

Notice that, in this case, the function f is continuous, but its derivative f' is not. The sequence $\langle s_k(f) \rangle$ of *partial sums* given by

$$s_k(f)(t) = \frac{1}{4} - 2 \sum_{n=0}^{k-1} \frac{\cos(2(2n+1)\pi t)}{\pi^2 (2n+1)^2}$$

converges *very fast* as the convergence is uniform. However, as the *square wave*

$$g(t) = \begin{cases} 1, & 0 \le t \le 1 \\ 0, & 1 < t \le 2 \end{cases}, \quad \text{with } f(t+2) = f(t), \ \forall \, t \in \mathbb{R},$$

has the Fourier series

$$\frac{2}{\pi} \sum_{n=0}^{\infty} \frac{\sin((2n+1)\pi t)}{(2n+1)},$$

which converges at each point of continuity of the function g. The convergence *at discontinuities* is not that good due to a general failure, known as the *Gibbs' phenomenon* (Example 10.14). Notice that, in this case, neither the function g nor its derivative g' is continuous.

10.3.1 Dirichlet Theorem

In 1829, Dirichlet proved that the Fourier series of a piecewise smooth (periodic) function converges pointwise at each point of continuity of its domain. This result is now known as the *Dirichlet theorem*. We know that the *pointwise convergence* is a weaker form of convergence, but it is very useful in many practical situations.

To prove our main theorem here, we need some preparation. First, notice that due to what has been said earlier about periodic extension of a half-range function it is natural to work with functions in the space of *piecewise continuous* functions defined over a bounded interval. We can assume periodicity of such functions if the interval is *symmetrical* about the origin.

Recall that $PC[a, b]$ is the space of real-valued piecewise continuous functions and $PS[a, b]$ the space of real-valued piecewise smooth functions, defined over the bounded interval $[a, b]$. Notice that, for $f \in PS([a, b])$, the derivative f' may not exists at the points of *jump discontinuity*. Moreover, since

$$\int_{-\infty}^{\infty} |f(t)| dt = \int_{-\infty}^{\infty} f^+(t) dt + \int_{-\infty}^{\infty} f^-(t) dt,$$

where $f^+(t) = \max(f(t), 0)$ and $f^-(t) = \max(-f(t), 0)$, it follows that

$$\int_{-\infty}^{\infty} f^+(t)dt < \infty \quad \text{and} \quad \int_{-\infty}^{\infty} f^-(t)dt < \infty.$$

This is equivalent, in Lebesgue sense, to the integrability of f. So, every piecewise continuous function defined on a bounded interval is bounded, square integrable, and absolutely integrable.

Theorem 10.2 (Dirichlet Theorem) *Let $f \in L^1[a, a + 2\ell]$ be a 2ℓ-periodic function such that, for each $t \in (a, a + 2\ell)$, we have*

1. The limits $f(t^+) = \lim_{h \to 0} f(t + h)$ and $f(t^-) = \lim_{h \to 0} f(t - h)$ exist.

2. The one-side limits $f'(t^+) \lim_{h \to 0} f'(t + h)$ and $f'(t^-) \lim_{h \to 0} f'(t - h)$ exist.

Suppose the coefficients a_n and b_n are computed by using formulas (10.2.3) and (10.2.4). Then, the Fourier series (10.3.1) converges pointwise to the mean jump

$$\frac{f(t^+) + f(t^-)}{2}, \quad \text{for all } t \in (a, a + \ell).$$

Also, at the end points of the interval $[a, a + 2\ell]$, the series (10.3.1) converges to $(f(t^-) - f(t^+))/2$.

Notice that, in particular, Theorem 10.2 implies that $s(f)(t) \to f(t)$ for any f that is continuous and differentiable over $t \in (a, a + 2\ell)$. An outline proof of Theorem 10.2 is as follows: Putting in Eq. (10.2.1) the expressions for the coefficients a_n and b_n, we obtain

$$s_k(f)(t) = \frac{1}{2\ell} \int_a^{a+2\ell} f(x)dx + \frac{1}{\ell} \int_a^{a+2\ell} f(x) \left[\sum_{n=1}^k \cos(\omega_n(t - x)) \right] dx$$

$$= \frac{1}{\ell} \int_a^{a+2\ell} f(x) \left[\frac{1}{2} + \sum_{n=1}^k \cos(\omega_n(t - x)) \right] dx$$

$$= \frac{1}{\ell} \int_a^{a+2\ell} f(x) D_k(t - x) \, dx, \tag{10.3.3}$$

where the expression

$$D_k(t - x) = \frac{1}{2} + \sum_{n=1}^k \cos(\omega_n(t - x)), \quad k \geq 1, \tag{10.3.4}$$

is known as the *Dirichlet kernel*. For any fixed $k \geq 1$, we may write

$$D_k(y) = \frac{1}{2} + \sum_{n=1}^{k} \cos(\omega_n y), \quad k \geq 1. \tag{10.3.5}$$

The main argument used in the proof given below is based upon the fact that although the function $D_k(y)$ has a huge spike when y is near zero, i.e., when x is close to t, the area under the spike for $D_k(y)$ is approximately 1. So, the integral in (10.3.3) approaches f, as $k \to \infty$. For values of x away from t, the function $D_k(x - t)$ is oscillatory, where as k approaches to ∞ these oscillations tend to cancel out in the integral on the right of the Eq. (10.3.3).

Proof In view of Remark 10.2, it suffices to prove the theorem for a piecewise smooth function $f : [-\pi, \pi] \to \mathbb{R}$, with $f(t + 2\pi) = f(t)$ for all $t \in \mathbb{R}$. In this case, the *Dirichlet kernel* is given by

$$D_k(y) = \frac{1}{2} + \sum_{n=1}^{k} \cos(n\,y), \quad k \geq 1, \tag{10.3.6}$$

so that the *k*th partial sums $s_k(f)$ of the Fourier series

$$\frac{a_0}{2} + \sum_{n=1}^{\infty} [a_n \cos(n\,t) + b_n \sin(n\,t)]$$

can be written as

$$s_k(f)(t) = \frac{1}{\pi} \int_{-\pi}^{\pi} f(x)\, D_k(t - x)\, dx. \tag{10.3.7}$$

Multiplying both sides of (10.3.6) by $\sin(y/2)$, $y \neq 0$, and using the relation $2 \sin \alpha \cos \beta = \sin(\alpha + \beta) - \sin(\alpha - \beta)$, we obtain

$$\sin\left(\frac{y}{2}\right) D_k(y) = \frac{1}{2} \left\{ \sin\left(\frac{y}{2}\right) + \sum_{n=1}^{k} \sin\left(\left(n + \frac{1}{2}\right)y\right) - \sin\left(\left(n - \frac{1}{2}\right)y\right) \right\}.$$

Since the sum appearing on the right side of this equation is *telescopic*, we obtain the following closed-form expression for the *Dirichlet's kernel*:

$$D_k(y) = \frac{\sin\left(\left(k + \frac{1}{2}\right)y\right)}{2\sin\left(\frac{y}{2}\right)}, \quad \text{for } y \neq 0. \tag{10.3.8}$$

Clearly, we have $D_k(0) = (2k + 1)/2$. Notice that the points $y = 2m\pi$ are only *removable discontinuities* of the function on the right side of the Eq. (10.3.8), so it holds for all $y \in \mathbb{R}$ subject to that limits are taken at points wherever it is not well-defined. Notice that D_k is a 2π-periodic, even function. Also, by definition, we have

$$\frac{1}{\pi} \int_{-\pi}^{\pi} D_k(y)dy = 1, \quad \text{for each} \quad k \geq 1, \tag{10.3.9}$$

so that we may write

$$f(t) - s_k(f)(t) = \frac{1}{\pi} \int_{-\pi}^{\pi} D_k(t-x)[f(t) - f(x)]\,dx. \tag{10.3.10}$$

By using the periodicity of D_k and f, from (10.3.9), it follows that

$$s_k(f)(t) = \frac{1}{\pi} \int_{-\pi}^{\pi} f(x)D_k(t-x)dx$$

$$= \frac{1}{\pi} \int_{-\pi+b}^{\pi+b} f(x)D_k(t-x)dx$$

$$= \frac{1}{\pi} \int_{-\pi}^{\pi} f(u+b)D_k(t-u-b)\,du$$

$$= \frac{1}{\pi} \int_{-\pi}^{\pi} f(t-x)D_k(x)dx. \tag{10.3.11}$$

A similar manipulation gives

$$s_k(f)(t) = \frac{1}{\pi} \int_{0}^{\pi} \left[f(t-x) + f(t+x)\right]D_k(x)dx. \tag{10.3.12}$$

Therefore, by taking

$$h(x) = \frac{f(t) - f(x)}{2\sin((t-x)/2)},$$

we obtain from relations (10.3.8) and (10.3.10) that

$$f(t) - s_k(f)(t) = \frac{1}{\pi} \int_{-\pi}^{\pi} \sin\left(\left(k+\frac{1}{2}\right)(t-x)\right) h(x)dx.$$

Now, for any fixed $t \in [-\pi, \pi]$, if f is differentiable at t then

$$\frac{f(t) - f(x)}{t - x} \in PC[-\pi, \pi] \Rightarrow h(x) \in PC[-\pi, \pi].$$

As said before, as $k \to \infty$, the oscillations of the sine function become large, and the difference between its contribution to the integral from the intervals where *it is positive* and where it is *negative* is almost zero. A rigorous proof of this last assertion follows from Lemma 10.1. Therefore,

$$|f(t) - s_k(f)(t)| \to 0 \quad \text{as} \quad k \to \infty, \ \forall\, t \in [-\pi, \pi].$$

Likewise, at a point of *jump discontinuity* $t = t_0$, we may use (10.3.12) to conclude that the assertion of the theorem is equivalent to the statement

$$\lim_{k \to \infty} \frac{1}{\pi} \int_0^\pi \left[f(t_0 - x) + f(t_0 + x) \right] D_k(x) dx = \frac{f(t_0^+) + f(t_0^-)}{2}.$$

However, by using the relations

$$\frac{1}{\pi} \int_0^\pi D_k(y) = \frac{1}{\pi} \int_{-\pi}^0 D_k(y) = \frac{1}{2}, \quad k \geq 1, \tag{10.3.13}$$

it is enough to prove the following equalities:

$$\lim_{k \to \infty} \int_0^\pi \left[f(t_0 \pm x) - f(t_0^\pm) \right] D_k(x) dx = 0,$$

which in integral form are given by

$$\lim_{k \to \infty} \frac{1}{\pi} \int_0^\pi \frac{f(t_0 \pm x) - f(t_0^\pm)}{2 \sin(x/2)} \sin\left(k + \frac{1}{2} \right) dx = 0.$$

Therefore, by using the fact (Exercise 10.12) that the function

$$g_\pm(x) := \frac{f(t_0 \pm x) - f(t_0^\pm)}{2 \sin(x/2)} \in L^1(0, \pi], \tag{10.3.14}$$

we may apply the generalised form of the Riemann-Lebesgue lemma (as in (10.2.15)), and the argument as above, to complete the proof of the Dirichlet theorem.

In particular, it follows that

$$f(t) = \frac{a_0}{2} + \sum_{n=1}^\infty \left[a_n \cos(nt) + b_n \sin(nt) \right] \tag{10.3.15}$$

at every point $t \in (-\pi, \pi)$, whenever both f and f' are continuous. We say that a function defined on a bounded interval has *Dirichlet's property* if Theorem 10.2 holds for the function. For example, a continuous function with *bounded variations* has the Dirichlet's property. The argument used in the above proof, together with Lemma 10.1, show that for a function $f : [-\pi, \pi] \to \mathbb{R}$ satisfying Dirichlet's property, the value of the Fourier series

$$s(f)(t) := \lim_{k \to \infty} \left[\frac{a_0}{2} + \sum_{n=1}^k \left[a_n \cos(n t) + b_n \sin(n t) \right] \right]$$

at a given point depends only on the *local values* of the function f near that point. This is known as the *localisation principle* (Exercise 10.11).

10.3.2 Gibbs Phenomenon

The behaviour at a *jump discontinuity* illustrated in the next two worked examples is known as *Gibbs phenomenon*, which says that the truncated Fourier series *overshoots* by about 9% of the *size of the jump* near a point of discontinuity.

Example 10.13 Consider the *step function*

$$h(t) = \begin{cases} 0 & \text{for } -\pi < t < 0 \\ 1 & \text{for } 0 < t < \pi \end{cases}, \quad \text{with } h(t + 2\pi) = h(t), \ \forall \, t \in \mathbb{R}.$$

Let t be a very small positive number. Now,

$$s_k(h)(t) = \frac{1}{\pi} \int_0^\pi D_k(x - t)\,dx = \frac{1}{\pi} \int_{t-\pi}^t D_k(y)\,dy. \tag{10.3.16}$$

We may use (10.3.13) to write

$$s_k(h)(t) = \frac{1}{2} + \frac{1}{\pi} \int_0^t D_k(y)\,dy - \frac{1}{\pi} \int_{-\pi}^{-\pi+t} D_k(y)\,dy. \tag{10.3.17}$$

For a fixed $a > 0$, take $t_n = a/[k + (1/2)]$. Since $D_k(-\pi) = (-1)^k/2$, for $-\pi \le y \le -\pi + t$, we have

$$|D_k(y)| < \frac{1}{2}, \quad \text{for small values of } t.$$

Therefore,

$$\left| \int_{-\pi}^{-\pi+t} D_k(y)\,dy \right| \le \frac{t_n}{2} < \frac{a}{2n}. \tag{10.3.18}$$

Also, since the function $(1/\sin(y/2)) - (2/y)$ is differentiable, and it vanishes at $y = 0$, for small values of t and

$$\widehat{D}_k(y) = \frac{\sin\left((k + \frac{1}{2})\, y\right)}{y},$$

we have

$$\left| \int_0^{t_n} D_k(y)\,dy - \int_0^{t_n} \widehat{D}_k(y)\,dy \right| < \frac{t_n}{2} < \frac{a}{2n}. \tag{10.3.19}$$

Notice that, by using the substitution $x = (k + 1/2)y$,

$$\int_0^{t_n} \widehat{D}_k(y)\, dy = \int_0^a \frac{\sin x}{x} = \mathrm{Si}(a), \qquad (10.3.20)$$

where $\mathrm{Si}(x) = \sin x / x$, which attains its *maximum* at $a = \pi$ and $\mathrm{Si}(\pi) \approx 1.85194.$, is known as the *Gibbs constant* It thus follows from the discussion above that

$$\lim_{k \to \infty} s_k(h)(t_n) = \frac{1}{2} + \frac{\mathrm{Si}(\pi)}{\pi} \approx 1.08949,$$

that is, in the limiting case, the partial sequence of the Fourier series of the function h overshoots its maximal value by 9%.

Example 10.14 Consider the *step function*

$$f(x) = \begin{cases} -1 & \text{for } -\pi < x < 0 \\ 1 & \text{for } 0 < x < \pi \end{cases}, \quad \text{with } f(x + 2\pi) = f(x), \ \forall\, x \in \mathbb{R}.$$

Then, as shown in Example 10.8 (with $a = 1$), the sequence $\langle s_k(f) \rangle$ of *partial sums* of the Fourier series of f is given by

$$s_k(f)(x) = \frac{4}{\pi} \sum_{n=0}^{k-1} \frac{\sin(2n+1)x}{2n+1}$$

$$= \frac{4}{\pi} \mathrm{Im} \left[\sum_{n=1}^{k-1} \frac{e^{i(2n+1)x}}{2n+1} \right], \quad \text{for all } x \in (0, \pi).$$

By assuming termwise differentiation is valid, we have

$$s_k'(f)(x) = \frac{4}{\pi} \mathrm{Im} \left[\sum_{n=1}^{k-1} i\, e^{i(2n+1)x} \right] = \frac{4}{\pi} \mathrm{Im} \left[i\, e^x \sum_{n=1}^{k-1} e^{i2nx} \right]$$

$$= \frac{4}{\pi} \mathrm{Im} \left[i\, e^x \frac{1 - e^{i2kx}}{1 - e^{i2x}} \right] = \frac{4}{\pi} \mathrm{Im} \left[i\, \frac{1 - e^{i2kx}}{e^{-x} - e^{ix}} \right]$$

$$= \frac{4}{\pi} \mathrm{Im} \left[i\, \frac{e^{i2kx} - 1}{e^{ix} - e^{-ix}} \right]$$

$$= \frac{2}{\pi} \mathrm{Im} \left[\frac{e^{i2kx} - 1}{\sin x} \right] = \frac{2}{\pi} \frac{\sin 2kx}{\sin x}.$$

Therefore, $s_k(f)(x)$ attains its maximum value at $x = \pi/2k$. Also, since we can write

$$s_k(f)(x) = \frac{2}{\pi} \int_0^x \frac{\sin 2ku}{\sin u}\, du = \frac{2}{\pi} \int_0^{2kx} \frac{\sin t}{2k \sin(t/2k)}\, dt,$$

it follows that

$$s_k(f)(x) \approx \frac{2}{\pi} \int_0^{2kx} \mathrm{Si}(t)dt, \quad \text{for } k \text{ sufficiently large.}$$

where $\mathrm{Si}(t) = \dfrac{\sin t}{t}$ is the *integral sine function*, with $\mathrm{Si}(\pi) = 1.85194$. Thus, the maximum value of $s_k(f)(x)$ at $x = \pi/2k$ is approximately 1.17898. Thus, as $k \to \infty$, the *overshoots* is 0.18, which is 9% of the jump at $x = 0$.

In the original article, Gibbs used the Fourier series of the sawtooth function as obtained in Example 10.7. Significance of this function is due to the fact that we can write any periodic function $f \in PS[-\pi, \pi)$, with $f(-\pi) = f(\pi)$, as the sum of a continuous piecewise smooth function and a finite linear combinations of sawtooth-like functions. So, by Theorem 10.5, Gibbs' phenomenon occurs only due to the second part of the sum. The related details are left as exercise for the reader.

10.3.3 Convergence in Mean

We prove here, if $f \in L^2[-\pi, \pi]$ is a periodic function of period 2π with $f(-\pi) = f(\pi)$, then $s(f) \to f$ with respect to L^2-*norm*. Recall that $f \in L^2[-\pi, \pi]$ imply that $f : [-\pi, \pi] \to \mathbb{R}$ is a Lebesgue measurable function such that

$$\|f\|^2 = \int_{-\pi}^{\pi} |f(x)|^2 dx < \infty, \tag{10.3.21}$$

The above assertion is a special case of the celebrated theorem due to Riesz and Fischer (Theorem 10.7). Let f be as above, and for a fixed k, consider a trigonometric polynomial given by

$$\sigma(t) := \frac{c_0}{2} + \sum_{n=1}^{k} [c_n \cos(nt) + d_n \sin(nt)], \tag{10.3.22}$$

such that $(f - \sigma) \geq 0$, where the coefficients c_n and d_n are some real numbers. Notice that, in general, we have

$$\int_{-\pi}^{\pi} (f - \sigma)^2 dt = \int_{-\pi}^{\pi} (f^2 - 2f\sigma + \sigma^2) dt. \tag{10.3.23}$$

Now, by using the *orthogonality relations*, we have

$$\int_{-\pi}^{\pi} \sigma(t)^2 dt \tag{10.3.24}$$

$$= \int_{-\pi}^{\pi} \left[\frac{c_0}{2} + \sum_{n=1}^{k} \left[c_n \cos(nt) + d_n \sin(nt) \right] \right]^2 dt$$

$$= \int_{-\pi}^{\pi} \left[\frac{c_0^2}{4} + \sum_{n=1}^{k} \left[c_n^2 \cos^2(nt) + d_n^2 \sin^2(nt) \right] \right] dt$$

$$= \frac{c_0}{4} \int_{-\pi}^{\pi} (1) dt + \sum_{n=1}^{k} \left(c_n \int_{-\pi}^{\pi} f(t) \cos(nt) dt + d_n \int_{-\pi}^{\pi} f(t) \sin(nt) dt \right)$$

$$= \frac{c_0^2}{4} \cdot 2\pi + \pi \sum_{n=1}^{k} \left(c_n^2 + d_n^2 \right)$$

$$= \pi \left[\frac{c_0^2}{2} + \sum_{n=1}^{k} \left(c_n^2 + d_n^2 \right) \right], \tag{10.3.25}$$

This is called the *Plancherel-Parseval Identity* for trigonometric polynomials. Similarly, by using Eqs. (10.2.6) and (10.2.6), it follows that

$$\int_{-\pi}^{\pi} f(t)\sigma(t) dt \tag{10.3.26}$$

$$= \int_{-\pi}^{\pi} f(t) \left[\frac{c_0}{2} + \sum_{n=1}^{k} [c_n \cos(nt) + d_n \sin(nt)] \right] dt$$

$$= \frac{c_0}{2} \int_{-\pi}^{\pi} f(t) dt + \sum_{n=1}^{k} \left(c_k \int_{-\pi}^{\pi} f(t) \cos(nt) dt + d_k \int_{-\pi}^{\pi} f(t) \sin(nt) dt \right)$$

$$= \frac{c_0}{2} \cdot \pi a_0 + \sum_{n=1}^{k} \left(c_k \cdot \pi a_k + d_k \cdot \pi b_k \right)$$

$$= \pi \left[\frac{1}{2} a_0 c_0 + \sum_{n=1}^{k} \left(a_n c_n + b_n d_n \right) \right]. \tag{10.3.27}$$

Putting the last two expressions in (10.3.23), a simple algebraic manipulation of the resulting expression gives

$$\int_{-\pi}^{\pi} (f - \sigma)^2 dt = \int_{-\pi}^{\pi} f^2 dt - 2 \int_{-\pi}^{\pi} f\sigma \, dt + \int_{-\pi}^{\pi} \sigma^2 dt$$

$$= \int_{-\pi}^{\pi} f^2 dt - \pi A + \pi B, \tag{10.3.28}$$

where we have

$$A = \left[\frac{a_0^2}{2} + \sum_{k=1}^{n} (a_k^2 + b_k^2) \right], \tag{10.3.29}$$

$$B = \left[\frac{(a_0 - c_0)^2}{2} + \sum_{k=1}^{n} \left[(a_k - c_k)^2 + (b_k - d_k)^2 \right] \right]. \tag{10.3.30}$$

Notice that only B depends on the trigonometric polynomial σ, and the left side integral in (10.3.28) has the **minimum value** only when $c_k = a_k$ and $d_k = b_k$, for $0 \le k \le n$. Further, if a square integrable function $f : [-\pi, \pi] \to \mathbb{R}$, with $f(t + 2\pi) = f(t)$ ($t \in \mathbb{R}$) is represented by the Fourier series (10.1.12), then $B = 0$, and so the positivity of L^2-norm proves that the following *Bessel's inequality* holds:

$$\frac{|a_0^2|}{2} + \sum_{n=1}^{\infty} \left(|a_n|^2 + |b_n|^2 \right) \le \frac{1}{\pi} \int_{-\pi}^{\pi} [f(t)]^2 dt. \tag{10.3.31}$$

This assertion also holds for the complex form of the Fourier series:

$$\sum_{n=-\infty}^{\infty} |c_n|^2 \le \frac{1}{2\pi} \int_{-\pi}^{\pi} [f(t)]^2 dt. \tag{10.3.32}$$

To prove the inequality (10.3.32), one may use the exponential form of the definition for the L^2-*norm*. In particular, it follows that all the following three series

$$\sum_{n=1}^{\infty} |a_n|^2, \quad \sum_{n=1}^{\infty} |b_n|^2, \quad \text{and} \quad \sum_{n=-\infty}^{\infty} |c_n|^2,$$

converge. Therefore, we have $a_n, b_n, c_n \to 0$, as $n \to \infty$. That is, the *absolute values of the Fourier coefficients of a function $f \in L^2[-\pi, \pi]$, with $f(t + 2\pi) = f(t)$ ($t \in \mathbb{R}$), are always arbitrarily small*. Recall that, by Riemann-Leibesgue Lemma, this assertion holds even for 2π-periodic functions $f \in L^1[-\pi, \pi]$. Notice that the equality holds in (10.3.31) for any $f \in L^2[-\pi, \pi]$:

$$\frac{1}{2\ell} \int_{-\pi}^{\pi} f(t)^2 dt = \frac{a_0^2}{4} + \frac{1}{2} \sum_{n=1}^{\infty} [a_n^2 + b_n^2], \tag{10.3.33}$$

where a_n and b_n are the Fourier coefficients of the function $f(t)$.

There are other forms of convergences that hold importance in some situations. For example, Féjer was the first to use Theorem 3.6 to prove *Cesáro summability* of Fourier series of an integrable 2π-periodic function $f : [-\pi, \pi] \to \mathbb{R}$ in terms of the sequence of functions

$$f_n(t) := \frac{1}{n+1} \left(\frac{\sin[(n+1)t/2]}{\sin(t/2)} \right),$$

known as the *Féjer kernels*. It is very instructive to study a connection between this and the Dirichlet's approach discussed above. Poisson used Corollary 3.3 to develop his version of convergence of complex form of Fourier series.

Consider the sequence of *exponential polynomials*

$$p_k(t) = \sum_{n=-k}^{k} c_k\, e^{i\,\omega_k\, t}, \quad \text{for } k \geq 0. \tag{10.3.34}$$

Then the proof of Theorem 10.1 can be modified, with $a = -\ell$, to show that if the series (10.2.11) converges uniformly to a function f then the coefficients c_n $(n \in \mathbb{Z})$ can be recovered from the series by using the orthogonality relation

$$\int_{-\ell}^{\ell} e^{i(n-k)x}\,dx = \begin{cases} 0, & \text{for } n \neq k \\ 2\ell, & \text{for } n = k \end{cases}.$$

The *complex version* of Theorem 10.2 states as follows.

Theorem 10.3 *Let $f : [a, a+\ell] \to \mathbb{R}$ be a 2ℓ-periodic function such that $f \in PS([a, a+\ell])$. If the coefficients c_n are given by the formula as in (10.2.11), then $c_n = O(|n|^{-1})$, and*

$$\lim_{k \to \infty} p_k(t) = \frac{f(t^+) + f(t^-)}{2}, \quad \text{for all } t \in (a, a+\ell).$$

The situation at the end points remains the same as in Theorem 10.2.

10.3.4 Uniform Convergence

Finally, we prove uniform convergence of Fourier series with respect to L^2-norm. Recall that it makes sense to talk about *complex Fourier series* of a function $f \in PC(\mathbb{T})$, with the sequence of polynomial $\langle p_n(t) \rangle$ as its nth *partial sum*, where $c_k = c_k(f)$ is given by (10.2.11), where $\mathbb{T}$ is the unit circle.

In this case, it follows easily that the *Dirichlet kernel* is given by

$$D_k(y) = \sum_{n=-k}^{k} e^{i\,n\,y} = \frac{e^{i(k+1)y} - e^{-iky}}{e^{iy} - 1}, \quad \text{for } k \geq 0,$$

and the function $g_{\pm}(x) \in PS(\mathbb{T})$ is given by

$$g(x) = \begin{cases} \dfrac{f(t-x) - f(t^+)}{e^{ix} - 1}, & \text{if } -\pi < x < 0 \\[2mm] \dfrac{f(t-x) - f(t^-)}{e^{ix} - 1}, & \text{if } 0 < x \leq \pi \end{cases}.$$

Therefore, a proof of the following *complex version* of Dirichlet's theorem can be obtained by imitating the proof of Theorem 10.2.

Theorem 10.4 *Let $f \in PS(\mathbb{T})$, and suppose $c_n = c_n(f)$ are given by (10.2.11), with $\omega_n = n$. Then*

$$\lim_{n \to \infty} c_n = \lim_{n \to \infty} c_{-n} = 0,$$

and we have

$$\lim_{n \to \infty} p_n[f](t) = \frac{f(t^+) + f(t^-)}{2}, \quad \text{for all } t \in \mathbb{T}.$$

Notice that the first assertion follows from Lemma 10.1. In view of Theorem 10.2, 2π-periodic functions in $PS(\mathbb{T})$ are good for many applications. However, these are not good enough to meet all practical needs such as termwise differentiation and integration of the Fourier series. For example, differentiating the Fourier series of *sawtooth function* as obtained in Example 10.7, we get

$$1 = 2 \sum_{n=1}^{\infty} (-1)^{n+1} \cos nt, \quad -\pi < t < \pi,$$

which is certainly *false*. Notice that the series on the right is not even convergent. In fact, by Riemann-Lebesgue lemma, it can't even be a Fourier series of any function in $PC(\mathbb{T})$. As has been said before, such type of problem do not arise when the convergence is *uniform*. That is, if *sup metric* is used on the space of 2π-periodic *continuous* functions in $PS(\mathbb{T})$.

Theorem 10.5 *Every continuous, piecewise smooth function in $PS(\mathbb{T})$ can be represented by its Fourier series.*

Proof Let $f(t) \in PS(\mathbb{T})$ be continuous. Then, $f' \in PC(\mathbb{T})$ implies that

$$c_n(f') = \frac{1}{2\pi} \int_{-\pi}^{\pi} f'(x) e^{-inx} dx < \infty. \quad \text{for all } n \in \mathbb{Z}. \tag{10.3.35}$$

Integrating by parts, we obtain

$$c_n(f') = in \, c_n(f), \quad \text{for all } n \in \mathbb{Z}. \tag{10.3.36}$$

Similarly, we can obtain

$$a_n(f') = n \, a_n(f) \quad \text{and} \quad b_n(f') = -n \, b_n(f),$$

for all applicable n. Now, we have

$$\sum_{n=-\infty}^{\infty} |c_n(f)| = |c_0(f)| + \sum_{0\neq n\in\mathbb{Z}} |c_n(f)|$$

$$= |c_0(f)| + \sum_{0\neq n\in\mathbb{Z}} \left|\frac{c_n(f')}{in}\right|, \qquad (10.3.37)$$

$$\leq |c_0(f)| + \frac{1}{2}\sum_{0\neq n\in\mathbb{Z}} |c_n(f')|^2 + \frac{1}{2}\sum_{0\neq n\in\mathbb{Z}} \frac{1}{n^2}, \qquad (10.3.38)$$

where the second equality uses (10.3.36), and the last inequality follows by applying the following standard inequality to $a = c_n(f')$ and $b = (in)^{-1}$:

$$|ab| \leq \frac{|a|^2 + |b|^2}{2}, \quad \text{for } a, b \in \mathbb{C}.$$

The first sum in (10.3.38) converges by Bessel's inequality (10.3.32), and the second by usual p-test. We thus conclude that

$$\sum_{n=-\infty}^{\infty} |c_n(f)| < \infty. \qquad (10.3.39)$$

Finally, to conclude that proof, we prove that the sequence $\langle S_k(f) \rangle$ converges uniformly to f over $\mathbb{T}$. For, let $t \in \mathbb{T}$. Then

$$|S_k(f)(t) - f(t)| = \left| \sum_{n=-k}^{k} c_n(f)e^{int} - \sum_{n=-\infty}^{\infty} c_n(f)e^{int} \right|$$

$$= \left| \sum_{|n|>k} c_n(f)e^{int} \right|$$

$$\leq \sum_{|n|>k} |c_n(f)e^{int}| = \sum_{|n|>k} |c_n(f)|.$$

Therefore, by completeness property of $\mathbb{R}$, we obtain

$$0 \leq \sup_{t\in\mathbb{T}} |S_k(f)(t) - f(t)| \leq \sum_{|n|>k} |c_n(f)|,$$

which proves the assertion, because the right side series is convergent by (10.3.39).

Notice that (10.3.39) implies that

$$\sum_{n=-\infty}^{\infty} |c_n(f)e^{int}| < \infty.$$

So, for a continuous, piecewise smooth function in $PS(\mathbb{T})$, the Fourier series converges absolutely, as well. Further, for a continuous function $f \in PS(0, \ell)$, Theorem 10.5 applies only to the *even periodic extension* of the function $f(t)$. The problem with the *odd periodic extension* is that it is continuous on $\mathbb{R}$ if and only if $f(0^+) = f(\ell^-)$.

A similar remark applies to ℓ-*periodic extension* of a continuous, piecewise smooth function in $PS(\mathbb{T})$. However, in this case, the problems arise due to points of discontinuities introduced, and also slow rate of decay of the Fourier coefficient $b_n(f)$. The reader is encouraged to use worked examples given earlier to reflect upon these comments.

Recall that $L^2(\mathbb{T})$ denotes the space of *square integrable* real or complex-valued function defined on $\mathbb{T} = [-\pi, \pi)$, with L^2-norm given by (10.3.21):

$$\|f\|_2^2 := \int_{-\pi}^{\pi} |f(t)|^2 dt < \infty, \quad f \in L^2(\mathbb{T}).$$

Here, we discuss L^2- *convergence* of the Fourier series $s(f)$ for $f \in L^2(\mathbb{T})$. Notice that, since

$$e^{ikx} \in C(\mathbb{T}) \quad \Rightarrow \quad e^{ikx} \in L^2(\mathbb{T}),$$

so the product $f(t)e^{ikx} \in L^1(\mathbb{T})$. Thus, all the coefficients c_k in (10.3.34) are elements of the space $L^1(\mathbb{T})$. Hence, the polynomial $p_n(t) \in L^2(\mathbb{T})$, for all $n \in \mathbb{Z}$.

By Stone-Weierstrass theorem, the space of polynomials $p(t)$ defined over the interval $[-\pi, \pi]$, satisfying $p(-\pi) = p(\pi)$, is *dense* in $C[-\pi, \pi]$. As mentioned in the concluding part of the previous section, the 2π - *periodic extension* of any such polynomial lies in the class of continuous, piecewise smooth functions defined over the set $\mathbb{T}$, which we denote by $PSC(\mathbb{T})$. Certainly then, the space $PSC(\mathbb{T})$ is dense in $C(\mathbb{T})$. Also, we know that every $f \in L^2(\mathbb{T})$ can be approximated by a sequence of functions in the space $C(\mathbb{T})$. Notice that the function $f(t) = -\ln |2\sin(t/2)| \in L^2(\mathbb{T})$ is 2π - periodic, but $f \notin PC(\mathbb{T})$. It is instructive to obtain the Fourier series of this function. Does $p_n[f](t) \to f(t)$ for all t ?

Theorem 10.6 *For every $f \in L^2(\mathbb{T})$,*

$$\lim_{n\to\infty} \|p_n[f](t) - f(t)\|_2 = 0, \quad \text{for all } t \in \mathbb{T}. \tag{10.3.40}$$

The converse holds too.

Proof Let $f_n \in PSC(\mathbb{T})$ such that

$$\|f_n - f\|_2 \to 0, \quad \text{as } n \to \infty. \tag{10.3.41}$$

Also, by triangle inequality, we have

$$\|p_n[f] - f\|_2 \le \|p_n[f] - p_n[f_n]\|_2 + \|p_n[f_n] - f_n\|_2 + \|f_n - f\|_2.$$

However, since

$$\|p_n[f] - p_n[f_n]\|_2 = \|p_n[f - f_n]\|_2 \le \|f - f_n\|_2,$$

we obtain

$$\|p_n[f] - f\|_2 \le \|p_n[f_n] - f_n\|_2 + 2\|f_n - f\|_2.$$

Now, we know that $p_n[f_n]$ converges uniformly to f_n for each $n \in \mathbb{N}$, by Theorem 10.5, and so with respect to L^2-norm as well. Hence, (10.3.41) proves (10.3.40).

Conversely, since we can write

$$f = p_n[f] + (f - p_n[f]),$$

so if $p_n[f] \to f$ with respect to L^2-norm, then from the implication

$$p_n[f] \in L^2(\mathbb{T}) \quad \Rightarrow \quad f - p_n[f] \in L^2(\mathbb{T}), \quad \text{for large } n,$$

it follows that $f \in L^2(\mathbb{T})$.

In particular, we obtain L^2—convergence of the sequence $\langle p_n[f] \rangle$ for a 2π-periodic function $f(t)$ with $f, f' \in PC(\mathbb{T})$.

Theorem 10.7 (Riesz-Fischer, 1907) *If $\{\varphi_n\}$ is an orthonormal system in $L^2(\mathbb{T})$ and $\{c_n\}$ is a sequence in $\ell^2(\mathbb{C})$, then the convergence of the series $\sum_{-\infty}^{\infty} c_n^2$ is a necessary and sufficient condition for the existence of a function f such that*

$$\int_{-\pi}^{\pi} f(x)\varphi_n(t)\,dt = c_n, \quad \text{for every } n. \tag{10.3.42}$$

In particular, for $\varphi_n = (1/\pi)e^{-2nt}$, it follows that the series (10.2.11) converges uniformly to f, if $f \in L^2(\mathbb{T})$, and the coefficients c_n are given by Eq. (10.3.42).

10.4 An Application to Infinite Series

Fourier series provides a useful tool to find the sum of important infinite series. The basic idea is to *identify* an appropriate continuously differentiable 2ℓ - periodic function $f(t)$ defined over the interval $[-\ell, \ell]$ so that Theorem 10.2 applies. Then, for $t \in (-\ell, \ell)$, we can write

$$\frac{f(t^+) + f(t^-)}{2} = \frac{a_0}{2} + \sum_{n=1}^{\infty}[a_n \cos(\omega_n t) + b_n \sin(\omega_n t)], \tag{10.4.1}$$

where the Fourier coefficients a_n, b_n are given by Eqs. (10.2.3) and (10.2.4). The sum of given infinite series is subsequently obtained by evaluating the Fourier series of the function $f(t)$ at a suitable point of continuity $t = t_0 \in (-\ell, \ell)$.

Example 10.15 By Theorem 10.2, the Fourier series of 2π - periodic function $f(t)$ obtained in Example 10.6 gives at $t = \pi$

$$\frac{f(\pi^+) + f(\pi^-)}{2} = \frac{(-\pi + \pi^2) + (\pi + \pi^2)}{2} = \pi^2.$$

In particular, it follows that

$$\sum_1^\infty \frac{1}{n^2} = \frac{\pi^2}{6},$$

which is known as the *Basel theorem*. Further, by taking $t = 0$, we obtain

$$\sum_{n=1}^\infty \frac{(-1)^{n-1}}{n^2} = \frac{\pi^2}{12}.$$

Example 10.16 By Example 10.7, we know that the Fourier sine series of 2π - periodic function $f(t) = t$, $-\pi < t \leq \pi$, is given by

$$t = 2 \sum_{n=1}^\infty \frac{(-1)^{n+1}}{n} \sin nt \quad (-\pi < t < \pi).$$

In particular, by taking $t = \pi/2$, we obtain the *Gregory series*

$$\frac{\pi}{4} = 1 - \frac{1}{3} + \frac{1}{5} - \frac{1}{7} + \cdots.$$

On the other hand, the Fourier cosine series of 2π-periodic function $f(t) = |t|$, $t \in [-\pi, \pi]$, obtained in Example 10.9 is given by

$$\frac{\pi}{2} - \frac{4}{\pi} \sum_{k=1}^\infty \frac{\cos(2k-1)t}{(2k-1)^2}.$$

In particular, by taking $t = 0$, we obtain

$$\sum_{k=1}^\infty \frac{1}{(2k-1)^2} = \frac{\pi^2}{8}.$$

Example 10.17 For any $f \in L^2([-\pi, \pi])$, notice that equality holds in (10.3.31):

$$\frac{1}{2\ell} \int_{-\pi}^{\pi} f(t)^2 dt = \frac{a_0^2}{4} + \frac{1}{2} \sum_{n=1}^{\infty} [a_n^2 + b_n^2], \qquad (10.4.2)$$

where a_n and b_n are the Fourier coefficients of the function f. By applying this to the function $f(t) = t$, $t \in (0, 2)$, and by Example 10.11, we have

$$\frac{1}{4} \int_0^2 x^2 dx = 1 + \frac{32}{\pi^4} \sum_{n=1}^{\infty} \frac{1}{(2n-1)^4},$$

so that

$$\sum_{n=1}^{\infty} \frac{1}{(2n-1)^4} = \frac{\pi^4}{96}. \qquad (10.4.3)$$

Notice that

$$S = \sum_{n=1}^{\infty} \frac{1}{n^4} \implies \sum_{n=1}^{\infty} \frac{1}{(2n)^4} = \frac{1}{2^4} \sum_{n=1}^{\infty} \frac{1}{n^4} = \frac{S}{16}.$$

Thus, by using (10.4.3), we obtain

$$S = \sum_{n=1}^{\infty} \frac{1}{n^4} = \sum_{n=1}^{\infty} \frac{1}{(2n-1)^4} + \sum_{n=1}^{\infty} \frac{1}{(2n)^4}$$

$$= \frac{\pi^4}{96} + \frac{S}{16}$$

$$\implies S = \sum_{n=1}^{\infty} \frac{1}{n^4} = \frac{\pi^4}{90}.$$

Notice that the infinite series in the above example is a special case of the *Riemann zeta function* $\zeta(s)$ defined as an absolutely convergent series given by

$$\zeta(s) := \sum_{n=1}^{\infty} \frac{1}{n^s}, \quad s \in \mathbb{C}, \quad \text{with } \mathrm{Re}(s) > 1,$$

which can extended to $\mathbb{C} \setminus \{s = 1\}$ by meromorphic continuation. Most notable conjecture about $\zeta(s)$ is *Riemann Hypothesis*: *Other than the even negative integers, all solutions of the equation $\zeta(s) = 0$ have their real part on the line $x = 1/2$.*

Exercises 10

10.1 Compute the Fourier series of the *half-sine wave* function given by

$$f(t) = \begin{cases} 0, & \text{if } -\pi < t < 0 \\ \sin t, & \text{if } 0 < t < \pi \end{cases}$$

10.2 Compute the Fourier series of the function $f : [-2, 2] \to \mathbb{R}$ given by $f(x) = e^{-x}$, viewed in natural manner as a 4-periodic function.

10.3 Compute the Fourier series of the function $f : [0, 2\pi] \to \mathbb{R}$ given by $f(x) = e^{-x}$, with $f(x + 2\pi) = f(x)$ for all $x \in \mathbb{R}$.

10.4 Expand $f(x) = \dfrac{(\pi - x)^2}{4}$ in a Fourier series on the interval $(0, 2\pi)$ and obtain the value for $\displaystyle\sum_{0}^{\infty} \dfrac{1}{n^2}$.

10.5 Obtain the Fourier series expansion of $f(x) = \cos(ax)$ on the interval $(-\pi, \pi)$.

10.6 Compute the Fourier series of the function $f : [-\pi, \pi] \to \mathbb{R}$ given by $f(x) = x \cos x$, with $f(x + 2\pi) = f(x)$ for all $x \in \mathbb{R}$.

10.7 Expand f in a Fourier series where

$$f(x) = \begin{cases} -\pi & -\pi < x < 0 \\ x & 0 < x < \pi. \end{cases}$$

Also find the sum of the Fourier series at the point of discontinuity.

10.8 Find a Fourier series with period 3 to represent $f(x) = 2x - x^2$ in the range $(0, 3)$. Also, obtain the Fourier series with period 2 to represent the same function in the range $(0, 2)$.

10.9 Discuss the pointwise, uniform, and L^2-convergence of the Fourier sine series of each one of the following functions on the interval $[0, \pi]$: (a) $f(x) = \pi x - x^2$; (b) $g(x) = x^2 + 1$; (c) $h(x) = x^3$.

10.10 Verify Lemma 10.1 for (i) $f = \chi_{(a,b)}$, with $(a, b) \subset [-\pi, \pi]$; (ii) $f = \sum_{k=1}^{n} \chi_{(a_k, b_k)}$; and, (iii) $f = \sum_{k=1}^{n} \chi_{M_k}$, where each $M_k \subset (-\pi, \pi)$ is a measurable set.

10.11 Let $f : [-\pi, \pi] \to \mathbb{R}$, with $f(t + 2\pi) = f(t)$ for all $t \in \mathbb{R}$, be an integrable function such that for some $\epsilon > 0$, $f(t) = 0$, for all $t \in (t_0 - \epsilon, t_0 + \epsilon)$. Prove that $S(f)(t_0) = 0$. Hence deduce that if another such function $g : [-\pi, \pi] \to \mathbb{R}$ such that $f(t) = g(t)$, for all $t \in (t_0 - \epsilon, t_0 + \epsilon)$, then either both limits do not exist or $S(f)(t_0) = S(g)(t_0)$, if any one of these exists.

10.12 Prove the assertion (10.3.14).

References

1. A. Razdan, V. Ravichandran, *Fundamentals of Partial Differential Equations with Applications* (Springer, to appear)
2. D.C. Champeney, *A Handbook of Fourier Theorems* (Cambridge University Press, New York, 1987)

Appendix
Mathematical Logic

Every statement in modern mathematics is expressed in terms of sets and functions, and the *language* used to communicate the same is provided by the mathematical logic. In layman terms, the following four types of sentences are used to communicate about *anything* in our daily life:

1. A declaration (or a declarative sentence) is a phrase such as "*the capital of India is New Delhi*";
2. An *interrogative sentence* (or a question) is used to request information such as "*what is the height of himalaya ?*";
3. An exclamation (or an *exclamatory sentence*) is generally a more emphatic form of phrase such as "*what a wonderful joke this is !*";
4. A command (or an *imperative sentence*) is used to make a demand or a request such as "*do your discrete mathematics homework*".

Clearly, only a declarative sentence can be classified as true or false.

Definition A.1 We say a (declarative) sentence is a **proposition** if it is either *true* (T) or *false* (F).

For *propositional calculus*, every proposition is a variable, and we usually write $p, q, r, \ldots$ to denote propositions. The two possible values (T) or (F) that a proposition p can take are called the *truth values*. A *universally true* proposition is called a **tautology**, and a *universally false* proposition is called a **contradiction**. For example, the proposition

$$p : \text{ the sum of an odd integer and an even integer is an odd integer}$$

is a tautology. We usually write a tautology as **T**. On the other hand, the proposition

$$q : \text{ there is no integral solution to the equation } x^2 + y^2 = z^2$$

is a contradiction. We usually write a contradiction as **F**.

© The Editor(s) (if applicable) and The Author(s), under exclusive license to Springer Nature Singapore Pte Ltd. 2022
A. K. Razdan and V. Ravichandran, *Fundamentals of Analysis with Applications*,
https://doi.org/10.1007/978-981-16-8383-1

We form a *compound proposition* by using **connectives** such as *"and"*, *"or"*, *"not"*, *"if ..., then ..."*, etc. A proposition that involves no connective is said to *atomic*. The following *four* connectives are fundamental to propositional calculus in the sense that the rest of all the connectives can be expressed in terms of these four.

1. The *disjunction* of propositions p and q, denoted by $p \vee q$, is the proposition *"p or q"* with the truth value F if and only if both p and q have the truth value F. In algebraic terms, a disjunction corresponds to the *sum*;
2. The *conjunction* of propositions p and q, denoted by $p \wedge q$, is the proposition *"p and q"* with the truth value T if and only if both p and q have the truth value T. In algebraic terms, a conjunction corresponds to the *product*;
3. The negation of a proposition is the proposition with the truth value opposite to that of the given proposition. We write $\neg$ to denote *"not"*. So, for any proposition p, $\neg p$ is the proposition whose truth value is the opposite to the truth value of p. In algebraic terms, $\neg$ corresponds to an involutory *unary operation*;
4. For propositions p and q, the *conditional connective*, denoted by $p \Rightarrow q$, is the proposition *"if p, then q"* with the truth value F if and only if p has the truth value T and q has the truth value F. In the conditional statement $p \Rightarrow q$, we call p a *hypothesis* (or *antecendent*) and q a *conclusion* (or *consequent*).

As an illustration, we write the next sentence as a logical expression.

You can access the internet from campus only if you are a computer science student or you are not a freshman.

Notice that we can also write this sentence as follows.

If you are a computer science student or you are not a freshman, then you can access the internet from campus.

Let p, q, and r denote respectively the following three propositions:

$$p : \text{"you are a computer science student"},$$
$$q : \text{"you are a fresher"},$$
$$r : \text{"you can access the internet from campus"}.$$

Then, the given sentence is conditional, with the hypothesis $p \vee \neg q$, and the conclusion r. So, the logical expression for the above sentence is $(p \vee \neg q) \Rightarrow r$. Notice that its negation is the proposition

$$\neg((p \vee \neg q) \Rightarrow r).$$

A *biconditional statement* "*p if and only if q*", denoted by $p \Leftrightarrow q$, is the proposition whose truth value is T if and only if both p and q have the same truth value. The propositions $p \Rightarrow q$ and $p \Leftrightarrow q$ are usually expressed as follows:

$p \Rightarrow q$	$p \Leftrightarrow q$
if p, (then) q	p if and only if q
p implies q	p is necessary and sufficient for q
p only if q	If p, (then) q and conversely
q is necessary for p	
p is sufficient for q	

Notice that the compound propositions such as $p \vee q$ and $(p \wedge q) \Rightarrow r$ have no truth values unless some specific propositions are substituted for the *variables* $p, q, r, \ldots$. In general, the truth values of a compound proposition depend on the truth values of its atomic propositions. A tabular representation of the truth values of a compound proposition for different choices of truth values of its atomic propositions is called a *truth table*. We usually apply the following algorithm to obtain a new proposition or a *statement formula*:

1. Any propositional variable is a formula; **T** and **F** are also formulas.
2. If p and q are formulas, then $(\neg p)$, $(p \vee q)$, $(p \wedge q)$, $(p \Rightarrow q)$, and $(p \Leftrightarrow q)$ are also formulas.
3. Any expression obtained by applying the above rule (finite number of times) is a formula.

Definition A.2 A formula obtained by above procedure is called a **well-formed formula**.

As our main concern always is to get a well-formed formula, we may use the terminology *formula* for it. For example, the following logical expressions

$$((p \vee q) \Rightarrow r), \quad ((p \wedge q) \Leftrightarrow (\neg p))$$

are formulas. It is not necessary to use parenthesis always, so we may omit some of the parenthesis to write a formula in a simpler form. For example, we may write the above two formulas simply as

$$(p \vee q) \Rightarrow r, \quad (p \wedge q) \Leftrightarrow \neg p.$$

In doing so, we follow the convention that the negation affects as little as possible. For example, in $\neg p \vee (q \vee r)$, the negation is applied only to p. If we want to apply the negation to all the terms, then we must write $\neg(p \vee (q \vee r))$. For example, the logical expressions such as

$$\neg((\neg p) \vee (q \vee r)), \qquad \neg(p \vee q) \vee \neg(\neg p \vee \neg q),$$

and

$$(p \vee q) \rightarrow (p \wedge \neg q), \quad (p \Rightarrow q) \Rightarrow (q \Rightarrow p)$$

are well-formed formulas.

As before, a propositional formula that is true always irrespective of the truth values of the individual variables is called a *tautology*. For example, the proposition $p \vee \neg p$ is a tautology, and so is $(p \vee q) \vee \neg p$. Also, $(p \Rightarrow q) \Leftrightarrow (\neg p \vee q)$ is a tautology. Notice that a formula is a tautology if the last column of its truth table has only T. On the other hand, a propositional formula that is always false irrespective of the truth values of the individual variables is called a *contradiction*. Notice that a formula is a contradiction if the last column of its truth table has only F. For example, the proposition $p \wedge \neg p$ has the truth value F for all choices of truth values of p and q. Also, $q \wedge (p \wedge \neg q)$ is a contradiction. A formula that is neither a tautology nor a contradiction is called a *satisfiable formula*.

Definition A.3 We say a proposition p (logically) **implies** q if and only if $p \Rightarrow q$ is a tautology. In this case, we write $p \rightarrow q$. Further, we say two propositions p and q are (logically) **equivalent** if and only if $p \Leftrightarrow q$ is a tautology. In this case, we write $p \leftrightarrow q$.

Said differently, two propositions are *logically equivalent* if they have the same truth values irrespective of the choices for the truth values of the individual/atomic propositions may take. We can use $\leftrightarrow$ to define an equivalence relation on the collection of all propositions. Notice that $p \leftrightarrow q$ if and only if $p \Rightarrow q$ and $q \Rightarrow p$. In general, we can prove the equivalences and implications by constructing the truth tables or by using other equivalences.

Definition A.4 For a conditional statement $p \Rightarrow q$, the proposition $q \Rightarrow p$ is called its *converse*; the proposition $\neg p \Rightarrow \neg q$ is called the inverse; and, the proposition $\neg q \Rightarrow \neg p$ is called its *Contrapositive*.

Notice that, for every pair p, q, the conditional proposition $p \Rightarrow q$ and its contrapositive $\neg q \Rightarrow \neg p$ are equivalent; also, the converse $q \Rightarrow p$ and inverse $\neg p \Rightarrow \neg q$ are equivalent. The detachment law (or *modus ponens*) states that

$$(p \Rightarrow q) \wedge p \Rightarrow q.$$

The contrapositive (or *modus tollens*) states that

$$(p \rightarrow q) \wedge \neg q \Rightarrow \neg p.$$

The chain rule (or *hypothetical syllogism*) states that

$$(p \rightarrow q) \wedge (q \rightarrow r) \Rightarrow (p \rightarrow r).$$

It follows easily that $p \Rightarrow (p \vee q)$. In general, we can prove implications by using truth tables or by using *known equivalences*. Recall that $p \rightarrow q$ if $p \Rightarrow q$ is a tautology. Now, when p is a conjunction of the propositions $p_1, p_2, \ldots, p_n$ and q (logically) follows from $p_1, \ldots, p_n$, we write

$$(p_1 \wedge p_2 \wedge \cdots \wedge p_n) \Rightarrow q,$$

where $p_1, p_2, \ldots, p_n$ are called the *premises* of the (valid) conclusion q. We can also show that p follows from $p_1, \ldots, p_n$ by constructing the truth table for the expression $(p_1 \wedge p_2 \wedge \cdots \wedge p_n) \Rightarrow q$.

Theorem A.1 *If $p_1, p_2, \ldots, p_n$ and p imply q, then $p_1, p_2, \ldots, p_n$ imply $p \to q$.*

Proof As $(p_1 \wedge \cdots \wedge p_n \wedge p) \Rightarrow q$, so

$$(p_1 \wedge \cdots \wedge p_n \wedge p) \Rightarrow q$$

is a tautology. We also have that the following equivalence holds:

$$(r \wedge s) \to t \quad \leftrightarrow \quad r \to (s \to t).$$

It, thus, follows that

$$(p_1 \wedge \cdots \wedge p_n) \Rightarrow (p \to q)$$

is a tautology, and hence

$$(p_1 \wedge \cdots \wedge p_n) \to (p \to q),$$

as asserted. □

It follows easily that q is not a valid conclusion drawn from premises $p \to q, \neg p$. However, $\neg p$ is a valid conclusion drawn from the premises $p \to q, \neg(p \wedge q)$. It is convenient to use truth tables to verify these assertions. However, when the number of variables in premises is large, the method of truth table gets tedious. In such situations, we use *rules of inference*.

Theory of Inference

We give here a brief introduction of the *theory of inference* for propositional calculus. In general, the following *Rules of inferences* are applied to show that a valid conclusion follows from a number of premises:

1. Rule P: A hypothesis or premise can be introduced at any point in the derivation.
2. Rule T: A formula r can be introduced at any point in the derivation if r is implied by one or more of preceding formulas in the derivation.
3. Rule CP: If we can derive s from r and a set of premises $p_1, \ldots, p_n$, then we can derive $r \to s$ from the set of premises $p_1, \ldots, p_n$ alone.

The last rule says that if we have to derive a conditional conclusion $r \to s$ from premises $p_1, \ldots, p_n$, then it is enough to show that the hypothesis r of the conditional conclusion and the premises $p_1, \ldots, p_n$ implies s. This follows because

$$
\begin{aligned}
p \to (r \to s) \quad &\leftrightarrow \quad \neg p \vee (\neg r \vee s) \\
&\leftrightarrow \quad (\neg p \vee \neg r) \vee s \\
&\leftrightarrow \quad \neg (p \wedge r) \vee s \\
&\leftrightarrow \quad (p \wedge r) \to s.
\end{aligned}
$$

This method of obtaining the conclusion starting with the premises and applying the rules of inferences is called the method of *direct proof*.

In an axiomatic system, a *proof* is a logically valid premise consisting of certain *axioms of the theory* or a tautology or a true implication $p \Rightarrow q$, where each is called a *logical argument*. It is mostly possible to use axioms alone to derive a *direct proof* of statements. For example, in Chap. 2, we used the five Peano's axioms to construct the set of real numbers $\mathbb{R}$, starting with the set of natural numbers $\mathbb{N}$ as being provided as an *axiomatic system*. Subsequently, all the properties of $\mathbb{R}$ are derived from the properties of the set $\mathbb{N}$, using basic rules of logic and deductive reasoning. Further, since

$$
\begin{aligned}
(p \wedge \neg q) \to \mathbf{F} \quad &\leftrightarrow \quad \neg (p \wedge \neg q) \vee \mathbf{F} \\
&\leftrightarrow \quad \neg p \vee q \\
&\leftrightarrow \quad p \to q,
\end{aligned}
$$

we see that $p \wedge \neg q \Rightarrow \mathbf{F}$ is equivalent to $p \Rightarrow q$. This gives another method of proof, called *indirect method* or the *method of contradiction* (or *reductio ad absurdum*): To prove that $p_1, \ldots, p_n$ implies q, show that $p_1, \ldots, p_n, \neg q$ implies a contradiction. In such case, the propositions $p_1, \ldots, p_n, \neg q$ are *inconsistent*.

Definition A.5 We say premises $p_1, \ldots, p_n$ are **inconsistent** if $p_1 \wedge \cdots \wedge p_n \to \mathbf{F}$.

Notice that $p \wedge \neg p \to \mathbf{F}$, always. For an illustration, we show that $p \to q, q \to r$ and p implies r. A direct proof is as given below.

(1)	$p \to q$	Rule P
(2)	$q \to r$	Rule P
(3)	$p \to r$	(1), (2), chain rule
(4)	p	Rule P
(5)	r	modus ponens, (3), (4)

We next give an indirect proof of this implication:

(1)	$\neg r$	negated conclusion
(2)	$q \to r$	Rule P
(3)	$\neg q$	modus tollens, (1), (2)
(4)	$p \to q$	Rule P

(5)	$\neg p$	modus tollens, (3), (4)
(6)	p	Rule P
(7)	$p \wedge \neg p$	(5), (6), $p \wedge \neg p$ is a Contradiction

□

We now show that $p \rightarrow q$ and $\neg q$ implies $\neg p$. In fact, the result is the *modus tollens*. We give a direct proof of it below.

(1)	$p \rightarrow q$	Rule P
(2)	$\neg q \rightarrow \neg p$	(1), $p \rightarrow q \leftrightarrow \neg q \rightarrow \neg p$
(3)	$\neg q$	Rule P
(4)	$\neg p$	modus ponens, (2), (3)

We now give an indirect proof of this implication:

(1)	p	negated conclusion
(2)	$p \rightarrow q$	Rule P
(3)	q	modus ponens, (1), (2)
(4)	$\neg q$	premise
(5)	$q \wedge \neg q$	(3), (4), $q \wedge \neg q$ is a Contradiction

□

Likewise, we can show that $\neg(p \wedge \neg q)$, $\neg q \vee r$ and $\neg r$ implies $\neg p$. In the following examples, we give proof of an implication where the conclusion is conditional.

Example A.1 We show that $p \rightarrow q$, $\neg p \rightarrow r$, $r \rightarrow s$ implies $\neg q \rightarrow s$. We first derive the conditional conclusion directly.

(1)	$\neg p \rightarrow r$	premise
(2)	$r \rightarrow s$	premise
(3)	$\neg p \rightarrow s$	chain rule, (1), (2)
(4)	$p \rightarrow q$	premise
(5)	$\neg q \rightarrow \neg p$	contrapositive, (4)
(6)	$\neg q \rightarrow s$	chain rule, (5), (3)

The following is proof where the premise of the conditional conclusion is used as an additional premise, and we have derived the conclusion of the conditional conclusion.

(1)	$p \rightarrow q$	premise
(2)	$\neg q$	premise
(3)	$\neg p$	modus tollens, (1), (2)
(4)	$\neg p \rightarrow r$	premise
(5)	r	modus ponens, (3), (4)
(6)	$r \rightarrow s$	premise
(7)	s	modus ponens, (5), (6)

□

The next illustration gives an example of propositions that are inconsistent.

Example A.2 We show that the premises $p \to q$, $p \to r$, $q \to \neg r$, p are inconsistent. We prove this by showing that $p \to q$, $p \to r$, $q \to \neg r$, p implies a contradiction $p \wedge \neg p$.

(1)	p	premise
(2)	$p \to q$	premise
(3)	q	(1), (2), modus ponens
(4)	$q \to \neg r$	premise
(5)	$\neg r$	(3), (4), modus ponens
(6)	$p \to r$	premise
(7)	$\neg r \to \neg p$	(6), contrapositive
(8)	$\neg p$	(5), (7), modus ponens
(9)	$p \wedge \neg p$	(1), (8)

□

Predicate Calculus

In *predicate calculus*, atomic statements are analysed using *predicates*, which describe a common property of several propositions. Notice that the proposition "the number 5 is not divisible by 3" can be written as "it is not true that the number 5 is divisible by 3". In all finite number of proposition, with the phrase "is divisible by 3" is common, and this common property is called a predicate. By denoting this predicate by D, and writing "x *is divisible by* 3" as $D(x)$, we can symbolise any such statement as $D(x)$, called a *predicate formula*. More precisely, it is a single place predicate,[1] where x is a variable for the predicate formula $D(x)$. For example, the proposition $D(6)$ is an assertion about the number 6, and the number 6 is the subject of the proposition. So, the predicate is what we assert or say about the subject. In $D(x)$, the subject x is a variable while in $D(6)$ the subject is a constant.

In general, if A is a predicate, then "x is A" is written as $A(x)$, and it is called a *predicate formula*. We cannot assign truth values for $A(x)$ when x is a variable, and, thus, they are not propositions. Therefore, $A(x)$ is a proposition only when the subject x is a constant. If s is some specific subject, then $A(s)$ is a proposition, called a *substitution instance* (or *instantiation*) of $A(x)$. We can combine different predicates using connectives. For example, consider the proposition: $D(x)$ - x is divisible by 2 and $G(x, y)$ "x is greater than y". Then $D(x) \wedge G(x, y)$ gives the statement "x is divisible by 2 and x is greater than y". For the predicate formula "$P(x) : x$ is a prime number", we can find the truth value of $P(2)$, $P(5)$, and $P(1729)$. The universe of discourse is the set of all integers. As 2 and 5 are prime numbers, $P(2)$ and $P(5)$ are true whereas $P(1729)$ is false.

Definition A.6 An expression consisting of a predicate symbol and a single variable (such as $A(x)$, $B(x)$, $C(y)$) is called a **simple statement function**. The statement

---

[1] An **n- place predicate** is a predicate requiring n subjects.

obtained by replacing the variable by the name of a subject is called a **substitution instance** of the statement function.

A statement function is not a statement because it cannot be assigned any truth value. However, the substitution instance of the statement function is a statement (in the propositional calculus). For example, let H be the predicate "is a mortal" and a the name "Aryabhatta" and b the name "Baskara". Then $H(x)$ denotes "x is a mortal". $H(a)$, $H(b)$ are substitution instances of the statement function $H(x)$. Note that $H(a)$, $H(b)$ are statements while $H(x)$ is not a statement.

Definition A.7 An expression consisting of a predicate symbol and n variables is called a **simple statement function of n variables**. The statement obtained by replacing each of the n variables by the corresponding names of the subjects is called a **substitution instance** of the statement function of n variables.

For example, let A be the predicate "is older than" and h the name "G. H. Hardy", r the name "S. Ramanujan". The notation $A(x, y)$ is used to denote the statement function "x is older than y". It is a statement function of two variables. Also $A(h, r)$ and $A(r, h)$ are both substitution instances of the statement function $A(x, y)$. The truth value of $A(h, r)$ is true while the truth value of $A(r, h)$ is false.

Definition A.8 An expression consisting of simple statement functions (of one or more variables) connected by logical connectives is called a **compound statement formula**.

For example, let M denotes the predicate "is a man", and N denotes the predicate "is an Indian mathematician". Then $M(x)$ and $N(x)$ stand for the statement functions "x is a man" and "x is an Indian mathematician" respectively. Let h denotes the name "G. H. Hardy", r the name "S. Ramanujan". The statements "Ramanujan is an Indian mathematician" and "Hardy is not an Indian mathematician" translate into

$$M(r) \wedge N(r) \quad \text{and} \quad M(h) \wedge \neg N(h),$$

respectively. The statement functions

$$M(x) \wedge N(x) \quad \text{and} \quad M(x) \wedge \neg N(x)$$

are compound statement functions.

Definition A.9 The **existential quantification** of a predicate formula $P(x)$ is the proposition, denoted by

$$(\exists x)P(x),$$

that is true if $P(a)$ is true for at least one subject a. The existential quantifier symbol $\exists$ is read "**there exists**" or "**there is**".

For example, consider the proposition "Some real numbers are rational". We can rewrite this proposition as: "For some x, x is a real number and x is rational". Now, we can symbolise the proposition using the following statement functions:

$R(x)$: x is a real number
$Q(x)$: x is rational.

Then the given proposition is symbolised as

$$(\exists x)(R(x) \wedge Q(x)).$$

Definition A.10 The **universal quantification** of a predicate formula $P(x)$ is the proposition, denoted by

$$(\forall x)P(x)$$

or simply by

$$(x)P(x),$$

that is true if $P(a)$ is true for all subject a. The universal quantifier symbol $\forall$ is to be read as **for all, for each**, or **for every**.

For example, consider the proposition "Every Monday I go to library". We can rewrite this proposition as: "For every x, if x is a Monday, then x is when I go to library". Now, we can symbolise the proposition using the following statement functions:

$M(x)$: x is a Monday
$L(x)$: x is when I go to library.

Then the given proposition is symbolised as

$$(x)(M(x) \rightarrow L(x)).$$

Let P denote an n-place predicate and $x_1, x_2, \ldots, x_n$ be the individual variables, then $P(x_1, x_2, \ldots, x_n)$ is an n-place predicate formula. The predicate formula $P(x_1, x_2, \ldots, x_n)$ is an *atomic formula* of the predicate calculus. For example, the following are examples of atomic formula of predicate calculus:

$$A, \quad Q(x), \quad R(y), \quad M(x, y), \quad A(x, y, z).$$

Definition A.11 The formulas obtained by the following procedure are called **well-formed formulas**:

1. Any atomic formula is a well-formed formula.
2. If A and B are formulas, then $(\neg A)$, $(A \vee B)$, $(A \wedge B)$, $(A \Rightarrow B)$ and $A \Leftrightarrow B)$ are also well-formed formulas.
3. If A is a well-formed formula and x is any variable, then $(x)A$ and $(\exists x)A$ are well-formed formulas.

4. Any expression obtained by applying the above rules (any number of times) is a well-formed formula.

For example, the predicate formula $(x)(P(x) \vee Q(x)) \wedge R(x)$ is a well-formed formula but not a statement in the predicate calculus while $(x)(P(x) \wedge Q(x)) \wedge (\exists x)S(x)$ is a well-formed formula and also a statement in the predicate calculus.

Definition A.12 The scope of a quantifier is the predicate to which it applies. A variable x in a predicate formula is a **bound occurrence** if it lies inside the scope of the quantifier (x) or $(\exists x)$. Otherwise it is a **free occurrence**.

For example, in the formula

$$(\exists x)(P(x) \vee Q(x)) \wedge R(x),$$

the scope of the existential quantifier is the formula $P(x) \vee Q(x)$, and the occurrence of x in $P(x) \vee Q(x)$ is a bound occurrence. The variable x in $R(x)$ is a free occurrence. Further, in the predicate formula

$$(x)P(x) \vee (Q(x) \wedge R(x)),$$

the scope of the universal quantifier is $P(x)$, and the variable x in $Q(x) \wedge R(x)$ is a free occurrence. Thus, the occurrence of the variable x in $P(x)$ has no relationship to the free occurrence of the variable x of $Q(x) \wedge R(x)$. In fact, the predicate formula

$$(x)(P(x) \vee (Q(x) \wedge R(x))$$

can be written as

$$(y)(P(y) \vee Q(y)) \wedge (Q(x) \wedge R(x)).$$

The variable y is a bound variable while the variable x is a free variable.

If the subjects are taken only from a set U, then the set U is called the **universe of discourse** or **universe** or **domain**.

Example A.3 Here, we express the following statement using quantifiers: "There exists an integer which is a square root of an integer". Let S denotes the predicate "is a square root of an integer". Since the statement "There exists an integer which is a square root of an integer" can be written as "For some x, x is a square root of an integer", the symbolisation of it is

$$(\exists x)S(x).$$

In this case, we assume that the universe is the set of integers. If we do not assume that the universe is the set of integers, then the symbolisation is given by using an additional predicate. Let I denote the predicate "an integer" so that $I(x)$ denote the formula "x is an integer". The given statement is symbolised by

$$(\exists x)(I(x) \wedge S(x)).$$

□

Example A.4 Here, we express the following statement using quantifiers: "Some integers are not square of any integers". Let S denotes the predicate "a square root of an integer". Since the statement "Some integers are not square of any integers" can be written as "For some x, it is not the case that x is a square of an integer", the symbolisation of it is

$$(\exists x)(\neg S(x)).$$

In this case, we assume that the universe is the set of integers. If we do not assume that the universe is the set of integers, then the symbolisation is given by using an additional predicate. Let I denotes the predicate "an integer" so that $I(x)$ denotes the formula "x is an integer". The statement is symbolised by

$$(\exists x)(I(x) \wedge \neg S(x)).$$

□

Example A.5 Here, we symbolise the following statement with and without using the set of integers as the universe of discourse: "Given any positive integer, there is a greater positive integer." Let $P(x)$ denotes the statement function "x is a positive integer", and $G(x, y)$ the statement function "x is greater than y". The given statement can be written as "For all x, if x is a positive integer, then there is a y such that y is greater than x." Thus the symbolic form is

$$(x)(P(x) \rightarrow (\exists y > 0)G(y, x)).$$

In this case, we assumed that the universe is the set of all integers. If we do not assume that the universe is not the set of integers, then we can symbolise it by using another predicate. Let I denotes the predicate "an integer" so that $I(x)$ denotes the formula "x is an integer". In this case, the symbolic form is

$$(x)(P(x) \rightarrow (\exists y)(I(y) \wedge G(y, x))).$$

□

A proposition in the predicate logic quantified with an existential quantifier is analogues to the disjunction in the propositional logic. If the universe of discourse U is a finite set $\{x_1, x_2, \ldots, x_n\}$, then $(\exists x)P(x)$ is true if and only if $P(x_1) \vee P(x_2) \vee \cdots \vee P(x_n)$ is true. Thus

$$(\exists x)P(x) \Leftrightarrow P(x_1) \vee P(x_2) \vee \cdots \vee P(x_n).$$

Similarly a proposition in the predicate logic quantified with an universal quantifier is analogues to the conjunctions in propositional logic. The proposition $(x)P(x)$ is true if and only if $P(x_1) \wedge P(x_2) \wedge \cdots \wedge P(x_n)$ is true. Thus

$$(x)P(x) \Leftrightarrow P(x_1) \wedge P(x_2) \wedge \cdots \wedge P(x_n).$$

Example A.6 Let the universe of the discourse be the set $\{a, b, c, d\}$. Then the formula $(x)P(x)$ is same as $P(a) \wedge P(b) \wedge P(c) \wedge P(d)$. Similarly, the formula $(x)(\neg P(x)) \vee (x)P(x)$ is the formula

$$[\neg P(a) \wedge \neg P(b) \wedge \neg P(c) \wedge \neg P(d)] \vee [P(a) \wedge P(b) \wedge P(c) \wedge P(d)].$$

Lemma A.1 *If the universe of discourse is finite, then*

$$\neg[(\exists x)P(x)] \Leftrightarrow (x)[\neg P(x)].$$

Proof Let the universe of discourse

$$U = \{x_1, x_2, \ldots, x_n\}$$

be finite. By using the De Morgan's law of propositional calculus, we have

$$\begin{aligned}
\neg[(\exists x)P(x)] &\Leftrightarrow \neg[P(x_1) \vee P(x_2) \vee \cdots \vee P(x_n)] \\
&\Leftrightarrow \neg P(x_1) \wedge \neg P(x_2) \wedge \cdots \wedge \neg P(x_n) \\
&\Leftrightarrow (x)[\neg P(x)].
\end{aligned}$$

$\square$

Lemma A.2 *If the universe of discourse is finite, then*

$$\neg[(x)P(x)] \Leftrightarrow (\exists x)[\neg P(x)].$$

Proof Let the universe of discourse

$$U = \{x_1, x_2, \ldots, x_n\}.$$

be finite. By using the De Morgan's law of propositional calculus, we have

$$\begin{aligned}
\neg[(x)P(x)] &\Leftrightarrow \neg[P(x_1) \wedge P(x_2) \wedge \cdots \wedge P(x_n)] \\
&\Leftrightarrow \neg P(x_1) \vee \neg P(x_2) \vee \cdots \vee \neg P(x_n) \\
&\Leftrightarrow (\exists x)[\neg P(x).]
\end{aligned}$$

$\square$

Definition A.13 Since $(x)P(x)$ is true if and only if $P(x)$ is true for every x in the universe of discourse, we get the following implication:

$$(x)P(x) \Rightarrow P(y).$$

It should be noted that $P(y)$ is a proposition and y is any subject from the universe of discourse. The above implication is called as the **rule of universal specification** or US for short.

Definition A.14 Similarly if $P(x)$ is true for all subjects from the universe of discourse, then we can conclude that $(x)P(x)$ is true. Thus, we have

$$P(x) \Rightarrow (y)P(y).$$

This rule is called the **rule of universal generalisation** or UG for short.

It should be noted that the rule of universal generalisation can be applied only when $P(x)$ is true for all x in the universe of discourse.

Definition A.15 If $P(x)$ is true for at least one subject a in the universe of discourse, then we can make the following conclusion:

$$P(a) \Rightarrow (\exists x)P(x).$$

This rule is known as the **rule of existential generalsation** or EG for short.

Definition A.16 If $(\exists x)P(x)$ is true, then $P(x)$ is true for some a in the universe of discourse, we get the following implication:

$$(\exists x)P(x) \Rightarrow P(a).$$

This rule is called as the **rule of existential specification** or ES for short.

Note that a variable cannot be generalised universally if it was obtained by an application of the Rule ES. Let P and Q be predicates and A be a proposition (without any variable), then the following equivalences are valid:

$$(i)\neg[(\exists x)P(x)] \Leftrightarrow (x)[\neg P(x)]$$
$$(ii)\neg[(x)P(x)] \Leftrightarrow (\exists x)[\neg P(x)]$$
$$(iii)(\forall x)P(x) \wedge (\forall x)Q(x) \Leftrightarrow (\forall x)[P(x) \wedge Q(x)]$$
$$(iv)(\exists x)P(x) \vee (\exists x)Q(x) \Leftrightarrow (\exists x)[P(x) \vee Q(x)]$$
$$(v)(\forall x)P(x) \vee (\forall x)Q(x) \Leftrightarrow (\forall x)(\forall y)[P(x) \vee Q(y)]$$
$$(vi)(\exists x)P(x) \wedge (\exists x)Q(x) \Leftrightarrow (\exists x)(\exists y)[P(x) \wedge Q(y)]$$
$$(vii)(\forall x)P(x) \wedge (\exists x)Q(x) \Leftrightarrow (\forall x)(\exists y)[P(x) \wedge Q(y)]$$
$$(viii)(\forall x)P(x) \vee (\exists x)Q(x) \Leftrightarrow (\forall x)(\exists y)[P(x) \vee Q(y)]$$

$$(ix) A \lor (\forall x) P(x) \Leftrightarrow (\forall x)[A \lor P(x)]$$
$$(x) A \lor (\exists x) P(x) \Leftrightarrow (\exists x)[A \lor P(x)]$$
$$(xi) A \land (\forall x) P(x) \Leftrightarrow (\forall x)[A \land P(x)]$$
$$(xii) A \land (\exists x) P(x) \Leftrightarrow (\exists x)[A \land P(x)].$$

Index

© The Editor(s) (if applicable) and The Author(s), under exclusive license to Springer
Nature Singapore Pte Ltd. 2022
A. K. Razdan and V. Ravichandran, *Fundamentals of Analysis with Applications*,
https://doi.org/10.1007/978-981-16-8383-1

Printed in the United States
by Baker & Taylor Publisher Services

Towards Real-World Neurorobotics:
Integrated Neuromorphic Visual Attention*

Samantha V. Adams, Alexander D. Rast, Cameron Patterson, Francesco Galluppi,
Kevin Brohan, José-Antonio Pérez-Carrasco, Thomas Wennekers,
Steve Furber, and Angelo Cangelosi

[1] Plymouth University, Plymouth, UK
[2] School of Computer Science, University of Manchester
Manchester, UK M13 9PL
{rasta}@cs.man.ac.uk,
{steve.furber}@manchester.ac.uk,
{samantha.adams,A.Cangelosi,thomas.wennekers}@plymouth.ac.uk
http://www.cs.manchester.ac.uk/apt
http://www.plymouth.ac.uk

Abstract. Neuromorphic hardware and cognitive robots seem like an obvious fit,
yet progress to date has been frustrated by a lack of tangible progress in achieving
useful real-world behaviour. System limitations: the simple and usually proprietary nature of neuromorphic and robotic platforms, have often been the fundamental barrier. Here we present an integration of a mature "neuromimetic" chip,
SpiNNaker, with the humanoid iCub robot using a direct AER - address-event
representation - interface that overcomes the need for complex proprietary protocols by sending information as UDP-encoded spikes over an Ethernet link. Using
an existing neural model devised for visual object selection, we enable the robot
to perform a real-world task: fixating attention upon a selected stimulus. Results
demonstrate the effectiveness of interface and model in being able to control the
robot towards stimulus-specific object selection. Using SpiNNaker as an embeddable neuromorphic device illustrates the importance of two design features in a
prospective neurorobot: *universal configurability* that allows the chip to be con-
formed to the requirements of the robot rather than the other way 'round, and *standard interfaces* that eliminate difficult low-level issues of connectors, cabling,
signal voltages, and protocols. While this study is only a building block towards
that goal, the iCub-SpiNNaker system demonstrates a path towards meaningful
behaviour in robots controlled by neural network chips.

Keywords: cognitive, robotics, attention, neuromorphic.

* Alexander Rast and Steve Furber are with the School of Computer Science, The University of
Manchester, Manchester, UK (email: rasta@cs.man.ac.uk). Samantha Adams, Thomas
Wennekers and Angelo Cangelosi are with Plymouth University, Plymouth, UK. Francesco
Galluppi is with the Institut de la Vision, Paris, France. José-Antonio Pérez-Carrasco is with
the Departemento de Teoría de la Señal y Comunicaciones, Universidad de Sevilla, Seville,
Spain. Cameron Patterson is with Supplemango Networks Ltd., Aberdeen, UK.

C.K. Loo et al. (Eds.): ICONIP 2014, Part III, LNCS 8836, pp. 563–570, 2014.

1   Introduction: The Need for Practical Neuromorphic Robotics

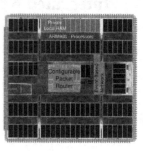

(a) Real iCub   (b) Simulated iCub   (c) SpiNNaker chip

Fig. 1. The iCub Robot and SpiNNaker chip

Neural networks seem like an obvious fit for robots. Indeed, behavioural roboticists assert, with some justification, that *embodiment matters*: it is not enough, to achieve meaningful behaviour, to implement a neural network as an abstract "disembodied brain" operating on synthetic stimuli in an artificial environment [11]. Within neuro-robotics itself, advances in understanding of the neurobiology have many researchers suggesting that neural models more closely matching the biology [1] may more clearly reveal the computational principles necessary for cognitive robotics while illuminating human (and animal) brain function. Critically, some neurobiological experiments suggest that spike-based signalling is important at the cognitive/behavioural level as well as in learning [12]. If this is true perhaps spike-based neurorobots can embody behavioural features that seem intractably hard without neural hardware [7]. However, until now, in practice most neurorobotic systems, e.g. [3] have simulated the neural component on an external host PC due to internal hardware constraints or incompatible interfaces. Both the hardware [4] and the robotic systems [2] have now reached a point of maturity where *integrated* neurorobots able to demonstrate effective behaviour in nontrivial real-world scenarios are within reach. Our aim is to create such a system in a way that both fulfils the long-held promise of practical neurorobotics and illustrates their potential to act as a tool for neurobiological experimentation. Here we introduce a system integrating 2 mature platforms over a direct interface: the humanoid iCub robot (Figs. 1a, 1b) and the embeddable neuromimetic SpiNNaker chip (Fig. 1c) to solve a behaviourally relevant task: goal-directed attentional selection.

2   Test Materials and Methods

There are 3 main components to our test system: the iCub robot and its associated support systems, the SpiNNaker system with its AER interface, and the network model.

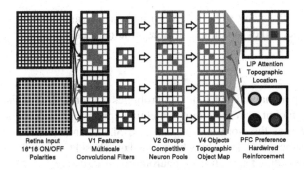

Fig. 2. The test model. The input retina layer is a real or simulated visual field taken either from the preprocessed robot imaging system or from a software image generator. Each of layers V1, V2, V4, and PFC are separated into 4 orientations per layer. Layer LIP merges orientations via a winner-take-all.

2.1 Neural Model

As a test network, we use the attentional model of F. Galluppi, K. Brohan, et al. in [5]. We chose this network because it exhibits realistic real-world behaviour in a non-trivial task while remaining simple, proven, and scalable. The network (Fig. 2) has 6 layers roughly corresponding to selected brain areas. We retained the original network parameters and sizes: 16×16 neuron visual field, 10×10 image maps in 4 orientations for both V1 and V2, and 5×5 location fields for V4, PFC, and LIP. However we made the following modifications/additions to the network:

1. We sharpened the winner-take-all filtering in the LIP of the original model.
2. We extended the input stimulus to allow testing with either real or synthetic sources.
3. We inserted an automatic parameter scaling module in the PyNN script.
4. We added an option to enable STDP learning between the V2 and V4 layers.

2.2 SpiNNaker

The SpiNNaker chip (Fig. 1c) is a universal neural network platform designed for real-time simulation with an array of programmable cores operating in parallel over a configurable asynchronous multicast interconnect. While the typical neuromorphic device implements a fixed model, SpiNNaker can be easily programmed by users with a wide range of different models. Inter-processor communications occur exclusively through Address-Event Representation (AER) spikes: small packets carrying only the address of the neuron that spiked (and possibly a 32-bit data payload).

To interface to SpiNNaker, external devices (such as the robot) send spikes rather than pass data structures directly. This conveniently and data-efficiently abstracts away internal processing particulars on both sides of the interface. SpiNNaker supports the AEtheRnet general-purpose direct AER interface [10], transmitting up to 256 spikes per frame as 32-bit words via UDP over an Ethernet connection. We enhanced the original AEtheRnet interface with additional support for multiple (up to 6) input and output devices (spike sources and spike receivers) and an internal software router to allow multicast of output spikes to multiple devices simultaneously.

2.3 iCub

iCub is a flagship humanoid developmental robotics platform [8]. See Fig. 1. There has been a limited amount of research implementing spiking neural networks for control of the iCub, such as the work of Bouganis and Shanahan [2] and the iSpike library developed by Gamez, et al. [6]. These have relied on a host PC to run the neural network whereas our aim is direct execution on neuromorphic hardware.

We use YARP (Yet Another Robot Platform) for a communications protocol and also Aquila - an easy-to-use, high-performance, modular and scalable software architecture for cognitive robotics [9]. In particular we use the Tracker module for extraction of objects from the scene and basic image processing, transforming the 240x320 RGB raw image from a single iCub camera into a downsampled 16x16 image of black and white pixels which are then converted to spikes by mapping "ON" pixels to spike outputs. We also use the iCubMotor module which converts image coordinates into head motor movements to enable the iCub to look at a location corresponding to a point in a 2D image. We then configured the iCub as a virtual AEtheRnet device with a bidirectional link which maps iCub camera input to the two input layers and receives spikes from the LIP output layer.

3   Results

We ran 4 experiments. Experiment I was a simulation using synthetic visual inputs to develop an initial weight tuning for the network. Experiments II, III and IV tested the ability of the network (with optimised weights, STDP off), and the iCub robot, to locate and attend to the location of a preferred object when it was the only object present and when two objects (one preferred and one aversive) were present in the scene.

3.1 Experiment I

To explore the weight parameters of the network, we first ran a set of preliminary tests using the synthetic stimulation model noted in 2.1. The stimulus (3a) was a pair of black horizontal and vertical bars with slow upwards and rightwards motion respectively. We successively tuned the spike outputs for each of layers V1, V2, V4, and LIP to respond just below its continuous spiking threshold - at the point of maximum sensitivity (Fig. 3). As can be seen in Fig. 3f, when correctly tuned the output LIP layer was able, with synthetic inputs, to narrow its output (area of visual fixation) to a single neuron.

3.2 Experiment II

In this experiment we tested the ability of the robot to attend to single stimuli (either a horizontal or vertical object). Figure 4 shows the raw and preprocessed input. Figure 5 shows the LIP maps (fixation frequency by location) produced. The maps show spike

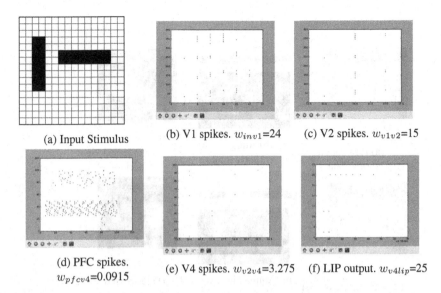

(a) Input Stimulus   (b) V1 spikes. w_{inv1}=24   (c) V2 spikes. w_{v1v2}=15

(d) PFC spikes. w_{pfcv4}=0.0915   (e) V4 spikes. w_{v2v4}=3.275   (f) LIP output. w_{v4lip}=25

Fig. 3. Synthetic input and spike raster plots for each layer. The x-axis is time. Neuron numbers are depicted on the y-axis in ascending order, by orientation first, then by row, then by column. For each layer other than LIP, each orientation is $\frac{1}{4}$ of the population. The LIP layer has only one combined orientation. Neuron 18 thus corresponds to the neuron at row 4, column 4 in the output field.

count over a 1 second run (normalised so that values lie between 0.0 and 1.0) with lighter coloured areas indicating higher saliency. In both cases the network is able to determine whether the object to left or to the right is the preferred object.

3.3 Experiment III

In this experiment we tested the ability of the robot to attend to one object when both were present in the scene. Figures 4g and 4h show the input stimulus in this case. Figure 5c shows the LIP saliency map (attentional preference) produced when the horizontal object is preferred and the vertical object is aversive and Fig. 5d the LIP map when the vertical object is preferred and the horizontal is aversive. In both scenarios, although there is activity produced for both objects, the network is able to determine which one is the preferred object and thus the more active location. Table 1 compares the estimated mean firing rates in both scenarios for the V2 and V4 layers and shows the effect of the biasing of V4 by the PFC. In the V2 layer there is very little discrimination between the preferred and aversive stimuli but in V4 the activity of the preferred stimulus is amplified and that of the aversive stimulus is suppressed.

3.4 Experiment IV

In this experiment we ran 20 repeats of Experiment III, 10 of each with the preferred orientation set as horizontal and vertical respectively. Figure 4i shows the raw camera

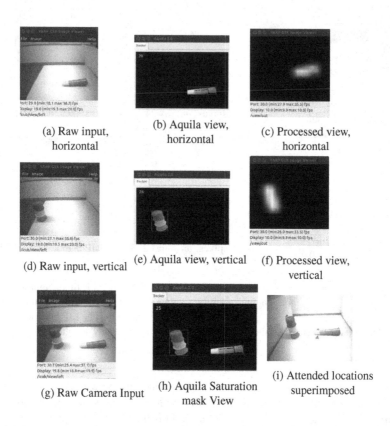

(a) Raw input, horizontal

(b) Aquila view, horizontal

(c) Processed view, horizontal

(d) Raw input, vertical

(e) Aquila view, vertical

(f) Processed view, vertical

(g) Raw Camera Input

(h) Aquila Saturation mask View

(i) Attended locations superimposed

Fig. 4. iCub camera view and Aquila saturation masks for objects taken separately and together. Downsampled black and white images for the horizontal and vertical objects (top). Superimposed attended location map (bottom right).

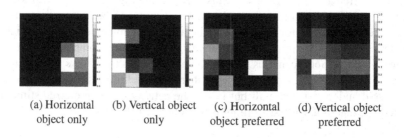

(a) Horizontal object only

(b) Vertical object only

(c) Horizontal object preferred

(d) Vertical object preferred

Fig. 5. LIP saliency maps for Experiment II (single stimuli) and Experiment III (dual stimuli)

Table 1. Estimated mean firing rates (Hz) in V2 and V4 for dual stimuli

Preferred	Aversive	V2 (unbiased)		V4 (biased)	
		P	A	P	A
Vertical	Horizontal	8.04	7.63	55.48	4.92
Horizontal	Vertical	8.34	7.83	23.0	4.48

view with a selection of the locations that were attended to superimposed as red squares. Of the 20 runs 14 resulted in the correct object being selected. Nearly equal activation in both areas of the LIP map caused the wrong object to be selected in 6 cases.

4   Discussion: The Test of Real-World Applications

Our system significantly expands the model complexity that can be embedded in a neurorobot. Some issues remain to be addressed. *Scalability:* The small network we used allows only a broad distinction between 2 objects; we are scaling it to sizes with finer discrimination. *Interfaces:* Translating from the frame-based camera was rather artificial. Experiments with a neuromorphic version of the iCub show that native AER is clearly the preferred interface. *Learning:* Disabling STDP during tests clarified interface aspects at the expense of cognitive aspects. SpiNNaker-only trials of the model with STDP on have run successfully and will be tested on the iCub. *Context:* Using vision alone tends to lead to a fixed context but future plans include expanded top-down processing in the PFC and the addition of bio-inspired auditory processing to assist with learning the preferred stimulus. When these systems are in place we hope to be in a position to run a genuinely practical cognitive test. We propose several key criteria about what kind of application would demonstrate achievement of practical neurorobotics:

Adaptive Behaviour: The application should require the robot to synthesise behaviours that have not been specified imperatively.

Intrinsic Unpredictability: The environment should generate unpredictable events whose impact on behaviour is more than perturbative.

Multidimensional Cognition: Tasks should involve the integration of multiple sensory and response modes.

Real-World Applicability: Behaviour should be meaningful in a real human context.

5   Conclusions

We have integrated neuromorphic computing and humanoid robotics to perform a cognitive task that, while equally feasible with traditional robotics techniques, can here be achieved with a model scalable to a variety of contexts. In possibly nonstationary situations, where decisions may need to be made faster than their consequences can be predicted, classical offline machine learning and Bayesian approaches may fall short. A cognitive system using configurable neural hardware with standard interfaces, running biologically derived models, offers perhaps the critical margin of flexibility. Our experiments point the way to one of the goals of cognitive robotics: self-directed robots able to respond adaptively and appropriately rather than imperatively to the combination of unexpected events and indeterminate consequences characteristic of the real world.

Acknowledgements. This work was supported under EPSRC Grant EP/J004561/1 (BABEL). The SpiNNaker project is supported by EPSRC Grant EP/G015740/1, and by industry partners ARM, Silistix, and Thales.

References

1. Arbib, M., Metta, G., van der Smagt, P.: Neurorobotics: from vision to action. In: Springer Handbook of Robotics, pp. 1453–1480. Springer (2008)
2. Bouganis, A., Shanahan, M.: Training a spiking neural network to control a 4-DOF robotic arm based on spike timing-dependent plasticity. In: Proc. 2010 Int'l Joint Conf. Neural Networks (IJCNN 2010), pp. 4104–4111 (2010)
3. Chersi, F.: Learning Through Imitation: a Biological Approach to Robotics. IEEE Trans. Autonomous Mental Development 4(3), 204–214 (2012)
4. Furber, S.B., Lester, D.R., Plana, L.A., Garside, J.D., Painkras, E., Temple, S., Brown, A.D.: Overview of the SpiNNaker system architecture. IEEE Trans. Computers PP(99) (2012)
5. Galluppi, F., Brohan, K., Davidson, S., Serrano-Gotarredona, T., Carrasco, J.-A.P., Linares-Barranco, B., Furber, S.: A Real-Time, Event-Driven Neuromorphic System for Goal-Directed Attentional Selection. In: Huang, T., Zeng, Z., Li, C., Leung, C.S. (eds.) ICONIP 2012, Part II. LNCS, vol. 7664, pp. 226–233. Springer, Heidelberg (2012)
6. Gamez, D., Fidjeland, A., Lazdins, E.: iSpike: a spiking neural interface for the iCub robot. Bioinspiration and Biomimetics 7(2) (2012)
7. Kuniyoshi, Y., Berthouze, L.: Neural learning of embodied interaction dynamics. Neural Networks 11(7-8), 1259–1276 (1998)
8. Metta, G., Sandini, G., Vernon, D., Natale, L., Nori, F.: The iCub humanoid robot: an open platform for research in embodied cognition. In: Proc. IEEE Workshop on Performance Metrics for Intelligent Systems, PerMIS 2008 (2008)
9. Peniak, M., Morse, A., Cangelosi, A.: Aquila 2.0 software architecture for cognitive robotics. In: Proc. IEEE Third Joint International Conference on Development and Learning and Epigenetic Robotics, ICDL (2013)
10. Rast, A., Partzsch, J., Mayr, C., Schemmel, J., Hartmann, S., Plana, L.A., Temple, S., Lester, D.R., Schüffny, R., Furber, S.: A Location-Independent Direct Link Neuromorphic Interface. In: Proc. 2013 Int'l Joint Conf. Neural Networks (IJCNN 2013), pp. 1967–1974 (2013)
11. Scheier, C., Pfeifer, R., Kuniyoshi, Y.: Embedded neural networks: exploring constraints. Neural Networks 11(7-8), 1551–1569 (1998)
12. Shmiel, T., Drori, R., Shmiel, O., Ben-Shaul, Y., Nadasdy, Z., Shemesh, M., Teicher, M., Abeles, M.: Temporally precise cortical firing patterns are associated with distinct action segments. The Journal of Neurophysiology 96, 2645–2652 (2006)

On the Role of Working Memory in Trading-Off Skills and Situation Awareness in Sudoku

George Leu, Jiangjun Tang, and Hussein Abbass

School of Engineering and Information Technology
University of New South Wales, ADFA, Canberra, Australia
{G.Leu,J.Tang,H.Abbass}@adfa.edu.au

Abstract. Working memory accounts for the ability of humans to perform cognitive processing, by handling both the representation of information (the mental picture forming the situation awareness) and the space required for processing these information (skill processing). The more complex the skills are, the more processing space they require, the less space becomes available for storage of information. This interplay between situation awareness and skills is critical in many applications. Theoretically, it is less understood in cognition and neuroscience. In the meantime, and practically, it is vital when analysing the mental processes involved in safety-critical domains.

In this paper, we use the Sudoku game as a vehicle to study this trade-off. This game combines two features that are present during a user interaction with a software in many safety critical domains: scanning for information and processing of information. We use a society of agents for investigating how this trade-off influences player's proficiency.

Keywords: cognitive challenge, scanning skills, working memory.

1 Introduction

We investigate the trade-off between the processing and storage functions of working memory in Sudoku, one of the most popular puzzles. During a game, players visually scan the Sudoku grid in a continuous manner searching for cells containing information that can be propagated throughout the grid to narrow down the degrees of freedom of empty cells. These scanning skills are an integral part of the skill set required for solving the Sudoku puzzle, and the players learn gradually how to use them properly.

Since our interest is to understand the cognitive functions and working memory within Sudoku, we only consider players playing the game without note-taking abilities. External note-taking extends players' working memory with an external holder of information. This type of game play is not suitable for this investigation because it does not give us the right insight into the player's cognitive capacity.

Depending on the complexity of the scanning patterns used for improving situation awareness, the scanning activity can become strenuous. This is because

C.K. Loo et al. (Eds.): ICONIP 2014, Part III, LNCS 8836, pp. 571–578, 2014.

the player must use the working memory for handling both the current scanning task (space for executing skills), and the information resulted from scanning (situation awareness). Between the two functions of the working memory, there is always a trade-off [3], given its finite capacity. Understanding how these two functions interact would help us in designing training and education programs that are cognitively plausible and that account for this trade-off.

The rest of this paper is structured as follows: Section 2 presents background information on working memory, followed by the methodology in Section 3, results in Section 4, then conclusions are drawn in Section 5.

2   Background on Working Memory

Early memory models emphasized one of the two functions of working memory. On one side, the multi-store model [1], suggested that memory was a series of stores, i.e. the sensory memory, the short-term memory, and the long term memory, focusing on the storage function of memory. On the other side, the "levels of processing" model [9] concentrates on the processes involved in memory. It considers memory as a consequence of the depth of information processing, with no clear distinction between short-term memory, long-term memory or other stores.

Later, Baddeley et al. [2] introduced the concept of Working Memory (WM) and showed that short-term memory is more than just one simple store. WM is still short-term memory, but instead of all information going into one single store, there are different subsystems with different functions (e.g. visual, auditory, etc.). The resource-sharing approach on memory adds on Baddeley's work and emphasizes the "storage versus processing" paradigm. Case et al. [7] consider that WM accounts for the processing resources of an individual, and is the sum of a storage space and an operating space. They show that the more complex the current processing task is, the more operating space it uses, leaving less space for information storage. This was further supported by other studies which assumed that human performance in various cognitive tasks is strongly related to working memory [12,3].

Another aspect involved in the study of working memory is the forgetting mechanism associated with its normal operation. The displacement theory [1] assumes the working memory is a first-in-first-out (FIFO) queue with limited capacity, and explains how the most recent information stored in the memory is easier to be recalled (recency effect). The trace decay theory [14] assumes that the events between learning and recall have no effect on recall, the essential influencing factor being the period of time the information has been retained. The interference theory challenges the decay concepts, and attributes forgetting to the existence of interference [8,5]. A review of the forgetting theories can be found in [15].

From a computational perspective, numerous studies proposed various combinations of working memory models and theories of forgetting, in order to instantiate plausible behaviors. Two major approaches have been proposed over the years: the localist and the distributed representation of working memory. In the localist approach, the memory is seen from an item-unit perspective, in which the informational items are stored in corresponding units situated in certain locations of the memory. The Competitive Queueing [6] and the Primacy [13] models propose time-related displacement and displacement only, respectively, whereas the Start-End [10] model proposes a multi-level displacement-based approach. The distributed approach sees an informational item as distributed throughout the memory space rather than as in the item-unit equality view. Thus, in the distributed approaches, the focus shifts from the localist practice of retrieving items from their corresponding locations, to identifying patterns of activation which recompose informational items from their parts distributed through multiple layers of interconnected units. A displacement-based distributed approach was proposed by Lewandowsky and colleagues [11] based on the Theory of Distributed Associative Memory. Later, models such as OSCAR [4] and SIMPLE [5] considered interference-based implementations in which the working memory is hierarchical in structure, oscillatory in time, and contextually activated.

3   Methodology

From the Sudoku puzzle perspective the scanning skills are forming situation awareness picture required to solve the Sudoku game, which is further stored in the WM. The scanning pattern is based on Sudoku cells belonging to the grid. This suggests a localist approach for the WM, in which each step in the scanning pattern produces a scanned item corresponding to a memory unit. We consider that the items are stored in a FIFO queue, as in the displacement theory.

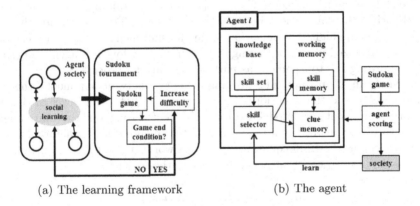

(a) The learning framework   (b) The agent

Fig. 1. Methodological approach

Figure 1 presents the social learning framework and the internal structure of an agent. The agents in the society are endowed with fixed working memory and adaptive ability to choose the scanning skills to be loaded into memory and get executed. Their resultant Sudoku proficiency is tested in a tournament with several rounds of increasing difficulty. In each round one game is proposed, which all agents try to solve. If no agent is able to successfully fill the grid at the end of the game in the first attempt, the agents learn socially from each-other how to adapt their skill selection towards a better proficiency, and the game is replayed in the new conditions. The learning process continues until at least one agent becomes proficient enough to complete the grid or until learning does not bring any more improvement. The tournament continues with the next round where a game with increased difficulty is proposed and the process is repeated.

3.1   The Agent

The Scanning Skill Set: Each agent can choose from a set of scanning skills, constant over the whole society. The complexity of a skill can be related to the amount of information that must be stored in the working memory to describe the scanning pattern. The skill set shown below presents the number of scanned cells and the number of memory units needed for storing them. Since we adopt a localist approach on working memory, the two numbers are equal. Each skill is named as COL for scanning a column, ROW for scanning a row, or BOX for scanning a box. The subsequent digit in the name represents the number of dimension of this scanning activity; thus, the size of the storage required to perform this scanning task.

Skill code	1	2	3	4	5	6	7	8	9	10
Skill name	ROW3	ROW5	ROW7	ROW9	COL3	COL5	COL7	COL9	BOX5	BOX9
Grid cells	3	5	7	9	3	5	7	9	5	9
Mem. units	3	5	7	9	3	5	7	9	5	9

The skill selector is a vector $V = V(v_1, \ldots, v_{10})$ containing the weights associated with each skill in the skill set, where $v_i \in [0, 1]$. The agent loads the first m skills with the highest weight that fit in the skill memory M_s. The selection vector is initialized at the beginning of the simulation for each agent, then it is updated during the tournament as part of the learning process.

The Working Memory: The working memory has two components: the skill memory (M_s), and the situation awareness memory (M_c). The sizes of the two memory components are predefined for each agent throughout the tournament, and they differ from one agent to another, but their sum is identical for all agents in the society, as shown in equation 1. In other words, the total working memory space is maintained constant for each agent in the society.

$$M = M_i = M_{s_i} + M_{c_i}, \forall i = 1, n; \tag{1}$$

The skill memory stores the representation of the skills selected by the skill selector. At the end of each game or round, the skill memory is erased and then reloaded with the new skills selected by the skill selector as a result of the learning process. The new skills are to be used in the next game if the game must be replayed or in the next round if a new round must start. Thus, the skill memory is rewritten through incremental social learning, each time a game is replayed in a round of the tournament.

The situation awareness memory stores the results of the scanning process and is modelled from a displacement theory perspective, as a FIFO queue of size M_c. In this queue, the information discovered by applying the skills currently stored in skill memory are stored in continuation of those stored at previous step. Thus, older information, which exceeds the queue size is pushed out and lost.

3.2   The Tournament

Playing a Game: During a game an agent scans the Sudoku grid in order to form its situation awareness picture which allows the propagation of Sudoku constraints/rules. The agent visits each empty cell of the grid and applies to it the selected skills. After the agent applies the skills on all empty cells and propagates the Sudoku rules, a score is given to the game reflecting on agent's proficiency.

Scoring a Game: Agents receive scores based on the remaining degrees of freedom of empty cells in the grid after a fixed number of allowed steps. The degree of freedom for an empty cell, f_c, is the number of possible candidates found after propagation of domains. If the Sudoku grid is complete at the end of the game, there is no degree of freedom left. If the grid still has empty cells, the degrees of freedom in each empty cell are added, generating $f_i = \sum f_c$, the total degree of freedom for agent i. The performance of an agent is inversely proportional to f_i: the less degrees of freedom remaining at the end of a game, the better the agent performs. Thus, the score s_i is defined in Equation 2, where f_i is the total degrees of freedom for agent i, and f_{max} is the maximum total degrees of freedom over the society.

$$s_i = 1 - \frac{f_i}{f_{max}}; \qquad (2)$$

Learning: At the end of a game, with or without completion of the grid, each agent updates its skill selection vector using the experience of the other agents. The update can be viewed from two perspectives, which we combine: the current agent searches for the agent with the highest score and most similar memory. Thus, the agents in the society learn from other agents with as similar memory (cognitive capacity) as possible, and score as high as possible.

First, we define a similarity metric D_{ij} for working memory of agents, as in Equation 3, where M_{c_i} and M_{c_j} are the awareness memory size of agents i (current agent) and j (an agent from the society) respectively, and $M_{c_{max}}$ and $M_{c_{min}}$ are the maximum and minimum sizes of the awareness memory. If $i = j$ then the memory similarity of an agent to itself $D_{ii} = 1$.

$$D_{ij} = 1 - \left| \frac{M_{c_i} - M_{c_j}}{M_{c_{max}} - M_{c_{min}}} \right|; \qquad (3)$$

Then, we couple the memory similarity metric with the score, as in Equation 4, where F_{ij} is the fitness between current agent i and an agent j from the society.

The agent j corresponding to the max F_{ij} will be the agent in the society which is best fit for participating in agent i learning process. We note that the fitness F_{ii} of an agent to itself equals its score in the Sudoku game, since $D_{ii} = 1$.

$$F_{ij} = s_j \times D_{ij}; \tag{4}$$

The amount of participation of agent j in agent i's learning is given by the actual value of the maximum fitness coefficient, hence, the update function becomes as in Equation 5 and applies when $F_{ij} > F_{ii}$.

$$v_i(t+1) = (1 - F_{ij})v_i(t) + F_{ij}v_j(t); \tag{5}$$

3.3   The Experimental Setup

The skill selection vector is initialized with random weights for each agent at the beginning of tournament. Ten sets of experiments are run with different seeds.

The total working memory M for each agent in the society is 54 units, corresponding to 54 digits associated with the Sudoku puzzle. Within the total memory, the skill memory M_s and the situation awareness memory M_c can be tuned in the range between 9 and 45. The agents in the society are endowed with a situation awareness memory ranging between 9 to 45 progressively with an increment of 2. The skill memory follows the opposite variation pattern. Consequently, the society has 19 agents.

The tournament consists of nine rounds of increasing level of difficulty, where difficulty is associated with the initial number of non-empty cells existent in the grid. In this study, grids with 76, 74, 71, 67, 62, 56, 49, 41 and 32 initial non-empty cells are used for the 9 tournament rounds in that order.

4   Results and Discussion

Figure 2(a), shows that skill 8 (COL9) is the most used by the agent society throughout the Sudoku tournament, followed by skill 1 (ROW5). However, a skill being chosen may not imply it produces high scores in Sudoku games. We test this by investigating what are the skills most used for obtaining the highest scores. Figure 2(b) demonstrates that indeed skill 8 and skill 1 are also the most effective scanning skills, given the experimental setup used in this paper. We understand that skill 8 is the most effective skill, followed by skill 1.

Further, we select the skill with highest proficiency (skill 8) and investigate for which size of the M_c it is most used. This is equivalent to searching which agent used this skill most, since the agents differentiate from each other through the ratio between skill and situation awareness memory. Figure 2(c) shows that the highest effective scanning skill is used more by agents with high amount of memory reserved for situation awareness storage, and less used by other agents.

We continue to investigate the overall influence on performance of the scanning skills and situation awareness components, in order to see if skills (and

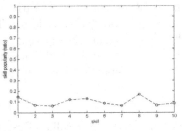

(a) The popular skills over the tournament

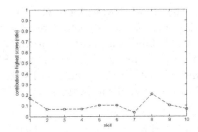

(b) The effective skills over the tournament

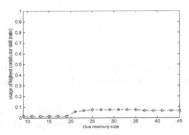

(c) Usage of most effective skill over all memory setups

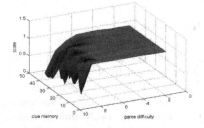

(d) Performance throughout the tournament for all memory setups

Fig. 2. Skills over the tournament

subsequently the processing function) prevail over situation awareness (the storage function) or otherwise. Figure 2(d) displays the score at the end of the games for all agents (clue memory sizes) and all difficulty levels throughout the tournament. For low difficulty games, the score is always maximum, with the games being completed regardless of the size of memory or skill complexity. As the game difficulty increases, the difference in score between agents with low and high situation awareness memory becomes significant. The score drops significantly for agents with high situation awareness memory, with the drop starting early in the tournament at difficulty 5. Recalling that high M_c leaves a low amount of working memory to be used for loading skills (M_s), only limited amount, number and/or complexity, of skills can be used. On the other hand, results show that the performance is less affected in the opposite situation, when the situation awareness memory is low. This suggests that the scarcity of skills can jeopardize entirely the ability of an agent to complete the game, whereas severe limitation of situation awareness memory still allows a certain level of performance. We conclude that the presence of scanning skills (inherent ability to process) prevails the storage of situation awareness (ability to store) in the working memory.

5 Conclusions

We investigated the trade-off between processing and storage functions of working memory in Sudoku. We used a society of agents capable of learning from each-other how to effectively use their existing skills in conjunction with their working memory in order to solve Sudoku games of various difficulty levels. The most used skills, and most importantly the effective skills, i.e. the ones that contribute to acquiring high scores in a Sudoku, have been established. The main finding is that, the scanning skills tend to be more important than the space available for situation awareness.

Acknowledgement. This project is supported by the Australian Research Council Discovery Grant DP140102590, entitled "Challenging systems to discover vulnerabilities using computational red teaming".

References

1. Atkinson, R.C., Shiffrin, R.M.: Human memory: A proposed system and its control processes. In: The psychology of learning and motivation, vol. 2, pp. 89–195. Academic Press, New York (1968)
2. Baddeley, A.D., Hitch, G.: Working memory. In: Bower, G.H. (ed.) The psychology of learning and motivation: Advances in research and theory, vol. 8, pp. 47–89. Academic Press, New York (1974)
3. Baddeley, A., Chincotta, D., Adlam, A.: Working memory and the control of action: evidence from task switching. J. of Exper. Psyc.: General 130(4), 641 (2001)
4. Brown, G.D.A., Preece, T.: Oscillator-based memory for serial order. Psyc. Rev. 107(1), 127–181 (2000)
5. Brown, G.D.A., Neath, I., Chater, N.: A temporal ratio model of memory. Psyc. Rev. 114(3), 539–576 (2007)
6. Burgess, N., Hitch, G.: Computational models of working memory: putting long-term memory into context. Trends in Cog. Sc. 9(11), 535–541 (2005)
7. Case, R., Kurland, D.M., Goldberg, J.: Operational efficiency and the growth of short-term memory span. J. of Exper. Child Psyc. 33(3), 386–404 (1982)
8. Chandler, C.C.: Specific retroactive interference in modified recognition tests: Evidence for an unknown cause of interference. J. of Exper. Psyc. 15, 256–265 (1989)
9. Craik, F.I.M., Lockhart, R.S.: Levels of processing: A framework for memory research. J. of Verbal Learning and Verbal behavior 11, 671–684 (1972)
10. Henson, R.N.A.: Short-term memory for serial order: the start-end model. Cog. Psyc. 36, 73–137 (1998)
11. Lewandowsky: Redintegration and response suppression in serial recall: A dynamic network model. Int. J. of Psyc. 34, 434–446 (1999)
12. Lovett, M.C., Reder, L.M., Lebiere, C.: Modeling working memory in a unified architecture. In: Models of Working Memory: Mechanisms of Active Maintenance and Executive Control, pp. 135–182. Cambridge University Press, NY
13. Page, M.P.A., Norris, D.: The primcay model: A new model of immediate serial recall. Psyc. Rev. 105, 761–781 (1998)
14. Towse, J., Hitch, G.: Is there a relationship between task demand and storage space in tests of working memory capacity? Q. J. of Exp. Psyc. 48, 108–124 (1995)
15. Wixted, J.T.: The psychology and neuroscience of forgetting. Annual Rev. of Psyc. 55(1), 235–269 (2004)

GA-Tetris Bot: Evolving a Better Tetris Gameplay Using Adaptive Evaluation Scheme

Somnuk Phon-Amnuaisuk

Media Informatics Special Interest Group,
School of Computing & Informatics
Institut Teknologi Brunei,
Brunei Darussalam
somnuk.phonamnuaisuk@itb.edu.bn

Abstract. Genetic Algorithm (GA) is employed to evolve a solution for any given tetromino sequence. In contrast to previous works in this area where an evolutionary strategy was employed to evolve weights (i.e., preferences) of predefined evaluation functions which then were used to determine players' actions, we directly evolve the actions. Each chromosome represents a plausible gameplay strategy and its fitness is evaluated by simulating the game and rating the gameplay quality using two fitness evaluation approaches: evaluating the whole board at once and evaluating local parts of the board in which they will be expanded to the whole board as the evolution progresses. We compare the results of these two evaluation tactics and also compare the evolved gameplay with actual human gameplay.

Keywords: GA-Tetris, Evolving game playing behaviours, Adaptive evaluations.

1 Introduction

All search strategies, unless it is a pure random search, explore available information (e.g., gradient information, heuristic information) to guide the search. Evolutionary Computing (EC) is a population-based search technique that stochastically explores the state-space without gradient information. EC explores the state-space using different heuristics derived from the knowledge content inherited in the population [1,2,3].

This paper investigates an application of adaptive evaluation in GAs [2]. An adaptive evaluation may be obtained through modifying specific parameters such as mutation rate, crossover rate or modifying weights of various objective functions according to some properties observed from the population. We investigate the performance of GAs in searching for an optimal sequence of actions that would place 50 tetrominoes on the Tetris board such that there are minimum unfilled tiles left. This work offers unique contributions on the following points. Firstly, that the GA is employed to find the playing strategy of a given tetrominoe sequence. This is different from previous works in this area where the GA

C.K. Loo et al. (Eds.): ICONIP 2014, Part III, LNCS 8836, pp. 579–586, 2014.

was employed to find the percentage contributions of different objective functions [4,5]. The information encoded in the chromosome in this work is a sequence of a player's actions while the chromosomes in [4,5] encoded the weight of predefined objective functions. Secondly, we investigate the use of GAs to evolve a sequence of actions using two different fitness evaluation styles, i.e., evaluating the whole chromosome at once and partially evaluating the chromosome and gradually adapting the size of the evaluation region.

The rest of the materials in the paper are organized into the following sections: Section 2: Related works; Section 3: Problem formulation; Section 4: Experimental results & Discussion; and section 5: Conclusion.

2   Related Works

Tetris is a casual game created by Alexey Pajitnov. The game is played on a 20×10 tiles board. There are seven tetrominoes in a standard Tetris game. Each tetrominoe is a pattern formed by four connected tiles which are often referred to, according to their shapes, as $\{I, J, L, O, T, S, Z\}$. The goal of the game is to place each given tetrominoe on the board such that the horizontal line is completely filled so that the line can be removed. Else, newly added Tetriminoes will stack up and will lead to game over if a new tetrominoe piece cannot be placed on the board.

Although Tetris has a simple gameplay, the game has been analysed by [6] and was classified as an NP complete problem. The Tetris game has been a popular game for AI researchers in the past decades. Early works in this area focused on finding powerful heuristics to guide search in the traditional graph search paradigm. Many useful objective functions derived from an analysis of board properties have been proposed e.g., height and maximum height of columns, maximum depth of a well, etc., (a good summary was provided in [7]). These handcrafted heuristic and rules are employed by all top performance Tetris controllers today.

Assigning different preferences (i.e., weights) to different objective functions affects the gameplay strategy. The state-space of all possible weight combinations is quite huge. Hence, many researchers applied evolutionary computation approach to find the optimal weights for the objective functions. To the best of our knowledge, all previous works reported in the literature worked on assigning weights to high-level heuristic functions and not on the actual actions, for examples, see [7,8,9].

We found two previous works that employed GA to evolve the weights. In [4], GA the researchers applied various weight combination e.g., linear combination, exponential combination, etc., and suggested that non-linear combination of weight could produce a better performance. In [5], it was shown that the evolutionary strategy (ES) successfully learned the weights and the weights learned by ES were quite similar to those hand tuned by humans.

3   Problem Formulation

In this work, we employ GA to evolve a player's actions when given a sequence of tetrominoes. The aim is to find a sequence of actions that would optimally place the tetrominoes so that the horizontal lines are completely filled. We have decided to limit the number of tetrominoes to 50 pieces [1]. Hence each individual in the population represents a sequence of 50 actions.

3.1   Evolving Tetris Playing Actions

Let us formally define a board States $s \in \mathcal{S}$, let s be a unique arrangement of tetrominoes on the Tetris board area. In this implementation, the state $s \in \{0, 1\}^{20 \times 10}$ is represented as a binary matrix, where the entry 1 denotes a filled tile and the entry 0 denotes an empty tile.

A population of chromosomes $\mathcal{C}$ is a $m \times n$ matrix where m denotes the length of the sequence of actions and n denotes the number of chromosome in the population. Each chromosome c_n represents a sequence of actions $[a_1, a_2, ..., a_m]$. An action a is expressed as a tuple (r, x) where $r \in \{1, 2, 3, 4\}$ indicates the desired rotated angle of $90(r-1)$ degrees; and $x \in \{1, 2, ..., 10\}$ indicates the left most position of the tetrominoe with reference to the x-axis of the board. Hence, each chromosome represents a different playing strategy of a given sequence of tetrominoes. The fitness of each chromosome is evaluated by simulating the Tetris gameplay. The chromosomes with lesser unfilled tiles are considered better and are desired to be reproduced in the next generation.

Before describing the evolutionary process implemented in this work, let us describe the assumptions made about the Tetris gameplay in this implementaion:

1. For each game, a sequence of 50 tetrominoes are generated randomly by uniformly picking them from the set $\{I, J, L, O, T, S, Z\}$.
2. For each tetrominoe τ, only one action (a combination of rotation and translation) is allowed.
3. No row eliminations is carried out during the game play. This means that a perfect game will require only 50 tetrominoes to fill all 200 tiles.

Given a sequence of tetrominoes $\mathcal{T} = [\tau_1, \tau_2, ..., \tau_m]$, the perfect gameplay is the gameplay with a sequence of actions $\mathcal{A} = [a_1, a_2, ..., a_m]$ that yields an optimal board state s i.e., the board with no holes (all tiles are filled).

Figure 1 graphically summarises the evolutionary process implemented here. Each GA population evolves a specific gameplay for a random generated tetrominoe sequence. Two different fitness evaluation strategies f_1 and f_2 were used (see details in subsection 3.2). The pseudo code below highlights the evolutionary process employed in this implementation.

---

[1] Only 50 tetrominoes are needed to fill the 20×10 board.

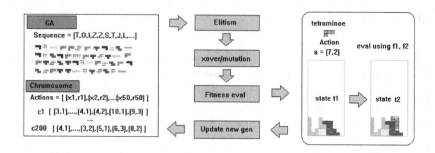

Fig. 1. A graphical summary of the GA-Tetris system

function GA-TETRIS(C, FITNESS-FN) **returns** c
// C : is an $m \times n$ matrix represent a set of individual $\{c_1, ..., c_n\}$
// Each $c_n = [a_1, ..., a_m]$ where $a = (r, x)$; $r \in \{0, 90, 180, 270\}$; $x \in \{1, 2, ..., 10\}$;
repeat
    $newC \leftarrow \phi$
    evaluate C according to FITNESS-FN
    add top 20% individuals to $newC$
    while $size(newC) < size(C)$ **do**
        $c_x \leftarrow$ SELECTION(C), according to its fitness ranking
        $c_y \leftarrow$ RANDOM-SELECTION(C)
        $c_z \leftarrow$ CROSSOVER(c_x, c_y)
        $c_z \leftarrow$ MUTATE(c_z)
        if c_z is well-formed // actions can be carried out on a Tetris board
        then $newC$.APPEND(c_z)
        else $c_z \leftarrow$ REPAIR(c_z) **and** $newC$.APPEND(c_z)
    $C \leftarrow newC$
until termination criteria are met
return the best individual $c \in C$, according to FITNESS-FN

3.2   Reproduction

Fitness Functions. A perfect game in our implementation only happens when all 50 tetrominoes fill the 200 tiles exactly (i.e., a board $s \in \{0, 1\}^{20 \times 10}$). This is very hard, even for a human player and is only feasible for specific sequences. A good gameplay can be determined by the number of tiles filled after gameover. The objective of our problem formulation is to maximise the filled tiles (or minimise the unfilled tiles). Two styles of fitness functions are implemented in our experiments: (i) we count all the unfilled tiles (see Eq 1) and (ii) count all the unreachable unfilled tiles in a desired region which grows as generation progresses (see Eq 2). Unreachable tiles refer to unfilled tiles that cannot be filled since they are not accessible as the game progresses. Counting unfilled tiles is simple: by subtracting filled tiles to 200 (total tiles), we get the value:

$$f_1 = 200 - \sum_{ij} s(i, j) \tag{1}$$

where $s(i,j)$ denotes the value of the entry (i,j) in the Tetris board.

In order to place more emphasis at the bottom region of the board in the early stage of the game and to gradually increase the region to a full board toward the end of the game, let us describe the process using these parameters: h, $step$ and w. The h denotes the number of rows of the Tetris board, $step$ denotes the position along the row of the Tetris board as the GA generation, n, increases toward the maximum generation (here set at 500):

$$step = h(maxGeneration - n)/maxGeneration$$

The weight w denotes the importance of row i and is defined as

$$w = e^\alpha \text{ where } \alpha = \begin{cases} 0 \text{ if } i > step, \text{ else} \\ (10 \times i)/step - 10 \end{cases}$$

$$f_2 = \sum_i \left(10 - w \sum_j s(i,j) \right) \tag{2}$$

Elitism and Selection Process. It has been decided that the best 10 % of the population (according to the fitness functions) will be selected and will continue to the next generation without going through the crossover and mutation process. The rest of the population will be selected in pair: c_x will be picked according to its rank in the whole population and c_y will be randomly selected from the population.

Crossover and Mutation. A standard one point crossover is applied to c_x and c_y. This produces two offsprings and the better offspring c_z is retained for the next generation. Then a one point mutation is randomly applied to the chromosome c_z. The process continues until all the termination criteria are met or the GA has reached the maximum generation.

4   Experimental Results and Discussion

Experimental Design. One hundred fifty tetrominoe sequences, each with 50 tetriminoes uniformly sampled from the set $\{I, J, L, O, T, S, Z\}$, are employed in this study. For each sequence, two versions of fitness evaluations (see subsection 3.2) are employed by two different GA populations. Each GA population evolves for the maximum of 500 generations to find the best playing strategy for a given sequence using two different fitness evaluation approaches. The following parameters: 200 chromosomes, 10% elitism scheme, one point crossover, and one point mutation are common to both GA populations.

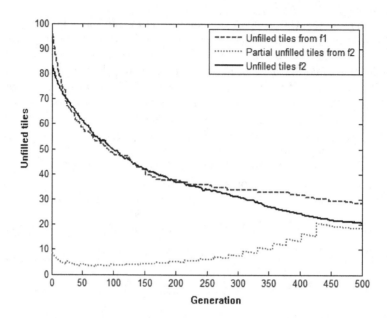

Fig. 2. A plot showing average values of unfilled tiles from evaluation scheme f_1 and f_2. The values are averaged over 150 sequences.

Results and Discussion. In our view, evaluating the whole board at once (approach taken in f_1) is a harder problem than partially evaluating part of the board then gradually expanding the area to a full board (approach taken in f_2). It is more difficult because f_1 attempts to solve a bigger problem, while f_2 attempts a smaller problem and gradually increases the size of the problem. This implicitly ensures that a complete solution is built up from good local solutions. One may argue that this common scenario is usually dealt with using appropriate objective functions. That is generally true but the approach also introduces more variables into the system e.g., the assignment of weighted contributions of available objective functions. It is also a challenging knowledge engineering task to devise appropriate objective functions for the problem. Hence, adapting the size of the problem could be one domain dependent strategy that can be universally applied in an evolutionary approach.

Figure 2 shows the average values of unfilled tiles observed in the GA populations over the whole evolution process. The *partial unfilled tiles from f_2* plots the number of unreachable tiles. The number of unreachable tiles increases as the evolution progresses since the evaluation area increases (i.e., the problem is getting more difficult). The *unfilled tiles from f_1* and *unfilled tiles from f_2* show the number of unfilled tiles. In both evaluation approaches, the number of unfilled tiles start at around 80-100 in the first generation. After about 250 generations, both fitness evaluations depart and it is conclusive that the adaptive evaluation approach yields a better gameplay. The instances of both boards are shown in Figure 3. The information of the tetrominoe sequence and actions is given below.

Sequence	OTLTILJZTTJTTZOLSTOTILOTIOOSIJILJLLOISTLZSOOJZZIIO
	GA with f1
x translation	47181426158894623679684711941351851947 2158
Rotation	13334442111322143124213442133334132422 2143
	GA with f2
x translation	6891895138861364236434336 1A9618274198651 3413696
Rotation	1343424424224232232222244244241242112242 14243
	Human
x translation	1364319477944771227367496997132511 39A55178637
Rotation	1342242413242212221124142112242244412142 12111

It is interesting to compare the gameplays of both evolution strategies to a human's. We have asked a student to play the same sequences [2]. The output from an adaptive evaluation are comparable to a human's gameplay (see Figure 3). It should be highlighted that the human has played the same sequence and has been exposed to the same information available to the GA program. For each tetrominoe, a human player has unlimited time to analyse the board before deciding on the action.

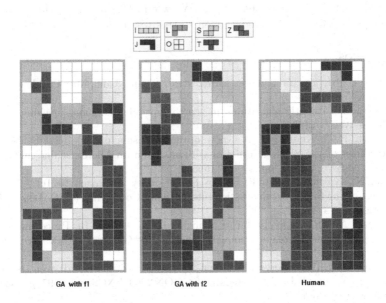

Fig. 3. An instant of a gameplay by GAs (left: with f_1 evaluation scheme, middle: with f_2 evaluation scheme) and by a human (right)

---

[2] Sequences are randomly taken from the pool of 150 sequences employed in this study.

5 Conclusion

Population-based approach commonly encodes a candidate solution as an individual in the population. This coding style offers a strength in dealing with long-range dependency among local groups in a solution since the solution is examined on a global scale. However, formulating appropriate fitness functions to handle complex dependency among local and global groups is not always possible. This work highlights this common issue observed in evolutionary strategy through the evolution of a Tetris gameplay.

Measuring the fitness of a solution based on the number of unfilled tiles is logical but the measure only gives us the global property of the board. It is always a challenge to create a set of objective functions that could provide good evaluations of both local and global properties at the same time. Alternatively, one may construct a solution from easy to hard. This is quite similar in spirit to the reductionism paradigm. In this work, we show two different fitness evaluation approaches: the whole solution is evaluated in a global scale in the first approach while only a partial solution is evaluated and gradually built up in the second approach. Our experiments support the merit of an adaptive evaluation approach.

Acknowledgments. We wish to thank anonymous reviewers for their comments, which help improve this paper. We would like to thank the GSR office for their partial financial support given to this research.

References

1. Perl, J.: Heuristics: Intelligence Search Strategies for Computer Problem Solving. Addoson-Wesley (1984)
2. Goldberg, D.: Genetic algorithms in search, optimization, and machine learning. Addoson-Wesley (1989)
3. Eberhart, R.C., Kennedy, J.: A new optimizer using particle swarm theory. In: Proceedings of the Sixth International Symposium on Micromachine and Human Science, Nagoya, Japan, pp. 39–43 (1995)
4. Böhm, N., Kókai, G., Mandl, S.: An Evolutionary Approach to Tetris. In: Proceedings of the Sixth Metaheuristics International Conference (MIC 2005), Vienna, Austria, pp. 1–6 (2005)
5. Boumaza, A.: On the evolution of artificial Tetris players. In: Proceedings of the IEEE Symposium on Computational Intelligence and Games (CIG 2009), pp. 387–393 (2009)
6. Demaine, E.D., Hohenberger, S., Liben-Nowell, D.: Tetris is hard, even to approximate. In: Warnow, T.J., Zhu, B. (eds.) COCOON 2003. LNCS, vol. 2697, pp. 351–363. Springer, Heidelberg (2003)
7. Thiery, C., Scherrer, B.: Building controllers for Tetris. International Computer Game Association Journal 32, 3–11 (2009)
8. Carr, D.: Adapting reinforcement learning to Tetris BSc (Honours) Thesis, Rhodes University (2005)
9. Langenhoven, L., van Heerden, W.S., Engelbrecht, A.P.: Swarm tetris: Applying particle swarm optimisation to Tetris. In: Proceedings of the IEEE Congress on Evolutionary Computation (CEC 2010), Barcelona, pp. 1–8 (2010)

An Effectiveness of Model-Based Development with User Model in Consideration of Human

Yoshinobu Akimoto, Eri Sato-Shimokawara,
Yasunari Fujimoto, and Toru Yamaguchi

Tokyo Metropolitan University, 6-6 Asahigaoka, Hino, Tokyo 191-0065, Japan
akimoto-yoshinobu@sd.tmu.ac.jp,
{eri,yfujimoto,yamachan}@tmu.ac.jp

Abstract. In this paper, we proved an effectiveness of a model-based development method with user model which can operate a product as an agency of human in the model-based simulation. In this method, the user model is generated on the basis of the user profile which is defined to clarify the features of the product's user. This method is effective to evaluate the products which is used by various users with different features, and is also effective to evaluate the products with the risks of damaging the human. And this method does not require any hardware sample production for evaluation. In order to investigate the effectiveness of this method by comparing the results between the simulations and the experiments, we applied this method to the development of an approach function and an emergency stop function of a service robot.

Keywords: model-based development, user model, robot, simulation.

1   Introduction

There are many previous studies regarding the user model related issues [1, 2, 3]. The VERITAS Project in EU had researched the user model of the elderly person and handicapped person. And in 2012 as a result, The VERITAS Project published a PDF document titled as "White Paper: Virtual User Modelling Public Document".

We have been studying the Persona which is one of user profiling methods [4, 5], and a user modeling method which can be used with the model-based development (MBD). As members of a committee organized by Information-Technology Promotion Agency, Japan (IPA), we reported a new idea of MBD with user model [6], which consists of a user profiling, a user modeling and a simulation. Because the user model behaves as an agency of the actual human in simulations, the MBD with user model is effective to evaluate the functions operated by various users with different features, and is effective also to evaluate the functions with the risk of damaging the user.

In this paper, we proved an effectiveness of the MBD with user model, by applying to a development of an approach function and an emergency stop function for the walker's safety. There are many previous studies regarding the collision avoidance for robots [7, 8, 9]. But we focused on behaviors of walker since the risk of collision is

C.K. Loo et al. (Eds.): ICONIP 2014, Part III, LNCS 8836, pp. 587–595, 2014.

high when the walker is careless. And there are many previous studies regarding the robot approaching a human [10]. But we focus on the personal space which is a key condition for the users to feel at ease when a robot approaches.

We adopted the Robot Technology Middleware [11] (RT Middleware) as a MBD tool, because the RT Middleware is a middleware to develop robots, and can be used as a simulator and also has a capability to define user models. And we adopted the Persona as a profiling method for identifying the features of each user as a base of user model.

2   MBD with User Model and System Configuration

This chapter shows a process of the MBD with user model, a configuration of the robot system, a process of an automatic stop sub-function for the approach function and the emergency stop function, and evaluation criteria.

2.1   Process of MBD with User Model

The actual system consists of a service robot and a human. And the system for the MBD with user model in this paper consists of a device model, a device scenario, a user model, a user scenario, and an MBD tool. The features of user model are defined on the basis of the user profile defined as Persona. And, the user scenario is defined on the basis of the user's behaviors defined as Persona. The functions of the device model are defined on the basis of the system requirements for the device. And, the device scenario is defined on the basis of the procedures required to the device. Therefore the user model and the user scenario could be defined to be separated from the device model and the device scenario. As a result, variety sets of a user model and a user scenario can be applied to each set of a device model and a device scenario.

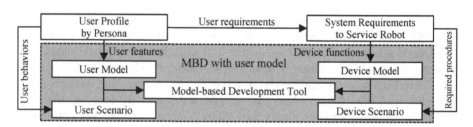

Fig. 1. Relationship of elements in MBD with user model

2.2   User Profiling, User Modeling and Component for User Model

We investigated the conditions under which users feel the anxiety and discomfort in the approach of a robot, and as a result, we focused on the size of personal space different for each user. And then, we investigated the conditions of walkers who have big risks to collide with a robot, and as a result, we focused on the careless walkers watching a smartphone while walking.

Then, in the human-readable form by Persona, we defined a user profile of persons requesting an approach to a robot, and a user profile of walkers having risks to collide with the robot. And then, we defined user models and user scenarios on the basis of the features and the behaviors described in the user profiles as Persona. And then, for the simulation on the RT Middleware, we developed components of a user and a walker based on the models and scenarios.

2.3   Configuration of Actual System and Model-Based System

The actual system consists of a human and a robot system. And the system for the simulation consists of a user model and a model of robot system.

The robot system consists of a drive function, a warning function, sensing functions, and a robot controller. The drive function is implemented by a drive unit or a drive unit model. The warning function is implemented by a caution unit. The sensing functions are implemented by a LRF (laser range finder) unit, a Kinect unit and a sensor manager. The LRF unit is a device for finding obstacles. A Kinect unit is a device for analyzing the attentiveness of a walker, and also for analyzing the gesture commands by human. And the robot controller manages these functions to achieve the automatic stop sub-function. The user model is implemented by combining a model of a human and two sensor units. Fig.2 shows a configuration of the actual system for the automatic stop sub-function for an emergency stop and an approach function.

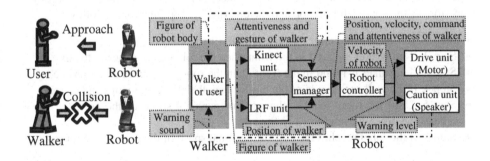

Fig. 2. Configuration of actual system for automatic stop sub-function

Fig.3 shows the configuration of system for simulation by the MBD with user model.

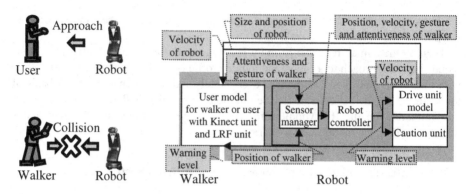

Fig. 3. Configuration of system for simulation of automatic stop sub-function

2.4   Process of Automatic Stop for Emergency Stop and Approach

We assumed that the personal space was one of key factors for the approach function, because a human felt uncomfortable when its personal space was invaded. And we assumed that the attentiveness of the walker was one of key factors for the emergency stop function, because the risk of the collision with a careless walker was much higher than the risk with a careful walker. The attentiveness of a walker decreases by its parallel behaviors at walking. Therefore, we developed an algorithm to calculate the border line on which a robot stops and warns a walker, according to the level of attentiveness of walker. And we focused on an operation of smartphone during walking, as a parallel behavior which is a cause decreasing the walker's attentiveness.

Then based on the above assumption, we defined a process of the automatic stop sub-function for the approach function and the emergency stop function, as follows.

- Step 1. Sensor units monitor obstacles in the surroundings.
- Step 2. If no walker is found in the front, then the robot approaches to the user.
- Step 3. If a walker is found in the front, then the robot calculates a border line.
- Step 4. If the walker is inside of the border line, then the robot stops and warns.
- Step 5. Until the robot reaches the edge of personal space of the user, the robot repeats the process from Step 1 to Step 4.
- Step 6. If the robot reaches the edge of personal space of the user, the robot stops.

The walker is either a walking person or a standing person, and has risks to collide with the robot. In order to calculate the border line, following values are required.

- Attentiveness of human: A rate value estimated by the behavior of a human, which is acquired by sensor units. For example, if the human watches a smartphone when walking, the attentiveness rate is assumed to be low.
- Relative position of obstacle: A relative position of an obstacle which is acquired by sensor units.
- Absolute velocity of obstacle: A velocity which is calculated by a velocity of the robot and a time-series data of the relative position of obstacle.

- Personal space of human: An occupied space by a human. This personal area is specified for each user. For the walker, a fixed value is applied.

2.5   Evaluation Criteria for Simulation and Experiment

We defined the evaluation criteria as follows. In this paper, we describes the evaluation criteria only in the x-direction, but the process of the automatic stop is designed in two-dimension. The suffix x means the direction which the robot and the human are facing.

The HA is an attentiveness rate to estimate a required time for a human to stop, after hearing a warning from a robot, and it is estimated based on a behavior of human.

$$HA = max\left(\frac{t_{human\,best}}{t_{human\,worst}}, min(HA_{max}, HA_{sensor})\right) \tag{1}$$

The $t_{human\,best}$ is the best reflection time (seconds) of the walker to the stop from the recognition of a warning, and the $t_{human\,worst}$ is the worst reflection time (seconds) of the walker to the stop from the recognition of a warning. The HA_{max} is the maximum attentiveness rate of the walker which is 1.0. And the HA_{sensor} is the attentiveness rate of the walker which is acquired by sensors installed on the actual robot. The $t_{human\,stop}$ is a required time for a walker to the stop from the recognition of a warning.

$$t_{human\,stop} = \frac{t_{human\,best}}{HA} \tag{2}$$

$RD_x{}^{border}$ is a distance in the x-direction where the robot stops and warns a walker.

$$RD_x^{border} = HD_x^{personal\,space} + v_x^{human} * t_{human\,stop} \tag{3}$$

The $v_x{}^{human}$ is an absolute velocity (millimeter per second) of the walker, which is calculated by a velocity of the robot and a time-series data of the relative position of obstacle. The $HD_x{}^{personal\,space}$ is a distance (millimeter) in the x-direction of a personal space. And the HD_x is a distance (millimeter) in the x-direction to the robot from the human at the time when the human stops.

However, if the walker does not travel (that is, $v_x{}^{human} = 0$), then $HA = 1.0$, $t_{human\,stop} = t_{human\,best}$, and $RD_x{}^{border} = HD_x{}^{personal\,space}$.

In this paper, we assumed that $t_{human\,best} = 1.0$ (second), $t_{human\,worst} = 5.0$ (second) and $HA_{maximum} = 1.0$.

3   Simulations and Experiments

This chapter shows the system configuration and the results of simulations by the MBD with user model, the system configuration and the results of experiments with a human and an actual robot system, and the discussion.

3.1   Configuration and Results of Simulation of Approach Function

Based on the Fig.3, we developed a system of the approach function for the simulation on the RT Middleware, as shown in Fig.4. In this simulation, the values of 700, 900 or 1,100 was set to the $HD_x^{personal\ space}$. And the Table 1 shows the results of the simulation.

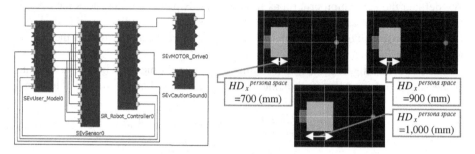

Fig. 4. Configuration for the simulation with user model

Table 1. Results of simulation of approach function

$HD_x^{persona\ space}$ (mm)	HA	v_x^{human} (mm/sec)	$t_{human\ stop}$ (sec)	RD_x^{border} (mm)	HD_x (mm)
700	1.0	0.0	0.0	700	695
900	1.0	0.0	0.0	900	887
1100	1.0	0.0	0.0	1100	1094

3.2   Configuration and Results of Experiment of Approach Function

Based on the Fig.2, we developed an actual system for the approach function on the RT Middleware, as shown in Fig.5. In this experiment, the value of 700, 900 or 1,100 was set to the $HD_x^{personal\ space}$. And the Table 2 shows the results of the experiment.

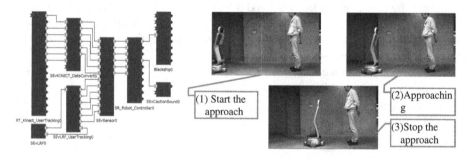

Fig. 5. Configuration for experiment with actual human

Table 2. Results of experiment of approach function

$HD_x^{persona\ space}$ (mm)	HA	v_x^{human} (mm/sec)	$t_{human\ stop}$ (sec)	RD_x^{border} (mm)	HD_x (mm)
700	1.0	0.0	0.0	700	686
900	1.0	0.0	0.0	900	899
1100	1.0	0.0	0.0	1100	1080

3.3   Configuration and Results of Simulation of Emergency Stop Function

The system configuration of the emergency stop function for the simulation on the RT Middleware was the same configuration with the approach function shown in Fig.4. In this simulation, the value of 500 or 1,000 was set to the v_x^{human}, the value of 1.0 or 0.5 was set to the HA and the value of 700 was set to the $HD_x^{personal\ space}$. And the Table 3 shows the results of the simulation.

Table 3. Results of simulation of emergency stop function

v_x^{human} (mm/s)	HA	$t_{human\ stop}$ (sec)	$HD_x^{persona\ space}$ (mm)	RD_x^{border} (mm)
500	1.0	1.0	700	1200
500	0.5	2.0	700	1700
1000	1.0	1.0	700	1700
1000	0.5	2.0	700	2700

3.4   Configuration and Results of Experiment of Emergency Stop Function

The system configuration of the emergency stop function for the experiment on the RT Middleware was the same configuration with the approach function shown in Fig.5. In this experiment, the v_x^{human} and the HA were acquired by the sensor, and the value of 700 was set to the $HD_x^{personal\ space}$. And the Table 4 shows the results of the experiment.

Table 4. Results of experiment of emergency stop function

v_x^{human} (mm/s)	HA	$t_{human\ stop}$ (sec)	$HD_x^{persona\ space}$ (mm)	RD_x^{border} (mm)
801	1.0	1.0	700	1501
386	0.5	2.0	700	1472
1150	1.0	1.0	700	1850
718	0.5	2.0	700	2135

3.5   Discussion

In this study, we applied the MBD with user model to develop the approach function and the emergency stop function, on the basis of the automatic stop sub-function. In this study, we applied two types of user model. One was the agent of a user whom the robot approached to give services. And the other was the agent of a walker who is the obstacle of the robot to approach to the user.

In the section 3.1, we evaluated the approach function by the simulation on the RT Middleware with three user models with different size of personal space, and, accomplished the process of the approach function without actual robot and actual user. And then in the section 3.2, we verified the approach function of the actual robot by the experiments with actual human, without risks. And we also recognized the differences of tension by the differences of the distance where the robot stopped.

In the section 3.3, we evaluated the emergency stop function by the simulation on the RT Middleware with four user models with different velocities and different attentiveness, and, accomplished the process of the emergency stop function without actual robot and actual walker. And then in the section 3.4, we verified the emergency stop function of the actual robot by experiments with actual walker, without risks.

And then, by comparing the results of the simulations and the experiments, we recognized that the user model could work as an agent of the human. And also, we recognized that if the evaluation by simulation could achieve the expected results, then the experiment would be completed with low risk.

4   Conclusion

We applied the method of the MBD with user model to the development of the approach function and the emergency stop function, which are implemented on the basis of the automatic stop sub-function. And, we evaluated the approach function and the emergency stop function, by simulations with user models. And, we verified the actual system with actual human, without risks by experiments. And then, by comparing the results of the simulations with the results of the experiment, the similarity of both results are certified. By these results, we proved the effectiveness of the MBD with user model.

References

1. Biswas, P., Langdon, P., Robinson, P.: Designing inclusive interfaces through user modelling and simulation. International Journal of Human Computer Interaction 28(1) (2011)
2. Carmagnola, F., Cena, F., Gena, C.: User model interoperability: a survey. User Modeling And User-Adapted Interaction 21(3), 285–331 (2011)
3. Moschonas, P., Paliokas, I., Tzovaras, D.: A Novel Accessibility Assessment Framework for the Elderly: Evaluation in a Case Study on Office Design (2014), http://rehab-workshop.org/papers/rehab2014_submission_28.pdf
4. Akimoto, Y., Shimokawara, E.S., Fujimoto, Y., Yamaguchi, T.: User model by Persona and the application of the user behavior model by sensor network to the service robot. In: Proc. 31st Annual Conference of the Robotics Society of Japan (IRH) (2013) (in Japanese)
5. Takai, S., Moriguchi, S., Browne, J., Miyazaki, H., Namamura, J., Shirane, H., Sasaki, H., Uehira, T., Hisanabe, Y., Hirota, T., Akimoto, Y., Kataoka, T.: Practice for Persona Marketing, pp. 138-159 (2014) (in Japanese) ISBN: 978-4-532-31930-4
6. Information-Technology Promotion Agency, Japan (IPA): Activity report, of the Model-based development Technical Group. IPA, pp. 6-28 (2011), http://www.ipa.go.jp/files/000026871.pdf (in Japanese)
7. Hennes, D., Claes, D., Meeussen, W., Tuyls, K.: Multi-robot collision avoidance with localization uncertainty. In: Proc. 11th International Conference on Autonomous Agents and Multiagent Systems (AAMAS), pp. 147–154 (2012)
8. Pelechano, N., Allbeck, J.M., Badler, N.I.: Controlling individual agents in high-density crowd simulation. In: Proc. 2007 ACM SIGGRAPH/Eurographics Symposium on Computer Animation (SCA 2007), pp. 99–108 (2007)
9. Guy, J.S., Lin, M.C., Manocha, D.: Modeling collision avoidance behavior for virtual humans. In: Proc. 9th International Conference on Autonomous Agents and Multiagent Systems (AAMAS), pp. 575–582 (2010)

10. Woods, S.N., Walters, M.L., Koay, K.L., Dautenhahn, K.: Methodological Issues in HRI: A Comparison of Live and Video-Based Methods in Robot to Human Approach Direction Trials. In: Proc. 15th IEEE International Symposium on Robot and Human Interactive Communication (ROMAN 2006), pp. 51–58 (2006)
11. Ando, N., Suehiro, T., Kitagaki, K., Kotoku, T., Yoon, W.K.: RT-middleware: distributed component middleware for RT (robot technology). In: Proc. 2005 IEEE/RSJ International Conference on Intelligent Robots and Systems (IROS 2005), pp. 3933–3938 (2005)

Dynamic Programming for Guided Gene Transfer in Bacterial Memetic Algorithm

Tiong Yew Tang[1], Simon Egerton[1], János Botzheim[2], and Naoyuki Kubota[2]

[1] School of Information Technology, Monash University Malaysia, Jalan Lagoon Selatan, 47500 Bandar Sunway, Selangor Darul Ehsan, Malaysia
{tang.tiong.yew,simon.egerton}@monash.edu

[2] Graduate School of System Design, Tokyo Metropolitan University, 6-6 Asahigaoka, Hino, Tokyo, 191-0065, Japan
{botzheim,kubota}@tmu.ac.jp

Abstract. Evolutionary Computation (EC) approaches are known to empirically solve NP-hard optimisation problems. However, the genetic operators in these approaches have yet to be fully investigated and exploited for further improvements. Hence, we propose a novel genetic operator called Dynamic Programming Gene Transfer (DPGT) operator to improve the existing gene transfer operator in the Bacterial Memetic Algorithm (BMA). DPGT integrates dynamic programming based edit distance comparisons during gene transfer operator in BMA. DPGT operator enforces good gene transfers between individuals by conducting edit distance checks before transferring the genes. We tested the DPGT operator in an artificial learning agent ant's perception-action problem. The experimental results revealed that DPGT gained overall improvements of training accuracy without any significant impact to the training processing time.

Keywords: Dynamic Programming, Edit Distance, Evolutionary Computation, Bacterial Evolution, Memetic Algorithm, Perception-Action.

1 Introduction

Evolutionary Computation (EC) approaches are commonly inspired from nature's adaption process and modelled these phenomena into computational optimisation algorithms [5]. In this research, we model an artificial learning agent ant's perception-action problem by using EC strategy. In order for an artificial learning agent ant to survive in its environment, it is important that both perception-action learning errors and learning computational time to remain as low as possible. For example, if the ant perceived a threat from its environment by its limited insect vision [7], the ant will need to react to the situation with the correct action such as escaping to mitigate the situation. Therefore, this research aim is to improve the EC optimisation performance without significantly impacting the overall perception-action learning time.

In this paper, we propose a novel gene transfer operator called *Dynamic Programming Gene Transfer (DPGT)*. DPGT operator as its name suggests

C.K. Loo et al. (Eds.): ICONIP 2014, Part III, LNCS 8836, pp. 596–603, 2014.

integrates an efficient string similarity comparison algorithm called *Dynamic Programming (DP)* [1] algorithm to calculate the similarity of two individual's bacterial chromosome elements. The DPGT operator checks the source's and destination's bacterium chromosome elements similarities with *Levenshtein Distance* (or *Edit Distance*) [8] calculation before transferring the elements. Hence, by transferring different gene elements to the destination gene elements (known as good gene transfer) it will improve the overall optimisation performance. On the other hand, when transferring similar gene elements it could cause poor optimisation performance (or bad gene transfer). In nature, we can observe this bad gene transfer phenomena when closely related breeding in a population will cause *inbreeding depression* [6]. Interestingly, dynamic programming edit distance calculation has been proposed in genome similarity comparison in bioinformatics research field [12]. Therefore, it is very intuitive to integrate dynamic programming edit distance calculation in EC's gene transfer operator.

Memetic algorithms combine global and local searches in their optimisation process. An example of a local search method is the Levenberg-Marquardt (LM) method [10]. In this research, we improved the gene transfer operator of a memetic algorithm known as Bacterial Memetic Algorithm (BMA) [3]. BMA was applied to many different applications such as fuzzy logic rules extractions [3], path planning problem [2], and fuzzy neural network optimisation problems [4]. We apply DPGT operator to improve BMA's existing gene transfer operator for fuzzy logic rules optimisation application [3]. We conduct experimental testing on the proposed DPGT operator to an artificial learning agent ant's perception-action problem. This problem was inspired by the ant's limited vision ability [7] where it's visual perception is limited to pixels resolution. The contributions of this paper are: 1) the proposed DPGT operator can achieve overall best performance in terms of Mean Square Error (MSE) bacterial evolution generations when it is compared to the benchmark. 2) DPGT operator is able to converge faster (achieved lower MSE) in shorter (or early) bacterial evolution generations when it is compared to the benchmark. 3) DPGT operator's execution time does not differ much from the benchmark.

2   Bacterial Memetic Algorithm

In order to achieve good quasi-optimal solution for a given optimisation problem, BMA utilised population based stochastic optimisation memetic approach that

---

Algorithm 1. Bacterial Memetic Algorithm

1: **procedure** BMA
2:     Initial bacterial population
3:     $generation \leftarrow 0$
4:     **while** $generation \neq N_{gen}$ **do**
5:         Bacterial mutation applied to each bacterium
6:         Local search for each baterium
7:         Gene transfer for the population
8:         $generation \leftarrow generation + 1$

combined global and local search (Refer to Algorithm 1). In case of global optimal search, BMA performs the bacterial operators that are the bacterial mutation and the gene transfer operators. The purpose of bacterial mutation operator is the optimisation of the bacterium's chromosome.

The BMA applied the Levenberg-Marquardt (LM) [10] method in its local search technique that was initially introduced in Botzheim et al. [3]. BMA begins with the generation of a random initial population that consist of N_{bac} bacterial individuals. Then, BMA repeats a loop until the number of generation N_{gen} that is the terminating condition is reached. BMA performs the bacterial mutation, LM local search, and gene transfer operators. Bacterial mutation operator initialises N_{clones} number of clones of a bacterium. These clones are then subjected to random changes in their genes, except one clone. When the mutations are complete, all the clones are evaluated, and the one with the best fitness value replaces the other clones. The total number of gene elements that are modified is determined by the length of mutation segment (L_{ms}) parameter of the algorithm. In the local search step, for each bacterium the LM algorithm is performed until a certain number of iterations (N_{iter}). The two parameters that determine the iteration terminal condition are τ and the initial bravery factor, α.

The last operator in BMA is the gene transfer operator that performs horizontal gene transfer. The horizontal gene transfer refers to copy of genes information from good bacterium individuals to the bad ones. In order to achieve this, the population is divided into two portions according to the fitness values. The two portions are the superior portion and the inferior portion of the population. The number of infections N_{inf} in a generation and infection segment length L_{is} that get transferred in each operation are predetermined by the parameter settings.

3   Artificial Learning Agent Ant's Perception-Action Problem

In this research we investigate the artificial learning agent ant's perception-action problem. The property of optimisation performance improvement without significantly impacting the processing time is very important in this problem. The reason

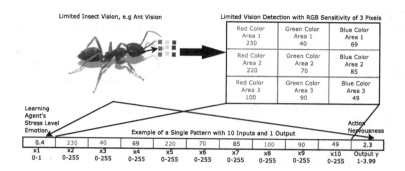

Fig. 1. The artificial learning agent ant's perception-action problem

is, that the agent's survival depends on the quick and correct learned actions. Figure 1 illustrates the problem. The artificial learning agent ant's perception is defined as the combination of Red-Green-Blue (RGB) pixels. The ant's vision is limited to RGB pixels representation of 3×3 dimensions which is inspired from limited insect vision [7]. Figure 2 depicts the Mamdani [9] fuzzy inference system process for the artificial learning agent ant's perception-action problem. The agent's stress emotion is also modelled in the data as a real value between 0 and 1. The bacterium's chromosome elements are modelled by the ant's vision RGB pixels in integer values between 0 to 255. The proposed DPGT operator optimises the membership function scope of the bacterium's chromosome elements in Fuzzy Rule Base Extraction (FRBE) application [3].

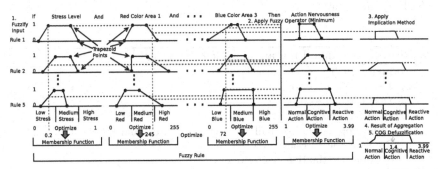

Fig. 2. Mamdani fuzzy inference system process with Center of Gravity (COG) defuzzification for an artificial learning agent ant's perception-action problem

4   Dynamic Programming Gene Transfer Algorithm

Nawa and Furuhashi [11] introduced the *gene transfer* operator in their Bacterial Evolutionary Algorithm (BEA) approach. In this work, we introduce the dynamic programming algorithm to calculate the edit distance similarity between source bacterium and destination bacterium chromosome elements known as *Dynamic Programming Gene Transfer (DPGT)* presented in Algorithm 2.

DPGT operator performs edit distance gene similarity comparisons before transferring genes from a superior individual bacterium to the inferior individual bacterium's chromosome elements in the population. If the current edit distance is larger than the average edit distance threshold *averageEDThreshold* of the population, only then the source bacterium's chromosome elements will be transferred to the destination bacterium within the number of similarity search count N_{search}. The real numbers in bacterium's chromosome elements are converted into string characters with string decimal point precision. For example, the minimum edit distance similarity between 2 pairs of real values (3.34, 0.25) and (3.334, 0.2) are converted to string characters (334025) and (333402). In this case the Hamming distance is 4 which means that it failed to capture similarity between these real values pairs. On the other hand, minimum edit distance calculation is at 2 which is able to determine the suitability to perform gene transfer depending on the source and destination bacterium chromosome elements similarity. It is intuitive to refrain the gene transfer if source and destination bacterium chromosome elements are similar for a better gene transfer result. Thus, the training performance can be improved by reducing the bad gene transfers (similar source and destination bacterium chromosome elements gene transfer). In addition to that, DPGT operator applied efficient edit distance calculation so that additional gene similarity checks will not impose significant processing cost to the original operator performance but still improve overall optimisation performance. The edit distance calculation with dynamic programming algorithm is very efficient because dynamic programming is known to have trade memory space with time property. In this work, we extended the BMA's gene transfer step (See Algorithm 1 step 7) with our proposed DPGT operator and applied it in FRBE application [3].

Algorithm 2. Dynamic Programming Gene Transfer

1: **procedure** DPGT
2:     $editDistanceCount \leftarrow 0$
3:     $editDistanceSum \leftarrow 0$
4:     **for** q:=1,2 ... N_{inf} **do**
5:         Order Population into half as Superior (Source) and Inferior (Destination)
6:         Random Source Bacterium
7:         Random Destination Bacterium
8:         **for** r:=1,2 ... L_{is} **do**
9:             $assigned \leftarrow False$
10:            $searchCount \leftarrow 0$
11:            **while** *Not assigned And searchCount* < N_{search} **do**
12:                Random Select Source Bacterium Elements
13:                Random Select Destination Bacterium Elements
14:                Concatenate All Selected Source Bacterium Elements to String a
15:                Concatenate All Selected Destination Bacterium Elements to String b
16:                $d_{m,n} \leftarrow minEditDistanceCalc(a, b)$
17:                $editDistanceCount \leftarrow editDistanceCount + 1$
18:                $editDistanceSum \leftarrow editDistanceSum + d_{m,n}$
19:                $averageEDThreshold \leftarrow editDistanceSum/editDistanceCount$
20:                **if** $d_{m,n} \geq averageEDThreshold$ **then**
21:                    *Assign Source Bacterium Elements to Destination Elements*
22:                    $assigned \leftarrow True$
23:                $searchCount \leftarrow searchCount + 1$

Let $d_{i,j}$ denote the edit distance matrix with dimensions $i \times j$ between string a and string b. Finally, c is defined as the cost function. The cost for delete, insert and substitution are the same with $c_{del,ins,sub} = 1$. The $minEditDistanceCalc$ function in Algorithm 2 performs the following steps: First, the algorithm initialises the first row and column with Equations (1) and (2). Then, the algorithm continues to fill up the entire remaining empty cells in the $d_{i,j}$ matrix with Equation (3).

$$d_{i,0} = \sum_{k=1}^{i} c_{del}, \quad for\ 1 \le i \le m \tag{1}$$

$$d_{0,j} = \sum_{k=1}^{j} c_{ins}, \quad for\ 1 \le j \le n \tag{2}$$

For $1 \le i \le m$, $1 \le j \le n$:

$$d_{i,j} = \begin{cases} d_{i-1,j-1} & \text{if } a_j = b_i, \\ min \begin{cases} d_{i-1,j} + c_{del} \\ d_{i,j-1} + c_{ins} \\ d_{i-1,j-1} + c_{sub} \end{cases} & \text{otherwise.} \end{cases} \tag{3}$$

5   Experimental Simulations

The experiment was conducted on a MacBook Pro machine with 2.8 GHz Intel Core i7 processor and 16 GB 1600 MHz DDR3 RAM. The machine's operating system is Mac OS X 10.9.4 and DPGT approach was implemented in C++ language execution environment. The total number of input features X is 10 and the number of input patterns is 50 for both training and test dataset. The number of fuzzy rules N_{fuz} is set to 5. The maximum search count N_{search} is set to 5. String decimal point precision is set to 2 decimal point. The BMA parameters are presented in Table 1. The mutation unit and the infection unit are set to membership function.

Table 1. Parameter setting for the proposed algorithm

N_{gen}	N_{bac}	N_{clones}	L_{ms}	N_{inf}	L_{is}	N_{iter}	τ	α
20	3	4	1	5	1	5	0.0001	1

The experimental result for the *best* bacterium in terms of Mean Square Error (MSE) is shown in Fig. 3a. Next, the experimental result for the *population average* is shown in Fig. 3b. Both experimental results' MSE values were calculated based on average of 10 trials. These experiments were conducted in the perspective of comparison between DPGT and benchmark operators. The proposed DPGT operator achieved overall lower MSE for both the best bacterium and the population average sections (See Fig. 3a and Fig. 3b). In addition to that, the proposed DPGT operator and its best bacterium can achieve lower MSE in earlier generations than the benchmark, thus the proposed method converges faster (See Fig. 3a). We observe this phenomenon as the good gene transfer between bacterial individuals are enforced by the DPGT operator that derived the

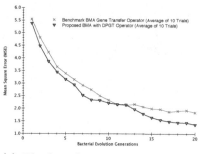
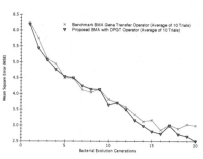

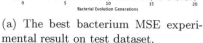

(a) The best bacterium MSE experimental result on test dataset.   (b) The population average MSE experimental result on test dataset.

Fig. 3. Experimental results

Table 2. Comparison

Experiment	GT	DPGT	Difference
Best MSE based on the average of 10 trials	1.843	1.369	34.62%
Population average MSE based on the average of 10 trials	2.988	2.502	19.42%
Best trial's best bacterium's MSE	1.460	1.133	28.86%
Best trial's population average MSE	2.425	1.924	26.04%

optimisation performance gain. Please refer to Table 2 for the experiment result comparison between original GT and DPGT operator.

The benchmark operator's average computational time based on 10 trials is 39.1 seconds and that of the DPGT operator is 41.7 seconds. Therefore, the DPGT operator has only an additional 6.65% computation cost over benchmark GT operator. Therefore, the proposed approach does not have significant impact to overall processing time. The time-complexity for the Algorithm 2 is $O(N_{bac} \times log(N_{bac}) \times N_{inf} \times L_{is} \times |a| \times |b| \times N_{search})$. On the other hand, the time-complexity for the original gene transfer operator is $O(N_{bac} \times log(N_{bac}) \times N_{inf} \times L_{is})$. The current limitations of DPGT approach is that the string decimal point precision parameter had to be manually configured and it is problem dependent.

6   Conclusion and Future Work

We proposed a novel DPGT operator that effectively addressed the artificial learning agent ant's perception-action problem. The proposed DPGT operator achieved overall best optimisation MSE performance results. Furthermore, the proposed DPGT operator also provided faster solution convergence in earlier bacterial evolution generations for its best bacterial individual. These performance gains were achieved without significantly impacting the training processing time.

In future work, we will try the DP approach in the bacterial mutation operator as well. We also intend to investigate the possibility to use DPGT operator on other problem domains. We will also consider the automatic parameter setting for string decimal point precision parameter.

References

1. Bellman, R.: Dynamic programming and Lagrange multipliers. Proceedings of the National Academy of Sciences of the United States of America 42(10) (1956)
2. Botzheim, J., Toda, Y., Kubota, N.: Path planning in probabilistic environment by bacterial memetic algorithm. In: Watanabe, T., et al. (eds.) Intelligent Interactive Multimedia: Systems & Services. Smart Innovation, Systems and Technologies, vol. 14, pp. 439–448. Springer, Heidelberg (2012)
3. Botzheim, J., Cabrita, C., Kóczy, L.T., Ruano, A.: Fuzzy rule extraction by bacterial memetic algorithms. Int. Journal of Intelligent Systems 24(3), 312–339 (2009)
4. Botzheim, J., Földesi, P.: Novel calculation of fuzzy exponent in the sigmoid functions for fuzzy neural networks. Neurocomputing 129, 458–466 (2014)
5. Eiben, A.E., Schoenauer, M.: Evolutionary computing. Information Processing Letters 82(1), 1–6 (2002)
6. Jiménez, J.A., Hughes, K.A., Alaks, G., Graham, L., Lacy, R.C.: An experimental study of inbreeding depression in a natural habitat. Science 266(5183), 271–273 (1994)
7. Land, M.F.: Visual acuity in insects. Annual review of entomology 42(1), 147–177 (1997)
8. Levenshtein, V.: Binary Codes Capable of Correcting Deletions, Insertions and Reversals. Soviet Physics Doklady, vol. 10 (1966)
9. Mamdani, E.H.: Application of fuzzy algorithms for control of simple dynamic plant. Proceedings of the Institution of Electrical Engineers, vol. 121 (1974)
10. Marquardt, D.W.: An algorithm for least-squares estimation of nonlinear parameters. J. of the Society for Industrial & Applied Mathematics 11(2), 431–441 (1963)
11. Nawa, N.E., Furuhashi, T.: Fuzzy system parameters discovery by bacterial evolutionary algorithm. IEEE Transactions on Fuzzy Systems 7(5), 608–616 (1999)
12. Sankoff, D.: Edit distance for genome comparison based on non-local operations. In: Proc. of the 3rd Annual Symposium on Combinatorial Pattern Matching, pp. 121–135 (1992)

Topological Gaussian Adaptive Resonance Associative Memory with Fuzzy Motion Planning for Place Navigation

Wei Hong Chin and Chu Kiong Loo

Faculty of Computer Science and Information Technology, University of Malaya,
Kuala Lumpur, Malaysia

Abstract. This paper presents a neural network for online topological map construction inspired by beta oscillations and hippocampal place cell learning. The memory layer represents the hippocampus, the input layer represents the entorhinal and the ρ is the orientation system. In our proposed method, nodes in the topological map represent place cells (robot location), while edges connect nodes and store robot action (i.e. orientation, direction). It comprises two layers: input and memory. The input layer collects sensory information and incrementally clusters the obtained information into a set of topological nodes. In the memory layer, the clustered information is used as a topological map where nodes are associated with actions. The advantages of the proposed method are: 1) it does not require high-level cognitive processes and prior knowledge to make it work in a natural environment; 2) it can process multiple sensory sources simultaneously in continuous space, which is crucial for real-world robot navigation; and 3) it is an incremental and unsupervised learning method. Thus, we integrate our Topological Gaussian Adaptive Resonance Associative Memory (TGARAM) with fuzzy motion planning to constitutes a basis for place navigation with little or no human intervention. Finally, the proposed method was validated using several standardized benchmark datasets and implemented to a real robot.

Keywords: Topological map, Adaptive Resonance Theory, Navigation, unsupervised learning.

1 Introduction

Autonomous mobile robots are able to move in a given environment and can perform desired tasks and navigate in unstructured environments with little or no human intervention.

A human-friendly autonomous guided robot knows at least a little about where it is and how to reach a particular location in order to achieve certain goals. Building the representation of map is crucial to autonomous navigation

C.K. Loo et al. (Eds.): ICONIP 2014, Part III, LNCS 8836, pp. 604–611, 2014.
© Springer International Publishing Switzerland 2014

and therefore a reliable map not only improves maintenance, but also map-based localization and path planning in any environment.

In mobile robotics, representations of the world are grouped into metric maps, topological maps and hybrid models that combine both metric and topological information [1]. In the metric mapping framework, the environment is represented as a set of objects with coordinates in a 2D space. The construction of the map is based on a grid occupancy or feature map approach [2]. However, this approach is limited by the high computational cost of feature matching. Pure metric map approaches are susceptible to inaccuracies in both map-making and the dead reckoning abilities of the robot.

In the topological framework, the environment is represented by a set of distinct places [3] and how a robot travels from one place to another. Places are defined by information gathered from sensors placed in the environment, which is stored in nodes. Some of the robot's odometry information (gathered while it travels from one place to another) is stored in the links of the map.

The first advantage of topological maps is that they do not require a metric sensor model to convert sensor data into a 2D frame of reference [4]. Furthermore, memorizing the environment as a set of places generates a discrete spatial layout that is suitable for higher-level processes, which may be very natural and similar to the places defined by humans (such as corridors and rooms). Therefore, a topological map mimics the way in which a human memorizes a map.

Another area of research has focused on emulating the biological systems thought to be the basis for mapping and navigation in animals. Early work with rats identified place cells in their hippocampus that appeared to respond to the animal's spatial location [5]. Other research has discovered that beta oscillations occur during the learning of hippocampal place cell receptive fields in novel environments [6]. Beta oscillations explain how place cells may be learned as spatially selective categories from the feedback between the entorhinal cortex and the hippocampus.

In this paper, we continue our previous work [8] by integrating Fuzzy Motion Planning from Yau-Zen Chang [7] to constitute a basis for robot place navigation in dynamic environment. The proposed learning model can simultaneously build and maintain the topological map and works in natural environments. The metric map does not need to be constructed in advance and because it is based on a simple mathematical model, our proposed method requires less computational power than any of the other map building methods mentioned above. Lastly, robot can perform fuzzy motion path planning based on the topological map. We utilized the fuzzy motion path planning because it can detect and avoid obstacles continuously, thus it can be used in dynamic environment.

The remainder of this paper is organized as follows. Section 2 introduces the construction of the online topological map using TGARAM. The experimental results of map building and place navigation are summarized in Section 3.2 and 3.3 and discussed in Section 4. Finally, Section 5 present some conclusions.

2   Proposed Method

2.1   System Architecture

To build the topological map, we divided the system into two layers, as shown in Figure 1.

In the robot navigation process, we divided the process into two steps, exploration step and path/action planning step.

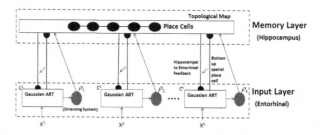

Fig. 1. TGARAM architecture and learning: The memory layer represents the hippocampus, the input layer represents the entorhinal and the ρ is the orientation system. Top-down attentive matching and mismatch-mediated reset (beta oscillations), which are triggered by the orientation system.

2.2   Building a Topological Map Online with TGARAM

Each node contains an input vector, V encoded from robot's odometer. It is defined as a Gaussian distribution, with mean, μ_j, standard deviation, σ_j in each dimension, and a priori probability as shown in equation 1. Such node definition is based solely on the robot's perceptual capacities and does not rely on a human definition of what a place is supposed to be. Equation 2, 3, 4 and 5 is used for node detection and recognition. Details of TGARAM can be refer to our previous publications [8].

$$G_j = \exp\left(-\frac{1}{2}\sum_{i=1}^{M}\left(\frac{x_i - \mu_{ji}}{\sigma_{ji}}\right)^2\right) \tag{1}$$

If the training involves K sensory channels, the match value, G_J for each node is:

$$
\begin{aligned}
G_j &= \sum_{k=1}^{K}\alpha^k[G_j^k] \\
&= \sum_{k=1}^{K}\alpha^k\left[\exp\left(-\frac{1}{2}\sum_{i=1}^{M}\left(\frac{x_i^k - \mu_{ji}^k}{\sigma_{ji}^k}\right)^2\right)\right]
\end{aligned} \tag{2}
$$

$$n_J = n_J + 1 \tag{3}$$

$$\mu_{Ji}^k = (1 - n_J^{-1})\mu_{Ji}^k + n_J^{-1}x_i^k \tag{4}$$

$$\sigma_{Ji}^k = \begin{cases} \sqrt{(1 - n_J^{-1})\sigma^2{}_{Ji}^k + n_J^{-1}(x_i^k - \mu_{Ji}^k)^2} & \text{if } n_J > 1 \\ \gamma & \text{otherwise} \end{cases} \tag{5}$$

2.3 Fuzzy Motion Planning

We modified and adapted the fuzzy motion planning strategy from Yau-Zen Chang [7] to enable robot navigation in any indoor environment.

Obtaining of the laser range finder is recorded in terms of the scanning angle, θ, and the distance between the obstacle and the robot, d_o . The target angle, denoted by ϕ, is the angle between the robot heading direction and the goal direction, viewed from the robot. This target angle is always obtained from the edges of topological map created by TGARAM.

The relation between the robot and a target node is represented as a target membership function, $\mu_{Target}(\theta)$:

$$\mu_{Target}(\theta) = \begin{cases} 0.1 + 0.9 \cdot \frac{\theta + 122}{\phi + 122} & \text{for } -120° \leq \theta < \theta_1 \\ 0.1 + 0.9 \cdot \frac{\theta - 122}{\phi - 122} & \text{for } \theta_1 \leq \theta < 120° \\ 0.1 & \text{for } \theta < -120° \text{ or } \theta > 120° \end{cases} \tag{6}$$

Next, the obstacle membership function is defined as $\mu_{Obs}(\theta)$. Instead of considering the size of the robot directly, each detected obstacle is amplified accordingly:

$$\mu_{Obs}(\theta) = \begin{cases} \exp[-0.5 \cdot (d_o - (r_c - d_s + \triangle_m))] & \text{when } \theta_1 \leq \theta \leq \theta_2 \\ 0 & \text{otherwise} \end{cases} \tag{7}$$

Where

d_o: distance between the obstacle and the robot.

r_C: offset based on size of the vehicle, usually defined as the radius of the smallest sphere that can cover the robot.

d_S: extra safety distance for tolerance of non-detectable obstacle elongations.

$\triangle_m$: maximum mobile distance during the sampling interval, which is 0.1 second for URG-04LX-UG01, of the laser range finder.

The traversable membership function, $\mu_{Traversable}(\theta)$, is merely the fuzzy complement of the obstacle membership function, $\mu_{Obs}(\theta)$:

$$\mu_{Traversable}(\theta) = 1 - \mu_{Obs}(\theta) \tag{8}$$

In our approach, the heading membership function, $\mu_{Heading}(\theta)$ is defined as the product of the traversable membership function, $\mu_{Traversable}(\theta)$ and the target membership function, $\mu_{Target}(\theta)$.

$$\mu_{Heading}(\theta) = \mu_{Traversable}(\theta) \cdot \mu_{Target}(\theta) \tag{9}$$

At each movement, the robot is commanded to move in the average direction that is calculated from the grade of $\mu_{Heading}(\theta)$.

3   Real Robot Implementation

We have conducted the following experiments using robot named as H20 equipped with laser range finder and vision-landmark base indoor localization sensor as shown in Figure 2.The experimental place as shown in Figure 3 is a meeting & lounge room, moderately populated during the day and empty at night. The size of the environment is 10 x 4 meters. The environment was by no means static, with moving people, re-arrangement of furniture and equipment and changing door states.

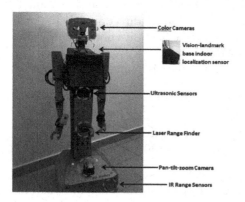

Fig. 2. Robot named H20 equipped with various sensors and cameras

Fig. 3. Photo of the environment

3.1   Experimental Design

Parameters for the robot system and algorithms were set as shown in Table 1. These parameter values have been obtained from enormous indoor experimentation and are usable for most typical indoor environments.

Table 1. TGARAM and fuzzy motion path planning parameters

TGARAM Parameter	Value	Fuzzy Motion Parameter	Value
γ	4	r_c	0.5
ρ_1, ρ_2	0.7	d_s	0.2
α_1, α_2	0.5	$\triangle_m$	0.1
e_{max}	0.5		

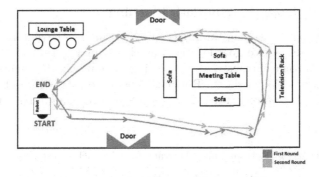

Fig. 4. H20 Navigation Path

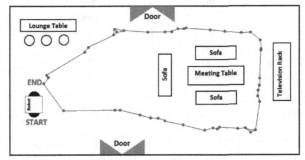

Fig. 5. Topological map for the first run

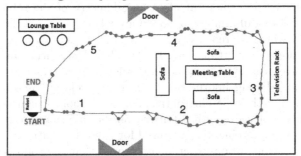

Fig. 6. Topological map for the second run

3.2   Results: Topological Map Construction

Figure 4 shows the robot navigation path. the nodes are plotted as a blue color circle at the (x, y) coordinates in the stored position. Linked nodes are joined with a black line. The topological map has only 55 nodes in the first run as shown in Figure 5. In the second run, the map is updated and has 76 nodes as shown in Figure 6. The map is not absolutely accurate in a Cartesian sense, but need not be for effective navigation.

3.3   Results: Fuzzy Motion Path Planning

Fuzzy motion path planning was assessed by examining the goal-reaching rate. We randomly select five goal locations as shown in Figure 4 for verification. During place navigation, obstacles are placing randomly in the environment and pedestrians are walking around. Robot is capable to avoid these obstacles and navigate to goal locations. Table 2 shows the goal reaching rates for each goal location.

Table 2. Path planning accuracy rate for each goal location

Location	Trials	Success	Accuracy
1	20	20	100%
2	20	18	90%
3	20	18	90%
4	20	17	85%
5	20	19	95%

4   Discussion

We have shown that the TGARAM algorithm we developed is able to construct a topological map from scratch using nearly unprocessed sensory information about the environment. The TGARAM training process not only takes into account sensor measurements but also the odometry of the robot. Therefore, TGARAM is able to disambiguate locations where the sensorial information is very similar, which overcomes problem of online detection and recognition of topological nodes. Unfortunately, the TGARAM algorithm relies on access to comprehensive sensory information in order to generate and maintain the topological map. Missing information as well as irrelevant information would significantly degrade the reliability of the map. The topological map is structured into a set of linked locations with particular sensory information, which provides the flexibility and maintainability required for robot navigation.

The robot implementation experiments proved that the topological map enables robot to perform place navigation. The robot first obtained the start and goal destination on the map. Next, the robot starts moving to the destination goal using fuzzy motion planning. During robot moving, pedestrians are randomly moving in the environment. Robot is avoiding pedestrians and using the topological map information to continue moving to destination points. The experiment results show that the successful rate of goal reaching is approximately 90%.

5   Conclusion

In summary, we proposed a biologically-inspired online learning method for topological map building. It is an incremental and unsupervised learning method and able to produce a consistent and stable topological map in natural environments without prior knowledge of the environment. Thus, TGARAM is integrated with fuzzy motion planning enables robot to perform place navigation in dynamic environment. Therefore, TGARAM constitutes a basis for robot navigation in any environment.

In future work, image features should be used as input for our proposed learning method to improve the online detection and recognition and also increase the reliability of the topological map.

References

1. Tomatis, N., Nourbakhsh, I.R., Siegwart, R.: Hybrid simultaneous localization and map building: a natural integration of topological and metric. Robotics and Autonomous Systems 44, 3–14 (2003)
2. Leivas, G. Botelho, S. Drews, P. Figueiredo, M.: Sensor fusion based on multi-self-organizing maps for SLAM. In: Multisensor Fusion and Integration for Intelligent Systems (MFI) (2010)
3. Kuipers, B., Byun, Y.-T.: A Robot Exploration and Mapping Strategy Based on a Semantic Hierarchy of Spatial Representations. Journal of Robotics and Autonomous Systems 8, 47–63 (1991)
4. David, F.: Map-based navigation in mobile robots: I. A review of localization strategies: Cognitive Systems Research 4, 243–282 (2003)
5. O'Keefe, J., Dostrovsky, J.: The hippocampus as a spatial map. Preliminary Evidence from Unit Activity in the Freely-Moving Rat: Brain Research 34, 171–175 (1971)
6. Berke, J., Hetrick, V., Breck, J., Greene, R.: Transient 23-30 Hz oscillations in mouse hippocampus during exploration of novel environments. Hippocampus 18, 519–529 (2008)
7. Chang, Y.-Z., Huang, R.-P., Chang, Y.-P.: A Simple Fuzzy Motion Planning Strategy for Autonomous Mobile Robots. In: Industrial Electronics Society, pp. 477–482 (2007)
8. Chin, W.H., Loo, C.K.: Biologically Inspired Topological Gaussian ARAM for Robot Navigation. In: Robot, Vision and Signal Processing (RVSP) (2013)

Anomaly Based Intrusion Detection through Temporal Classification

Shih Yin Ooi[*], Shing Chiang Tan, and Wooi Ping Cheah

Faculty of Information Science and Technology, Multimedia University
Jalan Ayer Keroh Lama, 75450 Melaka, Malaysia
{syooi,sctan,wpcheah}@mmu.edu.my

Abstract. Many machine learning techniques have been used to classify anomaly-based network intrusion data, encompassing from single classifier to hybrid or ensemble classifiers. A nonlinear temporal data classification is proposed in this work, namely Temporal-J48, where the historical connection records are used to classify the attack or predict the unseen attack. With its tree-based architecture, the implementation is relatively simple. The classification information is readable through the generated temporal rules. The proposed classifier is tested on 1999 KDD Cup Intrusion Detection dataset from UCI Machine Learning Repository. Promising results are reported for denial-of-service (DOS) and probing attack types.

Keywords: anomaly-based intrusion detection, machine learning, temporal classification, temporal decision tree, temporal sequences.

1 Introduction

With the rapid development of networking and interoperation in public networks, the volumes as well as the sophistication of computer network security threats have been significantly increased. Referring to the Table 1 on the compiled data from National Vulnerability Database (NVD) [1], almost 13 new vulnerabilities were reported per day in year 2013, which is the highest number for the past five years.

Table 1. Reported Vulnerabilities by NVD from 2009-2013

Year	Number of Reported Vulnerabilities
2009	4783
2010	4258
2011	3532
2012	4347
2013	4794

---

[*] Corresponding author.

C.K. Loo et al. (Eds.): ICONIP 2014, Part III, LNCS 8836, pp. 612–619, 2014.
© Springer International Publishing Switzerland 2014

Many researchers mitigated the aforementioned threats through various prevention methods, including but not limited to, authentication (i.e.: examining the complexity of passwords, identification and verification through smart devices, or biometrics), reducing the programming and configuration errors, protecting the information through encryption (i.e.: cryptography), improving the firewall architecture design, etc. Apart from these prevention methods, detection methods are also very essential to provide another layer of security mechanism on modern systems. Intrusion detection system (IDS) is one of these kinds. IDS can be defined as a system to detect any illegal or suspicious attempts to penetrate the confidentiality, integrity, and availability of the network resources. It works by monitoring all events occurring in a network (network-based IDS) or even in a single machine (host-based IDS). When an abnormal activity is observed, IDS will issue an alarm, as well as logging the report.

In general, IDS can be categorized into two categories: (1) misuse detection (or better known as signature detection), and (2) anomaly detection. Misuse detection was widely used over the decades, especially in the commercialized IDS. It is very reliable in terms of, it never miss in detecting the intrusions which their behavior patterns (or better known as "signatures") have been filed in its signature database. However, the failure rate for it to detect a "zero-day attack" is extremely high. In other words, the reliability of misuse detection IDS is heavily depending on its signature database (pattern matching approach), and lacking the ability to discover and predict new attacks. On the other hand, anomaly detection IDS is able to automatically build a profile of normal event behaviors, and to flag the abnormal event behaviors when they are deviated from the normal one. This can be done through machine learning approach, i.e.: event classification (normal vs. abnormal) as well as prediction of new attacks. However, there are so many variations of attack patterns, and it is almost impossible to build a "perfect" learning model. Thus, the reliability issues are still an open research question to address. In this work, we propose a new nonlinear temporal classifier through the decision tree approach in modeling the anomaly detection IDS.

The organization of this paper is as follows. In Section 2, some related literature reviews on the anomaly detection methods are presented. In Section 3, the methodology of this work is justified. In Section 4, some experimental results and benchmark comparisons on the proposed method are discussed. Lastly, Section 5 concluded our work.

2   Related Works

Many machine learning techniques have been applied in anomaly detection systems [2]. Some researchers adopted single learning techniques such as neural networks [3], support vector machines (SVM) [4] and decision tree [5]. Some claimed that single classifier itself is a weak learner, thus, they used hybrid and ensemble techniques [6]. The winner [7] and the first runner-up [8] for KDD Classification Cup 1999 were using hybrid techniques. They [7] combined the usage of bagging and boosting, and managed to record the highest overall classification accuracy in the competition; while [8] used an ensemble method of decision forest – Kernel Miner. Later, [9] proposed to reduce the data dimension by reducing the number of features (attributes) and rules based on association rules. [9] reported that this method is effective in reducing false alarm. [10] proposed a hybrid model (fuzzy inference system) based on neuro-fuzzy classifiers and genetic algorithm (ESC-IDS), and they managed to optimize the prediction by using small number of training instances. [11] presented a novel anomaly detection technique (with the combination usage of k-means clustering, Naïve Bayes feature selection and C4.5 decision tree) especially to detect the "zero-day attack". The reported results were tested on their independent database. Among the aforementioned techniques, SVM is most widely exploited by the researchers, i.e.: [12] proposed a hybrid detection system by using hierarchical clustering as feature selector, and SVM as classifier, while [13] mined the intrusion data through the combination usage of SVM and ant colony networks. However, the nature of SVM is not fit to large-scale data set, thus rendering its usage on real application. Most of these classifiers are presented with high performance accuracy. However, to our best knowledge, there are no research teams exploited on this field by considering temporal order of the connection records. Most of the network attacks may not happened in a single connection, but they could be a sequential attacks, or could be appeared in multiple connection records, i.e. denial-of-service (DOS), and probing. Therefore, we see the potential to apply the temporal framework in this context.

3   Methodology

3.1   Temporal Tree

In the temporalization procedure, a temporalized record can be formed by merging consecutive connection records in a network intrusion dataset by using a time window of w. In other words, the connection records are flatten to form a new connection record with w time steps (w observations). A time label will be set for the same attributes (features) from the subsequent connection records, so that they start at 1 and end in w. The assumption with bigger window values indicates a longer delay in events' effects (i.e.: when the real attack happens). By arranging the connection records in such way, the non-temporal and ordinary classifiers such as C4.5 [14] can be used [15].

In this work, the temporal tree is modified based on the latest research version of C4.5 approach, which is better known as C4.8 and implemented as J48 in Weka package [16]. Instead of using the corrected gain ratio criteria as in the original work [17], we modify the splitting criteria in such a way that: if a segment of data is missing in one of the two temporal series, we will ignore that piece during information gain calculation. The rationale of doing this is that, we can avoid the node to have an unknown value during classification.

The best-performed window size is subject to the network intrusion dataset or it can be determined by the IDS expert. If there is no temporal effect, then the window size will set to 1. However, due to the lack of industry experts in our case, we let the algorithm run recursively on the different window sizes (start from $w = 1$ until $w = $ max) until the error rates stop growing, the window size at this stage will be deemed as the most appropriate window size. The chosen window size implies the expected number of time steps for a previous connection record to have effect on current connection record's traffic. Ideally, by merging the subsequent connection records into a single one, the possible causes and effects can be brought together. To understand how the temporalization algorithm works, consider a dataset with 4 temporally consecutive records, each with 3 attributes. The last attribute is the decision attribute. By flatten these records using a window size of 2, the temporalized records can be generated as shown in Table 2.

Table 2. An example of temporalization

Original Records (w=1)			Temporalized Records (w=2)					
x	y	Z	$x(t1)$	$y(t1)$	$z(t1)$	$x(t2)$	$y(t2)$	$z(t2)$
0	0	0	0	0	0	1	0	0
1	0	0	1	0	0	1	1	1
1	1	1	1	1	1	0	0	0
0	0	0						

For J48 (or C4.8), the last variable (z in this case) will always be deemed as decision attribute. One of the rules that can be extracted from the above example (for original records) would be: if $\{(x = 1) \text{ AND } (y = 1)\}$ then $z = 1$. To adopt this in temporal fashion, we modified the J48 algorithm in such a way that, all attributes must carry along their respective time label based on the selected window size, w. This is to ensure that during the step of choosing the best condition attribute for splitting, the time criterion (temporal order) will also be considered. For an instance, if a condition attribute with time label $t1$ is used at a node, then its children can only use the condition attributes with a time label $t2 >= t1$. In this way, the temporal rule extracted will be: if $\{(x_{t1} = 1) \text{ AND } (y_{t1} = 1) \text{ AND } (z_{t1} = 1) \text{ AND } (x_{t2} = 0) \text{ AND } (y_{t2} = 0)\}$ then $z_{t2} = 0$, where z_{t2} is the decision attribute.

Temporal Tree Algorithm
Input: D, a network intrusion dataset;
                    Maximum window size, w_{max}.
Learning scheme: Temporal-J48.
Output: a temporal decision tree with a set of temporal rules.

Learning Procedure:
    1.   **Generate** a new training dataset, D_{train} from a given dataset, D.
    2.   **for** $w = 1$ to max **do**
    3.   **If**: all the cases in D_{train} belong to the same class or D_{train} is small, label the leaf of the tree with the most frequent class in D_{train} and their respective time-label based on w.
            Else:
            Choose a test based on a single attribute with two or more outcomes. Make this test the root of the tree with one branch for each outcome of the test, partition D_{train} into corresponding subsets, D_{train1}, D_{train2} ... according to the

outcome for each case, and apply the same procedure recursively to each subset.

4. **Rank** the possible test by using the information gain and
the default gain ratio (considering both suitability and
temporal sequence). The default gain ratio is used to divide the information gain by the information from the
test outcomes. All missing values will be ignored.

5. **Determine** the upper limit of the binomial probability by
using a user-specified confidence or using the default
value of 0.25.

6. **Select** a subset of temporal rules for each class, and
choose the default class.

7. **Derive** the training model.

8. **Classify** each example in D_{train} by using the derived training model and record the performance accuracy.

9. **endfor**

10. **Choose** the best window size with highest performance accuracy, w_{best}.

Testing Procedure:

11. **Generate** a new testing dataset, D_{test} from a given dataset,
D.

12. **Set** window size = w_{best} from step 10.

13. **Classify** the test or unseen examples using the derived
training model from learning procedure.

The discovery of temporal relations within the selected w time steps for a network
intrusion dataset has been made possible and easily interpretable through the generated temporal rules. The transparent rules are very useful for the field experts to identify
and interpret the trends and patterns from the historical connection records.

4   Experimental Evaluation

To evaluate the proposed method, we use the *1999 KDD Cup Intrusion Detection*
dataset (it is the only benchmark dataset for IDS despite its obsolete attack types for
modern network attacks) from UCI Machine Learning Repository [18], which recorded a wide variety of intrusions simulated in a typical U.S. Air Force LAN (local area
network) environment. The dataset contains a set of training and testing dataset re-
spectively: (i) training set with 4,898,431 connection records (972,781 normal |
3,925,650 attack) collected in 7 weeks, and (ii) testing set with 311,029 connection
records (60,593 normal | 250,436 attack) collected in 2 weeks; both with 41 attributes.
Attack types are further labeled into four main categories, which are (i) DOS, i.e.:
synchronization (SYN) flood, (ii) R2L, i.e.: password guessing, (iii) U2R, i.e.: buffer
overflow, and (iv) probing, i.e.: port scanning. Several specific attack types in the
testing dataset are intentionally excluded from the training dataset. This is to make the
identification task more challenging and realistic, i.e.: to determine how well the proposed algorithm can identify a new attack. Each connection record represents a sequence of TCP packets starting and ending within the well-defined times and protocol. This dataset is discrete and continuous.

In this work, all training data and testing data are utilized. To examine the robustness of the proposed classification algorithm, we intentionally do not remove any attributes (41 in this case). The connection records are sorted based on the destination host, thus it is possible to observe if there is any temporal relationship among the connection records (i.e. an attack targeted on a host is caused immediately, or the effect could be possibly delayed?).

4.1 Experimental Results

According to the prediction results as shown in Fig. 1, the robustness of the classifier based on all 41 attributes is considered promising. By using the proposed algorithm of Temporal-J48, the result for window size = 1 (no temporal value) is better than window size > 1 in the context of normal traffic, U2R and R2L, indicating there could possibly no temporal orders exist in these three traffic types. However, we observed that the predictive accuracy for DOS and probing are improving when the window size is > 1. The accuracy slightly reduce when $w = 2$, and 3, this is because when the window size is > 1 (when temporal value is considered), the tree tends to become smaller and this leads to the increasing of the tree's error rate. However, we observed that the error rates are reduced while moving from window size of 4 to 6 (when more windows are considered). This is very interesting because it implies that there are relationships between the traffic of current connection and the traffic of previous connections. The performance accuracy of Temporal-J48 on DOS (when $w_{best} = 6$) and probing (when $w_{best} = 7$) are even higher than an ordinary J48 (when $w = 1$). Both of these scenarios substantiated our argument on the delayed temporal effects, which is makes sense as in DOS can be caused by SYN packets flooding, and probing can be caused by multiple scanning attempts (both to be reflected by multiple connection records, and not in a single connection record).

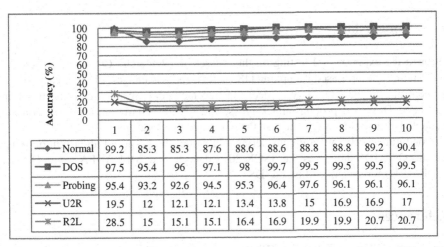

	1	2	3	4	5	6	7	8	9	10
Normal	99.2	85.3	85.3	87.6	88.6	88.6	88.8	88.8	89.2	90.4
DOS	97.5	95.4	96	97.1	98	99.7	99.5	99.5	99.5	99.5
Probing	95.4	93.2	92.6	94.5	95.3	96.4	97.6	96.1	96.1	96.1
U2R	19.5	12	12.1	12.1	13.4	13.8	15	16.9	16.9	17
R2L	28.5	15	15.1	15.1	16.4	16.9	19.9	19.9	20.7	20.7

Fig. 1. Prediction results of Temporal-J48 for normal, DOS, Probing, U2R, and R2L traffic types on windows size, $w = 1, 2, 3, 4, 5, 6, 7, 8, 9,$ and 10 respectively

However, due to the insufficient data for U2R (52 connection records) and R2L (1,126 connection records) in the training dataset, the performances of Temporal-J48 on these two attacks are relatively poor. Unlike the DOS and probing attack types, R2L and U2R attack types normally occur within a single connection. This implying that there is no temporal patterns exist among their connection records. Thus, Temporal-J48 could not outperform in these cases.

4.2   Performance Comparison

In order to validate the robustness of Temporal-J48 in the network intrusion classification, we compared it against another five research works as shown in Table 3. All of them were using the same experimental set-up as ours, i.e.: by utilizing the full set of original training and testing data. The results from the proposed method are very encouraging. As can be seen from the Table 3, Temporal-J48 can perform better than others when dealing with DOS and Probing.

Table 3. Identification accuracy of Temporal-J48, hybrid method of Hierarchical Clustering & SVM, ESC-IDS, Association Rules, KDD99 Winner and KDD99 Runner-Up for 1999 KDD Cup Intrusion Detection dataset

	Temporal-J48	Hierarchical Clustering + SVM [12]	ESC-IDS [10]	Association Rules [9]	KDD99 Winner [7]	KDD99 Runner-Up [8]
Normal	99.2	99.3	98.2	99.5	99.5	99.4
DOS	99.7	99.5	99.5	96.8	97.1	97.5
Probing	97.6	97.5	84.1	74.9	83.3	84.5
U2R	19.5	19.7	14.1	3.8	13.2	11.8
R2L	28.5	28.8	31.5	7.9	8.4	7.3

5   Conclusion

Through the experimental testing on the dataset of *1999 KDD Cup Intrusion Detection* dataset, the proposed Temporal-J48 spells three advantages: (1) it is easy to implement due to the simplicity of decision tree methodology, (2) higher accuracy in predicting DOS and probing attack types, and (3) all classification information are readable by the field experts through the generated temporal rules.

For future works, we are considering evaluating the proposed algorithm on other recent public IDS databases, i.e. ISCX 2012 dataset, and 2014 ADFA Linux Dataset (ADFA-LD) for system call based IDS.

References

1. National Vulnerability Database [NVD]: http://nvd.nist.gov
2. Tsai, C.-F., Hsu, Y.-F., Lin, C.-Y., Lin, W.-Y.: Intrusion Detection by Machine Learning: A Review. Expert Systems with Application 36, 11994–12000 (2009)

3. Joo, D., Hong, T., Han, I.: The Neural Network Models for IDS Based on the Asymmetric Costs of False Negative Errors and False Positive Errors. Expert Systems with Application 25, 69–75 (2003)
4. Zhang, Z., Shen, H.: Application of Online-Training SVMs for Real-Time Intrusion Detection with Different Considerations. Computer Communications 28, 1428–1442 (2005)
5. Stein, G., Chen, B., Wu, A.S., Hua, K.A.: Decision Tree Classifier for Network Intrusion Detection with GA-Based Feature Selection. In: Proceedings of the 43rd Annual Southeast Regional Conference, vol. 2, pp. 136–141 (2005)
6. Peddabachigari, S., Abraham, A., Grosan, C., Thomas, J.: Modeling Intrusion Detection System Using Hybrid Intelligent Systems. Journal of Network and Computer Applications 30, 114–132 (2007)
7. Pfahringer, B.: Winning the KDD99 Classification Cup: Bagged Boosting. KDD 1999 1(2), 65–66 (2000)
8. Levin, I.: KDD-99 Classifier Learning Contest LLSoft's Results Overview. SIGKDD Explorations 1(2), 67–75 (2000)
9. Xuren, W., Famei, H., Rongsheng, X.: Modeling Intrusion Detection System by Discovering Association Rule in Rough Set Theory Framework. In: Proceedings of the International Conference on Computational Intelligence for Modeling Control and Automation, and International Conference on Intelligent Agents, Web Technologies and Internet Commerce (CIMCA-IAWTIC), p. 24 (2006)
10. Toosi, A.N., Kahani, M.: A New Approach to Intrusion Detection Based on an Evolutionary Soft Computing Model Using Neuro-Fuzzy Classifiers. Computer Communications 30, 2201–2212 (2007)
11. Louvieris, P., Clewley, N., Liu, X.: Effects-Based Feature Identification for Network Intrusion Detection. Neurocomputing 121, 265–273 (2013)
12. Horng, S.-J., Su, M.-Y., Chen, Y.-H., Kao, T.-W., Chen, R.-J., Lai, J.-L., Perkasa, C.D.: A Novel Intrusion Detection System Based On Hierarchical Clustering and Support Vector Machines. Expert Systems with Applications 38, 306–313 (2011)
13. Feng, W., Zhang, Q., Hu, G., Huang, J.X.: Mining Network Data for Intrusion Detection through Combining SVMs with Ant Colony Networks. Future Generation Computer Systems 37, 127–140 (2014)
14. Quinlan, J.R.: Induction of Decision Trees. Machine Learning 1(1), 81–106 (1986)
15. Karimi, K., Hamilton, H.J.: Temporal Rules and Temporal Decision Trees: A C4.5 Approach. Technical Report CS-2001-02, Department of Computer Science, University of Regina, Canada (2001)
16. Hall, M.A., Frank, E., Holmes, G., Pfahringer, B., Reutemann, P., Ian, H.W.: The WEKA Data Mining Software: An Update. SIGKDD Explorations 11(1) (2009)
17. Quinlan, J.R.: Unknown Attribute Values in Induction. In: Segre, A. (ed.) Proceedings of the 6th International Machine Learning Workshop Cornell. Morgan Kaufmann (1989)
18. Bache, K., Lichman, M.: UCI Machine Learning Repository. University of California, School of Information and Computer Science, Irvine, CA (2013),
   http://archive.ics.uci.edu/ml

Threshold Visual Secret Sharing Based on Boolean Operations and Random Grids

Xuehu Yan, Shen Wang*, and Xiamu Niu

School of Computer Science and Technology,
Harbin Institute of Technology, 150080 Harbin China
`shen.wang@hit.edu.cn`

Abstract. Visual secret sharing (VSS) by random grids (RG) has gained much attention since it can avoid the pixel expansion problem without codebook design. However, all of the reported RG-based VSS schemes fail to fit normal color representation method of digital images, which leads to additional computational time to flip the bits to fit with digital applications. In this paper, the same color representation method of VSS as normal digital images is exploited. Based on the same color representation, a RG-based VSS based on Boolean operations is given. Experimental results show the effectiveness of the exploited color representation and extended scheme.

Keywords: Visual secret sharing, Boolean operations, Random grid, Color representation.

1 Introduction

Secret image sharing techniques are important in protecting secret images in addition to traditional cryptography. It has attracted more attention to scientists and engineers. Visual secret sharing (VSS) is an efficient branch of secret sharing techniques, since VSS reveals the secret based on human visual system (HVS) without cryptographic computation. In (k, n) threshold-based VSS scheme, a binary secret image is shared by generating n noise-like shadow images. Any k or more noise-like shadow images are superposed to obtain the secret image based on HVS. Random grid (RG)-based VSS has no the pixel expansion and codebook design, hence it is a primary method of VSS. The basic model of RG-based VSS for binary secret images was introduced by Kafri and Keren [1]. This model forms the basic structure for various VSS schemes proposed in the literature. The secret image is generated into two noise-like RGs (shadow images or share images) that have the same size as the original secret image. Unfortunately, Kafri and Keren's model only supports (2, 2) threshold. Various efforts [2] were proposed to address the issue of Kafri and Keren's VSS, (k, n) threshold is still not satisfied. Chen et al. [3] proposed a (k, n) RG-based VSS based on Boolean operations (XOR and OR operations). Inspired Chen et al.'s method [3], some approaches [4]

---

* Corresponding author.

C.K. Loo et al. (Eds.): ICONIP 2014, Part III, LNCS 8836, pp. 620–627, 2014.

were given to improve the visual quality. However, the reported RG-based VSS suffers from different color representation method of normal digital images [5]. For digital images and common digital processing software in digital application, such as binary BMP and Matlab, the normal color representation is "0" denotes black or opaque and "1" denotes white or transparent. While in reported RG-based VSS, "1" denotes black pixel, "0" denotes white pixel, which will lead to additional computation for flipping, reversing or complementing operations in digital applications, i.e., $0 \rightarrow 1$ or $1 \rightarrow 0$.

Motivated by exploiting more and better properties for RG-based VSS schemes, in this paper, the same color representation method of VSS as digital images is exploited, and the recovery method is also based on stacking (Boolean AND operation) and HVS. Based on the same color representation, a new threshold VSS scheme based on Boolean operations (XOR and AND operations) and RG is extended through applying the same color representation method as digital images that other previous schemes fail to have. The extended scheme has several properties such as (k, n) threshold, simple computation (stacking recovery), alternative order of shadow images in recovery, no codebook design, and avoiding the the pixel expansion. Experimental results show the effectiveness of the exploited color representation method and extended RG-based VSS.

The rest of the paper is organized as follows. Section 2 introduces the preliminary techniques. The exploited color representation method and extended scheme are presented in Section 3. Section 4 is devoted to experimental results. Finally, Section 5 concludes the paper.

2   Preliminary Techniques

This section gives preliminaries of a well-known RG-based [3] VSS. In what follows, "$\oplus$" "&" and "$\otimes$" denote Boolean XOR, AND and OR operation. The binary secret image denoted as S with pixel value $S(i, j), 1 \leq i \leq M, 1 \leq j \leq N$, is shared among n ($n \geq 2, n \in Z^+$) shares, and the reconstructed secret image S' is reconstructed from t ($1 \leq t \leq n, t \in Z^+$) shares. The generation and recovery phases of RG-based [3] VSS are described as follows. Here "1" denotes black pixel, "0" denotes white pixel.

Step 1: Randomly generate 1 RG SC_1

Step 2: Compute SC_2 as in Eq. (1), where $\overline{SC_1(i, j)}$ is a bit-wise complementary operation.

Recovery: $S' = SC_1 \otimes SC_2$ as in Eq. (2), where $\otimes$ denotes the Boolean OR operation. If a certain secret pixel of $S(i, j)$ is 1, the recovery result $SC_1 \otimes SC_2 = 1$ is always black. If a certain secret pixel of $S(i, j)$ is 0, the recovery result $SC_1 \otimes SC_2 = SC_1(i, j) \otimes SC_1(i, j)$ has half chance to be black or white since SC_1 are generated randomly

$$SC_2(i, j) = \begin{cases} SC_1(i, j) & if\ S = 0 \\ \overline{SC_1(i, j)} & if\ S = 1 \end{cases} \tag{1}$$

$$S'(i,j) = SC_1(i,j) \otimes SC_2(i,j) =$$
$$\begin{cases} SC_1(i,j) \otimes SC_1(i,j) & if \ S(i,j) = 0 \\ SC_1(i,j) \otimes \overline{SC_1(i,j)} = 1 & if \ S(i,j) = 1 \end{cases} \tag{2}$$

The same approach can be extended to (k,n) threshold scheme by applying the above process repeatedly for the first k bits and generating the last $n - k$ bits randomly. However, the color representation of the scheme is different from normal color representation method of digital images, which is not convenient in digital applications

In fact, in the generation phase, Eq. (1) is equal to Eq. (3).

$$sc_2 = sc_1 \oplus s \quad or \quad s = sc_1 \oplus sc_2 \tag{3}$$

Since if $s = 0 \Rightarrow sc_2 = sc_1 \oplus 0 \Rightarrow sc_2 = sc_1$, and if $s = 1 \Rightarrow sc_2 = sc_1 \oplus 1 \Rightarrow$ $sc_2 = \overline{sc_1}$. And the equations will be used in the extended scheme The same approach could be extended to

$$s = sc_1 \oplus sc_2 \oplus \cdots \oplus sc_k \tag{4}$$

In the recovery phase, in a similar way, the recovered secret bit by stacking any $t(k \leq t \leq n)$ bits of shadow images satisfies to:

$$s = sc_1 \otimes sc_2 \otimes \cdots \otimes sc_t \tag{5}$$

3   The Exploited Color Representation and Extended Scheme

In this section, we exploit the same color representation method as digital images and give a threshold RG-based VSS scheme applying the exploited color representation based on Boolean operations (XOR and AND operations). In the extended scheme, the binary secret image is shared among n shadow images RG, while the secret image is recovered from t $(2 \leq t \leq n)$ shadow images. Here "1" denotes white pixel, "0" denotes black pixel, which is the same as normal color representation method of digital images.

3.1   Exploited Color Representation

We first give an $(2,2)$ threshold scheme as an example to show the idea of the exploited color representation. The generation and recovery phases are described as follows. The recovery is also based on stacking or HVS with no cryptographic operation.

Step 1: Randomly generate 1 RG SC_1.

Step 2: Compute SC_2 as in Eq. (6), where $\overline{SC_1(i,j)}$ is a bit-wise complementary operation.

Recovery: $S' = SC_1 \& SC_2$ as in Eq. (7). If a certain secret pixel of $S(i,j)$ is 0, the recovery result $SC_1 \& SC_2 = 0$ is always black. If a certain secret pixel of $S(i,j)$ is 1, the recovery result $SC_1 \& SC_2 = SC_1(i,j) \& SC_1(i,j)$ has half chance to be black or white since SC_1 are generated randomly.

$$SC_2(i,j) = \begin{cases} SC_1(i,j) & if\ S(i,j) = 1 \\ \overline{SC_1(i,j)} & if\ S(i,j) = 0 \end{cases} \tag{6}$$

$$S'(i,j) = SC_1(i,j)\&SC_2(i,j) =$$
$$\begin{cases} SC_1(i,j)\&SC_1(i,j) & if\ S(i,j) = 1 \\ SC_1(i,j)\&\overline{SC_1(i,j)} = 0 & if\ S(i,j) = 0 \end{cases} \tag{7}$$

Here, we explain why the recovery of above example is also stacking or HVS.

Visual cryptography is a secret sharing technique that allows a "visual" reconstruction of the secret. That is, participants have to be able to simply stack the shares (printed on transparencies) to recover the secret. And the HVS could be performed after the stacking operation with no cryptographic operation.

From the generation and recovery phases of traditional (2, 2) RG-based VSS [3], we can present the idea of RG-based (2, 2) VSS in Figure 1, a certain pixel of the secret image is generated into a white or black subpixel in each of the two shadow images. Here "1" denotes black pixel, "0" denotes white pixel. The subpixels are randomly selected from the two columns tabulated under the certain secret pixel, which lead to certain secret pixel that is encoded into two subpixels with the same probabilities (50%). Based on this, an individual shadow image gives no information about the secret image. When the subpixels are stacked, the opaque (black) pixels will cover the transparent (white) pixels. The black secret pixel will be decoded into black pixel, and the white secret pixel into white pixel or black pixel with the same probabilities (50%). Different stacking results and probabilities will lead to the contrast. The secret could be revealed by HVS when contrast is greater than 0. Thus, if the secret image could be recovered with contrast by stacking the two shadow images together, then, the secret image will be revealed by HVS with no cryptographic operation.

Remark that, the stacking operation results of traditional (2, 2) RG-based VSS are the same as Boolean OR operation of 0 or 1. Thus, in traditional VSS, the stacking is corresponding to Boolean OR operation.

The idea of the proposed (2, 2) RG-based VSS is presented in Figure 2, where "1" denotes white pixel, "0" denotes black pixel, which are the same as color representation method of digital images. When stacking the two shadow images, the opaque (black) pixels will also cover the transparent (white) pixels. The black secret pixel will be decoded into black pixel, and the white secret pixel into white pixel or black pixel with the same probabilities (50%), thus the contrast is also introduced. So, the secret image will be revealed by HVS with no cryptographic operation.

While, the stacking operation results of the proposed (2, 2) RG-based VSS are the same as Boolean AND operation of 0 or 1. Thus, the stacking operation in the extended scheme is corresponding to Boolean AND operation.

Based on the above discussion, both traditional (2, 2) RG-based VSS and the proposed (2, 2) RG-based VSS are based on stacking. The extended scheme uses the AND operation for the recovery the recovery also can be performed by the stacking or HVS. Hence the proposed (2, 2) RG-based VSS in fact is visual.

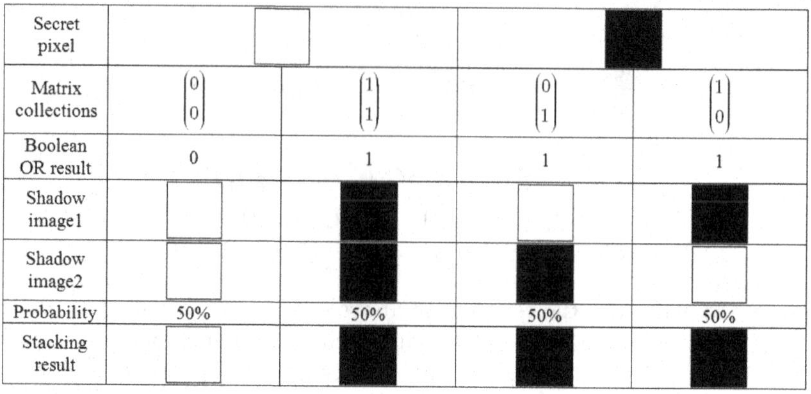

Secret pixel	(white)		(black)	
Matrix collections	$\binom{0}{0}$	$\binom{1}{1}$	$\binom{0}{1}$	$\binom{1}{0}$
Boolean OR result	0	1	1	1
Shadow image1				
Shadow image2				
Probability	50%	50%	50%	50%
Stacking result				

Fig. 1. The idea of traditional (2, 2) RG-based VSS

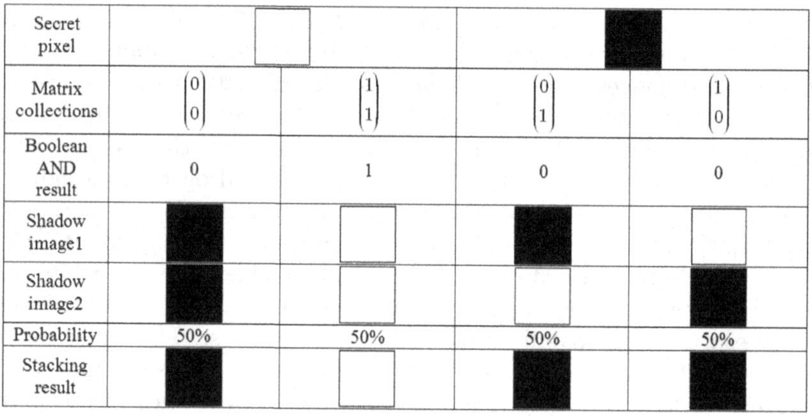

Secret pixel	(white)		(black)	
Matrix collections	$\binom{0}{0}$	$\binom{1}{1}$	$\binom{0}{1}$	$\binom{1}{0}$
Boolean AND result	0	1	0	0
Shadow image1				
Shadow image2				
Probability	50%	50%	50%	50%
Stacking result				

Fig. 2. The idea of the exploited (2, 2) RG-based VSS

3.2 Extended Scheme

Based on the above exploited color representation and scheme in [3], a (k, n) RG-based VSS with the same color representation as digital images is extended. The shadow image generation algorithmic steps are described in Algorithm 1.

The secret recovery of the extended scheme is also based on stacking (&) or HVS, while the color representation is the same as normal digital images.

Algorithm 1. The proposed (k, n) threshold scheme
Input: A $M \times N$ binary secret image S, the threshold parameters (k, n) **Output**: n shadow images $SC_1, SC_2, \cdots SC_n$
Step 1: For each position $(i, j) \in \{(i, j)\|1 \le i \le M, 1 \le j \le N\}$, repeat Steps 2-4.   **Step 2:** Compute $b_1, b_2, \cdots b_k$ one by one repeatedly using Eq. (6).   **Step 3:** Randomly select $b_{k+1}, b_{k+2}, \cdots b_n$ from $\{0, 1\}$.   **Step 4:** Randomly rearrangement $b_1, b_2, \cdots b_n$ to $SC_1(i, j), SC_2(i, j), \cdots SC_n(i, j)$.   **Step 5:** Output the n shadow images $SC_1, SC_2, \cdots SC_n$

Based on the steps above, we can find that the extended scheme is the same as [3] except the basic (2, 2) generation method. Thus, the color representation method and the recovery method are different. In addition, different color representation will lead to that, the Eq. (4) will not satisfy in the extended scheme. We give the explaination as follows.

In [3], "0" denotes white pixel, "1" denotes black pixel. The first k bits before rearrangement have the relationship in Eq. (4) of the generation phase, the stacking t bits have the relationship in Eq. (5) of the recovery phase.

In the generation phase of the extended scheme, we have:

$$\bar{s} = \overline{sc_1 \oplus sc_2 \oplus \cdots sc_{k-2} \oplus sc_{k-1} \oplus sc_k}$$
$$= sc_1 \oplus sc_2 \oplus \cdots sc_{k-2} \oplus sc_{k-1} \oplus \overline{sc_k}. \tag{8}$$

Where $\overline{sc_i}$ is a bit-wise complementary operation.

If k is even, then $k - 2$ is even, thus Eq. (8) is equivalent to $\bar{s} = \overline{sc_1} \oplus \overline{sc_2} \oplus \cdots \overline{sc_{k-3}} \oplus \overline{sc_{k-2}} \oplus sc_{k-1} \oplus \overline{sc_k}$, hence $\bar{s}! = \overline{sc_1} \oplus \overline{sc_2} \oplus \cdots \overline{sc_{k-3}} \oplus \overline{sc_{k-2}} \oplus \overline{sc_{k-1}} \oplus \overline{sc_k}$ and $s = \overline{sc_1} \oplus \overline{sc_2} \oplus \cdots \overline{sc_{k-3}} \oplus \overline{sc_{k-2}} \oplus \overline{sc_{k-1}} \oplus sc_k$

If k is odd, then $k - 1$ is even, thus Eq. (8) is equivalent to $\bar{s} = \overline{sc_1} \oplus \overline{sc_2} \oplus \cdots \overline{sc_{k-3}} \oplus \overline{sc_{k-2}} \oplus \overline{sc_{k-1}} \oplus \overline{sc_k}$

In the recovery phase of the extended scheme, we have:

$$\bar{s} = \overline{sc_1 \otimes sc_2 \otimes \cdots sc_{t-2} \otimes sc_{t-1} \otimes sc_t}$$
$$= \overline{sc_1} \& \overline{sc_2} \& \cdots \overline{sc_{t-2}} \& \overline{sc_{t-1}} \& \overline{sc_t} \tag{9}$$

Where $\overline{sc_i}$ is a bit-wise complementary operation, where $\&$ denotes the Boolean AND operation

While in Eq. (6) and Eq. (7) of the extended scheme, "1" denotes white pixel, "0" denotes black pixel.

Hence, we have: in the generation phase,

$$s = \begin{cases} sc_1 \oplus sc_2 \oplus \cdots \oplus sc_k & if \ k \in 2Z^+ - 1 \\ \overline{sc_1 \oplus sc_2 \oplus \cdots \oplus sc_k} & if \ k \in 2Z^+ \end{cases} \tag{10}$$

In the recovery phase,

$$s' = sc_1 \& sc_2 \& \cdots \& sc_t \tag{11}$$

Based on the above discussion, the color representation method of the extended scheme is different from [3], and the same as color representation method of digital images. Furthermore, the extended scheme has the same generation and recovery results except for the classified discussion of threshold k. The contrast and security of the extended scheme is the same as [3]. Hence, the extended scheme is secure.

4   Experimental Results

Herein, we conduct experiments and analysis to evaluate the effectiveness of the proposed scheme. In the experiments, original binary secret image as shown in Figure 3 (a) is used as the binary secret image, with standard size 512×512, which also can be applied for color images [3]. In our experiments, $(2, 3)$ (i.e. $k = 2$, n $=3$) threshold with secret image is used to do the test of the extended scheme.

Figure 3 (b-d) show the 3 shadow images SC_1, SC_2 and SC_3, which are random noise-like. Figure 3 (e-h) show the recovered binary secret image with 2 or 3 shadow images based on & operation, from which the secret image1 could be recognized. And the visual quality of Figure 3 (h) with 3 shadow images is better than with 2 shadow images. Although some contrast loss occurs, the revealed image is clearly identified.

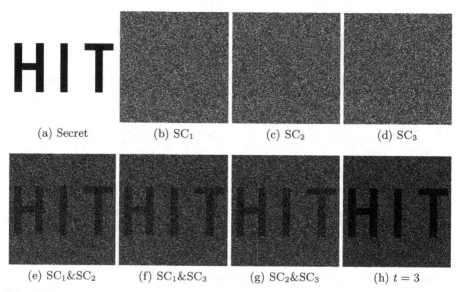

(a) Secret   (b) SC_1 (c) SC_2 (d) SC_3

(e) SC_1&SC_2 (f) SC_1&SC_3 (g) SC_2&SC_3 (h) $t = 3$

Fig. 3. Experimental example of the extended $(2, 3)$ threshold scheme for binary secret image1

From the image illustrations, the shadow images are random noise-like, when $t(k \leq t \leq n)$ shadow images are stacked (Boolean AND operation) the secret image could be recognized. While $k - 1$ shadow images are collected there is no information of the secret image could be recognized, which shows the security of the extended scheme.

5   Conclusion

This paper exploited normal color representation to be the same as digital images in RG-based VSS. In addition, based on the exploited color representation, we extended a VSS scheme based on Boolean operations and Random grid (RG). The new scheme applied XOR and AND (stacking) operations into the generation phase of RG-based VSS scheme and realized some functionalities in VSS such as (k, n) threshold, simple computation (stacking recovery), alternative order of shadow images in recovery, no codebook design and avoiding the the pixel expansion problem.

Acknowledgement. The authors wish to thank the anonymous reviewers for their suggestions to improve this paper. This work is supported by the National Natural Science Foundation of China (Grant Number: 61100187, 61301099, 61361166006).

References

1. Kafri, O., Keren, E.: Encryption of pictures and shapes by random grids. Optics Letters 12(6), 377–379 (1987)
2. Shyu, S.J.: Image encryption by random grids. Pattern Recognition 40(3), 1014–1031 (2007)
3. Chen, T.H., Tsao, K.H.: Threshold visual secret sharing by random grids. Journal of Systems and Software 84(7), 1197–1208 (2011)
4. Yan, X., Wang, S., Niu, X.: Threshold construction from specific cases in visual cryptography without the pixel expansion. Signal Processing (2014), doi:10.1016/j.sigpro.2014.06.011
5. Wang, D., Zhang, L., Ma, N., Li, X.: Two secret sharing schemese based on boolean operations. Pattern Recognit. 40(10), 2776–2785 (2007)

Wavelet Based SDA for Face Recognition

Goh Fan Ling[1], Pang Ying Han[1], Liew Yee Ping[1], Ooi Shih Yin[1],
and Loo Chu Kiong[2]

[1] Faculty of Information Science and Technology, Multimedia University, Malaysia
{goh.fan.ling10,yhpang,syooi}@mmu.edu.my,
lyping8@yahoo.com
[2] University Malaya, Malaysia
ckloo.um@gmail.com

Abstract. Semi-supervised discriminant analysis (SDA) is a popular semi-supervise learning technique for limited labelled training sample problem in face recognition. However, SDA resides in the illumination variations and noise of the face features. Hence, SDA exposes the illumination variations and noise when constructing the optimal projection. It could affect the projection, leading to poor performance. In this paper, an enhanced SDA, namely Wavelet SDA, is proposed. This proposed technique is to resolve the problem of intra-class variations due to illumination variations and noise on image data. The robustness of the proposed technique is evaluated using three well-known face databases, i.e. ORL, FERET and FRGC. Empirical results validated the good effects of wavelet transform on SDA, leading to better recognition performance.

Keywords: face recognition, semi-supervised, wavelet transform.

1   Introduction

In the fields of pattern recognition and machine learning, dimensionality reduction has being explored by researchers since the past decades [1]. Dimension reduction is not only important for curse of dimensionality, it is also meaningful underlying features of the data [2]. These are numerous techniques have been proposed for dimensionality reduction. Among them, Principal Component Analysis (PCA) [3] and Linear Discriminant Analysis (LDA) [1] are the most popular techniques. PCA is an unsupervised technique which works in maximizing the scatter of all projected samples [4]. On the other hand, LDA is a supervised technique which aims to search for the projection axes through maximizing the between-class scatter and minimizing the within-class scatter of the data [5].

Even though, LDA is outperformed PCA but the system will be degraded significantly when there are not enough training samples relative to the number of dimensions [6]. In this case, semi-supervised learning works in the way where only a small number of labelled data is used along with a large amount of unlabelled data to produce a suitable function that could well present the property of the data [7]. In real world application, labelled data is usually very limited, as a result it could cause the

C.K. Loo et al. (Eds.): ICONIP 2014, Part III, LNCS 8836, pp. 628–635, 2014.
© Springer International Publishing Switzerland 2014

supervised data learning becomes difficult. In addition, considerable amount of human resources or skills are needed in labelling huge amounts of data. On the other hand, unlabelled data could be easily obtained in large quantities and less costly price such as obtain image or dataset from website.

However, semi-supervised data learning technique is sensitive to illumination variations and noise on face images which is still a central problem in face recognition [8]. To tackle this problem, wavelet is a powerful tool to remove these additive variations. Wavelet transform offers analysis filter bank that is able to decompose the face image into smooth variations in low frequencies and the detail information in high frequencies.

Generally, an image is composed of low frequency components (smooth colour variations) and high frequency components (detail information) [9]. Upon refining an image, high frequency components is added to give detailed image. Hence, low frequency component demands more importance compared with high frequency components. According to [10], pose and scale of a face can affect intensity manifold globally which is low frequency spectrum. Thus, only low frequency spectrum (smooth variations) is sufficient for recognition and representation of image can be expressed at different resolutions [10, 11]. Based on this idea, we combine wavelet transform and SDA, namely Wavelet SDA, to remove high frequency components induced by noise and addition variation, while revealing informative features on face image with limited labelling data.

2　Wavelet Transform

The basic idea of wavelet transform is to decompose a signal into different scales on different resolution levels [12]. It is a signal analysis used efficiently for noise reduction and image compression to obtain a new image while signal frequency varies over time. Wavelet transform is one of a most excellent device to solve problem for study area of the low frequency and high frequency. The wavelet functions of a signal $f(a)$ can be obtained by a convolution of signal with a family of real orthonormal basis, $\psi_{xy}(a)$ as defined below:

$$\left(W_\psi f(a)\right)(x,y) = |x|^{1/2} \int_{-\infty}^{\infty} f(a)\psi\left[\frac{a-y}{x}\right] dx$$

$$f(a)\epsilon\, L^2(\Re) \tag{1}$$

where $x, y \in \Re$ and $x \neq 0$ are the dilation parameter and the translation parameter respectively.

3　Semi-supervised Discriminant Analysis

Semi-supervised discriminant analysis (SDA) is a learning technique which utilizes both unlabelled and limited labelled data [6]. SDA merges regularization strategies [13] and graph based semi-supervised learning algorithms [14] into LDA structure such that the labelled data points are utilized to suppose discriminant structure and the unlabelled data points are to calculate approximately the fundamental geometric structure of the data. The SDA functions as defined below:

$$v_{opt} = \max_{v} \frac{v^T S_b v}{v^T (S_t + \alpha X L X^t) v} \qquad (2)$$

where $X = [x_1, x_2, ..., x_m]$, S_b is the between-class scatter, S_t is the total scatter matrix, and $L = D - S$ is the Laplacian matrix [15]. The D is diagonal matrix and its entries are row or column sum of S, $D_{ii} = \Sigma_j S_{ij}$ while S_{ij} is the corresponding weight matrix. For the simple minded case, S_{ij} could be binary weight that defined as:

$$S_{ij} = \begin{cases} 1, & \text{if } x_i \in N_k(x_j) \text{ or } x_j \in N_k(x_i) \\ 0, & \text{otherwise} \end{cases} \qquad (3)$$

where $N_k(x_i)$ indicates the set of k-nearest neighbor of x_i to model the relationship between nearby data point. Detailed formulation of SDA could be referred to [7].

4   Experimental Results and Discussions

4.1   Databases and Classifier

In this experiment, three publicly available databases are adopted to evaluate performance of the proposed technique. These databases are (1) Olivetti Research Ltd. (ORL) [16], (2) Face Recognition Technology (FERET) [17], and (3) Face Recognition Grand Challenge (FRGC) [18]. The databases will be divided into two sets: training set and testing set. Training set is used to adjust the weights during the learning process as well as construct projection directions; while testing set is used to measure the recognition performance. There are plentiful classifiers could be applied such as k-nearest neighbors (KNN) [19], support vector machines (SVM) [20], linear discriminant analysis (LDA) and etc. However, nearest neighbour classifier with Euclidean metric is adopted in this study for sack of simplicity. Since the proposed method mainly deals with illumination variations and noise problem, ORL database was chosen as the base for parameter k tuning because all the images are against a dark homogeneous background with the subject is taking in different times, varying the lighting, facial expressions and facial details.

4.2   Parameter Setting on SDA

These are several parameters affect the result in the proposed technique. To obtain the optimal result, there are an experiment have been tested based on nearest neighbour classifier with Euclidean metric and ORL database. In this section are conducted using different weight mode of SDA with different neighbour mode setting based on the ORL databases for the idea of determining the order of neighbor parameter value(k) that optimally describe face features.

These are two types of SDA weight modes S_{ij} (eq (3)) have been tested, which are Binary and Heat Kernel weight modes. The recognition performance of the weight modes is shown in Table 1. From Table 1, we observe that Heat Kernel with k value 2 performs the best recognition rate which is 92%.

Table 1. Comparative weight modes S_{ij} with different k values in SDA

k	Binary	HeatKernel
1	91.5	91.5
2	91.5	92.0
3	91.0	91.0
4	89.5	90.0

4.3 Parameter Setting on Wavelets

In addition, the performance of different subbands of wavelet transform is also assessed on ORL database. Table 2 records the performance of the subbands of wavelet, i.e. approximation, vertical, horizontal and diagonal subbands. The approximation subband achieves a prime result while the other frequency subbands execute unsatisfying recognition rate. From the results obtained in Section 4.2 and 4.3, the approximation subband of wavelet and SDA weight mode of heat kernel with k = 2 are selected for the subsequent experiments in this study.

Table 2. Performance Comparison on wavelet subbands

Subbands	Recognition rate
Approximation, cA	92.5
Vertical, cV	41.0
Horizontal, cH	57.0
Diagonal, cD	11.5

Next, the following experiment is conducted for determining the optimal wavelet filter(s). Wavelets of Haar, Daubechies, Symmlets, Coiflets and BiorSplines filters are considered in this study. The recognition performances of these filters on ORL database are shown in Table 3. From the result, it is observed that SDA with Daubechies wavelet filter order 2, Symmlets wavelet filter order 2 and Coiflets wavelet filter order 2 in level 2 perform the best recognition rate with 95.5%.

Hence, the performances of the proposed techniques SDA with Daubechies wavelet filter order 2, SDA with Symmlets wavelet filter order 2 and SDA with Coiflets wavelet filter order 2 in level 2 are further assessed on FERET and FRGC databases. From Table 4 below, the average error obtained on FERET and FRGC databases is about 66%. Figure 1 shows the comparative result in between different databases.

Table 3. Comparative result of recognition rate with different wavelet filters

Wavelet Filter	Decomposition	
	Level 1	Level 2
haar	92.5	95.0
db1	92.5	95.0
db2	92.0	95.5
db3	92.5	94.0
sym2	92.0	95.5
sym3	92.5	94.0
sym4	94.0	93.5
coif1	94.0	95.5
coif2	94.5	94.0
coif3	93.5	94.5
bior1.1	92.5	95.0
bior1.3	93.5	94.5
bior1.5	93.0	95.0

Table 4. Average result on other databases

Techniques	Recognition rates (%)		Average
	FERET	FRGC	recognition rate (%)
db2+SDA	62.2	71.5	66.9
sym2+SDA	62.2	71.5	66.9
coif1+SDA	61.3	69.7	65.5

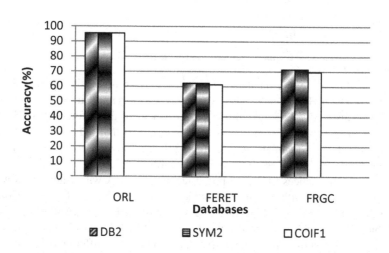

Fig. 1. Comparison Result between Optimal Wavelets on Diffrent Databases

4.4 Performance Comparison with other Techniques: ONE Labelled Training Data

In this experiment, ONE labelled and four unlabelled training images per subject are considered for data learning in the training stage. The performances of the proposed technique with other dimensionality reduction techniques, such as PCA and SDA, are presented. LDA is excluded because LDA needs more than 2 initial training samples per subject for projection construction. From Table 5, it is observed that the proposed techniques averagely outperform PCA and SDA. Here, we deduce that wavelet SDA is able to extract significant features for recognition.

Table 5. Compare with other technique with 1 labelled and 4 unlabelled training data

Technique / Database	FERET	FRGC	Average(%)
PCA	53.0	63.9	58.5
SDA	66.1	51.3	58.7
Wavelet db2+ SDA	62.2	71.5	**66.9**
Wavelet sym2 +SDA	62.2	71.5	**66.9**
Wavelet coif1+ SDA	61.3	69.7	**65.5**

4.5 Results on Wavelet SDA: TWO Labelled Training Data

In this experiment, TWO labelled and three unlabelled training images per subject are considered for data learning. The performances of the proposed technique with other dimensionality reduction techniques, such as PCA, LDA and SDA, are presented. The proposed techniques demonstrate superior performance to the other techniques, as shown in Table 6. This again validates the superiority of the integration of wavelet and SDA in extracting informative and discriminative features that are significant to discriminate the data.

Table 6. Compare with other technique with 2 labelled and 3 unlabelled training data

Technique / Database	FERET	FRGC	Average(%)
PCA	53.0	63.9	58.5
LDA	59.7	50.3	55.0
SDA	65.9	56.9	61.4
Wavelet db2+ SDA	62.9	74.9	**68.9**
Wavelet sym2 +SDA	62.9	74.9	**68.9**
Wavelet coif1+ SDA	62.3	72.0	**67.2**

5 Conclusion

This work studies the integration of the wavelet transform and SDA in the face recognition system. Wavelet performs well in noise reduction and image compression while SDA is effective in extracting informative features from data, especially when the labelled training samples are limited. Wavelet SDA achieves better recognition performance. Its superior performance to other feature extraction algorithms is observed as per discussed in section 4 with different scenario.

Acknowledgement. The authors would like to thank MMU-UM Collaboration and FRGS (#MMUE/140020) for the financial support.

References

1. Fukunaga, K.: Introduction to statistical pattern recognition. Academic Press (1990)
2. Carreira-Perpinan, M.A.: A review of dimension reduction techniques.: Department of Computer Science. University of Sheffield. Tech. Rep. CS-96-09, 1-69 (1997)
3. Jolliffe, I.: Principal component analysis. John Wiley & Sons, Ltd. (2005)
4. Turk, M., Pentland, A.: Eigenfaces for recognition. Journal of Cognitive Neuroscience 3(1), 71–86 (1991)
5. Lone, M.A., Zakariya, S.M., Ali, R.: Automatic Face Recognition System by Combining Four Individual Algorithms. In: 2011 International Conference on Computational Intelligence and Communication Networks (CICN), pp. 222–226. IEEE (2011)
6. Cai, D., He, X., Han, J.: Semi-supervised discriminant analysis. In: IEEE 11th International Conference on Computer Vision, ICCV 2007, pp. 1–7. IEEE (2007)
7. Zhu, X., Lafferty, J., Rosenfeld, R.: Semi-supervised learning with graphs (Doctoral dissertation, Carnegie Mellon University, Language Technologies Institute, School of Computer Science) (2005)
8. Qiao, L., Zhang, L., Chen, S.: An empirical study of two typical locality preserving linear discriminant analysis methods. Neurocomputing 73(10), 1587–1594 (2010)
9. Al Muhit, A., Islam, M.S., Othman, M.: VLSI implementation of discrete wavelet transform (DWT) for image compression. In: 2nd International Conference on Autonomous Robots and Agents, vol. 4(4), pp. 421–433 (2004)
10. Lai, J.H., Yuen, P.C., Feng, G.C.: Face recognition using holistic Fourier invariant features. Pattern Recognition 34(1), 95–109 (2001)
11. Naik, S., Patel, N.: Single Image Super Resolution in Spatial and Wavelet Domain. International Journal of Multimedia & Its Applications 5(4) (2013)
12. Graps, A.: An introduction to wavelets. IEEE Computational Science & Engineering 2(2), 50–61 (1995)
13. Friedman, J.H.: Regularized discriminant analysis. Journal of the American Statistical Association 84(405), 165–175 (1989)
14. Belkin, M., Niyogi, P., Sindhwani, V.: Manifold regularization: A geometric framework for learning from labeled and unlabeled examples. The Journal of Machine Learning Research 7, 2399–2434 (2006)
15. Chung, F.R.: Spectral graph theory. In: CBMS Regional Conference Series in Mathematics, vol. 92 (1996)
16. ORL face Database. AT&T Laboratories, Cambridge, U. K, http://www.uk.research.att.com/facedatabase.html
17. Phillips, P.J., Moon, H., Rizvi, S.A., Rauss, P.J.: The FERET evaluation methodology for face-recognition algorithms. IEEE Transactions on Pattern Analysis and Machine Intelligence 22(10), 1090–1104 (2000)

18. Phillips, P.J., Flynn, P.J., Scruggs, T., Bowyer, K.W., Chang, J., Hoffman, K., Worek, W.: Overview of the face recognition grand challenge. In: IEEE Computer Society Conference on Computer Vision and Pattern Recognition, CVPR 2005, vol. 1, pp. 947–954. IEEE (2005)
19. Larose, D.T.: k-Nearest Neighbor Algorithm. Discovering Knowledge in Data: An Introduction to Data Mining, pp. 90–106 (2005)
20. Steinwart, I., Christmann, A.: Support vector machines. Springer (2008)

Essential Visual Cryptographic Scheme with Different Importance of Shares

Xuehu Yan[1], Shen Wang[1,*], Xiamu Niu[1], and Ching-Nung Yang[2]

[1] School of Computer Science and Technology, Harbin Institute of Technology,
150080 Harbin China
[2] Department of CSIE, National Dong Hwa University, Hualien 974, Taiwan
shen.wang@hit.edu.cn

Abstract. Essential secret image sharing scheme allows some participants own special privileges with different importance of shares. The beauty of visual cryptographic scheme (VCS) is that its decoding is based on stacking and human visual system (HVS) without cryptographic computation. In this paper, for the first time essential and non-essential VCS (ENVCS) is introduced based on pre-existed (k, n) VCS. In the proposed (k_0, n_0, k, n) ENVCS, we generate the secret image into n shares which are classified into n_0 essential shares and $n - n_0$ non-essential shares. In the decoding phase, in order to reveal secret we should collect at least k shares, among which there are at least k_0 essential shares. Experiments are conducted to evaluate the security and efficiency of the proposed scheme.

Keywords: Secret sharing, Visual cryptographic scheme, Essential Secret sharing, Essential visual cryptographic scheme.

1 Introduction

Secret sharing encrypts the user data into different secret shadows (also called shares or shadow images) and distributes them to multiple participants, which has attracted more attention from scientist and engineers. Shamir's polynomial-based scheme [1–5] and visual cryptographic scheme (VCS) [6–8], are the primary branches in secret sharing.

A (k, n) -threshold secret sharing scheme was first proposed by Shamir in 1979 [1] through encrypting the secret into the constant coefficient of a random $(k-1)$-degree polynomial. The secret image can be perfectly reconstructed using Lagranges interpolation. Inspired by Shamir's scheme, Thien and Lin [2] reduced share size $1/k$ times to the secret image utilizing all coefficients of the polynomial for embedding secret. The advantage of Shamir's polynomial-based scheme [1–5] is the secret can be recovered losslessly. Although Shamir's polynomial-based scheme only needs k shares for reconstructing the distortion-less secret image, while it requires more complicated computations, i.e., Lagrange interpolations, for decoding and known order of shares.

---

[*] Corresponding author.

C.K. Loo et al. (Eds.): ICONIP 2014, Part III, LNCS 8836, pp. 636–643, 2014.
© Springer International Publishing Switzerland 2014

Naor and Shamir [7] first proposed the threshold-based VCS. In their scheme, a secret image is generated into n random shares which separately reveals nothing about the secret other than the secret size. The n shares are then printed onto transparencies and distributed to n associated participants. The secret image can be visually revealed based on human visual system (HVS) and probability by stacking any k or more shares. While less than k shares give no clue about the secret, even if infinite computational power is available. Whereas, traditional VCS has the limitation of the pixel expansion [9]. The pixel expansion will increase storage and transmission bandwidth. In order to remove the pixel expansion, probabilistic VCSs [10–12] and random grids (RG)-based VCSs [13–15] were proposed. Main properties for VCS are simple reconstructed method and alternative order of the shadow images. Simple reconstructed method means that the decryption of secret image is completely based on HVS without any cryptographic computation.

Some participants require special privileges because of their importance or status in some applications, hence some shares may be more important than others [5]. However, most of the above secret image sharing schemes have the same importance of shares. Aiming to solve this problem, recently Li et al. [5] proposed the (k_0, n_0, k, n) essential secret image sharing scheme. In their scheme, all n shadows are classified into n_0 essential shadows and $n - n_0$ non-essential shadows. The (k_0, n_0, k, n)-scheme needs k shadows including at least k_0 essential shadows, when reconstructing the secret image. Unfortunately, Li et al.'s scheme [5] suffers from more complicated computations and known order of shadow images for decoding, since their decoding method is based on Lagrange interpolations.

In this paper, for the first time we introduce (k_0, n_0, k, n) essential and non-essential VCS (ENVCS) with stacking decryption, i.e., lower complicated computations without any cryptographic computation, and known order of shares for decoding. In the generation phase of our (k_0, n_0, k, n) ENVCS, based on any adopting (k, n) VCS without the pixel expansion, the secret image is encoded into n shares which are divided into n_0 essential shares and $n - n_0$ non-essential shares. In the recovery phase, to reveal secret one should collect at least k shares, among which there are at least k_0 essential shares. Experimental results and analyses demonstrate the effectiveness and security of the proposed scheme.

The rest of the paper is organized as follows. Section 2 introduces the preliminary techniques as the basis for the proposed scheme. In Section 3, the proposed scheme is presented in detail. Section 4 is devoted to experimental results. Finally, Section 5 concludes this paper.

2   Review of the Related Work

Any (k, n) VCS without the pixel expansion can be applied as the input (k, n) VCS in the proposed scheme. One RG-based (k, n) VCS in [14] will be adopted in the proposed scheme. In what follows, symbols $\oplus$ and $\otimes$ denote the Boolean XOR and OR operations, respectively. $\overline{b}$ is a bit-wise complementary operation of a bit b. The binary secret image denoted as S with pixel value $S(i,j), 1 \leq i \leq M, 1 \leq j \leq N$, is shared among n ($n \geq 2, n \in Z^+$) shares, and the reconstructed secret image S' is reconstructed from t ($1 \leq t \leq n, t \in Z^+$) shares. Here '0' denotes white pixel, '1' denotes black pixel.

As the basis of the proposed scheme, herein a (k,n) RG-based VCS [14] is described as follows:

Wu and Sun's (k, n) RG-based VSS
Input: A $M \times N$ binary secret image S , the threshold parameters (k,n)
Output: n shadow images $SC_1, SC_2, \cdots SC_n$
Step 1: For each position $(i,j) \in \{(i,j)\|1 \leq i \leq M, 1 \leq j \leq N\}$, repeat Steps 2-6
Step 2: Select $b_1, b_2, \cdots b_k \in \{0,1\}$ randomly.
Step 3: If $S(i,j) = b_1 \oplus b_2 \cdots \oplus b_k$, go to Step 5; else go to Step 4
Step 4: Randomly select $p \in \{1, 2, \cdots, k\}$ flip $b_p = \overline{b_p}$ (that is $0 \to 1$ or $1 \to 0$).
Step 5: Set $b_{k+1} = b_k, b_{k+2} = b_k, \cdots b_n = b_k$
Step 6: Randomly rearrange $b_1, b_2, \cdots b_n$ to $SC_1(i,j), SC_2(i,j), \cdots SC_n(i,j)$
Step 7: Output the n shadow images $SC_1, SC_2, \cdots SC_n$

However, traditional (k,n) VCSs have the same importance of shares, which will restrict the applications. In order to solve this problem, we will present an ENVCS for case (k_0, n_0, k, n).

3   The Proposed Scheme

In the section, we first introduce the definition of (k_0, n_0, k, n) ENVCS, then a (k_0, n_0, k, n) ENVCS is proposed in detail.

3.1   Definition of (k_0, n_0, k, n) ENVCS

We denote all the participants as $\mathcal{V} = \{1, 2, \cdots n\}$, among which the first n_0 participants are essential denoted as E and the last $n - n_0$ participants are non-essential denoted as N. An access structure is denoted by $(\Gamma_{\text{Qual}}, \Gamma_{\text{Forb}})$, where Γ_{Qual} (resp. Γ_{Forb}) denotes the superset of qualified subsets (resp. the superset of forbidden subsets), $\Gamma_{\text{Qual}} \cap \Gamma_{\text{Forb}} = \emptyset$, $\Gamma_{\text{Qual}} \cup \Gamma_{\text{Forb}} = 2^{\mathcal{V}}$, and

$$\Gamma_{\text{Qual}} = \{B \subseteq \mathcal{V} \wedge B_E \subseteq E \wedge B_N \subseteq N : |B| \geq k \wedge |B_E| \geq k_0\} \text{ and}$$
$$\Gamma_{\text{Forb}} = \{B \subseteq \mathcal{V} \wedge B_E \subseteq E \wedge B_N \subseteq N : |B| < k \vee |B_E| < k_0\}$$

Definition of (k_0, n_0, k, n) ENVCS is introduced as follows.

Definition 1 ((k_0, n_0, k, n) **ENVCS**): A VCS is (k_0, n_0, k, n) ENVCS If the VCS satisfies :

1) Contrast condition: If $X = \{i_1, i_2, \cdots i_p\} \in \Gamma_{\text{Qual}}$, then we have $\alpha > 0$.
2) Security condition: If $F = \{i_1, i_2, \cdots i_p\} \in \Gamma_{\text{Forb}}$, then we have $\alpha = 0$.
The above two conditions imply:

1. The secret image can be revealed by any qualified subset of shares, i.e., at least k shares and among which at least k_0 essential shares.
2. Any information of the secret image cannot be obtained by any forbidden subset of shares , i.e., less than k shares or less than k_0 essential shares, other than the size of the secret image.

3.2   The Proposed Scheme

Prior to show the detail of the proposed scheme, the shares generation design concept of the proposed scheme is illustrated in Fig. 1.

Any pre-existed (k_x, n_x) VCS without the pixel expansion can be adopted in the design of the proposed scheme, where k_x, n_x denote the threshold parameters in the adopted original (k_x, n_x) VCS without the pixel expansion and $k_x, n_x \in Z^+, n_x \geq k_x \geq 2$.

The shares generation phase is divided into two steps. In the first step, the binary secret image S is encoded into $n_0 + 1$ essential shares denoted as $SC_1, SC_2, \cdots SC_{n_0}, \widetilde{SC}_{n_0+1}$ by the original $(k_0 + 1, n_0 + 1)$ VCS, where $\widetilde{SC}_{n_0+1}$ indicates the temporary share which will be utilized to generate the non-essential shares. In the second step, the temporary share $\widetilde{SC}_{n_0+1}$ is utilized to generate the $n - n_0$ non-essential shares by $(k - k_0, n - n_0)$ VCS. Thus, there are to-tally n_0 essential shares and $n - n_0$ non-essential shares. If one wants to recover the the binary secret image S, one should collect at least $k_0 + 1$ shares from $SC_1, SC_2, \cdots SC_{n_0}, \widetilde{SC}_{n_0+1}$ because of $(k_0 + 1, n_0 + 1)$ threshold, among which share $\widetilde{SC}_{n_0+1}$ can be recovered when more than $k - k_0$ non-essential shares are collected from $SC_{n_0+1}, \cdots SC_{n-1}, SC_n$ because of $(k - k_0, n - n_0)$ threshold. As a result, one should collect at leat k shares including at leat k_0 essential shares to recover the secret, i.e., (k_0, n_0, k, n) threshold will be achieved.

Based on the above design concept, the corresponding algorithmic steps are described in Algorithm 1 detailedly. The upper part, and bottom part in Fig. 1 corresponds to Steps 1 and 2 in Algorithm 1, respectively.

Algorithm 1. The proposed (k_0, n_0, k, n) ENVCS.
Input: A $M \times N$ binary secret image S, the threshold parameters (k_0, n_0, k, n) and a pre-existed original (k_x, n_x) VCS without the pixel expansion
Output: n shares $SC_1, SC_2, \cdots SC_n$.
Step 1: Utilize the original $(k_0 + 1, n_0 + 1)$ VCS to encode the binary secret image S into $n_0 + 1$ essential shares denoted as $SC_1, SC_2, \cdots SC_{n_0}, \widetilde{SC}_{n_0+1}$
Step 2: Utilize the original $(k - k_0, n - n_0)$ VCS to encode the temporary share $\widetilde{SC}_{n_0+1}$ into $n - n_0$ non-essential shares denoted as $SC_{n_0+1}, \cdots SC_{n-1}, SC_n$
Step 3: Output the n shares $SC_1, SC_2, \cdots SC_n$.

We can reconstruct the secret image by stacking shadows when sufficient and satisfied shares are collected.

We remark that in Step 2 of the proposed scheme, a different original $(k - k_0, n - n_0)$ VCS can be adopted as well.

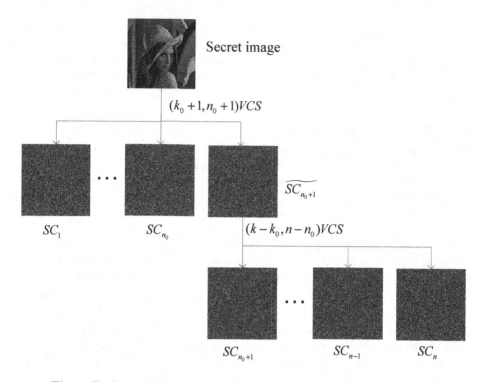

Fig. 1. Shadow images generation design concept of the proposed scheme

4   Experimental Results and Comparison

In this section, we conducted experiments and analyses to evaluate the effectiveness of the proposed scheme. In the experiments, original binary secret image as shown in Fig. 2 (a),of size 512×512, is used as the secret image to test the efficiency of the proposed scheme.

4.1   Image Illustration

In our experiments, $(2, 2, 4, 5)$ (i.e., $k_0 = 2, n_0 = 2, k = 4, n = 5$) threshold is used to test the proposed scheme, where the first t qualified or forbidden shares are used as examples for saving pages.

Fig. 2(b-c) show the two essential shares SC_1 and SC_2, which are random noise-like. Fig. 2(d-f) show the three non-essential shares SC_3, SC_4 and SC_5, which are also random noise-like. The five shares' contrast defined in [16] is -0.0027, -0.0012, -0.0024, 3.8070e-004 and -6.2170e-005, respectively. Fig. 2 (g-h) show the recovered binary secret image with any qualified subset of shares, from which the secret image can be revealed. Fig. 2 (i-l) show the recovered binary secret image with any forbidden shares, from which the secret image cannot be recognized.

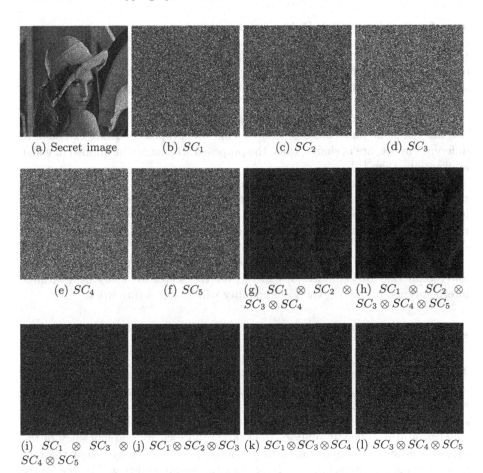

(a) Secret image   (b) SC_1 (c) SC_2 (d) SC_3

(e) SC_4 (f) SC_5 (g) $SC_1 \otimes SC_2 \otimes$ (h) $SC_1 \otimes SC_2 \otimes$
$SC_3 \otimes SC_4$ $SC_3 \otimes SC_4 \otimes SC_5$

(i) $SC_1 \otimes SC_3 \otimes$ (j) $SC_1 \otimes SC_2 \otimes SC_3$ (k) $SC_1 \otimes SC_3 \otimes SC_4$ (l) $SC_3 \otimes SC_4 \otimes SC_5$
$SC_4 \otimes SC_5$

Fig. 2. Experimental example of the proposed $(2, 2, 4, 5)$ scheme. (a) secret image; (b)-(c) two essential shares; (d)-(f) three non-essential shares; (g)-(h) stacking results by qualified shares; (i)-(l) stacking results by forbidden shares.

From the results shown in Fig. 2:

- The shares are noise-like, hence every single share gives no clue about the secret and there is no cross interference of secret image in the shares for the proposed scheme.

- When $t < k$ shares or less than k_0 essential shares are collected , there is no information of the secret image could be recognized, which shows the security of the proposed scheme.

- When $t(k \leq t \leq n)$ shares including k_0 or more essential shares are collected, the secret image is recognized by HVS.

- The proposed scheme achieves (k_0, n_0, k, n) threshold.

4.2 Comparisons with Related Schemes

In the subsection, we compare the proposed scheme with other related methods especially [14] and [5], since the proposed scheme is reconstructed from the scheme in [14] and the scheme in [5] is the first secret sharing scheme which has the feature of (k_0, n_0, k, n) threshold.

[14] and the proposed scheme both have the feature of (k_0, n_0, k, n) threshold. [14] has the same importance of shares, while the proposed scheme can achieve different importance of shares, thus, the proposed scheme can be applied in wider applications than [14].

[5] and the proposed scheme both require no codebook design without the pixel expansion. [5] has more complicated computations because of Lagrange interpolations for decoding requiring known order of shadow images, while the proposed scheme can decode the secret by HVS with free or alternative order of shares.

Visual quality of the reconstructed secret image of the proposed scheme is lower than that of [5], since [5] can recover the secret losslessly by Lagrange interpolations. Improving the visual quality will be our future work.

5 Conclusion

An essential and non-essential VCS (ENVCS) for case (k_0, n_0, k, n) is proposed in this paper, which can reconstruct the secret image by human visual system without cryptographic computation. At least k shares including at least k_0 essential shares should be stacked to reveal secret. We have performed several experimental results and analyses to evaluate the security and efficiency of the proposed scheme. Comparisons with previous approaches suggest that the proposed scheme has several merits. Improving the visual quality will be the future work.

Acknowledgement. The authors wish to thank the anonymous reviewers for their suggestions to improve this paper. This work is supported by the National Natural Science Foundation of China (Grant Number: 61100187, 61301099, 61361166006).

References

1. Shamir, A.: How to share a secret. Communications of the ACM 22(11), 612–613 (1979)
2. Thien, C.C., Lin, J.C.: Secret image sharing. Computers & Graphics 26(5), 765–770 (2002)
3. Lin, S.J., Lin, J.C.: Vcpss: A two-in-one two-decoding-options image sharing method combining visual cryptography (vc) and polynomial-style sharing (pss) approaches. Pattern Recognition 40(12), 3652–3666 (2007)

4. Yang, C.N., Ciou, C.B.: Image secret sharing method with two-decoding-options: Lossless recovery and previewing capability. Image and Vision Computing 28(12), 1600–1610 (2010)
5. Li, P., Yang, C.N., Wu, C.C., Kong, Q., Ma, Y.: Essential secret image sharing scheme with different importance of shadows. Journal of Visual Communication and Image Representation 24(7), 1106–1114 (2013)
6. Wang, Z., Arce, G.R., Di Crescenzo, G.: Halftone visual cryptography via error diffusion. IEEE Trans. Inf. Forensics Security 4(3), 383–396 (2009)
7. Naor, M., Shamir, A.: Visual cryptography. In: De Santis, A. (ed.) EUROCRYPT 1994. LNCS, vol. 950, pp. 1–12. Springer, Heidelberg (1995)
8. Yan, X., Wang, S., Niu, X.: Threshold construction from specific cases in visual cryptography without the pixel expansion. Signal Processing (2014), doi:10.1016/j.sigpro.2014.06.011
9. Weir, J., Yan, W.: A comprehensive study of visual cryptography. In: Shi, Y.Q. (ed.) Transactions on DHMS V. LNCS, vol. 6010, pp. 70–105. Springer, Heidelberg (2010)
10. Kuwakado, H., Tanaka, H.: Image size invariant visual cryptography. IEICE Transactions on Fundamentals of Electronics, Communications and Computer Sciences 82(10), 2172–2177 (1999)
11. Yang, C.N.: New visual secret sharing schemes using probabilistic method. Pattern Recognit. Lett. 25(4), 481–494 (2004)
12. Cimato, S., De Prisco, R., De Santis, A.: Probabilistic visual cryptography schemes. The Computer Journal 49(1), 97–107 (2006)
13. Kafri, O., Keren, E.: Encryption of pictures and shapes by random grids. Optics Letters 12(6), 377–379 (1987)
14. Wu, X., Sun, W.: Improving the visual quality of random grid-based visual secret sharing. Signal Processing 93(5), 977–995 (2013)
15. Guo, T., Liu, F., Wu, C.: Threshold visual secret sharing by random grids with improved contrast. Journal of Systems and Software 86(8), 2094–2109 (2013)
16. Shyu, S.J.: Image encryption by random grids. Pattern Recognition 40(3), 1014–1031 (2007)

Improved Biohashing Method
Based on Most Intensive Histogram Block Location

Munalih Ahmad Syarif[1,*], Thian Song Ong[1], Andrew Beng Jin Teoh[2],
and Connie Tee[1]

[1] Faculty of Information Science and Technology, Multimedia University,
Jalan Ayer Keroh Lama, 75450 Melaka
asyarifm@yahoo.co.id, {tsong,tee.connie}@mmu.edu.my
[2] School of Electrical and Electronic Engineering, Yonsei University, South Korea
bjteoh@yonsei.ac.kr

Abstract. Biohashing is a promising cancellable biometrics method. However, it suffers from a problem known as 'stolen token scenario'. The performance of the biometric system drops significantly if the Biohashing private token is stolen. To solve this problem, this paper proposes a new method termed as Most Intensive Histogram Block Location (MIBL) to extract additional information of the p-th best gradient magnitude. Experimental analysis shows that the proposed method is able to solve the stolen token problem with error equal rates as low as 1.46% and 7.27% when the stolen token scenario occurred for both FVC2002 DB1 and DB2 respectively.

Keywords: Biohashing, Cancellable Biometrics, MIBL.

1 Introduction

Biometrics is a promising solution to identify or verify the identity of a person using physiological or behavioural characteristics. However, there are several concerns including: biometrics cannot be revoked or cancelled; biometric traits are permanently associated with a user; a compromised biometric is forever compromised; all applications using the same biometrics will be compromised at once; and cross-matching can be used to track individuals [1].

Cancellable biometrics is proposed to overcome the mentioned problems [1] by enabling the cancellation and replacement of the biometric template. An additional information is given to the user in the form of a random key (K). The transformation function (F) uses K as the parameter to generate the transformed biometric template (F(T,K)) based on the user's biometric template (T). F is designed in such a way that it is hard to recover the original biometric data based on the transformed template. Moreover, to ensure privacy protection, the transformed template is stored in the database instead of the original biometrics and is used for matching. During the matching process, the user needs to provide K and query biometric data (Q) into the same

---

* Corresponding author.

C.K. Loo et al. (Eds.): ICONIP 2014, Part III, LNCS 8836, pp. 644–652, 2014.

transformation function (F) to generate the transformed query template (F(Q,K)) and compare it with the stored transformed template (F(T,K)) (see Fig.1). Regenerating a new template based on the same biometric data can be achieved easily by replacing the old K with a new one.

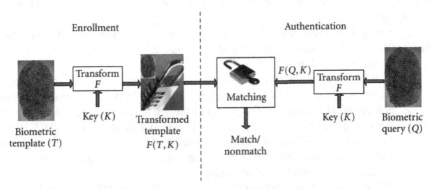

Fig. 1. Cancellable Biometrics Diagram Blocks [2]

Biohashing is one of the cancellable biometric template protection schemes [3]. First, the biometric features are extracted from the biometric data and represented in a vector format, $\Gamma \in \mathfrak{R}^l$ where l is the feature length of Γ. Then, a random private token is generated in the form of k orthonormal pseudo random vectors, $\{r_{1i} \in \mathfrak{R}^l | i = 1, \dots, k\}$ and $k \leq l$. Then the inner product between Γ and r_{1i} was performed. The result is converted to a k-bit string by thresholding. The value below or equal to a preset threshold will be converted to 0 and 1 otherwise. Normally the threshold value is set to 0. As a cancellable biometric method, the main advantage of Biohashing is the ability to revoke the existing biometric template and replace it with a newly generated template. Biohashing has been tested with several biometric traits for fingerprint [3], face [4] and palmprint [5]. Moreover, Biohashing is also able to solve the privacy issue, since the transformed biometric template is stored instead of the original biometric data.

However, Biohashing suffers substantial performance degradation in the stolen-token scenario, i.e. when an "impostor" B steals the Biohashing private token of A and tries to authenticate as A. In other words, a biometric system which implements Biohashing might recognize an imposter as an authorized user if the imposter steals the token from an authorized user. In this context, Lumini and Nanni [6] proposed several solutions to solve the Biohashing stolen-token problem with promising result of low EER.

2   Proposed Method

In this paper, an improved Biohashing method termed as the Most Intensive Block Location (MIBL) for fingerprint biometrics is proposed to address the stolen token problem. The solution extracts an additional information which is the p-th MIBL from the fingerprint image. The p-th MIBL can be obtained by taking into account the location of the histogram block which contains the p-th best gradient magnitude value. The location is unique for each user. The histogram bin recorded as the biometric features are the histogram bins which belong to blocks which are considered as the p-th MIBL, otherwise the features will be converted to zero. This idea is inspired by the work of Fuksis et al. [7].

The proposed solution can be divided into three main steps: MIBL extraction, MIBL training, and features filtering based on trained MIBL.

a) MIBL Extraction.

At first, Histogram of Oriented Gradient (HOG) is applied to extract the fingerprint features, f. The fingerprint image (fp) is divided into n number of histogram blocks to extract the fingerprint histogram features. In each block, the gradient magnitude $(g(1), g(2), ..., g(n))$ and orientation $(\theta(1), \theta(2), ..., \theta(n))$ are computed and both of them are represented by h number of histogram bin. The n rectangular cells and h histogram bins are concatenated to form a nh - dimensional fingerprint feature vector, f.

The g values of each blocks then passed to function $M(p, g)$ which records the p-th largest g value $(G(1), G(2), ..., G(p))$ in descending order, $G(1) > G(2) > \cdots > G(p)$.

$$G = M(p, g) \tag{1}$$

The blocks with g value bigger than or equal $G(p)$ are considered as the most intensive blocks and labelled as '1' otherwise will be labelled as '0'. Here the number of p can be adjusted where it should be greater than 0 and less than n $(0 < p < n)$. Fig. 2 shows the sample of a fingerprint image which is divided into 12 histogram blocks and p is set to 5. Most intensive blocks are located in blocks number 1, 4, 5, 8 and 11 so all of these blocks are labelled by 1 and 0 for the rest. As a result, the MIBL, denoted as v tained is 100110010010.

$$v = \begin{cases} 1, & g \geq G(p) \\ 0, & g < G(p) \end{cases} \tag{2}$$

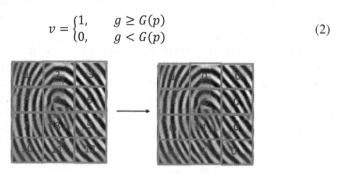

Fig. 2. Block Diagram of the Proposed Method

b) Training the Most Intensive Block Location

The w numbers of v for each person are trained in order to get more accurate MIBL information. The training process performed by applying the following summation function for w sample of v.

$$s = \sum_{i=1}^{w} v_i \qquad (3)$$

Then the summation result (s) and q are passed to function $M(q,s)$ which records the q-th largest s value ($S(1), S(2), \ldots, S(q)$) in descending order. The blocks with s value bigger than or equal $S(q)$ are considered as the trained most intensive blocks and labelled as '1' otherwise will be labelled as '0'. Here the number of q can be adjusted where it should be greater than 0 and less than p ($0 < q \leq p$). Below is the sample of q is equal to 5.

$$
\begin{array}{ll}
v_1 & : 100110010010 \\
v_2 & : 110010010010 \\
\\
\vdots \\
\\
v_w & : 110010010010 \\
s & : 530251050050
\end{array}
$$

$$S = M(q,s) \qquad (4)$$

$$v^t = \begin{cases} 1, & s \geq S(q) \\ 0, & s < S(q) \end{cases} \qquad (5)$$

Trained MIBL, v^t $\qquad$: 110010010010

c) **Feature Filtering based on the Trained MIBL**

Feature filtering is performed after MIBL training stage. In this stage f is filtered based on v^t. The filtering process is implemented as below. The f features of block is multiplied by the v^t which is belong to that blocks.

$$F = \sum_{j=1}^{n} \sum_{d}^{jh} f(d, 1) \times v^t(j) \qquad (6)$$

$$d = \big((j-1)h\big) + 1 \qquad (7)$$

where F is the result of features filtering process and d are the initiate number of a feature in a block. With the aid of private token, F is then passed to Biohashing process in order to obtain final secured template. Fig. 3 (a) and (b) show the illustration of enrolment and verification process of a fingerprint recognition system that applies MIBL + Biohashing.

3   Experimental Analysis

The experiments were performed on FVC2002 DB1 and DB2 fingerprint databases[8]. Both datasets consist of 100 persons with 8 different samples for each person. The device used to capture the fingerprint images was "TouchView II" produced by Identix. The size of the fingerprint images were 388 x 374 pixels. In this experiment, the region of interest (ROI) of the fingerprints were cropped to a size of 175 x 175 pixels.

The cropped fingerprint images were enhanced with the method proposed by Chikkerur et al. [9]. The results of the experiments were presented using Equal Error Rate (EER). The performance of HOG for FVC2002 DB1 was EER 16.06% and FVC2002 DB2 was EER 31.22% when Euclidean Distance is used as the Distance Matcher.

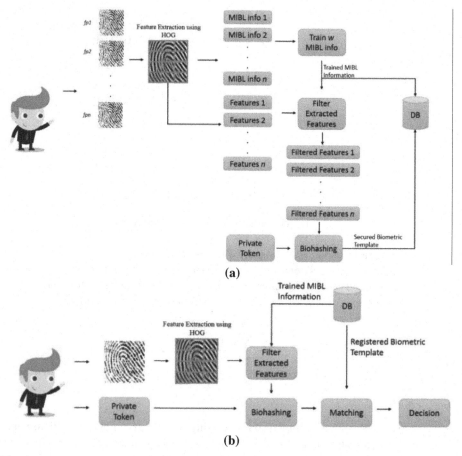

Fig. 3. Illustration of Enrolment (a) and Verification (b) Process of Fingerprint Biometric System which apply MIBL + Biohashing

3.1 Performance Comparison Using Different Values of MIBL Parameters p and q.

This experiment searches the best combination of MIBL parameters, p and q. p is the number of the most intensive block location before the MIBL training. Meanwhile q

is the number of the most intensive block location after the MIBL training. For this experiment the values of p used were 10, 15 and 20. Meanwhile for q the value were set to 3, 5, 7 and 9. The number of samples used for training MIBL was set to 3. Table 1 shows that the best result obtained when the value of p and q were set to 20 and 5, respectively for all databases. The best result was EER 0.0047% and EER 0.4245% for FVC2002 DB1 and DB2, respectively.

Table 1. Result of Experiment using Different p and q Values

Database	p	q	EER (%)
FVC2002 DB1	10	3	0.01
		5	0.0094
		7	0.0096
		9	0.01
	15	3	0.0071
		5	0.006
		7	0.0071
		9	0.0083
	20	3	0.0052
		5	**0.0047**
		7	0.0055
		9	0.0069
FVC2002 DB2	10	3	0.9
		5	0.7504
		7	0.7639
		9	0.78
	15	3	0.67
		5	0.6097
		7	0.6625
		9	0.7062
	20	3	0.4339
		5	**0.4245**
		7	0.5399
		9	0.6079

3.2 Performance Comparison between the Proposed Solution and the Existing Methods

The proposed MIBL + Biohashing was compared with the original Biohashing method and the existing methods proposed by Lumini and Nanni [6]. The experiment was performed under two different conditions, with and without stolen token condition (normal condition). Table 2 shows that the proposed method was able to improve the original Biohashing performance both in the normal condition and stolen token condition. The proposed method was able to produce the best EER at 1.46% and 7.27% for both FVC2002 DB1 and DB2 respectively under stolen-token condition.

Fig. 4 and Fig. 5 shows the genuine and imposter score for Biohashing and proposed method in stolen token condition for both FVC2002 DB1 and DB2 respectively. Fig. 6 (a) and (b) are the Receiver Operating Characteristic (ROC) curve of the experiments which are performed on FVC2002 DB1 and DB2.

Table 2. Result of Comparison between the Proposed Method and the Existing Methods

Method	Database	Biohashing Stolen Token (Yes/No)	EER (%)
Biohashing [3]	FVC2002 DB1	Yes	16.69
		No	0.02
	FVC2002 DB2	Yes	31.63
		No	0.52
MIBL + Biohashing (proposed)	FVC2002 DB1	Yes	**1.46**
		No	**0.0047**
	FVC2002 DB2	Yes	**7.27**
		No	**0.42**
Normalization + τ Variation [6]	FVC2002 DB1	Yes	10
		No	0.6
	FVC2002 DB2	Yes	8.9
		No	0.5
Normalization + Space Augmentation [6]	FVC2002 DB1	Yes	8
		No	0.4
	FVC2002 DB2	Yes	7.5
		No	0.2
Normalization + Features Permutation [6]	FVC2002 DB1	Yes	8.2
		No	0.4
	FVC2002 DB2	Yes	8
		No	0.2

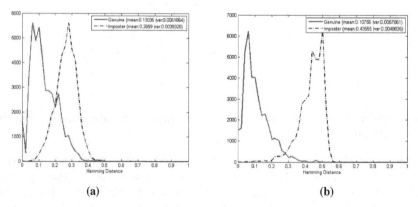

(a)   (b)

Fig. 4. (a) Biohashing Genuine and Imposter Score (b) MIBL+Biohashing Genuine and Imposter Score. (Stolen Token Condition, FVC2002 DB1).

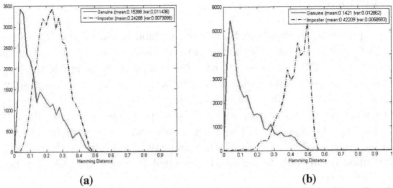

(a)   (b)

Fig. 5. (a) Biohashing Genuine and Imposter Score (b) MIBL+Biohashing Genuine and Imposter Score. (Stolen Token Condition, FVC2002 DB2).

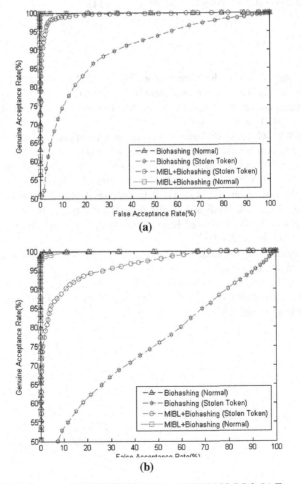

(a)

(b)

Fig. 6. ROC Curves of FVC2002 DB1 (a) and FVC2002 DB2 (b) Experiment

4   Conclusion

This paper proposes an improved Biohashing method based on the Most Intensive Block Location information. The method extracts additional information termed as the Most Intensive Block Location (MIBL) information from the biometric images in order to solve the Biohashing stolen token problem. Experimental results show that the proposed method is able to yield promising solution for the Biohashing stolen token problem. Moreover, the proposed method outperforms the state-of-the-art algorithms, which is able to produce EERs equal 1.46% and 7.27% when the stolen token scenario occurred for both FVC2002 DB1 and DB2, respectively. Meanwhile, the original method is only able to produce EERs equal 16.69% and 31.63% for the same datasets.

Acknowledgement. This work is supported by Fundamental Research Grant Scheme (FRGS) of Malaysia (Grant Number: MMUE/120124).

References

1. Ratha, N., Connel, J., Bolle, R.: Cancellable Biometrics: A Case Study in Fingerprints. In: The 18th International Conference on Pattern Recognition (2006)
2. Jain, A.K., Nandakumar, K., Nagar, A.: Biometric Template Security. EURASIP Journal on Advances in Signal Processing (2008)
3. Teoh, A., Ngo, D., Goh, A.: Biohashing: two factor authentication featuring fingerprint data and tokenised random number. Pattern Recognition (2004)
4. Ngo, D.C.L., Teoh, A.B.J., Goh, A.: Eigenspace-Based Face Hashing. In: Zhang, D., Jain, A.K. (eds.) ICBA 2004. LNCS, vol. 3072, pp. 195–199. Springer, Heidelberg (2004)
5. Connie, T., Teoh, A., Goh, M., Ngo, D.: Palmhashing: A Novel Approach for Dual-Factor Authentication. Pattern Analysis and Applications 7, 255–268 (2004)
6. Lumini, A., Nanni, L.: An Improved Biohashing for Human Authentication. Pattern Recognition 40(3), 1057–1065 (2007)
7. Fuksis, R., Kadikis, A., Greitans, M.: Biohashing and Fusion of Palmprint and Palm Vein Biometric Data. IEEE (2011)
8. FVC2002 (2002), http://bias.csr.unibo.it/fvc2002/
9. Chikkerur, S., Cartwright, A.N., Govindaraju, V.: Fingerprint Enhancement using STFT Analysis. Pattern Recognition 40, 198–211 (2007)

2.5D Face Recognition under Tensor Manifold Metrics

Lee-Ying Chong[1,*], Andrew Beng Jin Teoh[2], Thian-Song Ong[1],
and Siew-Chin Chong[1]

[1]Faculty of Information Science and Technology, Multimedia University, Malaysia
{lychong,tsong,chong.siew.chin}@mmu.edu.my
[2]School of Electrical and Electronic Engineering, Yonsei University, South Korea
bjteoh@yonsei.ac.kr

Abstract. Gabor-based region covariance matrix (GRCM) is a very flexible
face descriptor where it allows different combination of features to be fused to
construct a covariance matrix. GRCM resides on Tensor manifold where the
computation of geodesic distance between two points requires the consideration
of geometry characteristics of the manifold. Affine Invariant Riemannian Metric (AIRM) is the most widely used geodesic distance metric. However, it is
computationally heavy. This paper investigates several geodesic distance metrics on Tensor manifold to find out the alternative speedy method for 2.5D face
recognition using GRCM. Besides, we propose a feature-level fusion for 2.5D
partial and 2D data to enhance the recognition performance.

Keywords: 2.5D face recognition, GRCM, Tensor manifold, feature-level
fusion, geodesic distance metrics.

1 Introduction

Face recognition drew a heavy publicity in recent years due to the maturity of evolving technologies. The face recognition system mainly focuses on utilizing the 2D face
images. However, human face is a 3D object where the 2D projection could not able
to capture the face geometric information adequately. Besides, the facial appearance
greatly varied depends on the environment factors (such as illumination conditions,
head orientations) and it is sensitive to the facial expressions and cosmetics [1].

The limitations of 2D facial data can be overcome by utilizing 3D facial data. Yet,
3D face modelling is impractical due to high cost of 3D data acquisition in which it
involves the use of more than one 3D sensors and complex processing of matching
2.5D partial face data into a 3D model. The 2.5D partial face data is a simplified 3D
(x,y,z) surface representation that contains at most one depth value (z direction) for
every point in the (x, y) plane [2]. The 2.5D partial face data explicitly represent the
3D shape, which is invariant under the change of color and independent to any illumination conditions [3].

The emergence of 2.5D face recognition overcomes the drawbacks of 2D face recognition and improves the efficiency in 3D face recognition. The usage of 2.5D partial

---

* Corresponding author.

C.K. Loo et al. (Eds.): ICONIP 2014, Part III, LNCS 8836, pp. 653–660, 2014.
© Springer International Publishing Switzerland 2014

data is much more reliable compared to 3D due to the ability of capturing the 2.5D partial face data in a very short time. It also greatly improves the accuracy of the recognition performance.

Gabor-based region covariance matrix (GRCM) becomes a popular face descriptor recently. The innovation of GRCM is stemming from the introduction of region covariance matrix (RCM) by Porikli et al. [4] and Tuzel et al. [5]. RCM is a type of matrix-form feature descriptor which consists of covariance matrices computed from image statistics. Direct use of RCM in face recognition is not preferable as reported in [5] and [6]. Pang et al. [7] proposed GRCM for 2D face recognition, an approach to overcome the discrimination limitation of RCM, by embedding Gabor features in the construction of region covariance. GRCM is illumination invariant compared to conventional face recognizer that is sensitive to the illumination.

The covariance matrix of GRCM does not lie on Euclidean space. Thus, common Euclidean distance metric is not suitable to be used for distance matching in GRCM. Instead, GRCM resides on Tensor manifold, a type of topological space which locally similar with Euclidean space [8]. Each covariance matrix in the GRCM uniquely corresponds to a point in the Tensor manifold. The distance between two GRCMs, which always search for the shortest curve, is defined as *geodesic distance*. Affine Invariant Riemannian Metric (AIRM), the most widely used metric in computing the geodesic distance, is computationally heavy [9][10].

Based on the above-mentioned problems in 2D and 3D face recognition, objectives of this paper have been outlined as follows:

- To propose GRCM as a novel 2.5D face descriptor, which is resulted from the fusion of range and texture images in feature level.
- To study the performance of several geodesic distance metrics on Tensor manifold in order to find out the best alternative for 2.5D face recognition. The result serves as guideline to further enhance complicated classifiers in 2.5D face recognition.

2   Methodology in 2.5D Face Recognition

There are five steps involved in the proposed 2.5D face recognition, namely data preprocessing, fusion, Gabor feature extraction, GRCM computation and matching as illustrated in Fig 1.

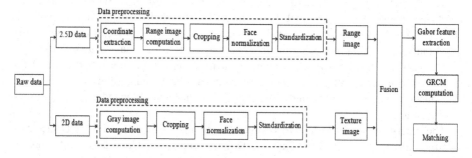

Fig. 1. Block diagram of proposed 2.5D face recognition

2.1   Data Preprosessing

There are many type of representation of 2.5D partial data, such as point cloud, polygon mesh, range image and facial curve [11][12][13]. Range image is selected because the structure of data is easy to be utilized and processed. Each point of the range image is associated with depth value which is nearest to the 3D sensor.

The processes involve in the data preprocessing for 2.5D partial data are coordinate extraction, range image computation, cropping, face normalization and standardization. The coordinates of x, y and z are extracted from the raw data by filtering unrelated point. Then, the range image is computed by extracting the z value through the (x,y) plane. The cropping is done manually and it generates the normalized face based on the location of eyes and mouth. Standardisation is used to rescaling the data to have zero mean and unit variance. Data preprocessing for 2D data consists of gray image computation, cropping, face normalization and standardization. The 2D texture data is converted into gray image and manual cropping is performed. The normalized face is produced based on the location of eyes and mouth. Standardisation is applied for zero mean and unit variance purpose.

2.2   Fusion

Each subject consists of two type of images; 2.5D partial image and 2D image. Since fused features are able to offer more discriminative result, feature-level fusion is done to combine the feature from range image and texture image. Two different fusion schemes are implemented. Fusion 1 is the concatenation between range image and texture image whereas Fusion 2 is integration of range image and texture image using the sum rule.

2.3   Gabor Feature Extraction

Gabor kernel is the product of elliptical Gaussian and a complex plane wave [14]. It is defined as:

$$\varphi_{u,v}(z) = \frac{\|k_{u,v}\|^2}{\sigma^2} e^{\frac{\|k_{u,v}\|^2 \|z\|^2}{2\sigma^2}} \left(e^{ik_{u,v}z} - e^{-\sigma^2/2}\right) \tag{1}$$

such that u and v represent the orientation and scale of the Gabor kernel. The wave vector $k_{u,v}$ is defined as $k_{u,v} = k_v e^{i\emptyset_u}$ where $k_v = k_{max}/f_v$ and $\emptyset_u = \pi u/8$. k_{max} is the maximum frequency and f_v is the spacing factor between the kernels in frequency domain. In this paper, we construct the Gabor kernel by taking 8 orientations ($u \in \{0...7\}$) and 5 scales ($v \in \{0...4\}$). This produces Gabor feature through the convolution between image **I** with pixel location (x, y) and Gabor kernel as illustrated in Equation 2 below:

$$g_{u,v}(x,y) = |\mathbf{I}(x,y) * \varphi_{u,v}(x,y)| \tag{2}$$

where $g_{u,v}(x,y)$ are the Gabor representations of image at orientation u and scale v.

2.4   Gabor-Based RCM

Let $\mathbf{I}$ represents an image with size h x w, where h and w are the height and width of the image region, respectively. The total number of pixels from the image region is, $n = hw$. A mapping function ϕ that used to extract d-dimensional vector z from a pixel at position (x, y) of $\mathbf{I}$ is defined as $\phi(\mathbf{I}, x, y) = \{z_i \in \mathfrak{R}^d \mid i=1\ldots n\}$. The d x d dimensional covariance matrix is then generated as below:

$$C = \frac{1}{n}\sum_{i=1}^{n}(z_i - \mu)(z_i - \mu)^T \tag{3}$$

where μ is the mean of vector z.

Mapping function is then further enhanced with the additional data such as pixel location, intensity component and Gabor feature to construct vector z in the form of $\phi(\mathbf{I}, x, y) = \begin{bmatrix} x\ y\ I(x,y) g_{0,0}(x,y)\ g_{0,1}(x,y) \ldots g_{7,4}(x,y) \end{bmatrix}$. By substituting the mapping function into Equation 3, Gabor-based covariance matrix, $G \in \mathfrak{R}^{43 \times 43}$ is computed. The dimensionality of G is 3 + 8 x 5 = 43, resulting from the additional data and Gabor feature. Table 1 below states three possible mapping features used to develop the covariance; GRCM1 includes the pixel locations, GRCM2 covers the pixel locations and intensity component, and GRCM3 excludes both pixel locations and intensity component.

Table 1. Feature Mapping Functions used in Experiment

Method	$\emptyset$ $(\mathbf{I}, x, y)$
GRCM1	$[x\ y\ g_{00}(x,y)\ g_{01}(x,y) \ldots g_{74}(x,y)]$
GRCM2	$[x\ y\ \mathbf{I}(x,y)\ g_{00}(x,y)\ g_{01}(x,y) \ldots g_{74}(x,y)]$
GRCM3	$[g_{00}(x,y)\ g_{01}(x,y) \ldots g_{74}(x,y)]$

2.5   Distance Matching

Each image is then further divided into five regions – whole face, left-half, right-half, upper-half, lower-half portion as displayed in Fig 2. Each of this region produce Gabor-based covariance matrix. Thus, five corresponding Gabor-based covariance matrices $\{G^k \mid k = 1 \ldots 5\}$ are computed.

Fig. 2. Five regions of face image

Since GRCM is a symmetric positive definite (SPD) matrix which does not lie on a Euclidean space. Thus, it is necessary to find a proper distance metric for measuring two GRCMs. To compute the distance metric between two GRCMs, M and N are hence denoted as:

$$\text{dist}(M, N) = \sum_{i=0}^{5} p(G_i^M, G_i^N) - \max_j(p(G_j^M, G_j^N)) \tag{4}$$

where $p(\cdot, \cdot)$ can be one of the distance measures listed in Table 2.

Table 2. Distance metrics used in the experiment

Name	Equation	Remarks
Euclidean Distance	$p_E(M, N) = \|M - N\|$	Euclidean distance is used to measure the distance between two points that connected by a straight path in Euclidean space. It is used here to serve as a baseline for comparison purpose.
Cholesky Distance (CHOL)	$p_E(M, N) = \|\|L_M - L_N\|\|_F$	Cholesky distance is a re-parameterisation measure that decomposes each GRCM uniquely into a product of a lower triangular matrix and its transpose, such as $M = L_M L_M^T$ and $N = L_N L_N^T$.
Kullback-Leibler Divergence Metric (KLDM)	$p_E(M, N)$ $= \sqrt{\dfrac{1}{2}Tr(M^{-1}N + N^{-1}M - 2I)}$	Kullback-Leibler Divergence Metric belongs to a family of information theoretic approach, which is used to measure the distance between two probability distributions.
LogDet Divergence (LD)	$p_E(M, N)$ $= \log\left\|\dfrac{M - N}{2}\right\| - \dfrac{1}{2}\log\|MN\|$ where $\|\cdot\|$ is matrix determinant.	LogDet Divergence (LD) is derived from family of information theoretic approach. It is a type of matrix divergence which measures the distance between two GRCMs.
Affine Invariant Riemannian Metric (AIRM)	$p_E(M, N) = \sqrt{\displaystyle\sum_{i=1}^{5} \ln^2\lambda_i(M, N)}$ where $\lambda_1, \ldots, \lambda_5$ are the eigenvalues of M and N.	Affine Invariant Riemannian Metric (AIRM) was proposed by Xavier et al. [15] for similarity measure on the tensor manifold. The affine-invariant computation involves the use of eigenvalues decomposition, logarithms, exponentials and square roots.
Log-Euclidean Riemann Metric (LERM)	$p_E(M, N)$ $= \|\|\text{Log}(M) - \text{Log}(N)\|\|_F$ where $\text{Log}(\cdot)$ is the matrix logarithm.	Log-Euclidean Riemann Metric (LERM) utilises the Euclidean metrics in the domain of logarithms in measuring the distance between two GRCMs.

3   Experiments and Discussion

The experiments are conducted on a subset of the FRGC version 2.0 dataset [16]. The experiments include randomly selected 8 samples of 2.5D range images and corresponding 2D texture image for each subject with total of 4064 images. Both range image and texture image are normalised to have size of 61 x 73. The parameter of Gabor kernel are set as $k_{max} = \pi/2, f_v = \sqrt{2}$, $\sigma = \pi$ and it produces Gabor wavelet with size of 40 x 40. Fours samples from image are set as training samples and the rest are used as testing samples. Two different fusion schemes have been studied.

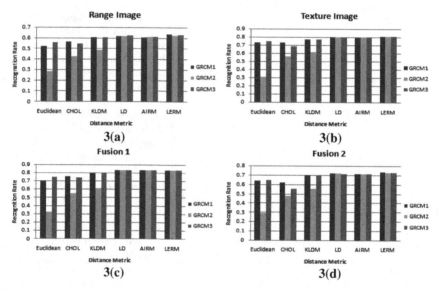

Fig. 3.   (a), (b), (c), (d).Experiment result of range image, texture image, fusion 1 and fusion 2 in term of recognition rate

The recognition performance of range image, texture image, fusion 1 and fusion 2 is presented in Fig 3. It can be seen that fusion 1 exhibits the best recognition rate among all images used in this experiment. Range image is unable to compete with texture image in term of recognition rate as shown in Fig 3. This is due to the simple preprocessing process that does not include the registration and alignment process for range image. However, two fusions are performing well than range image. These results suggest that the used of fusion improves the result of 2.5D face recognition.

We also compare the performances of geodesic distance metrics outlined in section 2.5. The results are shown in Fig 3 and Fig 4. Among the distance metrics, Euclidean distance obtains relatively low recognition rate in all cases. These directly prove that Euclidean distance is not suitable to be used for GRCM matching. CHOL achieves the fastest computation time as displayed in Fig 4 because the matrix decomposition only involves the product of lower triangular matrix and its transpose. Unfortunately, it exhibits second lower recognition rate due to the nature of CHOL where the Cholesky distance is not a true geodesic distance [17].

KDLM suffers from slowest computation time due to the high computation cost of KL distance as demonstrated in Fig 4. LERM is the distance metric that closely related to AIRM. From Fig 3 and Fig 4, LERM performs better than AIRM in term of recognition rate but LERM requires more computation time than AIRM. LD performs almost as well as LERM, and LD computes faster than LERM.

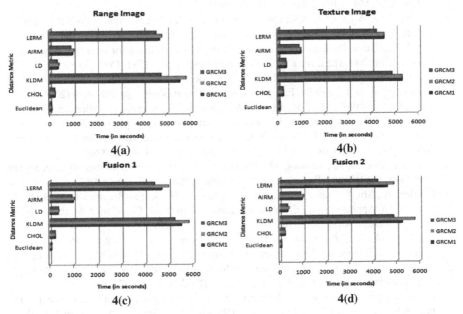

Fig. 4. (a), (b), (c), (d).Experiment result of range image, texture image, fusion 1 and fusion 2 in term of computation time

4   Conclusion

In this paper, we propose GRCM for 2.5D face recognition. From experimental result, fusion of range image and texture image is able to increase the recognition performance. We present a comparative study for several distance metrics through the use of GRCM in 2.5D face recognition. LD and LERM demonstrate the best performance in term of recognition rate in the chosen database. Based on the computation time, LD is chosen to be a good choice as distance metric in 2.5D face recognition.

Acknowledgements. This research was supported by Basic Science Research Program through the National Research Foundation of Korea (NRF) funded by the Ministry of Science, ICT and Future Planning (2013006574) and Fundamental Research Grant Scheme (FRGS) of Malaysia under grants MMUE/140021.

References

1. Bronstein, A., Bronstein, M., Kimmel, R.: Expression-Invariant 3D Face Recognition. In: Kittler, J., Nixon, M.S. (eds.) AVBPA 2003. LNCS, vol. 2688, pp. 62–70. Springer, Heidelberg (2003)
2. Lu, X., Jain, A.K., Colbry, D.: Matching 2.5D Face Scans to 3D Models. IEEE Trans. on Pattern Anal. Mach. Intell. 28(1), 31–43 (2006)
3. Liu, P., Woo, W.L., Dlay, S.S.: One Colored Image Based 2.5D Human Face Reconstruction. In: 17th European Signal Processing Conference, Scotland, pp. 2584–2588 (2009)
4. Porikli, F., Tuzel, O.: Fast Construction of Covariance Matrices for Arbitrary Size Image Windows. In: 2006 IEEE Inter. Conf. Image Processing, pp. 1581–1584 (2006)
5. Tuzel, O., Porikli, F., Meer, P.: Region Covariance: A Fast Descriptor for Detection and Classification. In: Proc of 9th European Conf. on Comp. Vision, pp. 589–600 (2006)
6. Tuzel, O., Porikli, F., Meer, P.: Human Detection via Classification on Riemannian Manifolds. In: IEEE Conf. Computer Vision and Pattern Recognition, pp. 1–8 (2007)
7. Pang, Y., Yuan, Y., Li, X.: Gabor-Based Region Covariance Matrices for Face Recognition. IEEE Trans. on Circuits Syst. Video Technol. 18(7), 989–993 (2008)
8. Tosato, D., Spera, M., Cristani, M., Murino, V.: Characterizing Humans on Riemannian Manifolds. IEEE Trans. Pattern Anal. Mach. Intell. 35(8), 1972–1984 (2013)
9. Cherian, A., Sra, S., Banerjee, A., Papanikolopoulos, N.: Jensen-Bregman LogDet Divergence with Application to Efficient Similarity Search for Covariance Matrices. IEEE Trans. on Pattern Anal. Mach. Intell. 35(9), 2161–2174 (2013)
10. Alavi, A., Wiliem, A., Zhao, K., Lovell, B.C., Sanderson, C.: Random Projections on Manifolds of Symmetric Positive Definite Matrices for Image Classification. CoRR (2014)
11. Fabry, T., Smeets, D., Vandermeulen, D.: Surface Representation for 3D Face Recognition. In: Oravec, M. (ed.) Face Recognition, pp. 273–294. InTech (2010)
12. Lu, X., Colbry, D., Jain, A.K.: Matching 2.5D Scans for Face Recognition. In: Zhang, D., Jain, A.K. (eds.) ICBA 2004. LNCS, vol. 3072, pp. 30–36. Springer, Heidelberg (2004)
13. Samir, C., Srivastava, A., Daoudi, M.: Three-Dimensional Face Recognition Using Shapes of Facial Curves. IEEE Trans. on 28(11), 1858–1863 (2006)
14. Lee, T.S.: Image Representation using 2D Gabor Wavelets. IEEE Trans. on Pattern Anal. Mach. Intell. 18(10), 959–971 (1996)
15. Pennec, X., Fillard, P., Ayache, N.: A Riemannian Framework for Tensor Computing. Int. J. Comput. Vis. 66(1), 41–66 (2006)
16. Phillips, P.J., Flynn, P.J., Scruggs, T., Bowyer, K.W., Chang, J., Hoffman, K., Marques, J., Min, J., Worek, W.: Overview of The Face Recognition Grand Challenge. In: IEEE Computer Society Conference, vol. 1, pp. 947–954 (2005)
17. Jayasumana, S., Hartley, R., Salzmann, M., Li, H., Harandi, M.: Kernel Methods on the Riemannian Manifold of Symmetric Positive Definite Matrices. In: IEEE Conference, pp. 73–80 (2013)

Predicting Mobile Subscriber's Behaviour from Contextual Information Extraction: SMS Data

Ayesha Javed Butt[1], Naveed Anwer Butt[2,*], Rabia Ghias Butt[3],
and Muhammad Touseef Ikram[4]

[1]Faculty of Computing and Information Technology, University of Gujrat, Gujrat, Pakistan
[2]Institute of Space Technology, Islamabad, Pakistan
{12031719-005,naveed}@uog.edu.pk, Rabiag7@gmail.com,
touseef.ikram@ist.edu.pk

Abstract. Mobile phones are the most significant way of communication in this era and SMS is the extensively used mobile service around the globe. This extensive use of SMS puts oil into fire of competition between different telecom companies. Only the one who manages to make subscribers believe "We care for you" can survive, obviously keeping company's own benefits and profits in mind. This research paper presents a technique to effectively design and develop smart SMS packages by using contextual information of SMS data. It also deals with the prediction of multiple patterns about subscriber's behaviour by using subscribers demographic and timing information.

Keywords: SMS, Mobile phone, Information Extraction, Contextual Information, Customer Satisfaction, Data Mining.

1 Introduction

The first and the foremost goal of a business are to increase profits. Every telecom company aims to increase and retain its patrons/subscribers in order to enhance company's market share. Therefore, a tough competition exists in the market. Only the best can survive and the best is the one who knows the client i.e. who knows how to attract the client by fulfilling their desires and needs. Now a day's conversion to another service provider is only a single click away. Facing with this threat, companies should be equipped and armed with the most efficient and effective methods of examining their client's behaviour. In the recent years, the trend of mobile phone subscribers' is towards SMS service. Recently, a survey showed that the most widely used mobile service is text messaging with 72% active SMS users out of all mobile phone users worldwide [1].

So the best area to target by a telecom company is the SMS [4] i.e. the company can attract the most subscribers by offering smart packages. Most of the companies till now are offering packages based on their resource management but it would be economical to consider contextual information while designing SMS packages.

---

* Corresponding author.

C.K. Loo et al. (Eds.): ICONIP 2014, Part III, LNCS 8836, pp. 661–672, 2014.

1.1   Contextual Information Extraction of SMS Data

SMS is a mobile phone service for sending short text messages from one phone to another mobile phone [2] and Package is the container that carries something for storage/transportation [3]. So SMS package is a container that contains desired choices for subscriber. Contextual information is defined as "information that is relevant to and understanding of the text". Contextual Information Extraction is the IE of the contextual data i.e. time, frequency of SMS, gender, age etc. to model user trends.

2   Literature Review

In order to cover the whole topic in detail for previous works, we divided the topic into four categories i.e. Context information, Information extraction, Contextual Information extraction and SMS trends.

2.1   Context Information

Contextual information is defined as "information relevant to understanding of the text/ data". Therefore the contextual information extraction of SMS data means to identify the kind of people using SMS packages, their satisfaction level, motives, timing and frequency. User's interest modelling is important to effectively anticipate and fulfil user needs and context is important for this purpose [6]. Peter Bailey and Liwei Chen (2009) checked the effectiveness of five different sources of contextual information using them as basis for user's interest modelling. They concluded that the overlapping of multiple context variables provides better results as compared to a single context variable [6]. Tom Heath, Enrico Motta and Martin Dzbor (2005) also declared the same opinion i.e. the search engines need to use different aspects of user context in order to support user's online tasks and surfing. These context aspects may include user's social network, experiences, preferences, current location, services they use and the third party whom user trusts etc [7].

2.2   Information Extraction

Information extraction [5] has been applied to various types of data. Richard Cooper and Sajjad Ali (2006) applied IE techniques on textual data, SMS text data was extracted by identifying linguistic structures in the messages with the goal to update data repository. The approach was to find a matching pattern for the sentence, data extraction and use of that data to update templates appropriate for that pattern [8]. Yonatan Autmann, Ronen, Benjamin, Yair and Jonathan were of opinion that although visual and typographic information is very important, most of the IE tools ignore such information and process only text. So, they presented an IE system based wholly and solely on documents' visual characteristics. They proved that for specific field such as publication date and author names, visual information alone is sufficient to provide high accuracy [9]. Chia-Hui Chang, Moheb Ramzy, Mohammed Kayed and Shaalan studied and compared major IE tools based on three factors; the automation degree, the techniques and the task domain. The first factor is used to measure the degree of automation of IE systems; the second one is used to classify IE systems based on the techniques used and the third factor answers the question why an IE system fails to handle some Web sites of particular structures [10].

2.3   Contextual Information Extraction

Contextual Information extraction is the IE of contextual data i.e. in terms of time, frequency, gender and age etc. Javier Tamames and Victore de Lorenzo (2010) proposed a text mining system for extracting contextual information automatically. They have developed a complete ontology for the annotation of a biological sample [11]. Jay Chen, Brendan Linn, Lakshmi narayanan Subramanian (2009) described the details of SMSFind which is a search engine based on SMS.The client sends a query to a number of server mobiles which processes the search query on a search engine; the results are then distilled to only 140 bytes. The results showed that SMSFind gives a correct response of 22 out of 31 queries [12]. Khaled Amailef, je Lu and Jun Ma presented   an algorithm with in MERS applications. The purpose of the algorithm is to extract and aggregate information from text messages received via mobile phones sent by the user in emergency situations. Results of the algorithm may assist decision makers to respond in emergency situations [20]. Duan yu chen, chia hsun chen, yung sheng and jun jhe wang (2011) proposed a dynamic visual intelligent system to detect and analyse the taillights of vehicles and extract corresponding contextual information. The results of their experiments showed that under different traffic conditions and lighting, their proposed system can extract critical contextual information effectively [21].

2.4   SMS Trends

\To consider subscribers' SMS trends, we considered three major factors i.e. frequency, age and customer satisfaction.

Frequency. Generally seven factors i.e. convenience, enjoyment, personal communication, escape, economical reasons, social involvement and public expression affect SMS usage frequency. T.Ramayah, Yulihasri, Amlus Ibrahim and NorzalilaJamaludin (2006) predicted SMS usage among university students using technology acceptance model (TAM). The two factors PU (perceived usefulness) and PEU (perceived ease of use) are considered the key determinants of the actual system usage proposed by TAM. They determined that the usage and acceptance of any technology or system perceived usefulness was more essential [13]. Whereas Ian Phau, Min Teah (2009) conducted a research to find the motives for using SMS and how these motives influence SMS usage frequency and attitudes towards SMS promotions/advertising. They also tried to find if there is any positive relationship b/w SMS promotions and its usage [14].

Age Group. The number of text messages sent every day has increased as compared to the number of calls and the text messaging trend is increasing alarmingly. A survey carried out in US showed that the kids below age 12 send 428 messages per month on average. Nor Shahriza Abdul Karim and Ishaq Oyebisi Oyefolahan (2009) explored the differences of mobile use in terms of age, gender and occupation. In their study, appropriation pattern, the relationship between age, occupation and gender status and appropriation variables are discussed. For entertainment or leisure, the younger the age, the more probability of engaging the mobile phone to gratify the purpose [15]. Vilama Balkrishman and Paul H.P. Yellow (2007) have investigated the effect of age and gender on mobile phone users' texting satisfaction, text entry factors being a focus of study. The younger users were found to be more satisfied with the navigation and learning while texting than users in their thirties. Males have the capability of understanding and learning technical things faster than the females, thus making them to be more satisfied than females. Users' Satisfaction declines as age increases as well [16].

Customer Satisfaction. A direct relationship exists b/w number of subscribers' of a network and their satisfaction level i.e. greater customer satisfaction means more customers. Two factors that generally influence customer satisfaction are price fairness and SMS services offered. Ishfaq Ahmed, Zafar Ahmad, Muhammad, Muhammad Musarrat Nawaz, Zeeshan Shaukat, Ahmad Usman and Sarfraz Ahmed (May 2010) presented in their paper that the cost required to retain customers is much less than the cost required making new customers. The results indicate that the satisfaction has a significant relationship with the customer retention. Similarly, gender has a significant relationship with satisfaction and intentions to retain as customer [17]. Muzammil Hanif, Sehrish Hafeez and Adnan Riaz (2010) conducted a research to determine the factors that affect customer satisfaction. Customer satisfaction and price fairness has strong and direct relationship; if customer services are good then customers are willing to pay more for the product or service and if price fairness exists then it would add to customer satisfaction [18].

3   Methodology

To deal with the data increasing day by day, several processes for automated information retrieval have been made. Association rule mining, one of these information extraction/ rule generation technique is used for this work.

3.1   Data Collection

The data was collected from NUS (National University of Singapore) website [22]. Two databases were available, one was English database and other was Chinese database. The corpus of databases was available in SQL and XML formats. Both formats had same fields. English database contained 38,725 records that were last updated on 31 August 2011. Chinese database contained 21,444 records. The databases contained 19 fields i.e. sender, receiver, send date, collect_time, collect_method, content, native, country, age, input_method, experience, frequency, phone_model, collector, gender, Smartphone, lang, city, and id.

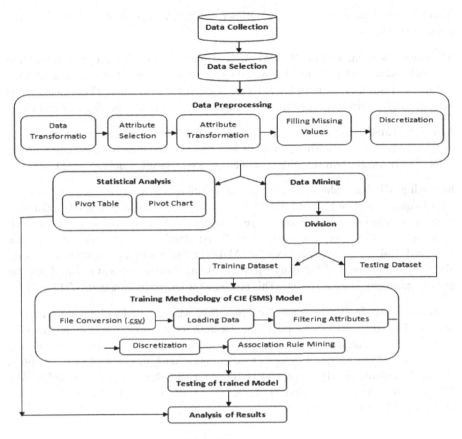

Fig. 1. Methodology

3.2   Data Selection

We selected English database stored in SQL format. This selected database has been used lately as the basic data set for model building.

3.3   Data Pre-processing

Raw data are seldom used for data mining. Many data mining methods such as prediction require data set in particular format to produce efficient results. Therefore we may need to perform various transformations on the data; this is the most critical step and is called pre-processing. Steps that are taken to preprocess the data are as follow:

Data Transformation. The data was not in proper format. So the most important task was to arrange it according to the needs of the model building. To perform statistical analysis, we needed the database in .xlxs format. But database was in .txt format. Moreover, it was not compatible with SQL Server 2010 so we modified it to the correct syntax.

After properly compiling the data in SQL Server 2010, we exported the database in Microsoft Excel format.

Attribute Selection. Out of 19 fields, we took out only 10 fields that were relevant to our work and discarded the rest because those were irrelevant with the context of SMS and more of all, they do not play any role in behavior prediction. These fields were: Sender, receiver, send date, country, age, experience, frequency, gender, city and id.

Attribute transformation. Sender and receiver fields of database contained encrypted data for security reasons. This encrypted data was not readable so we replaced the encrypted entries with short readable identities i.e. send1, send2, send3…. Rec1, Rec2…

Removing Missing Values. We replaced the "unknown" value in each cell with the most frequent value of that cell. E.g.in age column, cells which contained "unknown" value, was replaced by most frequent age group i.e. 21-25. Similarly, in experience column, "unknown" values were replaced by "5-10 years". In gender column, cells that contained "unknown" were replaced by "Male" and in frequency; "unknown" was replaced by "more than 10 SMS daily". In City column, "unknown" was replaced with the most frequent city of each country. This procedure is known as "Imputation" [26][27].

Discretization. In the "send_time" column, time was recorded with the difference of seconds. But we needed categorical time duration for the sake of statistical Analysis, so we split the column "send_time" into two columns i.e. "send_time_hour" and "send_Date". Then we make groups of the column "send_time_hour" with the option "Group" available in MS Excel. So resulting groups that we got were 0(12A.M)-3(A.M), 4(A.M)-7(A.M), 8(A.M)-11(A.M), 12(P.M)-15(3 P.M), 16(4 P.M)-19 (7 P.M) and 20(8 P.M)-23(11 P.M).

3.4   Statistical Analysis

To analyse subscriber's trend toward SMS and their SMS usage patterns, we conducted a statistical analysis with the help of pivot table and pivot chart in MS Excel. In our case, the features that we considered are Age Group vs. Frequency, Gender vs. Frequency, Gender and Age vs. Frequency, Country vs. Frequency, Country and Gender vs. Frequency, Experience and Gender vs. Frequency, Country and city vs. Frequency, Time vs. Frequency, Country and Age vs. Frequency, Experience vs. Frequency and Age and experience vs. Frequency.

3.5   Data Mining

The weka (Waikato environment for knowledge Analysis) tool is selected for predictive model building to train data. Weka is a complete and comprehensive tool bench for machine learning and data mining. This is particularly strong in the area of classification. Association rules, clustering algorithms and regression have also been implemented in this software. In this phase, following steps were taken out: The records in excel format were statistically analysed with the help of pivot table along with its pivot charts. The pivot table and pivot charts give us a summarized data and visualization of patterns found in database. From these patterns, we were able to derive out rules that show subscriber's SMS trends.

Age Group Vs. Frequency

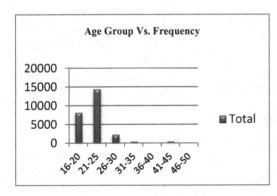

Fig. 2. Age Group vs. Frequency

By inspecting the graph, we can see the trend that users having age 21-25 send maximum messages.

Rule# 1:

If (age >16 && age <25) then frequency=5-10 smsdaily

Else if (age>25) then frequency=(less than 1 sms daily‖ 1 -2 sms daily)

Else frequency= (2-5 smsdaily‖ more than 10 sms daily‖more than 50 sms daily).

Gender Vs. Frequency

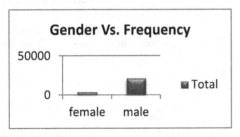

Fig. 3. Gender vs. Frequency

We can observe the trend that, males send more message than females. To increase this trend in female as well, Company can introduce better packages to females.

Rule # 2:

If (gender= male) then frequency=5-10 sms daily

Else frequency= =(less than 1 sms daily‖ 1 -2 sms daily)

Country Vs. Frequency

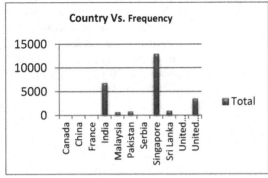

Fig. 4. Country vs. Frequency

Rule # 3:

If (country=Singapore) then frequency=5-10 smsdaily

Else if ((country=Serbia) ‖ (country=France) ‖ (country=United Kingdom)) then Frequency= (2-5 smsdaily‖ more than 10 sms daily‖more than 50 sms daily)

Else frequency= (2-5 smsdaily‖ more than 10 sms daily‖more than 50 sms daily)

Experience and Gender Vs. Frequency

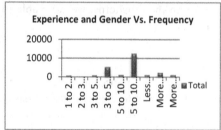

Fig. 5. Experience and Gender vs. Frequency

Country and Gender vs. Frequency

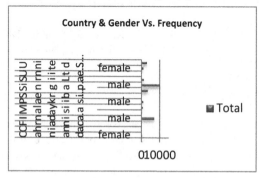

Fig. 6. Country and Gender vs. Frequency

Time vs. Frequency

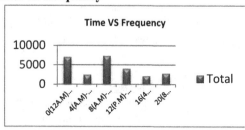

Fig. 7. Time Vs Frequency

Rule # 4:

If (gender=male && (exp>5yrs && exp <10yrs)) then frequency=5-10 smsdaily

Else if (gender=male && (exp>2yrs && exp <3yrs)) then frequency=(less than 1 sms daily‖ 1 -2 sms daily)

Else frequency=(2-5 smsdaily‖ more than 10 sms daily‖more than 50 sms daily)

Rule # 5:

If (country= Singapore && gender=male) then frequency=5-10 smsdaily

Else if ((Country= Serbia ‖ country = France ‖ country= United Kingdom) && gender=male) then frequency=(less than 1 sms daily‖ 1 -2 sms daily)

Else frequency=(2-5 smsdaily‖ more than 10 sms daily‖more than 50 sms daily).

Rule # 6:

If ((time>= 12 am && time <= 3am) && (time>= 8am && time >=11 am)) then frequency=5-10 smsdaily

Else if ((time>= 4 pm && time <= 7pm) then frequency= (less than 1 sms daily‖ 1 -2 sms daily)

Else frequency=(2-5 smsdaily‖ more than 10 sms daily‖more than 50 sms daily).

Country & age Vs Frequency:

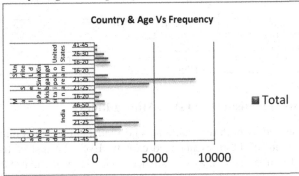

Fig. 8. Country and age vs. Frequency

Rule # 7:

If (country= Singapore && (age > 16 && age <25)) then frequency= 5-10 sms daily
Else if (country=India ‖ country=United Kingdom) && (age>36 && age<40) then frequency=(less than 1 sms daily‖ 1 -2 sms daily) Else frequency = (2-5 smsdaily‖ more than 10 sms daily‖more than 50 sms daily).

Age & Experience Vs Frequency:

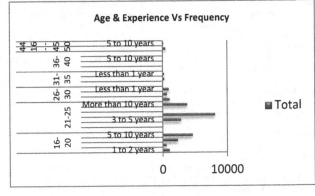

Fig. 9. Age and Experience vs. Frequency

Rule # 8:

If((age>16 && age <25) && (exp>5 years && exp < 10 years)) then frequency = 5-10 sms daily
Else if ((age>31 && age <40) && (exp>3 years && exp < 5 years)) then frequency =(less than 1 sms daily‖ 1 -2 sms daily)
Else frequency = (2-5 smsdaily‖ more than 10 sms daily‖more than 50 sms daily).

3.6 Association Rule Mining

Weka was used to generate rules using Association Rule Mining and varying matrices. We created 100 rules out of which some are mentioned below;

Table 1. Resultant Rules of Association Rule Mining

1. country=India 5061 ==> gender=male 5061 conf:(1)	8. sender=send90 3282 ==> age=16-20 3282 conf:(1)
2. age=16-20 frequency=5 to 10 SMS daily 4617 ==> gender=male 4617 conf:(1)	9. sender=send90 3282 ==> experience=5 to 10 years 3282 conf:(1)
3. city=Adam Park 4121 ==> country=Singapore 4121 conf:(1)	10. sender=send90 3282 ==> frequency=5 to 10 SMS daily 3282 conf:(1)

4. country=Singapore frequency=5 to 10 SMS daily gender=male 4012 ==> experience=5 to 10 years 4012 conf:(1)	11. sender=send90 3282 ==> gender=male 3282 conf:(1)
5. country=Singapore experience=5 to 10 years frequency=5 to 10 SMS daily 4012 ==> gender=male 4012 conf:(1)	12. sender=send90 age=16-20 3282 ==> country=Singapore 3282 conf:(1)
6. country=Singapore age=16-20 3645 ==> experience=5 to 10 years 3645 conf:(1)	13. sender=send90 country=Singapore 3282 ==> age=16-20 3282 conf:(1)

3.7   Comparative Analysis of Statistical and Data Mining Rules

The comparison of rules generated by both the techniques i.e. statistical and data mining techniques showed that 7 out of 11 rules matched exactly. Even the rest of the data mining rules had almost similar pattern as in statistical rules. It means that data mining rules support our statistical analysis. In the next step, we will check the precision, recall and accuracy of our rules.

Table 2.   Comparative Analysis of Rules

Statistical Rules	Association Rules
If (age >16 && age <25) then frequency=5-10 smsdaily Else if (age>25) then frequency=(less than 1 sms daily‖ 1 -2 sms daily) Else frequency= (2-5 smsdaily‖ more than 10 sms daily‖more than 50 sms daily)	age=16-20 3282 ==> frequency=5 to 10 SMS daily 3282 conf:(1)
If (gender= male) then frequency=5-10 sms daily Else frequency= =(less than 1 sms daily‖ 1 -2 sms daily)	gender=male 3282 ==> frequency=5 to 10 SMS daily 3282 conf:(1)
If (country=Singapore) then frequency=5-10 smsdaily Else if ((country=Serbia) ‖ (country=France) ‖ (country=United Kingdom)) then Frequency=(2-5 smsdaily‖ more than 10 sms daily‖more than 50 sms daily) Else frequency=(2-5 smsdaily‖ more than 10 sms daily‖more than 50 sms daily)	country=Singapore 3282 ==> frequency=5 to 10 SMS daily 3282 conf:(1)
If (country= Singapore && gender=male) then frequency=5-10 smsdaily Else if ((Country= Serbia ‖ country = France ‖ country= United Kingdom) && gender=male) then frequency=(less than 1 sms daily‖ 1 -2 sms daily) Else frequency=(2-5 smsdaily‖ more than 10 sms daily‖more than 50 sms daily)	country=Singapore gender=male 3282 ==> frequency=5 to 10 SMS daily 3282 conf:(1)
If (exp > 5 && exp < 10) then frequency= 5-10 sms daily Else if (exp > 2 && exp < 3) then frequency = (less than 1 sms daily‖ 1 -2 sms daily) Else if (exp > 1 && exp <2) ‖ (exp>3 && exp <5) ‖ (exp >10) ‖ (exp<1) = (2-5 smsdaily‖ more than 10 sms daily‖more than 50 sms daily)	experience=5 to 10 years 3282 ==> frequency=5 to 10 SMS daily 3282 conf:(1)
If (country= Singapore && (age > 16 && age <25)) then frequency= 5-10 sms daily Else if (country=India ‖ country=United Kingdom) && (age>36 && age<40) then frequency=(less than 1 sms daily‖ 1 -2 sms daily) Else frequency = (2-5 smsdaily‖ more than 10 sms daily‖more than 50 sms daily).	country=Singapore age=16-20 3282 ==> frequency=5 to 10 SMS daily 3282 conf:(1)
If ((age>16 && age <25) && (exp>5 years && exp < 10 years)) then frequency = 5-10 sms daily Else if ((age>31 && age <40) && (exp>3 years && exp < 5 years)) then frequency =(less than 1 sms daily‖ 1 -2 sms daily) Else frequency = (2-5 smsdaily‖ more than 10 sms daily‖more than 50 sms daily)	age=16-20 experience=5 to 10 years 3282 ==> frequency=5 to 10 SMS daily 3282 conf:(1)

4   Conclusion

In the nutshell, we have successfully extracted behaviour patterns from the context database of users. Moreover, we have successfully predicted customer's future trends towards SMS packages. It was observed that both data mining and statistical rules support each other.

Future Work

- Scope of the work can be extended by using real time data or more user context attributes.
- If the real time data is available, then we can also suggest that which particular customer a package should be suggested. In this way, a company does not need to broadcast a message to all its subscribers rather it can broadcast it to an automatically generated list of subscribers. This will save company's expenditure. In future, we can implement this research work with the help of AI techniques; it will help to automate package development procedure.

References

[1] http://www.smsfeedback.com.au/facts.htm
[2] http://www.webopedia.com/TERM/S/SMS.html
[3] http://www.thefreedictionary.com/package
[4] Zerfos, P., Meng, X., Wong, S.H.Y., Samanta, V., Lu, S.: A study of short message service of a nationwide cellular network. In: Proc. of ACM SIGCOMM Internet Measurement Conference (IMC), Rio de Janeiro (2006)
[5] Sarawagi, S.: Information Extraction. Foundations and Trends in Databases 1(3) (2008)
[6] White, R.W., Bailey, P., Chen, L.: Predicting User Interests from Contextual Information. In: Proceedings of the 32nd International ACM SIGIR Conference on Research and Development in Information Retrieval (2009)
[7] Heath, T., Motta, E., Dzbor, M.: Uses of Contextual Information to Support Online Tasks. In: 1st AKT Doctoral Symposium, Milton Keynes, UK, June 14-16 (2005)
[8] Cooper, R., Ali, S.: Extracting Data from Personal Text Messages (2006)
[9] Aumann, Y., Feldman, R., Liberzon, Y., Rosenfeld, B., Schler, J.: Visual Information Extraction. Knowledge and Information Systems (2006)
[10] Chang, C.-H., Kayed, M., Girgis, M.R., Shaalan, K.: Survey of Web Information Extraction Systems. IEEE Transactions on Knowledge and Data Engineering, TKDE-0475-1104.R3 (2006)
[11] Tamames, J., de Lorenzo, V.: EnvMine: A text mining systemfor the automatic extraction of contextual information (2010)
[12] Chen, J., Linn, B., Subramanian, L.N.: SMS-Based Contextual Web Search. In: Proceedings of the 1st ACM Workshop on Networking, Systems, and Applications for Mobile Handhelds (2009)
[13] Ramayah, T., Yulihasri, E., Ibrahim, A., Jamaludin, N.: Predicting Short Message Service (SMS) Usage among University Students using the Technology Acceptance Model (TAM). In: 15th International Conference on Management of Technology (IAMOT 2006), Beijing, P.R. China, May 22-26 (2006)

[14] Phau, I., Teah, M.: Young consumers' motives for using SMS and perceptions towards SMS advertising. Direct Marketing: An International Journal 3(2), 97–108 (2009)

[15] Karim, N.S.A., Oyefolahan, I.O.: Mobile Phone Appropriation: Exploring Differences in terms of Age, Gender and Occupation. In: The 6th International Conference on Information Technology and Applications, ICITA 2009 (2009)

[16] Balkrishman, V., Yellow, P.H.P.: Texting satisfaction: does age and gender make a difference. International Journal of Computer Science and Security (2007)

[17] Ahmed, I., Nawaz, M.M., Ahmad, Z., Shaukat, M.Z., Usman, A., Ahmed, S.: Impact of Demographical Factors and Extent of SMS usage on Customer Satisfaction and Retention. Interdiscilinary Journal of Contemporary Research in Business (2010)

[18] Hanif, M., Hafeez, S., Riaz, A.: Factors Affecting Customer Satisfaction. International Research Journal of Finance & Economics (60), 44 (2010)

[19] Amailef, K., Lu, J., Ma, J.: Text Information Extraction and aggregation in a mobile based emergency response system (2009)

[20] Chen, D.-Y., Wang, J.-J., Chen, C.-H., Chen, Y.-S.: Video based intelligent vehicle contextual information extraction for night conditions

[21] http://wing.comp.nus.edu.sg:8080/SMSCorpus/sql.jsp

[22] Insignt into Data Mining (theory and Practice) Book

[23] Saar-Tsechansky, M., Provost, F.: Handling Missing values when applying classification models. Journal of Machine Learning Research (2007)

[24] Allison, P.D.: Multiple imputation for missing data: A cautionary tale. Sociological Methods of Research 28(3), 301–309 (2000)

Discovering Plain-Text-Described Services Based on Ontology Learning

Hai Dong[1], Farookh Khadeer Hussain[2], and Athman Bouguettaya[1]

[1] School of Computer Science and Information Technology, RMIT University, Australia
{hai.dong,athman.bouguettaya}@rmit.edu.au
[2] Decision Support and e-Service Intelligence Lab, Centre for Quantum Computation and Intelligent Systems, School of Software, University of Technology Sydney, Australia
farookh.hussain@uts.edu.au

Abstract. In this paper, we present an approach to efficiently discover domain-specific services that are described by plain text over the Internet. Plain-text-described service advertisements account for the vast majority of service advertisements over the Internet, but current research rarely focuses on this area. To address this issue, we design a domain-ontology-based approach for automatic plain-text-described service discovery. This approach incorporates a plain-text-described service ontology for standard service description, a plain-text-described service discovery framework for domain-relevant service discovery and ontology learning, and a machine-learning-based model for ontology-based service functionality annotation. The experimental results show that this approach is able to efficiently discover more relevant plain-text-described services than other approaches.

Keywords: service computing, service discovery, Web mining, ontology learning, focused crawler.

1 Introduction

The IT industry is now in a period of transformation. In the enterprise market, sales of ready-made software or hardware products have gradually been replaced by the provision of customized IT services, aimed at individual problem solving. In large IT companies, such as IBM and HP, sales of enterprise services generate substantial revenue[1,2]. Service computing is a new research field which studies the science and technology underlying the popularity of this emerging industry and spans multiple research areas[3].

Service discovery is an important research topic covered by service computing, which aims to search for and discover services for service users (human or software

---

[1] http://www.ibm.com/investor/4q12/press.phtml
[2] http://h30261.www3.hp.com/phoenix.zhtml?c=71087&
p=irol-newsarticle&ID=1760639
[3] http://conferences.computer.org/scc/2014/

C.K. Loo et al. (Eds.): ICONIP 2014, Part III, LNCS 8836, pp. 673–681, 2014.
© Springer International Publishing Switzerland 2014

agents) [1]. Traditional service discovery research mostly focuses on services described in standard languages, such as Web Service Description Language[4] and Web Application Description Language[5]. Since these languages provide unified notation systems to denote the descriptive parameters of functionality and the attributes of services, service discovery becomes a straightforward matching process between the annotated parameters of service requests and service descriptions. These languages undoubtedly ease the operational complexity of service discovery.

Most services on the Internet, especially business services, are unfortunately described in plain text. Service providers write and publish service descriptions on their websites or in online business directories. Unlike services described in standard formats, a major difference of services described in plain text is that they do not have standard notation systems. The functionality and attributes of these services are embedded in ambiguous plain-text descriptions. Domain knowledge is therefore required to mine the service information from these descriptions. Ontology, as a means of specific domain knowledge representation [2], can be utilized for plain-text-described service discovery. Nevertheless, a natural gap exists between domain-expert-defined ontologies and service-provider-published service descriptions. That is, the terms for defining service ontological concepts might be inconsistent with the terms for describing services. The impact of this inconsistency of terms in the plain-text-described service discovery process is that the originally used ontologies are incapable of completely capturing the functionality and attributes of the services from Web information. This eventually results in a substantial number of domain-relevant services being undiscovered.

In our previous research [3, 4], we targeted the plain-text-described service discovery problem. We proposed a novel ontology-learning-based focused crawling approach that enables ontological vocabulary learning from Web information. This approach shows a preliminary improvement on existing approaches in the retrieval of relevant plain-text-described services. However, the experimental results show that the improvement is still inadequate when the substantial proportion of non-retrieved relevant services is taken into account.

In this paper, we propose a more efficient approach for plain-text-described service information discovery. This approach incorporates 1) an enhanced ontology for annotating the functionality and attributes of plain-text-described services; 2) an improved framework for plain-text-described service discovery and ontology learning; and 3) a more efficient matching model for ontological-concept-based service functionality annotation.

The rest of the paper is organized as follows: in Section 2 we briefly review the previous work in the areas of plain-text-described service information discovery and ontology-learning-based focused crawling; in Section 3 we present the plain-text-described ontology; in Section 4 we present the plain-text-described service discovery framework; in Section 5 we deliver the matching model; in Section 6 we reveal the evaluation details of the approach, and in Section 7 we summarize our work and plan future research.

---

[4] http://www.w3.org/TR/wsdl
[5] http://www.w3.org/Submission/wadl/

2   Related Work

In this section, we briefly review the relevant research work in the areas of plain-text-described service discovery and ontology-learning-based focused crawling.

Although service discovery is a hot research topic in service computing, little work has been conducted in the area of plain-text-described service discovery. It is only recently that this research topic has attracted the attention of researchers. Noor et al. [5] worked on the topic of cloud service discovery. The results of their survey showed that up to 98% of cloud services are described in plain text. Hence, they designed an ontology-based crawler engine prototype – CSCE for cloud service discovery. CSCE uses concepts of a prebuilt Cloud Service Ontology as keywords to retrieve the URLs of cloud service Web pages from the Internet. The retrieved URLs are sent to a Cloud Service Verifier to verify their activation and a Cloud Service Validator to evaluate their validity. The information about the active and valid cloud services is extracted and stored in a database. Compared to our research, this work does not address the issue of real-time ontology update in the crawling process.

Ontology-learning-based focused crawling is a subfield of ontology-based focused crawling. This type of crawler uses ontologies for topic-specific crawling while refining ontologies in the crawling process [4]. Zheng et al. [6] proposed such a crawler based on a Feedforward Neural Network (FNN) model. The main principle of the crawler is to use the trained FNN model to determine the relatedness between Web documents and topical ontologies based on the number of occurrences of ontological concepts in the documents. Su et al. [7] presented a reinforcement-learning-based crawler. Given a topical ontology, the relevance score between a Web document and the ontology is the sum of the occurrences of ontological concepts in the document multiplied by the weights of those concepts. Reinforcement learning [10] is utilized to train the weight of a concept by calculating the probability of the relevant Web documents and the conditional probability of the relevant Web documents where the concept occurs. Phuc et al. [8] proposed a Support Vector Machine (SVM)-based ontology-learning crawler. Analogously, ontological concepts are used to retrieve Web documents. A SVM model [11] is used to determine the relatedness between Web documents and the ontology, which is trained based on the *tf-idf* values of selected tokens in relevant documents.

One common limitation of the above studies is that they only attempt to learn the associations between ontologies and Web documents. None of these approaches tries to *"learn"* new knowledge from crawled Web information, which could lead to Web documents with new relevant terms being misidentified as non-relevant. This motivates us to design a novel crawling framework that will enable ontological vocabulary learning. Our previous work [3, 4] show preliminary progress in retrieving a greater number of relevant service documents. In this project, we will present a novel approach that aims to further enhance this progress.

3   Plain-Text-Described Service Ontology

In this section, we present a Plain-Text-Described Service Ontology for describing the functionality and attributes of plain-text-described services. In fact, most plain-text-described services exist in the context of business services. Hence, this ontology is extended from the Web Ontology Language for Web Services (OWL-S)[6] and adapted to the context of generic business services. As shown in Fig. 1, the service ontology comprises two major parts – a Service Ontology and a Service Profile Ontology.

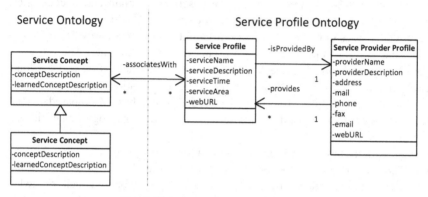

Fig. 1. Plain-Text-Described Service Ontology

- The Service Ontology represents the knowledge in a specific business service domain, in which each concept (i.e., Service Concept) is used to denote the functionality of a service subdomain. The concepts are associated with *class/subclass* relations whereby the upper concept is the generalization of its associated lower concepts. Each concept follows a standard schema which consists of three properties:
  - *conceptDescription* is a datatype property defined by domain experts which refers to a brief description of the functionality of a service subdomain. A concept may contain more than one *conceptDescription* property;
  - *learnedConceptDescription* is a datatype property learned from the Internet by our approach presented in Section 4, the purpose of which is also to describe the functionality of a service subdomain. A concept may contain more than one *learnedConceptDescription* property;
  - *associatesWith* is an object property which associates a relevant Service Concept and an instance of Service Profile.
- The Service Profile Ontology represents the two types of schema for describing the functionality and attributes of a real-world service, namely
  - Service Profile is the descriptive schema of a service, the datatype properties of which include
    ○ *serviceName* – the name of the service;
    ○ *serviceDescription* – the functional description of the service;

---

[6] http://www.w3.org/Submission/2004/SUBM-OWL-S-20041122/

- o *serviceTime* – the weekly time at which the service is available;
- o *serviceArea* – the area in which the service is available;
- o *webURL* – URL of the Web page at which the service can be found.
- Service Provider Profile is the descriptive schema of a service provider, the datatype properties of which include
  - o *providerName* – the name of the service provider;
  - o *providerDescription* – the description of the service provider;
  - o *address/mail/phone/fax/email* – the contact details of the service provider;
  - o *webURL* – URL of the Web page at which the service provider can be found.
- *Provides/isProvidedBy* is a pair of inverse object properties that links a service and a service provider to indicate their provision relationship. A service provider may provide more than one service.

The services discovered from the Internet are stored as instances of the Service Profile Ontology for standard description and are linked to relevant Service Concepts for functionality annotation.

4   System Architecture

In this section, we present a novel framework for plain-text-described service discovery (Fig. 2). Most of the functions and algorithms offer improvements to our previous work [3, 4].

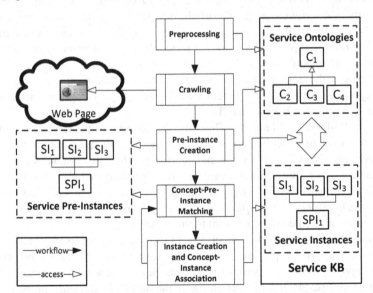

Fig. 2. System Architecture of the Proposed Plain-Text-Described Service Discovery Framework

As shown in Fig. 2, the framework comprises a Service Knowledge Base and five primary functions. The Service Knowledge Base stores service ontologies and service instances. The latter are instances of the Service Profile Ontology, automatically created by the proposed framework. The five primary functions are:

- Preprocessing. This function searches and weighs the key terms contained in the *conceptDescription* properties of the Service Concepts before the Crawling function starts. This search function is realized by using WordNet[7] to implement tokenization, part-of-speech tagging, nonsense word filtering, and stemming for all the terms contained in the *conceptDescription* properties of a Service Ontology. The weighting function is achieved by searching the synonyms of key terms using WordNet and calculating the specificity of the terms using Inverse Document Frequency. Therefore, the weight of a term t occurring in a *conceptDescription* property CD of an ontology O is

$$\underset{(t \in CD) \cap (\forall CD \in O)}{w(t) =} \quad \log \frac{\{|C| \, | \forall C \in O\}}{\{|C_\alpha| \, | (s \in CD_\beta) \cap (CD_\beta \in C_\alpha) \cap (\forall C_\alpha \in O) \cap (\forall s \in \delta(t))\}} \tag{1}$$

where $|C|$ is the number of concepts and $\delta(t)$ is the synset of the term t.
- Crawling. This function downloads Web pages by means of a pre-assigned URL list. The URL assignment is realized according to user-defined crawling policies.
- Pre-Instance Creation. This function creates service and service provider pre-instances according to the Service Profile Ontology. Property values of the pre-instances are extracted from the crawled Web pages, and the extraction is realized by observing the common patterns in the HTML/XHTML codes of the Web pages. The *serviceDescription* properties of the service pre-instances are processed by WordNet through tokenization, part-of-speech tagging, nonsense word filtering, and stemming. If a processed term in a *serviceDescription* property can retrieve its identical counterpart from the key term list obtained in the Preprocessing function, the weight of the term is equal to the weight of its counterpart; otherwise the term is assigned with a maximum weight $\{\log|C| \, | \forall C \in O\}$. The pre-instances are not stored in the Service Knowledge Base.
- Concept-Pre-Instance Matching. This function determines the relevance between a concept and a pair of service and service provider pre-instances in two steps:
  — Direct matching. If the service and service provider pre-instances do not have identical counterparts in the Service Knowledge Base, and if the *serviceDescription* property of the service pre-instance is contained in the *conceptDescription* or *learnedConceptDescription* properties of the concept, the pair of pre-instances is deemed to be relevant to the concept.
  — Syntactic matching. If the Direct Matching fails, the processed *serviceDescription* property of the pre-instance is syntactically matched with the preprocessed *conceptDescription* properties of the concept in a Concept-Pre-Instance Matching Model (explained in Section 5). If the *serviceDescription* property of the pre-instance matches any *conceptDescription* property of the concept according

---

[7] http://wordnet.princeton.edu/

to the model, the concept and the pair of pre-instances are determined as being relevant; in addition, the *serviceDescription* property is *"learned"* by the concept as a new *learnedConceptDescription* property.

- Instance Creation and Concept-Instance Association. If the result of the Concept-Pre-Instance Matching is positive, the pair of service and service provider instances is stored in the Service Knowledge Base and associated with the relevant concept. The Concept-Pre-Instance Matching is invoked iteratively until the relevance between all Service Concepts in the Service Knowledge Base and the pair of service and service provider instances is determined.

5   Concept-Pre-Instance Matching Model

To determine the relevance between a *conceptDescription* property (*CD*) of a concept and a *serviceDescription* property (*SD*) of a pre-instance, we design a Support Vector Machine (SVM) classifier. This classifier provides a binary classification function (relevant/non-relevant), which is characterized by a hyperplane in a given feature space. For a detailed explanation, let $X = [0, 1]^3$ be the feature space with feature vectors $x_i = (sim_{Dice}(CD,SD), sim_{Jaccard}(CD,SD), sim_{Overlap}(CD,SD))$, in which the features represent three types of matching result between a pair of *CD* and *SD* shown below:

$$sim_{Dice}(CD, SD) = \frac{2 \times \sum_{(\forall u \in \delta(CD)) \cap (\forall u \in SD)} w(u)}{\sum_{\forall t \in CD} w(t) + \sum_{\forall s \in SD} w(s)} \tag{2}$$

$$sim_{Jaccard}(CD, SD) = \frac{\sum_{(\forall u \in \delta(CD)) \cap (\forall u \in SD)} w(u)}{\sum_{\forall t \in CD} w(t) + \sum_{\forall s \in SD} w(s) - \sum_{(\forall u \in \delta(CD)) \cap (\forall u \in SD)} w(u)} \tag{3}$$

$$sim_{Overlap}(CD, SD) = \frac{\sum_{(\forall u \in \delta(CD)) \cap (\forall u \in SD)} w(u)}{\min\left(\sum_{\forall t \in CD} w(t), \sum_{\forall s \in SD} w(s) \right)} \tag{4}$$

where $w(t)$ is the weight of a term t and $\delta(CD)$ is the synset of the terms in *CD*.

The y_i value of the training dataset equals -1 for a pair of non-relevant *conceptDescription* property and *serviceDescription* property, and 1 for a relevant case. As usual, relevance sets in the test collections are subjectively defined by domain experts. Eventually, the input to the SVM classifier is a set of training dataset $\{(x_1, y_1), ..., (x_m, y_m)\}$ with $x_i \in X$ and $y_i \in \{-1, 1\}$.

The result of running the SVM classifier on such input is a hyperplane, which separates training dataset in the feature space as precisely as possible while the distance of the nearest points of each category is maximized to avoid biased categorization.

6   Evaluation

To evaluate the proposed plain-text-described service discovery approach, we compare the performance of this approach with our previous approach – SOF [4] on precision and recall. Both approaches enable the ontological vocabulary learning paradigm. We also use a non-vocabulary-learning approach – Su et al.'s approach [7] – as a baseline for this evaluation.

The evaluation is run in a computer with Intel i5 CPU@1.9G HZ, 8G Memory and Windows 7. The Plain-Text-Described Service Ontology is built in Protégé-OWL[8], in which the Service Ontology is extended from our previously designed transport service ontology [9]. This is a four-tier ontology that consists of 304 concepts. The SVM model is implemented using LibSVM[9]. For the purpose of evaluation, we use the proposed crawler to download 4400 transport-related Web pages from Australian Yellowpages[10] and 10000 transport-related Web pages from Kompass[11]. We use the former as the training dataset and the latter as the testing dataset.

As shown in Fig. 3, the improvement of our approach (labeled as DJO) on recall compared to the SOF is outstanding, in addition to achieving close performance on precision. Since the new SVM classifier determines the relevance between a *conceptDescription* property of a service concept and a *serviceDescription* property of a service instance from three distinct perspectives, more instances are correctly annotated with relevant concepts, and more relevant terms, i.e., *serviceDescription* properties, are learned by the ontology, in contrast to our previous approach. This result preliminarily proves the capability of the proposed approach to discover more domain-relevant plain-text-described services than other approaches.

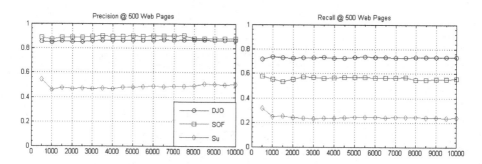

Fig. 3. Performance Comparison on Precision and Recall

7   Conclusion and Future Work

In this paper, we presented a novel approach for discovering plain-text-described services. Three basic components are contained in the approach: 1) a plain-text-described

---

[8] http://protege.stanford.edu/
[9] http://www.csie.ntu.edu.tw/~cjlin/libsvm/
[10] http://www.yellowpages.com.au/
[11] http://au.kompass.com/

service ontology extended from OWL-S for annotating the functionality and attributes of plain-text-described services; 2) a plain-text-described discovery framework for automatically discovering plain-text-described services while enriching the vocabulary of ontological concepts; and 3) a machine learning model for automatically annotating the functionality of plain-text-described services with relevant ontological concepts. To evaluate this approach, we compared its performance with our previous work and an existing approach. The experimental result preliminarily proves the efficiency of the proposed approach in discovering relevant plain-text-described services.

In future work we will target 1) the application of this approach in the domain of cloud services; and 2) the design of novel plain-text-described service discovery approaches using ontological relationship learning methods.

References

1. Dong, H., Hussain, F.K., Chang, E.: Semantic Web Service Matchmakers: State of the Art and Challenges. Concurrency Computat. Pract. Exper. 25, 961–988 (2013)
2. Gruber, T.R.: A Translation Approach to Portable Ontology Specifications. Knowledge Acquisition 5, 199–220 (1993)
3. Dong, H., Hussain, F.K., Chang, E.: Ontology-Learning-Based Focused Crawling for Online Service Advertising Information Discovery and Classification. In: Liu, C., Ludwig, H., Toumani, F., Yu, Q. (eds.) Service Oriented Computing. LNCS, vol. 7636, pp. 591–598. Springer, Heidelberg (2012)
4. Dong, H., Hussain, F.K.: SOF: A Semi-Supervised Ontology-Learning-Based Focused Crawler. Concurrency Computat. Pract. Exper. 25, 1755–1770 (2013)
5. Noor, T.H., Sheng, Q.Z., Alfazi, A., Ngu, A.H.H., Law, J.: CSCE: A Crawler Engine for Cloud Services Discovery on the World Wide Web. In: ICWS 2013, pp. 443–450. IEEE, New York (2013)
6. Zheng, H.-T., Kang, B.-Y., Kim, H.-G.: An Ontology-Based Approach to Learnable Focused Crawling. Inform. Sciences 178, 4512–4522 (2008)
7. Su, C., Gao, Y., Yang, J., Luo, B.: An Efficient Adaptive Focused Crawler based on Ontology Learning. In: HIS 2005, pp. 73–78. IEEE, New York (2005)
8. Phuc, H.L., Gauch, S., Qiang, W.: Ontology-Based Focused Crawling. In: eKNOW 2009. IEEE, New York (2009)
9. Dong, H., Hussain, F.K.: Focused Crawling for Automatic Service Discovery, Annotation, and Classification in Industrial Digital Ecosystems. IEEE Trans. Ind. Electron. 58, 2106–2116 (2011)
10. Rennie, J., McCallum, A.: Using Reinforcement Learning to Spider the Web Efficiently. In: ICML 1999, pp. 335–343. Morgan Kaufmann Publishers Inc., San Francisco (1999)
11. Cortes, C., Vapnik, V.: Support-Vector Networks. Machine Learning 20, 273–297 (1995)

A Fuzzy VSM-Based Approach
for Semantic Service Retrieval

Supannada Chotipant[1], Farookh Khadeer Hussain[1], Hai Dong[2],
and Omar Khadeer Hussain[3]

[1]Decision Support and e-Service Intelligence Lab (DeSI)
Quantum Computation and Intelligent Systems (QCIS) and School of Software
University of Technology Sydney (UTS), Sydney, NSW, Australia
Supannada.Chotipant@student.uts.edu.au,
Farookh.Hussain@uts.edu.au
[2] School of Computer Science and Information Technology, RMIT University, Australia
Hai.Dong@rmit.edu.au
[3]School of Business
University of New South Wales, ADFA, Canberra BC
ACT, 2601, Australia
O.Hussain@adfa.edu.au

Abstract. A vast number of business services have been published on the Web in an attempt to achieve cost reductions and satisfy user demand. Service retrieval consequently plays an important role, but unfortunately existing research focuses on crisp service retrieval techniques which are unsuitable for vague real world information. In this paper, we propose a new fuzzy service retrieval approach which consists of two modules: service annotation and service retrieval. Related service concepts for a given query are semantically retrieved, following which services that are annotated to those concepts are retrieved. The degree of retrieval of the retrieval module and the similarity between a service, a concept, and a query are fuzzy. Our experiment shows that the proposed approach performs better than a non-fuzzy approach on Recall measure.

Keywords: Fuzzy Service Retrieval, Semantic Service Annotation, Semantic Service Retrieval.

1 Introduction

Nowadays, a huge number of business services have been published on the Internet because companies desire to reduce costs and easily access their customers. Annotating services semantically enables machines to understand the purpose of services and can further assist in intelligent and precise service retrieval, selection and composition. Nowadays, meanings are manually ascribed to most services by service providers. Although this makes the results more acceptable, it is time-consuming, and there are difficulties in dealing with online tasks. The automation of service annotation is therefore desirable, but unfortunately no service annotation technique for service technology exists. Research about Web service annotation [1,2,3,4,5,6] concerns

C.K. Loo et al. (Eds.): ICONIP 2014, Part III, LNCS 8836, pp. 682–689, 2014.
© Springer International Publishing Switzerland 2014

semi-automated systems which suggest annotations to service providers. While [1], [6] are term-based annotation approaches, the others [2,3,4,5] are ontology-based annotation approaches. However, these approaches focus on only crisp annotation. To link the parameters of service descriptions to concepts of ontology is fuzzy. One parameter can be linked to several concepts with different degrees of relevance. Moreover, those works focus only on Web services. In this paper, we attempt to annotate business services to service concepts by using fuzzy variables.

Understanding the semantics of a query is the other significant factor for improving service retrieval performance. Zhai et al [7] focus on semantic query expansion by using ontology. In contrast, in this paper, we apply the ECBR algorithm [8] to semantically retrieve relevant services by using the synonyms of a query. Moreover, the intention of our approach is to overcome the issue of vague information, such as the similarity between a service and a concept, the similarity between a query and a concept, and the retrieval degree of a service. Therefore, we apply Fuzzy logic to the semantic service retrieval, which has not been done previously in the existing literature. Although there are some papers relating to fuzzy service retrieval [7], [9,10,11], their methods are based on fuzzy ontology, instead of crisp ontology which is used in this paper. Moreover, the existing work [9], [11] focus on retrieving Web services, while we focus on retrieving business service information on the Web.

The rest of this paper is organized as follows. In Section 2, we introduce the design of our semantic service retrieval system. The experiments and results are provided in Section 3, and the work is concluded in Section 4.

2   System Design

In this paper, we propose a Fuzzy VSM-based approach for a semantic service retrieval system, to enable users to search services based on service concepts.

The overall system architecture of the semantic service retrieval system is shown in Fig.1. The system consists of three main components, namely the service knowledge base, service annotation module and service retrieval module. The service knowledge base stores information related to services. It contains service ontology and service description entity (SDE) metadata. The service annotation module automatically annotates a SDE to relevant service concepts in the domain specific service ontology. The service retrieval module retrieves SDEs which are relevant to a user's query.

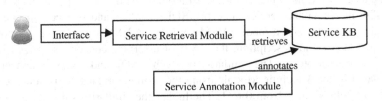

Fig. 1. Overall system architecture of the semantic service retrieval system

2.1   Service Knowledge Base

As mentioned above, the service knowledge base stores domain specific service ontology and SDE metadata. In this paper, we apply the service knowledge base from
the work by Dong, Hussain and Chang [8]. The SDEs are easily linked to service
concepts according to the categories of those services. In this paper, we propose a
new service annotation method, which is presented in Section 2.2.

Service Ontology

The service ontology is a conceptualization of the services offered. Its structure is
separated into four layers, namely the abstract concept layer, the service subdomain
concept layer, the abstract service concept layer, and the actual service layer. Each
service concept consists of a service concept name and service concept description.
Only service concepts in the actual service layer are linked to SDE metadata.

Service Description Entity (SDE)

The SDE metadata is the information about the actual business services which may be
relevant to more than one service concept. In this paper, the SDE metadata consists of
five properties, namely linked concepts, service provider name, provider address,
provider contact details, and SDE description. For example, the SDE metadata "Airline Agents Bookings" is assigned to "Virgin Airlines", "131 Fortitude Valley QLD,
4006 Australia", "Phone : 13 6789", and "Low Fares, Great Service." as service provider, provider address, provider contact details, and SDE description respectively.

2.2   Service Annotation Module

The service annotation module is a pre-processing element of the service retrieval
system. The main purpose of this module is to automatically link SDE metadata to
related service concepts. For example, the SDE called "Airline Agents Bookings" is
connected to the service concept called "Airline_Booking". This enables the system to
semantically retrieve the services associated with the relevant service concepts.

We present two approaches to SDE annotation, namely non-fuzzy and fuzzy service annotation respectively. The workflow of both approaches is presented in Fig. 2.
The input data are SDE metadata and service ontology, while the output data are links
between SDE metadata and service concepts. SDE metadata and service concepts are
represented by a vector space model (VSM). While the VSM of a SDE is generated
from the SDE name and description, the VSM of a service concept is created from
service concept descriptions. The similarity between a SDE and a service concept is
then calculated in the Matching Module using the cosine similarity between vectors.
The similarity values range from 0 to 1. The higher the similarity value is, the more
closely a SDE relates to a service concept. A value of 0 means that a SDE and a service concept are totally different. On the other hand, a value of 1 means that a SDE
and a service concept are the same. The SDE-Service Concept Linking Module will

subsequently link a SDE to a service concept if they are related. There is a difference in this step between a non-fuzzy system and a fuzzy system. If the SDE-Concept similarity value is greater than the link threshold (LT) in a non-fuzzy system, a SDE will be linked to a concept. In contrast, a SDE in a fuzzy system will be linked only if the similarity is greater than 0.

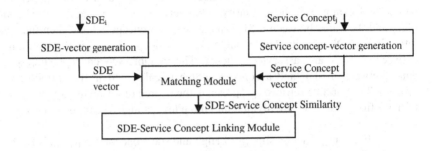

Fig. 2. System architecture of the service annotation module

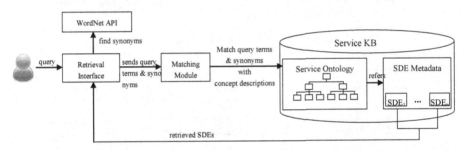

Fig. 3. System architecture of the service retrieval module

2.3   Service Retrieval Module

The service retrieval module fetches SDEs which are relevant to a user's query. For example, when a user sends the query "Flight booking service" to the system, the module retrieves related SDEs, such as "Airline Agents Bookings", which is provided by Air Niugini. The architecture of the service retrieval module is shown in Fig. 3.

First, the module receives a query via the retrieval interface. It extracts a set of separating terms from the query and removes the stop words. The retrieval interface then sends each query term to the WordNet API which returns the synonyms of the received term. The retrieval interface sends a set of query terms and their synonyms to the matching module, and we add those synonyms to the query terms. The matching module computes the similarity value between the set of query terms and each service concept in the service ontology. The processes of this module for non-fuzzy and fuzzy service retrieval systems are different. For the non-fuzzy system, arbitrary service concepts whose similarity values are greater than the retrieval threshold (RT) are selected, and the module refers to SDEs that are linked to the selected service concepts in the previous step. For the fuzzy system, the matching module applies Fuzzy logic to retrieve the relevant SDEs. The fuzzy rules and fuzzy membership functions are defined in Fig.4 and Fig.5 respectively. The fuzzy rules in this paper are quite simple. For example, if the similarity value between a query and a service concept is high, and the value between a SDE and a service concept is high, then the degree of retrieving a SDE is high. The variable query_concept_sim$_{(Q,C)}$ is the similarity value between a query and a service concept. The variable sde_concept_sim$_{(SDE,C)}$ is the value between a SDE and a concept, which is calculated in the service annotation process. The variable retrieve$_{(SDE)}$ presents the degree of retrieving a SDE. The values of these fuzzy variables are divided into three levels: high, medium, and low.

IF query_concept_sim$_{(Q,C)}$ is High and sde_concept_sim$_{(SDE,C)}$ is High
        THEN retrieve$_{(SDE)}$ is High.
IF query_concept_sim$_{(Q,C)}$ is Medium and sde_concept_sim$_{(SDE,C)}$ is Medium
        THEN retrieve$_{(SDE)}$ is Medium.
IF query_concept_sim$_{(Q,C)}$ is Low and sde_concept_sim$_{(SDE,C)}$ is Low
        THEN retrieve$_{(SDE)}$ is Low.

Fig. 4. Fuzzy rules for service retrieval system

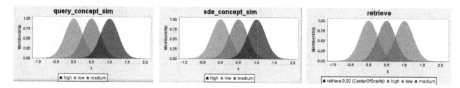

Fig. 5. Membership functions for service retrieval system

Query-Concept Similarity

We apply an extended case-based reasoning (ECBR) algorithm [8] to compute the similarity value between a set of query terms and a service concept. The main concept of the ECBR algorithm is to compare a set of query terms with the descriptions of a service concept. As previously mentioned, a service concept may contain many service concept descriptions. Therefore, the module will first compute the similarity values between query terms and each service concept description. Then, their maximum similarity values will be returned.

The similarity value for a service concept description is the summation of the matching value between each term and that concept description. We normalize the summation with the length of that concept description. If a query term from a user appears in the description, a value of 1 will be added to the matching value. On the other hand, if a synonym of a query term appears, a value of 0.5 will be added instead.

3   Experiment and Result

In this section, we compare the performance of the service retrieval system based on the non-fuzzy VSM model with the system based on the fuzzy VSM model in a Transportation domain ontology (TO) defined in [8]. We apply four performance measures from the area of information retrieval [12], namely Precision, Recall, Harmonic Mean, and Fallout Rate.

3.1   Experiment and Results

To evaluate the non-fuzzy system, we test the performance by setting the retrieval threshold (RT) for the ECBR algorithm as 0.8 and setting the link threshold (LT) in the service annotation module from 0.1 to 0.9 with an increment of 0.1. The performance of non-fuzzy service annotation and retrieval are shown in Table 1 and Table 2 respectively. For the fuzzy system, we set the RT as 0.8, and the retrieval degree (RD), the result of firing fuzzy rules, from 0.1 to 0.9 with an increment of 0.1. The performance of the fuzzy service annotation and retrieval are presented in Table 3 and Table 4 respectively. Note that we set the value of the RT as 0.8 because this value gives us the best performance values in term of Precision and Recall.

Comparing the Performance of Non-fuzzy with Fuzzy Service Annotation
The Precision values of non-fuzzy service annotation in Tables 1 and 3 are greater with every LT than those in the fuzzy system. It should be noted that all measure values (Precision, Recall, Harmonic Mean and Fallout) are the same values with different levels of RD, because the system links all SDEs that relate to a concept, even if that relationship is low. The performance of the fuzzy service annotation is equal to the performance of non-fuzzy annotation with LT value 0. Consequently, the Recall value of the fuzzy system is greater than all the Recall values of the non-fuzzy system. However, the non-fuzzy system performs better on Harmonic Mean and Fallout.

Table 1. The Performance of Non-Fuzzy Service Annotation for the TO, RT = 0.8

Link Threshold (LT)	Precision	Recall	Harmonic Mean	Fallout
0.1	17.43%	99.85%	11.01%	5.91%
0.2	20.34%	98.81%	12.86%	4.50%
0.3	27.14%	97.10%	16.20%	2.95%
0.4	34.08%	90.25%	18.01%	2.01%
0.5	42.08%	78.51%	21.13%	1.06%
0.6	53.95%	67.93%	24.18%	0.52%
0.7	59.72%	55.89%	24.48%	0.16%
0.8	56.28%	36.29%	19.34%	0.06%
0.9	54.98%	28.80%	16.71%	0.01%

Table 2. The Performance of Non-Fuzzy Service Retrieval for the TO, RT = 0.8

Link Threshold (LT)	Precision	Recall	Harmonic Mean	Fallout
0.1	67.59%	100.00%	37.67%	3.44%
0.2	70.23%	99.35%	38.91%	2.61%
0.3	78.83%	94.45%	41.49%	1.45%
0.4	87.95%	84.71%	41.77%	0.91%
0.5	82.15%	61.71%	33.04%	0.58%
0.6	69.55%	43.55%	25.94%	0.20%
0.7	75.29%	37.66%	22.91%	0.12%
0.8	60.00%	18.53%	12.44%	0.20%
0.9	62.50%	19.57%	12.29%	0.00%

Table 3. The Performance of Fuzzy Service Annotation for the TO, RT = 0.8

Retrieval Degree (RD)	Precision	Recall	Harmonic Mean	Fallout
All RDs	16.59%	100.00%	10.48%	6.48%

Table 4. The Performance of Fuzzy Service Retrieval for the TO, RT = 0.8

Retrieval Degree (RD)	Precision	Recall	Harmonic Mean	Fallout
0.1	67.38%	100.00%	37.58%	3.53%
0.2	67.38%	100.00%	37.58%	3.53%
0.3	67.38%	100.00%	37.58%	3.53%
0.4	67.38%	100.00%	37.58%	3.53%
0.5	67.56%	100.00%	37.67%	3.41%
0.6	87.34%	87.76%	42.36%	1.03%
0.7	79.30%	69.74%	35.57%	0.77%
0.8	78.17%	53.37%	29.61%	0.39%
0.9	57.84%	18.54%	11.66%	0.18%

Comparing the Performance of Non-fuzzy with Fuzzy Service Retrieval

Because the processes of retrieving the services for non-fuzzy and fuzzy systems are quite different, we will compare them with their maximum Precision values and we will focus on the performance of a non-fuzzy system with the LT value 0.4 and a fuzzy system with the RD value 0.6. Both non-fuzzy and fuzzy service retrieval systems perform well on Precision measure with values of 87.95% and 87.34% respectively. We observe that the Precision value of the fuzzy retrieval system is still high, although the value of the fuzzy annotation system is very low. This means that the Precision values of fuzzy retrieval system do not depend on those of fuzzy annotation system. With the same Precision, it can be seen that the fuzzy system performs slightly better than the non-fuzzy system – around 3% on the Recall measure. This is because fuzzy annotation links all related SDEs to a concept, which non-fuzzy annotation fails to do. As a result, the Fallout Rate of the fuzzy system is a marginally higher because it retrieves more non-relevant services.

4   Conclusion

In this paper, we proposed a Fuzzy VSM-based approach for semantic service retrieval. This approach is divided into two main modules: an annotation module and a retrieval module. The purpose of the annotation module is to prepare the Service Knowledge Base for the retrieval module. In this step, SDEs are automatically linked to relevant service concepts by comparing the similarity between their VSM-based vectors. The retrieval module semantically retrieves relevant service concepts using the ECBR algorithm and then retrieves SDEs that relate to selected concepts using Fuzzy technique. The experiment demonstrates that fuzzy service retrieval performs better on Recall than non-fuzzy service retrieval, but that both non-fuzzy and fuzzy retrieval perform well on Precision and Harmonic Mean.

References

1. Stavropoulos, T.G., Vrakas, D., Vlahavas, I.: Iridescent: A tool for rapid semantic annotation of web service descriptions. In: Proceedings of the 3rd International Conference on Web Intelligence, Mining and Semantics, Madrid, Spain, pp. 1–9. ACM (2013)
2. Scicluna, J., Blank, C., Steinmetz, N., Simperl, E.: Crowd Sourcing Web Service Annotations. In: 2012 AAAI Spring Symposium Series (2012)
3. Patil, A.A., Oundhakar, S.A., Sheth, A.P., Verma, K.: Meteor-s web service annotation framework. In: Proceedings of the 13th International Conference on World Wide Web, pp. 553–562. ACM, New York (2004)
4. Hui, W., Zhiyong, F., Shizhan, C., Jinna, X., Yang, S.: Constructing Service Network via Classification and Annotation. In: 2010 Fifth IEEE International Symposium on Service Oriented System Engineering (SOSE), pp. 69–73 (2010)
5. Bo, J., Zhiyuan, L.: A New Algorithm for Semantic Web Service Matching. Journal of Software (1796217X) 8, 351–356 (2013)
6. Meyer, H., Weske, M.: Light-Weight Semantic Service Annotations Through Tagging. In: Dan, A., Lamersdorf, W. (eds.) ICSOC 2006. LNCS, vol. 4294, pp. 465–470. Springer, Heidelberg (2006)
7. Zhai, J., Cao, Y., Chen, Y.: Semantic information retrieval based on fuzzy ontology for intelligent transportation systems. In: IEEE International Conference on Systems, Man and Cybernetics, SMC 2008, pp. 2321–2326. IEEE (2008)
8. Dong, H., Hussain, F.K., Chang, E.: A service search engine for the industrial digital ecosystems. IEEE Transactions on Industrial Electronics 58, 2183–2196 (2011)
9. de Castilho, R.E., Gurevych, I.: Semantic Service Retrieval Based on Natural Language Querying and Semantic Similarity. In: ICSC, pp. 173–176 (2011)
10. Madkour, M., Maach, A., Driss, E., Hasbi, A.: Fuzzy-based approach for context-aware service retrieval. In: 2012 Second International Conference on Innovative Computing Technology (INTECH), pp. 396–401 (2012)
11. Zhao, W., Peng, L., Zhang, J., Tian, C.: A fuzzy service matching algorithm based on Bloom filter. In: 2012 IEEE 14th International Conference on Communication Technology (ICCT), pp. 945–950. IEEE (2012)
12. Baeza-Yates, R., Ribeiro-Neto, B.: Modern Information Retrieval. Addison-Wesley Longman, Boston (1999)

Maintaining Trust in Cloud Computing
through SLA Monitoring

Walayat Hussain[1], Farookh Khadeer Hussain[1], and Omar Khadeer Hussain[2]

[1] School of Software, Centre for Quantum Computation and Intelligent Systems,
University of Technology Sydney, Sydney, New South Wales 2007, Australia
[2] School of Business, University of New South Wales Canberra
walayat.hussain@student.uts.edu.au, farookh.hussain@uts.edu.au,
o.hussain@adfa.edu.au

Abstract. Maintaining trust in cloud computing is a significant challenge due to the dynamic nature of cloud computing and the fragility of trust. Trust can be established by conducting successful transactions and meeting all the parameters of the Service Level Agreement (SLA) drawn up between two interacting parties. Trust can be maintained by continuous monitoring of these predefined SLA parameters. There are number of commentaries on SLA monitoring that describe different frameworks for the proactive or reactive detection of SLA violations. The aim of this research is to present an overview of the literature and make a comparative analysis of SLA monitoring in respect of trust maintenance in cloud computing.

Keywords: Service level agreement monitoring, cloud computing, maintaining trust, cloud monitoring, SLA monitoring metrics, proactive SLA monitoring, hierarchical self-monitoring.

1   Introduction

Contemporary technologies create great opportunities for multiple online users to connect to the network simultaneously, but they also present a number of challenges. One of the key challenges is trust management [1]. The need for effective trust management is inevitable, given the large number of service providers and service consumers, and it is very difficult to monitor disreputable activity by consumers or violations of agreed service commitments by   providers [2].

Cloud computing is an emerging and popular new technology in parallel computing, thanks to the accessibility of resources irrespective of a user's location, timing or platform [3]. As a result of the increased number of providers, however, several challenges arise for cloud consumers such as data security, efficiency, trustworthiness and the reliability of the provider.

The cloud computing Service Level Agreement (SLA) is a mutual agreement between service provider and service consumer that describes such factors as Quality of Service parameters, mutually agreed services, service deliverability, transaction credits and penalties, based on which credibility and trustworthiness can be measured.

C.K. Loo et al. (Eds.): ICONIP 2014, Part III, LNCS 8836, pp. 690–697, 2014.

This paper attempts to integrate these findings by addressing the following research questions:

- What are the safeguards in the SLA monitoring framework that guarantee the trustworthiness of the cloud provider?
- What approaches have been suggested to maintain a trusting relationship between cloud consumer and cloud provider?

The structure of the paper is organized as follows. Section 2 presents our proposed classification of SLA monitoring in cloud computing, which is based on the functionality and working attributes of each approach. Section 4 provides insights from the findings and a comparative analysis of all the approaches covered in this study. Section 4 concludes the paper.

2   Classification of SLA Monitoring Scheme

The credibility of the provider, trust feedback from consumers, and assurance that the terms of the SLA will be met are the key drivers for the successful adaptation and growth of cloud computing [4] and for understanding the fundamental limitations and invariants associated with the adaptation of cloud computing to include features such as SLA monitoring. In this work, we present a classification of approaches that contribute to the issue of maintaining trust in cloud computing by monitoring SLAs. We have divided these approaches into four classes, based on their functionality and working attributes. These classes are:

- Self-manageable case-based reasoning approach
- SLA-based trust model approach
- Broker-based approach
- Workflow composition and reputation-based approach

2.1   Self-manageable Case-Based Reasoning Approach

The self-manageable approach provides service providers with the opportunity to detect and prevent any possible threat before it affects end users. [5]proposed a self-manageable monitoring mechanism in which low level hardware resource metrics are plotted to high level SLA parameters to detect possible future SLA violation threats and invoke an enactor component to undertake remedial action to avoid violation. A user-defined mapping rule is stored in the repository, and monitoring agents are responsible for measuring the resource metrics. A run-time monitor accesses the mapped metric repository and uses mapped SLA values to check the status of the deployed services and compare the mapped SLA metrics with the threshold threshold. If any violation is identified, the enactor component is alerted to the possible threat. Low level metric conversion is used in hierarchical monitoring and combined with a bottom-up approach for the propagation of SLA violation threats [6]. The propagation of SLA violation continued till specific layer which is able to perform suggested operation. The holistic SLA validation framework proposed by Haq et al. [7] used LoM2HiS [5] for SLA violation detection, LAYSI [6] for the bottom-up propagation of violations, and the rule-based SLA aggregation method [8] to identify the reasons for such violations and to impose penalty for violation. [9] proposed CASViD architecture which monitors and detects SLA violations at application level. To detect violations, CASViD finds the effective measurement interval to identify resource consumption by each application. Effective measurement is conducted by sampling time intervals and checking the applications for each interval. If the utility of the current time interval is greater from previous interval then the current interval is set as an effective measurement interval. The process continues till end.

The hierarchical self-healing SLA approach was proposed by [10]. Based on the hierarchical nature of cloud, SLA monitors its own parameters by itself. Each layer has multiple resources or providers. Each upper layer is dependent on the lower layer. Response time and throughput are used as measurement attributes. The monitoring function in each SLA continuously monitors the attributes based on the metric. When any violation is detected, then SLA first tries to prevent it by switching to another resource in that layer, otherwise it informs the SLA in the upper layer. This helps to prevent violations before they affect the end user.

[11] proposed a self-adaptive monitoring mechanism that monitors both the application and infrastructure layer and triggers *on-the-fly* reconfiguration functionality which enables the system to monitor during runtime. The monitoring mechanism consists of six components arranged in three layers of cloud. All the hardware level information concerning the execution of the virtual machine unit is sent from IaaS to the global repository in PaaS. All high level parameters are first collected into the local SaaS repository and then sent to the global repository in PaaS. Self-adaption allows both the hardware and software monitoring components to readjust resources or monitor time intervals.

2.2   SLA-Based Trust Model Approach

Trust in the provider, trust representation, and the criteria for trust calculation are three issues which are always of concern to the consumer, and need to be addressed in any business. Proactive performance monitoring was proposed by [12] which introduced a third party agent. [13] presented an effective QOS monitoring technique in which they proposed two techniques i.e. state monitoring and derived monitoring to monitor the trust of the provider and the representation of that trust. The authors proposed a dynamic trust calculation method based on Markov Chain theory and formulated conditions of steady state, un-steady state or failure state. Trust value is calculated at regular intervals, and when a provider attains peak level of trust, then 'extra' trust is considered to be a surplus which can be used when there is failure, without affecting the trust value.

Given the nature of cloud, a consumer can request services at any time. Scheduling the request is a difficult job, but if priorities are set in the SLA then the performance of both the cloud consumer and the cloud provider can be improved. [14] proposed a scheduling scheme using SLA by defining the priorities of requests in the SLA.

The trust monitor component acts as a third party agent. When a dishonest action is detected, the trust monitor considers it an intrusion and reports it immediately to the scheduler, also notifying both the provider and the consumer.

[2] proposed a model that consists of a SLA agent, cloud consumer model, cloud service directory and cloud provider. Each provider advertises their services in a cloud service directory to assist consumers to find a suitable cloud provider. When a cloud consumer query for a related provider from cloud service directory, a list of providers is obtained this is submitted to the trust management system for scrutiny. The list of trusted cloud providers is sent to the SLA agent with the service level objectives. When a cloud consumer submits a request for service, the ID of the provider and complete detail of the SLA are released to the consumer. If the consumer accepts the agreement, the transaction will be finalized and communication with the chosen provider will commence.

[3] proposed a cloud service registry and discovery (CSRD) model which acts as a monitoring agent between a cloud consumer and a provider. The trust of a provider is calculated according to feedback from credible service providers and credible service consumers. The credibility of a service provider is calculated by the length of time the provider has provided services divided by total number of services offered and the credibility of the service consumer is calculated by length of time services have been consumed. Trust is measured dynamically by standard deviation which is inversely proportional to trust.

[15] proposed that trust between a provider and a consumer can be maintained by monitoring trust at each layer. Monitoring should be done periodically, and can be evaluated by complex formula. Trusted third party monitors conduct communication between consumer and provider; however, they cannot determine the internal state of either the consumer or the provider. The trust module on the provider side has access to the internal state of the provider and can deal with any violation by itself. Although a trust module on the consumer side is not very effective, it can nevertheless be used to create trust for an assured provider.

2.3   Broker-Based Approach

A cloud service broker responsible for SLA negotiation using SaaS provisioning was proposed by [16]. Multi-attribute negotiation allows concurrent negotiating between two parties on multiple issues. The cloud service broker is responsible for delivering customized services to the cloud consumer. The service provider measures its Quality of Service by collecting data at predefined intervals, and if there is a decrease in the agreed level of QOS, the service provider may allocate further resources to meet its SLA obligations.

Multilevel management and monitoring of SLAs in a federated cloud environment was proposed by [17]. Monitoring of the SLA is achieved by retrieving SLA metrics from a different layer, checking the current SLA parameters and comparing them with the SLA metrics. The monitoring agent has services which are responsible for monitoring the SLA periodically and assesses performance against respective thresholds.

Intercloud computing and a cloud service brokerage was proposed by [18]. SLA monitoring starts when the SLA manager receives a service request from a consumer; the SLA manager translates the SLA and a service request is sent to the deployment manager to arrange the requested service. The deployment manager then forwards the request to the appropriate Intercloud gateway for the creation of the service. The consumer can request agreed QOS metrics based on the service ID, as well as metrics which can monitor the SLA. The consumer can terminate the service once the agreement has been completed and all resources have been released.

[19] proposed a cloud service broker portal which is a single entry point for the cloud service broker, cloud service provider, and cloud service consumer. It interacts with a unique interface designed for each stakeholder. Cloud service portals have a brokerage Application Programming Interface that is responsible for integrating various cloud service providers into the cloud service broker portal.

2.4   Reputation-Based and workflow Composition Approach

A reputation-based system approach was proposed by [20] that assists cloud consumers to select the most trustworthy and reliable cloud provider. It evaluates the reliability of reputation. To overcome biased evaluation, the authors proposed an IP monitoring mechanism. Management service is responsible to manage services by selective violation method to violate selective SLA to lower the monetary impact of penalties. [21] suggested the cancellation of service instances that have low priority and penalties. Unused resources which were reserved by the consumer are assigned to other consumers who are in need of those resources.

A fuzzy logic approach was used by [22]. The selection of a provider is dependent on recommendations by other users. Credible recommending users receive reputation requests from a third party and, based on the previous record of the provider, which is stored in the information repository of the Recommending user, reply with the trust value. The third parties SLA monitoring components aggregates all the reputation values from all recommending users and calculates the final reputation value of the service provider. The third party SLA monitoring component accesses the runtime SLA parameter and compares it with the threshold to identify the probability of failure.

A violation detection model was proposed by [23] which considers the utility function to measure the level of satisfaction for quality and control charts. The proposed model is comprised of three parts and considers four criteria for measurement. The Western Electric rule is used for SLA violation detection.

3   Comparative Analysis of Proposed Approaches

In the previous section, we presented our classification based on the functionality and working attributes of each approach. To select a better approach, we present a brief comparison of the different classes and candidate approaches in each class. This comparison is based on the monitoring approach, algorithm, SLA management and post procedure of SLA violation expressed in Table 1.

Table 1. Comparative analysis of monitoring SLA mechanism

Source	Domain	Monitoring Approach	Framework	Approach (host + broker)	Predict future SLA violation	Algorithm	Procedure after SLA violation
[5]	Cloud IaaS	Self-manageable	LoM2HiS + FoSII	Host + Broker	Yes based on threat threshold	Proactive	Not defined
[24]	Cloud IaaS	Self-manageable	DeSVi = FoSII + LoM2HiS	Host + Broker	Yes based on threat threshold	Proactive	Not defined
[6]	Cloud IaaS	Self-manageable	LAYSI= LoM2HiS + Threat propagation	Negotiator + Broker	Yes based on threat threshold	Proactive. Threat propagation	Self handle or propagated to upper layer.
[7]	Cloud IaaS	Self-manageable	Holistic SLA validation= LoM2HiS + LAYSI + Rule based	Negotiator + Broker	Yes based on threat threshold	Proactive. Threat propagation. Penalty enforcement	Propagate to upper layer. Renegotiate or abort service.
[9]	Cloud SaaS	Self-manageable	CASViD	Negotiator	Yes based on threat threshold	Proactive. Threat threshold.	Calculate SLA violation penalties
[25]	Cloud.	Self-manageable	HS-SLA	Hierarchical self monitoring	No	Reactive. Violation propagation and prevention.	Propagate to upper layer.
[11]	Cloud IaaS, SaaS	Self-adaptive	Multi layer monitoring	Platform monitoring	No	Reactive	Not defined
[13]	Cloud IaaS	Trust model	Dynamic trust model	Third party broker	Yes based on Markov Chain model	Proactive. Markov Chain	Not defined
[14]	Cloud SaaS	Trust model	SLA-based scheduling	Broker	No	Reactive	Not defined
[2]	Cloud SaaS	Trust model	SLA-based trust model	cloud service directory	No	Reactive	Not defined
[3]	Cloud	Trust model	CSRD	Broker	No	Reactive	Not defined
[16]	Cloud SaaS	Broker-based	Multi-attribute negotiation model	Broker	No	Proactive by collecting data at predefined intervals	Not defined
[17]	Cloud	Broker-based	Holistic SLA management model	Broker	No	Reactive	Not defined
[18]	Cloud IaaS	Broker-based	Generic cloud broker	Broker	No	Reactive	Not defined
[20]	Cloud SaaS	Reputation-based	Reputation system	Broker	No	Reactive	Not defined
[22]	Cloud	Reputation-based	TP SLA monitor	Broker	No	Reactive	Not defined

4　Conclusion

A service level agreement is a document which defines all the service level objectives and business norms and methods. Trust is dynamic and fragile in nature, and it is very difficult to maintain trust in cloud computing. One method of maintaining trust in cloud computing is real-time monitoring of SLA, to ensure that interacting parties fulfill all the service level objectives predefined in the SLA document. In this paper, we have described state-of-the art SLA monitoring frameworks in cloud computing. We divided our work into four groups based on functionality and working attributes, and made a comparative analysis of all these approaches.

References

1. Resnick, P., Zeckhauser, R.: Trust among strangers in Internet transactions: Empirical analysis of eBay's reputation system. Advances in Applied Microeconomics 11, 127–157 (2002)
2. Alhamad, M., Dillon, T., Chang, E.: Sla-based trust model for cloud computing. In: 2010 13th International Conference on Network-Based Information Systems (NBiS). IEEE (2010)
3. Muchahari, M.K., Sinha, S.K.: A New Trust Management Architecture for Cloud Computing Environment. In: 2012 International Symposium on Cloud and Services Computing (ISCOS). IEEE (2012)
4. Almathami, M.: Service level agreement (SLA)-based risk analysis in cloud computing environments, p. 91. Ann Arbor, Rochester Institute of Technology (2012)
5. Emeakaroha, V.C., et al.: Low level Metrics to High level SLAs-LoM2HiS framework: Bridging the gap between monitored metrics and SLA parameters in cloud environments. In: HPCS (2010)
6. Brandic, I., et al.: Laysi: A layered approach for sla-violation propagation in self-manageable cloud infrastructures. In: 2010 IEEE 34th Annual Computer Software and Applications Conference Workshops (COMPSACW). IEEE (2010)
7. Haq, I.U., Brandic, I., Schikuta, E.: Sla validation in layered cloud infrastructures. In: Altmann, J., Rana, O.F. (eds.) GECON 2010. LNCS, vol. 6296, pp. 153–164. Springer, Heidelberg (2010)
8. Haq, I.U., et al.: Rule-based workflow validation of hierarchical service level agreements. In: Workshops at the Grid and Pervasive Computing Conference, GPC 2009. IEEE (2009)
9. Emeakaroha, V.C., et al.: Casvid: Application level monitoring for sla violation detection in clouds. In: 2012 IEEE 36th Annual Computer Software and Applications Conference (COMPSAC). IEEE (2012)
10. Mosallanejad, A., et al.: HS-SLA: A Hierarchical Self-Healing SLA Model for Cloud Computing. In: The Second International Conference on Informatics Engineering & Information Science (ICIEIS 2013). The Society of Digital Information and Wireless Communication (2013)
11. Katsaros, G., et al.: A Self-adaptive hierarchical monitoring mechanism for Clouds. Journal of Systems and Software 85(5), 1029–1041 (2012)
12. Fachrunnisa, O., Hussain, F.K.: A methodology for maintaining trust in industrial digital ecosystems. IEEE Transactions on Industrial Electronics 60(3), 1042–1058 (2013)

13. Chandrasekar, A., Chandrasekar, K., Mahadevan, M., Varalakshmi, P.: QoS monitoring and dynamic trust establishment in the cloud. In: Li, R., Cao, J., Bourgeois, J. (eds.) GPC 2012. LNCS, vol. 7296, pp. 289–301. Springer, Heidelberg (2012)

14. Daniel, D., Lovesum, S.: A novel approach for scheduling service request in cloud with trust monitor. In: 2011 International Conference on Signal Processing, Communication, Computing and Networking Technologies (ICSCCN). IEEE (2011)

15. Quillinan, T.B., et al.: Negotiation and monitoring of service level agreements. In: Grids and Service-Oriented Architectures for Service Level Agreements, pp. 167–176. Springer (2010)

16. Badidi, E.: A Cloud Service Broker for SLA-based SaaS provisioning. In: 2013 International Conference on Information Society (i-Society). IEEE (2013)

17. Falasi, A.A., Serhani, M.A., Dssouli, R.: A Model for Multi-levels SLA Monitoring in Federated Cloud Environment. In: 2013 IEEE 10th International Conference on Autonomic and Trusted Computing (UIC/ATC) Ubiquitous Intelligence and Computing. IEEE (2013)

18. Jrad, F., Tao, J., Streit, A.: SLA based Service Brokering in Intercloud Environments. In: CLOSER (2012)

19. Lee, J., et al.: Cloud Service Broker Portal: Main entry point for multi-cloud service providers and consumers. In: 2014 16th International Conference on Advanced Communication Technology (ICACT), pp. 1108–1112. IEEE (2014)

20. Wang, M., et al.: A conceptual platform of SLA in cloud computing. In: 2011 IEEE Ninth International Conference on Dependable, Autonomic and Secure Computing (DASC). IEEE (2011)

21. Schulz, F.: Towards measuring the degree of fulfillment of service level agreements. In: 2010 Third International Conference on Information and Computing (ICIC). IEEE (2010)

22. Hammadi, A.M., Hussain, O.: A framework for SLA assurance in cloud computing. In: 2012 26th International Conference on Advanced Information Networking and Applications Workshops (WAINA). IEEE (2012)

23. Sun, Y., et al.: SLA detective control model for workflow composition of cloud services. In: 2013 IEEE 17th International Conference on Computer Supported Cooperative Work in Design (CSCWD). IEEE (2013)

24. Emeakaroha, V.C., et al.: Towards autonomic detection of sla violations in cloud infrastructures. Future Generation Computer Systems 28(7), 1017–1029 (2012)

25. Mosallanejad, A., Atan, R.: HA-SLA: A Hierarchical Autonomic SLA Model for SLA Monitoring in Cloud Computing. Journal of Software Engineering and Applications 6(3B), 114–117 (2013)

Author Index